The Safe Handling of Chemicals in Industry

P. A. CARSON M.Sc., Ph.D., A.M.C.T., C. Chem., F.R.S.C., F.I.O.S.H
Safety Liaison and Quality Assurance Manager, Unilever Research, Port Sunlight

C. J. MUMFORD B.Sc., Ph.D., D.Sc., C. Eng., M.I. Chem. E.
Reader in Chemical Engineering, University of Aston

Volume 3

Longman Group Limited
Longman House, Burnt Mill, Harlow
Essex CM20 2JE, England
and Associated Companies throughout the world

*Copublished in the United States with
John Wiley & Sons, Inc., 605 Third Avenue, New York
NY 10158*

© P. A. Carson and C. J. Mumford 1996

All rights reserved; no part of this publication may be reproduced, stored in any retrieval system, or transmitted in any form or by any means, electronic, mechanical, photocopying, recording, or otherwise without either the prior written permission of the Publishers or a licence permitting restricted copying in the United Kingdom issued by the Copyright Licensing Agency Ltd., 90 Tottenham Court Road, London W1P 9HE.

First published 1996

British Library Cataloguing in Publication Data
A catalogue entry for this title is available from the British Library.

ISBN 0-582-06307-8

Library of Congress Cataloging-in-Publication data
(Revised for vol. 3)
Carson, P. A. (Phillip A.), 1944–
   The safe handling of chemicals in industry
   Includes bibliographies and index.
   1. Hazardous substances – Safety measures.
I. Mumford, C. J.   I. Title.
T55.3.H3C37 1988   604.7   88–13137
ISBN 0–470–20886–4

Set in Linotron 202 10/12 pt Times
Printed and bound by Bookcraft (Bath) Ltd

Contents to Volumes I and II

Volume I
1	Introduction	1
2	General occupational safety	9
3	Physico-chemical principles and safety	75
4	Flammable materials handling	122
5	Toxicology	185
6	Control measures for handling toxic substances	269
7	Ambient air analysis for hazardous substances	338
8	Radioactive chemicals	444
9	Safety with chemical engineering operations	463

Volume II
10	Hazardous chemical processes	491
11	Safety in laboratories and pilot plants	533
12	Safety in chemical-process plant design and operation	653
13	Safety in marketing and transportation of chemicals	769
14	Waste disposal	821
15	Control of hazards from large-scale installations	864
16	Legislative controls	915
17	Management of safety and loss prevention	966
18	Chemical safety information	1031
	Index	1071

Contents

Preface	vi
Acknowledgements	ix
List of Abbreviations	x

Volume III

Further Considerations and Developments

19	Physico-chemical principles and safety – II	1091
20	Process development and pilot planting	1142
21	The environmental impact of chemicals	1205
22	Environmental pollution control	1256

Flammable and Explosive Hazards and their Control

23	Storage, handling, utilisation and transport of liquefied petroleum gases	1313
24	Spontaneous combustion and pyrophoric chemicals	1357
25	Dust explosions	1416
26	Chemicals that may explode	1473
27	Fire protection practice	1507

Toxic Hazards and their Control

28	Control of substances hazardous to health	1571
29	Toxic solids	1642
30	Toxic gases and vapours	1699
31	Toxic liquids	1761
32	Microbiological hazards and preventative measures	1786
	Index	1831

Preface

Throughout the twentieth century, chemicals have had a significant impact on society's improved physical and economic well-being. Output from the chemical industry provides the raw materials for a wide range of important processes and manufacturing activities including agriculture, food, textiles, paper and printing, pharmaceuticals, cosmetics and toiletries, the motor trade, plastics and rubber, electronics, paint, carpets and flooring, etc. Such consumer goods have been instrumental in enhancing standards of living, preventing disease, helping energy conservation and minimising labour. Chemicals are therefore encountered not only in the chemical and allied process industries, but also in other sectors including construction, engineering, offices, hospitals, farms, the service industries, academic laboratories, and the home.

However, unless properly controlled, many chemicals can prove to be hazardous. Their manufacture, storage, use and transportation are therefore governed by legislation and industry codes. Modern-day socio-economic, scientific and technological challenges are geared to further exploitation of chemicals and chemical products. At the same time society demands ever-increasing levels of health, safety and environmental standards, necessitating continuous programmes for monitoring of safety and health performance, which nowadays must extend beyond the workforce to customers, to the general public and to the environment. Devising safe plant or chemical operations such that no significant danger exists even when mishaps occur requires heightened awareness of the hazardous properties of chemicals, together with an appreciation of the measures for their safe handling.

Volumes 1 and 2 of this book, therefore, describe techniques for the recognition of the hazards associated with the manufacture, handling, storage, transport and disposal of chemical substances, from small-scale laboratory operations through to the R&D, pilot plant and production stages and eventual transportation from plant to customer. Areas and topics covered include a background to reactive, flammable, explosive, toxic, corrosive, radioactive and carcinogenic hazards, plus practical guidance on handling procedures, environmental monitoring, safe design and operation of plant, laboratory safety, major hazard installations, and waste disposal. Legal aspects, management responsibilities and sources of information are also discussed.

Volumes 3 and 4 provide more detailed guidance on the application of these general principles by reference to selected examples. Additional material is included on controlling the environmental impact of chemicals, controlling microbiological hazards, chemical hazards in the consumer market, emergency planning, the role of ergonomics and communication in accident causation and prevention, and the economics of safety and loss prevention.

Brief case histories are again included to illustrate some incidents or chemical exposures which have arisen in the past, and to show how they were dealt with and, where appropriate, what technical and/or management measures were taken to prevent recurrence. Although not an original approach this has proved effective; it was noted in 1886 that[1]

> ... nothing was so instructive to the younger Members of the Profession, as records of accidents in large works, and of the means employed in repairing the damage. A faithful account of those accidents, and of the means by which the consequences were met, was really more valuable than a description of the most successful works. The older Engineers derived their most useful store of experience from the observations of those casualties which had occurred to their own and other works, ...

The principles, chemical engineering, waste disposal, transport and administrative practices are not restricted to any particular country. Therefore, in the main, the recommended procedures and precautions are not legislation-based. For this reason the numerous items of legislation repealed, updated or introduced since publication of the earlier volumes are generally not referred to in Volume 3 since – whilst they represent important changes in law – they do not affect the general principles. Examples within the UK include the Chemicals (Hazard Information and Packaging for Supply) Regulations 1994, the Pressure Systems and Transportable Gas Containers Regulations 1989, and the '6 pack'[2]. Reference is however made to selected key legislation with a particular impact upon the subject under discussion, e.g. the Control of Substances Hazardous to Health Regulations 1988 and the Environmental Protection Act 1990 which are considered in some detail. Important selected European Directives are mentioned briefly in Chapter 28 to illustrate the trends. Other relevant legislation, e.g. the UK Planning (Hazardous Substances) Regulations 1992 – which requires the holder of dangerous substances to obtain a hazardous substances consent for any site at which it is intended to hold a bulk quantity of any of 71 substances above a controlled quantity, where appropriate – are discussed in Volume 4.

References
1. Stephenson, R., quoted in Petroski, H., *To Engineer is Human*, St. Martin's Press, New York, 1986, 224.
2. Provision and Use of Work Equipment Regulations 1992; Workplace (Health, Safety and Welfare) Regulations 1992; Management of Health and Safety at Work Regulations 1992; Manual Handling Operations Regulations 1992; Personal Protective Equipment at Work Regulations 1992, and the Health and Safety (Display Screen Equipment) Regulations 1992.

Acknowledgements

We should like to express our appreciation to Dr. John Bond for reading through a first draft of this volume and providing many useful comments and suggestions for improvement. Thanks are also due to Lithium Corporation of Europe (a company of FMC Corporation), and in particular Mr. R. M. Patel, for constructive comments on the sections dealing with the pyrophoric hazards, and industrial control measures, relating to butyl lithium.

P. A. C.
C. J. M.

Explosion and fireball that can be generated from an LPG cylinder in the back of a closed van (Courtesy Health and Safety Executive)

List of Abbreviations

ACGIH	American Conference of Governmental Industrial Hygienists
AIT	autoignition temperature
AZDN	azo-diisobutyronitrile
BCME	bischloromethyl ether
BLEVE	boiling liquid expanding vapour explosion
BOD	biological oxygen demand
BSE	bovine spongiform encephalopathy
BSI	British Standards Institute
CEN	Comité Européen de Normalisation
CENELEC	Comité Européen de Normalisation Electrotechnique
CFC	chlorofluorocarbon
CIA	Chemical Industries Association (U.K.)
COD	chemical oxygen demand
COSHH	Control of Substances Hazardous to Health Regulations 1988 (U.K.)
CTS	constant temperature stability
DBCP	dibromochloropropane
DFP	diisopropyl fluorophosphate
DMAC	dimethylacetamide
DSC	differential scanning calorimetry
DTA	differential thermal analysis
ECL	exposure control limit
EGDN	ethyleneglycol dinitrate
EPA	Environmental Protection Agency (U.S.)
ESD	electrostatic discharge
FDA	Food and Drugs Administration (U.S.)
GU	glycine units

HASAWA	Health and Safety at Work etc. Act 1974 (U.K.)
HAZOPS	Hazard and Operability Study
HEPA	high efficiency particulate air
HSC	Health and Safety Commission (U.K.)
HSE	Health and Safety Executive (U.K.)
IBA	iminobisacetonitrile
Ig	immunoglobulin
IR	infrared
ISO	International Standards Organisation
LEL	lower explosive limit
LNG	liquefied natural gas
LPG	liquefied petroleum gas
MCS	multiple chemical sensitivity
MEK	methyl ethyl ketone
MEL	maximum exposure limit
MIC	methyl isocyanate
MIE	minimum ignition energy
NFPA	National Fire Protection Association (U.S.)
NIOSH	National Institute for Occupational Safety and Health (U.S.)
ODMS	oligo(dimethylsiloxane)
OEL	Occupational Exposure Limit
OES	Occupational Exposure Standard
OSHA	Occupational Safety and Health Administration (U.S.)
PAH	polyaromatic hydrocarbons
PbB	level of lead in blood
PBN	phenyl-β-naphthylamine
PbU	level of lead in urine
PCB	polychlorinated biphenyl
PDMS	poly(dimethylsiloxane)
PMMA	poly(methyl methacrylate)
PPE	personal protective equipment
PTFE	poly(tetrafluoroethylene)
PVC	poly(vinyl chloride)
RTFR	reference temperature for filling ratio
SINOBS	sodium isononanoyl oxybenzene sulphanate
TCDD	2,3,7,8-tetrachlorodibenzo-p-dioxin

TDI	toluene diisocyanate
TEL	tetraethyl lead
TGIC	triglycidyl isocyanate
THF	tetrahydrofuran
TLV	threshold limit value
TNT	trinitrotoluene
TON	threshold odour number
TWA	time-weighted average
UEL	upper explosive limit
UV	ultraviolet
UVCE	unconfined vapour cloud explosion
VC, VCM	vinyl chloride monomer
VDU	visual display unit
VOC	volatile organic chemicals

CHAPTER 19

Physico-chemical principles and safety – II

Some background as to how basic physics and physical chemistry determine the safety, or otherwise, of a particular situation is provided in Chapter 3. Since an appreciation of these fundamental principles is so valuable, and the lessons are not constrained by industry or process application, further examples relevant to the safe handling of chemicals are included in this chapter.

Overpressure through chemical reactions evolving gaseous products

Explosions can arise when the design specification of plant is exceeded as a result of direct gas overpressure or hydraulic overloading.

> In May 1978 an explosion occurred at a liquid petroleum gas (LPG) storage tank farm of a refinery in the USA, causing seven deaths and ten injuries.[1] The cause was rupture of an isobutane storage sphere which had been overfilled and overpressurised due to failure of its level monitoring equipment.

Catastrophic failure can also result from plant being subjected to conditions which affect its integrity, thereby reducing its ability to withstand pressure even below design specifications; e.g. heated, unwetted parts of vessels can become weakened and burst to produce a BLEVE (page 877).

Perhaps less well appreciated is the insidious danger from excessive pressure that can arise indirectly, e.g. from *in situ* generation of gas from 'unanticipated' chemical reactions.

Many liquid phase or heterogeneous solid–liquid or gas–liquid reactions result in gaseous products or by-products. These may be flammable as in the following cases, and as discussed further on page 492.

> Detailed instructions issued to a student required him to use an aqueous solution of 30 g sodium hydroxide. By mistake he used 30 g sodium hydride dispersion, which reacted violently with water evolving heat and hydrogen gas.[2] The gas caught fire but was extinguished with a CO_2 extinguisher. Fortunately the student sustained only a slight hand burn and there was no other damage.

The sodium hydride, which was available for a subsequent experiment, was a commercial product; the bottle bore a warning of the hazard on contact with water but this was not visible from the side showing the name of the compound. (The very different physical appearances of the chemicals should possibly have been distinguishable to an experienced assistant.)

A firm marketed a cleaning fluid, 'pluperfect liquid', which evolved hydrogen in contact with cast iron. A firm of repairers, under the instruction of shipowners, used the liquid to clean a condenser on a ship. An explosion occurred, resulting in damage to the ship, when a workman carrying a naked flame approached the condenser. (Adequate warning of the flammable gas generation potential was not given in the instructions for use and the suppliers were held to be liable.)[3]

Alternatively the products may be toxic, as discussed on page 494.

Arsine gas, a powerful haemolytic poison, may be evolved from the reaction of water or moisture on calcium arsenide present as an impurity in ferrosilicon, in the purification of tin, and in the process of wetting aluminium and phosphate dross. Alternatively it may be generated by the action of water on soots in flues of furnaces in which metals contaminated with arsenic have been processed.[4]

The reactions may be initiated because of the ingress of air or moisture, or because of the presence of contaminants. For example, sodium hydrosulphite is very sensitive to water; thus small amounts of water (even sweat), or contact with a highly humid atmosphere in the presence of air, may cause a fire or ignite nearby combustible materials.

An operator opened a full drum of sodium hydrosulphite and removed a small quantity. On his return just over five hours later the contents of the drum were ablaze.[5] Possible causes of moisture ingress included the fit of the drum lid, moisture in the storage area, splashing of water into the area, or pick-up from the scoop or lid. (Incidentally, since the chemical provides its own oxygen, carbon dioxide and dry extinguishers are not effective fire extinguishing agents. Deluging with water is the recommended method; large amounts of sulphur dioxide are liberated so self-contained breathing units should be worn.)

It is often not realised that pressures sufficient to rupture metal containers, or vessels, can be produced by gas generation from relatively small volumes of reactants.

Some five gallon pails were rinsed into a drum containing solvents, water and approximately 1000 ppm of an active ingredient; the drum was then sealed. One pail had contained lumps of soda ash which settled to the bottom of the drum and were isolated from the acid water by an organic layer. Carbon dioxide liberation by chemical reaction was therefore gradual but after two days the internal pressure blew the top off the drum splattering liquid all over the area.[6]

$$Na_2CO_3 + 2H^+ \rightarrow CO_2 + H_2O + 2Na^+$$

The problem of overpressurisation appears to be more significant on a laboratory scale or with portable containers, presumably because pressure relief is not generally provided. However, the principles are not scale-dependent. Some of the more common causes are considered below.

Reaction with water

As described on page 492, many chemicals react vigorously with traces of water and this can result in overpressurisation of substantially engineered equipment or pipework, or more often of drums or other portable containers.

In the incident involving reaction of thionyl chloride with water in a 2.5 cm steel hose described on page 493 the expansion ratio due to formation of SO_2 and HCl was of the order of 3700 to 1. Calculations showed that one teaspoonful of water would, on reaction, result in a pressure >190 atmospheres in the hose.[7]

The bottom blew off a drum of (presumably organic) isocyanate blend and the drum dented a structural beam. A small amount of water in the drum prior to filling was not detected by a flashlight inspection. The resulting reaction generated carbon dioxide. It was calculated that, dependent on fill volume, only 20–150 g of water were required to generate the 50–75 psig at which rupture occurred; as little as 2 g would result in bulging as shown in Fig. 19.1.[6]

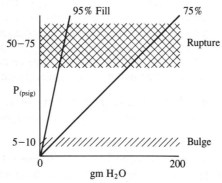

Fig. 19.1 Pressure associated with CO_2 gas generation in an isocyanate drum containing water

$$RNCO + H_2O \longrightarrow RNH_2 + CO_2$$

A similar reaction involving catalytic and exothermic hydrolysis of methyl isocyanate (CH_3NCO) led to a runaway with the formation of carbon dioxide and vaporisation of the boiling isocyanate, in the Bhopal incident (page 883).

The measures to avoid such incidents are described elsewhere in the text and include minimising stocks, proper inspection, preparation and sealing of equipment. Even then, however, the 'engineering' of any processes may also have a bearing on safe processing.

> 600 kg of 'dry' xylene was charged to a reactor. The moisture content had been checked using the Karl Fischer method but, due to inadequate agitation, there was a layer of water beneath it. When almost 300 kg of phosphorus oxychloride was added a violent reaction generated an internal gas pressure of 2 bar causing the relief valve to lift. A number of joints leaked and hydrogen chloride vented into the building.[8]

$$POCl_3 + 3H_2O \rightarrow 3HCl + H_3PO_4$$

Overpressurisation due to the reaction of residual phosphorus oxychloride and water in a drum was exemplified on page 493.

Because of this risk, detailed procedures have been established for washing out drums contaminated with phosphorus oxychloride[9] and also for drums contaminated with phosphorus trichloride.[9,10]

> The potential hazard in washing non-returnable phosphorus trichloride drums will be considerably reduced if it is arranged that only the minimum amount of phosphorus trichloride is left in them after use. However, proper supervision is essential.[9]
>
> Operatives must wear full protective clothing and goggles or a visor. They should avoid breathing PCl_3 fumes and a suitable respirator should be available. The drum is checked as black and labelled phosphorus trichloride. All three plugs are removed from the drum and it is inverted on a tray so that the PCl_3 can drain from the end plug-holes. (This operation is performed away from buildings and people.) Thorough draining is required so that only the minimum of PCl_3, about 100 ml, remains in the drum. The tray is then removed. The drum is kept inverted and supported clear of the ground, e.g. on housebricks. A thoroughly-emptied hosepipe of about 2.5 cm bore is placed in the side hole and fixed so that it will not be displaced when water flow is started. Care is essential to avoid putting a small quantity of water in the drum.
>
> The operator should move at least 10 m from the drum and preferably behind shelter before fully turning on the water supply to the hose. A flowrate of 10 gallons per minute is adequate; it is advisable to check that this is achievable beforehand.

These types of precautions can clearly be applied by analogy, with appropriate modifications, to the washing-out of other containers.

In the foregoing examples the reactions with water were all known to be rapid. Sometimes the hazard stems from a lack of reaction kinetic data or misunderstanding of it.

A 30% aqueous solution of KOCN was prepared in a 50 gal glass-lined steel reactor and heated to 50 °C for three hours to dissolve all the solid. The solution was then stored in a 55 gal HDPE drum, the bung removed and a valve inserted. On hearing a rumble inside the drum the operator vented the container by opening the valve. The drum began to bulge and within minutes it burst at its seam spraying the area and propelling the drum cover 3 m in the air. Fortunately no-one was injured.[11]

The nature of KCNO hydrolysis is dependent on KCNO concentration, temperature and pH and whether the system is open or closed, e.g.

$$NCO^- (aq) + 2H^+ (aq) + H_2O (l) \rightarrow NH_4^+ (aq) + CO_2 (aq)$$

$$KCNO (s) + 2HCl (aq) + H_2O (l) \rightarrow KCl (aq) + NH_4Cl (aq) + CO_2 (g)$$

$$OCN^- + 2H_2O \xrightarrow{HO^-} NH_3 + HCO_3^-$$

(Whether open or closed can affect, e.g. gas concentration and pH.)

Studies indicated that even under isothermal conditions, once underway the hydrolysis rates accelerated, peaking in 30 h at 25 °C and in 6.7 h at 40 °C. The gas pressure in the drum would have exceeded 25 psi after 1 h at 50 °C. Following the runaway the potential final gas pressure would be 470 psi at 118 °C. In addition the drum would be subjected to expansion forces from the swelling liquid.

Reaction due to contaminants

Reactions with, or initiated by, traces of contaminants have been a common cause of overpressurisation incidents. Such contaminants may inadvertently be present in storage containers.

After several months in storage a sealed bottle of a mixture of unused potassium dichromate and sulphuric acid exploded. The possibility was proposed that failure was caused by carbon dioxide pressure arising from trace contamination by carbon compounds.[12,13]

A diazonium compound in sulphuric acid had been left standing for 45 hours at ambient temperature in a closed, jacketed, enamel-lined vessel. The agitator was started. After 25 minutes the pressure was 2.8 kg/cm^2 and rising fast and the temperature was 160 °C. The operator immediately opened the vent valve and turned on cooling water to the jacket but, within a minute, the chargehole cover was blown off and the contents of the vessel blown out. Diazo solutions of this type are stable at ambient temperatures and the mixture must be heated to at least 115 °C before decomposition becomes spontaneous. At higher temperatures decomposition becomes violent. For diazo solutions contaminated with as little as 3% water, however, decomposition becomes spontaneous at 100 °C. The accident was attributed to rain water having entered the vessel during a violent thunderstorm via a vent pipe, forming an unmixed layer of water on top of the contents. On starting the stirrer the heat of mixing was sufficient to raise the temperature of the contaminated diazo solution to above 100 °C and a violent, spontaneous decomposition occurred.[14]

Note that diazonium salts such as benzene diazonium sulphate are converted to the phenol and nitrogen gas when boiled or steam distilled under acid conditions:

$$C_6H_5-N_2HSO_4 + H_2O \rightarrow C_6H_5OH + N_2 + H_2SO_4$$

Alternatively, contaminants may be present in transfer lines,[15] e.g. as in the case of the flexible metal transfer hose referred to earlier, and in the following case.

When an old acid line was dismantled, as the second joint was opened acid was ejected under pressure and splashed two workmen. The pipeline had sustained acid attack causing a build-up of gas pressure in some sections and blockage with sludge in others.

Alternatively, blockages may result in underpressurisation hazards.

Incidents have also arisen due to catalysis of a chemical reaction by traces of iron from the storage/transport vessel. (This was a factor in the Bhopal incident described on pages 883 and 1094.) Some runaway explosions result not from *generation* of gas but because of pressure rise of trapped gas (such as air in the following case history) due to the increase in temperature and the expansion of the liquid/solid reactants and products.

A pressure-relief valve lifted on a rail tank car containing acid-washed methacrylic acid. The area was secured and a remote fire monitor was set up to knock-down the discharging vapour; adjacent tank cars were removed. The relief valve closed several hours later but the water spray was left on covering the dome. The tank car exploded 20 hours after the incident was noted.[16] The likely cause was contained polymerisation of the methacrylic acid due to a combination of iron contamination from corrosion of the stainless steel tank by a lower acid layer normally present in acid-washed methacrylic acid (previously phenolic resin-lined tanks were used) and a reduced hydroquinone inhibitor level.

Sometimes reaction is promoted by impurities arising from previous use of a liquid or solution, i.e. when returned to prolonged storage.

A decalcifying liquid consisted of a mixture of 10% nitric acid and formalin. After use it was returned to a closed bottle. Interaction promoted by dissolved impurities resulted in overpressurisation after a few hours.[17]

Clearly, then, on a production scale particular attention has to be given to the storage of recycled, intermediate or 'waste' liquid streams.

Electrolytic corrosion

One hazard from electrolytic corrosion stems from the generation of internal pressure in ever-weakening closed equipment (page 103).

> A 200 litre galvanised steel drum was filled with a mixture of water, methanol and butyric acid which had been neutralised to pH 9.8 with ammonia. During the night it burst spontaneously; the bottom was blown off and the drum was projected onto the loading ramp. The cause was dissolution of the zinc coating with evolution of hydrogen which occurs under alkaline conditions, or under neutral conditions in the presence of ammonium salts.[18]

$$Zn + 2H_2O \xrightarrow{OH^-} Zn(OH)_2 + H_2$$

> During training an apprentice was instructed to erect some pipework for demonstration purposes. He required a blank flange and decided to use one from redundant piping. When he attempted to remove it he was sprayed in the face with acid. The redundant section had been in service with concentrated sulphuric acid and internal corrosion had generated hydrogen which was under pressure.[19]

$$Fe + H_2SO_4 \rightarrow FeSO_4 + H_2$$

Sometimes a combination of hydrolysis and corrosion has been involved:

> The presence of moisture in a sealed iron drum containing diethyl sulphate resulted in hydrolysis to produce sulphuric acid. This acid corroded the drum generating hydrogen, the pressure of which caused it to rupture.[20]

Reaction with construction materials

The possibility of gas evolution due to reaction with construction materials may be easier to foresee. For example, nitric acid in contact with copper will generate nitric oxide, one proposed cause of the failure of a canned pump in which 89 per cent nitric acid may have penetrated the sealed copper windings of the motor,[21] although this was subsequently challenged.[22]

$$3Cu + 2HNO_3 \rightarrow 3CuO + 2NO + H_2O$$

However, problems may arise only if there is some inherent fault in, e.g. a container:

> Winchester bottles of fuming nitric acid may develop internal pressure if the inert liner to the plastic cap is faulty or is missing. The mechanism is oxidation of the plastic to gaseous carbon dioxide and reduction of the acid to nitrous fumes; this gas pressure may cause failure.[23]

Decomposition

Slow decomposition may also result in pressure generation.

Slow decomposition in storage of 98–100% formic acid results in the liberation of carbon monoxide gas. This has caused sealed glass containers to rupture.[24] With no leakage of gas, a full 2.5 litre bottle would develop a pressure in excess of 7 bar during one year at 25 °C.

$$HCO_2H \rightarrow CO + H_2O$$

Hydrogen peroxide is commercially available in a variety of strengths, all of which undergo continuous exothermic decomposition to water and oxygen:

$$H_2O_2 \text{ (l)} = H_2O \text{ (l)} + O_2 \text{ (g)} \; [-23.5 \text{ kcal/g mol}]$$
$$H_2O_2 \text{ (l)} = H_2O \text{ (g)} + O_2 \text{ (g)} \; [-13.0 \text{ kcal/g mol}]$$

Although the rate of isothermal decomposition of commercial hydrogen peroxide is well under 1 per cent per year, the amount of oxygen evolved can be significant, as illustrated by Table 19.1.[25]

Table 19.1 Generation of gas on H_2O_2 decomposition*

Solution strength (wt.% H_2O_2)	Isothermal decomposition: volume of oxygen	Adiabatic decomposition volume of oxygen and steam
100	512.0	6643
80	378.8	3870
60	263.3	1672
40	163.5	932
20	76	296
10	36.6	45

* Volume of oxygen evolved isothermally, and volume of total gas (oxygen plus steam) evolved adiabatically, per unit volume of hydrogen peroxide solution at 1 atm and initial temperature of 20 °C.

A screw-capped Winchester bottle containing 35% hydrogen peroxide solution exploded after two years due to internal pressure from liberated oxygen.[26]

Twenty tons of 70 per cent hydrogen peroxide decomposing at the slow rate of 0.1 per cent per year would still produce 13 litres of gas per day, which in a sealed tank, 95 per cent full, could produce a pressure rise of 1 atm in about two months. The rate of gas evolution increases with temperature and a range of catalysts. Each 10-degree rise in temperature increases the decomposition rate by a factor of 2.3. The homogeneous decomposition could also increase by an order of magnitude in the presence of contaminants such as vanadates, chromates, tungstates, molybdates, bromides, iodides and salts of iron and copper, at the ppm level. Decomposition is also catalysed by active surfaces such as copper, mild steel, palladium, platinum, etc. Because the heat of reaction may not be easily dissipated, the material may boil and generate vast volumes of steam and oxygen. Hence considerable pressure build-up could occur in just a few days. Data on the adiabatic decomposition are also included in Table 19.1.

Heterogeneous decomposition is a surface phenomenon, so small plant items such as pumps, pipes and valves are particularly vulnerable because of the high surface-to-volume ratio. Furthermore, because pipelines conveying liquids usually have no free gas space, rapid overpressurisation can occur in stationary systems. Accelerated decomposition of peroxide may arise from the heat generated when running centrifugal pumps against closed heads (i.e. without flow). As demonstrated by Table 19.2[25] a relatively high percentage of incidents involving hydrogen peroxide are caused by overpressurisation due to inadequate care, housekeeping and lack of venting.

Table 19.2 Causes of hydrogen peroxide accidents (some incidents fall into more than one category)

Cause	Per cent
Spillage	38
Rapid decomposition	36
Fire	34
Pressure burst	13
Runaway reaction	12
H_2O_2/organic vapour-phase explosion	14
Condensed-phase explosion	5

The appropriate precautions with materials of limited stability therefore include the following.
- The selection of materials of construction compatible with compound use, and which have been properly cleaned and passivated.
- Adequate date-labelling and inventory control.
- Provision of automatic pressure relief valves on pipes and vents on tanks (one study recommends a minimum vent area of 200 cm^2 per ton of 100 per cent hydrogen peroxide).
- High standards of housekeeping (e.g. to avoid contamination).
- Immediate dilution of spillages and prevention of accumulation in pits, drains, etc.

The thermal degradation of otherwise stable materials can also lead to a dangerous build-up of gases.

> Three men were repairing an extruder used in production of polyethylene film. As they loosened flange bolts on the extruder head, the housing suddenly burst with a loud bang. The men suffered burns and grazing but fortunately escaped more severe injuries.[27]

When one of the heater elements failed on the extruder outlet the equipment jammed. Simultaneously, a second heater element increased output to compensate for the failure of the defective element. The explosion was attributed to the build-up of ethylene; the thermal degradation of the polyethylene above 260 °C produces a range of saturated and unsaturated hydrocarbons of varying chain length down to C_2.[28]

Self-initiated reaction

Reactions which may be self-initiated may also result in pressure generations.

A chemical polishing solution consisted of one volume of nitric acid, one volume of hydrofluoric acid and two volumes of glycerol. After storage in a sealed plastic container for four hours oxidation of the glycerol generated sufficient gas to overpressurise and rupture the container.[29]

Some months after a biocide dosing unit was installed on a cooling water unit at a research establishment one of the dosing tanks ruptured due to overpressurisation. The water was found to have contained approximately 14% of ethylene glycol, as an antifreeze, and it was subsequently demonstrated that after a period of several days this mixture would react with the biocide releasing bromine gas.[30]

In May 1979 a vessel ruptured explosively during preparation of a catalyst, killing two operatives and injuring fifty. Damage to property amounted to $2 million.[31]

The multi-stage process involved chlorination of powdered iron suspended in an agitated, partially-chlorinated organic solvent. In the first stage solvent was heated to 80 °C and the metal powder added. Chlorine was then bled-in so as to maintain a nominal pressure until the temperature reached 140 °C. The chlorine supply was then adjusted so as to maintain the temperature, and the reactor cooled externally to remove heat of reaction. Upon completion the resulting ferric chloride solution was transferred to storage. Pressure was controlled by venting to a scrubber and pressure safety valves were provided. In the later stages of the preparation of one batch of catalyst, pressure control problems were encountered and the temperature continued to rise despite full cooling. Pressure also rose even when the chlorine supply was shut off. Temperature alarms were activated at 150–155 °C and pressure alarms at 22 psi (150 kPa) when the temperature had reached 194 °C. The vent header pressure alarm sounded one minute later, indicating pressure relief discharge at 75 psi (517 kPa); the explosion then occurred.

Metal thinning suggested reactor failure at 800 psi (5500 kPa) compared to 75 psi design pressure. Safety tests indicated a gassy exotherm attributed to the (presumably Friedel–Crafts type) reaction of the —CCl_3 moiety with an H atom in an unchlorinated position of the aromatic ring of the solvent, and the formation of gaseous hydrogen chloride, thus

$$2 \text{ Ar}-\underset{\underset{\text{Cl}}{|}}{\overset{\overset{\text{Cl}}{|}}{\text{C}}}-\text{Cl} \longrightarrow \text{Ar}-\overset{\overset{\text{Cl}}{|}}{\underset{\underset{\text{Cl}}{|}}{\text{C}}}-\text{Ar}-\overset{\overset{\text{Cl}}{|}}{\underset{\underset{\text{Cl}}{|}}{\text{C}}}-\text{Cl} + \text{HCl}$$

Studies also confirmed that the cooling capacity was inadequate to cope with this exothermic reaction.

Overpressurisation of containers is particularly prone to occur with process wastes, because the process control may be less rigorous than with raw materials and products, and can have serious consequences if no pressure relief – however rudimentary – is provided.

Stencilled instructions on a waste drum containing flammable monomers said to leave the bung loose. However, it was not loose and when exothermic self-polymerisation occurred the internal pressure blew off the lid, damaging some plastic piping. The area was filled with toxic flammable vapours and the building had to be evacuated.[6] Subsequently all drums were properly labelled to specify which chemicals may be added to them. Some drums are colour-coded and segregation is preferred to avoid 'simple mistakes', e.g. due to misreading a label. Access to waste drums is limited to trained personnel and consideration given to relief bungs with proper ventilation.

Many unstable materials can be stored safely for extended periods provided the temperature is controlled.

Pure pyruvic acid is stable on a long-term basis if properly refrigerated, and if light and air are excluded. Internal pressurisation of a bottle of analytical grade acid, due to carbon dioxide gas, followed storage in a laboratory at 25 °C and caused it to burst.[32] This may be accelerated by enzymic catalysis, i.e. from ingress of airborne yeasts.

$$CH_3COCO_2H + [O] \rightarrow CH_3CO_2H + CO_2$$

The appropriate storage temperature, and the reliability of the coolant supply, are obvious considerations on a plant scale – as indeed is failure of power to the refrigerator on a laboratory scale.

Many more examples of overpressurisation could be cited. One list of gas evolution cases, including combustion incidents, has been published in ref. 33. However, as was emphasised in Chapter 10, reliance on lists of incompatible chemicals, or experience, or 'memory' is not a proper substitute for a systematic hazard evaluation in all such cases. Reference to the literature and data bases noted on page 1047, or in special cases use of the procedure outlined in Chapter 20 is recommended.

The need for such care is exemplified by the example quoted on page 770 involving the explosive properties of boron tribromide on contact with water. The defendant company who marketed the chemical for industrial use did not know of this hazard which was not referred to in a standard work on the industrial hazards of chemicals, nor in three other modern works they had apparently consulted. However, the property had been described in scientific literature dating from 1878 and it was held that research with the exercise of reasonable care would have revealed it.[34]

In the end, however, an unforeseeable chain of events may still result, albeit rarely, in an explosion. In the following case the stability of the reactions was investigated in accordance with the norms ten years previously; the effect was only demonstrated using more sophisticated techniques after the accident.

A reactor exploded due to an unexpected exothermic reaction during the manufacture of 2,4-difluoronitrobenzene from 2,4-dichloronitrobenzene by reaction with potassium fluoride in the presence of dimethylacetamide (DMAC) as solvent.

Halex reaction

2,4-dichloronitrobenzene + 2 KF $\xrightarrow{\text{DMAC}}$ 2,4-difluoronitrobenzene + 2 KCl

The plant was partially destroyed, with missiles and blast damage extending to 500 m and secondary fires. Six operators were injured and one subsequently died.[35,36]

The runaway was caused by contamination of the recycled solvent with acetic acid, formed by reaction of the solvent with water which passed into a tank in which the reaction product was stored prior to distillation.

$$CH_3CONMe_2 + H_2O \rightleftharpoons CH_3COOH + HNMe_2$$
(DMAC)

(The azeotrope of DMAC and acetic acid has the same boiling point as DMAC and hence acetic acid was recycled with the recovered solvent.)

Water contamination on previous occasions had been removed at the start of the batch distillation but this time the amount was greater so that it formed a separated layer, favouring acetic acid production. Moreover, although the equilibrium in the acetic acid formation equation (shown above) is well to the left, any unconverted raw material acts as a scavenger and by removal of the dimethylamine formed with the acetic acid moves the equilibrium to the right.

2,4-difluoronitrobenzene + HNMe$_2$ $\longrightarrow$ (2-NMe$_2$-4-F-nitrobenzene) + HF

In the runaway reaction itself potassium acetate is formed initially by reaction of potassium fluoride and acetic acid; this reacts with the 2,4-dichloronitrobenzene to form acetoxychloronitrobenzene which is unstable under the reaction conditions and reacts further to give, e.g. ketene, carbon dioxide, polyaryl ethers and tars.

Generation and dispersion of droplets

Many chemical engineering and physical processes result in droplet dispersion and these may be in the respirable range, i.e. aerosols with drops <7 μm. A general indication of the drop sizes associated with different processes is given in Fig. 19.2.

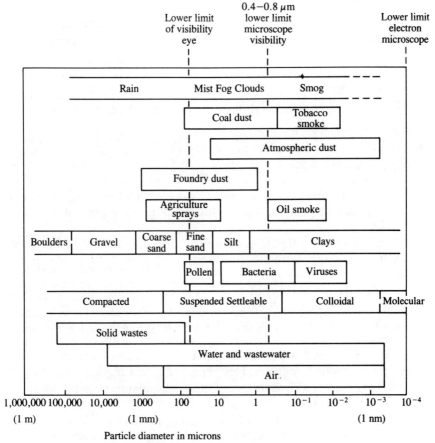

Fig. 19.2 Typical particle size ranges

If the drops are of solutions rather than pure liquids, evaporation will result in nuclei of suspended and dissolved solids. The smaller the solid content of the initial solution the smaller the resultant nuclei for a given drop size. The consequences of smaller nuclei are:
- An increased probability of dispersion
- Deeper penetration into the respiratory tract on inhalation (page 192)

With flammable liquids the risk of fire or explosion is also increased by atomisation.

A mist of kerosene which has a flash point of 38 °C may be ignited at temperatures as low as −17 °C. (The effect of increased surface area in heterogeneous reactions is discussed on page 115.)

Therefore mist and spray generation into the working environment is generally to be avoided.

Laboratory processes

Processes which generate aerosols during laboratory procedures are, as noted on page 1798, of particular relevance to microbiological techniques.

In the approximately 80% of cases of recorded human laboratory infections not associated with an accident, unsuspected aerosol generation whilst performing standard laboratory procedures has been identified as a factor.[37]

The main processes involved are summarised below (after ref. 38).
- The bursting of a liquid film in bubbles, froth, across bottle mouths and pipette tips and in platinum loops. Droplet generation is enhanced by an air pressure differential across the film. Aerosol dispersion by such film-bursting is possibly the most common cause of laboratory infection and, whilst it can be reduced by careful handling, may present a serious hazard.
- Mixing of gas and liquid in shake cultures, fermenters or test tubes results in frothing. With increased energy input, drops may be expelled directly into the atmosphere.
- Forceful ejection of the contents of a pipette or syringe already containing gas bubbles inevitably results in droplet dispersion. Similarly, forced injection of gas into a liquid, or liquid into a gas, will generate an aerosol.
- Vibration, e.g. the twanging of a hypodermic needle or a platinum loop, results in droplet dispersion. Hence proper care is required in their use. Other apparatus, such as the use of ultrasonic mixers, requires appropriate safety measures.
- Impact of a droplet on a liquid surface may produce an aerosol, as may disintegration due to falling under gravity. (Fall onto an adsorbent surface may prevent this; hence covering an open bench with lint or similar

material moistened with a germicide is recommended for some microbiological work.[37])
- Separation of two moist surfaces produces a filament which disrupts into droplets, e.g. as in withdrawal of the plunger in a syringe barrel or a stopper from a tube.
- Centrifugal force, e.g. as in a laboratory centrifuge, may disperse an aerosol. General precautions with laboratory centrifuges are summarised on page 582, but where toxic materials are involved in centrifuges operated under vacuum, the hermetically sealed rotors and sealed buckets should only be opened in a safety cabinet, and an air filter is required between the bowl and vacuum pump to avoid contamination of pump oil and redispersion as mist.[37] For microbiological work all other types of centrifuge should be operated in ventilated cabinets, or in ventilated rooms with adequate means for decontamination and respiratory protection for the operator.[37]
- 'Sizzling', e.g. when a platinum loop or inoculating tool is flamed or plunged hot into liquid, produces a measurable aerosol. Use of a safety cabinet is recommended.

Production operations

Aerosol generation may occur during production operations because of the following mechanisms.
- Atomisation is deliberately used in some humidification operations, in gas scrubbing and in spray drying. Either pressure nozzles, rotary atomisers or twin fluid atomisers may be used but in most cases the spray is contained within the appropriate equipment. Similarly mists produced in diesel engines or crankcases are normally contained. Escape of aerosol is, however, inevitable to some extent in spray painting, spraying of adhesive, and high pressure jet cleaning operations. Properly designed booths with local exhaust ventilation backed up by appropriate protective clothing and respiratory protection are therefore essential.
- Leakage of liquids transferred under pressure, e.g. from flange joints or pump glands, may occur as an aerosol. Enclosure of flanges may therefore be desirable, e.g. by shielding, if concentrated acids are transferred by pumping, or when transfering sensitising agents.
- Misting commonly occurs on condensation of a vapour, in gas–liquid contacting operations, e.g. distillation and gas absorption, and in evaporation. Various types of demister are available to prevent the emission of such aerosols into the environment, e.g. from atmospheric vents.
- Liquid dispersion by propellants, e.g. using hand-held aerosol cans, inevitably produces a localised concentration of aerosol. The nature of the hazard is then directly related to the liquid being applied; for example, silicone lubricants pose an eye hazard.
- Electrolytic processes may generate mists due to the bursting of minute

gas bubbles at a liquid–air interface. Hence, for example, special precautions possibly involving the use of a foam blanket but certainly including lip extraction are required on chromium-plating tanks.

Fractionation due to vaporisation or condensation

As discussed earlier, condensation can result in fractionation of a mixture of components. For example, any material with a lower boiling point than oxygen may cause fractionation of air (page 645).

> Some pork scratchings were cooled using liquid nitrogen to facilitate mincing. A mixture containing 70% oxygen (boiling point −183 °C) condensed onto the pork. A source of ignition, probably friction in the mincer, caused the pork and oxygen to explode, killing two men.[39]

A further example of oxygen condensation is given on page 1171.

Vaporisation can result in fractionation in an analogous manner. For example, substituted vinyl monomers to which an inhibitor (e.g. hydroquinone, hydroquinone monomethyl ether, tertiary butyl catechol or phenothiazine) is normally added to prevent polymerisation in storage and transport have much higher vapour pressures than the inhibitor. Hence vapour from the monomer will be denuded of inhibitor when condensed out. Vents can thus become blocked by monomer condensing and polymerising in cold weather. Flame arrestors are therefore not recommended on tanks containing substituted vinyl monomers.[40] Similarly some monomers can crystallise at ambient temperature; when this occurs the inhibitor may be excluded from the crystals. On thawing a liquid layer may be formed which is inhibitor-free and subject to low temperature initiation of polymerisation.

> A road tanker containing 13.6 tonnes of glacial acrylic acid burst open with explosive violence. Gelatinous polymer above its ignition temperature was dispersed, blazing, over a 150 m semi-circle. The explosion is believed to have been initiated by thawing of a layer of crystallised acrylic acid by warm water, used to maintain the tank above the freezing point of 14 °C. Crystallisation of the acrylic acid during loading occurred without its attendant inhibitor, normally present to prevent monomer polymerisation. A liquid layer without inhibitor formed on thawing in the tanker; polymerisation started in this layer and consumed the inhibitor in the bulk leading eventually to tank rupture.[40]

The physico-chemical principle is simply that the composition of the vapour above a miscible liquid mixture, or the composition of solid crystallised from a solution, will always differ from the initial feed. This can have important consequences related to chemical reaction, toxicological or flammable hazards. For example, many violent incidents which arose when liquid air was used as a cryogenic liquid were due to the increased content of residual liquid oxygen produced by fractional evaporation during storage. Liquid nitrogen is recommended as a safer coolant but condensation of atmospheric oxygen should be avoided during use.

Liquid–solid phase changes

The change in volume associated with a phase change from solid to liquid was referred to on page 84. However, the manner in which this change occurs at a specific temperature (or over a temperature range) dependent on the composition of the substance, and the fact that for many common chemicals the temperatures are near to ambient temperatures, can also have significant safety implications.

Solidification problems

In general all equipment and pipes intended to handle as a liquid substances which are solid at ambient temperatures should be heated, e.g. by steam tracing or electrically. It may be that such heating is only necessary to prevent freezing in cold weather; e.g. phosphorus oxychloride (page 1094) has a freezing point of 1.22 °C and all pipes are normally steam traced. In special situations, e.g. with molten white phosphorus, concentric pipes may be used with hot water flowing through the outer annulus.

> On a pilot plant in which vegetable oil was hydrogenated both the raw material and product had melting points of 20 °C. Samples had to be taken from a line joining two converters operating at 270 °C and 250 bar. The sample was transferred via a small bore pipe into a pot by opening an isolation valve and cracking open a fine adjustment valve; with both valves closed the pot was then drained into a can. The sample line and pot were heated; the pot had a 2.5 cm bore, 2.75 m long, unheated vertical vent pipe.
> After two weeks a sample could not be obtained. An attempt was therefore made to clear a suspected solids blockage by opening the isolation valve and pot drain valve and gradually opening the fine adjustment valve. As this was in progress the vent line which, like the pot, was not clamped moved backwards – presumably initiated by a sudden flow of gas as a blockage cleared. The pipe hit the chargehand involved on the head, causing fatal injuries.[41]
> The similarities between this incident and that described on page 90 have been highlighted; they apparently occurred in the same factory at locations only 200 m apart, but this one occurred twenty years earlier.[41]

More commonly blockages may create overpressurisation hazards, as in the case described on page 90 in which an unheated flametrap blocked, or underpressurisation hazards.

> A molten cyanotic material was delivered around a factory in a lagged, steam-heated tanker. To discharge it nitrogen pressure was applied to blow material up a dip pipe into a catchpot from whence it was pumped to storage; once pumping was established the nitrogen was turned-off and the tanker vented through a separate branch to atmosphere.[42]

One cold January night persistent blockages of crystallised material occurred and the nitrogen inlet/venting arrangement had to be dismantled several times to steam them clear. When pumping was achieved the system was left in operation for an hour until it was noticed that the level in the storage vessel had stopped rising; the tanker was then found to have collapsed and buckled. The cause was simply blockage of the vent line and pressure/vacuum relief valve on the tanker by sublimed product; the discharge pump then created a vacuum in the tanker which it was not designed to cope with.

In all cases where blockages must be freed there should be proper isolation (page 978), appropriate protective clothing, particularly head and face protection, and a procedure which mitigates the effects of possible pressure release, e.g. when breaking a joint by slackening the nuts furthest away first and springing apart the joint faces.

Thermal expansion or contraction of solids

Hazardous situations which may be associated with the volume changes of liquids with temperature were summarised on page 99 and with gas on page 95. The analogous phenomenon of thermal expansion or contraction of solids, e.g. with construction materials, is a matter of elementary knowledge yet the magnitude of its effect has often been overlooked. It is, of course, a property of value in assembly and construction operations, where the heating or cooling procedures themselves may introduce quite different risks.

In two separate incidents men involved in 'sweating' tiller heads onto the rudder stock in the steering engine compartments of ships using open flames were hospitalised due to the inhalation of 'nitrous fumes'.[43]

Some of the risks created are indeed barely credible.

It was necessary to fit a shaft into a bearing in a confined space. Since it was a tight fit the fitters decided to cool the shaft and heat the bearings, so demonstrating a good understanding of the topic under discussion. Unfortunately they elected to cool the shaft by pouring LPG onto it and to heat the bearing with an oxyacetylene torch.[44]

Coefficients of expansion

For a given material the amount of its linear expansion, or contraction, in one direction of measurement is directly proportional to the change in temperature and its original linear size. The linear size includes every linear dimension, e.g. length, diameter, circumference. Thus

$$\text{Change in length} = \alpha L \Delta t$$

where α = thermal coefficient of linear expansion;
L = original linear measurement;

Δt = change in temperature.

Coefficients for a range of materials of construction are given in Table 19.3.

Table 19.3 Thermal coefficients of linear expansion

Material	Coefficient of expansion per °C
Aluminium	0.000 025
Brass	0.000 018
Brass (Admiralty)	0.000 021
Bronze	0.000 018
Copper	0.000 017
Glass (ordinary)	0.000 008 5
Glass (Pyrex)	0.000 003
Invar	0.000 000 9
Iron (cast)	0.000 011
Lead	0.000 029
Platinum	0.000 009
Silver	0.000 021
Steel (mild)	0.000 012
Tin	0.000 025
Zinc	0.000 029

Changes in normal service temperature

Because of the differences in the coefficient of thermal expansion between different construction materials, or if a construction material is rigidly constrained, it is important for the designer to allow for all possible changes in temperature.

In 1974 a 137 m plastic gas main laid in a street alongside a hotel in Freemont, Nebraska, was affixed to metal gas mains at each end using compression couplings. In the winter, when the ground temperature fell, it contracted and almost pulled out of the couplings. In the next winter, which was colder, it contracted about 7.6 cm and came out, allowing gas to leak into the hotel basement where it exploded.[45]

Numerous examples of pipe support failures resulting from thermal expansion are given in ref. 51. Furthermore, problems can also arise with freely supported piping.

A 2 cm branch was fitted to the underside of a 25 cm pipeline carrying oil to 300 °C. The pipe rested on a girder such that the branch was approximately 13 cm from it. Expansion in service brought the branch into contact with the girder and knocked it off.

Therefore thermal movements of piping or equipment may be taken care of by some combination of

- expansion bellows[46]
- expansion joints
- designed expansion loops or offset legs
- careful positioning of equipment to utilise the inherent flexibility within pipework
- accommodation of the stresses by control of expansion direction via supports, guides, anchors, piles, etc.[47]
- decreased pipe wall thicknesses to reduce thermal forces and moments.

Thermal movements may, of course, also arise inadvertently during construction:

> In order that it could be painted, scaffolding was erected around a distillation column whilst it was still hot. On cooling down the scaffolding distorted.[34]

Thermal expansion of liquids

The propensity for thermal expansion of a liquid to cause build-up of pressure when sealed within piping or equipment was noted on page 99. The main sources of heat for thermal expansion include steam tracing, solar radiation, adjacent process fluid (e.g. a heat exchanger blocked in on the cold side but not on the hot side), and fire. Even a 16.6 °C rise in temperature may raise the pressure within enclosed plant above the design pressure; for example, the associated pressure increases with some common liquids are summarised in Table 19.4.[16]

Table 19.4 Pressure increase of liquids due to thermal expansion on 16.6 °C temperature rise

Liquid	P (psia)
Acetic acid	3200
Acetone	3260
Aniline	5190
Benzene	3860
n-Butyl alcohol	2590
Carbon tetrachloride	3310
Methyl alcohol	3900
Petroleum (s.g. = 0.8467)	2340
Toluene	3340
Water	1100

Calculation enables the relief volume and potential pressure increase to be calculated for any situation.[48] In general a thermal relief valve is not considered necessary if

- the equipment or pipeline cannot be blocked in and heated during normal operations;
- the liquid temperature is controlled by electrically heated tracing provided with a temperature cutout;

- the line is normally in service and can be vented and drained for shutdown;
- the line is above ground and <30 m in length (because the relief volume is small, any increase in volume would be offset by allowable leakage through the block valves);
- one end of the line is blocked in by a check valve (when leakage is considered sufficient).

If protection is necessary against overpressurisation the methods available are, in decreasing order of reliability:
- installation of an adequately sized relief valve;
- adding an open bypass around one of the block valves, provided some degree of backflow is permissible;
- placing a check valve around one of the block valves, provided leakage through the check valve can be tolerated;
- establishing a procedure which ensures that liquid is drained before the equipment or piping is blocked in.

Any relief valve should preferably discharge into a tank. Any flammable liquid at a temperature higher than its flash point, or any liquid at a temperature above 580 °C, or any toxic liquid, should discharge into a closed system, e.g. a special vessel or a flare header.

Chemical reactions: conversion versus time

The rates of chemical reactions are dependent on both the concentration of each reactant and the reaction temperature, as explained on page 91. Intentional catalytic effects, or unexpected catalytic effects due to contaminants, are also important in effectively reducing the relevant activation energy E (referred to on pages 93, 1095 and 1171).

From an excellent review of UK industrial incidents involving thermal runaways the principal chemical processes were identified as those in Table 19.5.[49] The prime causes of these incidents were related to process chemistry or plant design and operation as shown in Table 19.6.

A guide to the technique for evaluating chemical reaction hazards and interpretation of the resulting data with respect to process operation and plant design is given in ref. 50.

The main precautions for preventing thermal runaways include control of temperature, mixing of reagents in solvents, containment, venting and monitoring with provision in emergencies of pressure relief, dumping/quenching with compatible medium (e.g. with water where appropriate) or inhibition (e.g. with free-radical scavengers such as tertiary butyl catechol to halt polymerisation of certain monomers).

Effect of concentration

The Law of Mass Action governs the effect of reactant proportions and concentrations upon reaction rate; viz. the rate is proportional to the active

Table 19.5 Number of incidents per specified chemical process* from analysis of thermal runaway incidents[49]

Chemical process	Number of incidents
Polymerisation (including condensations)	64
Nitration	15
Sulphonation	13
Hydrolysis	10
Salt formation	8
Halogenation (chlorination and bromination)	8
Alkylation using Friedel-Crafts synthesis	5
Amination	4
Diazotisation	4
Oxidation	2
Esterification	1
Total	134

* Due to lack of information it was possible to identify the chemical processes involved in only 134 of the 189 incidents.

masses of the reacting substances. Hence unintended, or ill-considered, changes in the proportion and/or concentration of reactants can result in an uncontrollable reaction.[33] Examples are given on pages 497, 512 and 1145.

Effect of time and temperature

Equation 3.10 exemplifies how the rate of conversion is crucially dependent upon temperature. Thus at ambient temperature many reactions are quite slow.

> A porter sustained a very serious hand injury whilst cleaning a reagent bottle containing a white isopropyl ether precipitate. The bottle exploded resulting in a traumatic amputation of one finger, one-half or another finger, and multiple fractures of the left hand.[51] An attempt had earlier been made to dissolve the precipitate over a two-day period in an area not normally used by the porter for washing glassware. It was several days later when it exploded, having been collected by the porter.

Other types of reactions, such as the following, have a significant rate at ambient temperature.
- Catalysed reactions.

> Whilst disposing of waste solvents used for chromatography, chloroform was added to a residue bottle containing other solvents including acetone. There was a strong exothermic reaction and a few seconds later the bottle exploded with considerable violence. Flying glass injured two people.[52] (It was subsequently ascertained that chloroform will undergo a highly exothermic condensation with acetone to form 1,1,1-trichloro-2-hydroxy-2-methyl propane.

Table 19.6 The prime causes which led to overheating and eventual thermal runaway* from analysis of thermal runaway incidents[49]

PROCESS CHEMISTRY

Reaction chemistry and thermochemistry

34 of the incidents are attributable to little or no study, research or development being done beforehand, with the result:

No appreciation of the heat of reaction on which to base cooling requirements for the reactor (scale up).	8
The product mixture decomposed.	7
Unstable and shock-sensitive by-products were produced.	6
All reagents were added simultaneously at the start of the reaction, when semi-batch reaction would have been appropriate.	4
Unintended oxidation occurred instead of nitration.	3
The reaction was carried out with reactants at too high a concentration.	2
The reaction was carried out at too low a temperature resulting in accumulation of reactants and subsequent en-masse reaction.	1
The reaction accelerated due to:	
catalysis by materials of construction of the reactor.	1
unsuspected autocatalysis.	1
a phase change of the product (to the vapour state).	1
Total	34 (20%)

Raw material quality control

15 of the incidents are attributable to the use of out-of-specification materials:

Water contamination.	9
Other impurities.	5
Changed specification; an inhibitor should have been used on start of new supply but this change was not recorded in instructions.	1
Total	15 (9%)

PLANT DESIGN AND OPERATION

Temperature control

Failure to control steam pressure or duration of steam heating (includes one case of improper use of steam to unblock vessel outlet, causing product to decompose).	6
Probe wrongly positioned to monitor reaction temperature.	6
Failure of temperature control system (leading, for example, to cooling water being automatically shut off; heating oil overheating; steam valve remaining open).	7
Loss of cooling water (not monitored) (reactor 3, condenser 2).	5
Error in reading of thermometer or chart recorder.	4
Failure to provide sufficient separation distance between reactor and adjacent hot plant.	2
Too rapid heating at reaction initiation.	1
Thermocouple coated with polymer giving slow response.	1
Total	32 (19%)

Agitation

Inadequate stirrer specification.	4
Mechanical failure, for example, when stirrer blades sheared off due to solidification of the 'heel' from the previous batch. Although an overload switch was fitted, the motor was too powerful for the paddle securing bolts.	3
Operator either failed to switch on agitator or switched it on too late, causing accumulation followed by en-masse reaction.	6

Table 19.6 (continued)

Loss of power supply.	2
Agitator stopped by operator to make an addition (localised high concentration caused liquor to boil and erupt).	2
Total	17 (10%)

Mischarging of reactants or catalysts

Overcharging. (Includes 2 cases of overcharging a catalyst and 1 where the metering device was faulty. In 5 cases the total volume of the reaction mixture was incorrect and the cooling capacity of the reactor was inadequate. In the other 7 cases the reaction mixture contained the wrong proportions of reactants.)	12
Too rapid addition.	8
Wrong sequence of addition.	4
Wrong material.	5
Undercharging.	3
Improper control (use of hosepipe).	2
Addition too slow.	1
Total	35 (21%)

Maintenance

Equipment leaks (scrubber 1, valves 3, cooling pipes/jacket 3).	7
Blockages (vent pipes 2, transfer pipes 3, separator 1).	6
Condenser solvent-locked because valve in reflux return line was closed following shutdown for maintenance.	3
Residues from previous batch.	2
Water in transfer lines (including one case of water siphoning from quench tank).	3
In situ replacement of closures (cracked sight-glass 1, cover plate 1) during course of reaction.	2
Unauthorised modifications.	1
Loss of instrument air supply.	1
Total	25 (15%)

Human factors

Operator failed to follow written instructions.	4
Product run off before completion.	3
Deviations caused by poor communications at times of staff changeover (change of shift, holiday, sickness).	3
Product filtered at wrong stage of process.	1
Total	11 (6%)

* Due to lack of information it was possible to identify the prime causes for only 169 of the 189 incidents.

$$CH_3.CO.CH_3 + CHCl_3 \rightarrow (CH_3)_2C\begin{smallmatrix}OH\\CCl_3\end{smallmatrix}$$

The reaction is base-catalysed but basic substances may have been present from the chromatographic materials. Also acetone in contact with alumina will condense to yield mesityl oxide and phorone, which may have been involved in the explosion.) Clearly chloroform and acetone, which are common laboratory reagents, should be kept apart and separate arrangements should be made for disposal of residues.

$$\text{CH}_3.\text{C}.\text{CH}_3 \xrightarrow{\text{Al}_2\text{O}_3} \underset{\text{Mesityl oxide}}{(\text{CH}_3)_2\text{C}=\text{CH}.\text{COCH}_3} +$$

$$\underset{\text{Phorone}}{(\text{CH}_3)_2\text{C}=\text{CH}.\text{COCH}=\text{C}(\text{CH}_3)_2}$$

- Oxidations. As mentioned on page 158 it is crucially important that a stream of oxygen at high pressure does not come into contact with fittings or connections contaminated with oil or grease. Otherwise an explosion may result from spontaneous combustion.

  An oxygen cylinder was thought to be nearly empty. The regulator was therefore adjusted to its fully open position and the oxygen pipe connected to the works high-pressure oxygen line. An explosion occurred as the connection was made. Subsequent investigation revealed that during painting of the workshop some oil-bound paint had dripped inside the oxygen pipeline; this constituted sufficient oil to cause an explosion in the presence of oxygen.[53]

- Self-polymerisation of certain monomers.

  Butadiene must be stored at low temperatures to retard the formation of dimers even in the presence of an antioxidant. An increase in dimer content from 0.08% to the permissible limit of 0.2% will take >4 months at −3 °C but only 3 days at 50 °C. (An oxygen-free atmosphere must also be maintained in the butadiene tank, to prevent the formation of unstable peroxides, and the liquid circulated to prevent stratification and hence polymerisation.)[54]

Elevation of temperature above ambient introduces consequential hazards.

An explosion occurred in a bag filter positioned after a rotary dryer in a recirculating system for drying a powdered ABS-type polymer. An ensuing fire was extinguished after 20 minutes but the filter, surrounding insulation and electrical wiring and equipment were damaged. If the polymer is held for a long time at an elevated temperature it may undergo self-heating to ignition and this was deduced to be the most likely cause here.[55] (The plant was provided with safety features, of the type summarised on page 1462, to minimise damage from a dust explosion. Explosion venting of the dust collectors to NFPA 68 relieved effectively and an explosion suppression system, using water, prevented spread of fire or explosion to other parts of the system. Nevertheless recommendations were made to improve the effectiveness of the safety systems.)

Prior to the incident described on page 1095 the operating procedure was for KOCN solutions to be prepared and kept below 30 °C even though no hazards had at that time been identified with any particular temperature.[11]

The cover was blown off one drum of humid 1,4-benzoquinone due to internal pressure. Two other drums heated up spontaneously. It was subsequently determined that self-heating which commenced in the range 40–50 °C could lead to spontaneous decomposition at 60–70 °C and that impurities caused a marked reduction in thermal stability.[56]

The contents of a sealed glass reaction tube exploded violently whilst being heated to the reaction temperature of 110 °C in an oil bath. It had been half-filled with 2.7 mole (180 cm^3) of acrylonitrile together with a small quantity of initiator for the polymerisation. A thermal explosion occurred before the final temperature was reached. (The experiment represented a considerable scale-up and the risk had not been correctly assessed.)[57]

As discussed on page 93, any exothermic reaction which has a normal temperature dependence can run away, leading to a thermal explosion. Where the heat of reaction is transmitted through the reactant to the walls solely by conduction and these are kept at a constant temperature, e.g. as with a solid reactant in a metal container in a well-stirred oil bath, the critical condition for a thermal explosion depends upon the shape of the reacting material. For a cylinder, theory indicates that explosion will occur when a dimensionless quantity S for the specific system exceeds 2.0. S is defined as

$$S = \frac{vqa^2E}{\lambda RT^2}$$

where v = maximum rate of reaction at the container temperature, mole cm^{-3} s^{-1};
q = reaction exothermicity, cal mole^{-1};
a = half-thickness (or radius) of reactant, cm;
E = activation energy, cal mole^{-1};
λ = thermal conductivity of reaction mixture, cal cm^{-1} s^{-1} K^{-1};
R = gas constant, cal mole^{-1} K^{-1};
T = vessel temperature, K.

For the acrylonitrile polymerisation the rate at 110 °C was estimated as approximately 2.5×10^{-6} mole cm^{-3} s^{-1}. Taking $q = 20\,000$ cal mole^{-1}, $E = 17\,000$ cal mole^{-1}, $a = 1.6$ cm and estimating $\lambda = 2.5 \times 10^{-4}$ cal cm^{-1} s^{-1} K^{-1} gives $S = 30$, which greatly exceeds the critical value for a cylinder.

It is recommended that where calculations of S are used to assess safe conditions, values of 0.1 should not be exceeded,[57] S_{crit} for a slab theoretically being 0.9. Where the heat is not readily removed from the reactant the critical value of S is even lower (e.g. if the container is of low thermal conductivity material, stands in air, or is surrounded by insulating material).

Batch versus continuous processing

Numerous designs of chemical reactor are in commercial use in which operation may be batch, continuous, semi-batch or semi-continuous.[58,59]

Batch operation is flexible since each reactor can be used to produce different products, and the ability to identify a specific batch may be important for quality assurance. Alternatively batch reactors may be preferred because the interval between batches facilitates cleaning and ensures that no deleterious intermediates accumulate. They are often preferred for new, untried, reaction processes which can be subsequently converted to continuous operation. The latter is eventually adopted in the majority of large-scale chemical manufacturing processes for a combination of economic and safety advantages, as exemplied in Table 19.7.

Table 19.7 Safety considerations in continuous verses batch processing

Continuous process:
- Smaller quantities of materials are held up in the reactor system. The fire/explosion or toxic release hazard is reduced in scale.
- Any hazardous intermediate products may be consumed as fast as they are produced hence minimising their hazard.
- Steady-state operation aids automatic control. The probability of operator error is also reduced.
- The reactor is less subject to cyclical fluctuations in pressure and temperature.

Batch process:
- Reactors may be isolated from one another. Spread of fire can hence be minimised by the use of small, isolated parallel units.
- Careful analytical control can be applied to each batch of raw materials, material in process, and products.
- Identification of materials sources and process conditions assists in quality assurance.

The physico-chemistry of fire

Fire is an uncontrolled chemical change which evolves heat and is generally accompanied by light (flames).

Combustion occurs through chain reactions generally described in four groups:
- Initiation, in which fuel compounds are decomposed by heat, or other means, to yield active radicals.
- Propagation, in which the radicals react with more fuel molecules or with oxygen to give an equal number of radicals.
- Branching, in which the radicals react with other fuel molecules or oxygen to give a greater number of radicals.
- Quenching, in which radicals react with inert species, or with vessel walls, and are 'killed'. They can then no longer propagate the reaction or contribute to branching reactions.

Lack of control is associated with much more of a material being exposed to heat than is actually taking part in the reaction at any instant. Through a transfer process, usually dominated by heat transfer, more of the exposed material changes into a condition which allows it to take part in the reaction.

In the great majority of situations the reaction involves the oxidation of organic materials by atmospheric air or oxygen and the physical change is the generation of flammable vapour from the solid, or liquid, fuel. There are, however, exceptions, e.g. the oxidation of metals or hydrogen (page 1368), acetylene (pages 1113 and 1136), and the decomposition of explosives (page 1475).

Flame

Flame often characterises fires of organic material in air. For this to occur with an ignition source the concentration must be within the limits of flammability discussed on page 131.

Flames may be[60]

- Pre-mixed flames, when a large volume of mixture is present within the flammable range and the rate of burning depends on the rate at which flame propagates through the mixture. The usual effect is a rise in pressure and an explosion. The speed of flame propagation depends on the nature of the fuel and upon the temperature, composition, pressure and quiescence or turbulence of the mixture. Burned gases behind the flame will expand in relation to the temperature reached in the combustion process. If the gases cannot escape the flame is pushed forward at a correspondingly greater rate; the gas mixture in front of the flame is also moved forward. This results in the development of turbulence in front of the flame which increases its speed of travel, because it causes mixing of hot combustion products, which would normally flow away behind the combustion front, with unburned gas ahead of the combustion front. Turbulence generated due to the presence of obstructions or the existence of movement in a gas prior to ignition can also cause flame acceleration.
- Diffusion flames, when fuel and air are fed separately to a zone, mix to attain a flammable mixture, and burn. The effect is a continuing fire. Under free-burning conditions the rate of fuel supply controls the rate of burning; under 'air-starved' conditions it will depend upon the rate of air supply.

Rate of flame travel

The rate of flame travel can be several orders greater in a turbulent gas than in a quiescent gas. In either laminar or turbulent flame propagation the energy exchange to bring the unburned material into a state of reaction is by a process of heat transfer. Such combustion processes are termed 'deflagrations'. However, a 'detonation' can be established, i.e. a shock wave of sufficient power to heat the reacting mixture to a temperature at which the reaction rate is sufficient to drive the shock wave. Flame speeds in a gaseous detonation are of the order of 1000 m/s but one will not generally develop unless the gases are confined by a long duct.[60]

The movement of flame over large distances may, of course, also result in a widespread escalation of fire, e.g. as described for dust explosions (page 1416).

The size of a diffusion flame is that necessary for, at least, sufficient air to feed into the flame by natural convection processes associated with flame buoyancy to burn the vapour emitted from the fuel. Therefore for a given amount of fuel vapour burned in a compartmented fire the flames which move beneath a ceiling are much larger than those which move upwards from a fuel source. Hence the time required for the flames to reach the ceiling is a critical factor in the fire history. The rate of fuel supply is controlled by heat transfer between the flames and the fuel. For large flames this is predominantly by radiation. For small flames, and for solid elements within a large flame, convective heat transfer is important. The upward flame velocity and flame temperatures are also influenced by the flame buoyancy.[60]

The rate of spread of fire along a fuel surface depends on the rate of heat transfer to the surface, in order to raise it to a temperature that produces sufficient flammable vapour to burn. In general the lower this temperature the more rapid will be the spread of fire across the surface. Heat transfer to the neighbouring fuel includes radiation from the flame and, with flammable liquids, convection within the fuel. With a cellulosic solid fuel, flame spread will occur when the fuel has been heated to 300 °C and for horizontal beds of fuel, e.g. layers of vegetation, cribs, etc., the main mechanism of heat transfer is radiation through the fuel matrix.[60]

The probable rate of flame spread is an important consideration when selecting fire protection measures. In cellulosic fuels flame will spread upwards at metres per minute but horizontally at only centimetres per minute. By comparison with a liquid having a flash point below ambient, the rate of surface flame spread is of the order of metres per second.

Ignition

Ignition of flammable gases or vapours with air requires a source of energy to raise a finite mass of the mixture to a temperature at which the heat produced by the rate of reaction exceeds the rate of heat loss. This concept is illustrated in Fig. 4.6. (If a chain branching mechanism of ignition predominates, the rate of branching must exceed that of chain terminations.) Flame will then propagate away from the finite mass into the complete flammable mixture.

All the sources listed on pages 141–57 and discussed further on page 1512 *et seq*. can act as a local source of energy. The total amount of energy required will depend upon the size and shape of the source and upon the nature of the flammable mixture.

> A fitter inadvertently removed a thermowell from a fuel oil line at a refinery power station. Fuel oil at 10 bar pressure and a temperature of 90 °C was discharged directly onto a steam header at 515 °C and ignited immediately. The ensuing fire lasted for thirty minutes and caused serious damage to four boilers.[61]

With ignition of packed solids the combustion reaction is between the solid and air present within the packing. Thus, as discussed for spontaneous combustion on page 1361, the important factors are the ambient temperature at the surface of the stack and the stack size; these control the rate of heat loss and thus determine whether the material inside the stack will reach the critical temperature for ignition. Therefore unless sealed from extraneous air, or subjected to cooling, any stack of flammable materials will eventually ignite if it is large enough.

A variant of this is the case of heating of a solid, or liquid, fuel to produce a sufficient quantity of flammable vapours to become ignited.[60] An example of this is the ignition of paper by a lighted match.

Extinction

As described on page 168, a fire may be extinguished by removal of any one corner of the fire triangle. Most often it is achieved by removal of the fuel vapour that burns. Hence the predominant action of water on ordinary fires is to cool the burning material below 300 °C so that insufficient vapour enters the flame to support combusion. Below a critical rate of water addition (depending on the conditions of heat transfer between the flames and the fuel, and between the fuel and applied water) heat removal is insufficiently rapid and the fire will continue.

Whilst fires in flammable liquids of high flash point can similarly be extinguished by a water spray it is usual to use water foam which, as discussed on page 169, forms a blanket over the surface and prevents the flammable vapour reaching the air.

Flames may also be extinguished by changing the fuel/air mixture to a condition under which the flame will not propagate.[62] If all the fuel and oxygen in a mixture reacts to the maximum possible extent, and heat loss is excluded, the combustion products attain the 'adiabatic flame temperature'. Hence a lower limit of fuel concentration exists below which flame will not propagate through a mixture; this is associated with a certain adiabatic flame temperature, about 1500 K for many flammable vapours. So if a heat-absorbing additive is dispersed uniformly in the flammable vapour–air mixture to reduce the adiabatic flame temperature to the value for the lower limit then, generally, the mixture ceases to propagate flame. Suitable 'additives' include carbon dioxide, nitrogen, and inert powders, e.g. limestone or talc. Alternatively 'inhibitors' and vaporising liquids containing bromine and dry powders containing alkali metals can extinguish flame even if insufficient is present to reduce the adiabatic flame temperature to that at the lower flammability limit.

Combustion reactions leading to explosions

In combustion the rate of energy release is balanced by the rate of dissipation and, as illustrated in Fig. 19.3(a), a limiting reaction rate is reached.

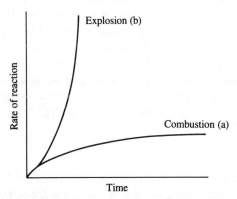

Fig. 19.3 Comparison between (a) combustion and (b) explosion characteristics

However, if within a confined vessel, or building, the products and heat of combustion are not removed, the temperature and (since the volume is constant) the pressure continues to rise. The rate of reaction is accelerated and an explosion follows, as illustrated in Fig. 19.3(b).

Confinement, as distinct from complete and effective containment, invariably results in serious consequences. If the confinement is fairly close, and the fuel–air mixture is approaching stoichiometry, pressures sufficient to rupture the container (i.e. vessel or building) are readily generated.[63] Such confined vapour cloud explosions are discussed on page 875.

The ratio of the pressure attained in an explosion in a closed vessel to the initial pressure can be estimated, since the volume of the reactants and products will be constant. Hence, for example, considering combustion of a mixture of 5 per cent methane in air at an initial temperature of 300 K,[19]

$$CH_4 + 3.98O_2 + 15.02N_2 \rightarrow CO_2 + 2H_2O \text{ (g)} \\ + 1.98O_2 + 15.02N_2$$

The total enthalpy of the products at 300 K can be calculated as follows.

Product	No. of moles, n	Enthalpy at 300 K, kcal	Enthalpy $\times n$, kcal
CO_2	1	1.658	1.658
H_2O	2	1.786	3.572
O_2	1.98	1.986	3.93
N_2	15.02	1.485	22.30
	20.00		31.46

For this reaction the heat of combustion is 191.8 kcal. Hence, since

$$\text{Heat of combustion} = \begin{bmatrix} \text{Enthalpy of} \\ \text{products at} \\ \text{adiabatic} \\ \text{flame temp.} \end{bmatrix} - \begin{bmatrix} \text{Enthalpy} \\ \text{of products} \\ \text{at 300 K} \end{bmatrix}$$

the enthalpy of the products at the adiabatic flame temperature is 191.8 + 31.46 = 223.3 kcal. Reference to enthalpy tables enables the adiabatic flame temperature to be calculated as 1818 K. The ratio of the pressures before and after the explosion is

$$\frac{P_e}{P_i} = \frac{T_e}{T_i} \times \frac{N_e}{N_i}$$

where T is the absolute temperature, subscript e refers to conditions after the explosion and i refers to the initial condition. Hence $P_e = 6.1 P_i$, and the vessel will be subjected to about six times the initial pressure.

If the mixture of fuel and air is of limited volume and unconfined when ignition occurs then the result is a deflagration or 'soft' explosion. Anyone close to the incident may suffer flash burns, and scorching, or the initiation of secondary fires, may also occur. Such an unconfined rapid combustion will produce no significant overpressures. However, large-scale incidents, especially with some degree of containment, may result in Unconfined Vapour Cloud Explosions (page 873).

Latent heat of vaporisation/condensation

The specific latent heat of vaporisation of a material is simply the quantity of heat, e.g. in kJ/kg, required to change unit mass of it from a liquid to the vapour phase with no associated change in temperature. This heat is absorbed on vaporisation so that residual liquid or the surroundings are cooled. Alternatively an equivalent quantity of heat must be removed to bring about condensation.

Thus when the pressure above a liquefied gas is reduced and a proportion of the liquid evaporates, as described in Chapter 3, the remaining liquid is cooled in the process. This absorption of heat, and consequent lowering in temperature, may have a serious effect on associated heat transfer media or upon the strength of materials of construction.

The temperature of liquid propylene is −47 °C at atmospheric pressure. When a plant containing liquid propylene was shut down the flow of cooling water to a heat exchanger was turned off. As the propylene pressure was reduced its temperature dropped and water retained in the tubes of the exchanger froze; this resulted in fracture of seven bolts in the floating head. On start-up propylene entered the cooling water system and a 15 cm line ruptured under the pressure. The subsequent gas escape was ignited at a furnace source 40 m away, causing a serious fire. (The recommended procedure was to keep water flowing through the tubes during depressurisation, in which case it would not have frozen provided the operation took longer than 10 minutes.)

Figure 19.4 illustrates a double-walled tank in which liquid ethylene was stored

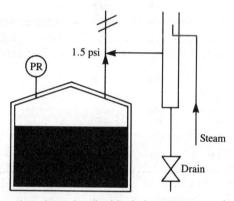

Fig. 19.4 Steam-purged stack serving liquid ethylene storage tank

at −100 °C. Off-gas was refrigerated and returned to the tank. When the refrigeration unit was shut down, and the pressure rose, gas was vented via a relief valve which led to the stack. A flow of steam was provided up the stack to disperse the cold ethylene vapour.

On one occasion condensate was frozen due to chilling by the cold ethylene, the stack was blocked, and internal pressure resulted in the tank splitting near its base.[64] (The mind-set of the designers who introduced this solution, after construction had started, to prevent cold, heavy ethylene vapour descending to ground level is discussed in ref. 36. There is little doubt that the eventuality would have been identified by the type of HAZOP study summarised on page 741 or some other formal analysis.)

During an upset on a plant separating hydrogen and light hydrocarbons at subzero temperatures, liquefied product at sub-zero temperature entered a carbon steel section of a common flare line. The metal became brittle and, due to uneven stresses, ruptured releasing a large quantity of hydrocarbons.

Critical temperatures of gases and cryogens

Every gas has a critical temperature above which it cannot be liquefied by the application of pressure alone. The critical pressure is that which is required to liquefy it at this critical temperature. Examples are given in Table 19.8. This has important practical consequences.

Table 19.8 Chemicals commonly stored as liquefied gases

Gas	Boiling point at 1 bar abs. (°C)	Liquid density at boiling point (kg/m³)	Volume ratio of gas (1 bar abs., 20 °C) to liquid (at boiling point)	Vapour pressure at 38 °C (bar abs.)	Critical temperature (°C)
Can be stored without refrigeration					
Ethylene oxide	11	883	425	2.7	195
n-Butane	0	602	242	3.6	152
Butadiene (1,3)	−4	650	280	4.1	152
Butylene (α)	−6	626	262	4.3	146
Isobutane	−12	595	241	5.0	135
Ammonia	−33	682	962	14.6	133
Propane	−42	582	315	13.0	97
Propylene	−48	614	347	15.7	92
Require refrigeration					
Ethane	−89	546	436	—	32
Ethylene	−104	568	487	—	9
Methane (LNG)	−162	424	637	—	−82
Oxygen	−183	1140	860	—	−119
Nitrogen	−196	808	696	—	−147

Liquefied gas storage

Liquefied gas may be stored fully-refrigerated, with the liquid at its bubble point at near-atmospheric pressure; fully-pressurised, i.e. at ambient temperature; or semi-refrigerated, with the temperature below ambient but the vapour pressure above atmospheric. Of the gases listed in Table 19.8 as commonly being stored as liquids,[54]

- those whose critical temperatures are below ambient require refrigeration;
- the remainder can be liquefied using pressure only.

Liquefied gas cylinders

Another consequence of the critical temperature phenomenon is that, provided the temperature remains constant, the pressure within any cylinder containing liquefied gas will remain constant as gas is drawn off.

More liquid simply evaporates to make up for the gas discharged until all the liquid is vaporised. Thus the quantity of gas in a cylinder cannot be deduced from the internal pressure.

Conversely, with a true gas, i.e. a vapour above its critical temperature, the quantity of gas in a cylinder is indicated by the pressure.

Liquefied gas cylinders are never completely filled, to allow for expansion of the liquid and its ultimate conversion to vapour if heated. Otherwise increase in temperature above the critical temperature could result in a dangerous pressure rise and possibly an explosion as described on page 99. Filling ratios vary from gas to gas, e.g. from 0.5 for ammonia to 1.25 for sulphur dioxide and 1.46 for sulphur hexafluoride.

Cryogens

The use of cryogenics, i.e. low-temperature technology, in laboratories is described on page 644. Industrial applications arise in food processing, rocket propulsion, microbiology, electronics, medicine, and metal working. Such low temperatures are primarily achieved by the liquefaction of gases as exemplified by Table 19.9.

Table 19.9 Physical properties of cryogenic liquids

Property	Units	Oxygen	Nitrogen	Argon	Carbon dioxide
Molecular weight		32	28	40	44
Boiling point (at atmospheric pressure)	°C	−183	−196	−186	−78
Freezing point	°C	−219	−210	−190	—
Density of liquid (at atmospheric pressure)	kg/m^3	1141	807	1394	1562 (solid)
Density of vapour (at NBP)	kg/m^3	4.43	4.59	5.70	2.90
Density of dry gas at 15 °C and at atmospheric pressure	kg/m^3	1.34	1.17	1.67	1.86
Latent heat of vaporisation at NBP and atmospheric pressure	kJ/kg	214	199	163	151
Expansion ratio (liquid to gas at 15 °C and atmospheric pressure)		842	682	822	538*
Volume per cent in dry air	%	20.95	78.09	0.93	0.03

* From liquid CO_2 at 21 bar, −18 °C.
NBP = normal boiling point.

Certain gases, e.g. those used as an inert medium to reduce the oxygen content of atmospheres containing flammable gas or vapour (page 132), are often shipped and stored as cryogenic liquid for convenience and economy. The main hazards with cryogens stem from the following factors.

- Low temperature. If the materials come into contact with the body, they can cause severe tissue burns. Whilst inadvertent contact due to splashing should be avoidable, the low viscosity of, e.g. liquid nitrogen will allow it to penetrate woven or porous clothing material much more rapidly than water. Flesh may stick fast to cold uninsulated pipes or vessels and tear on attempting to withdraw it. Low temperatures may also cause failure of service materials due to embrittlement; metals can become sensitive to fracture by shock.

A 30 bar carbon steel buffer tank ruptured into twenty pieces during filling with cold nitrogen gas. The gas was supplied from vaporisation of liquid nitrogen in an ambient heat exchanger which was not provided with a temperature control device or low-temperature trip.[65]

- Asphyxiation (except for oxygen). If the cryogen evaporates in a confined space it may result in oxygen deficiency.
- Catastrophic failure of containers. Evaporation of a cryogen may cause pressure build-up within the vessel beyond its safe working pressure. For example, pressures up to 280 000 kPa (40 600 psi) can develop from heating liquid nitrogen to ambient temperature in a confined space.
- Fog formation. Spillage of a liquid cryogen will produce a fog, due to condensation of water vapour in the surrounding air, which can restrict visibility.

A seventeen-vehicle collision was caused by reduced visibility when fog from a spillage of liquid nitrogen drifted over an interstate highway. Four people were killed and fourteen injured.[66]

- Flammability (e.g. hydrogen, acetylene), toxicity (e.g. carbon dioxide, fluorine), or chemical reactivity (fluorine, oxygen).

A summary of general precautions is given in Table 19.10. The cryogens encountered in the greatest volume include oxygen, nitrogen, argon and carbon dioxide. Their physical properties are summarised in Table 19.9 and further details are given in ref. 67.

Liquid oxygen

Liquid oxygen is pale blue, slightly heavier than water and magnetic. It is non-flammable but contact with a reducing agent can result in an explosion.

Gaseous oxygen does not burn but supports combusion of most elements (page 1106). Upon vaporisation liquid oxygen can produce an atmosphere which enhances fire risk (page 157). It may cause certain substances normally considered to be non-combustible, e.g. carbon steel, to inflame. In addition to the general precautions set out in Table 19.10 the following are relevant to the prevention of fires and explosions.

Table 19.10 General precautions with cryogenic materials

- Obtain authoritative advice from the supplier.
- Select storage/service materials and joints with care, allowing for the reduction in ductility at cryogenic temperatures.
- Provide special relief devices as appropriate.
- Materials of construction must be scrupulously clean, free of grease, etc.
- Use labelled, insulated containers designed for cryogens, i.e. capable of withstanding rapid changes and extreme differences in temperature; fill them slowly to minimise thermal shock. Keep capped when not in use and check venting. Glass Dewar flasks for small-scale storage should be in outer metal containers and any exposed glass taped to prevent glass fragments flying in the event of fracture/implosion. Large-scale storage containers are usually of metal and equipped with pressure-relief systems.
- In the event of a fault developing (as indicated by high boil-off rates or external frost), cease using the equipment.
- Provide a high level of general ventilation taking cognisance of density and volume of gas likely to develop. Initially gases will slump but those less dense than air (e.g. hydrogen, helium) will eventually rise. Do not dispose of liquid in confined areas.
- Prevent contamination of fuel by oxidant gases/liquids.
- With flammable gases eliminate all ignition sources. Possibly provide additional high and low level ventilation; background gas detectors to alarm, e.g. at 40% of the Lower Explosive Limit.
- With toxic gases, possibly provide additional local ventilation; monitors connected to alarms; appropriate air-fed respirators. (The flammable/toxic gas detectors may be linked to automatic shutdown instrumentation.)
- Limit access to storage areas to authorised staff with knowledge of the hazards, positions of valves, switches, etc. Display emergency procedures.
- Wear face shields and impervious dry gloves, preferably insulated and loose fitting.
- Wear protective clothing which avoids the possibility of cryogenic liquid becoming trapped near the skin. Avoid turnups and pockets, and wear trousers over boots, not tucked in.
- Remove bracelets, rings, watches, etc. to avoid potential traps of cryogen against skin.
- Prior to entry into large tanks containing inert medium, ensure pipes to the tank from cryogen storage are blanked off or positively closed off and purge with air and check oxygen levels. Follow the requirements for entry into confined spaces (page 981).
- First-aid measures include:
  – move casualties becoming dizzy or losing consciousness into fresh air and provide artificial respiration if breathing stops. Obtain medical attention;
  – in the event of 'frostbite' do not rub the affected area but immerse rapidly in warm water and maintain general body warmth. Seek medical aid.
- Ensure staff are trained in the hazards and precautions for both normal operations and emergencies.

- Prohibit smoking or other means of ignition in the area.
- Avoid contact with flammable materials (including solvents, paper, oil, grease, wood, clothing) and reducing agents. Thus oil or grease must not be used on oxygen equipment.
- Purge oxygen equipment with oil-free nitrogen or oil-free air prior to repairs.
- Post warning signs.
- In the event of fire evacuate the area and if possible shut off oxygen supply. Extinguish with water spray unless electrical equipment is involved when carbon dioxide extinguishers should be used.

Liquid nitrogen and argon

Liquid nitrogen is colourless and odourless, slightly lighter than water and non-magnetic. It does not produce toxic or irritating vapours. Liquid argon is also colourless and odourless but significantly heavier than water. Gaseous nitrogen is colourless, odourless and tasteless, slightly soluble in water and a poor conductor of heat. It does not burn or support combustion, nor readily react with other elements. It does, however, combine with some of the more active metals, e.g. calcium, sodium and magnesium, to form nitrides. Gaseous argon is also colourless, odourless and tasteless, very inert and does not support combustion.

The main hazard from using these gases stems from their asphyxiant nature. Thus in confined, unventilated spaces small leakages of liquid can generate sufficient volumes of gas to deplete the oxygen content to below life-supporting concentrations when personnel can become unconscious without warning symptoms. Gas build-up can occur when a room is closed overnight. Thus when using gaseous or liquid nitrogen, particularly in large quantities, the areas must be well ventilated to avoid the formation of an oxygen-deficient atmosphere. Nitrogen should never be released or vented in enclosed areas or buildings with inadequate ventilation; preferably it should be vented into the open air well away from areas frequented by personnel.[65]

Although gaseous nitrogen is slightly less dense than air at the same temperature (page 95), liquid nitrogen and cold nitrogen vapour are heavier than air and can therefore accumulate in pits, trenches or other low-lying areas. Moreover the fog formed in the vicinity of a large spill may contain oxygen concentrations appreciably lower than that of air, resulting in a local asphyxiation hazard.[65]

Because the boiling points of these cryogenic liquids are lower than that of oxygen, they can on exposure to air cause oxygen to condense preferentially, resulting in similar hazards to liquid oxygen (pages 645, 1106 and 1171).

Liquid carbon dioxide

Liquid carbon dioxide is usually stored under 20 bar pressure at -18 °C. Compression and cooling of the gas between the temperature limits at the 'triple point' and the 'critical point' will cause it to liquefy. The triple point is the pressure–temperature combination at which carbon dioxide can exist simultaneously as gas, liquid and solid. Above the critical temperature point of 31 °C it is impossible to liquefy the gas by increasing the pressure above the critical pressure of 73 bar. Reduction in the temperature and pressure of liquid below the triple point causes the liquid to disappear, leaving only gas and solid. (Solid carbon dioxide is also available for cryogenic work and at -78 °C the solid sublimes at atmospheric pressure.) Liquid carbon dioxide produces a colourless, dense, non-flammable vapour with a slightly pungent odour and characteristic acid 'taste'. Physical properties and the effect of temperature on vapour pressure are given in ref. 67.

The eight-hour time-weighted average (TWA) OEL for carbon dioxide is 0.5 per cent; at higher levels life may be threatened by extended exposure. The following considerations therefore supplement those listed in Table 19.10.

- Ensure operator exposure is below the hygiene standard. For environmental monitoring, because of its toxicity, a CO_2 analyser must be used as distinct from simply relying on checks of oxygen levels.
- When arranging ventilation remember that the density of carbon dioxide gas is greater than air.
- Ensure that pipework and control systems are adequate to cope with the higher pressures associated with storage and conveyance of carbon dioxide compared with those with most other cryogenic liquids.

Stored pressure energy and compressed gases

Any gas stored under pressure represents an energy source, i.e. if vessel failure occurs energy will be given out as the gas expands rapidly down to virtually atmospheric pressure. Elastic strain energy in the walls is also given out but this is normally very small.[68]

The energy content of a perfect gas may be calculated as follows:[69]

$$W' = P_o V \ln \frac{P_o}{P_a}$$

where W' = energy content (joules);
V = vessel volume (m^3);
P_o = internal gas pressure (MPa);
P_a = atmospheric pressure (MPa).

That high inventories of such energy can result in a major hazard was noted on page 881. Even failure of a small piece of equipment may lead to a serious explosion producing missiles as described on page 90.

A pressure not registered on an indicator may be sufficient to cause an accident.

An 8 m^3 vessel had been tested under nitrogen pressure for leaks. A man was instructed to blow off the pressure and change the top end of the closure. He opened the bottom valve and released the pressure until the pressure gauges on the top and bottom indicated zero and he could not feel gas blowing out with his hand. He next removed eight nuts which secured the top fitting and inserted a wedge to loosen it. The fitting, weighing 385 kg, was suddenly blown 7.6 m into the air.

The pressure gauges were calibrated up to 4000 psig, so that a pressure of 15 psig would not show up on them. Hence a pressure sufficient to blow off the vessel top was too small to be indicated. In addition, the vent line used to blow down the pressure was only 1.5 cm in diameter and was partially filled with loose debris.

Therefore when testing boilers or similar installations hydraulic testing is advisable, with care being taken to ensure that all air is expelled before sealing the equipment.

The diesel engine of a heavy vehicle had been overhauled and the water jacket refilled. The practice to detect leakages at the gaskets was to seal up the filler hole and other orifices and to screw a Schraeder-type valve into the jacket; compressed air at approximately 80 psi was then introduced through the valve and any visual signs of leaks noted. The compressed air line had just been connected when the side plate of the cylinder block blew out and struck a fitter's apprentice in the chest, injuring his ribs.[70] Previously it had been usual to apply a tyre foot-pump to the valve which can also be dangerous.

Pneumatic testing should be used only as a last resort on vessels which it is either impracticable to fill with liquid or undesirable for process reasons, e.g. reactors lined with firebrick or containing liquid-sensitive catalyst.[69] Measures to be taken then include (after ref. 69):
- Ensuring no-one would be injured if an explosion occurred, e.g. by siting the vessel in a blast pit or, as is done with gas cylinders, behind adequate retaining walls with entry prohibited during testing.
- Careful pre-inspection of the vessel, e.g. with radiographic or other appropriate non-destructive testing of welds.
- Avoidance of pneumatic testing of vessels constructed from materials prone to brittle fracture.
- Reduction of the internal air space in the vessel with metal or hardwood cores, hence reducing W'.
- Adequate methods for sealing vessel openings.
- Incorporation of appropriate reducing valves, pressure gauges and pressure relief valves in the test rig.

A proper procedure, given in ref. 69, should also be followed during hydraulic testing since this has also occasionally resulted in some spectacular failures:[71]

A 1.7 m diameter by 16 m long pressure vessel designed to operate at 350 bar failed during testing at the fabricators. A brittle-type failure occurred at 345 bar gauge pressure and four large pieces were ejected; one piece weighing 2 tonnes went through the workshop and travelled almost 50 m.[72]

This failure occurred in the winter and it is recommended that pressure tests be carried out above the ductile–brittle transition temperature for the grade of steel used. In this particular case, however, the vessel was stress-relieved at too low a temperature.

In service, as discussed on pages 697–701, reliable overpressure protection venting in a safe manner (e.g. via a flame arrestor, flare stack, knockout drum, scrubbing tower or seal pot as appropriate) should be inspected and maintained.

Relief facilities on pressurised process plants should be designed to cope with the largest single emergency relief case for all operating states of the plant. Also where instrumentation contributes to pressure protection a high-integrity installation should be used.[73]

A major explosion occurred on a Hydrocracker Unit (shown in Fig. 19.5)

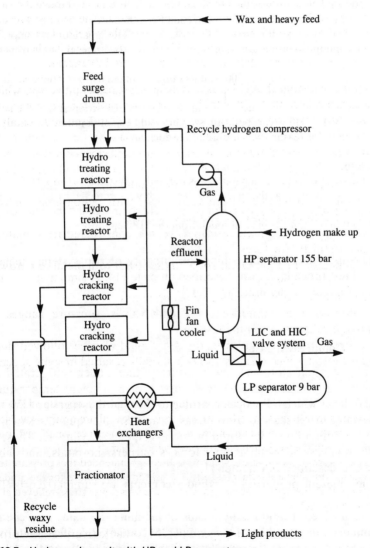

Fig. 19.5 Hydrocracker unit with HP and LP separators

which seriously damaged plant and equipment and completely disintegrated the Low Pressure (LP) Separator Vessel. Debris was widely scattered, with some weighing several tonnes being propelled up to 1 km, in some instances off site. A crane driver was killed.

The unit was being prepared for recommissioning and the unit reactor section was on gas circulation at 140 bar. The incident resulted from severe overpressure of the LP Separator designed to operate at 10 bar which caused it to disintegrate, at a pressure probably in the range 40–65 bar, with release of the contents, primarily hydrogen. The explosive forces were reported as equivalent to 90 kg of TNT. Ignition of the hydrogen cloud produced a fireball with some blast effects.[73,74]

The LP Separator was directly linked to the high pressure reactor circuit via a 200 mm nominal bore liquid connection from the base of the High Pressure (HP) Separator. At the time of the explosion the HP Separator had emptied of liquid and the overpressurisation was caused by rapid transfer of gas from the HP Separator into the LP Separator via the level control which was probably opened between 40% and 100% of its range on manual. Two independent extra low-level switches were provided on the HP Separator to protect against a falling level; when activated individually these switches tripped shut the level control valve. However, both trip switches had been disconnected several years previously for reasons which were never documented. Alarm functions had been retained but both associated visual alarms were inoperable due to bulb failure; audible alarms were operable.

Off-gas from the LP Separator was normally transferred to an ADIP unit but this was valved off at the time. This Separator was protected by a single relief valve sized to cope with failure of the pressure controller and the external fire condition but not gas breakthrough from the HP Separator.

(The importance of maintaining the integrity of trip systems, including formal test procedures, was described on page 714. A similar situation has been examined in detail in ref. 75.)

Compressed gases

Gases are often stored at low pressure, either under refrigerated conditions, e.g. cryogens as already discussed, or at ambient temperature in 'gasholders', which telescope according to the quantity of gas and are fitted with water or oil seals to prevent gas escape. Smaller quantities of gas at high pressure are usually stored in bottle-shaped gas cylinders. These find widespread use in welding, fuel for gas burners, hospitals, and laboratories, as described in Chapter 11.

The inherent properties of cylinder contents are summarized in Table 11.29.

Primary identification is by means of labelling the name and chemical formulae on the shoulder of the cylinder. Secondary identification is by use of ground colours on the cylinder body and colour bands on the cylinder shoulder to denote the nature of the gas as exemplified by Plate 2 in Volume 2 (Chapter 13). (The full scheme is given in BS 349: 1973.)

Table 19.11 summarises general guidance for handling compressed gases. Selected common, hazardous compressed gases are discussed below; chlorine, hydrogen sulphide and ammonia are dealt with in Chapter 29.

Acetylene

Acetylene has a high chemical reactivity, and is utilised in the synthesis of vinyl ethers, vinyl acetylene, trichloro- and tetrachloro-ethylene, and now less commonly in the manufacture of vinyl chloride, vinyl acetate, acrylonitrile, etc., in oxyacetylene cutting and welding, and as a fuel for atomic absorption instruments.

Acetylene is a simple asphyxiant and anaesthetic. Pure acetylene is a colourless, highly flammable gas with an ethereal odour. Material of commercial purity has an odour of garlic. Its physical properties are given in ref. 67.

In the free state acetylene can decompose violently, e.g. above 9 psig undissolved (free) acetylene will begin to dissociate and revert to its constituent elements. This is an exothermic process which can result in explosions of great violence. Therefore acetylene is transported in acetone contained in a porous material inside the cylinder. Voids in the porous substance can result from settling, e.g. if the cylinder is stored horizontally, or through damage to the cylinder in the form of denting. Voids may enable acetylene to decompose, e.g. on initiation by mechanical shock from dropping the cylinder. Figure 19.6 illustrates the rise in cylinder pressure with temperature. Normally, acetylene cylinders are fitted with a fusible metal plug which melts at about 100 °C.

Acetylene can form metal acetylides, e.g. copper or silver acetylide, which on drying become highly explosive; service materials hence require careful selection. Explosive acetylides may be formed on copper, or brasses containing more than 50 per cent copper, when exposed to acetylene-containing atmospheres.

Conditions favouring acetylide formation in copper equipment, e.g. in the incident described on pages 1135 and 1150, are:[76]

- More than 50 per cent copper in the construction material. The higher the copper content the easier the acetylide forms.
- Corrosion of the copper surface, e.g. mineral acids, caustic solutions and ammoniacal solutions all encourage acetylide formation.
- Moderate temperatures, from 10–50 °C, are probably necessary for active material formation, although decomposition can occur at any temperature.
- The probability of active acetylide formation increases the higher the concentration of acetylene present. However, a few parts per million of acetylene may be sufficient.

Table 19.11 General precautions with compressed gases

Consult the supplier for data on the specification, properties, handling advice and suitable service materials for individual gases.

Storage
- Segregate according to hazard.
- Stores should be adequately ventilated and, ideally, located outside but provide adequate protection from the weather. Fire-proof partitions/barriers may be used to separate cylinders.
- Store away from sources of heat and ignition.
- In laboratories cylinders should be restricted to those gases in use. Specially designed compartments with partitions may be required to protect people from the effects of an explosion. Cognisance must be taken of emergency exits, steam or hot water systems, the proximity of other processes, etc. Consider the possibility of dense gases accumulating in drains, basements, cable ducts, lift shafts, etc.
- Protect from mechanical damage.
- All cylinders should be properly labelled and colour-coded (e.g. BS 349).
- Full and empty cylinders should not be grouped together.
- Stock should be used in rotation: first in, first out.
- Access to the stores should be restricted to authorised staff.
- 'No smoking' and other relevant warning signs should be displayed as appropriate.
- Ensure all staff are fully conversant with the correct procedures when using pressure regulators. For cylinders without handwheel valves, the correct cylinder valve keys should be kept readily available, e.g. on the valve. Only use such keys. Do not extend handles or keys to permit greater leverage; do not use excessive force, e.g. hammering, when opening or closing valves or when connecting or disconnecting fittings. The pressure regulator must be fully closed before opening the cylinder valve. This valve can then be opened slowly until the regulator gauge indicates the cylinder pressure but should not be opened wider than necessary. The pressure regulator can then be opened to give the required delivery pressure. If a cylinder is not in use, or is being moved, the cylinder valve must be shut. When a cylinder has been connected the valve should be opened with the regulator closed; joints should then be tested with soap-detergent solution.
- Clearly and permanently mark pressure gauges for use on oxygen. Do not contaminate them with oil or grease nor use them for other duties.
- Cylinders which cannot be properly recognised should not be used; do not rely on colour code alone.
- Never try to refill cylinders.
- Never use compressed gas to blow away dust and dirt.
- Provide permanent brazed or welded pipelines from the cylinders to near the points of gas use. Select pipe materials suitable for the gas and its application. Any flexible piping used should be protected against physical damage. Rubber or plastic connections should not be used from cylinders containing toxic gases.
- On acetylene service, use only approved fittings and regulators. Avoid any possibility of it coming into contact with copper, copper-rich alloys or silver-rich alloys. (In the UK use at a pressure greater than 600 mbar g must be notified to HM Explosives Inspectorate for advice on appropriate standards.)
- On carbon dioxide service rapid withdrawal of gas may result in plugging by solid CO_2. Close the valve, if possible, to allow the metal to warm up; this will avoid a sudden gas discharge.
- Replace caps or guards on cylinder valves when not in use and for return to the supplier.
- Test and inspect cylinders regularly in accordance with current legislation.
- Design and manage cylinder stores in accordance with suppliers' recommendations.
- Appropriate personal protection should be worn when entering any store.
- Inspect condition of cylinders regularly, especially those containing hazardous or corrosive gases.

Use
- Transport gases in specially designed trolleys and use eye protection, stout gloves (preferably textile or leather) and protective footwear.

Table 19.11 (continued)

- Do not roll or drop cylinders off the back of wagons and never lift cylinders by the cap.
- Depending upon the length of pipe run, preferably locate cylinders outside. (For hazardous gases valves may be required in the laboratory to turn off the main supply at the cylinder remotely in the event of an emergency.)
- Site cylinders so that they cannot become part of an electrical circuit.
- Securely clamp, or otherwise firmly hold in position, cylinders on installation. (Unless otherwise specified, cylinders containing liquefied or dissolved gases must be used upright.)
- Avoid subjecting cylinders containing liquid to excessive heat.
- Fit approved cylinder pressure regulators, selected to give a maximum pressure on the reduced side of the valve commensurate with the required delivery pressure. (The regulator and all fittings upstream of it must be able to withstand at least the maximum cylinder pressure.)
- In-line flame arrestors should be fitted for flammable gases and ignition sources eliminated.
- Use compatible pipe fittings. (Flammable gas cylinders have valves with left-hand threads; cylinders for oxygen and non-flammable gases, except occasionally helium, have valves with right-hand threads. Certain liquefied gas cylinders have two supply lines, one for gas and one for liquid, dependent on cylinder position.)
- Do not use oil, grease or jointing compounds on any fittings for compressed gas cylinders.
- Fit an excess flow valve to the outlet of a regulator, selected to allow the maximum required gas flow.
- Respirators and face protection, etc., will be required when changing regulators on cylinders of toxic gases.
- Turn off gas supply at the cylinder at the end of each day's use.
- Consider the need for gas detection/alarms, e.g. for hazardous gases left in use out of normal hours.
- When gases are flammable ensure ignition sources are eliminated.
- Periodic checks include:
  - ensure no gas discharges when gauge reading is zero;
  - ensure reading on the gauge does not increase as the regulator valve is closed;
  - 'crawl' due to wear on the regulator valve and seat assembly;
  - ensure no leak between cylinder and regulator;
  - overhaul regulators on a 3–6-monthly basis for corrosive gases and annually for others.
- Staff should be trained in the hazards and correct handling precautions.

(In the UK numerous measures are encompassed by the 'Pressure Systems and Transportable Gas Containers Regulations 1985' and the ACoP, 'Safety of transportable gas containers'.)

A makeshift handle of rubber-covered electric cable was used in the effluent pit of an acetylene plant. Copper acetylide was formed, due to reaction with residual acetylene. The acetylide detonated when disturbed and triggered an explosion of acetylene in air.[77]

Silver and mercury and their salts also form unstable acetylides. Hence all heavy metals should be excluded from contact with acetylene.

In addition to the general precautions for compressed gases in Table 19.11, the following control measures are advisable with acetylene.

- Avoid the use of free acetylene at pressures above 9 psig unless special safety features are employed.
- Only store and use cylinders in an upright position.
- Always store reserves separate from oxygen cylinders.

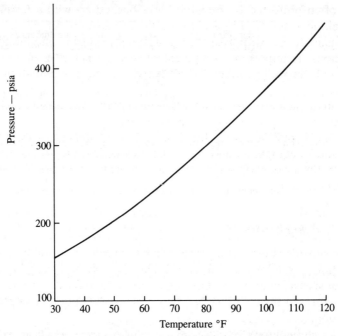

Fig. 19.6 Increase in acetylene cylinder pressure with temperature

- Eliminate means of accidental ignition in the area and provide adequate ventilation.
- Consult local regulations for use of this gas.
- Close the valves on 'empty' cylinders to prevent evaporation of acetone.
- Close cylinder valve before shutting off regulator to permit gas to bleed from regulator.
- When used, e.g. for welding, care is needed to prevent the careless use of flame which could fuse the metal safety plug in the cylinder.
- In the event of fire issuing from the cylinder, close the gas supply if it is safe to do so and evacuate the area.
- Consider the need for detection/alarm systems. In any event check periodically for leaks with, e.g. soap solution, *never* with a naked flame.

Hydrogen

Hydrogen is used for the hydrogenation of oils and fats, and in metallurgical processes, metal welding/cutting, ammonia synthesis, and petroleum refining. It is the lightest gas known. It is colourless and odourless, only slightly soluble in water but readily soluble in hydrocarbons. Hydrogen is non-toxic but can act as an asphyxiant. It is usually shipped at 2000 psig at

21 °C, often protected by frangible discs backed up with a fusible metal plug melting at 100 °C. Physical properties are given in ref. 67.

The main danger with hydrogen is of fire or explosion and the following precautions are important to supplement those in Table 19.11.

- Use only in well-ventilated conditions to avoid accumulation at high levels.
- Eliminate means of accidental ignition to prevent accidents such as that illustrated by the case history on page 638.
- Use only explosion-proof electrical equipment and spark-proof tools.
- Ground all equipment and lines used with hydrogen.
- Check for leaks with soapy water and consider the need for automatic detection/alarms.

Sulphur dioxide

Sulphur dioxide is used as a preservative for beer, wine and meats; in the production of sulphites and hydrosulphites; in solvent extraction of lubricating oils; as a general bleaching agent for oils and foods; in sulphite pulp manufacture; in the cellulose and paper industries and in disinfection and fumigation.

It is a non-flammable, colourless gas which is twice as dense as air, and slightly soluble in water forming sulphurous acid. It is readily liquefied as a gas under its own vapour pressure of about 35 psig at 21 °C. Its physical properties are listed in ref. 67 and the effects of low concentrations upon human health in Table 21.12. Cylinders tend to be protected against overpressurisation by metal plugs melting at about 85 °C.

Gaseous sulphur dioxide is highly irritant and practically irrespirable. Its effects are summarised in Table 19.12. It can be detected at about 3.5 ppm

Table 19.12 Effects of exposures to sulphur dioxide

Concentration (ppm)	Response
0.5–0.8	Minimum odour threshold
3	Sulphur-like odour detectable
6–12	Immediate irritation of nose and throat
20	Reversible damage to respiratory system
>20	Eye irritation
	Tendency to pulmonary oedema and eventually respiratory paralysis
10 000	Irritation of moist skin within a few minutes

and the irritating effects would preclude anyone from suffering prolonged exposure at high concentrations unless unconscious.

Liquid sulphur dioxide may cause eye and skin burns resulting from the freezing effects upon evaporation. Dry sulphur dioxide is non-corrosive to common materials of construction except zinc. The presence of moisture renders the environment corrosive.

The following precautions supplement those in Table 19.11.
- Use in well-ventilated areas.
- Wear eye/face protection, approved footwear and rubber gloves.
- Showers and eye wash facilities and respiratory protection should be conveniently located for emergencies.
- Insert traps in the line to avoid liquid suckback into the cylinder.
- Check for leaks with soap solution, aqueous ammonia, or colour indicator tubes.

Water hammer

Water hammer, often referred to as surge pressure, may occur either
- when the flow of a liquid in a pipeline is stopped suddenly, e.g. by a valve being closed rapidly;
- when slugs of liquid within a gas line are transferred by gas flow. This may arise if condensate accumulates in a steam main.

A 25 cm diameter steam main ruptured suddenly soon after being brought back into service. Several workers were injured.[78]

The line was up to operating pressure, 40 bar gauge, but there was no flow in it. The steam trap had been isolated because it was leaking. An attempt was made to remove condensate via a bypass valve, shown in Fig. 19.7, but steam

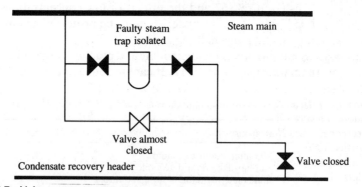

Fig. 19.7 Valve arrangement on steam main which ruptured due to water hammer

entered the condensate header so the line was isolated. This allowed condensate to accumulate in the steam main.

A 2 cm valve was opened from the main allowing steam to flow; movement of condensate caused the main to fracture.

References

1. Anon., *Loss Prevention Bulletin*, 1987(077), 11.
2. Anon., *Safety News*, Universities' Safety Association, April 1980, **13**, 13.

3. Anglo-Celtic Shipping Co. Ltd. v Elliott and Jeffery, 1926, 42, T.L.R. 297.
4. H.M. Factory Inspectorate, *Arsine Health and Safety Precautions*, Technical Data Note 6 (Rev.), Dept. of Employment, 1971.
5. Anon., *Loss Prevention Bulletin*, 1990(091), 22.
6. Capraro, M. A. & Strickland, J. J., *Plant/Operations Progress*, 1989, **8**(4), 189.
7. Manufacturing Chemists Assoc., Case History 1808.
8. Anon., *Loss Prevention Bulletin*, 1975(03), 3.
9. Albright & Wilson Ltd., *Loss Prevention Bulletin*, 1977(013), 21.
10. Anon., *Loss Prevention Bulletin*, 1977(012), 11.
11. Pinsky, M. L., Vickery, T. P. & Freeman, K. P., *J. Loss Prev. Process Ind.*, 1990 (Oct), **3**, 345.
12. Bryson, W. R., *Chem. Brit.*, 1975, **11**, 377.
13. Pitt, M. J., *Chem. Brit.*, 1975, **11**, 456.0.
14. Jennings, A. J. D., *Courses on Process Safety – Theory and Practice*. I. Chem E. (University of Durham), 11–16 July 1982.
15. Yoshida, T., *Safety of Reactive Chemicals*, Elsevier, Amsterdam, 1987, 57.
16. Wong, W. Y., *Chem. Eng.*, 1989 (May), 137–140.
17. Pirie, J. C. *et al.*, *Chem. Brit.*, 1979, **15**, 11.
18. Anon., *Loss Prevention Bulletin*, 1979(027), 74.
19. Jennings, A. J. D., *The Chem. Engr.*, 1974 (Oct), **190**, 637.
20. Siebeneicher, K., *Angew Chem.*, 1934, **47**, 105.
21. Anon., *Loss Prevention Bulletin*, 1987(074), 23.
22. Anon., *Loss Prevention Bulletin*, 1988(080), 31.
23. Bretherick, L., *CIHSC Chem. Safety Summ.*, 1978, **49**(157), 3.
24. Anon., *BCISC Quarterly Safety Summary*, 1973, **44**, 18.
25. Mackenzie, J., *Chem. Engineering*, 1990 (June), 84.
26. Clark, M. C. *et al.*, *Chem. Ind.*, 1974, 113.
27. Anon., *Loss Prevention Bulletin*, 1986(071), 31.
28. Grassie, N., in Polymer Handbook. Brandrup, J. & Immergut, E. H. (eds), John Wiley & Sons, New York, 3rd edn., 1989. Section II, p. 365.
29. Buck, R. H., *J. Electrochem. Soc.*, 1966, **113**, 1352.
30. Mansfield, D., *Loss Prevention Bulletin*, 1990(094), 25.
31. De Haven, E. S. & Dietsche, T. J., *Plant/Operations Progress*, 1990, **9**(2), 131.
32. Anon., *Sichere Chemiarbeit*, 1977, **29**, 87.
33. Bretherick, L., *Handbook of Reactive Chemical Hazards* (4th edn), Butterworths, 1990.
34. Vacwell Engineering Co. Ltd. v. B.D.H. Chemicals, 1969, **3**, All E.R. 1681.
35. Kletz, T. A. & Redman, J., *The Chem. Engineer*, 28 Feb. 1991, 15.
36. Kletz, T. A., *Loss Prevention Bulletin*, 1991(100), 21.
37. Darlow, H. M., Safety in the Microbiology Laboratory; An Introduction in '*Safety in Microbiology*', Shepton, D. A., & Board, R. G. (eds), Academic Press, New York, 1972.
38. Sulkin, S. E., *Bact. Rev.*, 1961, **25**, 203.
39. Health & Safety Executive, *Health & Safety Research*, 1979, 72–76.
40. Bond, J., *Loss Prevention Bulletin*, 1985(065), 20.
41. Kletz, T. A., *Lessons from Disasters – How Organisations have no Memory*, Institution of Chemical Engineers, 1993.
42. Anon., *Chemical Safety Summary*, Chemical Industries Association, 1984, **55**, 469.

43. Morley, R. & Silk, S. J., *Ann. Occup. Hyg.*, 1970, **13**, 101.
44. Cloe, W. W., *Selected Occupational Fatalities related to Fire and/or Explosion in Confined Workspaces as found in reports of OSHA Fatality/Catastrophy Investigation Report OSHA/RP-82/002*, U.S. Dept. of Labour, April 1982.
45. Vervalin, C. K. (ed.), *Fire Protection Manual for Hydrocarbon Processing Plants* (3rd edn.), Gulf Publishing, Houston, TX, 1985, p. 95.
46. B.S. 3351, *Piping Systems for Petroleum Refineries and Petrochemical Plants*, British Standards Institution, 1971.
47. Hancock, B., *Safety and Loss Prevention in the Chemical and Oil Process Industries*, Instn. of Chem. Eng., Symp. Series, 1989, **120**, 589.
48. *Recommended Practice for the Design and Installation of Pressure Relieving Systems in Refineries, Part 1 – Design*; Appendix C. API RP-520, American Petroleum Institute, Washington, DC., 3rd edn., 1967.
49. (a) Barton, J. A. & Nolan, P. F., *Hazards X: Process Safety in Fine and Speciality Chemical Plants*, I. Chem. E. Symp. Ser. No. 115: **3** (1989).
    (b) Pantony, M. F., Scilly, N. F. & Bowton, J. A., *Plant/Operations Progress*, 1989, **8**(2), 113.
50. *Guidelines for Chemical Reactivity Evaluation and Application to Process Design*, A.I. Chem. E./CCPS, New York, 1993.
51. MCA Case History 1607, in *Quarterly Safety Summary of British Chemical Industries Safety Council*, Jan.–Mar. 1970, 5.
52. King, H. K., *Chemistry in Britain*, 1970, **6**, 231.
53. Anon., *Industrial Accident Prevention Bulletin*, Royal Soc. for the Prevention of Accidents, 1958, **280**, 26.
54. Dharmadhikeri, S. & Heck, G., *The Chemical Engineer*, 27 June 1991(499), 17.
55. Skinner, S. J., *Plant/Operations Progress*, 1989, **8**(4), 211.
56. Anon., *Loss Prevention Bulletin*, 1977(013), 4.
57. Taylor, B. J., *Safety News*, Universities Safety Association, No. 1 (January 1972), 5.
58. Carberry, J. J., *Chemical and Catalytic Reaction Engineering*, McGraw-Hill, New York, 1976.
59. Lee, H. H., *Heterogeneous Reactor Design*, Butterworth, 1985.
60. Rasbach, D. J., *The Chem. Eng.*, November 1970, CE 385.
61. Anon., *Loss Prevention Bulletin*, 1986(069), 33.
62. Craven, A. D., *Loss Prevention*, 1975, **9**, 60.
63. Lunn, G., *Venting Gas and Dust Explosions – A Review, Institution of Chemical Engineers*, 1984.
64. Kletz, T. A., *Over or Under-pressuring of Vessels, Accidents Illustrated*, Case History 23, ICI, UK, November 1970.
65. Hempseed, J. W., *Loss Prevention Bulletin*, 1991(097), 1.
66. Report, *The State*, Columbia, South Carolina, 11 September 1988.
67. Carson, P. A. & Mumford, C. J., *Hazardous Chemicals Handbook*, Newnes-Heinemann, 1994.
68. Manning, W. R. D. & Lebrow, S., *High Pressure Engineering*, Leonard Books, 1971, 356.
69. Dooner, R. & Marshall, V. C., *Loss Prevention Bulletin*, 1989(086), 5.
70. Anon., *Industrial Accident Prevention Bulletin*, Royal Soc. Prevention of Accidents, 1968, **26**(280), 25.

71. Kletz, T. A., *The Chem. Eng.*, December 1987(443), 44.
72. Anon., *British Welding Research Association Bulletin*, June 1966, **7**(6), 149.
73. Anon., *Loss Prevention Bulletin*, 1989(089), 13–17.
74. Health & Safety Executive, *The Fires and Explosion at BP Oil (Grangemouth) Refinery Ltd. – A Report of the investigations by the Health and Safety Executive into the fires and explosion at Grangemouth and Dalmeny, Scotland, 13 March, 22 March and 11 June 1987*, HMSO, 1989.
75. Kletz, T. A. & Lawley, H. G., *High Risk Safety Technology*. Green, A. E. (ed.), John Wiley, New York, 1982, 317.
76. Bond, J., *I. Chem. E. Symp. Series*, **102**, 37.
77. Assoc. Brit. Chem. Info., *Quarterly Safety Summary*, 1946, **17**, 24.
78. *Explosion from a Steam Line*, Report of Preliminary Inquiry No. 3471, London, HMSO, 1975.

Fig. 19.8 Low Pressure Separator End section projected 75 m in the explosion described on p. 1131.

CHAPTER 20

Process development and pilot planting

Costs of new product or process development are high, typically tens of millions of dollars for speciality chemicals. Stages involved in getting a product onto the market place include exploratory bench research, applied research including semi-tech and pilot plant trials, development usually in conjunction with an operating unit, and finally full-scale production for commercial exploitation.

The safety and commercial risks of process development and pilot planting have recently been reviewed.[1] Whilst many of the dangers associated with operating pilot plants are similar to those for bench-scale operations and even full-scale production, some unique hazards exist as described on pages 646–49. Safety testing is mentioned briefly in Chapters 11 and 13. The present chapter explains in more detail the requirements for a planned programme of acquiring data for new projects and how to design and run pilot plants safely.

Process development

Experience dictates that because of the complexity of process development a team approach is essential. Individuals must be identified with specific functional responsibilities, and formal systems are needed for documenting all safety data and hazard assessments, and for authorising progress from one stage to the next. Safety issues encountered with production can often be traced to

- the lack of any strategy for safety assessment during process development which has been reactive rather than proactive-based, e.g. with actions triggered by incidents, and data collated in an *ad hoc* rather than a systematic manner
- safety assessments conducted at the wrong time, e.g. after, rather than before, capital proposals are authorised
- frequent changes in project leadership
- the lack of any clear organisation or lines of responsibility

- no team approach or obvious 'ownership' for the project
- teams comprising the wrong people, often inadequately-qualified to decide on the data requirements or interpretation, or of too-low a grade within the organisation to initiate (or stop) work or to transcend inter-departmental boundaries
- politics and culture differences between research laboratories within the same company
- inadequate funding for safety studies
- last minute, and often unrecorded/unauthorised, modifications to plant design, process or formulations

Typical stages in bringing a new chemical process, or product, to successful commercial production are summarised in Fig. 20.1.[2] During process development it is usual to cover all aspects related to safety in the design and operation of the full-scale plant, embracing normal variations in operating parameters, non-specific fault conditions, and all conceivable abnormal situations including emergencies. How this may be done is illustrated in the figure. Such a structured procedure serves:

- to assist those carrying out exploratory, predevelopment or development work in laboratories to foresee avenues likely to lead to hazardous conditions;
- to indicate what additional tests may be necessary to establish fully the hazards of a new chemical or process.

A simple desk study is performed at the initital exploratory stage and consideration is given to side-reactions and comparisons with known hazardous chemicals. This can provide some early warning of a particular hazard, or indicate a preferred process route. Reference to literature or database information on unusual reactions, or accidents, may also be recorded at this stage.

At the predevelopment stage proposals begin to be formulated for production; this may introduce other hazards associated with the production methods. The requirements for additional testing will then become apparent and may influence the selection of process route. How much testing should be carried out at each of the stages of development of a specific chemical process is clearly chemical-property-, process-, and scale-related. However, describing many of the likely hazards enables an appropriate choice to be made, so most hazards should have been identified by the end of the development stage.

A record of the analyses made, and of the testing selected and performed, should be documented and retained for appropriate Health and Safety Reviews.

The evaluation procedure involves three stages of implementation, although in practice these overlap. They are
- a desk study at the exploratory, or early-development, stages
- a laboratory study at the exploratory, or early-development, stages
- a hazard study at the process route and definition development stage.

1144 Further considerations and developments

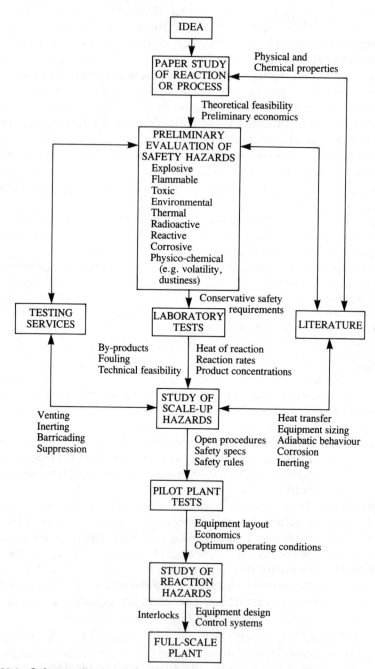

Fig. 20.1 Safety studies in process development

Desk study

Main reaction consideration

Examination starts by considering the basic chemical reaction and questions systematically the chemistry to assess what unplanned products could be produced. However, whilst the reaction kinetics can usually be well established, some catalytic reactions in both organic and inorganic chemistry can behave in bizarre and unruly ways.[3] Products are then compared with similar products to assess their likely explosive, flammable, toxic or environmental hazards.

In any event, as the following example illustrates, there is no substitute for a thorough survey of the literature on the chemical reaction hazards.

> An autoclave which was in use for the ethoxylation of a sugar derivative using ethylene oxide was destroyed by a violent explosion. The probable cause was the explosive decomposition of ethylene oxide vapour initiated by the temperature and pressure conditions in the autoclave. This hazard was not appreciated, so the method and scale of the proposed reactions were inappropriate and the temperature control was inadequate for the highly exothermic reaction.[4]
>
> It was found that some texts made no mention of the explosive nature of ethylene oxide vapour but others described previous autoclave explosions.[5,6] At that time manufacturers' warning labels often referred only to the flammability and toxicity of ethylene oxide, whereas even the liquid may undergo explosive self-polymerisation particularly in the presence of alcohols, amines, mineral or organic acids and bases (see Chapter 26).

Guidewords, adapted from HAZOPS terminology, may be used when considering the chemistry. For example,

None The reaction is carried out with *none* of one necessary chemical present, i.e. one chemical not added for some reason. Each chemical and the catalyst has to be considered in turn.

More or Less *More* or *less* of each necessary chemical is added to the reaction, i.e. greater or lower concentrations. This also applies to a catalyst. *More* or *less* temperature is also considered.

As Well As Contaminants are added to the chemical reaction *as well as* the normal chemicals involved. This can involve steam, water, Dowtherm, heat transfer salt, etc., as well as the product reacting further with a raw material.

> The first step in a chemical process involved dissolution of cyanuric chloride in acetone in a reactor. This normally involved no reaction but contamination of the acetone with wet acetone reduced its strength to 40%. The high water content led to an exothermic reaction resulting in excessive pressure which blew off a manway cover; acetone vapour was released and ignited externally.[7]

The 'As Well As' search is particularly relevant if trace contaminants may catalyse a reaction, such as decomposition.

> In the production of benzyl formate equal volumes of benzyl alcohol and phosgene were reacted in an excess of toluene at 12 °C to 16 °C. On one occasion the glass-lined kettle ruptured due to an internal explosion during vacuum distillation of the toluene. Corrosion of a ferrous alloy valve by the phosgene had provided ferric ions which catalysed decomposition of the ester.[8]

Change   Change in physical condition, e.g. frothing, misting; change in chemical condition, etc.

Instead of   One of the chemicals is substituted.

Even with the relatively small quantities of chemicals used in the laboratory changes between chemicals used for a similar function can significantly increase the hazard potential:

> Potassium dichromate and potassium permanganate are both familiar oxidising agents. As noted on page 278 chromic acid solution, a mixture of potassium dichromate and sulphuric acid, was commonly used as an agent for cleaning glassware; an explosion occurred when potassium permanganate was substituted for chromic acid.[9] The difference in reactivity is exemplified by, e.g. if 2.5 g of each chemical is mixed separately with 2.5 g of ethylene glycol the potassium permanganate mixture will ignite after 10 seconds but no ignition or heat release is evident with the potassium dichromate mixture.

Reactions in processing

Further reactions may occur during subsequent processing of the chemical. These have to be considered to determine whether any new hazards, or any unacceptable chemicals, are formed.

Reaction hazards

Chemical reaction hazards may stem from detonation/deflagration, that precludes manufacture in standard chemical plants, or problems with temperature or pressure. They may be particularly scale-dependent and include the following.

- Exotherms resulting from the reaction or decomposition.[10]

  > Exotherms of reactions become increasingly important in scale-up because, as explained on page 94, the heat transfer area-to-volume ratio decreases. Hence, a system which appears to exhibit a moderate laboratory exotherm may show a strong temperature rise on a pilot plant and could result in a runaway reaction on a commercial scale.[11]

- Gas evolution resulting from inherent characteristics of materials, e.g. raw materials used, products, products from side reactions, and residues from the reaction (refer to page 1091).

In the manufacture of ethyl polysilicate, silicon tetrachloride and industrial methylated spirits were run continuously from header tanks into an agitated steam-jacketed reactor. The initial rapid endothermic reaction was followed by further hydrolysis and condensation reactions.[12] Towards the end of one batch, explosive evolution of hydrogen chloride gas, albeit with an estimated energy release of <0.23 kg TNT, blew the cover plus agitator motor and gearbox off the reactor. The cause was unrevealed failure of the agitation which was exacerbated by the arrangement of feed pipes, with the denser silicon tetrachloride being introduced to the bottom of the reactor.

All hazards require assessment whether arising from the materials, contaminants, combinations, test procedures or changes in reaction conditions. Even when handling reactive substances in dilute form there may be possible hazards due to the concentration of explosive materials in a reaction system.

Dilute aqueous hydrogen peroxide is generally considered safe to handle provided simple precautions are taken (page 1099). Nevertheless, an explosion occurred during pretreatment of waste water with dilute hydrogen peroxide. An explosion also occurred during an experimental oxidation of an amine using dilute hydrogen peroxide. The explanation in both cases was believed to involve the production of explosive by-products which, following separation and concentration, exploded.[9]

Conditions have to be identified which are unfavourable to the conduct of each reaction; these include foaming, separation into layers, sudden boiling, spontaneous polymerisation, pulsation of liquids, cavitation processes, physical explosions, self-ignition as a result of adiabatic compression or the formation of explosive or toxic by-products, etc.[13] Catalytic effects of materials of construction or adsorbents also require assessment.

A 500 g bottle containing approximately 300 ml of vinyl ethyl ether had exceeded its expiry date by some 18 months. In preparation for its disposal it was filled with Vermiculite (exfoliated mica), to adsorb the liquid – to minimise evaporation and spillage in the event of container damage. Two or three days later it exploded.

Two reaction routes were possible, either a free-radical-initiated peroxidation process or a cation-induced rapid polymerisation. The latter was considered more likely, since explosions involving ether peroxide normally occur only after evaporation has taken place and when, e.g. a peroxide-encrusted bottle closure is disturbed or a bottle is subjected to shock. The original supply of vinyl ethyl ether had been stabilised, as supplied, with 0.1% ethyl aniline indicating its instability to acids. The exfoliated mica would have presented a large, possibly catalytic, surface area with numerous sites which could have promoted a cation-induced polymerisation.

Choice of chemistry

Each of the several possible routes to the chemical product should be examined to avoid the following.

- Particularly hazardous chemicals or intermediates.[14–16]

Benzidene and 2-naphthylamine were formerly used in the manufacture of dyes; alternative products were developed to avoid use of these carcinogens.[17]

Explosives manufacture involving nitration processes using nitric acid/sulphuric acid mixtures is potentially difficult and hazardous. Recent development trials have shown that nitration of rubbery explosive fuels can be performed cleanly and cheaply using dinitrogen pentoxide and that the fuel can be modified such that less solid oxidant is required in the mixture comprising an explosive/propellant.[18]

To manufacture carbaryl at Bhopal (page 883), α-naphthol and methylamine were reacted together to produce methyl isocyanate which was then reacted with phosgene. In an alternative process, using the same raw materials, α-naphthol and phosgene are reacted together to produce a chloroformate ester which is then reacted with methylamine to give carbaryl. No methyl isocyanate is formed in the latter process[19] although, clearly, phosgene is still involved.

- High temperature.
- High pressure.
- Large inventories of hazardous chemicals.

MDI and TDI are predominantly manufactured using phosgene but the quantities stored have been substantially reduced; no liquid phosgene is stored and phosgene gas flows directly from production to consumption.[15]

Literature survey

The standard texts[14–16] are useful sources to identify hazards. Several databases may also be consulted to determine whether any accidents have been reported with the particular chemicals involved.

A short time after the majority of the monoethanolamine had been pumped out of a tank, steam was admitted to a steam coil in the base to reduce the viscosity of the remainder. The paint on the side of the tank subsequently blistered due to the heat. When the steam was shut off, the tank imploded.[20]

A complex trisethanolamine iron, previously reported in the literature,[21] had formed on the steam coil.

$$\left[Fe \begin{pmatrix} NH_2-CH_2 \\ | \\ O-CH_2 \end{pmatrix}_3 \right] \cdot 3\,H_2O$$

Trisethanolamine iron

This complex decomposed at 110 °C, when the steam was on, to produce pyrophoric iron. The burning iron ignited some monoethanolamine, producing carbon dioxide and probably consuming all the oxygen in the tank atmosphere. The tank vent was blocked and when the gases cooled down, and some carbon dioxide dissolved in the remaining monoethanolamine, an implosion followed.

The following incident serves to illustrate that this is not an isolated example of an accident caused by complex formation.[22]

Two low-level dust fires and explosions occurred within three weeks of each other in a rotary dryer on an adipic acid plant. Although there was no serious damage the potential for injury, and for material or business loss was significant.

Inspection of the physico-chemical properties (Table 20.1) highlighted no obvious hazard in relation to the process.

Table 20.1 Physical properties of adipic acid

Melting point	152 °C (305 °F)
Solubility in water	2.5 wt.% at 25 °C (77 °F)
Specific gravity	1.34 at 18 °C (64.4 °F)
Boiling point	333 °C (626 °F) at 760 mm Hg (decomposes)
Flash point	196 °C (385 °F)
Autoignition temperature	420 °C (788 °F)

Examination of the dryer indicated that the fires started in the heated zone. There was no obvious source of ignition. Charred deposits and adipic acid were analysed by scanning electron microscopy, atomic absorption spectroscopy, thermal analysis or accelerating calorimetry. Data suggested that the incidents probably resulted from formation of an adipic acid–iron complex which decomposes as associated adipic acid, producing cyclopentanone, water and carbon dioxide. In the dryer the slow oxidation of adipic acid in contact with the catalyst causes sufficient adiabatic self-heating to result in ignition/ explosion. Removal of iron from the system prevents complex formation and hence the potential for ignition.

An ethereal solution of *N*-methyl-*N*-nitrosotoluene-4-sulphonamide was employed in a standard method to prepare diazomethane. In order to recover the reagent from a quantity of unused ethereal solution, which contained about 50 g of reagent, it was evaporated in a steam bath under reduced pressure in a clean round-bottomed 500 ml flask. Near the end of the evaporation the mixture contained crystals suspended in about 50 ml of ether; evaporation was continued and after 2 to 3 minutes an explosion occurred.[23]

The presence of peroxide in the ether was discounted since the diethyl ether used was from a freshly-opened bottle. The cause of the explosion was deduced to have been local overheating of crystals of reagent on the sides of the flask above the liquid surface. (In recrystallising *N*-methyl-*N*-nitrosotoluene-4-sulphonamide the temperature should not exceed 45 °C whereas in the case described the crystals would be exposed to temperatures approaching 100 °C.)

Laboratory study – exploratory/early development stage

Explosibility considerations

Within the UK it is necessary to meet the requirements of the Explosives Acts 1875 and 1923 and of S.1(i)c. of the Health and Safety at Work etc.

Act 1974. Moreover, under the European Directive for Major Accident Hazards, explosive substances which may explode under the effect of flame or which are more sensitive to shocks or friction than dinitrobenzene have to be notified.

The recommended procedures to decide whether a material may come under these requirements is as follows.
- Examine the chemical for particular groupings.
- Calculate the oxygen balance of the reactive or unstable molecule.
- Check the reaction of the material in a small-scale laboratory test.
- Check the chemical for decomposition.
- Submit the chemical for Health and Safety Executive Tests for explosibility.

These techniques are discussed in more detail in Chapter 11. Alternative legislation and appropriate tests will be applicable outside the UK but it is recommended that at least the above approach should be followed. Even then, however, extreme care is necessary in extrapolating the results of small-scale tests since some materials considered to be 'non-explosive' may become explosive if they are present in very large quantities due to the large-scale initiation effect.

> Following the explosion involving a 1:2 ratio double salt of ammonium sulphate and ammonium nitrate referred to on page 880, it proved difficult to explode this type of double salt in tests.[9] (Cases involving sodium chlorate are also described on pages 162 and 164.)

Atomic group characterisation

Chemicals containing the atomic groupings listed in Table 20.2 are known, by experience, to be unstable or explosive.[14]

Examples include copper acetylide, azo compounds, and a range of peroxy compounds.

> The shell of a methanol product condenser ruptured whilst one of the water channels was being removed in preparation for inspection and seven men sustained minor burns. This was a horizontal fixed tube heat exchanger constructed of copper. The shell (process) side had been steamed-out and left open to atmosphere for several hours over a period of three days.[20]
>
> The explosion emitted a blue-green flame and a black deposit subsequently found in the shell contained 9% cuprous acetylide. The explosion was concluded to have been due to copper acetylide decomposition, the source of acetylene being a minor process stream containing 300 ppm acetylene which was used intermittently.

Whilst the hazards associated with these chemicals may be partly mitigated by small-scale working, they remain serious even with laboratory quantities.

Table 20.2 Atomic groupings characterising explosive compounds

Bond groupings	Class
—C≡C—	Acetylenic compounds
—C≡C—metal	Metal acetylides
—C≡C—X	Haloacetylene derivatives
$\overset{N=N}{\underset{\diagdown}{\overset{\diagup}{C}}}$	Diazirines
CN$_2$	Diazo compounds
≥C—N=O	Nitroso compounds
≥C—NO$_2$	Nitroalkanes, C-nitro and polynitroaryl compounds
$\overset{NO_2}{\underset{NO_2}{C}}$	Polynitroalkyl compounds
≥C—O—N=O	Acyl or aklyl nitrites
≥C—O—NO$_2$	Acyl or alkyl nitrates
≥C—C≤ with O bridge	1,2-Epoxides
>C=N—O—metal	Metal fulminates or aci-nitro salts
$\overset{NO_2}{\underset{NO_2}{-C-F}}$	Fluorodinitromethyl compounds
>N—metal	N-Metal derivatives
>N—N=O	N-Nitroso compounds
>N—NO$_2$	N-Nitro compounds
≥C—N=N—C≤	Azo compounds
≥C—N=N—O—C≤	≤Arenediazoates
≥C—N=N—S—C≤	Arenediazo aryl sulphides
≥C—N=N—O—N=N—C≤	Bis-arenediazo oxides
≥C—N=N—S—N=N—C≤	Bis-arenediazo sulphides
≥C—N=N—N(R)—C≤	Trizaenes (R=H, —CN, —OH, —NO)
—N=N—N=N—	High-nitrogen compounds, tetrazoles
≥C—O—O—H ≥C.CO.OOH	Alkylhydroperoxides, peroxyacids
≥C—O—O—C≤ C.CO.OOR	Peroxides (cyclic, diacyl, dialkyl), peroxyesters

1152 *Further considerations and developments*

Table 20.2 (continued)

Bond groupings	Class
—O—O—metal	Metal peroxides, peroxoacid salts
—O—O—non-metal	Peroxoacids
N→Cr—O$_2$	Amminechromium peroxo- complexes
—N$_3$	Azides (acyl, halogen, non-metal, organic)
>C—N$_2$$^+O^-$	Arenediazoniumolates
≧C—N$_2$$^+S^-$	Diazonium sulphides and derivatives, 'xanthates'
N$^+$—HZ$^-$	Hydrazinium salts, oxosalts of nitrogenous bases
—N$^+$—OH Z$^-$	Hydroxylammonium salts
≧C—N$_2$$^+$ Z$^-$	Diazonium carboxylates or salts
(N—metal)$^+$ Z$^-$	Amminemetal oxosalts
Ar—metal—X X—Ar—metal	Halo-arylmetals
N—X	Halogen azides, *N*-halogen compounds, *N*-haloimides
—NF$_2$	Difluoroamino compounds, *N,N,N*-trifluoroalkyl compounds
—O—X	Alkyl perchlorates, chlorite salts, halogen oxides, hypohalites, perchloric acid, perchloryl compounds

1,3-Benzodithiolylium perchlorate was being prepared in accordance with a published method. The structure was given as:

During drying in a dessicator a severe explosion occurred and badly injured a postgraduate student. The detonation was caused when the dessicator lid, which had a ground glass flange, was being removed; a little of the compound had been deposited on the flange.

The preparation had been reported as safe but, although an eight-fold scale-up was involved, it was subsequently considered extremely dangerous.[24] (Extreme caution should be taken with all work including perchlorates, including ionic perchlorates which were believed to be less hazardous to handle.)

An organic azide compound was safely prepared up to the final step. Whilst shields were not in place, nor goggles worn, the final dried product was transferred from a funnel into a permanent container. An explosion occurred; the funnel was propelled into a wall, nearby glassware was shattered and the chemist was injured by glass fragments.[25]

It follows that, as discussed in Chapter 11, the integrity of laboratory apparatus is equally as important as that of rigidly constructed plant – particularly since, as emphasised throughout this discussion, although the scale is smaller the proximity of staff and their physical involvement in operations tends to be more problematic.

> An explosion occurred, spraying a technician in the face with molten sodium hydroxide, whilst hydrazine was being dried with sodium hydroxide pellets under vacuum. A leak of one of the taps in the system may have allowed air to enter.[24] Storage of solvents, particularly ethers, under poor conditions for over-long periods has resulted in many laboratory accidents due to peroxide formation with diisopropyl ether being most notorious. (Methods have thus been devised to control peroxide hazards in the use of ethers.[26] See also pages 601 and 602.)

Oxygen balance

As discussed further in Chapter 26, an explosive is a substance which will decompose to give gaseous products. It is often one which can use its own oxygen to develop a large volume of gas; the degree of availability is then an indication of the explosive potential.

The reaction is assumed to take the form

$$C_xH_yO_z + \left(x + \frac{y}{4} - \frac{z}{2}\right) O_2 \rightarrow xCO_2 + \frac{y}{2}H_2O$$

Hence, the lack of oxygen in the molecule required for this reaction, and thus the percentage measure of oxygen demand for the stoichiometric combustion is:

$$\frac{-1600 \left(2x + \frac{y}{2} - z\right)}{\text{Molecular weight of product}}$$

This is termed the 'oxygen balance' and typical values are given in Table 20.3.

Table 20.3 Typical 'oxygen balance' values

Glyceryl trinitrate	+ 4
Trinitrobenzene	− 56
Trinitrotoluene	− 74
Dinitrobenzene	− 95
Dinitrotoluene	−114
Acetic anhydride	−126
Dinitro m-xylene	−146
Nitrobenzene	−163
t-Butyl cumyl peroxide	−256
Propane	−364

If the oxygen balance is $>+240$ or <-160, the energy hazard potential is considered to be low. If the balance is between $+240$ and $+120$ or between -160 and -80 then the energy potential is considered to be medium; if it is between -80 and $+120$ the energy hazard potential is high. If the oxygen balance is >-200 (i.e. less $-$ve, more $+$ve) then other tests should be considered to screen the material for deflagrating and detonating properties.

The oxygen balance should not be considered in isolation but in conjunction with other factors, e.g. the presence, or absence, of reactive groups such as in Table 20.2. Common sense should prevail since, e.g. sodium bicarbonate has an oxygen balance of $+28.6$ whereas that of glyceryl trinitrate is $+4$. Similarly the oxygen balance of oxalic acid is -18 but it is not an explosive.

Small-scale laboratory test

As a preliminary to any testing for flash point, flammability limits, etc., or of powders which may be subject to drying operations, simple heating, deflagration and impact tests should be performed on products and by-products. These tests are recommended:
- to confirm the non-explosive nature of compounds
- to obtain visual information on the effects of heating compounds.

Further laboratory tests, if required, are described in ref. 27.

Decomposition

A number of chemicals, e.g. hydrazides, sometimes termed blowing agents, decompose at low temperature producing large volumes of gas, e.g. nitrogen and steam. A selection of such chemicals is listed in Table 20.4.[27a] Equipment for these products requires special design and a knowledge is required of activators, decomposition rates and temperatures.

A 1.8 m³ feed hopper which contained a hydrazide blowing agent ruptured due to overpressurisation. This hopper was provided with three bottles of Halon to suppress any dust explosion and this eventuality was discounted on good evidence. However, the hydrazide had a decomposition temperature of 150 °C; decomposition was exothermic and generated vast amounts of nitrogen and water vapour. This was deduced to have occurred, possibly initiated by friction and with the decomposition temperature reduced by traces of other hydrazide products used in the hopper. The exotherm initiated the explosion suppression system which added to the volume of gas resulting in overpressurisation[20] – see Chapter 19.

Health and Safety Executive (HSE) explosibility tests

If the chemical structure indicates that the chemical may have deflagrating or detonating explosive properties and the oxygen balance is >-200 then in the UK the chemical has to be submitted to the HSE for full testing.

Table 20.4 Substances with a high rate of decomposition

(a) The following substances are not explosives but decompose rapidly to produce large volumes of gas. They are substances not classified as deflagrating or detonating explosives but exhibit violent decomposition when subject to heat:

Material	Trauzel Lead Block value ($cm^3 gm^{-1}$)	Combustion properties
1,8-Bis(dinitrophenoxy)4,5-dinitroanthraquinone	18.5	Combustion propagates fully and fast with flame
100% Dinitrosopentamethylene tetramine	18.5	As above
2,4-Dinitroaniline	17.5	As above
2-Amino-3,5-dinitrothiophene	13–17.5	As above
1,5-Bis(dinitrophenoxy)4,8-dinitroanthraquinone	10.5	As above
2-Formylamino-3,5-dinitrothiophene	8	As above
2-Acetylamino-3,5-dinitrothiophene	7	As above
2-Anisidine nitrate	6	As above
80% DNPT	2.5	As above
6-Nitro-1-diazo-2-naphtholsulphonic acid	5	Combustion propagates fully and fast by smouldering
Ammonium nitrate	23	Local decomposition – no propagation of the decomposition
2-Bromo-4,6-dinitroaniline	17.5	As above
6-Bromo-2,4-dinitroaniline	17	As above
2-Chloro-4,6-dinitroaniline	16.5	As above

(b) The following substance exhibits a high rate of decomposition without combustion when exposed to heat and certain initiators:

Material	Decomposition temperature (°C)	Property
p,p'-Oxybis(benzenesulphonyl hydrazide)	150	Decomposes and propagates

The tests for explosibility are as outlined on page 621:
- Ciba Geigy Burning and Ignition Tests
- ICI Train Firing, Ignition Tube and Small Scale Thermal Stability Tests.

Further information is given in refs. 28 and 29.

Consideration of thermal hazards

Thermal hazards give rise to high temperature and, through this, to decomposition and gas evolution which may then result in a high pressure. Adequate cooling and venting to control thermal hazards depends upon the establishment of appropriate data.

> Interest in the evaluation of the energy hazards of unstable substances was increased in Japan when in 1980 several hundred kilograms of a pharmaceutical

intermediate 5CT (5-chloro-1,2,3-thiadiazole) exploded for no apparent reason. Two persons were killed and 13 injured.[9]

Experimental techniques for thermal stability classification – The experimental techniques for classification of thermal stability (page 621 *et seq.*) are as follows:

Technique	Condition of interest
Differential thermal methods (DTA, DSC)	Presence of exotherm
Constant temperature stability (CTS)	Absence of exotherm for 2 hours
Heating under confinement	Temperature and pressure change
Large-scale holding tests including adiabatic calorimetry	Thermal mass effect

Thermodynamic data – Data of importance[30–34] are, as exemplified in Chapter 10,
- heats of combustion of raw materials and products
- heats of reaction, including side reactions
- maximum enthalpy of decomposition of all raw materials and products
- heat of formation of raw materials and products
- heat of polymerisation.

The CHETAH computer program (page 620) is available to calculate various thermodynamic parameters including oxygen balance, heat of combustion and maximum enthalpy from the parameters on an overall hazard rating. The use of this program assists in identification of those chemicals within which there is a potential for energy release. Further information on reactive chemical screening is given in Chapter 11 and ref. 35.

Accelerating rate calorimetry – Accelerating rate calorimetry is used to hold a sample in an adiabatic condition once an exothermic reaction is detected and then to observe its progress over a period of time.

> A process involved reaction of an isocyanate with an oxine to produce a blocked isocyanate. No problem was encountered in laboratory tests but reactive chemical screening tests were run before pilot-plant scale-up. During a DSC test a large exotherm was detected very close to the operating temperature. An accelerating rate calorimetry test was used to determine heat duty and venting requirements and lower temperatures chosen for process operation.[35]

Thermal decomposition – Some powders may exhibit exothermic activity in bulk.[36]

Reaction Hazard Index

The Reaction Hazard Index links the Arrhenius activation energy with the energy generated by the decomposition reaction. The latter is replaced by the maximum adiabatic temperature reached by the products of a decomposition reaction. Typical data are given in Fig. 20.2.

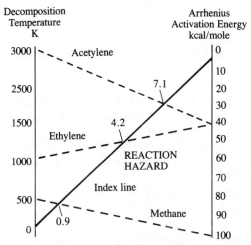

Fig. 20.2 Reaction Hazard Index data for methane, ethylene and acetylene

Values of the Reaction Hazard Index and the National Fire Protection Association Reactivity Rating[15] provide some indication of the hazard of handling the chemical.

More detailed guidance on chemical reaction hazard analysis is given in ref. 2.

Flammability considerations

The next factor to be considered is flammability.[37] Whether solid, liquid or gas it is important to establish flammability limits (and, for liquids, the flash point), sensitivity to ignition, and the potential violence of any ensuing fire or explosion. For solids the most important initial test is that for Fire and Explosion Classification, required in the UK under the Factories Act 1961 Part II S.31. For liquids the most important test is for flash point, required under various statutory provisions.

Flammability tests: dusts or powders – Many statutory authorities classify powders according to whether or not the solid will ignite and propagate a flame when suspended in air. The test is performed on powder <1400 μm diameter. In the UK the classification is
- Group (a), dusts which ignite and propagate flame
- Group (b), dusts which do not ignite and propagate a flame.

If the solid does not propagate a flame, even when in the form of a fine dust, no further flammability testing is required. If a flame is propagated

then the following tests may be necessary depending on design work requirements:[28, 36–41]
- Explosible Concentration
- Minimum Ignition Temperature
- Minimum Ignition Energy
- Maximum Permissible Oxygen Concentration to prevent ignition
- Maximum Explosion Pressure and Rate of Pressure Rise.

Spontaneous combustion in solid materials – Many solid products can catch fire as a result of self-heating, a particular problem when bagged material is stored in a warehouse. Tests, and further references, are given in ref. 42.

Flammability tests for liquids – Flash point determination for liquids[43] is required in order to classify them under various items of legislation as summarised in Table 20.5.

Furthermore, in the UK the CIMAH Regulations have certain requirements[44] for
- Flammable gas
  Substances which in the gaseous state at normal pressure and mixed with air become flammable and the boiling point of which at normal pressure is <20 °C.
- Highly flammable liquids
  Substances which have a flash point <21 °C and the boiling point of which at normal pressure is >20 °C.
- Flammable liquids
  Substances which have a flash point <55 °C and which remain liquid under pressure, where particular processing conditions, such as high pressure and high temperature, may create major accident hazards.

Tests of Flammable Limits,[45, 46] Minimum Ignition Energy, Auto-ignition Temperature[47] and Flame Speed may also be required.

Toxicity considerations[48–53]

For the purpose of toxicity hazard evaluation one recommended procedure is to assign each substance to one of the Hazard Categories listed in Table 20.6.[54] More generally, however, toxic substances are classified as:

Very toxic — A substance which if it is inhaled or ingested or it penetrates the skin may involve extremely serious acute or chronic health risks and even death.

Toxic — A substance which if it is inhaled or ingested or it penetrates the skin may involve serious acute or chronic health risks or even death.

Harmful — A substance which if it is inhaled or ingested or it penetrates the skin may involve limited health risks.

Table 20.5 Legislation including classification by flash point

Requirements	Classification	Flash point (°C) (closed cup)
United Nations	Inflammable Liquids: Class 3.1 Class 3.2	 <23 23 to 60.5
IMO/Blue Book (Shipping)	Inflammable Liquids: Class 3.1 Class 3.2 Class 3.3	 < minus 18 minus 18 to < 23 23 to 61
ICAO/IATA (Air Transport)	Flammable Liquid	≯60.5
ADR/RID (International Road/Rail)	Inflammable Liquids – Not miscible or only partially miscible with water 3.1° ⎱ (≯30% solids) 3.2° ⎰ (>30% solids) 3.3° 3.4° Inflammable Liquids – Miscible in all proportions with water 3.5°	 <21 21 to 55 >55 to 100 <21
EC Dangerous Substances Directive (Packaging and Labelling of Dangerous Substances Regulations, Council Directive 67/548/EEC as adopted for 12th time 1st March 1991	Extremely Flammable Highly Flammable Flammable	<0 0 to <21 21 to 55
Substances defined by the Petroleum (Consolidation) Act 1928	Highly Inflammable	<23
Highly Flammable Liquids and Liquefied Petroleum Gases Regulations 1972	Highly Flammable	<32
Dangerous Substances (Conveyance by Road in Road Tankers and Tank Containers) Regulations 1981	Flammable Liquid	<55
European Model Code of Safe Practice in the Storage of Petroleum Products. Part 1. Operation	Class 0 Liquefied Petroleum Gas (LPG) Class I Class II Class III Unclassified	 <21 21–55 55–100 >100
Postal Regulations (UK)	Inflammable Liquids	≯60.5

Corrigendum   Corrosive   A substance which may on contact with living tissues destroy them.

Irritant   A non-corrosive substance which, through immediate, prolonged or repeated contact with the skin or mucous membrane, can cause inflammation.

Carcinogenic   A substance which if it is inhaled or ingested or it

Table 20.6 Guidelines for toxic hazard categories

EXTREME HAZARD	Substances of known or suspected exceptional toxicity (e.g. carcinogens)
HIGH HAZARD	All substances whose toxicity exceeds that of the medium hazard category, except for those known or believed to be so highly toxic as to merit special precautions (i.e. those in the 'extreme' category)
MEDIUM HAZARD	Substances meeting criteria for CPL* classification as 'Harmful' or 'Irritant'
LOW HAZARD	Substances not matching criteria for CPL* classification as 'Harmful' or 'Irritant'

* CPL = the Classification, Packaging and Labelling Regulations 1984.‡
1. The toxicity considered should be that of the substance or mixture handled, including any impurities.
2. Substances may have other properties (e.g. flammability) which may call for additional precautions.
3. The above general guidance may need to be supplemented by developing additional criteria with the help of expert toxicological advice.
‡ Replaced in September 1993 by The Chemicals (Hazard Information and Packaging) Regulations.

	penetrates the skin may induce cancer in man or increase its incidence.
Teratogenic	A substance which if it is inhaled or ingested or it penetrates the skin may involve a risk of subsequent non-hereditable birth defects in offspring.
Mutagenic	A substance which if it is inhaled or ingested or it penetrates the skin may involve a risk of hereditable genetic defects.

The criteria for the first three categories are given in Table 20.7.

Table 20.7 Criteria for Very Toxic, Toxic and Harmful categories

Category	Median Lethal Dose (LD_{50})		Median Lethal Concentration (LC_{50})
	Absorbed orally in rat (mg/kg)	Absorbed percutaneously in rat or rabbit (mg/kg)	Absorbed by inhalation in rat (mg/litre) (4 hours)
Very toxic	≤25	≤50	≤0.5
Toxic	>25–200	>5–400	>0.5–2
Harmful	>200–2000	>400–2000	>2–20

If facts show that for the purposes of classification it is inadvisable to use LD_{50} and LC_{50} values as a principal basis because the substances produce other effects, the substance shall be classified according to the magnitude of these effects. In fact, as explained on page 252, animal test data must be interpreted with caution since different animal species and humans may react differently to a specific exposure. Moreover in risk assessment the potential routes for, and the rapidity of, entry into the body are obviously important factors.

> The single dose LD_{50} (monkeys) for 2,3,7,8-tetrachlorodibenzo-*p*-dioxin [i.e. TCDD, page 510] is 50 µg/kg. The comparable value for hydrogen cyanide is 500 µg/kg. However the risk posed by the latter is enhanced considerably by its ability to enter the body.[54a]

Toxic chemicals may be considered under two groupings:
- Acutely toxic, which have an immediate effect upon a person
- Chronically toxic, which have an effect after prolonged exposure to small quantities.

A chronic poison may, of course, become an acute poison if exposure occurs to a sufficient quantity of material. All chemicals with a sufficiently high vapour pressure may act as asphyxiants by excluding oxygen, i.e. to cause oxygen deficiency (page 242).

In planning any chemical reaction at the laboratory stage consideration has to be given to the potential for the production of toxic chemicals either by first intent, as a by-product or as a result of a subsidiary reaction.

Chemicals affecting different target organs are discussed in Chapter 5.

Acutely toxic chemicals

One medical classification for poisons is:

Medical classification	Overall effect
Nerve agents	Lethal via nervous system
Lung damaging agents	Lethal via choking
Blood-affecting agents	Lethal via blood system
Vesicant agents	Damaging – blisters
Psychomimatic agents	Incapacitating – mental
Vomiting agents	Incapacitating – physical
Sensory irritants	Lachrymatory, etc.
Miscellaneous agents	

Nerve agents are based on the phosphonofluoridic acid structure. These and related chemical structures which must therefore be avoided completely are:

$$\begin{array}{ccc}
R_1\text{—O}\diagdown\;\diagup\text{O} & R_1\text{—O}\diagdown\;\diagup\text{O} & R_1\text{—O}\diagdown\;\diagup\text{O} \\
P & P & P \\
\diagup\;\diagdown & \diagup\;\diagdown & \diagup\;\diagdown \\
R_2 \quad F & R_2N \quad CN & R_2 \quad SR_3
\end{array}$$

where R_1, R_2 and R_3 are organic groups.

> A fitter in a pilot plant for nerve gas production was instructed to enter a storage cubicle without wearing protective clothing or a gas mask. He noticed a

small drop hanging from a Sarin GB gas-bearing pipe. His chest and eyes became painful and subsequently his health deteriorated until he suffered from shortness of breath, dizziness, general weakness and depression. Some years later most of these symptoms were confirmed as due to exposure to an organic phosphorus compound.[54b]

Lung-damaging agents do not have particular chemical structures but some of the best known include chlorine (page 1730) and phosgene (page 1734). Many other gases also cause choking.

Blood-affecting agents replace oxygen in the blood system and are therefore transmitted to other parts of the body, often with serious consequences. The best known examples are acetone cyanhydrin (which readily decomposes to acetone and hydrogen cyanide), amyl nitrite, carbon monoxide, hydrogen cyanide, methyl dichloroarsine, nickel carbonyl and phenol.

Vesicant agents cause blistering of the skin. They are based on sulphur or nitrogen mustards having the general formulae:

$$R(CH_2CH_2Cl)_2$$

where R is either sulphur or CH_3N.

Other chemicals with similar effects are beta-chlorovinyl-dichloroarsine (Lewisite), $ClCH:AsCl_2$; and phosgene oxine, $Cl_2C=N-OH$.

Psychomimetic agents (which primarily affect the nervous system) include: *LSD* (lysergic acid diethylamide) – a well-known hallucinogen;

LSD

Bz (3-quinuclidinyl benzilate), which has similar effects to LSD. A film of its use on a soldier on guard duty showed that, instead of challenging a stranger, he forgot the relevant password and let the stranger pass whilst he pondered on what to do next;

BZ

CS (*o*-chlorobenzylidene malonitrile), which is extremely irritating and acts on exposed sensory nerve endings (e.g. eyes and respiratory system);

$$\text{CS structure: 2-chlorobenzylidene malonitrile, } C_6H_4Cl\text{-}CH=C(CN)_2$$

Sensory irritants principally affect the eyes, causing watering. They include ethyl iodoacetate, xylylene bromide, alpha-bromobenzyl cyanide and 2-chloroacetophenone. Other compounds such as diphenylcyanoarsine can cause sneezing.

Miscellaneous agents include the fluor acetates and fluor citrates.

An alternative taxonomy with illustrative examples is given in Chapter 5.

Acutely toxic chemicals, carcinogens or sensitisers should, in general, be avoided at the planning stage of any chemical reaction; if they are encountered their properties must be fully investigated. Authority to proceed is then a decision for the highest level of technical management. At this stage data also need to be established for setting in-house hygiene standards for new materials.

Under COSHH Regulations the following chemicals must not be manufactured in the UK for any purpose:[55] 2-naphthylamine, benzidine, 4-aminobiphenyl, 4-nitrodiphenyl, their salts and any substance containing any of those compounds, except as the by-product of a chemical reaction, in any other substance in a total concentration not >1 per cent. Other materials are prohibited in identified processes (e.g. white phosphorus in matches).

Health surveillance is necessary for the chemicals listed in Table 30.14.

Environmental considerations

Environmental considerations and their control are discussed in Chapters 21 and 22, respectively.

Two types of chemicals are particularly harmful to the environment:[56-58]
- those that are persistent in the environment
- those that bioaccumulate in the cell structure.

A 'black-list' of 129 chemicals harmful with respect to the environment is provided in List 1 of Council Directive 76/464/EEC, reproduced as Table 20.8.

Radioactivity considerations

Specialist expertise should be sought if radioactive chemicals are involved (Chapter 8).

Table 20.8 Chemicals listed as being potentially harmful to the environment

LIST I OF FAMILIES AND GROUPS OF SUBSTANCES

List I contains the individual substances which belong to the families and groups of substances enumerated below, with the exception of those which are considered inappropriate to List I on the basis of a low risk of toxicity, persistence and bioaccumulation.

Such substances which with regard to toxicity, persistence and bioaccumulation are appropriate to List II are to be classed in List II.

1. Organohalogen compounds and substances which may form such compounds in the aquatic environment.
2. Organophosphorus compounds.
3. Organotin compounds.
4. Substances which possess carcinogenic, mutagenic or teratogenic properties in or via the aquatic environment. Where certain substances in List II are carcinogenic, mutagenic or teratogenic, they are included in category 4 of this list.
5. Mercury and its compounds.
6. Cadmium and its compounds.
7. Mineral oils and hydrocarbons.
8. Cyanides.

LIST II OF FAMILIES AND GROUPS OF SUBSTANCES

List II contains the individual substances and the categories of substances belonging to the families and groups of substances listed below which could have a harmful effect on groundwater.

1. The following metalloids and metals and their compounds:
   1. Zinc
   2. Copper
   3. Nickel
   4. Chromium
   5. Lead
   6. Selenium
   7. Arsenic
   8. Antimony
   9. Molybdenum
   10. Titanium
   11. Tin
   12. Barium
   13. Beryllium
   14. Boron
   15. Uranium
   16. Vanadium
   17. Cobalt
   18. Thallium
   19. Tellurium
   20. Silver
2. Biocides and their derivatives not appearing in List I.
3. Substances which have a deleterious effect on the taste and/or odour of groundwater, and compounds liable to cause the formation of such substances in such water and to render it unfit for human consumption.
4. Toxic or persistent organic compounds of silicon, and substances which may cause the formation of such compounds in water, excluding those which are biologically harmless or are rapidly converted in water into harmless substances.
5. Inorganic compounds of phosphorus and elemental phosphorus.
6. Fluorides.
7. Ammonia and nitrites.

List of substances which could belong to List I of Council Directive 76/464/EEC

*** Substances which are the subject of a proposal or a communication to the Council.
** Substances which have been or are being studied.
* Substances to be studied next.

CAS number (Chemical Abstract Service)

1.	***	309-00-2	Aldrin
2.		95-85-2	2-Amino-4-chlorophenol
3.		120-12-7	Anthracene
4.	**	7440-38-2	Arsenic and its mineral compounds
5.		2642-71-9	Azinphos-ethyl

Table 20.8 (continued)

6.		86-50-0	Azinphos-methyl
7.	**	71-43-2	Benzene
8.	**	92-87-5	Benzidine
9.		100-44-7	Benzyl chloride (alpha-chlorotoluene)
10.		98-87-3	Benzylidene chloride (alpha, alpha-dichlorotoluene)
11.		92-52-4	Biphenyl
12.	***	7440-43-9	Cadmium and its compounds
13.	**	56-23-5	Carbon tetrachloride
14.		302-17-0	Chloral hydrate
15.	***	57-74-9	Chlordane
16.		79-11-8	Chloroacetic acid
17.		95-51-2	2-Chloroaniline
18.		108-42-9	3-Chloroaniline
19.		106-47-8	4-Chloroaniline
20.	*	108-90-7	Chlorobenzene
21.		97-00-7	1-Chloro-2,4-dinitrobenzene
22.		107-07-3	2-Chloroethanol
23.	**	67-66-3	Chloroform
24.		59-50-7	4-Chloro-3-methylphenol
25.		90-13-1	1-Chloronaphthalene
26.			Chloronaphthalenes (technical mixture)
27.		89-63-4	4-Chloro-2-nitroaniline
28.		89-21-4	1-Chloro-2-nitrobenzene
29.		88-73-3	1-Chloro-3-nitrobenzene
30.		121-73-3	1-Chloro-4-nitrobenzene
31.		89-59-8	4-Chloro-2-nitrotoluene
32.			Chloronitrotoluenes (other than 4-chloro-2-nitrotoluene)
33.		95-57-8	2-Chlorophenol
34.		108-43-0	3-Chlorophenol
35.		106-48-9	4-Chlorophenol
36.		126-99-8	Chloroprene (2-chlorobuta-1,3-diene)
37.		107-05-1	3-Chloropropene (allyl chloride)
38.		95-49-8	2-Chlorotoluene
39.		108-41-8	3-Chlorotoluene
40.		106-43-4	4-Chlorotoluene
41.			2-Chloro-p-toluidine
42.			Chlorotoluidines (other than 2-chloro-p-toluidine)
43.		56-72-4	Coumaphos
44.		108-77-0	Cyanuric chloride (2,4,6-trichloro-1,3,5-triazine)
45.		94-75-7	2,4-D (including 2,4-D-salts and 2,4-D-esters)
46.	**	50-29-3	DDT (including metabolites DDD and DDE)
47.		298-03-3	Demeton (including Demeton-o, Demeton-s, Demeton-s-methyl and Demeton-s-methylsulphone)
48.	*	106-93-4	1,2-Dibromethane
49.			Dibutyltin dichloride
50.			Dibutyltin oxide
51.			Dibutyltin salts (other than dibutyltin dichloride and dibutyltin oxide)
52.			Dichloroanilines
53.		95-50-1	1,2-Dichlorobenzene
54.		541-73-1	1,3-Dichlorobenzene
55.		106-46-7	1,4-Dichlorobenzene
56.			Dichlorobenzidines
57.		108-60-1	Dichlorodiisopropyl ether
58.	*	75-34-3	1,1-Dichloroethane
59.	*	107-06-2	1,2-Dichloroethane
60.	*	75-35-4	1,1-Dichloroethylene (vinylidene chloride)

Table 20.8 (continued)

61.	*	540-59-0	1,2-Dichloroethylene
62.	*	75-09-2	Dichloromethane
63.			Dichloronitrobenzenes
64.		120-83-2	2,4-Dichlorophenol
65.	*	78-87-5	1,2-Dichloropropane
66.		96-23-1	1,3-Dichloropropan-2-ol
67.		542-75-6	1,3-Dichloropropene
68.		78-88-6	2,3-Dichloropropene
69.		120-36-5	Dichlorprop
70.		62-73-7	Dichlorvos
71.	***	60-57-1	Dieldrin
72.		109-89-7	Diethylamine
73.		60-51-5	Dimethoate
74.		124-40-3	Dimethylamine
75.		298-04-4	Disulfoton
76.	**	115-29-7	Endosulfan
77.	***	72-20-8	Endrin
78.		106-89-8	Epichlorohydrin
79.		100-41-4	Ethylbenzene
80.		122-14-5	Fenitrothion
81.		55-38-9	Fenthion
82.	***	76-44-8	Heptachlor (including heptachlorepoxide)
83.	**	118-74-1	Hexachlorobenzene
84.	**	87-68-3	Hexachlorobutadiene
85.	**	608-73-1 58-89-9	Hexachlorocyclohexane (including all isomers and lindane)
86.		67-72-1	Hexachloroethane
87.		98-83-9	Isopropylbenzene
88.		330-55-2	Linuron
89.	*	121-75-5	Malathion
90.		94-74-6	MCPA
91.		93-65-2	Mecoprop
92.	***	7439-97-6	Mercury and its compounds
93.		10265-92-6	Methamidophos
94.		7786-34-7	Mevinphos
95.		1746-81-2	Monolinuron
96.		91-20-3	Naphthalene
97.		1113-02-6	Omethoate
98.		301-12-2	Oxydemeton-methyl
99.	**		PAH (with special reference to: 3,4-benzopyrene and 3,4-benzofluoranthene)
100.		56-38-2 298-00-0	Parathion (including parathion-methyl)
101.	**		PCB (including PCT)
102.	**	87-86-5	Pentachlorophenol
103.		14816-18-3	Phoxim
104.		709-98-8	Propanil
105.		1698-60-8	Pyrazon
106.		122-34-9	Simazine
107.		93-76-5	2,4,5-T (including 2,4,5-T salts and 2,4,5-T esters)
108.			Tetrabutyltin
109.		95-94-3	1,2,4,5-Tetrachlorobenzene
110.	*	79-34-5	1,1,2,2-Tetrachloroethane
111.	*	127-18-4	Tetrachloroethylene
112.		108-88-3	Toluene
113.		24017-47-8	Triazophos
114.		126-73-8	Tributyl phosphate

Table 20.8 (continued)

115.		Tributyltin oxide
116.	52-68-6	Trichlorfon
117. *		Trichlorobenzene (technical mixture)
118.	120-82-1	1,2,4-Trichlorobenzene
119. *	71-55-6	1,1,1-Trichloroethane
120. *	79-00-5	1,1,2-Trichloroethane
121. *	79-01-6	Trichloroethylene
122. **	95-95-4 88-06-2	Trichlorophenols
123.	76-13-1	1,1,2-Trichlorotrifluoroethane
124.	1582-09-8	Trifluralin
125.	900-95-8	Triphenyltin acetate (fentin acetate)
126.		Triphenyltin chloride (fentin chloride)
127.	76-87-9	Triphenyltin hydroxide (fentin hydroxide)
128.	75-01-4	Vinyl chloride (chloroethylene)
129.		Xylenes (technical mixture of isomers)

Testing and notification of new products

As discussed on page 929 those who manufacture or supply chemicals have a duty, arising in the UK from the Health and Safety at Work Act, to arrange for adequate testing and examination of them.

Many countries operate mandatory premanufacturing and premarketing notification schemes of which safety testing is the cornerstone. Within the European Community[59] Competent Authorities must be notified before new substances are supplied in the marketplace. The exceptions to this are:
- substances listed in EINECS
- medicinal products, narcotics and radioactive substances
- pesticides and fertilisers
- foodstuffs and feedingstuffs
- substances in transit under customs supervision
- waste substances
- munitions and explosives
- certain categories of polymer

The amount, and type, of data required in the regulatory submissions depends on the quantity supplied. The tonnage requirements are:
- Limited Announcement: <1 tonne per year
- The Base Set: >1 tonne per year
- Levels 1 and 2: >10 tonnes per year

Unlike the requirements for >1 tonne, member states have differing test requirements and formats for Limited Announcements which must be made in each member state in which the substance is supplied. The requirements for the UK are given in Table 20.9.

The test requirements for the Base Set (Level 0) are given in detail in Table 13.5 and summarised in Table 20.10. With the data the submission dossier should contain the test reports plus chemical specification, intended

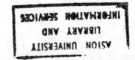

Table 20.9 UK guidance for Limited Announcements for the supply of new substances

Quantity	For supply to:	Tests
I: up to 1 kg p.a.	Public (including as part of a formulation)	Acute toxicity (oral, dermal or inhalation as appropriate to expected use(s)), skin irritation
II: 1 kg to 100 kg p.a.	Public (including as part of a formulation)	Category I + Ames test
III: 100 kg to 1 tonne p.a.	Public (including as part of a formulation)	Category II + eye irritation (unless skin irritation test positive), skin sensitisation
IV: up to 1 tonne p.a.	Within own factory or to skilled persons	As for category III
V: over 1 tonne for up to 1 year only (for 'commercial development')	Limited number of registered industrial users	As for category III

Some information on the ecotoxicity of the substances may also be desirable for category V and, if it seems likely that harmful amounts may reach the environment, for categories III and IV.

Table 20.10 Base Set tests for EC notification

Physico-chemical properties
Melting point
Boiling point
Relative density
Vapour pressure
Surface tension
Water solubility
Fat solubility
Partition coefficient
Flash point
Flammability
Explosivity
Autoflammability
Oxidising properties

Toxicological properties
Acute toxicity – by two routes of administration
Skin irritation
Eye irritation
Skin sensitisation
28-Day subacute toxicity
Ames test
In vitro metaphase analysis (or mouse micronucleus test)

Ecotoxicological properties
Acute toxicity to fish
Acute toxicity to *Daphnia*
Biodegradability
Hydrolysis (abiotic degradability)

use, immediate and longer-term marketing volumes, a declaration concerning any unfavourable effects of the substance, proposed classification and labelling, and recommended precautions for safe use.

The precise test requirements for Level 1 will be subject to the results of

Level 0 and the exposure patterns to people and the environment. Examples of Level 1 tests which *may* be required at 10 tonnes per year, or 50 tonnes total, and *will* be required at 100 tonnes per year, or 500 tonnes total, are given in Table 20.11.

Table 20.11 Level 1 tests for EC notification

Toxicological	Ecotoxicological
Fertility study (one generation)	Algal growth inhibition
Embryotoxicity (one species)	21-Day *Daphnia* toxicity
Subchronic/chronic toxicity (one species)	Prolonged toxicity to fish
Additional mutagenicity	Bioaccumulation in fish
	Prolonged biodegradation
	Effect on higher plants
	Effect on earthworms

Level 2 studies which will be required at 1000 tonnes per year, or 5000 tonnes total, unless the supplier can justify their exclusion to the Competent Authorities, are listed in Table 13.6 and summarised in Table 20.12.

Table 20.12 Level 2 tests for EC notification

Toxicological	Ecotoxicological
Chronic toxicity	Prolonged fish toxicity (including reproduction)
Carcinogenicity	
Fertility study (multi-generation)	Avian acute/subacute toxicity
Embryotoxicity (non-rodent)	Accumulation, degradation and mobility tests
Acute/subacute toxicity in second species	
Toxicokinetics	

Until safety data are available the substance should be labelled

'Caution – substance not yet fully tested'

All testing to support notification must be performed by specified methods[59] and in accordance with the principles of Good Laboratory Practice.[87] The detailed testing requirements are currently under review and publication of the 7th amendment to the Directive is imminent.

Hazard study – process route and definition development stage

During definition of the process for the chemical, each additional hazard which may arise from the selected route has to be considered and efforts made to minimise it. This may involve either additional work or selection of a new route.

Physical properties

Additional information is likely to be required on physical properties. The physical properties of all materials used and produced, as well as likely by-products, should be ascertained and recorded on an appropriate experimental chemical safety data sheet. Not all properties are necessarily required but in addition to these chemical properties and reactions, physiological effects, special protection, information, and data on materials of construction will normally be included. The headings for data sheets are summarised on page 794 together with the main properties which may be relevant (page 656). Some physical property data and calculation methods are given in refs. 11 and 60–63.

The acquisition of fundamental thermodynamic data is often essential for scientific scale-up.

> In the laboratory preparation of an organic chemical one of the raw materials was isolated from a reaction mixture using a typical batch distillation, i.e. a rectification section only. This was scaled-up directly to commercial plant operation. For 10 years it proved impracticable to recover the raw material below 10 to 15 wt.% in the bottoms; this was imagined to be due to complex formation and formation of an azeotrope in the reboiler. Eventually the selective volatility of the light raw material to the next heavy component was found to be 10 (compared with a value of 1.0 for an azeotrope). A simple McCabe–Thiele calculation showed that a top-fed continuous stripping column could reduce the light component to parts per million in the bottoms and yield an overhead stream containing 80% to 90% light component. This was then confirmed by running a miniplant distillation column.[64]

Process chemistry considerations

The process chemistry then needs to be examined with regard to all possible deviations. The possibilities are, of course, dependent on the specific chemicals used or produced but include the following.

Unplanned chemistry

A common feature of numerous explosions involving perchlorates is that they have been unexpected and the workers have been inadequately protected.

> Preparation was undertaken of a reaction mixture of dimethyl sulphoxide (1.5 g) and perchloric acid (3 g) in methylene chloride (5 ml). The dimethyl sulphoxide had been measured out using a 5 ml syringe. This syringe had been washed out in methylene chloride and was being used to withdraw the perchloric acid from the stock bottle when there was a violent explosion which shattered the syringe.[65] Similar transfers of perchloric acid had been made over several years without incident albeit usually using either acetonitrile or chloroform as solvent. The methylene chloride may have formed an explosive

mixture with perchloric acid with, or without, the presence of a small amount of dimethyl sulphoxide.

More commonly hazards may arise from circumstances such as the following.
- Unsuspected changes in feed composition.

Helium contaminated with air and/or nitrogen was returned to suppliers for purification by passage through a carbon bed which was cooled by vaporising liquid nitrogen, at -196 °C, in a jacket. The oxygen concentration in the feed was normally far less than that of nitrogen. On one occasion, however, it contained 1.3% nitrogen and 2.2% oxygen; the oxygen thus condensed out preferentially and the gas adsorbed on the top of the carbon bed was 85% oxygen. This carbon and oxygen mixture exploded causing extensive blast and missile damage.[66] It had not previously been appreciated that any change in feed composition could be hazardous. (Subsequently more rigorous analytical procedures were introduced and the carbon adsorbent was replaced with silica gel.)

- Peroxide formation. Examples of common peroxidisable chemicals and precautions necessary with them are summarised on page 600.
- Pyrophoric problems (page 1366).
- Polymerisation.

A two-compartment tank was used for styrene blending. One compartment was full; the other was empty and being steamed-out. After two days steaming a jet of styrene liquid, gas and polymer was ejected from the hatch of the other compartment and ignited. (Styrene is normally kept cool and inhibited to prevent polymer formation. If the temperature rises polymerisation can be initiated with considerable heat evolution. 30 °C is the maximum recommended storage temperature.)[67]

- Unplanned effects on materials of construction, reactions with gasket materials, reactions with insulation. For example, trace contaminants from construction materials may act as catalysts.

A 0.45 m^3 vessel was used to chlorinate an aromatic monomer dissolved in carbon tetrachloride at 50 °C. When only approximately 10% of the gaseous chlorine had been added to one batch the top of the vessel was blown off and polymer solution was ejected over the working area.[68]

Iron chlorides had entered the reactor via stainless steel feed lines. These had catalysed a very rapid side-reaction between the monomer and the chlorine, or HCl, evolving gas and producing a polymeric residue.

The suitability of the materials of construction to resist corrosion, temperature effects, etc., is also a factor.

Formaldehyde gas is released from formalin, for disinfection purposes, by the addition of limited quantities of potassium permanganate. When 0.1 litres of formalin were added in a plastic beaker, however, an extremely exothermic overnight reaction melted the beaker.[69]

Ethylene oxide leaked from a hairline crack on a distillation column and contaminated the mineral wool insulation on a level indicator. Here it reacted with moisture to form non-volatile polyethylene glycols. When the insulation was removed so that the indicator could be repaired, air leaked in and the glycols ignited, possibly due to thermal degradation to low flash-point materials that self-ignited. The fire heated the wall of the piping system in which ethylene oxide was static. The ethylene oxide decomposed and the decomposition travelled into the column which exploded, wrecking the plant and injuring five men. It is now recommended that non-absorbent insulation should be used for ethylene oxide service and that vessels containing ethylene oxide should be inerted with nitrogen, which is standard practice on storage vessels.[70]

Two people were killed and 350 injured in an explosion involving ammonium perchlorate in 1988.[71] This was manufactured for use as an oxidant in solid fuel rockets. The process comprised four batch steps:
(i) Electrolytic oxidation of sodium chloride to sodium chlorate in a series of cells:
$$NaCl + H_2O \rightarrow NaClO + H_2$$
$$3\ NaClO \rightarrow NaClO_3 + 2\ NaCl$$
(ii) Electrolytic oxidation of sodium chlorate to the perchlorate:
$$NaClO_3 + H_2O \rightarrow NaClO_4 + H_2$$
(iii) Production of crude ammonium perchlorate in agitated tanks:
$$NaClO_4 + NH_4Cl \rightarrow NH_4ClO_4 + NaCl$$
(iv) Crystallisation, filtration, drying, screening and blending of ammonium perchlorate to the required specification. Product was stored in 115 kg polyethylene drums or 2770 kg aluminium tote bins.

At about 11.30 a.m. flames were observed from a small batch dryer. Operators fought this fire using fire hoses. Several polythene drums of ammonium perchlorate and drums of floor sweepings were on fire as were the reinforced plastic walls of the building. The fire hoses then lost water pressure and at 11.51 a.m., shortly after the operators had left the building, an explosion occurred. Fire spread to a larger dryer building some 6 m away and ignited more drums of ammonium perchlorate. Operators in the plant and in an adjacent factory commenced site evacuation. At 11.53 a.m. a massive explosion, estimated at 108 tonnes TNT equivalent (see page 874), was caused by detonation of 2270 kg of perchlorate in a tote bin. This demolished the administration building and ruptured a 0.41 m natural gas pipeline. At 11.57 a.m. a special area of tote bins detonated, with an explosion estimated at 235 tonnes TNT equivalent. This demolished an adjacent factory and caused property damage up to 19 km away. An area of 8 km around the site was evacuated and the fire was left to burn out.

(Ignition of gas from a pipeline, or flame-cutting work in an area 25–30 m away from the batch dryer, were partly discounted as the initial cause. Contamination of the dryer insulation with ammonium perchlorate at the loading point was considered the most probable cause. This had apparently occurred twice in the previous year.)

In a reaction of an aromatic amine with a chloronitro compound, the formation of ferric chloride, which was known to catalyse exothermic side reactions, was prevented by using synthetic soda ash as an acid acceptor. After 20 years' successful operation the synthetic soda ash was replaced by natural soda ash. In the non-aqueous reaction medium the difference in crystallinity of natural soda ash rendered it less efficient as an acid acceptor. This allowed acid build-up and ferric chloride formed from reaction with the mild steel vessel. This catalysed exothermic side reactions which resulted in overpressurisation of the reactor and a serious explosion.[72] A weakness in the manhole closure caused the cover to blow off despite operation of a relief valve on the reactor which was designed for 100 psi (confinement tests subsequently indicated that the side reactions could develop pressures of 400–600 psi). Thrust of gases from the manhole propelled the vessel downwards, releasing gases into the building where combustion caused an explosion.[73]

Process physics considerations

Appropriate aspects of the process physics then need to be examined. These may include:
- static electrification potential
- maximum permissible oxygen concentration to prevent ignition
- decomposition rates and products
- maximum explosion pressure and rate of pressure rise in the 20 litre apparatus.

Batch reactor hazards

If batch reaction is involved then consideration has to be given to all the common possible hazards arising from maloperation.

A process involved reaction of chromium trioxide with hydrochloric acid and isopropanol to produce chromium(III) chloride. Because the batch reactor was not hot enough when the operator commenced isopropanol addition the reaction did not commence immediately. Unreacted isopropanol accumulated and when the reaction started decomposition occurred rapidly. The normal temperature was exceeded by 50 °C and the reactor was overpressured.[74]

Failure of agitation must also be taken into account.

A mixture containing a nitrite and an imine was hydrolysed in an exothermic reaction, carried out at 200 °C under pressure, to convert the imine to a ketone. During one batch the agitator failed and the reduced rate of heat transfer to the cooling coil – consequent upon the lowered heat transfer film coefficient on the process side (page 53) – allowed the temperature to rise slowly. Suddenly the temperature rose rapidly and the lid was blown off the reactor.

Under the slightly alkaline conditions in the hydrolyser the nitrite itself was hydrolysed at temperatures slightly higher than 200 °C. The appreciable increase in heats of reaction caused the exotherm.[75]

Insufficient mixing may also create a hazardous situation. A number of thermal runaways in batch and semi-batch processes have been traced to such problems. A third, less obvious, agitation problem is that of the potential for energy input from mechanical friction.

> After 33% of a nitric acid charge was added during the nitration of an intermediate the addition was discontinued. The reactant mass was agitated for 4 hours, with cooling applied, to allow for complete reactant consumption. The final shift before the weekend had finished so the cooling was shut off with the agitation left on; the temperature was approximately ambient. The temperature rose linearly over the next 35 hours reaching 80 °C by Sunday morning; that evening the batch erupted. Subsequently a series of adiabatic heat flow calorimeter experiments in which the energy input from agitation was measured confirmed that this had increased the batch temperature into its decomposition range and resulted in the incident.[64]

Failure of cooling or other services, or reaction with heat transfer media, are other scenarios.

Consideration of unit operations etc.

Hazards which may be associated with the chemical engineering operations in the process should be carefully analysed. Inventories of hazardous materials may be reduced by various strategies.[17]

> Following a decision by its suppliers to phase out delivery of bottled phosgene outside the company, an agrochemical company put in a skid-mounted phosgene plant to generate phosgene on demand. No liquefaction or intermediate storage is involved and its state-of-the-art phosgene storage facilities became redundant.[76]

The possible effects of contamination of materials upon the process, or upon materials of construction (e.g. presence of trace chlorides), should be assessed.

> A vacuum pump was used in a difficult pilot-plant distillation process. Contamination of the vacuum pump oil was anticipated; therefore DSC and ARC tests were performed on the mixtures. The vacuum pump diluted the reactive mixture, resulting in less heat evolution, but the oil did not catalyse or accelerate the decomposition so no process modifications were required.[35]

Uncontrolled overconcentration, catalysis, e.g. as described on page 1096, or overheating need to be guarded against.

> Minor upsets during the start-up of an ethylene oxide purification column resulted in a low level in a reboiler. Vapour above the liquid level was heated to 150 °C. A film of ethylene oxide polymer and iron oxide in the tubes catalysed a previously unknown reaction which resulted in the temperature rising to 500 °C. The ethylene oxide decomposed explosively resulting in extensive damage and one fatality.[77]

Storage and transport considerations

Consideration has to be given to potential hazards during storage and transportation, including, for example:
- effects of contamination.

> A drum of acrylonitrile on a flat-bed lorry exploded in a public car park dispersing polymer over a wide area. This had been the first drum filled in a batch via a flexible hose believed to have been contaminated with acid, having previously been used for 10% nitric acid. Ten days elapsed between filling the drum and its explosion due to initiation of polymerisation by the acid.[67]

- self-heating of stacked solid material.
- effect of solar heating or other changes in ambient temperature.
- separation of inhibitor.
- oxidative, or other, degradation.

Substitution of a less hazardous material for a hazardous product, as summarised on page 1736 will also, of course, result in improvements in the safety of shipping and associated storage. Recent examples[78] include transportation of
- ethylene dibromide instead of bromine
- ethyl benzene instead of ethylene
- chlorinated hydrocarbons instead of chlorine
- methanol instead of liquefied methane
- carbon tetrachloride instead of anhydrous hydrochloric acid (burning the CCl_4 with supplemental fuel produces HCl on demand at the user's site).

Evaluation of wastes is necessary both to control any environmental hazards and to ensure safety in the disposal operation, e.g. pre-treatment, collection, segregation, conveyance, treatment and disposal. The information required on properties is essentially similar to that summarised above but further guidance is given in Chapter 22.

> A 5 gallon plastic container used to store 'waste organic slops (monomerics)' ruptured violently whilst inside a laboratory fume hood. Some *t*-butylperoctoate had been added to it inadvertently and a runaway polymerisation started within 25 minutes. The capped container ruptured across its seam and, since the fume hood sash was open, polymer was ejected into the laboratory. The panels inside the hood were cracked and broken and many 20 ml bottles were blown out; some were broken and some contained cyanide.[79] (Apart from the need to keep the fume hood sash down when not in use, the appropriate precautions here require improved labelling of waste containers, detailing exactly which chemical wastes are allowed in each container, and segregation of non-combustible waste containers.)

It will be appreciated that the chemical engineering safety principles in the various treatment processes are not altered significantly merely because the batches or streams involved are 'waste'.

1176 Further considerations and developments

> Neutralisation of an acidic effluent was carried out by addition of chalk to a stirred tank. On one occasion the operator observed that the treated effluent was too acidic; the stirrer had actually stopped some time previously and unreacted chalk had settled.
>
> When the operator switched the stirrer back on, a rapid evolution of carbon dioxide blew off the manhole cover and lifted the bolted lid off the tank.[80]
>
> Detailed instructions as to the procedure to follow if the stirrer stopped were in the foreman's office but the operator was unaware of them.

However, common experience is that unanticipated reactions may occur in waste disposal, irrespective of the scale of operation.

> A fire hazard was discovered when discarding soda lime (which had been used to absorb hydrogen sulphide). When sulphide 'Sofnolite' had been discarded into waste bins containing moist paper wipes, charring or smouldering was produced in most cases; on one occasion use of a fire extinguisher was required. This was due to simultaneous interaction of the spent soda lime and air with a considerable exotherm, a 10 g conical heap heating from 20 °C to 120 °C in 1.5 ks.[81]
>
> Subsequently, to minimise the possibility of waste bin fires and the slow liberation of hydrogen sulphide all sulphided soda lime residues were saturated with water and sealed into a metal container for separate disposal.

Pilot planting

Some problems which may be encountered at the intermediate pilot-plant scale are summarised in Chapter 11. Certain hazards and precautions make pilot-plant work unique in terms of direct loss control, and particularly the hazards which may arise in subsequent commercial operation in the event of inadequate pilot studies.

Pilot plants vary considerably. Some are dedicated single-process operations, others are multi-purpose; their scale ranges from mini-plant operations to almost full commercial scale (see page 1181). The utilities may be dedicated or shared, as may operating and maintenance labour.

As pilot plants have grown in sophistication so too have costs risen in terms of control systems, instrumentation, equipment maintenance, environmental controls and worker protection. Typical costs of pilot plants are summarised in Table 20.13. Clearly this is a very expensive, and often time-consuming, development stage. However, unless the limits of operation are established on a small scale, 'learning' may not be transferred to the production scale with consequent higher costs.

> During the final phase of the manufacture of triglycidyl isocyanate (TGIC), a special epoxy resin, a batch decomposed with explosive violence when the excess of epichlorohydrin was distilled off. The upper part of a 2500 litre kettle was ruptured.[82]
>
> In the manufacturing process epichlorohydrin and isocyanuric acid are combined, as a first step, to form 1,3,5-tri-(2-hydroxy-3-chloropropyl)iso-

Table 20.13 Typical costs of pilot plants by size

Class	Characteristics	Cost range £×10³, 1988 (UK)
Bench-top and microunits	<1.5 m² area, typically located in a hood or on a bench top Small tubing throughout (3–6 mm dia.) Limited automation	7–35
Integrated pilot plant	10–25 m² area placed in open bay or overpressure cell Tubing and small (13–25 mm dia.) pipe Fairly heavily automated Feed and product systems generally included	35–17.5
Demonstration or prototype unit	>250 m² area, usually in a dedicated building or area Pipe used extensively Heavily automated Feed and product systems included	>350

cyanurate. In the second step the chlorohydrin compound is dehalogenated with sodium hydroxide to yield TGIC. On this occasion the content of hydrolysable chlorine was higher than usual, possibly because of unsatisfactory separation of organic and aqueous phases during washing operations, so that impurities were entrained. In any event the thermal stability of the reaction mass was drastically reduced by the high hydrolysable chlorine content.

Subsequently it was determined that the hydrolysable chlorine content, the epoxy equivalent and the pH should be within established limits before distillation could be permitted.

The hazards of, and precautions for, pilot-plant experiments parallel in many ways those of laboratories and production units. However, the differing purpose, coupled with the scale and intermittent or 'jobbing' mode of operation, introduce hazards requiring special consideration in pilot-plant design and management.

Purpose of pilot plants

The main aim of pilot-plant operations is to obtain data on process conditions and products to aid progression of successful projects from the laboratory to market place. The position of pilot planting in the general scheme of a research programme is shown in Table 20.14 and particular emphasis may be placed on any selection of the purposes listed in Table 20.15.[83]

Costly pilot-plant failures are generally attributable to:[84]
- Short-cutting the development process, i.e. piloting without adequate data or before equipment and process options have been properly screened. During scale-up from bench to pilot-plant temperature and concentration gradients and residence times often change. This will have

Table 20.14 Typical stages in a research programme

Step	Purpose
Origin of concept	Initiate ideas/concepts to be investigated
Basic research	Determine basic feasibility of concept
Preliminary economic evaluation	Determine if economic incentive is sufficient to proceed with investigation
Laboratory research	Develop basic data
Product evaluation	Evaluate suitability of product
Process development and preliminary engineering	Produce preliminary design of actual commercial plant, and resolve process questions as they arise
Pilot-plant study	Prove basic reliability of proposed process or process design
Demonstration unit	Demonstrate feasibility of process on mid-size scale
Prototype unit	Demonstrate feasibility of process on small-size commercial scale
Plant	Commercial production

Table 20.15 Purpose of pilot plants

- Solve scale-up problems
- Obtain design information
- Gain process know-how
- Gain operating experience and train staff
- Test reaction mechanism theories
- Obtain environmental and pollution abatement data
- Produce sample quantities for preliminary performance and safety testing
- Prove process theories
- Prove process reliability
- Improve on existing processes and products
- Convince management

implications for type, number and locations of vessels, type of heat transfer equipment and product-removal systems. The impact of temperature and concentration on possible runaway reactions is highlighted in Chapter 19.

- Design of the pilot plant without time to think through its operation, its ability to handle upset conditions, or its flexibility to meet objectives.

Hence numerous accidents have resulted from lack of complete process information on the operating ranges of a plant or pilot plant, or because of a failure by management to establish a mechanism for reviewing the available data at critical stages of project development, or because of inadequate studies to specify safe operating procedures for scale-up.

The inherent loss control problem with pilot plants arises from the use of new chemistry, testing of new technology, and working in new areas on an untried scale. For example, small temperature rises or bubbling in a laboratory apparatus may become significant exotherms or foaming as the scale is increased.

A complete package of data on the physical properties, fire and explosion characteristics, acute and chronic toxicity, and physico-chemical properties of all reactants, products and by-products, and on process kinetics and the effect of impurities on process and product safety, etc., is rarely available from earlier laboratory studies. This area requires increased attention in the UK with the introduction of the Control of Substances Hazardous to Health Regulations.

> A flammable liquid mixed with chloroform was distilled in a glass still heated by an electric isomantle in an electrically-flameproof area. Overpressure in the still resulted in a glass stopper blowing out and ignition of flammable vapour by the isomantle. It was assumed, wrongly, that the presence of chloroform would render the vapour non-flammable and that the heating element in the flameproof isomantle would not reach ignition temperature.[1]

The hazards from corrosivity, exothermic reactions, pyrophoric substances, detonatable mixtures, etc., may not have been fully assessed. There may indeed have been minor abnormalities which went unobserved in laboratory experiments, e.g. 'bumping' in laboratory flasks may have serious consequences in a 20 000 litre pilot-plant vessel.

Heat transfer may present different problems on scale-up from laboratory to pilot plant to production plant.

> An exothermic, liquid-phase reaction was performed batchwise in a 379 litre agitated pilot-plant reactor. This vessel was provided with a steam jacket which, for multi-purpose use, was equipped with isolation valves. At the end of the day the reaction was left to progress with the isolation valves closed. Thermal expansion of the water retained in the jacket occurred during the night due to the predictable temperature rise associated with the exotherm; this buckled the vessel inwards and fouled the agitator.[1]

Although in this case it would have made no difference, there is generally, as discussed on page 94, a larger surface area per unit volume available for heat transfer (and hence per unit of feed) on the smallest scale. Accordingly a lower ΔT may be required for heating in laboratory experiments. Thus, if thermally sensitive substrates are being heated, decomposition not observed in the laboratory may occur as scale is increased. Similarly, some corrosion rates are logarithmic with temperature; thus a few degrees' excursion on the pilot plant may lead to unforeseen results.

With highly exothermic reactions there is always the possibility of a runaway. It has been proposed that runaway conditions are possible if the adiabatic temperature rise is 200 °C.[85] Formulae and curves are provided from which the safe ΔT between reactants and coolants can be estimated; there is also a list of means for the effective control of highly exothermic processes. The complete thermal behaviour of a process must be understood to keep a reaction, and any possible side-reactions, under control.

A nitration reaction involving substituted benzoic acid had been run many times in the laboratory and once on a 50 gallon pilot scale.

$$(X)_n\text{-C}_6\text{H}_4\text{-CO}_2\text{H} + \text{HNO}_3 \xrightarrow{\text{H}_2\text{SO}_4} (X)_n\text{-C}_6\text{H}_3(\text{NO}_2)\text{-CO}_2\text{H} + \text{H}_2\text{O}$$

Fuming nitric acid was used as the nitronium ion source and fuming sulphuric acid as the solvent at 80 °C to 90 °C. On a 200 gallon scale a runaway occurred. The nitric acid–sulphuric acid mixture was being slowly added to the substituted benzoic acid when, a few minutes into the run, the temperature suddenly increased and the rupture disc blew; the dome gasket also ruptured as personnel evacuated the area.[86] Subsequent procedures for running reactions in the pilot plant were reviewed and a Standard Operating Procedure established. This set up a Batch Record Review Board to meet on a regular basis to review and approve the written Batch Record and required more information and instructions to be included in this document.

The evaluation of toxicological, ecotoxicological and physico-chemical properties may be required as an input into the safe design of production facilities, part of the consumer safety programme, or to satisfy legislation such as the Notification of New Substances Regulations 1982 in the UK. The data have to be generated in accordance with 'Good Laboratory Practice'.[87] Clearly when preparing material for such tests the samples should be of a consistent purity profile and preferably of a specification closely akin to that of the product as eventually marketed. Thus one purpose of a pilot-plant run may be to prepare a sufficiently large, single batch of material on which the complete safety test programme will be undertaken. The full risks at this stage of the development may be unknown and the precautions must reflect this lack of knowledge.

Hydrazinium diperchlorate was being investigated as a possible new oxidiser for solid rocket motors. In laboratory quantities, the only abnormal reaction observed was gassing due to inadvertent decomposition during drying. However, in the larger quantities handled in pilot plants, the material accidentally detonated with destructive force. Fortunately, good advanced planning dictated that, since the nature of the material had not been fully defined, the critical drying step should be undertaken at a remote station behind a barricade.[88]

Scale of operation

Scale-up ratios of a range of typical process systems are given in Table 20.16[89] and representative capacities for pilot plants in Table 20.17[1b].

As discussed in Chapter 11 standard laboratory equipment (e.g. fume cupboards, glove boxes, blast screens) generally enable hazardous ma-

Table 20.16 Typical scale-up ratios

System	Scale of operation (kg/h)		Scale-up ratio	
	Laboratory	Pilot plant	Laboratory to pilot plant	Pilot plant to commercial
Substantially gaseous (ammonia, methanol)	0.01–0.10	10–100	500–1000	200–1500
Gaseous reactants, liquid or solid products (sulphuric acid, urea, maleic anhydride)	0.01–0.2	10–100	200–500	100–500
Liquid and gaseous reactants, liquid products (benzene chlorination, oxidation)	0.01–0.2	1–30	100–500	100–500
Liquid reactants, solid or viscous liquid products (polymerisations, agricultural chemicals)	0.005–0.2	1–20	20–200	20–250
Solid reactants, solid products (phosphoric acid, cement, ore smelting)	0.10–1.0	10–200	10–100	10–200

Table 20.17 Representative capacities for pilot plants

Scale	Batch	Continuous
Lab bench	0.1 – 1000 g	10 – 100 g/h
Mini-pilot plant	1 – 100 kg	0.1 – 10 kg/h
Pilot plant	100 – 1000 kg	10 – 100 kg/h
Semi-works plant	1000 – 10000 kg	100 – 1000 kg/h

terials to be handled with relative ease at the bench. Special arrangements may be required to afford the same degree of protection to pilot-plant workers. Once the hazards are fully evaluated and business opportunities assessed, any special control measures for production will either be relatively easily justified or deemed unnecessary. However, more persuasive skills and enlightened management are helpful when proposing capital expenditure for safety reasons at the intermediate and more speculative pilot-plant stage of R&D.

The four basic pilot-plant layouts are:
- in a separate building
- in a containment (blast) cell
- in an open bay
- in a laboratory.

Location in a separate building eliminates the possibility of mutual interference between operations; special services, e.g. air conditioning, may be provided and an appropriate electrical classification adopted.[90] However, serving separate buildings may not be economical. Moreover, all intrinsic hazards still have to be catered for.

Total containment cells are very expensive.[90] Typically, therefore, cells have walls or roofs designed to vent the force of any explosion to a safe area. In one facility the offices, laboratory, control room, locker and washing areas are separated from the cells by 30 cm thick reinforced concrete walls and floors. The cell area is similarly constructed and the backs of the cells, which face a dirt and rock embankment, and their roof area are constructed of corrugated plastic sheeting as blow-out areas.[91] Cells are considered the safest operating arrangement provided all operations are remotely controlled and operators only enter the cell when the plant is shut down. Operating problems, or upsets, can be more effectively contained and addressed in such isolated areas which conventionally have provision for spill containment.[92] Greater risks may be acceptable if only equipment, not personnel, could be exposed to hazards. Installation within a cell is expensive, e.g. because of the instrumentation required for remote control and operating efficiency is limited by the restricted access.

Open bay location is the least expensive and is increasingly favoured, as risks have been reduced by more reliable instrumentation. The open layout has the advantage that operators are not working alone, or even out-of-sight. However, the arrangement may mean that a spillage or fire can readily spread from one plant to others.[90]

Laboratory-type pilot plants are a consequence of relatively recent reductions in size. Some pilot plants can actually fit inside an expanded laboratory hood, e.g. as illustrated in Fig. 6.3, with the advantage that operations can be conducted in a controlled, well-ventilated environment. The high rate of ventilation reduces considerably the hazards of handling toxic materials. However, such a facility is expensive to construct and operate.[90] A summary of the advantages and disadvantages of various area types is shown in Table 20.18.

In one company, in order to classify the hazards with regard to a reaction, a set of categories was developed to help fit the personnel protection required to the nature of the reaction:[92]

Category 1   Detonating reactions; operations with a high potential for detonation or massive rupture with a large and extremely rapid release.

Category 2   Ruptures and fires; operations with a high potential for generating large ruptures and large-scale fires.

Category 3   Leaks and small fires; operations with a relatively low potential for leaks. Expected, at worst, to produce only leaks or small fires.

Category 4   Low hazard operations; low-temperature and low-

Table 20.18 Pilot-plant arrangements

Area type	Intended use	Advantages	Disadvantages
Laboratory	General purpose, low risk, small-scale units	Readily available Temperature controlled Support equipment nearby Utilities readily available	Small-size unit only Prone to crowding Unsuited to high quantity, or high toxicity, of feeds or products Utility capacity limited Optimum layout usually not possible
Open bay	General purpose, moderate risk, larger-scale units	Optimum layout possible Handles larger units Large quantity of utilities usually available	Laboratory facilities may be some distance away Utilities may have to be run some distance
Overpressure cells	High risk or very hazardous experiments	Explosion and/or fire easily contained Operator safety maximised	Unit usually remote-controlled and more expensive to install Cell configuration frequently limits space and flexibility
High ventilation, face velocity >0.5 m/s	Toxic or flammable materials involved	Increases operator safety	Costly to install Space usually limited
Low ventilation, face velocity <0.5 m/s	General purpose, non-hazardous materials involved	Usually readily available	Precautions necessary to minimise build-up, or use, of unsuitable materials
Electrically classified*	Large amounts of flammable materials present at most times	Minimises danger of ignition leading to fire or explosion	Requires approved, more expensive equipment Prohibits or restricts certain operations
Electrically unclassified	Small amounts of flammable materials present at most times	More readily available User standard equipment	Possibility of flammable material introduction on a large scale must be considered Very expensive and time-consuming to convert to an electrically classified area later

* E.g., Class I, Division 2, Group D.

pressure operations shown by experience to have a low potential for either ruptures or fires.

A reaction category is established in the initial review of a new operation, before design work commences, and the personnel protection facilities are then set according to Table 20.19.[92] This determines whether the proposed pilot-plant location can provide the necessary protection.

If the category is not fixed by the specific nature of the reaction it can be determined from the operating pressure/temperature level via Fig. 20.3.[92]

In Table 20.19 'baffle' refers to the placement of instrument panels or sheet metal walls between equipment and the operator. Shields of 1.27 cm polycarbonate are used to provide missile protection, e.g. in front of groups of valves operating at high pressure.[88] The 0.64 cm steel walls of a sandbath provide protection against missiles from an exploding reactor; a barricade can be created by providing a spring-loaded relief head on the sandbath. The need for, and scope of, the protective devices are emphasised in the operating manual. The permissible degree of access into the operating equipment area is defined as follows:

1. Excluded access; personnel are not permitted into the enclosure area when a Category 1 operation (see above) is in progress.
2. Limited access; personnel are permitted in the enclosure area only for short periods to make adjustments to equipment when a Category 2 or 3 operation is in progress.
3. Unlimited access; no restriction is applied to Category 4 operations.

The scale of pilot-plant operations often requires considerable man-handling of kegs, sacks, drums, items of plant, etc. Overall, about 40 000 accidents are reported to the UK HSE each year as being caused by the handling, lifting and carrying of goods. Furthermore, manual transfer of materials tends to bring operators into close proximity with the chemical raw materials, solvents, etc.

The intermediate scale of pilot plant may introduce inherent engineering problems:

A glass-lined pilot-plant reactor was used, under pressure, for a heterogeneous acetylation reaction using acetic anhydride at high temperature. It was found impossible to discharge the product via the diaphragm valve connected to a stub-pipe on the reactor base because of solidification in the dead-end. (This was avoided on the production scale by use of a standard, rising-stem mushroom valve flush with the reactor base.)

Thus glass or other transparent media are more common than on a commercial scale, since there is often more incentive to provide 'windows' on the process. Similarly plastic may be used, for economy reasons, in preference to lined piping on corrosive duties. However, unless their limitations are appreciated, the desirability of using more convenient materials of

Table 20.19 Pilot-plant protection categories classification and requirements

Typical hazards	Category 1 Detonating reactions		Category 2 Rupture & fire hazards		Category 3 Leaks & small fires		Category 4 Low hazards	
	Hazard potential	Protection device	Hazard potential	Protection device	Hazard potential	Protection device	Hazard potential	Protection device
Detonation (internal)	High	Barricade Blast relief	Negligible		Negligible		Negligible	
Shrapnel	High	Barricade Shrapnel stop	Negligible		Negligible		Negligible	
Explosive atm. (external)	High	Barricade Blast relief Ventilation Min. cubicle size	Moderate	Baffle Ventilation Cubicle size	Negligible		Negligible	
Fire	High	Baffle	High	Baffle	Moderate	Baffle	Negligible	
Pressure leaks	Possible	Baffle (barricade)	Possible	Baffle	Possible	Baffle	Negligible	
Skin toxic	Possible	Baffle (barricade)	Possible	Baffle	Possible	Baffle	Possible	Baffle
Lung toxic	Possible	Ventilation	Possible	Ventilation	Possible	Ventilation	Possible	Ventilation
Permitted access to operating areas	Excluded		Limited		Limited		Unlimited	
Summary of protective devices		Barricade Blast relief Shrapnel stop Ventilation Min. cubic size		Baffle Ventilation Cubicle size		Baffle Ventilation required if lung toxic chemicals		Baffle required if skin toxic chemicals or adjacent operations (see baffle design) Ventilation required if lung toxic chemicals

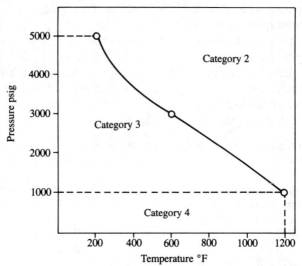

Fig. 20.3 Categorisation of pilot-plant hazard based on operating temperature and pressure[92]

construction on a pilot scale may create a hazard which would not be present with more robust materials on a production scale.

A 0.45 m diameter, 2 m high industrial-glass extraction column was fed with continuous aqueous phase by a pump capable of delivering 0.02 m^3s^{-1} against a head of 4 m. Whilst filling the column prior to commencing an extraction run, the top atmosphere vent/outlet valve was inadvertently left closed. The column was overpressurised and shattered. (The recommended working pressure of this diameter glass section was not significantly greater than the hydraulic head at the base of the column.)

A 10 cm diameter vertical glass pipe section was left pressurised overnight with 34 kN/m^2 (5 psig) nitrogen. The pipe had been installed out-of-line and was therefore, unevenly stressed. It failed, projecting slivers of glass throughout the pilot plant – fortunately whilst it was unoccupied. (Particular care, and appropriate shielding, is necessary with industrial glass pipework.)[1]

An operator applied steam, regulated to 30 psig, to a jacketed glass vessel in a fluid bed dryer system. The outlet was connected to a steam trap. The glass unit, which was not pressure rated and had no provision for pressure relief, exploded. It was not shielded, taped or netted but fortunately the operator was up on a ladder, out of line with the explosion.[93]

A subtle change in feed composition resulted in the introduction of minute quantities of solvent into a PVC plastic pipeline. The solvent degraded the plastic and a minor impact subsequently caused the line to fail and a spillage of material.[93]

Many devices with glass or plastic elements, e.g. rotameters, gauge glasses, tubular gauge columns and the front faces of pressure gauges, may be subject to fragmentation under pressure. Protective shields, with adequate pressure relief space between the shield and device, have been designed;[92] these have been tested by rupturing to check their effectiveness in protecting against flying glass, plastic or squirting of hot liquids.

The supply of feedstocks is often more difficult to cater for, or less well thought out, than on full-scale production.

> A cylinder which was heated to provide a supply of butadiene gas to a pilot plant exploded causing a fatality. Subsequent tests showed that the temperature of the liquid surface in the cylinder could reach 160 °C: at about 144 °C the liquid butadiene would have expanded to fill the cylinder and at 100 °C the Diels–Alder dimerisation to vinyl cyclohexene could produce a temperature excursion to over 555 °C.[94]

$$\text{Butadiene} \xrightarrow{\Delta} \text{Vinyl cyclohexene}$$

Operations almost inevitably result in many partial containers of raw materials, by-products, etc., all of which require proper storage. Special coding or labels may be applied to identify raw materials for a particular process since common storage for different projects, shared pallets, and similar chemical names and packaging can cause confusion.[93]

As discussed on page 100, any liquid near to its boiling point at a high pressure will 'flash' when let down to a reduced pressure, the saturation temperature for which is less than the temperature of the liquid.

> After a vessel had been 'boiled-out' using a large volume of water the operator began venting the steam which was at a positive pressure. Before venting was complete, he loosened the manhole cover; as a result copious quantities of steam and water blew out via the cover.[95]

Because of this phenomenon it is recognised that during process development the inventories which it is preferable to avoid are those of 'flashing' flammable or toxic liquids, i.e. liquids under pressure and above their atmospheric boiling points. In summary:[91]

> Liquids below their boiling points produce very little vapour; gases leak at a lower mass rate than liquids through an opening of a given size and are often dispersed by jet mixing. Flashing liquids leak at about the same rate as cooler liquids but then turn into a mixture of vapour and spray. The spray, if fine, is

just as flammable or toxic as the vapour and can be spread as easily by the wind.

Work on pilot plants can, however, be carried out safely when the correct procedures are adopted, and to rate pilot-plant operations as always likely to be more hazardous than laboratory or full-scale operation is too simplistic.

> An industrial chemist developed an allergic contact dermatitis affecting his hands whilst working continuously on a project involving the use of a diamino intermediate in a laboratory synthesis. He had worn wrist-length heavy-duty industrial rubber gloves when performing manual transfers of liquid and solid diamino compounds. Other chemists also developed rashes. At the pilot-plant stage as nearly a closed system as possible was employed and a 'no touch' technique was used when running laboratory checks on samples from reaction vessels. During full-scale production there were some additional episodes of dermatitis.[96] It may be that standards were relaxed for production since the problem appeared to have disappeared on the pilot plant.

This example demonstrates the need to consider longer-term occupational health hazards when planning for loss control.

Scale-up from laboratory to pilot plant also has implications with regard to exposure to emissions and for waste disposal and pollution control.

Mode of operation

Pilot plants are, like laboratories, normally housed indoors, thereby increasing the opportunity for build-up of dangerous concentrations of vapours, gases, dusts, fumes and mists. The common practice of manual control normally brings pilot-plant workers in close proximity to the process. Processing with reactions with poorly-defined rates and thermochemical characteristics may result in extreme and rapid overpressurisation of vessels and/or ancillary equipment. Building design may, therefore, incorporate discrete cubicles, with the provision of local exhaust ventilation, blow-out panels and remote operation and observation[97] as described earlier.

Pilot plants need a high degree of flexibility and often therefore comprise multi-purpose facilities. Depending on the management structure, unit operations may be grouped according to 'ownership' or 'responsibility' rather than as a reflection of the hazards involved. Plants are often constructed on a temporary basis and require regular modifications. This calls for careful management to ensure that safe systems of work operate, e.g. permits-to-work are required to prevent trial mock-ups and unauthorised modifications and maintenance by 'enthusiastic' scientists. Of the 2100 incidents in the chemical industry which were reported to the UK HSE between 1982 and 1985 about 500 cases involved process maintenance activities on pumps, pipes and valves, etc., and in 14% of the accidents investigated an unsafe system of work was being used.[98]

Manual operations

As noted ealier, pilot plants characteristically have more manual, less dedicated operations; in multi-purpose flexible facilities, for example, the same equipment may be adapted for use in a variety of short-term projects. This can result in misdirected material transfers.

> A mixture of highly flammable monomers was charged to a feed tank. The transfer hose was blown-down with nitrogen, hence pressurising the tank. When the hose was disconnected without venting the tank or isolating the dip pipe, monomers sprayed into the process area and over operators, who were fortunately properly attired with protective equipment.[93]

Multi-purpose operation

Multi-purpose use of sections of a pilot plant is common. This may result in safety problems associated with re-routing of process flows or services, adjustment of operating parameters, cleaning of equipment and pipework on changeover, e.g. the thorough removal of all residues. There is also a need for operating personnel to adapt to different processes and operations.

> A pilot plant contained a multi-purpose oven. Two drums of a temperature-sensitive material were placed in it at 45 °C but not properly logged-in. A different employee added a third drum and, without checking the labels on the drums, which indicated a 70 °C limit, or enquiring about them, raised the set point to 90 °C. Fortunately this was discovered several hours later before a catastrophic reaction was initiated.[93]

Variation of materials and amounts

It follows from multi-purpose use, and from changes in conditions during development of a process, that variations arise in the types and amounts of chemicals used in various operations. Detailed administrative systems are therefore advisable to avoid confusion.

> The wrong material was charged to a 350 gallon reactor because of confusion over chemical names. Three operators who were involved in moving and charging the material in drums failed to notice that it was not triethanolamine but triethylamine.[93]

Frequency of shut-down and start-up

Pilot plants, even those based on a continuous process, tend to operate on short runs resulting in extensive numbers of start-ups and shut-downs. Some indication of how potentially hazardous intermittent operation may be if inadequately controlled, is given by the fact that a study of the 100 largest losses in the hydrocarbon processing industry revealed that 22 per cent occurred during the relatively brief periods of start-up and shut-down.[99]

Batch operation

Unless specifically set up to mimic a continuous process, pilot plants tend, because of the limited scale, to be batch operated. Batch processes present their own hazards such as the need for more frequent cleaning or decontamination of reactors. Hence, as in all batch processes, hazards may arise from one or more of the following factors:

- Inadequate preparation for cleaning of reactors, other equipment, or ancillaries.

   The procedure for washing out a reactor after discharging a batch of aluminium chloride melt into water involved filling it with water. On one occasion, the wash water was added before the vessel was completely empty, no dipping or visual inspection having been performed. There was a violent reaction and the hydrogen chloride evolved ruptured the glass vent.[100]

   Organic peroxides in white oil were used as polymerisation initiators. These solutions were pumped into distribution tanks and fed from these to the plant via ring mains. The tanks were cooled by refrigerated glycol passing through internal coils because the peroxide was heat sensitive (the special hazards associated with peroxy compounds are noted on page 505). An outside Service Section was employed to clean the glycol system using hot citric acid solution. No clearance certificate was issued. Whilst four distribution tanks were empty a fifth contained 250 gallons of peroxide solution. Upon reaching a specific temperature this began to decompose violently. The rapid pressure increase fractured the lid holding-down bolts and it was blown off; the tank was forced downwards and buckled supporting steelwork and fractured a number of pipes. Fortunately the peroxide did not ignite.[101]

- Incorrect cleaning procedure.

   To clean a reactor an operator was instructed to fill it with water, heat it to its boiling point with the agitator on, stop heating and allow the contents to cool for 20 minutes, blow out the dip pipe with air and then empty the reactor. He apparently omitted to allow the requisite cooling period, and admission of air ejected boiling water out of an open manway resulting in 90% scalding of his body which subsequently proved fatal.[102]

- Incomplete removal of cleaning media.[103]

Added to these, the pilot plant may be the first opportunity to try out a cleaning procedure.

> A commercial cleaning solution was used to remove polymer build-up from a 50 gallon reactor. Some uncured material beneath the hard crust reacted with the solution causing a pressure rise. Pressure relief was impeded by the relief valve being blocked with polymer solids and pressure blew out a 5 cm glass elbow. Such an incident could have been anticipated using the tests summarised earlier.[93]

Safety in pilot-plant operation

The general considerations for the safe operation of plant and laboratory experiments are covered in Chapters 4, 6 and 11. The detailed precautions for a pilot plant are dictated by a combination of its intended use, its proposed location and the nature of the chemicals involved as discussed earlier. Management must also, however, devise special codes of practice for dealing with the peculiar problems identified for pilot-plant operations.

As with other activities discussed in previous chapters, management are responsible for safe design, construction, operation and maintenance of pilot plants and for the R&D programme. A team approach is essential for risk assessment and as a means of obtaining the commitment of all those involved. They must ensure that all relevant data are transferred from chemist to chemical engineer and to process engineer. Safety should also be formally reviewed at various critical stages of project development, i.e.
- at the planning stage
- at the design stage
- on commissioning equipment or plant
- planning a modification
- commissioning a modification
- routine operation, e.g. annual or more frequently.

Arising from the experimental nature of pilot-plant work coupled with their flexibility, management must provide adequate training, adopt a disciplined approach, and devise strict and explicit rules to prevent hazardous conditions arising. Any failure to achieve this can introduce psychological hazards associated with poor operating instructions, misunderstood instructions, use of improper equipment, crowded work-space, poor housekeeping, inattention and carelessness.

Omission of pilot planting

If a process or product is novel then, generally it is considered prudent to build and operate a pilot plant. Risks may otherwise be incurred in process development due to:
- Inability to predict the performance of equipment from first principles or laboratory data alone.

- Oversight or lack of knowledge relating to all the impurities or by-products which may accumulate in recycle or waste streams.
- Lack of knowledge of circumstances which may arise after extended periods of operation, e.g. fouling of heat transfer surfaces, accumulation of impurities.

There is also no opportunity for staff to be trained on the process prior to commissioning of a full-scale unit.

Designing a plant, or individual sections for specific unit operations, from pilot-plant data reduces the risk and indeed minimises the chance of a complete technical failure. Some generalisations regarding which unit operations require pilot-plant testing are summarised in Table 20.20 (after ref. 89). Excursions of temperature can lead to overpressurisation since, as

Table 20.20 Pilot-planting requirements of unit operations

Unit operation	Pilot-plant requirement	Notes
Distillation	Not usual. May be to obtain tray efficiency or HETP data	Foaming problems may arise unexpectedly
Fluid flow	Not usual for single phase flow. Normal for two phase flow, e.g. slurries which may cause plugging	Unpredictable behaviour may arise with polymer systems and non-Newtonian liquids
Reactors	Frequent	For homogeneous systems scale-up from laboratory data may be justified. Heterogeneous systems tend to be less easy to scale-up
Polymerisers	Normal	Geometric similarity needed
Crystallisation	Normal	Sensitive to impurities
Liquid–liquid extraction	Normal	Similar types of equipment to production-scale essential[105]
Solid–liquid extraction	Normal	Solids conveying is critical. Similar type of equipment to production-scale essential
Solids handling	Normal	Generally with vendor's equipment
Dryers	Normal	Similar type of equipment to production-scale essential.[106] Generally with vendor's equipment
Evaporation Boiling (reboilers/vaporisers), heat exchangers, condensers	Not usual for straightforward applications unless fouling is suspected	Special duties may require piloting, e.g. condensation when non-condensables are present

a rule-of-thumb, vapour pressures may double for every 20 °C rise in temperature.

Vendor testing

It is common practice for tests on equipment to be carried out, with the smallest unit size considered reliable, at the works of potential vendors. An important criterion is that the feed and processing conditions should be closely representative of those to be used at the full scale.[89] Therefore, an alternative may be to lease the equipment and install it in the pilot plant. Unit operations for which this is done include

Filtration	Flaking
Drying	Prilling
Solids conveying	Gas–solid separations
Crystallisation	Liquid–solid separations
Agitation	

If material is released for external tests it is essential that all relevant information relating to operating conditions and process hazards is supplied to the vendor, and that he understands it, or that the tests are carried out under strict supervision by the potential purchaser.

> Production of a herbicide was under consideration as a flaked product. Material and safety information were sent to an equipment vendor's test facilities. Prior to the arrival of the pilot-plant engineers the vendor 'tried out' some of the material. Adequate venting was not provided so that when some impurities in the herbicide degraded on heating evolution of toxic gas necessitated evacuation of the area. Subsequently tests with adequate venting yielded the requisite scale-up data and a production-scale flaker operated satisfactorily.

Special considerations may be appropriate for the pilot-plant manufacture of materials such as pharmaceuticals, toiletries, cosmetics and dental products which are to be used in human trials such as consumer tests and clinical studies to assure product quality, safety and efficacy. These products are often made according to Good Manufacturing Practice.[87,107]

All experiments require documentation and approval. In one large research establishment applications to carry out pilot-plant trials involving novel substances are submitted for clearance by the Head of the Laboratory. Another company makes use of approved construction and operating permits for all pilot plants. The construction permit is required prior to the start of any construction and the operating permit, which also serves for modifications in the operating procedures, as shown in Fig. 20.4, is required before the operation begins.[108]

As with many chemical process operations (page 990) control of modifications – however small they may appear – has a crucial impact on safety performance.

> Material from a new pilot-plant system was processed at a lower temperature than previous batches. Almost two weeks later a small retained sample

PERMIT TO OPERATE OR MODIFY PILOT PLANT OPERATING PROCEDURE
(ATTACH PAID AND OPERATING PROCEDURE)

Project no. _____ Submitted by _____
Pilot operating permit no. 0–8 _____ Date _____
Pilot construction permit no. C-8 _____ Department _____

SECTION A

Project title:

Nature of operation (initial/modification):

Location of equipment	Laboratory testing	In PP operation	Outside PP equipment	MPTF removal
Estimated time schedule				
Start				
Complete				

Type of operation (continuous/intermittent/time periods) _____
Scale of operation (size/rate) _____
Scaleup factor based on ORC experimental work _____
Experience at smaller scale (attach technical report and/or flowsheet) _____
Operating organization (names and home phone numbers for emergency contact)
Project leader _____
Technical personnel _____
Technicians _____

SECTION B

	Operating equip	Design	Prepared/(note)	Checked by	Date
Pressure limitations					
Temperature limitations					
Safety hazard data sheets					
First aid/antidote instructions					
Firefighting methods					
Hazards evaluation					
Adiabatic temp. & press. rise calc.					
Process description					
Start-up					
Shutdown procedure					
Emergency shutdown procedure					
Utility interruption procedure					
Equipment drawings					
Equipment field inspection					
Equipment, piping, electrical I.D.					
Prestart-up checklist					
Technical personnel training					
Operator training					
Data sheets					
Personnel protection reqt.					
Preventative maint. procedure					
Waste disposal (sol, liq, vap, silo.)					

Reviewed by	Date	Approved by	Date
Director Safety, Health & Envir.		Sponsored Department Director	
Sponsored Department Manager		Engineering Manager	
Supervisor, Facilities		Head, Project Services	
Safety Supervisor			
Pilot Plant Coordinator			

Fig. 20.4 Operation permit

exploded.[93] (All containers of the material were subsequently pressure checked, vented and monitored until being disposed of.)

Legislation is increasingly concerned with the need for employers to carry out formal assessments of risk. There may be a legal obligation for management to carry out a risk assessment and to document health monitoring exercises and checks on adequacy of safety precautions, e.g. in the UK under the COSHH Regulations 1988 discussed in Chapter 28.

As explained on page 648, pre-start-up safety check meetings should be held for each new project as this is usually the last opportunity to review the hazards and proposed precautions prior to the first trial run. The team should consider the process flowsheet, which must identify reactions, separations and auxiliary operations (such as material and energy transfer and conversion) needed to convert raw materials into finished products. They should also consider both normal operation and any foreseeable deviations, and the ultimate fate of the product including disposal, emergencies, spillages, etc., with a view to management drafting formal safe operating instructions.

A batch operation involved charging aldehyde to a solution of caustic, an aromatic solvent and a phenolic compound. Since he anticipated a problem from fumes the operator altered the sequence of addition, adding the aldehyde before the caustic. Whilst vacuum charging the solvent via a dip pipe the reactor contents exothermed; the pressure reached 25 psig and flammable solvent was splashed into the process area, and onto the operator, fortunately without igniting.

This incident would have been avoided by a proper safety review which would have revealed the potential fume problem, for remedial action, and cautioned against altering the order of addition.[93]

Other considerations such as machinery guarding, electricity, ergonomics, etc., should also be discussed. A checklist is given in Table 11.32.

Pilot plants are started up and shut down relatively frequently; detailed operating instructions are, therefore, just as essential as on a production facility. Some indication of issues to address when considering start-ups and shut-downs is illustrated by Table 12.7.

A good system of communications is essential, particularly when critical supplies of services are shared, and within the plant where multiple processes share common utilities.

Maintenance work at a site boiler house resulted in interruption of service water used as cooling on two pilot-plant reactors. Both reactors exothermed but overpressure was avoided by the installed safeguards.[93] That the pilot plant was serviced by this water system was not appreciated in the boiler house following changes in personnel.

Experimental design should ensure that adequate, good quality data are generated on all safety aspects of the chemicals and process to aid transfer first from laboratory to pilot plant and thence from pilot plant to production plant. Corrosion data often come from laboratory studies since difficulties can be encountered when relying on pilot-plant studies, as illustrated in Table 20.21.[109]

All accidents and incidents should be fully investigated and the underly-

1196　Further considerations and developments

Table 20.21 Some difficulties associated with pilot-plant corrosion tests

- Pilot plants tend to utilise highly corrosion-resistant materials.
- Emphasis in programme is on gathering process data, not testing materials. Corrosion impurities from the test samples can have undesirable process effects.
- Small size of equipment leads to difficulties in extrapolation to large-scale unit.
- Extreme variation in operating conditions may make corrosion tests less meaningful.
- Not all commercial plant operating conditions may be demonstrated in pilot plants.
- Short runs in pilot unit can lead to inaccurate corrosion data.
- Batch operation or semi-continuous operations of pilot plant can limit the exposure times for obtaining corrosion data.

ing cause established to prevent a recurrence and to ensure that costly errors are not passed on upon scale-up.

> An organic salt was prepared by a several-stage synthesis on the pilot plant and centrifuged. After acetone and water washing the resulting cake was separated.
>
> A 600 litre stainless steel vessel equipped with steam/water jacket, three fixed baffles, a turbine mixer and an extraction system was charged with water. With the mixer on, 46 kg of polyacrylic acid was added and the mixture heated to 70 °C. The salt was added manually over a period of 20 minutes. When the temperature was then raised to 80 °C, bubbles were observed on the liquid surface so heating was switched off and 6 kg of cold water was added. When a solution of sodium sulphate was introduced the reaction erupted violently with two-thirds of the contents boiling over. Fortunately, the operators managed to step clear.
>
> One hypothesis was that the accident resulted from the presence of acetone. Worst-case calculations based on analysis of the cake suggested that the acetone content would produce a mixture with a boiling point of 85 °C, i.e. in excess of the process temperature. Furthermore, laboratory experiments on the cake failed to reproduce a boil-over. The favoured theory was that the cause could be attributed to the inadvertent use of sodium carbonate rather than sodium sulphate, a hypothesis sustained by laboratory trials.
>
> No scientific conclusion could be drawn, so the company was left with a dilemma on scale-up; whether to set, perhaps unnecessarily stringent, specifications for acetone in the cake with consequential on-costs and implications for the purity of the organic salt, or to assume that the wrong ingredient had been added, an error not likely to be repeated on production.

Evaluation of ambient air concentrations of relevant materials (e.g. flammable and/or toxic gases, vapours, mists and dusts) provides valuable information on operator exposure and the safety of the particular pilot-plant run. Such evaluations also indicate the range of precautions required on scale-up to full production.

Protective clothing

The provision and use of protective clothing and equipment is, despite the smaller scale, equally as important as in production – particularly because of the proximity of chemicals and the probability of manual charging/ emptying/sampling.

> Whilst charging material from kegs using a dip-leg an operator took off his goggles to wipe them. His goggles were not over his face when he opened the next keg and unfortunately liquid from it splashed into his face and eyes.

This aspect is discussed further in Chapter 6.

Materials handling

Suitable storage is required to prevent stockpiling of raw materials and products on the plant. If unstable or degradable materials are stored the hazard may be compounded by low or irregular turnover rates. Thus an up-to-date inventory is preferable for seldom-used materials[93] and, as outlined on pages 600 and 1106, there should be a procedure for monitoring time-sensitive materials, e.g. peroxides and monomers. All materials should be properly labelled to identify, as a minimum requirement, the contents, hazards, date and ownership. Whereas fixed tanks are generally carefully designed and located and provided with appropriate protection, there may be a tendency for drum storage to grow in an unplanned manner. Standard outdoor drum storage recommendations should be followed.[110] Drums and carboys should be handled and stored as described on page 648.

> Two men sustained severe acid burns on their lower portions of their bodies and legs when a 9 litre bottle containing concentrated sulphuric acid exploded. They were engaged in the transfer of a mixture of silver sulphate and sulphuric acid from the bottle to a burette by applying nitrogen pressure from a 75 psig supply.

Compressed gases represent unique hazards associated with their chemical nature and the high pressures; these hazards and appropriate precautions are summarised in Chapter 11, page 627.

> As a test the contents of a hydrogen cylinder were released by suddenly opening the valve and ignited at the end of 15.34 m of 0.64 cm tubing by a flame.[111] The resulting flame envelope was >6 m in diameter and 10.7 m in length. The flame spalled concrete at 2.4 m below its centre even though it lasted only 15 seconds. (As a result a limiting orifice was subsequently installed just downstream of the regulator.)

All plant personnel should be trained in human kinetics to ensure the correct procedures are appreciated and adopted for manhandling goods. Table 2.4 provides an example of the correct lifting techniques.

Following the recent publication of a draft EEC Directive on the subject, the UK Health & Safety Commission have issued draft Regulations and Guidelines for handling loads at work.[112] If ratified, management will have a legal obligation to carry out, and keep up to date, formal risk assessments for all handling operations and to institute measures to prevent reasonably foreseeable injury to their employees from the handling of loads at work. The draft guidelines append the checklist reproduced here as Table 20.22 and further guidance on kinetic handling is given on pages 13–14.

Table 20.22 Checklist for employers – handling operations

- Have you assessed the handling tasks of your employees and whether they might cause injury?
- Have you made arrangements to monitor accidents and ill-health, to seek employees' and safety representatives' views, and to assess the effectiveness of any improved systems of work introduced?
- Can mechanical systems be introduced?
- Can the loads be improved? Can they be made smaller, lighter, more portable? Can handles be provided? Can you improve the external state of the loads, e.g. smooth off jagged edges? Can loads be marked to show how or where to hold them?
- Can handling aids be employed, such as trolleys, slides and chutes, or conveyors. Is personal protective equipment necessary? Is equipment properly maintained accessible?
- Can the workplace or task be redesigned to reduce bending, twisting, stretching, carrying distances, frequency of handling? Can jobs be rotated to avoid repetition and constant exertion? Is proper allowance made for rest pauses?
- Can the workplace be made safer by widening gangways, removing obstructions, keeping floors clean, providing proper lighting and temperature control?
- Has allowance been made for individual characteristics of the workforce?
- Is instruction and training necessary? If a training programme is introduced is its effectiveness being monitored?
- Do any of the tasks require special strength or fitness? If so, has this been properly evaluated and employees selected accordingly? Are the effects being monitored?

Detailed advice is also given in 'Guidance on Regulations – Manual Handling', HSE 1992.

Plant installation, maintenance and modification

Despite its limited scale and sometimes temporary nature, all equipment and pipework should be designed in accordance with established codes and standards.

A graphite rupture disc in a pilot-plant polymerisation reactor failed to burst and vent overpressure. Subsequent inspection revealed that several feet of water had accumulated in the vertical discharge pipe which vented directly to atmosphere with no bend to avoid rain water ingress.[1]

Any proposals to modify plants should be formalised and documented.

The work should be approved by relevant management and carried out under appropriate work permits. Pilot plants tend to have many joints per length of piping and are shut down, modified and cleaned relatively frequently. Furthermore, the typically small flowrates generally limit the amount of leakage which can be allowed without seriously affecting the accuracy of the data obtained. However, within the limited area of the pilot plant, the maximum leakage rate of any flammable or toxic materials must also be restricted so as not to result in potentially unsafe working conditions. Efficient and reliable leak testing is therefore essential.[113]

Maintenance work should be carefully supervised to avoid hazards being introduced inadvertently.

Problems were encountered with the discharge from a pilot-plant, batch-operated centrifuge containing an internal filter cloth. To check that it was running 'true', a fitter disarmed the microswitch on the lid and leaned over to observe rotation of the basket. Whilst so doing he accidently dropped a spanner into the basket necessitating a crash shut-down.[1]

Equipment and pipelines should be thoroughly decontaminated before maintenance personnel are allowed to work on them. Properly secured access equipment should be used and climbing on pipework, plant fittings, etc., should be prohibited. In addition to any protective gear required to guard against exposure to materials handled, personal protection should include appropriate foot- and head-wear. A strict standard of housekeeping is of paramount importance; hoses should be coiled and replaced immediately after use and never left trailing across part of the floor. Regular facility inspections are crucial as a means of monitoring compliance with company policy and codes.

In conclusion, in addition to the normal hazards encountered with laboratory and plant processes, pilot plants pose problems worthy of particular attention. These stem from the continuous modifications to plant and process, working at the frontiers of knowledge and difficulties associated with frequent manhandling of chemicals and items of plant. Throughout the laboratory and pilot-plant investigations data should be amassed on materials and the process to enable each stage of the development to proceed safely. The safety information required at the end of the R&D programme to support full-scale production, marketing and legal requirements is discussed in Chapters 12 and 13. Generation of sufficient high-quality data for pre-market clearance and notification may take several years to complete. Hence, unless the programme is started early on in the project, business opportunities may be lost. Inadequate data or inconclusive information will extend the approval procedure with consequent inefficient use of capital investment.

Much money was expended on developing a process to remove phthalate esters from PVC. However, because of the costs of meeting the environmental concerns, the project was abandoned.[114]

References

1. (a) Carson, P. A. & Mumford, C. J., *Loss Prevention Bulletin*. Institution of Chemical Engineers, 1989(089), 1; 1989(090), 3; (with Bond, J.) 1990(093), 5. (b) Jones, J. et al., *Chemical Engineering*, 1993 (April), 136.
2. *Guidance Notes on Chemical Reaction Hazard Analysis*. The Association of the British Pharmaceutical Industry, 1989.
3. Scot, S., *New Scientist*, 2 Dec. 1982, 53.
4. Universities Safety Association, *Safety News*, 9 May 1978, 12.
5. Bretherick, L., *Handbook of Reactive Chemical Hazards*, Butterworths, 1975.
6. Manufacturing Chemists Association, *Guide for Safety in the Chemical Laboratory*, 1972.
7. Jennings, A. J. D., *Chem. Eng.*, Oct. 1974, 637.
8. Anon., *Loss Prevention Bulletin*, Institution of Chemical Engineers, 1979(025), 1.2.
9. Yoshida, T., *Safety of Reactive Chemicals*, Elsevier, Oxford, 1987.
10. Boynton, D. E., Nichols, W. B. & Spurling, H. M., *Ind. Eng. Chem.*, 1959, **57**, 4.
11. Sebeko, Ju. N., Ivanov, A. V., Alehina, E. N. & Barmakova, A. A., Calculation of the upper concentration limits of flammability of the vapours of organic compounds and their derivatives in air. *Himiceskaja Promyslennost*, 1983, No. 10, 23–4, 20 ref. (in Russian).
12. Anon., *Loss Prevention Bulletin*, Institution of Chemical Engineers, 1979(025), 1.2.
13. Strizhevskii, I. I., *International Polymer Science and Technology*, 1988, **15**(12), T/51–3.
14. Bretherick, L., *Handbook of Reactive Chemical Hazards*, 3rd edn, Butterworths, 1986.
15. National Fire Protection Association, *Manual of Hazardous Chemical Reactions*, Section 491M, NFPA, Boston, MA.
16. Sax, N. I., *Dangerous Properties of Industrial Materials*, 5th edn, Van Nostrand Reinhold, New York, 1979.
17. Kletz, T. A., *Cheaper, Safer Plants or Wealth and Safety at Work*, Hazard Workshop Module, Inst. Chem. Eng., 1984.
18. Wright, P., *The Times*, 28 Sept. 1989.
19. Kletz, T. A., *Learning from Accidents in Industry*, Butterworths, 1988, 84.
20. Bond, J., *I. Chem. E. Symp. Series*, 1988, No. 102, 37.
21. Dixon, B. E. & Williams, R. A., *J. Soc. Chem. Ind.*, March 1950, 69.
22. Kaiser, M. A., *Plant/Operations Progress*, 1991, **10**(2), 101.
23. Anon., *Safety News*, Universities Safety Association, May 1972, No. 2, 3.
24. Anon., *Safety News*, Universities Safety Association, November 1978, 10, 11.
25. Kaufman, J., reported in *Hazardous and Toxic Materials Safe Handling and Disposal*, ed. H. W. Fawcett, 2nd edn, 1988, John Wiley & Sons, New York, 219.

26. Jackson, H. L. et al., *J. Chem. Educ.*, 1970, **47**, A175.
27. EEC Directive 84/449. The adaptation to technical progress for the sixth time. Council Directive 67/548/EEC on the approximation of laws, regulations and administrative provisions relating to the classification, packaging and labelling of dangerous substances, *Official Journal of the European Communities*, L251, 19 Sept. 1984.
27a. Bond, J. *Sources of Ignition – Flammability Characteristics of Chemicals and Products*, Butterworth-Heinemann, Oxford, 1991, Appendix 5.
28. Harper, D. J., Rogers, R. L. & Gibson, N., *Plant/Operations Progress*, July 1985, **4**(3).
29. Gibson, N. & Harper, D. J., *I. Chem. E. Symp. Series*, 1981, No. 68.
30. Benson, S. W. et al., *Chemical Reviews*, 1969, **69**, 279.
31. Stull, D. R., Westrum, E. F. & Sinke, G. C., *The Chemical Thermodynamics of Organic Compounds*, Wiley, New York, 1969.
32. Cox, J. D. & Pilcher, G., *Thermochemistry of Organic and Organometallic Compounds*, Academic Press, London, 1970.
33. Pedley, J. B. & Rylance, J., *N.P.L. Computer Analysed Thermochemical Data; Organic and Organometallic Compounds*, University of Sussex, 1977.
34. *Selected Values of Thermodynamic Properties*, US Bureau of Mines Standards, Technical Note 270/3, 1970.
35. Kohibrand, T., Reaction chemical screening for pilot plant safety, *Chem. Eng. Prog.*, April 1985.
36. Beever, P. F. & Thorne, P. F., *I. Chem. E. Symp. Series*, 1981, No. 68.
37. Methods for determining physico-chemical properties, *Official Journal of the European Communities*, L51, 19 Sept. 1985.
38. Butters, G., *Plastics Pneumatic Conveying and Bulk Storage*, Applied Science Publishers, London, 1981.
39. Field, P., *Handbook of Powder Technology, Vol. 4, Dust Explosions*, Elsevier Scientific Publishing, London, 1982.
40. Field, P., *Explosibility Assessment of Industrial Powders and Dusts*, HMSO, 1983.
41. Kravchenko, V. S. & Bondar, V. A., *Minimum Ignition Energy: Explosion Safety of Electrical Discharges and Frictional Sparks*, Part I, Chapter I, Moscow.
42. Spontaneous combustion in solid materials, *Fire Prevention*, No. 193, October 1985.
43. *IP Standards for Petroleum and its Products, Part I*, Hayden, London, 1980.
44. *The Control of Industrial Major Accident Hazard Regulations*, 1984.
45. Coward, H. F. & Jones, G. W., *US Bureau of Mines Bull.*, No. 503, 1952.
46. Zabetakis, M. G., *US Bureau of Mines Bull.*, No. 627, 1965.
47. *Auto-Ignition Temperature*, BS 4056, British Standards Institute, 1966.
48. *Medical Manual of Defence against Chemical Agents*, HMSO, London, 1972 (JSP 312) A/24/GEN/14392.
49. Inch, T. D., Chemical disarmament, *Chem. Brit.*, Aug. 1983.
50. *Toxic Hazard Assessment of Chemicals*, Royal Society of Chemistry, Cambridge.
51. Methods for determination of toxicity, *Official Journal of the European Communities*, L251, 19 Sept. 1984.

52. *Patty's Industrial Hygiene and Toxicology*, John Wiley & Sons, New York, 1985.
53. Blackburn, G. M. & Kellord, B., *Chem. & Ind.*, 15 Sept. 1986, 607.
54. Royal Society of Chemistry, *COSHH in Laboratories*, Royal Society of Chemistry, Cambridge, July 1989.
54a. Fawcett, H. H., *Dioxin (TCDD), Dibenzofurans and Related Compounds in Hazardous and Toxic Materials, Safe Handling and Disposal*, John Wiley, New York, 1984, Vol. 8, p. 219.
54b. Anon., *New Scientist*, 19 July 1979.
55. *The Control of Substances Hazardous to Health Regulations*, 1988.
56. Samullah, Y., *Prediction of the Environmental Fate of Chemicals*, Monitoring and Assessment Research Centre, London.
57. Methods for the determination of ecotoxity, *Official Journal of the European Communities*, L251, 19 Sept. 1984.
58. Dickson, K. L., Maki, A. W. & Brungs, W. A., *Fate and Effects of Sediment-bound Chemicals in Aquatic Systems*, Elsevier Science, 1987.
59. EC Directive 67/548/EEC and its 6th amendment 79/831 EEC.
60. Reid, R. C. & Sherwood, T. K., *The Properties of Gases and Liquids*, McGraw-Hill, New York, 1958.
61. Calcote, H. F., Gregory, C. A., Barnett, C. M. & Gilbert, R. B., *Ind. Eng. Chem.*, 1952, **44**(11), 2656.
62. Fujii, A. & Hermann, E. R., *Journal of Safety Research*, 1982, **13**(4).
63. Weissberger, A. (gen. ed.), *Vapour Pressure, Latent Heats and Specific Heats. Techniques of Organic Chemistry*, Vol. 1, Interscience, New York, 1959.
64. West, E. D., Gravenstone, G. W. & Hoppe, T. F., *Plant Operations Progress*, 1986, **5**(3), 142.
65. Anon., *Safety News*, Universities Safety Association, 9 May 1978, 7.
66. Hempseed, J., reported by Kletz, T. E., *Chem. Eng.*, January 1991, No. 488, 28.
67. Bond, J., *Loss Prevention Bulletin*, Institution of Chemical Engineers, 1985(065), 20.
68. Anon., *Loss Prevention Bulletin*, Institution of Chemical Engineers, 1977(012), 4.2.
69. Robinson, P. J., *Chem. & Ind.*, 1978, 723.
70. Kletz, T. A., *Chem. Eng.*, February 1990, 15.
71. Bond, J. (based on a report by the United Steelworks of America), *Loss Prevention Bulletin*, Institution of Chemical Engineers, June 1990(093), 1.
72. Russell, W. W., *Loss Prevention*, 1976, **10** (Chem. Eng. Prog. Tech. Manual), American Institution of Chemical Engineers, 80.
73. Manufacturing Chemists Association, Washington, DC, Case History 1911.
74. Anon., *Loss Prevention Bulletin*, Institution of Chemical Engineers, 1975(01), 3.4.
75. Anon., *Loss Prevention Bulletin*, Institution of Chemical Engineers, 1977(012), 4.5.
76. Anon., *Chem. Eng.*, 15 November 1990, 7.
77. Kletz, T. E., reported in *The Chem. Eng.*, 8 April 1993, 8.
78. Englund S. M., *Chem. Eng. Prog.*, March 1991, 85.
79. Anon., *Loss Prevention Bulletin*, Institution of Chemical Engineers, 1990(091) 15.

80. Anon., *Loss Prevention Bulletin*, Institution of Chemical Engineers, 1977(014), 1.1.1.
81. Bretherick, L., *Chemistry and Industry Weekly*, 11 September 1971.
82. Anon., *Loss Prevention Bulletin*, Institution of Chemical Engineers, 1977(013), 3.
83. Lowenstein, J. G., *Chem. Eng. Prog.*, September 1975, 51.
84. Pfeffer, H. A., Bhalla, S. K. & Dore, J. C., *Plant/Operations Progress*, April 1984, **3**(2), 98.
85. Boynton, D. E. W., Nichols, W. B. & Spurlin, H. M., *Ind. Eng. Chem.*, 1959, **51**, 489.
86. Brummel, R. N., *Plant/Operations Progress*, 1989, **8**(4), 228.
87. Carson, P. A. & Dent, N. J. (eds), *Good Laboratory and Clinical Practices*, Heinemann, 1990.
88. Graf, F. A., *Chem. Eng. Prog.*, 1967, **63**, 11, 67.
89. Bisio, A. & Kabel, R. L., *Scale-up of Chemical Processes*. John Wiley and Sons, New York, 1985.
90. Palluzi, R. P., *Chem. Eng.*, March 1990, 76.
91. Chaty, J. C., *Plant/Operations Progress*, 1985, **4**(2), 105.
92. Carr, J. W., *Chem. Eng. Prog.*, Sept. 1988, 52.
93. Capraro, M. A. & Strickland, J. H., *Plant/Operations Progress*, 1989, **8**(4), 189.
94. Bullock, C. J., Short course material, *Fundamentals of Process Safety*, Durham, 1985.
95. Jennings, A. J. D., *Chem. Eng.*, October, 1974, 637.
96. Rycroft, R. J. G., *Contact Dermatitis*, 1983, **9**, 453.
97. *AIChE Pilot Plant Safety Manual*, American Institute of Chemical Engineers, New York, 1972.
98. *Health and Safety Commission Newsletter*, 1994 (098), 4.
99. Garrison, W. G., *Hydrocarbon Processing*, 1988, **67**(9), 115.
100. Anon., *Loss Prevention Bulletin*, Institution of Chemical Engineers, 1980(035), 2.2.2.
101. Anon., *Loss Prevention Bulletin*, Institution of Chemical Engineers, 1977(014), 3.
102. Anon., *Loss Prevention Bulletin*, Institution of Chemical Engineers, 1980(035), 2.2.5.
103. Anon., *Loss Prevention Bulletin*, Institution of Chemical Engineers, 1977(014), 1.1.3.
104. Froment, G. F. & Bischoff, K. B., *Chemical Reactor Analysis and Design*, John Wiley, New York, 1979.
105. Hanson, C. (ed.), *Recent Advances in Liquid Extraction*, Pergamon, Oxford, 1971.
106. Majumdar, A. S. (ed.), *Handbook of Industrial Drying*, Marcel Dekker, New York, 1979.
107. Sharp, J. R. (ed.), *Guide to Good Pharmaceutical Manufacturing Practice*, HMSO, 1983.
108. Gundzik, R. M., *Chem. Eng. Prog.*, 1983 (Aug.), 29.
109. Krystow, P. E., in *Scale-up of Chemical Processes*, ed. A. Bisio, and R. L. Kabel, Ch. 16, John Wiley & Sons, New York, 1985.
110. Anon., *Loss Prevention Bulletin*, Institution of Chemical Engineers, 1974(028), 115.

111. Anon., *Loss Prevention Bulletin*, Institution of Chemical Engineers, 1981(041), 1.
112. Health & Safety Commission, Consultative Document: *Handling Loads at Work*, HMSO, 1988.
113. Carr, J. W., *Chem. Eng. Prog.*, 1988 (Sept.), 52.
114. Lederman, P. B., in *Scale-up of Chemical Processes*, ed. A. Bisio and R. L. Kabel, Ch. 17, John Wiley & Sons, New York, 1985.

CHAPTER 21

The environmental impact of chemicals

Introduction

A great cloud of smoke hangs continually over the town, and choking fumes assail the nose from various works. In the face of such an atmosphere it is not to be wondered at, that the trees and other green things refuse to grow.

Such was the town of Widnes in the mid-19th century following the introduction of a Leblanc Alkali works and before the recovery of the majority of the hydrogen chloride.[1] An early solution was to disperse the gaseous pollutants from tall chimneys. Of course, chimneys belching forth smoke are no longer considered a sign of affluence and successful industrial activity. Society undoubtedly benefits from the products of modern chemistry and technology but it is incumbent on manufacturers, processors and transporters of chemicals to be aware of the environmental impact of their products or waste, and to control the level of pollution to ever more stringent standards.

Whereas waste implies a lost resource with potential for recovery, pollution is considered as too much matter entering the environment via the air, water or land with consequential hazardous or obnoxious impact. However, as illustrated below, waste disposal can cause pollution and the two are inextricably linked. Hence 'pollution' and 'waste' are often used synonymously. UK industry and commerce generates in excess of 100 million tonnes of waste each year costing over £3 billion for disposal.[2]

> Alkali waste, comprising impure calcium sulphide, generated in the original Leblanc process was simply dumped in heaps near the factory. The piles were phosphorescent and presumably evolved hydrogen sulphide when acid rain fell on them. Some of these heaps – oxidised to calcium sulphate – still exist.[3] The source of the acid was hydrogen chloride vapour, which was said to be perceptible at a distance of six miles from the emitting works and to cause smarting eyes, irritating cough and difficulty in breathing within one or two miles.[5c]

> In 1894 a Mr Love began to build a canal linking the Niagara River with Lake Ontario but the project was abandoned because of cost. Half a century later

the 18-acre trench was used as a waste dump (page 854). Subsequent pollution resulted in the area becoming a Federal Disaster Zone; the school was closed, and it was recommended that pregnant women and small children should be evacuated, and a multi-million dollar drainage system was installed. An increased frequency of cancer and birth deformities in the area have been linked with the Love Canal. A $635 million lawsuit has been filed against the chemical company who were being sued by more than 100 000 individuals.[4a]

By June 1994 Occidental Chemical, the successors of the original chemical company, reached a settlement with New York State – 14 years after this litigation began and over 40 years after dumping of wastes into the landfill. The chronology was,[4b]

- 1942–53: Hooker Chemical & Plastics Corp. dump wastes in canal.
- 1953: Niagara Falls Board of Education buy canal area.
- 1968: Occidental Chemical purchases Hooker Chemical.
- 1977: Black, oily fluids ooze from ground in canal vicinity.
- 1978: New York State declares health emergency. State and federal governments evacuate residents, begin cleanup.
- 1979–80: U.S. and New York file lawsuits against OxyChem, Company sues U.S., New York, Niagara Falls and its Board of Education.
- 1988: Federal court finds OxyChem liable for cleanup costs.
- March 1994: Federal court rejects New York's claim for punitive damages against OxyChem.
- June 1994: New York and OxyChem settle suit.

(Under the settlement Oxychem will pay $98 million to New York State over 3 years and take over responsibility for pump-and-treatment operations.[4b] They still face about $200 million in federal remediation claims and some personal injury health claims from residents. However, some 60% of the injury claims were settled for $19 million.)

Following extensive pollution around the Denver site described on page 852 clean-up costs have been estimated at $5 billion and, in an unsuccessful claim against their insurers to reclaim the losses, the company have spent $50 million in bringing the claim and the insurers an equivalent amount in their defence.[5a]

It has been characteristic of problems with 'solid waste' disposal by landfilling that problems from leachate or gas generation are highlighted by subsequent land use.

The latter is exemplified by the Lekkerkerk experience summarised on page 1244.

Clearly these cases also illustrate the considerable financial penalties which can arise from uncontrolled waste disposal. The toxic effects of chemicals are discussed in detail in Chapter 6 while the control of exposure in the workplace is discussed in Chapter 7. The present chapter highlights the polluting effects of chemicals escaping into the natural environment with examples of how such incidents can arise. Methods for controlling such escapes are covered in Chapter 22.

Pollution and waste can be in the form of solid, liquid or gas, or combinations of these phases, as illustrated by Tables 21.1 and 21.2.[6] They may arise as a result of continuous fugitive emissions or sudden but more periodic episodes (see page 270) including spillages.

Severe pollution incidents may also follow fires, e.g. due to release of pyrolysis products, or of basic chemicals, or through contamination of firewater run-off.

> A fire in the State Office building in Binghampton, New York in 1981 caused major pollution due to the release of chemicals such as PCBs, dioxins and furans from transformer coolant throughout the 18-storey building. Detailed criticisms have been made of the emergency response.[7] In any event many people were affected, frightened or inconvenienced by the incident – ranging from janitors, who began cleaning up the day after the fire without training or protection, to people allowed access to the building's car park. No restrictions were placed on water release from the building, i.e. via flushed toilets.

Catastrophic and endemic exposure to chemicals in the environment may affect human health directly or affect the quality of life by damage to the ecosystem. Therefore, concern for the environment is, rightly, likely to be a continuing theme and this will have an increasing effect on the layout, design and operation of major hazard sites and the manner in which all chemicals are stored and processed. For example, there is a trend to replace floating roof or vented hydrocarbon liquid storage tanks by facilities with vapour recovery. However, care is required to avoid solution of an environmental problem increasing the potential for an accident to occur.

> On 21 August 1991 an explosion lifted a 230 tonne acrylonitrile tank off its base and projected it over four other tanks onto the forecourt of a terminal at Coode Island. A series of explosions and bund and tank fires followed. After about 3 hours the fires were almost completely extinguished but a further explosion occurred next day and a further serious incident developed. Only 13 of 45 storage tanks on the site remained undamaged. Whilst a common vapour recovery system was not the direct cause of this event, it did apparently allow propagation of flame to remote tanks.[8]

The aquatic environment

The world's waterways comprise vast expanses of oceans and seas, together with inland systems of rivers and lakes. Industrial contaminants of waterways are summarised in Table 21.3 (after ref. 6).

> Lake Baikal in the heart of Siberia with six or so towns and a few villages along its shoreline once provided breathtaking scenery and a sanctuary for rare inhabitants. Following the establishment of the timber industry, silt and rotting

Table 21.1 Types and forms of waste

Key:
- G Pollutant occurs as a gas
- L Pollutant occurs as a liquid
- S Pollutant occurs as a solid
- P Pollutant occurs in particulate form
- A Pollutant occurs in aqueous solution or suspension

Agriculture, horticulture

BOD waste (high and low)	L S A
Disinfectant	A
Pesticides, herbicides (see below)	P A
Fertilisers (see Inorganic chemicals)	A

Cement, bricks, lime

Chromium	L
Dust	S
Fluoride	G
Sulphur dioxide	G

Coal distillation, coal tar, coke ovens

Ammonia	G A
Aromatic hydrocarbons	G L
Combustion products	G
Cyanate	A
Cyanide	G A
Dust	S P
Fluoride	G A
Hydrocarbons (general)	G L A
Hydrogen sulphide	G
Phenols	L A
Polycyclic hydrocarbons	G
Sulphur dioxide	G
Tar	G L S
Thiocyanate	A

Construction, building, demolition

Combustion products	G P
Dust	S P
Metals	S
Rubble	S
Timber	S

Dry-cleaning

Chlorinated hydrocarbons	G L
Hydrocarbon solvents	G L
Bleach	A
Detergent	A
Phosphate	A
Sulphate	A

Electricity generation

Clinker	S
Combustion products	G
Cooling water	L
Pulverised fuel ash	S P
Sulphur dioxide	G

Explosives, pyrotechnics

Barium	S
Hydrocarbons	G L A
Lead	S
Manganese	S P
Mercury	L S P
Nitric acid	A

Table 21.1 (continued)

	Nitroglycerin	L A
	Phenol	S P
	Phosphorus	S
	Solvents	G L
	Strontium	S
	TNT	L A
Fibres, textiles		
	Bleach	A
	Cyanide	A
	Detergent	A
	Dyestuffs	A
	Grease	L S
	Oil	L A
	Resins	L A
	Silicones	A
	Speciality chemicals for fire-, rot- and waterproofing	L A
	Wax	P A
Food processing		
Animal	Abattoir waste	L S P A
	BOD waste (high and low)	A
	Disinfectants	A
	Grease	S A
	Oil	L A
Beverage	Alkali	A
	BOD waste (low)	A
	Carbon dioxide	G
	Cullet	S
	Detergent	A
Vegetable and fruit	Alkali	A
	Bleach	A
	BOD waste (high and low)	P A
	Oil	L
	Solvent	L A
	Wax	S P A
Inorganic chemicals		
Chloralkali	Brine	A
	Calcium chloride	A
	Mercury	L
Desalination	Brines	A
Fertiliser	Ammonia	G A
	Nitrates	S
	Oxides of nitrogen	A
	Phosphates	S
Glass and ceramic	Arsenic	A
	Barium	S
	Manganese	S
	Selenium	S
Hydrofluoric acid	Calcium sulphate	P A
Metals		
Extraction and refining	Acid mine waters	A
	Combustion products	G P
	Carbon monoxide	G
	Chloride	G S P
	Drosses	S
	Dust	S
	Fluoride	G S P

Table 21.1 (continued)

	Glass	S
	Spoil	S
	Sulphides	S P A
	Sulphur dioxide	G
Nitric acid	Ammonia	G A
	Oxides of nitrogen	A
Phosphoric acid	Calcium sulphate	P A
	Hydrofluoric acid	A
Pigments	Arsenic	S P
	Barium	S P
	Cadmium	S P
	Cobalt	S P
	Iron	S P A
	Lead	S P
	Manganese	S P
	Selenium	S P
	Titanium	S P
	Zinc	S P
Sulphuric acid	Sulphur dioxide	G P
Finishing/surface treatments		
Anodising	Chromium	A
Degreasing	Chlorinated hydrocarbons	G L
	Detergents	A
	Grease	L S
	Solvents	G L
Electroplating	Alkali	A
	Boron	A
	Cadmium	A
	Chromium	P A
	Copper	A
	Cyanide	A
	Detergent	A
	Fluoride	A
	Iron	A
	Nickel	A P
	Organic complexing agents	A
	Phosphate	A
	Precious metals	A
	Silver	A
	Sulphate	A
	Tin	A
	Zinc	A
Foundries	Dust	S P
	Sands	S
Machine shops	Oils	L A
	Oil absorbents	S
	Solvents	L G
	Swarf	S
	Synthetic coolants	L S A
Pickling	Acid	A
	Ferrous chloride	A
	Ferrous sulphate	A
	Hydrochloric acid	A
	Hydrofluoric acid	A
	Nitric acid	A
	Phosphoric acid	A
	Sulphuric acid	A

Table 21.1 (continued)

Pigments	see Inorganic chemicals	
Process/Engineering	Ammonia	G A
	Arsenic	G S A
	Cyanide	S A
	Emulsions	A
	Lubricating oils	L
	Phenols	L S A
	Soluble oils	A
	Thiocyanates	A
Products		
Batteries	Cadmium	S P
	Lead	S P
	Manganese	S
	Mercury	S
	Nickel	S
	Zinc	S
Catalysts	Cobalt	S
	Iron	S
	Manganese	S
	Mercury	S
	Nickel	S
	Organometallics	L S
	Platinum	S
	Silver	S
	Vanadium	S
Mining (excluding metals)	Spoil	S
	Dust	S P
Motor industry	Chromates	A
	Grease	L S
	Oil	L
	Paint	L S P
	Phosphates	S P A
	Solvents	L
Nuclear fuel and power	Radioactive substances	
	Radioisotopes	
Paint		
	Barium	S
	Cadmium	S
	Chromium	S
	Copper	S
	Lead	S
	Manganese	S
	Mercury	L S
	Selenium	S
	Solvents	G L
	Titanium	S
	Zinc	S
Paper, pulp		
	Bleach	A
	Chlorine	G A
	Copper	A
	Fibres	S P A
	Lignin	A
	Mercury	A
	Methanol	A
	Sulphides	A

1212 Further considerations and developments

Table 21.1 (continued)

	Sulphite liquor	A
	Titanium	S P A
	Wax	S P A
	Zinc	P A
Pesticides, herbicides		
	Arsenic	S P A
	Carbamates	S P A
	Chlorinated hydrocarbons	L S P A
	Copper	A
	Fluoride	G S A
	Lead	A
	Mercury	L A
	Organophosphorus compounds	P A
	Phenol	S A
	Polychlorinated biphenyls (PCB)	A
	Selenium	S
Petrochemicals		
Detergents	Boric acid	S A
	Phosphates	S A
	Sulphates	S A
Dyestuffs	Aniline	L A
	Chromium	L S P A
	Phenol	A
	Selenium	L S P A
General	Benzene	L
	Boric acid	A
	Chlorocarbons	G L A
	Fluorine	A
	Fluorocarbons	G L A
	Hydrocarbons	G L S P A
	Hydrochloric acid	A
	Hydrofluoric acid	A
	Phenol	L A
	Solvents	L
	Sulphuric acid	A
Miscellaneous	Polychlorinated biphenyls (PCB)	L
	Tetraethyl lead	G L P
Pharmaceuticals		
	Drug intermediates and residues	L S A
	Solvents	L A
Photography	Alkali	A
	Cyanide	A
	Mercury	A
	Phenols	A
	Silver	P A
	Thiosulphate	A
Polymeric, plastics, resins, rubber and fibres	Acid	A
	Alkali	A
	Asbestos	P S
	Cadmium	L S P A
	Cuprammonium compounds	A
	Detergent	A
	Dyestuffs	A
	Fibres	S P
	Formaldehyde	A
	Hydrocarbons	G L A
	Methanol	L A

Table 21.1 (continued)

	Phenols	L S A
	Phthalates	L A
	Polychlorinated biphenyls (PCB)	L A
	Solvents	L A
	Sulphide	A
	Urea	S A
	Wood flour	S P
	Zinc	A
Refineries		
	Alkali	A
	BOD waste	L P A
	Combustion products	G
	Emulsions	A
	Hydrocarbons	L A
	Mercaptans	G
	Mineral acids	A
	Phenol	L S P A
	Sulphides	G S P
	Sulphur	S P
	Tar	S P
Sewage treatment		
	Purified effluent	L
	Sewage sludge	S A
Tanneries		
	Arsenic	A
	Chromium	A
	Fibres	S P
	Hair	S P
	Lime	P A
	Sulphides	A
Textiles – see fibres		
Miscellaneous		
Electrical, electronics	Copper	S
	Mercury	L
	Precious metals	S
	Selenium	S
Vehicle exhaust	Aromatic hydrocarbons	G
	Lead	S P
	Nitrogen oxides	G
Water treatment	Calcium salts	A P
	Filtered solids	L P A

logs depleted the lake of much of its oxygen content. However, it was the millions of tons of effluent dumped each year from the many factories and chemical plants that developed on the coastline with no purification units that caused most ecological distress. Pollutants included sewage, alkali sludge, agrochemicals, and heavy metals such as mercury, zinc, tungsten and molybdenum. Some aquatic species have stopped spawning; for others the growth rate has slowed and they are less fertile. The average size of fish has halved. All but one of the test species of fish being monitored contain traces of DDT, PCBs and hexachlorocyclohexane, all banned but still produced in the former USSR. It is claimed that 'we are near the point where the process of negative change becomes irreversible'.[9]

Table 21.2 Common sources of pollutants

Key:	G	Pollutant occurs as a gas
	L	Pollutant occurs as a liquid
	S	Pollutant occurs as a solid
	P	Pollutant occurs in particulate form
	A	Pollutant occurs in aqueous solution or suspension

Chemical		Examples	Industrial source
Acids: Mineral	A	Hydrochloric acid Nitric acid Sulphuric acid	Pickling Chemical reagent By-products, petrochemicals
Organic	A	Acetic acid	Petrochemicals
Aldehydes	A	Acetaldehyde	Photochemical reaction in smog Petrochemicals
Alkali	P A	Sodium hydroxide Lime	Electroplating Beverage Photography Vegetable and fruit processing
Ammonia	G A	—	Coal distillation Nitric acid Urea and ammonium nitrate works
Aniline and related compounds	L A	—	Dyestuffs
Aromatic hydrocarbons	G L	Benzene Toluene	Coal tar Vehicle exhausts Petrochemicals Pesticides Herbicides
Arsenic	G S P A	Arsine Arsenous acid and salts	Pigment and dye Pesticide and herbicide Metallurgical processing of other metals Glass and ceramic Tanneries
Asbestos	S P	Chrysolite Amosite Crocidolite	Equipment and building; insulation Fillers in various industries Motor vehicle assembly Fabric manufacture Polymers, plastics
Carbon dioxide	G A		Combustion Fermentation
Carbon monoxide	G		Coke ovens Incomplete combustion Smelting Vehicle exhausts Metal extraction and refining

Table 21.2 (continued)

Chlorinated hydrocarbons				
Chemical	G L	Trichlorethylene 1,1,1-trichloroethane		Degreasing (engineering) Dry-cleaning Solvents
Pesticidal	L S	DDT BHC Aldrin Dieldrin		Pesticides Wood treatment
Chlorine and chlorides	G S P A			Chlorinated hydrocarbons Chloralkali Paper and pulp Petrochemicals Metal extraction and refining
Chromium and compounds	S P A	Chromic acid Sodium dichromate		Anodising Cement Dyes Electroplating Paint Tanneries
Cobalt and compounds	S P A	Cobalt oxide		Catalysts Fibres Paint Paper and pulp Pickling
Copper and compounds	S P A	Copper sulphate Copper pyrophosphate Cuprammonium compounds		Electroplating Electrical and electronics Etching Pesticides
Cyanate	A			Coal distillation Oxidation of cyanide
Cyanide	S P A	Sodium cyanide Copper cyanide		Heat treatment of metal Photography Coal distillation Electroplating Synthetic fibre
Disinfectants				Agriculture and horticulture Food processing
Fluorides	G S P A	Hydrogen fluoride Calcium fluoride		Cement Aluminium
Hydrocarbons, general				Coal distillation Petrochemicals Refineries
Iron and compounds	S P A	Iron oxide Ferrous chloride		Aluminium Electroplating Pickling Pigments Electronics Titanium dioxide

Table 21.2 (continued)

Lead and compounds	S P A	Lead oxide Tetraethyl lead (TEL)	Batteries Printing Vehicle exhausts Explosives and pyrotechnics Pesticides Paint Refineries Petrochemicals
Manganese and compounds			Catalyst Batteries Glass Paint Pyrotechnics
Meat wastes	S A		Meat processing and preparation Abattoirs Dairies Tanneries
Mercaptans	G		Refineries Coke ovens
Mercury			Herbicides
Organic	S A	Methyl mercury	Bacterial activity on inorganic mercury Pesticides
Inorganic	L S A	Mercurous chloride	Batteries Catalysts Cement Chloralkali Combustion of coal and oil Electrical and electronic Explosives Paints Paper and pulp Pesticides Pharmaceuticals Photographic Scientific instruments
Methanol	L A		Resins Paper
Nitrates	S P	Potassium nitrate	Metals heat treament Water treatment
Nitrogen oxides	G	Nitrogen dioxide	Combustion processes Electricity generation
Oil and soluble oil	L		Engineering Petrochemicals Refineries
Paraquat	L S		Herbicide

Table 21.2 (continued)

Pesticides (includes acaricides, avicides, bactericides, insecticides, molluscicides, nematocides, piscicides, rodenticides)		Chlorinated hydrocarbons (q.v.) Carbamates (q.v.) Organophosphorus compounds (q.v.)	
Pharmaceuticals	L S P A	Aspirin Penicillin	Pharmaceutical industry
Phenol and related compounds	S P A	Phenol Cresol	Photographic Coal distillation Dyestuffs Explosives Petrochemicals Pesticides Plastics Refineries
Phosphorus and compounds	S P A	Phosphoric acid	Boiler blowdown Corrosion protection Detergents Fertilisers Matches Metal finishing
Phthalates	L S P A	Dibutyl phthalate	Plasticiser (polymers)
Platinum and compounds	S P		Catalysts
Polychlorinated biphenyls (PCB)	L S A		Adhesives Lubricants and hydraulic fluids Pesticides Plasticiser in paint and polymers
Silicates	P		Cement Metal extraction and refining
Sulphur oxides	G	Sulphur dioxide Sulphur trioxide	Coal distillation Combustion of coal and heavy fuel oil Electricity generation
Tar	L		Refineries Coal distillation
Thiocyanate	A		Coal distillation
Tins and compounds	G S P A		Mining, tinplating
Titanium and compounds	S P A	Titanium dioxide	Astronautics Paper Paint
Vanadium and compounds	S P		Catalysts

Table 21.2 (continued)

Vegetable waste	L S P A	Breweries Cattle feed Natural rubber Starch Sugar refineries Vegetable and fruit processing and preparation
Wax	S	Fruit preserving Paper Refineries Textiles
Zinc and compounds	G S P A	Galvanising Electroplating Paper and pulp Synthetic fibres

(The former Soviet chemical industry is reported to be a serious source of ecological danger. One hundred chemical plants were shut down on environmental grounds in 1989 with lost capacity amounting to 1.2 Mt/a of fertiliser, 1.2 Mt/a of sulphuric acid, and 1 Mt/a of ammonia.[10])

It is convenient to divide aquatic pollutants into inorganic and organic sources.

Inorganic chemical pollutants

Metals

Metals occur in the natural aquatic environment. However, excessive concentrations above those normally encountered represent a pollution hazard. The most toxic compounds tend to be salts of heavy metals such as beryllium, cadmium, lead, mercury, nickel, silver, gold, chromium, zinc and copper.

Thus copper is a micro-nutrient found in natural waters at levels of <5µg/l but becomes toxic at elevated concentrations, e.g. as a result of mining activities and other industrial processes, e.g. smelting, electroplating and brass production. The toxicity appears to stem from the presence of Cu^{++} ions.

In one river, only 1–2 ppm of copper, from factory effluent, was found to have exterminated all animal life for 10 miles downstream and decimated the algae concentrations.[11]

Zinc, an essential trace element, is involved in nucleic acid syntheses and it occurs in many enzymes. It occurs widely in nature as the sulphide, carbonate and hydrated silica ores often in combination with other metals

Table 21.3 Common pollutants of aqueous effluents (after ref. 6)

Pollutant	Example
Alkali	Sodium hydroxide
	Potassium hydroxide
	Calcium oxide
	Calcium hydroxide
	Sodium, potassium and calcium carbonates
	Ammonia (*q.v.*)
Ammonia	—
Biocides	Fungicides, insecticides
Biodegradable waste	Food waste
	Organic chemicals
	Sewage
Boron, borates, fluoroborates	—
Bromine	—
Chloride	Sodium chloride
Chlorine	Hypochlorites
Chromic acid (hexavalent chromium)	Potassium dichromate
Cyanide	Copper cyanide
	Nickel cyanide
	Potassium cyanide
	Silver cyanide
	Sodium cyanide
	Zinc cyanide
Emulsified oil	'Suds' from machine coolants
Fluoride	—
Fluorine	—
Metal salts in alkaline solution	Cuprammonium complex
	Nickel and cobalt ammonia complex
	Cyanides (*q.v.*)
	Copper pyrophosphates
	Plumbites
	Zincates
Metal salts in acid solution	Most metals as acid salts, e.g. chloride, nitrate, sulphate
Mineral acid	Sulphuric acid
	Hydrochloric acid
	Hydrofluoric acid
	Nitric acid
	Sulphuric acid
Miscible/soluble organic material	Acetone
	Alcohols
	Organic acids
Non-metallic inorganic dissolved compounds	Arsenic
	Selenium
Oil, grease, wax and immiscible organics	Animal fat
	Chlorinated solvents
	Lubricating oil, fuel oil
Organometallic compounds	Catalysts
pH	Acids, alkalis
Pharmaceuticals	—
Phenols and related compounds	Phenol, cresol
Phosphate	Detergents, fertilisers
Sewage	—
Sulphate	Calcium sulphate
Sulphite liquor	Sodium sulphite
Suspended particles	Miscellaneous metals, organics

Table 21.3 (continued)

Pollutant	Example
Total dissolved solids	Carbonate Chloride Nitrate Phosphate Sulphate

such as iron and cadmium. Industrial uses include alloy manufacture (e.g. brass) and galvanising. Compounds such as the oxide, chloride, chromate and sulphide are widely used in other industries.

Chromium may be present in wastes from electroplating works or tanneries. The hexavalent form (chromate) is more phytotoxic than the trivalent form with soluble chromate concentrations as low as 0.5 µg/ml being significant[23]. The trivalent form is apparently more toxic to fish with concentrations for several species being in the range 0.2 to 5 µg/ml.[23]

Cadmium is usually found as an impurity in ores of metals such as zinc and copper; the key industrial use is in electroplating. Heavy metals are capable of bioaccumulation to biologically significant levels and they may be passed along the food chain at lethal levels: sub-lethal concentrations can affect the biochemistry, physiology, behaviour, and life cycles of living organisms.

The mass poisoning of humans as a result of contamination of the food chain by methyl mercury which had polluted the environment is referred to on page 200. In another episode involving agricultural organic mercury in 1960 there were at least 22 fatalities amongst 221 farmers in the Middle East who had consumed bread made from wheat dressed with a fungicide, Granosan M, which contained 7.7 per cent ethyl mercury sulphonanilide.[12] Mercury pollution of natural waters has also followed the use of slimicides containing mercury in pulp paper mills. More recently concern has been expressed in Japan that the mercury from discarded small batteries from widely distributed electronic devices may leach from garbage tips as dimethyl mercury, or a similar compound, and present the potential for pollution of water supplies.[13]

Lead is neither as toxic nor as bioavailable as many other heavy metals.[23] However, it is widely-used and, as discussed on page 1671, can accumulate in mammalian bone marrow.

The toxicity of these heavy metals to freshwater fish and other aquatic organisms and vegetation has been eloquently reviewed by Alabaster and Lloyd.[14] The toxicity of a given element is influenced by a range of factors such as test animal (species, age, size, etc.), pH, salinity, water hardness, temperature, dissolved oxygen concentration, etc. Table 21.4 summarises data for selected metals from one source;[15] because of the lack of test standardisation and the wide variety of species investigated, ranges of

Table 21.4 Aquatic toxicity of selected metals

Metal	TLm 96 (ppm)*
Arsenic (e.g. bromide or chloride)	10–1
Cadmium (e.g. chloride)	100–<1
Calcium (e.g. carbonate or chloride)	>1000
Mercury (e.g. bromide, chloride or sulphate)	<1
Tin (e.g. as chloride)	1000–100
Titanium (e.g. chloride)	1000–100

* The concentration that will kill 50% of the exposed organisms with 96 hours.

toxicities rather than a single dose are used to give an indication of the toxicity to aquatic life.

Aluminium exists in nature as insoluble material in rocks, soil, river bed sediments, etc. However, if solubilised by reduced pH, e.g. as a result of acid rain (see page 1222), aluminium compounds become remarkably toxic to many forms of aquatic life at concentrations as low as 0.1–1 mg/l. Aluminium sulphate and other aluminium compounds are widely used as coagulants in the treatment of public water supplies. These salts form an insoluble floc of hydrolysis product which entraps colloidal and suspended material in the raw water. The suspended floc is then filtered off but inefficient treatment can lead to breakthrough of floc into the distribution system and accidental overdosing can give rise to hazardous levels. Concern arises over aluminium because of its role in encephalopathy and high levels of aluminium have been found in brain tissue of patients dying from this disease, and in those who suffered from premature senile dementia – Alzheimer's disease. However, drinking water is thought to contribute a maximum of 5 per cent of the total dietary intake of aluminium, although high levels could arise from episodes of accidental overdosing.

> In July 1988, 20 tons of acidic aluminium sulphate (sufficient for several months) were accidentally tipped into the wrong tank (containing water supply for less than one day) at Lowermoor Treatment Works. The water supply to residents and tourists in the Camelford area of Cornwall hence became heavily contaminated: the levels of aluminium and sulphate ions were 600 and 24 times their respective acceptable levels for potable water. About 60 000 fish were killed in nearby rivers when the mains were flushed out; 20 000 local people suffered from acute effects, including a stinging sensation in their mouths when they drank the tap water, plus after-effects such as diarrhoea, skin rashes, sore throats, ulcers and chest pains. Some women reported that their hair turned green on washing. Chronic complaints included loss of memory and pain in the joints, and fear of the onset of Alzheimer's disease.
>
> It was two days after the incident, when staff at the water works noticed that stocks of aluminium sulphate were low and made contact with the supplier, that the mistake was realised. A decision was apparently taken not to tell the

public. The authority flushed-out its own system but failed to tell people to flush the poison from their domestic water tanks.[16]

Pollutants are often mixtures and whilst the combined effect for certain metals is additive (e.g. nickel and chromium) others have antagonistic effects (e.g. calcium on the toxicity of lead and aluminium) and others behave synergistically (e.g. zinc and copper or ammonia and zinc). Thus fathead minnows usually survive an 8-hour exposure to either 8 mg Zn/l or 0.2 mg Cu/l but most succumbed within this period in a mixture of the two metals at one-eighth of these concentrations, respectively.

Cyanides

Cyanides, e.g. from plating effluents, can be fatal to fish at concentrations less than 1 ppm, e.g. 0.04 ppm is fatal to trout.

Acid rain and lowered pH

Acid rain provides an example of the chronic effect of pollution on the acidity of the aquatic environment. Rain is normally very slightly acidic, i.e. about pH 6, but release of acidic oxides of sulphur and nitrogen into the atmosphere during the combustion of coal, oil and gas in the generation of electricity, and from industrial processes and vehicle exhausts, can increase the acidity of rain to pH 4. This in turn increases the acidity of the soil and freshwater systems with consequential adverse effects on the environment. Coniferous trees turn yellow, the needles fall off and the trees die. Aquatic life is also disturbed as organisms die, fail to reproduce, become deformed, behave abnormally (including trends to cannibalism), etc. A generalised summary of the effect of pH reduction is given in Table 21.5 (after refs. 17 and 18). Indirect effects include exposure of organisms

Table 21.5 Sensitivities of aquatic organisms to lowered pH

pH	Effects
6.0	Crustaceans, molluscs, etc., disappear. White moss increases.
5.8	Salmon, char, trout and roach die. Salamander eggs fail to hatch. Sensitive insects, phytoplankton and zooplankton die.
5.5	Whitefish and grayling die.
5.0	Perch and pike die.
4.5	Eels and brook trout die.
4.0	Crickets, frogs and northern spring peepers die.

to toxic metals leaching from the land or as a result of corrosion of plumbing systems. Clearly pH of waterways can also be acutely affected by direct discharges of acid effluent from industrial and mining operations.

During maintenance work on a wall, falling debris sliced a valve from a storage tank at a galvanising works releasing 4000 gallons of hydrochloric acid into the River Dibbin which became polluted over a 2-mile stretch. Workers at nearby factories were evacuated as the fire service controlled the venting-off of 18 tons of the corrosive chemical. The company was ordered to dam the river with limestone chippings to neutralise the acid and prevent its spread; children were warned not to play in the river, a local beauty spot. Some 23 000 gallons of contaminated water were removed. It was concluded that the accident could have been prevented by building a wall/embankment around the tank to protect it. Clean-up cost the company £6744 and they were fined £500.[19]

(The bunding of storage tanks, with valved drains where appropriate, is as noted on page 691, a common design feature which should be extended to include waste tanks; it may also be of value for drum storage areas.)

In addition to the direct effects of pH on living matter, an increase in pH is responsible for solubilising aluminium which, once in solution, is extremely toxic to fish. Drinking water acidified to pH 4 poses a health risk from contamination by aluminium and heavy metals, e.g. from metal pipes; copper solubility increases sharply below pH 5 and with increased temperature. (Concentrations of up to 45 mg/l Cu have been detected in acid water standing overnight in copper hot-water pipes; the WHO recommends a maximum Cu content of 1.5 mg/l.)

Nitrates and phosphates

Nitrates occur in nature and are essential constituents for the formation of chlorophyll and amino acids by plants. The use of nitrogen-containing fertilisers has resulted in enhanced crop yields. However, leaching of these materials from the land has resulted in increased concentrations in inland waterways and some sources of drinking water. Concern over the latter stems from the possible formation of methaemoglobinaemia in bottle-fed infants. There are also reports that nitrates may be converted to carcinogenic nitrosamines in the adult stomach. Therefore the European Health Standard for drinking water recommends a maximum nitrate concentration of 50 mg per litre.

Another important nutrient is phosphorus which if allowed to pollute waterways, e.g. by agricultural run-off or from use of domestic detergents, can stimulate algal growth in still or slow moving, shallow or stratified, warm water in the presence of sunlight. The financial consequences can be significant especially in water supply reservoirs. The algal blooms can block filters in water treatment works and when they decay degradation products may taint drinking water. The rapid growth of algae on the surface of water also reduces penetration of light and thereby inhibits photosynthesis of germinating plants leading to oxygen depletion with consequential loss of fish life.

Organic chemical pollutants

About two-thirds of the organic content of US watercourses originates from industrial waste. Some organic pollutants such as silage effluent, sewage treatment plant effluents, fats, etc. are themselves not necessarily toxic but provide food on which micro-organisms thrive; as a result they deplete the water of oxygen with fatal implications for higher organisms. (Chemicals such as sulphides, e.g. from tanning, and ferrous salts tend to oxidise in the water and thus deplete the oxygen content. Similarly eutrophication can lead to oxygen deficiency as described on page 1223, as can the formation of layers of oil on the water surface.)

Polychlorinated biphenyls (PCBs), pesticides, herbicides, and fungicides have the potential to persist in the environment. Examples include DDT, dieldrin, aldrin, endrin and lindane which are estimated to have half-lives of around 15–20 years. The materials also bio-accumulate in the food chain, e.g. one estimate suggests that the concentration factor for PCBs from aquatic to carnivorous mammals may be as high as 10 million.

PCBs were widely used between the 1930s and 1970s in electrical equipment, and in heat transfer fluids, hydraulic liquids, cutting oils, paint varnishes, adhesives, etc. From 1971 the sole manufacturers in the UK restricted sales for dielectric applications and production ceased soon after.[20] The long-term risks from exposure are largely unknown. PCBs have been associated with liver disease and chloracne and they are suspected human carcinogens.[21] Pollution by PCBs is particularly hazardous since concentrations as low as 0.001 ppm are fatal to some aquatic organisms. They undergo only 50 per cent degradation in 20 years and readily accumulate in, e.g. fish and shrimps, ending up in the food chain, and this source is believed to be the major contributor to the estimated human daily intake of 5–10 μg.[20] Clearly, therefore, any pollution of water supplies is serious and PCBs must be transported and disposed of with care, whether in concentrated or diluted form. Unfortunately careless disposal has contaminated soils and waters in most countries in which they were manufactured and used.

> A transporter loaded with PCBs began leaking as it was transported along a mountainous road. When informed of the leak, which was dissolving the asphalt road, the driver dumped the liquid onto an adjacent mountainside. Rain caused excessive penetration into the soil. Contaminated soil was eventually shipped in 11 000 drums for entombment in concrete vaults but not before impacting the public over an area of 52 square km (20 square miles), precipitating claims for clinical manifestations of PCB diseases in people and livestock. It is alleged that one person became mentally deficient from drinking PCB-contaminated water. The incident cost over $1 million in 1973 in clean-up costs and legal damages.[22]

> In 1976 strong antiseptic odours were detected in what was first thought to be an oil spill. PCBs were found in oil, water and sludge in old pits which had

been used for 30 years to dump waste transformer fluids. Because of falling dyke walls and porous soil the 26 500 m^3 (7 million gallons) of waste containing *ca*. 1.5 mg PCB/l threatened a major river. The entire contents of the dump had to be drained into safe areas for chemical and physical treatment to reduce the PCB content of the effluent to 1.0 μg/l prior to discharge into the river. Clean-up costs totalled $200 000.[22]

Phenols, e.g. from synthetic resin manufacture or from tar distillation, biodegrade if present at low concentrations. However, concentrations higher than 1–5 ppm may be fatal to fish and bacteria. Phenols also taint drinking water at very low levels, especially if chlorinated.

Clearly organics can pollute waterways as a result of natural usage (e.g. long-term leaching of pesticides from treated land) or as a result of acute events such as accidental spillage.

The cargo ship *Perentis* sank in very heavy weather on 13 March 1988 some 30 miles north-west of the Channel Islands. It was carrying several consignments of chemicals including 6 tonnes of lindane stowed in a container on deck and 1 tonne and 0.6 tonnes of permethrin and cypermethrin respectively stowed in drums in the cargo hold.[24] The lindane, a persistent organochlorine pesticide[25] classed by the US Environmental Protection Agency as an 'extremely hazardous substance', was contained in sealed polythene bags within fibreboard drums inside a 6.25 m freight container. The permethrin and cypermethrin were packed in 50 kg lots in sealed polythene bags in metal drums considered to have a minimum life of about 1 year in sea water before commencing to deteriorate.

After 2 days the lindane container and other deck cargo were located floating in the sea but while the container was being towed to Cherbourg in rough weather the towline parted and it sank in 70 m of water some 25 miles from port. (Five drums of the other pesticides were recovered from the seabed in early April but 27 were still missing.)

Octyl phenol and other materials, including drums of paraquat and diquat, were involved in a fire at Woodkirk in February 1982. These pesticides were released and, as a result of the large quantities of water used to fight the fire, were washed into nearby watercourses.[26]

A full-scale pollution alert was instigated by the local Water Authority. The water supply to the city of York was threatened at one stage. The dense black smoke from the fire was thought to be toxic so preparations were made for evacuation of local residents but, due to prevailing wind conditions, this proved unnecessary.

Following the Sandos accident in November 1986, referred to on page 911, approximately 10 000 m^3 of the copious amounts of water used for firefighting drained into the Rhine, polluted by chemicals. Some 300 firemen, police and workers who were on duty exhibited no adverse health effects. There were complaints of headaches, nausea, burning eyes and respiratory irritation from the public but no cases of serious illness.[27a] However, the Rhine sustained ecological damage for about 250 km. In a two-month mopping-up operation toxic substances deposited on the bed of the river near the Bisfelden power station were removed. Debris from the fire was packed into 250 truck bodies, 17 railway cars and over 6000 drums and stored at a separate site for future disposal.

Oil pollution

The effect of oil release into natural waters is to cause visible pollution since it floats on the surface (page 100). It tends to spread as extremely thin films which prevent diffusion of oxygen from the air and thus interfere with regeneration, to the detriment of water quality. Oil films >1 μm thickness, equivalent to an even spread of 500 gallons over one square mile have a marked effect on the rate of oxygen absorption.[27b] Large quantities of oil tend to coat the gills of fish, and hence affect their utilisation of dissolved oxygen, and to endanger water birds. 'Soluble' oils used as machine coolants may also cause troublesome, stable emulsions.

More is known about the toxic effects of oil on salt water systems than freshwater ecosystems even though chronic pollution of freshwaters with hydrocarbons is widespread. Water-soluble constituents can be toxic to marine life and some components, such as lead and polynuclear aromatics, can accumulate in tissues. Emulsifiers and dispersants used to clean up spillages are themselves not without toxic effects and the surface-active nature facilitates penetration of toxic substances. Physical properties of floating oil are a particular threat to higher vertebrates such as aquatic birds because contamination reduces buoyancy and insulation: ingestion can be lethal. At low levels crude oil distillates, waste from refineries and petrochemical plants can taint flesh, such as fish, to the human palate.

In 1983–86, 60 per cent of pollution incidents in Strathclyde which resulted in closure of the water supply intakes were due to spillage of petroleum products.[28] Sources of marine oil pollution include petrol and oil washed from roads, illegal discharges of engine oil, boat and irrigation pumps, leaks from storage tanks and spillages from land and sea transport systems. The ship/shore interface is particularly vulnerable:

> On 12 July 1990 about 10 tonnes of heavy crude oil were discharged into the River Mersey during the transfer of 125 000 tonnes of oil from a ship for refining. The ship was close to finishing discharge and was using a crude-oil hose to sluice down the inside walls of emptied tanks. A wrong valve was opened by mistake causing the spillage. A slick one kilometre in length formed and some 500 metres of foreshore became polluted.[29]

Underwater pipelines pose a threat unless protected and well maintained:

> On 19 August 1989 some 156 tonnes of crude oil leaked into the River Mersey from a six-inch fracture which had developed in a corroded 16-year-old underwater 12-inch-bore pipeline linking an oil terminal to a refinery 20 miles

away. A 20-mile stretch of the river became polluted and hundreds of birds were killed and thousands more affected. The company was prosecuted under the Control of Pollution Act 1974 and were fined £1m. In addition the company paid £1 400 000 towards the clean-up costs and in settlement of third-party claims.[30]

Whilst vast amounts of oil are, routinely, successfully transported around the world without incident, when they do occur accidents can have spectacular effects as illustrated by Table 21.6 (after refs. 31 and 32). There

Table 21.6 Illustrative large oil spills from tankers (after refs. 31 and 32)

Date	Location	Tanker	Spillage	
			Metric tonnes × 10^3	US Gal × 10^6
17.3.78	France	Amoco Cadiz	220	66
18.3.67	England	Torrey Canyon	117	36
19.12.72	Gulf of Oman	Sea Star	115	34
23.2.77	Pacific Ocean	Hawaiian Star	99	30
20.3.70	Baltic Sea	Othello	60–100	18–30
12.5.76	Spain	Urquiola	<88 (burned)	<26
29.1.75	Portugal	Jakob Maersk	<84 (burned)	<24
27.2.71	South Africa	Wafra	<63 (burned)	<19
9.8.74	Straits of Magellan	Motula	51	15
24.3.89	Alaska	Exxon Valdez	35	10

have been larger documented spills to date, viz. from the *Atlantic Express* off Tobago in 1979 (276 000 tons) and from the *Costello Belver* off South Africa in 1983 (256 000 tons).

On 16 March 1978 the *Amoco Cadiz* supertanker was wrecked only 1.5 miles off the Brittany coast with the loss of 223 000 tons of light crudes. Despite the nature of the oil with a 40% volatile fraction, and the effects of the winds which helped to dissipate much of the hydrocarbons, about one-third of the cargo was stranded along 300 km of shoreline. Salvage operations enabled one-fifth of the pollution to be removed from the coast and slicks were further dispersed using 2000 tons of chemicals. Nevertheless the cost of the salvage operations far exceeded the value of any reclaimed oil and the pollution took its toll in animals, shellfish and algae. The cost of this incident in terms of clean-up and compensation exceeded £20 million.[31,33]

Just after midnight on Good Friday, 1989 the *Exxon Valdez* strayed from its 10-mile-wide channel and hit a submerged, but well-marked reef, 22 miles south of the oil terminal at Valdez, Alaska. The vessel released one-fifth of its 53 million gallons cargo of crude oil into the environmentally sensitive waters of Prince William Sound. Despite attempts to clear the spillage using an armada of oil-recovery craft and battalions of anti-pollution workers, only 5% recovery was achieved. A huge slick covered 3000 square miles of unspoilt coastline, coating 300 miles of shore. In addition to the aesthetic effects, casualties included a significant proportion of the local population of 10 000 sea otters (which were particularly vulnerable), plus 200 bird species, fish, whales, seals and porpoises along with land-based mammals such as bears and deer which rely to some extent on the aquatic food chain. By May 20 000 carcasses had been collected.[32, 34-38]

It was estimated that cleaning up would take 3 years at a cost exceeding $200 million. Other losses include a more than $250 million claim by the fishing industry, $6 million to repair the tanker, and $30 million to cover the loss of cargo. The captain stood to face a hefty fine and prison sentence, and the company's prestigious reputation was destroyed. Some 5 years later the court proceedings continue. The first phase exposes the company to the possibility of punitive damages as high as $15 billion and additional compensation claims up to $1.5 billion. The captain was found to have been negligent and reckless. The case will be followed up with three additional phases. The tanker company voluntarily spent $2.2 billion on cleanup with another $1 billion in settlements.

The incident resulted from a catalogue of errors compounded by a slow response once the oil leaked from the ship. Criticisms included:
- At the time of the incident the vessel was captained by a man with a history of drunken driving convictions (he had failed to report two convictions for drunken driving when he renewed his master mariner's certificate) and with only an unlicensed third mate on the bridge.
- The failure to learn from previous similar incidents.
- Despite promises in the 1970s by the company to build tankers double-hulled for extra collision protection, the *Exxon Valdez* was built in 1986 with only a cost-saving single hull.
- The 24-hour twenty-man emergency response team established when the Valdez terminal opened in 1977 was disbanded in 1981 as an economy measure.
- The crew of the single emergency response barge at Valdez took hours to assemble, mainly because the barge had been taken out of service for repairs and the authorities had not been notified.
- It was 36 hours before the stricken vessel was surrounded by booms whereas contingency plans allowed only five hours.
- 48 hours elapsed before permission was granted to spray with dispersant, mainly because the company had not previously established the effectiveness of the dispersant under prevailing sea conditions and had to carry out tests. The application of dispersant was further hampered by rough seas and inadequate stocks of chemical at hand.

Similarly oil spills from sea rigs can be major hazards in safety and environmental terms.

The largest documented oil spill occurred on 3 June 1979 in Mexico Bay at the exploratory well known as Ixtoc-1.[39-41] During deep drilling beneath the sea bottom, mud, used to lubricate the drilling bit, stopped circulating. Against the advice of specialists the drill pipe was cautiously withdrawn and a section was partially unscrewed. Oil rushed out under a pressure of 6000 psi. The pipe threads jammed. A safety valve could not be screwed into place and blew out. Shears, designed to cut through the drill pipe, could not sever the thick drill collars which were all they could reach. Mud, oil and gas spewed uncontrolled from the well. The natural gas was ignited by pump motors resulting in several explosions. Human error and ineffective safety equipment caused the blow-out and none of the 'advanced' techniques for plugging the wells or recapping the oil worked satisfactorily. The gusher ran wild for nearly 10 months, fouling the Gulf of Mexico with about 140 million gallons of oil (i.e. more than twice that spilled from the *Amoco Cadiz* – see page 1227).

In total about 4–5% of the oil was recovered by mechanical means, 17% evaporated and 50% burned. However, 28% formed a slick. In nearly 12 days, 15 million gallons of oil spilled into the gulf and the slick covered an area 12 km wide by 128 km long, and by mid-July oil was impacting the Mexican shores. A month later it reached the US coastline some 600 miles from the source of the spill. The blow-out acutely affected the ecosystem in the Campeche Bay area: shrimp life was particularly threatened and fish and octopus catches in Texas were reportedly down by 50–70%.

Obstacles encountered during the clean-up included:
- Precise location of Ixtoc-1 was not known.
- Lack of lighting equipment prevented 24-hour operation.
- Oil/water recovery ratios were so low that the company pumped back into the sea.
- Depth of spilled oil was such that boats sucked oil into their water intakes.
- Changes in current direction shifted the plume emerging from the well, allowing oil to escape.
- Inadequate stockpiles of spill response equipment at hand (some 900 tons of dispersant were used).

The technical difficulties listed in Table 21.7 were also encountered. The

Table 21.7 Some technical problems encountered with the clean-up operation after the *Ixtoc-1* blow-out

Equipment	Problem	Cause
Skimming barrier	Flotation bags failed to inflate	Trip lines not connected
	Twist in barrier frame	Excessive boom sallying on deployment
	Barrier curtain parted	Vessel damage
	Barrier lines and rods broken	Excessive tow speed
	Main tension line parted	Vessel damage
	Skimming weirs plugged	Debris
	Foam pads carried away	High seas
	Unable to visually locate moor points	Oil coated identification pads
	Curtain fatigue	Weight of field-modified pump float
	Unable to locate frame and pump float moorings	Location and design
Pump float	Components sheared off	Under-design
	Excessive pressure on hydraulic supply	Fluid expansion in heat of sun
	Broken support bracket for hydraulic connection	Metal fatigue
	Suction hoses pulled loose	High seas
	Hydraulic pilot control valve failed	Unknown
	Extensive fibreglass damage	Personnel error

cost of the spillage in terms of clean-up, loss of oil rig, compensation, etc. was in excess of $600 million.

However, the environmental impact from spills can sometimes be overestimated:

> At 11.00 a.m. on 5 January 1993 the *MV Braer*, a 45 000 tonne bulk carrier, ran aground off rocks on Shetland's southern peninsula. At that stage 500 tonnes of bunker oil and half the cargo of 85 000 tonnes of light crude was believed to have leaked into the sea.[42a] Within hours several inland waterways had been sandbagged to exclude the oil and major roads were sanded to avoid airborne contamination making them slippery. Over the next few days the islands were hit by blizzards and gale-force winds and all recovery operations had to be suspended. However, this promoted natural dispersion and evaporation. Attacking the slick using aerial dispersion, however, proved controversial.
>
> Fortunately the slick contaminated only 30 miles of coastline and although oil concentrations reached 1000 times the background level of the west of Shetland they returned to normal as early as mid-February. Because of the speed of natural recovery processes the coastline appears to have made a surprising recovery.[42b]

'Physical' pollutants

Temperature

The temperature of any effluent, even cooling water, discharged into natural waters may cause a rise in temperature. This may have deleterious effects. The reduced solubility of oxygen with water temperature (page 81) is accompanied by an increased rate of oxygen consumption, because biochemical reactions proceed faster at higher temperatures (page 93).

> The 5-day BOD (biochemical oxygen demand) of sewage is approximately twice as great at 25 °C as at 5 °C.

Increased water temperature may also be deleterious to fish life with possible mortalities amongst sensitive species.

Suspended matter

Effluents commonly contain some suspended, insoluble matter. This may be organic, inorganic or a mixture of both depending upon the source. Their discharge into natural waters is disadvantageous because:
- penetration of light through the water is reduced so that photosynthesis, and hence self-purification is partially inhibited
- deposition on, and blanketing of, the bed may lead to destruction of plant and animal life
- they are an aesthetic nuisance.

In specific cases acid conditions may liberate toxic metals in a soluble state from suspended solids contaminated by metallic compounds; this could result in fish mortality.

Foam

The presence of foam on a water surface inhibits oxygen transfer. Moreover earlier detergent formulations, based on branched chain alkylbenzene sulphonates, resisted oxidation and therefore passed unchanged through sewage treatment works. Degradable formulations based on alkylbenzene sulphonates were consequently introduced in the 1960s. Foam in water may, however, also be caused by soaps, fish glue and saponins.

Turbidity

Turbidity of waste waters derives from the presence of colloidal matter or very finely-divided particulates. Such fine particles have negligible settling velocities (page 118) in water.

Whilst the greater the turbidity the greater the tendency is for pollution by a wastewater, although this may in fact be due to some 'inert' solid, e.g. clay, the absence of turbidity is not, of course, any indication of the absence of dissolved pollutants.

Tastes and odours in water supplies

Many 'natural' phenomena may result in a source of taste or odour in water. For example, decaying vegetation may produce a grassy, fishy or musty odour. Algae can cause offensive odours as they die off and some living algae cause taste and odour problems.[43] Moulds and actinomycetes may generate earthy, musty or mouldy tastes in waters, particularly in favourable conditions for growth, e.g. in water left standing in long pipe runs in warm buildings. Iron and sulphur bacteria produce deposits which release an offensive odour, in the latter case of hydrogen sulphide, on decomposition. However, many tastes and odours are also attributable to trace chemical contaminants.

Excessive sodium chloride imparts a brackish taste to water. Chlorides are present in nearly all natural waters; most combinations are with sodium, and to a lesser extent, with magnesium and calcium. An EC Directive and WHO 1983 Guidelines recommend a guide level of 25 mg/l Cl in drinking water, whereas a sensitive palate can detect 150 mg/l and concentrations above 250 mg/l impart a distinct salty taste.[43] Most rivers and lakes contain <50 mg/l; any marked increase above this is possibly indicative of pollution. An excessive amount of iron will impart a bitter taste to water.

Taste and odour problems can arise from industrial waste sources. Contamination by phenols is commonly experienced; in the presence of free chlorine these result in chlorophenols which have a pronounced taste. For example, an objectionable taste may arise from reaction of only 0.001 mg/l phenol with chlorine which is used for water treatment.

Air pollution

Air pollution can render an atmosphere unpleasant (e.g. see Tables 21.8[44] and 21.9[45]) and can also result in the erosion of structures. Building ma-

Table 21.8 Recognition odour thresholds in air (US Manufacturing Chemists Association, Inc.)

Compound	ppm (volume)	Compound	ppm (volume)
Acetaldehyde	0.21	Ethanol (synthetic)	10.0
Acetic acid	1.0	Ethyl acrylate	0.00047
Acetone	100.0	Ethyl mercaptan	0.001
Acrolein	0.21	Formaldehyde	1.0
Acrylonitrile	21.4	Hydrochloric acid gas	10.0
Allyl chloride	0.47	Methanol	100.0
Amine, dimethyl	0.047	Methyl chloride	> 10
Amine, monomethyl	0.021	Methylene chloride	214.0
Amine, trimethyl	0.00021	Methyl ethyl ketone	10.0
Ammonia	46.8	Methyl isobutyl ketone	0.47
Aniline	1.0	Methyl mercaptan	0.0021
Benzene	4.68	Methyl methacrylate	0.21
Benzyl chloride	0.047	Monochlorobenzene	0.21
Benzyl sulphide	0.0021	Nitrobenzene	0.0047
Bromine	0.047	Paracresol	0.001
Butyric acid	0.001	Paraxylene	0.47
Carbon disulphide	0.21	Perchloroethylene	4.68
Carbon tetrachloride (chlorination of CS_2)	21.4	Phenol	0.047
		Phosgene	1.0
Carbon tetrachloride (chlorination of CH_4)	100.0	Phosphine	0.021
		Pyridine	0.021
Chloral	0.047	Styrene (inhibited)	0.1
Chlorine	0.314	Styrene (uninhibited)	0.047
Dimethylacetamide	46.8	Sulphur dichloride	0.001
Dimethylformamide	100.0	Sulphur dioxide	0.47
Dimethyl sulphide	0.001	Toluene (from coke)	4.68
Diphenyl ether (perfume grade)	0.1	Toluene (from petroleum)	2.14
		Tolylene diisocyanate	2.14
Diphenyl sulphide	0.0047	Trichloroethylene	21.4

Note: Recognition levels in practice may be much lower than the above values, determined in low-odour air. Olfactory fatigue, synergism and the level/type of background odour are important factors.

terials such as steel, paint, plastics, limestone, etc., are at risk and the frequency at which structures need their protective coating replaced is estimated to cost billions of dollars annually. Air pollution can also alter microclimates and the chemistry of lakes, streams, rivers and soil with

Table 21.9 Air dilution odour threshold data on 214 industrial chemicals

Substance	Air odour threshold (ppm; v/v)	Substance	Air odour threshold (ppm; v/v)
Acetaldehyde	2000	Cyclohexene	0.18
Acetic acid	21	Cyclohexylamine	2.6
Acetic anhydride	39	Cyclopentadiene	1.9
Acetone	57	Decaborane	0.060
Acetonitrile	0.23	Diacetone alcohol	0.28
Acetylene	230	Diborane	2.5
Acrolein	0.61	o-Dichlorobenzene	0.30
Acrylic acid	110	p-Dichlorobenzene	0.18
Acrylonitrile	0.12	trans-1,2-Dichloroethylene	17
Allyl alcohol	1.8	β,β-Dichloroethyl ether	0.049
Allyl chloride	0.84	Dicyclopentadiene	0.0057
Ammonia	4.8	Diethanolamine	0.27
n-Amyl acetate	1800	Diethylamine	0.13
sec-Amyl acetate	61 000	Diethylaminoethanol	0.011
Aniline	1.9	Diethyl ketone	2.0
Arsine	0.10	Diisobutyl ketone	0.11
Benzene	0.85	Diisopropylamine	1.8
Benzyl chloride	23	N-Dimethylacetamide	47
Biphenyl	240	Dimethylamine	0.34
Bromine	2.0	N-Dimethylaniline	0.013
Bromoform	0.39	N-Dimethylformamide	2.2
1,3-Butadiene	640	1,1-Dimethylhydrazine	1.7
Butane	0.29	1,4-Dioxane	24
2-Butoxyethanol	250	Epichlorhydrin	0.93
n-Butyl acetate	390	Ethane	120 000
n-Butyl acrylate	290	Ethanolamine	2.6
n-Butyl alcohol	60	2-Ethoxyethanol	2.7
sec-Butyl alcohol	38	2-Ethoxyethyl acetate	0.056
tert-Butyl alcohol	2.1	Ethyl acetate	3.9
n-Butylamine	2.7	Ethyl acrylate	0.0012
n-Butyl lactate	0.71	Ethyl alcohol	84
n-Butyl mercaptan	510	Ethylamine	0.95
p-tert-Butyltoluene	2.0	Ethyl n-amyl ketone	6.0
Camphor	7.3	Ethyl benzene	2.3
Carbon dioxide	0.067	Ethyl bromide	3.1
Carbon disulphide	92	Ethyl chloride	4.2
Carbon monoxide	0.00050	Ethylene	290
Carbon tetrachloride	0.052	Ethylenediamine	1.0
Chlorine	3.2	Ethylene dichloride	88
Chlorine dioxide	0.011	Ethylene oxide	430
α-Chloroacetophenone	1.4	Ethylenimine	1.5
Chlorobenzene	110	Ethyl ether	8.9
Chlorobromomethane	0.50	Ethyl formate	31
Chloroform	0.12	Ethylidene norbornene	0.014
Chloropicrin	0.13	Ethyl mercaptan	0.00076
β-Chloroprene	0.68	N-Ethylmorpholine	1.4
o-Chlorotoluene	150	Ethyl silicate	17
m-Cresol	17 000	Fluorine	0.14
trans-Crotonaldehyde	17	Formaldehyde	0.83
Cumene	570	Formic acid	49
Cyclohexane	25	Furfural	0.078
Cyclohexanol	0.15	Furfuryl alcohol	8.0
Cyclohexanone	0.88	Halothane	33

Table 21.9 (continued)

Substance	Air odour threshold (ppm; v/v)	Substance	Air odour threshold (ppm; v/v)
Heptane	150	Methyl methacrylate	0.083
Hexachlorocyclopentadiene	0.030	Methyl n-propyl ketone	11
Hexachloroethane	0.15	α-Methyl styrene	0.29
Hexane	130	Morpholine	0.01
Hexylene glycol	50	Naphthalene	0.084
Hydrazine	3.7	Nickel carbonyl	0.30
Hydrogen bromide	2.0	Nitrobenzene	0.018
Hydrogen chloride	0.77	Nitroethane	2.1
Hydrogen cyanide	0.58	Nitrogen dioxide	0.39
Hydrogen fluoride	0.042	Nitromethane	3.5
Hydrogen selenide	0.30	1-Nitropropane	11
Hydrogen sulphide	0.0081	2-Nitropropane	70
Indene	0.015	m-Nitrotoluene	0.045
Iodoform	0.0050	Nonane	47
Isoamyl acetate	0.025	Octane	48
Isoamyl alcohol	0.042	Osmium tetroxide	0.0019
Isobutyl acetate	0.64	Oxygen difluoride	0.10
Isobutyl alcohol	1.6	Ozone	0.045
Isophorone	0.20	Pentaborane	0.96
Isopropyl acetate	2.7	Pentane	400
Isopropyl alcohol	22	Perchloroethylene	27
Isopropylamine	1.2	Phenol	0.040
Isopropyl ether	0.017	Phenyl ether	0.0012
Maleic anhydride	0.32	Phenyl mercaptan	0.00094
Mesityl oxide	0.45	Phosgene	0.90
2-Methoxyethanol	2.3	Phosphine	0.51
Methyl acetate	4.6	Phthalic anhydride	0.053
Methyl acrylate	0.0048	Propane	16 000
Methyl acrylonitrile	7.0	Propionic acid	0.16
Methyl alcohol	100	n-Propyl acetate	0.67
Methylamine	3.2	n-Propyl alcohol	2.6
Methyl n-amyl ketone	0.35	Propylene	76
N-Methylaniline	1.7	Propylene dichloride	0.25
Methyl n-butyl ketone	0.076	Propylene glycol 1-methyl ether	10
Methyl chloroform	120		
Methyl 2-cyanoacrylate	2.2	Propylene oxide	44
Methylcyclohexane	630	n-Propyl nitrate	50
cis-3-Methylcyclohexanol	500	Pyridine	0.17
Methylene chloride	250	Quinone	0.084
Methyl ethyl ketone	5.4	Styrene	0.32
Methyl formate	600	Sulphur dioxide	1.1
Methyl hydrazine	1.7	1,1,2,2-Tetrachloroethane	1.5
Methyl isoamyl ketone	0.012	Tetrahydrofuran	2.0
Methyl isobutyl carbinol	0.070	Toluene	2.9
Methyl isobutyl ketone	0.68	Toluene-2,4-diisocyanate	0.17
Methyl isocyanate	2.1	o-Toluidine	0.25
Methyl isopropyl ketone	1.9	1,2,4-Trichlorobenzene	1.4
Methyl mercaptan	0.0016	Trichloroethylene	28

Table 21.9 (continued)

Substance	Air odour threshold (ppm; v/v)	Substance	Air odour threshold (ppm; v/v)
Trichlorofluoromethane	5.0	n-Valeraldehyde	0.028
1,1,2-Trichloro-1,2,2-trifluoroethane	45	Vinyl acetate	0.50
		Vinyl chloride	3000
Triethylamine	0.48	Vinylidene chloride	190
Trimethylamine	0.00044	Vinyl toluene	10
1,3,5-Trimethylbenzene	0.55	m-Xylene	1.1
Trimethyl phosphite	0.00010	2,4-Xylidine	0.056

consequential adverse effects on the ecosystem. Air pollutants can also impair human health if excessive concentrations are present.

The primary air pollutants are carbon monoxide, particulates, hydrocarbons, oxides of nitrogen, and sulphur oxides. (Until relatively recently carbon dioxide, a major constituent of emissions from all fossil-fuelled combustion processes, was not considered as a pollutant; its role in the 'greenhouse effect' is now well-established.)

Carbon monoxide is, as discussed on page 825, an inevitable product of combustion of fossil fuels with restricted air supplies. Sulphurous compounds arise whenever sulphur is present.

In many situations a combination of pollutants are present so that as discussed on page 404, their additive, or other mixed, effects come into play.

In early December 1930 stagnant weather conditions and a cold mist developed along the Meuse River Valley, Belgium. Because of the lack of air movement, combustion products and other industrial emissions from factories and mills concentrated in the valley. Several thousand people became ill simultaneously from one end of the valley to the other and there were 60 additional deaths over and above the number in a normal three-day period for that area. Practically all deaths resulted from acute heart failure; coughing, shortness of breath, and vomiting were among the reported symptoms. Those at risk were the elderly, and people with pre-existing heart conditions and respiratory distress, although many young, healthy people also became ill. Wild and domestic animals were also noticeably affected. The major causes were suspected as sulphur dioxide and gaseous fluoride, in factory smoke.[46]

In October 1948 the winds died down in Donora, a small industrial town in a Pennsylvanian valley. For several days factory emissions accumulated in the still air. Almost half of the towns' inhabitants were stricken with severe eye, nose and throat irritation, chest pains and breathing difficulties. Twenty people died as a result of the incident. Those at greatest risk were again chronic bronchitics, asthmatics, and people with pre-existing heart disease. Animals also suffered. Whilst sulphur dioxide and zinc ammonium sulphate have been implicated as the cause, the main conclusion was that a mixture of toxic agents were responsible, none of which would have caused the problem alone.[46]

The effects of key pollutants on mammals and plants are summarised in Table 21.10. In addition to any toxic effects that can result from exposures

Table 21.10 Industrial sources of major atmospheric pollutants

Contaminant	Man-made source	Effects
Sulphur dioxide	Combustion of coal, oil and other sulphur-containing fuels; petroleum refining, metal smelting, paper-making	Vegetation damage, sensory and respiratory irritation, corrosion, discoloration of buildings
Hydrogen sulphide	Various chemical processes; oil wells, refineries; sewage treatment	Odours, toxic; crop damage and reduced yields
Carbon monoxide	Vehicle exhausts and other combustion processes	Adverse health effects
Carbon dioxide	Combustion processes	
Nitrogen oxides	High-temperature reaction between atmospheric nitrogen and oxygen, e.g. during combustion; by-product from manufacture of fertiliser	Adverse health effects, sensory irritants, reduced visibility, crop damage
Ammonia	Waste treatment	Odour, irritant
Fluorides	Aluminium smelting; manufacture of ceramics, fertiliser	Crop damage; HF has adverse health effects on cattle fed on contaminated food
Lead	Combustion of leaded petrol; solder, lead-containing paint; lead smelting	Adverse health effects
Mercury	Manufacture of certain chemicals, paper, paint; pesticides; fungicides	Adverse health effects
Volatile hydrocarbons	Motor vehicles, solvent processes, chemical industry	Vegetation damage (especially unsaturated hydrocarbons); some are irritants; adverse health effects
Particulates	Chemical processes; fuel combustion; construction; incineration processes; motor vehicles	Nuisance, adverse health effects

to unacceptable concentrations of these substances, emissions of many of the chemicals, in particular carbon dioxide and methane, together with fluorocarbons and oxides of nitrogen, lead to global warming, via the greenhouse effect.

Because there are established ways of preventing many pollutants from entering the atmosphere the socio-political pressures to reduce acute and chronic industrial air pollution episodes are significant.

The main sources of man-made atmospheric pollution given in Table 21.10 stem from combustion of coal, wood, oil and gas, and waste from chemical and process industries. Some materials which enter the atmosphere in smaller quantities may pose toxic or nuisance problems, e.g. mercaptans. Some environmental hazards arise from consumer goods as exemplified by dumping of domestic refrigerators containing chlorofluorocarbons, and the use of aerosol propellants.

Volatile organic chemicals (VOCs)

Volatile organic chemicals are emitted principally from industrial processes and from road transport. Some 2.7 million tonnes/annum were estimated to have been emitted into the atmosphere from the UK in 1988. The main sources, emission values in 1988 and expected emission values in 1999 (following adoption of the United National Economic Commission for Europe protocol on control of VOC emissions) are summarised in Table 21.11.[45a]

Odours

Odour problems may be caused by very low levels of some atmospheric pollutants. In the majority of cases the contaminant responsible is organic, often with small proportions of sulphur or nitrogen. The major causes of odour from sewage treatment and other common sources are hydrogen sulphide, skatole (C_9H_9N), methyl mercaptan, ethyl mercaptan, cadaverine and various amines. There are, however, a number of inorganic malodorous compounds.

Each odorant has a concentration in air, the olfactory or odour threshold, which is detectable. However, this varies from one person to another; moreover, continuous exposure may result in olfactory fatigue, when odour from a particular substance ceases to be noticed. Commonly occurring compounds known to cause odour in industrial processes at low concentrations include sulphides and mercaptans (thiols).

Recognition odour thresholds of a number of substances have been published in the USA by the Manufacturing Chemists Association, Inc. (see Table 21.8). (Measurement relies upon the presentation of static air samples containing the odorants to a panel of assessors trained in sensory evaluation. In obtaining the data in Table 21.8 the dilution medium was a low-odour background air and the odour threshold was recorded as the lowest concentration at which all members of the panel recognised the odour.)

More recently it has been established that the odour threshold determination by a worker under laboratory conditions can result in much lower

Table 21.11 Volatile organic compound emissions by source in the UK[45a]

Source	Emissions/k tonnes 1988	1999
Painting	278	186
Printing	42	28
Surface cleaning	46	19
Dry cleaning	11	7
Adhesives	58	30
Pharmaceuticals	40	28
Aerosols	86	96
Non-aerosol products	69	77
Agrochemicals	39	40
Seed oil extraction	10	3
Other solvent use	73	29
Oil production	191	209
Oil distribution	100	55
Refineries	69	45
Product distribution	128	58
Other oil industry	31	28
Chemical industry	200	76
Domestic combustion – coal	51	22
Domestic combustion – other	8	9
Industrial combustion	17	19
Other combustion	3	3
Baking	21	21
Alcoholic beverages	44	46
Animal by-products	1	0
Other food	20	12
Iron and steel	12	7
Waste disposal	70	37
Animal husbandry*	—	—
Other agriculture*	—	—
Gas leakage	34	45
Other miscellaneous	5	6
Petrol exhaust	644	410
Petrol evaporation	137	61
Diesel exhaust	167	117
Other transport	24	27
TOTAL	2734	1857

* No data available, source under review.
Note: Both high and low trends have been forecast for petrol exhaust, petrol evaporation and diesel exhaust. The high trend is used above to calculate the overall emission reduction.

levels than those at which odours are perceptible to a person concentrating on other tasks. Hence the threshold is misleading as an indication of the level at which the average process worker, concentrating on his normal duties, will become aware of an odour. Furthermore, typically 68 per cent of the population will vary in their detection response from 0.25 to 4 times the mean detection level.[44] A factor difference of 26 has been reported between laboratory conditions and test subjects whose attention was otherwise directed.[45] Collected odour threshold data are summarised in Table 21.9.[45] The differences between values in Tables 21.8 and 21.9 are striking.

Combustion products

Combustion processes are mainly associated with energy production such as electricity generation from oil or coal, heat from burning oil, gas, coal or wood, etc., or mechanical energy as in the case of the internal combustion engine. The effluent is usually exhausted to atmosphere as a gas or particulate aerosol (smoke) via a chimney or vent. Whilst carbon dioxide and water are the products of complete combustion of fossil fuels, oxides of sulphur and nitrogen, plus carbon monoxide and a wide range of hydrocarbons, are present in emissions as a result of impurities such as sulphur (e.g. as pyrites, sulphates, mercaptans, polysulphides, thiophenes) in the fuel, or because of incomplete combustion.

> As a result of high pressure over London in 1952, smoke from domestic coal fires mixed with the London fog to form a potent acid mist containing sulphuric acid, carbon monoxide and tar. There were 3000 deaths within a week of exposure to the pollutants and a further 1500 fatalities in the following weeks. Hospital admissions were up 40%. In a repeat incident in 1956, 1000 people died even though the problem lasted only 18 hours.[46] In neither incident are the acute effects of the exposures known.
>
> (In fact the report of the inquiry into the consequences of the November 1952 episode, when during 11 consecutive days of atmospheric inversion London was covered with dense fog and the concentrations of SO_2 and smoke rose to over 2000 and 3000 $\mu g/m^3$ respectively, led to the Clean Air Act 1956. A significant public health advantage was gained by requiring compliance with a memorandum on chimney heights albeit, as was subsequently recognised, by the transport of aerosols containing SO_2 to other countries, leading to complaints of acid rain.[47]

Carbon monoxide is the single most abundant toxic man-made air-pollutant with some 70 million tons vented into the atmosphere each year in the USA alone. Some 80 per cent of that present in urban atmospheres is believed to originate from motor vehicle exhausts.[48] Its toxicity is discussed in more detail in Chapter 30 whilst effects at varying concentrations are included on page 243. Carbon dioxide is not particularly toxic.

Sulphur oxides arise mainly from the combustion of inorganic sulphides and sulphur-bearing organic compounds present in oils and coals. In combustion gases the ratio of SO_2 to SO_3 is approximately 25:1 to 30:1,[49] but sulphur dioxide is slowly oxidised in the presence of moisture and sunlight in the atmosphere to the trioxide. With moisture this immediately produces sulphuric acid which contributes to the haze. Absorption onto particulates and neutralisation also occur, e.g. a complex of sulphuric oxides including metallic and ammonium sulphates may result to form an aerosol and solid agglomerates.

Sulphur oxides irritate the respiratory tract and may contribute to the onset of, or aggravate, asthma, emphysema and bronchitis. A summary of some observed health effects at differing levels of exposure is given in Table 21.12.

Table 21.12 Observed health effects of sulphur dioxide and particulates

Concentrations of sulphur dioxide			Concentrations of particulates		
$\mu g/m^3$	ppm	Measured as	$\mu g/m^3$	Measured as	Effect that may occur
1500	0.52	24-h average	—	—	Increased mortality
715	0.25	24-h mean	750	24-h mean	Increased daily death rate; a sharp rise in illness rates among bronchitics over age 54
300–500	0.11–0.19	24-h mean	Low	24-h mean	Increased hospital admissions of elderly respiratory disease cases; increased absenteeism among older workers
600	0.21	24-h mean	300	24-h mean	Symptoms of chronic lung disease cases accentuated
105–265	0.04–0.09	Annual mean	185	Annual mean	Increased frequency of respiratory symptoms and lung disease
120	0.05	Annual mean	100	Annual mean	Increased frequency and severity of respiratory diseases among school children
115	0.04	Annual mean	160	Annual mean	Increased mortality from bronchitis and lung cancer

The gases or their acid rain also affect plant and animal health. (Other industrial sources of atmospheric oxides of sulphur are sulphuric acid manufacture, e.g. from sulphur dioxide passing through the absorber tower, unabsorbed sulphur trioxide, particulate acid leaving the absorber and leaks in general; acid plant effluents or metallurgical ore sintering operations.) Oxides of nitrogen are also damaging to health and can also arise as a pollutant from nitric acid manufacture and in the removal of ammonia from coke oven gas.

In addition to the vast quantities of volatile hydrocarbons dispersed into the atmosphere annually from a variety of industrial sources, carcinogenic polycyclic hydrocarbons such as benzopyrene (see Chapter 6) are generated during the burning of coal and the production of coke. Hazardous emissions can also arise from inefficient combustion of toxic waste during incineration. Thus PCBs are hazardous because they are both notoriously difficult to dispose of and their incomplete combustion can lead to formation of dioxin. For this reason Her Majesty's Inspectorate of Pollution's interpretation of Best Practicable Means (see page 1259) for PCBs is a destruction and removal efficiency of 99.999 per cent. Similar standards are adopted by the US Environmental Protection Agency and by the German authorities,[50] and contravention can be costly:

> In 1990 a chemical company failed to comply with the federal regulations on disposing of PCBs at a plant in Chicago. They had failed to monitor stack emissions, had not operated air pollution control equipment in accordance with the company's Toxic Substances Control Act permit, and had not recorded the rate and quantity of solid PCB fed into the incinerator. The company was fined $3.75 million.[51]

(The hazard of aquatic pollution by PCBs was mentioned on page 1224.)

Particulates

Particulates constitute about 5 per cent of the weight of air pollutants, varying in size from 0.005 μm to about 100 μm, as illustrated in Fig. 19.2. Airborne solid matter larger than 76 μm is referred to as grit, whilst solid particles smaller than 76 μm are termed dust. Particles larger than 10 μm tend to be produced by mechanical processes, grinding, spraying, abrasion, the movement of vehicles and people, and from wind erosion and dispersion. Fumes are small airborne particles (generally less than 10 μm) arising from chemical reaction or condensation of vapour. Particles between 0.1 and 1 μm largely comprise combustion products and the products of photochemical reactions. Any below 0.1 μm are minute combustion products; they tend to form larger particles rapidly by sorption and nucleation of gas molecules and adhesion with other particles. Particulate pollutants include metal salts and oxides, sulphuric acid droplets, finely divided particles of silica and carbon, liquid sprays and mists, etc. The larger, heavier particles settle quickest and those around 2–5 μm represent the greatest hazard to human health (see page 193) and are the least effectively removed from the atmosphere by rain.

Pollution by asbestos is subject to strict control in most developed countries. Such pollution may cause mesothelioma which has in a few cases been found to arise many years, e.g. 20 to 30 years, after relatively light exposure to asbestos, particularly crocidolite (see page 221).

> Cases of mesothelioma have been reported amongst patients with no occupational exposure to asbestos but who lived in mining areas for one to two years as children[52] or who lived near to asbestos works.[53] However, no excess incidence occurred near the asbestos mines in Quebec which produced chrysotile asbestos.

Some particles are readily soluble in water, others, such as insoluble carbon, can complicate health effects by being contaminated with toxic substances, e.g. benzopyrene has been associated with soot particles.

Serious air pollution episodes associated with an emergency can, of course, result upon settlement of particulate emissions in land pollution as exemplified on page 1249. Fires involving buildings of asbestos-cement construction, or with asbestos-based insulation panels, may also cause problems.

> An asbestos pollution alert followed a fire at an Ordinance Depot in April 1988. A plume of smoke and dust rose high into the air and chrysotile asbestos particles from the burning roof material were carried westwards over a five to six mile radius. By 10 a.m. next day a 100-strong team of council workers wearing protective clothing was out collecting debris and washing down roads. (Of relevance to the discussion of emergency planning in Volume 4, the rapid emergency response was due, in part, to experience with a very similar fire five years earlier.[54])

Metals

Lead is a potential environmental problem as a result of such processes as lead smelting and use of lead-based solder or paint. However, the most important source of atmospheric contamination is from the exhausts of cars run on leaded petrol (see, e.g. page 249). The majority of the lead emitted is <1 μm in particle size and can therefore readily penetrate the respiratory system as far as the alveoli (see page 193), from which transfer takes place to the bloodstream.[48] However, one estimate was that only some 28 per cent of the total lead assimilated by the body can be derived from the atmosphere.

Mercury from industrial (e.g. paper, chemical and paint manufacture) and agricultural sources (pesticides and fungicides) escapes into the atmosphere as vapour or particulates. The nature of the compound and its form affects its toxicity. The hazards of metal fumes from, e.g. welding were discussed in Chapter 5. Fluoride compounds, e.g. from smelting iron and aluminium, can have disastrous effects on plants, animal health and humans.

> The first reported cases of neighbourhood fluorosis – with symptoms of gastric upsets, stiffness, and pains in the legs and joints – occurred in nine members of a farming family living near a fluoride-emitting ironstone works in South Lincolnshire. The duration of exposure was, with one exception, between 3 and 14 years.[55]

> Cadmium poisioning, due to emission of cadmium fumes from a lead–zinc mining operation, the fallout from which contaminated rice-paddy fields, was responsible for an outbreak of *itai itai* disease in Japan (see page 199).
> Research in the USA also demonstrated a relationship between cadmium in air and cardiovascular disease.[48]

Other metals have also been responsible for pollution episodes. Thus manganese has found use in the manufacture of alloys, dyestuffs, wood preservatives, textile bleaches, petroleum additives, etc. Cases of occupational poisoning have been observed in many industries following excessive exposure generally as a result of inhalation of dust. Target organs are the brain or lungs. A high incidence of lung problems has been reported in neighbouring populations around manganese works.

> The erection of an electrical plant for manganese smelting at Sauda in Norway was followed by a tenfold increase in the mortality rate for pneumonia in that area. A pall of smoke which overhung the town was found to contain particles less than 5 μm in size, of oxides of manganese.[56]

Chlorofluorocarbons

Ozone poses a health problem at the surface of the planet but is essential in the stratosphere to filter out mutagenic and carcinogenic UV radiation and thereby prevent it reaching the earth's surface. It is both produced and destroyed in the upper atmosphere by a series of free-radical reactions initiated by the absorption of solar UV radiation, thus:

$$O_2 \xrightarrow{h\nu} 2O\cdot$$

$$O\cdot + O_2 \xrightarrow{M} O_3 + M\cdot$$

$$O_3 \xrightarrow{h\nu} O\cdot + O_2$$

$$O\cdot + O_3 \longrightarrow 2O_2$$

where M = H, OH, NO, Cl or Br. The depletion of this UV shield represents a serious risk and of the many species capable of destroying the ozone layer, chlorofluorocarbons such as CFC-11 (trichlorofluoromethane) and CFC-12 (dichlorodifluoromethane) with half-lives of 77 and 139 years, respectively, represent the greatest danger:

$$CF_2Cl_2 \xrightarrow{h\nu} \cdot CF_2Cl + Cl\cdot$$

$$CFCl_3 \xrightarrow{h\nu} \cdot CFCl_2 + Cl\cdot$$

$$Cl\cdot + O_3 \longrightarrow ClO + O_2$$

$$ClO + O. \longrightarrow Cl. + O_2$$

$$\text{Net:} \quad O. + O_3 \longrightarrow 2O_2$$

Chlorofluorocarbons are low boiling, non-toxic, non-corrosive and non-flammable substances. Their main use is as refrigerants, aerosol propellants, solvents, and plastic foam blowing agents. By 1978 the above two CFCs were banned as aerosol propellants in some countries and in 1987 leading industrial nations signed the Montreal Protocol to control the use of a range of fluorocarbons listed in Table 21.13.

Table 21.13 Substances controlled by the Montreal Protocol

Compound	Number	Ozone depletion potential	Half-life (years)
$CFCl_3$	CFC11	1	77
CF_2Cl_2	CFC12	1	139
C_2F_3Cl	CFC113	0.8	92
$C_2F_4Cl_2$	CFC114	1	180
C_2F_5Cl	CFC115	0.6	380
CF_2BrCl	Halon 1211	2.7	12.5
CF_3Br	Halon 1301	11.4	101
$C_2F_4Br_2$	Halon 2402	5.6	unknown

Secondary air pollutants

Hydrocarbons, oxides of nitrogen and ozone react by photochemical processes to result in 'smog', as illustrated in Fig. 21.1. Petrol-engined vehicles are the primary source of hydrocarbons and oxides of nitrogen, but clearly any process involving loss of hydrocarbon volatiles, e.g. by evaporation, adds to their emissions, e.g. paint spraying, dispensing of petrol, industrial use of solvents. Oxides of nitrogen are formed in any high-temperature combustion process; hence there are multiple sources.

Land pollution

In the early 1970s the village of Lekkerkerk, Netherlands was built on reclaimed land but with the surface lower than the river. This produced an upward flow of ground water. In 1978 the first signs of land pollution were detected which included deterioration of plastic water pipes, polluted ground water, stunted plant growth and objectionable odours. Excavations revealed full and empty drums. Contaminated water entered the drinking water supply. It became necessary to evacuate most of the inhabitants and remove some 87 000 m^3 of contaminated soil and 1652 drums from beneath the jacked-up

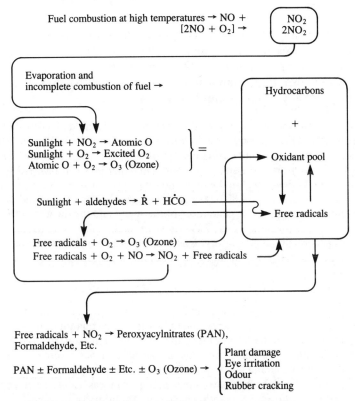

Fig. 21.1 The photochemistry of smog formation and its effects

buildings. These drums contained a wide range of hazardous chemicals including alkyd resins, epoxy resins, and paint solvents. The actual levels of soil contamination were 1000 mg/kg of toluene, 3000 mg/kg of lower boiling point hydrocarbons, 230 mg/kg of antimony, 97 mg/kg of cadmium, 8.2 mg/kg of mercury, 720 mg/kg of lead, and 1670 mg/kg of zinc. The clean-up operation was completed by January 1981 at a cost in excess of £30 million.[58,59]

Thus pollution of land can be harmful:
- to humans by virtue of acute and chronic effects resulting from ingestion of the soil, inhalation of vapours, smoke, or fume emitted from the soil, or dust, e.g. waste asbestos, created by wind or traffic movement. Those at risk include surveyors, workers investigating complaints or engaged in reclamation or using the land or sites nearby, or trespassers.
- to local fauna and flora by direct impairment of animal or plant health or by translocation of toxicant to edible matter. Thus ingestion of contaminated soil may poison some animals whilst in others no harm may be apparent but accumulation in milk fat may lead to elevated intake of the toxicant by those who drink the milk. Similarly levels of contaminant in the soil may reach levels so as to inhibit or prevent plant growth,

whereas other scenarios include concentration in edible fruit without direct phytotoxic effects but proving harmful to those who consume it. Those substances which are highly soluble tend to pose the greatest hazard by entering the plant via the root system.
- to water quality by contaminating ground water that lies beneath or within it and any surface water into which it drains:

> Some 30 million gallons of hazardous waste were dumped at a Californian site over a 15-year period. The rock was thought to be impermeable but proved to be porous and during heavy rain neighbouring land became polluted. As a result the site was closed and a dam was constructed. Within a year cracks developed in the dam and seepage occurred. Five years later the dam overflowed and 800 000 gallons of hazardous waste entered the public water supply. The generators of the waste and the owners of the dump were prosecuted by the EPA and over 5000 claims were submitted from individuals alleging injury. Clean-up costs amounted to $120 million with the possibility of a further $240 million compensation to injured Third Parties.[51]

> For 20 years ponds and pits around the Tucson Airport, Arizona, were used to store spent solvent and heavy-metal waste. Large quantities of waste liquid leaked into the environment and contaminated property and drinking water. Significant levels or carcinogenic trichloroethylene were detected in the drinking water and levels of leukaemia and testicular cancer in the area are abnormally high. A 20-year site clean-up plan was instigated at a cost of $35 million and over 800 plantiffs have sued for injury or damage.[4]

Amongst the many criteria for selecting waste disposal sites, soil condition is of paramount importance. Unless the permeability of the ground is such as to prevent leakage to underground water supplies within a 250-year period, the waste materials listed in Table 21.14 should not be deposited on land.

Table 21.14 Materials difficult to dispose of safely on land

1. Non-biodegradable hazardous materials
    Chlorinated hydrocarbons
    Slowly degradable materials
    Mineral oils, greases
2. Poisonous or persistent materials, including heavy metals
    Phenols
    Cyanides, including metal–cyanide complexes
    Drug residues
    Chromium, cadmium, lead, mercury, nickel, copper, zinc
    Rodenticide and pesticide residues
	3. Strongly acid or alkaline materials in quantities which would disturb the neutrality of the landfill
	    Hydrochloric, sulphuric, nitric, phosphoric, chromic, or hydrofluoric acid
	    Sodium hydroxide
	4. Non-aqueous liquid wastes where these would dissolve material used in sealing a site

- to building construction materials, particularly certain metals, cement and concrete as illustrated by Table 21.15 (after refs 46 and 60).
- from fire hazards, e.g. from evolution of methane by the anaerobic digestion of organic refuse or from contamination by spilled hydrocarbons.

Sources of pollution include:
- direct dumping of domestic and industrial solid waste. One estimate claims that hazardous waste disposed of on land amounts to 88 per cent flammable material, 72 per cent acid or caustic, and 96 per cent indisputably toxic. (Note: some materials fall into more than one category);
- excessive application of materials such as agricultural chemicals or polluted sewage;
- indirect contamination resulting from leaks or from leaching of hazardous components from liquid waste disposal sites or by fallout from atmospheric pollution episodes.

As described on pages 510, 698 and 892 the runaway explosion at the ICMESA plant, Seveso, in 1976, caused the release into the air of a mixture of toxic substances consisting of trichlorophenol and the hyper-poisonous tetrachlorodibenzo-p-dioxin (TCDD). In addition to causing external skin lesions such as chloracne in humans, the toxicant contaminated rivers, soil and the walls of the neighbouring houses; crops were destroyed and traces were detected in animals. Clean-up was both difficult and costly.

In the same year Italy experienced a second major industrial chemical disaster resulting in casualties and land pollution. A petrochemical plant located in an area between the town of Manfredonia and St Angelo Hill comprised a series of ammonia, urea, caprolactam and ammonium sulphate production facilities. Carbon dioxide from the ammonia process was absorbed with aqueous potassium carbonate solution activated by addition of $ca.$ 150 g/l arsenic trioxide at 100 °C and $ca.$ 30 atm pressure:

$$K_2CO_3 + CO_2 + H_2O \xrightarrow{As_2O_3} 2KHCO_3$$

At about 9.50 a.m. on Sunday, 26 September 1976, an estimated 10 tons of a mixture of potassium carbonate, potassium arsenite and potassium arsenate escaped from the plant following rupture of the upper part of the wash tower at a connection. Part of the contents were spilled onto the ground and part escaped to the atmosphere in the form of dust/aerosol which was carried by the wind and deposited over a 13 km^2 area of largely cultivated land including the extreme edge of Manfredonia. The incident led to acute concern, because of the intrinsic toxicity of arsenic and its compounds (see Chapter 5 for criteria used at the time), and because of the high solubility of potassium arsenite, which could have diffused deep into the ground and down to the water-bearing stratum at the time of autumn agricultural activities.

The day following the incident the mayors of Manfredonia and Monte S. Angelo met and a committee of experts was set up. Immediate specific

Table 21.15 Effects of major pollutants on materials

Chemical	Primary materials attacked	Typical damage
Carbon dioxide	Building stones, e.g. limestone	Deterioration
Sulphur dioxide	Metals	
	Ferrous metals	Corrosion
	Copper	Corrosion to copper sulphate (green)
	Aluminium	Corrosion to aluminium sulphate (white)
	Building materials (limestone, marble, slate, mortar)	Leaching, weakening
	Leather	Embrittlement, disintegration
	Paper	Embrittlement
	Textiles (natural and synthetic fabrics)	Reduced tensile strength, deterioration
Hydrogen sulphide	Metals	
	Silver	Tarnish
	Copper	Tarnish
	Paint	Leaded paint blackened due to formation of lead sulphide
Ozone	Rubber and elastomers	Cracking, weakening
	Textiles (natural and synthetic fabrics)	Weakening
	Dyes	Fading
Nitrogen oxides	Dyes	Fading
Hydrogen fluoride	Glass	Etches, opaques
Solid particulates (soot, tars)	Building materials	Soiling
	Painted surfaces	Soiling
	Textiles	Soiling
Acid water (pH < 6.5)	Cement and concrete	Slow distintegration
Ammonium salts	Cement and concrete	Causes slow to rapid distintegration; if cement is porous, corrosion of steel reinforcement may occur
Fats, animal or vegetable oils	Cement and concrete	May cause slow disintegration
Calcium chloride	Cement and concrete	May cause steel corrosion if concrete is porous or cracked
Calcium sulphate (and other sulphates)	Cement and concrete	Disintegrates concrete of inadequate sulphate resistance
Solvents	Cement and concrete	Some may cause changes resulting from penetration
Oxalic acid	Cement and concrete	Beneficial, reduces effects of CO_2, salt water, dilute acetic acid
Paraffin	Cement and concrete	Slow penetration, not harmful
Phenol (5% solution)	Cement and concrete	Slow disintegration
Caustic alkali	Cement and concrete	Not harmful up to 10–15% concentration

measures were instituted for the protection of human beings and animals; the emergency measures put into effect were closure of the plant, temporary evacuation of the population, suspension of work and commerce, prohibition of hunting, fishing and grazing, sequestration of farm animals (some 1500 chickens and rabbits had to be slaughtered), and destruction of fruit and vegetables. Monitoring stations were set up at sea for marine fauna and on the coast to check springs. The affected area was marked off and sampling and analytical methods were established. A medical advice centre was set up and requests sent out for suitable drugs to treat arsenic poisoning. Over 140 people were hospitalised for suspected arsenic poisoning. Further measures included health checks on workers and medical observations on the population of Manfredonia and the establishment of a clean-up team equipped with appropriate personal protection.

By 1 October the area had been divided into three zones and use was made of milk of lime and ferric sulphate for decontamination, by conversion of soluble trivalent arsenic (potassium arsenite) into insoluble pentavalent salt (iron or calcium arsenate), and of sodium hydrochlorite to wash the houses and roofs in northern Manfredonia. The soil within the factory site was stripped to a depth of about 30 cm. Urinary arsenic levels were monitored in workers: those with values above 100 μg/l were not admitted to the factory site and those with values in excess of 3000–4000 μg/l were admitted to hospital for observation. Not until 13 January 1977 did the Decontamination Committee declare the entire area affected by the incident fit for use. Follow-up medical checks continued until 22 December 1977. Alerted and sensitised by the Seveso accident, the administrative and sanitary authorities were prompt to act and fortunately there were no fatalities. However, the incident created much distress, was costly and took almost 4 months to clean up.[60]

That such extensive pollution can be difficult to clean up is further illustrated by the following case history of a more recent explosion at a Russian plant.

In mid-September 1990, an explosion at the beryllium unit of a nuclear fuel processing plant at the Ulba metallurgical works in the centre of the city of Ust-Kamenogorsk released a cloud of some 4000 kg of beryllium dust. This metal, which is used widely in the nuclear industry to emit α-particles when bombarded with neutrons (see Chapter 8), for cans and other accessories in nuclear reactors, and in the aerospace industry, is 250 times more toxic than arsenic. Effects of over-exposure include acute bronchitis and, in extreme cases, breathing difficulties, coughing, spitting blood, and skin and eye reactions.

The resulting dust cloud threatened the city's 300 000 inhabitants and caused extensive pollution. The entire Kazakhstan republic was eventually declared a disaster zone: orders were given to wash down roofs and streets and to replace sand in children's playgrounds. One analysis however, concluded that despite the use of high-powered hoses to spray buildings and roads, and the re-tarmacing of roads and removal of top soil, it would be virtually impossible to decontaminate the region completely, with consequential long-term damage to the water and food chain.[61]

Cases of neighbouring berylliosis had previously been reported amongst persons residing in the vicinities of beryllium production factories.[62,63]

Land contamination

Land may, as exemplified by the foregoing examples, become contaminated by chemicals processed, stored or dumped on the site. Such contamination may – in addition to its environmental impact – pose a health risk to workers on the site, those involved in building construction or engineering works, or the public, e.g. due to trespass.

The nature and degree of land contamination may only be determinable by detailed surveys culminating in soil surveys. However, any potential contamination and its nature may be deduced from historical information. Thus likely contaminants from particular industries are summarised in Table 21.16, although other contaminants may, of course, be present and other types of site may merit thorough investigation.

Sources of historical information include[64]
- old maps and plans from record offices and libraries
- local authorities' planning departments
- former employers and employees of companies who may have occupied the site
- local residents who may have detailed knowledge of past usage of the site and the immediate neighbouring sites
- local industrial archeological groups.

Preliminary inspection of the site may also reveal potential contamination. Indicators include[64]
- remains of tanks, pipes, culverts or other fixed plant
- disused storage areas, pits, etc.
- inconsistent vegetation growth, especially if it is poor or deformed
- surface materials, unusual contours or colours in the ground
- distinctive odours
- drums, containers, cans, skips, etc.
- discoloured and/or odoriferous water in sumps, pools, streams, etc.
- bubbling or frothing of standing water, possibly attributable to submerged gas pockets.

A soil sampling strategy may then be developed. In the case of construction work soil samples should be taken at least at the depth of any excavation. Consideration needs to be given to water seepage causing migration of contamination. Some guidelines for the interpretation of results from soil surveys are given in Table 21.17.[65]

Table 21.16 Soil contaminants likely to be associated with different industries[66]

Industry	Examples of sites	Likely contaminants
Chemicals	Acid/alkali works Dyeworks Fertilisers and pesticides Pharmaceuticals Paint works Wood treatment plants	Acids, alkalis, metals, solvents (e.g. toluene, benzene), phenols, specialised organic compounds
Petrochemicals	Oil refineries Tanks farms Fuel storage depots Tar distilleries	Hydrocarbons, phenols, acids, alkalis and asbestos
Metals	Iron and steel works Foundries and smelters Electroplating, anodising and galvanising works Engineering works Shipbuilding/shipbreaking Scrap reduction plants	Metals, especially Fe, Cu, Ni, Cr, Zn, Cd and Pb; asbestos
Energy	Gasworks Power stations	Combustible substances (e.g. coal and coke dust), phenols, cyanides, sulphur compounds, asbestos
Transport	Garages, vehicle builders Maintenance workshops Railway depots	Combustible substances, hydrocarbons, asbestos
Mineral extraction, land restoration (incuding waste disposal sites)	Mines and spoil heaps Pits and quarries Filled sites	Metals (e.g. Cu, Zn, Pb), gases (e.g. methane), leachates
Water supply and sewage treatment	Waterworks Sewage treatment plants	Metals (in sludges) Micro-organisms
Miscellaneous	Docks, wharfs and quays Tanneries Rubber works Military land	Metals, organic compounds, methane; toxic, flammable or explosive substances; micro-organisms, asbestos

Ubiquitous contaminants – hydrocarbons, polychlorinated biphenyls (PCBs), asbestos, sulphates and many metals used in paint pigments or coatings – may be present on almost any site

References

1. Campbell, W. A., *The Chemical Industry*, Longman, 1971, 39, quoting from the *Chemical Trades Journal*, 1889.
2. Ballard, J., *The Safety and Health Practitioner*, 1991, **9**(3), 37.
3. Marshall, V. C., *Env. Protection Bulletin*, 1989(001), 5.
4. a) Glaubinger, R. S. et al., *Chem. Eng.*, 1979, **22**, 86.
   b) Anon., *Chem. & Eng. News*, 1994, **72**(26), 4.
5. a) Madge, P., *The Safety and Health Practitioner*, 1989, **7**(5), 35.
   b) British Medical Association, *Hazardous Waste and Human Health*, Oxford University Press, 1991, Oxford.
   c) Gittins, L., *Chemistry in Britain*, August 1993, 684.

Table 21.17 Guidelines for classification of contaminated soils – suggested range of values (parts per million) on air-dried soils (except for pH)

Parameter	Typical values for uncontaminated soils	Slight contamination	Contaminated	Heavy contamination	Unusually heavy contamination
pH (acid)	6–7	5–6	4–5	2–4	<2
pH (alkaline)	7–8	8–9	9–10	10–12	12
Antimony	0–30	30–50	50–100	100–500	500
Arsenic	0–30	30–50	50–100	100–500	500
Cadmium	0–1	1–3	3–10	10–50	50
Chromium	0–100	100–200	200–500	500–2500	2500
Copper (available)	0–100	100–200	200–500	500–2500	2500
Lead	0–500	500–1000	1000–2000	2000–1.0%	1.0%
Lead (available)	0–200	200–500	500–1000	1000–5000	5000
Mercury	0–1	1–3	3–10	10–50	50
Nickel (available)	0–20	20–50	50–200	200–1000	1000
Zinc (available)	0–250	250–500	500–1000	1000–5000	5000
Zinc (equivalent)	0–250	250–500	500–2000	2000–1.0%	1.0%
Boron (available)	0–2	2–5	5–50	50–250	250
Selenium	0–1	1–3	3–10	10–50	50
Barium	0–500	500–1000	1000–2000	2000–1.0%	1.0%
Beryllium	0–5	5–10	10–20	20–50	50
Manganese	0–500	500–1000	1000–2000	2000–1.0%	1.0%
Vanadium	0–100	100–200	200–500	500–2500	2500
Magnesium	0–500	500–1000	1000–2000	2000–1.0%	1.0%
Sulphate	0–2000	2000–5000	5000–1.0%	1.0–5.0%	5.05
Sulphur (free)	0–100	100–500	500–1000	1000–5000	5000
Sulphide	0–10	10–20	20–100	100–500	500
Cyanide (free)	0–1	1–5	5–50	50–100	100
Cyanide	0–5	5–25	25–250	250–500	500
Ferricyanide	0–100	100–500	500–1000	1000–5000	5000
Thiocyanide	0–10	10–50	50–100	100–500	500
Coal tar	0–500	500–1000	1000–2000	2000–1.0%	2500
Phenol	0–2	2–5	5–50	50–250	1.0%
Toluene extract	0–5000	5000–1.0%	1.0–5.0%	5.0–25.0%	250
Cyclohexane extract	0–2000	2000–5000	5000–2.0%	2.0–10%	25.0%

6. Bridgwater, A. V. & Mumford, C. J., *Waste Recycling and Pollution Control Handbook*, George Godwin Ltd, 1979.
7. Clarke, L., *Acceptable Risk?*, University of California Press, 1989.
8. a) Weterings, R. A. P. M. & Van Eijndhoven, J. C. M., *Risk Analysis*, 1989, **9**(4), 473.
   b) Alexander, J., *Chem. Eng.*, 1992 (27 February).
9. Stewart, J. M., *New Scientist*, 1990 (June), 58.
10. Rose, J., *Chem. & Ind.*, 1990 (Nov.), 726.
11. Butcher, R. W., *J. Inst. Sewage Purification*, 1946, **2**, 92.
12. Jalili, M. A. & Abbasi, A. H., *Brit. J. Ind. Med.*, 1961, **18**, 303.
13. Anderson, A., *Nature*, 1984, **309**, 576.
14. Alabaster, J. S. & Lloyd, R., 'Fish Toxicity Testing Procedures' in *Water Quality Criteria for Freshwater Fish*, Butterworth-Heinemann, 1982.
15. The Registry of Toxic Effects of Chemical Substances, National Institute of Occupational Safety and Health, Cincinnati, OH.
16. Ghazi, P., *Observer*, 4 Feb. 1990; Brown, P., *New Scientist*, 4 Aug. 1990, 26.
17. Mason, C. F., Biological aspects of freshwater pollution, in *Pollution: Causes, Effects and Control*, 2nd edn, ed. R. M. Harrison, Royal Soc. Chem., Cambridge, 1990, 99.
18. Hare, F. K., Human environmental disturbances, in *Environmental Science and Engineering*, ed. J. G. Henry & G. W. Heinke, Prentice-Hall, Englewood Cliffs, NJ, 1989, 114.
19. Anon., *The Bebington News*, 5 Sept. 1990, 5; 5 Dec. 1990, 3.
20. Crittenden, B. D., Kolaczkowski, S. T. & Perera, S. P., *Env. Protection Bulletin*, 1990(004), 25.
21. IARC Monograph on the Evaluation of Carcinogenic Risk of Chemicals to Humans, Vol. 18, *PCBs*, WHO/IARC, Geneva, 1978.
22. Smith, A. J., PCBs, in *Hazardous Materials Spills Handbook*, ed. G. F. Bennett, F. S. Feates & I. Wilder, McGraw-Hill, New York, 1982.
23. Allowcy, B. J. & Ayres, D. C., *Chemical Principles of Environmental Pollution*, Blackie Academic & Professional, Glasgow, 1993, 256.
24. Anon., *Hazardous Cargo Bulletin*, 1989 (May), 68.
25. Lang, T. & Clutterbuck, C., *P is for Pesticides*, Ebury Press, 1991, p. 202.
26. Ryder, E. A., *Annales des Mines*, 1984 (August), 19.
27. a) Anon., *Loss Prevention Bulletin*, 1987(075), 11.
    b) Downing, A. L. & Truesdale, G. A., *J. Appl. Chem.*, 1955, **5**, 570.
28. Green, J. & Trett, M., *Fate and Effects of Oil in Freshwater*, Elsevier Applied Science Publishers, Oxford, 1989.
29. Anon., *Financial Times*, 13 July 1990, 6.
30. Anon., *The Times*, 24 Feb. 1990; *Daily Telegraph*, 24 Feb. 1990, and 25 May 1990.
31. Butler, J. N., *Ocean Industry*, 1978 (Oct.), 101.
32. a) Dayton, L., *New Scientist*, 8 April 1989, 20; 15 April 1989, 21; 6 May 1989, 24; 12 Aug. 1989, 23.
    b) Kirschner, E., *Chem. & Eng. News*, 1994 (June), 8.
33. Spooner, M., *Marine Pollution Bulletin*, 1978, **9**(11), 281; Laubier, L., *ibid.*, 285.
34. Anon., *The Economist*, 1 April 1989, 41.
35. Buchan, J., *Financial Times*, 3 April 1989, 3.

36. Exxon Corp., *ENDS Report* 171, 1989 (April), 11.
37. Nichols, A. B., *Journal WPCF*, 1989, **61**(7), 1176.
38. Crum, J. A., *ibid.*, 1186.
39. a) Anon., *New York Times*, 12 April 1980, 22.
    b) O'Brian, J. L., *Proc. 1981 Oil Spill Conference (Prevention, Behaviour, Control, Cleanup)*, Atlanta, GA, USA, 2–5 March 1981; American Petroleum Institute, Washington, DC, *National Strike Force Response Ixtoc 1 Blowout – Bay of Campeche*, 125–30.
40. Jernelov, A. & Linden, O., *Ambio*, 1981, 299.
41. Golob, R. S. & McShea, D. W., Implications of the Ixtoc 1 blowout and oil spill, presented at *Conf. on Petroleum and the Marine Environment*, Monaco, 1980, 743–59, Petro Mar 80 Euroccan, IPIECA, E&P Forum, UNEP.
42. a) Robbins, A., *Civil Protection*, **26**, 3.
    b) Ryan, S., *Sunday Times*, 6 June 1993, 2.5.
43. Twort, A. C., Law, F. M. & Crowley, F. W., *Water Supply*, 3rd edn, Edward Arnold, 1985.
44. Anon., Manufacturing Chemists Association, Washington, DC, USA.
45. Amoore, J. E. & Gautala, E., *J. Applied Toxicology*, 1983, **3**, 6.
    a) Department of the Environment, *Reducing emissions of volatile organic compounds (VOCs) and levels of ground level ozone: a UK strategy*, a consultation document, 1992.
46. Kupchella, C. E. & Hyland, M. C., *Environmental Science*, Allyn and Bacon, Boston, MA, 2nd edn, 1986, 283.
47. Warner, Sir F. W., *Environmental Protection Bulletin*, 1989(001), 5.
48. Butler, J. D., *Air Pollution Chemistry*, Academic Press, New York, 1979.
49. Chanlett, E. T., *Environmental Protection*, McGraw-Hill, New York, 1973.
50. Butcher, C., *Chem. Eng.*, 1990 (12 April), 27.
51. Office of Toxic Substances, US Environmental Protection Agency, *Chemicals-in-Progress Bulletin*, 1991, **12**(1), 12.
52. Hourihane, D. O'B. & McGauhey, W. T. E., *Portland Med. J.*, 1966, **42**, 613.
53. Dalguen, P., Dabbert, A. F. & Heinz, Y., *Prax. Pneumol.*, 1969, **23**, 548.
54. Anon., *Civil Protection*, 1988, **7**, 3.
55. Murray, M. M. & Wilson, D. C., *Lancet*, 1946, **2**, 821.
56. Hunter, D., The Diseases of Occupations, The English Universities Press, 5th edn, 1975.
57. Tilson, S., *Air Pollution*, Int. Sci. Technol., June 1965.
58. Brinkman, F. J. J., *Quality of Groundwater, Studies in Environmental Science* No. 17, eds. W. Duijvenbooden, P. Glasborgen & H. van Lelyveld, Amsterdam, 1981.
59. Finnecy, E. E. & Pearce, K. W., Land contamination and reclamation, in *Understanding Our Environment*, ed. R. E. Hester, Royal Soc. Chemistry, Cambridge, 1986.
60. a) Liberti, L. & Polemio, M., *J. Envir. Sci. Health*, 1981, **A16**(3), 297.
    b) Soleo, L. (ed.), Exposure to arsenic: five years after the Manfredonia accident, *La Med. Lavoro*, 1982, **73**, Suppl. No. 3, 262.
    c) Polemio, M. *et al.*, *Boll. Chimicio Un. Ital. Lab. RRov.*, 1980, **6**(2), 165.
61. Anon., *Sunday Times*, 30 Sept. 1990, 2.
62. Eisenbud, M., Wanta, B. S., Duston, C. *et al.*, *J. Ind. Hyg. Toxicol.*, 1949, **31**, 282.

63. Sussman, V. H., Lieben, J. & Cleland, J. G., *Amer. Ind. Hyg. Assoc. J.*, 1959, **20**, 504.
64. Collison, J., NEBOSH Diploma Course Notes, 1992, ATM Safety Ltd., U.K.
65. British Standards Institution, Draft for Development DD 175:1988, Code of Practice for the identification of potentially contaminated land and its investigation.

CHAPTER 22

Environmental pollution control

Chemical pollution comprises accidental periodic emissions, continuous effluent discharge and uncontrolled disposal of waste into the environment via air, land or water. Control of pollution is essential since, as illustrated in the previous chapter, it can prove to be
- damaging to public health and well-being, the environment, ecosystem, livestock, crops, and buildings/structures
- illegal
- wasteful of valuable resources
- expensive and technically difficult to rectify (e.g. to quantify effects, to clean up spillages and repair damage)
- harmful to company reputation

Chapter 14 briefly reviewed the general technical and administrative measures available for the control of pollution and more detailed information is provided in ref. 1. The present chapter amplifies on these approaches by reference to legislative, management and technical controls with examples of selected industrial applications.

Legislative controls

Regulations to control pollution and control hazardous waste vary considerably from country to country as illustrated by Table 22.1.[2] They represent one of the most complicated and ambitious areas of modern legislation. Details are beyond the scope of this general text but specific information is obtainable from the authoritative publications listed in the references.

US legislation

Important amongst US legislation are Regulations on Hazardous Wastes under The Resources Conservation and Recovery Act, The Toxic Substances Control Act, and The Comprehensive Environmental Response, Compensation and Liability Act (Superfund) (see page 840). These are explained in detail and their requirements compared in ref. 3.

Table 22.1 Legislation relating to hazardous wastes

	Belgium[a]		West Germany	Denmark	France	Netherlands	UK	USA
	Nat.	Fl.						
Legislation	+	+	+	+	+	+	+	+
Part of the general legislation relating to municipal wastes	–	+	+	–	+	–	+	+
Year of legislation	1974	1981	1972	1972	1976	1976	1974	1976
Year indicating hazardous wastes	1976	—	1977	1976	1977	1979	1980	1980
Implementation of legislation	+	–	+	+	+	+	+	+
Legislation relating to used oil	–	+	+	+	+	+	–	+
Separate enactment decree within main legislation (hazardous wastes) relating to used oil	–	+	–	+	+	+	–	+
Mainly with enactment of legislation relating to the official loading level	N	D	D	G	D	N	G	D
Special legislation for disposal and incineration of (hazardous) wastes at sea	–	+	+	+	+	+	+	+
'Cradle to grave' control for hazardous wastes	–	–	+	+	–	+	+	+[d]

	Belgium	West Germany	Denmark	France	Netherlands	UK	USA
The 'polluter pays principle'	+	+	+	+	+	+	+
State contribution to the costs of disposal[1]	O	O	+	+	O	–	–
Intentions regarding prevention/minimisation	+	+	+	+	+	+	+
Possibilities to force prevention/minimisation by legislation	+[2]	–	–	+/–	+/–	–	+/–
Intention to reuse/usefully apply	–	+/–[4]	+	+	+/–	+	+
Control possibilities for reuse/useful application in the legislation[3]							
Promoting reuse, useful applications, prevention and minimisation through:							
State establishment of a special authority charged with handling these matters	–	–	–	+[5]	–	–	–
State action to carry out and stimulate research by making financial provisions available	+	+/–	+/–	+/–	+	+	+
Financial support by the State in actual activities	–	–	+/–	+/–	–	–	–
Establishment of a waste exchange	–	+	–	+	+	–	+/–[4]

Table 22.1 continued

	Belgium	West Germany	Denmark	France	Netherlands	UK	USA
Public nuisance obligations (in which the government decides whether the responsibility for the wastes is taken over by the state when the producer reports their presence)	−						
Disposal facilities in the hands of the state	+/−[a]	+/−[c]	+	−	+/−[b]	−	−
Important state financial contribution towards the costs of disposal	−	−	+	+	+	−	−
Levies of taxes imposed on (disposal of) hazardous wastes	+	+	−	+	−	−	+/−[d]
The obligation of the state to set up a plan for the disposal of hazardous wastes	+	+	+	+	−	+	+

−, No contributions; O, relatively small contribution; +, relatively large contribution.

[1] Only in the Flemish region up to now.
[2] These means are enacted in principle in various national legislations, but rarely or never used in practice.
[3] Diverse regulations in the different countries/Federal States.
[4] A special institute exists in France for handling these matters (ANRED).

[a] In Belgium, a distinction must be made between national legislation and legislation concerning the Flemish region alone. This distinction is made in the table by the annotations Nat./Fl. Specific legislation relating to chemical wastes in the Wallonian region has not yet been enacted.
[b] The 'Deposit of Poisonous Waste Act' was withdrawn in 1972 and replaced by the 'Special Waste Regulations 1980'.
[c] N (National Federal); D (Federal State, Department, Province); G (Municipal, Regional).
[d] Reporting pro-formas (trip-ticket) are not submitted to EPA unless the producer, transporter or processor signals a deficiency in pro-forma flow. However, in many states there is an obligatory duty to report.

Environmental pollution control 1259

In the USA in 1991 the Environmental Protection Agency announced a negotiated consensus rule covering 149 of the 189 substances listed in the Clean Air Act and aimed at reducing fugitive emissions from industrial processes. Such emissions are said to account for about one-third of all emissions from the chemical industry. The rule will[4]
- make standards for pumps and valves more stringent;
- reduce the threshold level for leak repairs;
- subject flange and other pipe connectors to leak repairs;
- specify performance level;
- give incentives for continuous improvements in performance;
- impose penalties for failure to achieve performance levels.

UK legislation

Selected UK legislation relating to control of pollution and hazardous waste was mentioned on pages 837–9. Of the recent European legislation, key UK developments include
- The Control of Pollution Act (Special Waste Regulations) 1989.
- The Health and Safety (Emissions to the Atmosphere) (Amendment) Regulations 1989 which amended the definition of a number of scheduled works (asbestos, chemical incineration and electricity) and introduced new categories (glass and mineral fibre works, large combustion works, large glass works, and large paper pulp works).
- The Control of Industrial Air Pollution (Registration of Works) Regulations 1989 which expanded existing registration procedures and introduced the requirement to advertise applications.
- The Air Quality Standards Regulations 1989 to enshrine in UK law EC Directives on air quality standards for sulphur dioxide and particulate matter, lead and nitrogen dioxide.
- The Environmental Protection Act 1990. This completely reformed arrangements for controlling pollution at national and local levels and created a complete interlocking framework for pollution control. Intergrated Pollution Control (IPC) is a new approach with waste minimisation at its centre and a commitment to higher environmental standards. The concept of the Best Practicable Environmental Option (BPEO) is a statutory objective of IPC. The reform of waste disposal was designed to minimise waste and maximise recycling.

The Environmental Protection Act – Part I

Part I of the Act introduced an IPC system to control emissions to air, land or water for the most polluting industrial and similar processes. IPC is limited to prescribed processes (e.g. chemical, fuel and power, waste disposal, minerals, etc.) by prior authorisation. Authorisations are based upon the requirement for owners/controllers to prevent prescribed substances from being released or, where this is not practicable, to reduce the release to a minimum. Any residual releases must be rendered harmless. To achieve these aims operators have to use the Best Practicable Means Not Entailing Excessive Cost (BATNEEC).

This is to be supplemented by extending local authority control of air pollution to cover a second tier of less polluting processes.

Incineration – Certain incinerators are now subject to control under Part I of the Environmental Protection Act 1990 and the Environmental Protection (Prescribed Processes and Substances) Regulations 1991. Part A processes are subject to the IPC regime and Part B processes are subject to local authority air pollution control; the processes are listed in Table 22.2.

Table 22.2 Classification of incinerators for control under The Environmental Protection Act 1990

Part A processes include
a) incineration of any waste chemicals or waste plastic arising from the manufacture of any chemical or plastic
b) except where incidental to the burning of other wastes the incineration of any waste chemicals comprising (in elemental or compound form)

bromine	iodine	phosphorus
cadmium	lead	sulphur
chlorine	mercury	zinc
fluorine	nitrogen	

c) incineration of any other waste, including animal remains, in an incinerator with a capacity greater than 1 tonne/hour
d) burning of residues in metal containers used for transport or storage of chemicals prior to their re-use

Part B processes include
a) incinerators for the destruction of waste, including animal remains, in any other controlled incinerator
b) the cremation of human remains

Incinerators with a capacity of <50 kg/h are exempt except when used for the incineration of clinical waste (other than animal carcasses), sewage sludge, sewage screenings or municipal waste. The difference in control requirements is that whereas use of BATNEEC is applicable to both Part A and Part B processes, use of the Best Practicable Environmental Option (BPEO) is also required for the former.

The Environmental Protection Act – Parts II to VIII

Part II reformed waste disposal, handling and management. These provisions include new duties on producers of waste to ensure its safe disposal; stronger licensing powers for local authorities, and continuing responsibilities for licensees to monitor and maintain sites after closure. Section 34 introduced a statutory duty of care upon holders of waste, viz. all those

who manage or dispose of waste in any capacity. (These comprise any person who imports, produces, carries, keeps, treats or disposes of control of such waste.) The duty is to take all those measures which are applicable to that person in his capacity in relation to the waste and reasonable in the circumstances,
- to prevent unauthorised disposal,
- to prevent the escape of waste from the control of the holder or any other person,
- to ensure that the waste is transferred only to an authorised person and that a sufficient written description of the waste is given.

This requires use of the documentation shown in Fig. 22.1.

Part III re-enacted the law on statutory nuisance with changes to the Public Health Act 1936. Controls over offensive trades are transferred to Part I.

Part IV provided new measures to deal with litterers.

Part V contained amendments to the Radioactive Substances Act 1960; Part VI controls the use, import, containment or release of genetically modified organisms to the environment; and Part VII provides for the reorganisation of the Nature Conservancy Council and Countryside Commission.

Part VIII covers miscellaneous measures including new restrictions over the import, use, supply and storage of injurious substances; new controls over trade in waste; powers to obtain information about potentially harmful substances; public registers of land which may be contaminated; and amendments to planning legislation governing the siting for polluting controlled waters. (Water pollution is controlled by the Water Act 1989.)

The timetable for introduction is for key sections to be introduced up to 1998.

Code of Practice

In order to comply with the statutory duty of care in respect of waste the following procedures are advised in a Code of Practice.[5]
- Identify and describe the waste
    Is it controlled waste?
    What problems are associated with the waste and its disposal?
    What should be included in the description?
- Keep the waste safely
    Are containers secure for both storage and transit?
    What precautions have been taken to ensure security of waste?
- Transfer to the right person
    Will public authorities take the waste?
    Is the carrier registered, or exempt from registration (under The Control of Pollution (Amendment) Act 1989 and the Controlled

Section A — Description of Waste

1. Please describe the waste being transferred:

2. How is the waste contained?

    Loose ☐   Sacks ☐   Skip ☐   Drum ☐   Other ☐ → please describe:

3. What is the quantity of waste (number of sacks, weight etc):

Section B — Current holder of the waste

1. Full Name (BLOCK CAPITALS):

2. Name and address of Company:

3. Which of the following are you? (Please ✓ one or more boxes)

producer of the waste ☐	holder of waste disposal or waste management licence ☐ →	Licence number: Issued by:
importer of the waste ☐	exempt from requirement to have a waste disposal or waste management licence ☐ →	Give reason:
waste collection authority ☐	registered waste carrier ☐ →	Registration number: Issued by:
waste disposal authority (Scotland only) ☐	exempt from requirement to register ☐ →	Give reason:

Section C — Person collecting the waste

1. Full Name (BLOCK CAPITALS):

2. Name and address of Company:

3. Which of the following are you? (Please ✓ one or more boxes)

waste collection authority ☐	holder of waste disposal or waste management licence ☐ →	Licence number: Issued by:
waste disposal authority (Scotland only) ☐	exempt from requirement to have a waste disposal or waste management licence ☐ →	Give reason:
exporter ☐	registered waste carrier ☐ →	Registration number: Issued by:
	exempt from requirement to register ☐ →	Give reason:

Section D

1. Address of place of transfer/collection point:

2. Date of transfer: 3. Time(s) of transfer (for multiple consignments, give 'between' dates):

4. Name and address of broker who arranged this waste transfer (if applicable):

5. Signed: Signed:

Full Name: Full Name:
(BLOCK CAPITALS)   (BLOCK CAPITALS)
Representing: Representing:

Fig. 22.1 Controlled Waste Transfer Note, to meet the duty of care under S.34 of The Environmental Protection Act 1990

Waste (Registration of Carriers and Seizure of Vehicles) Regulations 1991)?
Is the waste manager properly licensed?

Discarded paint tins were reported on fire near some ponds. Paperwork discovered near the fire led the Waste Disposal Authority to a metal treatment business. They admitted having passed the waste to a scrap merchant and pleaded guilty under S.34 of the EPA to failing to use a registered waste carrier and failing to ensure that a transfer note accompanied the consignment.[6]

- Steps on receiving waste
  Who did the waste come from?
  Does the waste tally with its description?
  Are proper records being kept?
- Checking-up on disposal
  Report any suspicions to the Waste Disposal Authority
  Change carrier/contractor

Discharges of trade effluent to surface waters

Discharges of trade effluent, or sewage effluent, to 'controlled waters' (e.g. any inland waters but not sewers, landlocked lakes and reservoirs, any coastal or tidal waters within three miles of the territorial limit, underground or ground waters) require a consent from the National Rivers Authority under the Water Act 1989. Discharges through a pipe from land to the sea outside controlled waters, or from a building or plant to land or landlocked waters, are also covered.

Discharges of trade effluent to sewers

Discharge of trade effluent into a sewer requires a consent from the Sewerage Undertaker under a procedure set out in the Public Health (Drainage of Trade Premises) Act 1961, Control of Pollution Act 1974 and Water Act 1989. The Environmental Protection Act 1990 is also applicable to prescribed processes and substances.

A Trader is under a duty not to discharge trade effluent to a public sewer without the sewerage undertaker's consent – specifying the volume, structure and composition of the effluent – to comply in every respect with the consent conditions, and to have regard to safety. Guideline values considered realistic by one undertaker are summarised in Table 22.3.[7] If a particular trade effluent constitutes a high proportion of the flow to a treatment works, more stringent standards may be imposed to protect the treatment and disposal system.

Special provisions control specific dangerous substances likely to pass through the system into the natural environment. These 'Red List' substances are defined in the Trade Effluents (Prescribed Processes and Substances) Regulations 1989 and listed in Table 22.4.

An application for consent, or an Agreement to discharge trade effluent

Table 22.3 Standard conditions for more commonly occurring contaminants exemplified in Guidelines[7]

---

1. *Substances to be eliminated*
   The following matters shall be eliminated from the trade effluent before discharge into the sewers:
   (a) petroleum spirit
   (b) calcium carbide
   (c) carbon disulphide
   (d) mercury, cadmium and their compounds unless controlled by EC Authorisation
   (e) organohalogen compounds including pesticide residues and degreasing agents unless controlled by EC Authorities
   (f) any substances which either alone, or in combination with each other, or with any other matter lawfully present in the sewers would be likely to:
       (i) cause a nuisance or produce flammable, harmful or toxic vapours either in the sewers or at the sewage works of the Authority
       (ii) injure the sewers or interfere with the free flow of their contents or affect prejudicially the treatment and disposal of their contents or have injurious effects on the sewage treatment works to which it is conveyed or upon any treatment plant there
       (iii) be dangerous to or cause injury to any person working in the sewers or at the sewage treatment works
       (iv) affect prejudicially any watercourse, estuary or coastal water into which the treated effluent will eventually be discharged.

2. *General limitations*
   The following parameters should not normally exceed the values shown and may be less in appropriate circumstances:
   (a) sulphides, hydrosulphides, polysulphides and substances producing hydrogen sulphide on acidification — 1 mg/l
   (b) separable grease and oil in excess of — 100 mg/l
   (c) sulphates as SO_4 in excess of — 1000 mg/l
   (d) Toxic metals in excess either individually or in total, i.e. antimony, arsenic, beryllium, chromium, copper, lead, nickel, selenium, silver, tin, vanadium, zinc — 10 mg/l
   (e) cyanides and cyanogen compounds which produce hydrogen cyanide on acidification in excess of — 1 mg/l
   (f) no trade effluent shall be discharged which has a pH less than 6 or greater than 10
   (g) no trade effluent shall be discharged which has a temperature higher than 43.3 °C (110 °F).

(or any variation of an existing consent) must be referred to the Secretary of State if any substances listed in Schedule 1 in Table 22.5 will be present in a concentration greater than the background concentration. A similar procedure applies to any trade effluent derived from any of the processes listed in Schedule 2 in Table 22.5 if either asbestos or chloroform is present in a concentration greater than the background concentration.

Strict liability

The concept of 'strict liability' for pollution from a factory, or other industrial, site was noted briefly on page 918. So far as environmental law in the UK is concerned, environmental offences appear to be generally treated as

Table 22.4 UK 'Red List' of substances

Mercury
Cadmium
* gamma-Hexachlorocyclohexane (Lindane)
* DDT
* Pentachlorophenol (PCP)
Hexachlorobenzene (HCB)
Hexachlorobutadiene (HCBD)
* Aldrin
* Dieldrin
* Endrin
Chlorprene
3-Chlorotoluene
* Polychlorinated biphenyls (PCB)
* Triorganotin compounds
Triorganotin
* Dichlorvos
* Trifluralin
Chloroform
Carbon tetrachloride
1,2-Dichloroethane
Trichlorobenzene
* Azinphos-methyl
* Tenitrothion
* Malathion
* Endosulfan
* Atrazine
* Simazine

---

* Substances likely to enter the aquatic environment through a variety of indirect routes.

Table 22.5 Prescribed Substances and Processes, under the UK Trade Effluents (Prescribed Processes and Substances) Regulations 1989

Schedule 1 – Prescribed Substances
  Mercury and its compounds
  Cadmium and its compounds
  gamma-Hexachlorocyclohexane
  DDT
  Pentachlorophenol
  Hexachlorobenzene
  Aldrin
  Dieldrin

Schedule 2 – Prescribed Processes
Description of Process
  Any process for the production of chlorinated organic chemicals.
  Any process for the manufacture of paper pulp.
  Any industrial process in which cooling waters or effluents are chlorinated.
  Any process for the manufacture of asbestos cement.
  Any process for the manufacture of asbestos paper or board.

strict liability offences unless it is clear from the wording of the statute that they are not.[8] For example,

> Under the Control of Pollution Act 1974 any person who 'causes ... any poisonous, noxious or polluting matter to enter a stream' is strictly liable and guilty of an offence.[9]

Where such strict liability is imposed a Defendant incurs criminal liability although he was ignorant of those factors which rendered his conduct criminal, and even though his ignorance is not attributable to any fault or negligence on his part.

The meaning of 'cause' in the example, and how it gives rise to strict liability, is exemplified in a leading case[10] in which a company's conviction for causing polluting matter to enter a river or stream was upheld even though it was accepted that they had not been negligent.

> The defendant's paper manufacturing factory was located on the bank of a river. Effluent was fed to collecting tanks and its overflow was prevented by the use of two pumps; one pump was actuated automatically at a set level, the other was a manually actuated pump. One pump was generally adequate and they were inspected, cleaned and maintained every weekend. The intake to each pump was protected by a rose intended to filter-out foreign matter.
>
> On one occasion, despite both pumps being in operation, polluted water from the tanks overflowed into the river. A quantity of brambles, ferns and long leaves were found to have obstructed the pumps so that they failed to prevent an overflow. This had never happened before, and the company had taken reasonable steps to ensure that effluent did not escape into the river.
>
> Nevertheless the company were convicted of causing polluting matter to enter the river, i.e. they were deemed to be accountable for any pollution arising as a result of their manufacturing operations.[8]

However, if the effective cause of pollution was intervention by a third party, e.g. a trespasser, or depending on the circumstances an act of God, the Defendant would in all probability be not guilty if effluent from his process is released into a river. Thus liability is strict, not absolute.[8]

> A company with a fuel oil storage tank were found not guilty of having caused polluting matter to enter a stream on the basis that there was an intervening act of an unknown person who had opened a gate valve on the tank allowing oil to escape.[11]

In summary, if a company in the UK produces waste they will – in the absence of an intervening act of God or third party – be strictly liable if it escapes causing pollution. Thus companies are well advised to develop a rapport with the enforcing authorities, to take their advice and to heed their warnings. If after having worked with, and taken advice from, enforcing authorities there is an accidental escape of effluent or waste – provided it is not a major pollution incident – the risk of prosecution will usually be lessened.[8]

Furthermore a prerequisite of the recovery of damages under the rule in Rylands versus Fletcher (page 918) is foreseeability of harm or injury of the type complained of.[8a]

Contamination of water available for abstraction at a water company's borehole was caused by perchloroethylene used by a leather company in a process of degreasing pelts at its tanning works 1.3 miles away. Regular spillages of small quantities of perchloroethylene occurred onto the tannery floor up to 1976, and seeped into the ground and was conveyed in percolating water in the direction of the borehole. In 1976 the water company had bought the land in which the borehole was drilled and began pumping from an underground aquifer in 1979; the water quality was then in compliance with prevailing standards. Higher standards were imposed by the EC in 1980 and tests in 1983, using technology not previously available, detected perchloroethylene levels far exceeding the limits allowed in the Directive; as a result the water company had to close the borehole.

It was held that neither seepage of the perchloroethylene into the chalk aquifers below the tannery, nor that detectable quantities would be found down-catchment, were foreseeable by a reasonable supervisor employed by the leather company. Hence he could not have foreseen in or before 1976 that repeated spillage would lead to any environmental hazard or damage. As a result strict liability was not imposed upon the company.[8a]

Environmental assessment

Environmental assessment (EA), also termed environmental impact assessment (EIA), is a procedure to ensure that potentially significant environmental impacts are satisfactorily assessed – and taken into consideration – in the planning, design, authorisation and implementation of a proposed action.

A European Directive, which came into effect in July 1988,[12] requires EA to be carried out before consent is granted for a development for certain major projects likely to have significant effects on the environment. Within the UK the Directive is applied through the Town and Country Planning (Assessment of Environmental Effects) Regulations 1988 and formal guidance has been provided.[13] (The arrangements in Scotland are broadly similar, although there are some variations and EA is carried out under the Environmental Assessment (Scotland) Regulations 1988. Separate provision is made for Northern Ireland but the general principles are similar.) Further information on the procedures involved is given in ref. 14.

The steps that should normally be part of the assessment are:
- a decision that EA is necessary
- determining the coverage of the EA (scoping)
- preparing the EA report (environmental statement)
- reviewing the statement
- consultation and participation

1268 *Further considerations and developments*

- synthesising the findings of consultation
- decision-making
- monitoring and post-auditing[15]

Assessment is required in every case for Schedule 1 projects, and for Schedule 2 projects subject to certain criteria.[13]

The Environmental Statement is a published document which assesses the likely impact of the proposed development upon the environment. It has to contain:

- a description of the development, including information about the site, design, and size or scale of the development;
- the necessary data to identify and assess the main impacts that the development is likely to have on the environment;
- a description of the likely significant effects, direct and indirect, that the development is likely to have on human beings, flora, fauna, soil, water, air, climate, the landscape, the interaction between any of the foregoing, material assets, the cultural heritage;
- where significant adverse effects are identified, a description of the envisaged mitigation measures;
- a summary in non-technical language of the above information.

It may also include information relating to

- the physical characteristics and land-use requirements of the development during construction and operational phases;
- the main characteristics of production processes, including nature and quality of materials to be used;
- an estimation of type and quantity of expected residues and emissions when the development is in operation;
- an outline of the main alternatives studied and an indication of the reasons for choosing the development proposed;
- the likely significant direct and indirect effects on the environment which may result from natural resources, emission of pollutants, etc.;
- the forecasting methods used to assess the likely significance of the above effects;
- any difficulties, e.g. technical, encountered in compiling the information.[16]

The environmental safety of pesticides is now studied intensively before they are registered for commercial use. Hazard assessment is used to determine the nature and magnitude of hazards that may result from their release into the environment. It depends on the comparison of chemical concentrations that are expected to occur in the environment, the Expected Environmental Concentration (EEC), with those estimated to have no biological effects, the No Observed Effect Concentration (NOEC). Their relationship is shown in Fig. 22.2.

The majority of hazard assessment procedures involve a series of steps, which progress from relatively simple to more complicated laboratory tests and finally to field tests and environmental monitoring as illustrated in Fig.

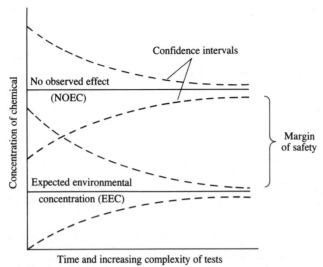

Fig. 22.2 Principle of environmental hazard assessment for pesticides

22.3.[17] The initial or screening stage is based largely on acute toxicity tests, physico-chemical properties and rates of chemical degradation.

As an example, cypermethrin is a broad-spectrum pyrethroid insecticide. Laboratory data relevant to risk assessment to fresh water environments are given in Table 22.6. These demonstrated that sorption of cypermethrin would play a dominant role in determining its environmental distribution. Although its water solubility is very low, its toxicity to aquatic organisms is very high so even very low concentrations in water are potentially toxic. Its very high octanol:water partition coefficient indicated a potential for bioaccumulation to relatively high levels. Laboratory and field studies were therefore undertaken to assess the environmental risks in more detail.[17]

The intermediate stage of assessment involves toxicity tests using a wider range of organisms, i.e. not only the obvious species of concern such as fish, birds and mammals, but also beneficial insects, spiders and crustaceans from a variety of terrestrial and aquatic habitats, tests to assess sublethal effects, bioaccumulation tests and studies on persistence and identification of metabolites in water, soil, plants, fish and mammals. A subsequent, advanced, stage of assessment may involve environmental studies in small field plots and outdoor ponds to assess the fate and effects of the chemicals under conditions closely simulating the natural environment.

Intensive environmental monitoring is also carried out during the early commercial marketing of new pesticides to check the soundness of the risk assessment.

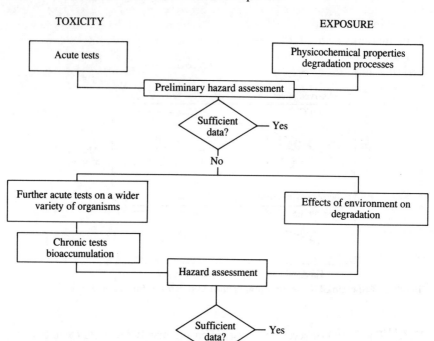

Fig. 22.3 Sequential scheme for hazard assessment for pesticides in the environment

Table 22.6 Cypermethrin laboratory data

Molecular weight	416.3
Vapour pressure at 70 °C	5.1×10^{-6} Pa
Solubility in water	<10 µg/l
Sorption	Octanol/water partition coefficient: 2×10^{-6}
	Sediment/water partition coefficient: 10^5
Volatilisation	Negligible
Phototransformation	Negligible
Hydrolysis	Significant
Biodegradation	Significant
Toxicity to fish (96 h LC_{50})	0.4–2.2 µg/l
Toxicity to sensitive species of aquatic invertebrates (24 h LC_{50})	0.05–5.0 µg/l

Management controls

Environmental protection is applied[12]
- to prevent loss in resources, performance and time

- to minimise disruption, errors and accidents
- to increase profitability by increasing efficiency, quality and harmony with the local community and with authorities.

To this end the person acting as an 'Environmental Manager' may be involved in all the aspects of plant operation summarised in Table 22.7.[18]

Table 22.7   Areas of Environmental Manager's involvement

- External affairs: the community, authorities, emergency services, public relations, etc.
- Security, safety and housekeeping: visible cleanliness, personnel interests
- Administration and records controls: records, inspections, inventories, audits
- Engineering and maintenance: pollution control equipment, requirements for maintenance, equipment monitoring, equipment performance
- Process engineering control and quality: process performance losses, yields improvement potentials, investment potential

External affairs

A company has to liaise with, and be sensitive to, the needs of the regulatory control agencies, the local community, local authorities and the local emergency services.

In its relationship with the community the company obviously needs to be aware of any development which could be adverse, e.g. a local nuisance or any current national concerns relating to the environment, specific pollutants or possible health effects from them. Two separate lines of communication are therefore desirable, one a formal link via senior line management and the second an informal link via the functional Environmental Manager. In the administration of external relations the keeping of proper records is essential since, with the frequency of staff turnover always likely to be greater than the turnover of local inhabitants, they serve as the institutional memory. To obtain independent early warning of adverse community reaction, and to be kept aware of the community's response to company initiatives, useful lines of communication may involve the workforce itself, retired employees, local action groups and elected local government representatives.

A good working relationship is necessary with local authorities, e.g. in the UK the Environmental Health Department and the Planning Department. With current trends in environmental legislation the role of such local agencies is likely to increase.[18]

As discussed in Chapter 17, good liaison with the emergency services is essential in order that they can deal efficiently with any in-plant incidents and to provide back-up for any incidents elsewhere in the locality.

Internal administration

Security, safety and housekeeping

Cleanliness and good housekeeping are essential elements of a good environmental protection programme. Therefore a system of work is desirable to ensure that spillages are cleaned up and leaks dealt with promptly, accumulations of waste or refuse are removed from accessible and inaccessible areas and a good standard of hygiene is aimed for. Improvements may be obtainable by engineering attention to persistent problems, e.g. leakages from pipelines, valves or seals; spillages from conveyors, bunkers or vehicle loading points; leakages from local extraction ductwork.

Administration, records and control

Full records of environmental protection data are useful for reference to assist with

- future pollution problems – complaints
- future developments/planning environmental studies
- diagnosis of in-plant problems
- diagnosis of areas for future development
- reliability data bank for future risk assessment
- loss prevention studies
- public relations

Formal data banks can be very valuable, provided they are well maintained. In any event records need to be adequate with regard to coverage, reliability, accuracy and retrievability. Summary reports are necessary for circulation to departmental management but care is necessary in their preparation, so as to avoid concealing significant events due to averaging or by loss of association.

A pollution inventory, which is merely a tabulation of emissions, discharges and wastes from the plant in various operating phases, e.g. as in Table 22.8, provides a basic data base for loss prevention and the costs of

Table 22.8 Details commonly tabulated in a Pollution Inventory

- Atmospheric emissions: continuous, occasional and fugitive emissions for normal operation start-up and shut-down
- Aqueous discharges: to sewer or direct discharge to river or sea, etc.
- Solid wastes: by type, origins and means of disposal
- Noise: usually handled by results of environmental noise surveys with identification of special features, tonal characteristics and intermittency
- Transport: especially if hazardous cargoes are involved

environmental protection. It should facilitate the identification of problem areas and those capable of improvement. The inventory should include data for design conditions and all the results from monitoring exercises;

thus it is a basic reference for on-going pollution audits. If no measurements exist for fugitive emissions then theoretical estimates are inserted, based upon standard emission factors; any engineering estimates inserted are clearly marked as such.

An audit may be conducted when the need, or a potential benefit, is identified. Once several audit results have been entered into the inventory the performance can be characterised, e.g. as erratic or consistently acceptable/unacceptable.

A typical Environmental Pollution Audit should include:
- Listing gaseous emissions, concentrations, smoke characteristics. Prevailing winds and exposed zones. Toxicity or nuisance potential. Effects of synergism or poor atmospheric dispersing conditions.
- Querying the adequacy of chimneys and stacks, scrubbers and particulates collection equipment (e.g. filters, cyclones, electrostatic precipitators).
- Listing effluents, their analyses and discharge from processes. Querying flammability, corrosivity, toxicity, miscibility, reactivity.
- Listing of 'solid' wastes, analyses and physical form (e.g. solid, slurry, suspension, sludge). 'Toxicity' and flammability.
- Bunding of storage areas, segregation of waste storage areas, security of landfill sites, etc.
- Monitoring (preferably continuous).

The importance of monitoring established systems for continued performance cannot be overemphasised. Monitoring can range from simple checks on management systems, e.g. inspection of returnable chemical containers (see Chapter 13) and on shipments for compliance with agreed procedures and quality standards:

> An intermediate bulk container (IBC) split during transport and spilt hydrochloric acid onto the roadway, resulting in a major call-out for police, the fire service and the emergency team of a local chemical firm.[19] The IBC was found to be three times older than its design life. Subsequently users of IBCs were warned that they must have a system for replacing old containers which become brittle and liable to split.

> A load of 74 drums arrived at a landfill site. Accompanying documentation indicated the consignment to be 'resins, fillers, pigments and paints'. One of the plastic containers disintegrated as it was being off-loaded prior to routine sampling. Subsequent analysis revealed that 50 drums contained liquid paint, 18 contained paint sludges, and six contained waste fluoroacetic acid, a highly toxic, corrosive, strong acid.[20]

to checks on levels of environmental pollution (e.g. methane production at landfill sites) by simple or sophisticated automated techniques:

> The Rhine, which rises in the Alps, collects a mixture of chemical effluent as it

meanders through Switzerland, Liechtenstein, Austria, France, Germany and the Netherlands. At Rotterdam the river annually dumps 10 million cubic metres of silt which must be removed and kept in a special dump since it is too contaminated to be deposited at sea. The Dutch national water authority installed a £1m self-checking, computerised monitoring system gathering data on water quality from instruments at seven stations along the 350 kilometres of river tributaries. The sensors check on chemical/mineral content, oxygen concentrations, temperature and pH. Data are analysed by a central computer which triggers alarms when values exceed thresholds. Management must acknowledge the alarm within a limited time to prevent diversion to another station and anti-pollution scientists are called in. The system directs alarms to the home telephone of operators outside normal working hours.[21]

Engineering and maintenance

If standards for environmental protection are set in-company at the minimum statutory requirements then pollution control may appear to involve only maintenance of the pollution control equipment. This avoids opportunities for preventing losses or increasing yield or performance and, in such cases, records of actual emissions or discharges tend to be minimal.[18]

Currently a more enlightened approach is for pollution control to determine maintenance needs, e.g. use of concepts of leak detection and repair.

Process engineering

The process engineering department may be considered the best location for line responsibilities for pollution control, since discharges or the production of waste may arise either from characteristics of the design of a unit of equipment or from the manner in which it is operated within a process.

Waste stream flows, and their compositions, are determined by the process. Furthermore the effluents from pollution control units are affected by both their efficiency and the feeds, i.e. the performance of upstream units. Problems are often associated with by-product lines due, for example, to the requirements for plant flexibility. These are best eliminated by considering the process as a whole, in an attempt to reduce upstream variability, since the design and installation of pollution control units which can cope with uncontrolled feed variability is an expensive option.[18]

The major advantage of achieving environmental protection from within the process itself, including close monitoring and control, is an economic one. If waste streams can be minimised at source then plant handling them is not subjected to additional loads above the specified duty and good performance may be attained more readily with reduced maintenance requirements.

Technical controls

A recommended approach to environmental control for new processes and new plants is to find ways of designing-out waste generation. Relevant process design considerations are summarised in Table 22.9.[22]

Table 22.9 'New' process or plant design: some considerations for minimising waste and environmental impact[22]

- Maximum chemical conversion efficiency to minimise intractable by-products
- Maximum recycle of by-products
- Maximum energy efficiency
- Minimum introduction of carrier gases
- Minimum use of solvents
- Avoidance of hazardous solvents
- Closed circuit transfers of volatile liquids
- Minimum inventories of hazardous materials
- Minimum isolation of intermediates
- Minimum use of process water
- Maximum raw material purity to minimise the formation of hazardous by-products
- Select equipment that produces minimal or no waste
- Maximum recovery of products and by-products

Waste generation can often also be significantly reduced by improving the efficiency of an existing production process, i.e. by improved operation and maintenance, material change and by equipment modifications.[23] Examples of improved operating procedures, i.e. measures to optimise the use of raw materials, are summarised in Table 22.10.[23] Substitution of a

Table 22.10 Operational changes leading to waste reduction[23]

- Reduce raw material and product loss due to leaks, spills, drag-out, off-specification process solution.
- Schedule production to reduce equipment cleaning, e.g. formulate light to dark paint so vats do not have to be cleaned out between batches.
- Inspect parts before they are processed to reduce number of rejects.
- Consolidate types of equipment or chemicals to reduce quantity and variety of waste.
- Improve cleaning procedures to reduce generation of dilute mixed waste, e.g. using dry clean-up techniques, mechanical wall wipers or squeegees, and 'pigs' or compressed gas to clean pipes and increasing drain time.
- Segregate wastes to increase recoverability.
- Optimise operational parameters (temperature, pressure, reaction time, concentration, chemicals) to reduce by-product or waste generation.
- Develop employee training procedures on waste reduction.
- Evaluate the need for each operational step. Eliminate steps that are unnecessary.
- Collect spilled or leaked material for reuse.
- Maintain a strict preventive maintenance programme (e.g. on motors, pumps, machines to minimise oil, hydraulic fluid losses).
- Redesign production lines to generate less waste (e.g. paint waste can be minimised by using high volume, low pressure paint guns or electrostatic paint equipment).

hazardous chemical in either a product formulation or a production process by a less hazardous material brings benefits both in terms of occupational health, as discussed on page 1623 *et seq.*, and in reducing the amount of hazardous waste generation in formulation, production and end-use.

> Aircraft manufacturers in the USA have switched to lower volatility substances exempt from the Resource Conservation and Recovery Act regulations (see page 1256), e.g. trichlorotrifluoroethane and TCA instead of MEK, for cleaning preassembly and subassembly components prior to bonding.[28a]

Other examples include,[28a]

- Use of phosphate-based corrosion inhibitors instead of chromates in water cooling towers.
- Use of plastic media blasting in place of solvent stripping for paint removal.
- Replace the use of solvents by detergents or treatment via chemical reactions (e.g. reacting chelating agents with soils to form soluble complexes).
- Use of a terpene–water emulsion instead of chlorofluorocarbons in certain electronic degreasing operations.

Waste may also be reduced by proper inventory management, i.e. by preparing an inventory of all raw materials and only purchasing the minimum amounts required so avoiding the need to dispose of unused raw materials as hazardous waste.[28a] Considerations include,

- Avoidance, when possible, of the expiry of the shelf-life of a substance before it is all consumed. A 'first-in, first-out' programme will assist in this.
- Avoidance of purchasing materials no longer in use.
- Purchasing materials in refillable containers (e.g. to avoid excessive waste-water generation in container washing and the disposal of unwanted containers).
- Purchasing materials in the container size which is most compatible with normal use (e.g. to avoid degradation of 'opened' but unused material resulting in a disposal requirement).

Process modifications can often reduce waste generation by more efficient use of raw materials, by reducing the quantities of reject/off-specification products, and by reducing inherent 'process losses'. Examples are given in Table 22.11.[23]

Experience suggests that once plant has been installed and is operational opportunities for significant modification are likely to be limited.

> Aluminium is obtained by electrolysis of aluminium fluoride/fluorspar in carbon-lined pots. Fluoride is liberated in both particulate and gaseous form. Emissions from pots are contained and extracted to bag filters prior to exhausting to atmosphere via 120 m stacks. The bag collectors were effective for removing particulate matter from discharges but ineffective for gaseous fluoride. Considerable modifications were required including incorporation of

Table 22.11 Examples of process modification for waste reduction[23]

Process step	Technique
Chemical reaction	Optimise reaction variables and improve process controls.
	Optimise reactant-addition method.
	Eliminate use of toxic catalysts.
	Improve reactor design.
Filtration and washing	Eliminate or reduce use of filter aids and disposal filters.
	Drain filter before opening.
	Use countercurrent washing.
	Recycle spent washwater.
	Maximise sludge dewatering.
Parts cleaning	Enclose all solvent cleaning units.
	Use refrigerated freeboard on vapour degreaser units.
	Improve parts draining before and after cleaning.
	Use mechanical cleaning devices.
	Use plastic-bead blasting.
Surface finishing	Prolong process bath life by removing contaminants.
	Redesign part racks to reduce drag-out.
	Reuse rinse water.
	Install spray or fog nozzle rinse systems.
	Properly design and operate all rinse tanks.
	Install drag-out recovery tanks.
	Install rinse water flow control valves.
	Intall drip racks and drainboards.

dry scrubbing using flowing alumina powder to absorb fluoride. Pilot studies indicated a 30-fold reduction in gaseous fluoride emissions.[24]

Thus pollution problems need to be considered at the design stage of projects and risks from novel processes should be carefully assessed at the laboratory or pilot-plant stage to avoid difficulties on scale-up (Chapter 20). Technically and economically it is often preferable to reduce the generation of waste than to manage it.[25]

The Threshold Limit Value for toluene in air was 100 ppm for which the corresponding equilibrium concentration in water, explained on page 81, is 2.0 mg/l at 25 °C. Limits on toluene concentrations in trade effluent should ensure a concentration less than this, possibly resulting in a Consent Limit of <5.0 mg/l compared with the solubility of toluene in water of 516 mg/l at 20 °C. Thus elimination of contamination may be a more practicable approach than toluene removal.[7] (The OEL is now 50 ppm as an 8 h TWA value.)

As mentioned on page 81 the concentration of a solute in a solvent near the vapour–liquid interface can be calculated for dilute solutions using Henry's Law:

$$K_i = y_i/x_i = H_i/p_{\text{total}}$$

where

y_i/x_i = mole fraction of component i in the vapour phase over its mole fraction in the liquid phase
H_i = Henry's constant (Table 22.12).

Table 22.12 Henry's Constant for selected organic compounds[26]

Compound	Henry's Constant (atm per mole fraction)
Acetone	2376
Benzene	3092
Biphenyl	4743
Butane	5095
Carbon dioxide	1216
Carbon disulphide	1067
Carbon monoxide	6343
Carbon tetrachloride	1634
Ethane	2679
Hexane	7173
Toluene	3531
Styrene	1463

For water containing 10 ppm of CCl_4 on a molar basis, at 1 atm pressure the level of CCl_4 in the atmosphere is calculated as follows. From Table 22.12 CCl_4 has an H value of 1634, therefore

$$K_{CCl_4} = 1634/1 = 1634$$
$$y_{CCl_4} = K_{CCl_4} \times x_{CCl_4} = 1634 \times 10 = 16\,340 \text{ ppm}.$$

In much the same way as it is safest to substitute toxic materials by less hazardous ones to reduce the risk when handling chemicals in industry (Chapter 28), substitution can also reduce risk from hazardous waste as illustrated by Table 22.13.[27] Recovery can also provide an attractive option using technologies exemplified by Table 22.14.[27]

Solid waste

Solid waste is generated in vast quantities by industry. Much of it is recycled or dumped as illustrated by Fig. 22.4.

Industrial waste dumping at sea was ended in the UK in 1992 and the dumping of sewage sludge is to be terminated in 1998.[28]

In the UK landfill is the predominant disposal route for chemical wastes by co-disposal with household refuse.[29] Since the objective is for attenuation processes within the landfill to produce relatively non-polluting products, the composition of the chemical waste must be known before acceptance at a site; therefore a strict monitoring regime is required. Some

Table 22.13 End-product substitutes for reduction of hazardous waste

Product	Use	Ratio of waste* to original product	Available substitute	Ratio of waste* to substitute product
Asbestos	Pipe	1.09	Iron Clay PVC	0.1 phenol, cyanides 0.05 fluorides 0.04 VCM manufacture + 1.0 PVC pipe
	Friction products (brake linings)	1.0+ manufacturing waste	Glass fibre Steel wool Mineral wools Carbon fibre Sintered metals Cement	0
	Insulation	1.0+ manufacturing	Cellulose fibre	0.2
PCBs	Electrical transformers	1.0	Oil-filled transformers Open-air-cooled transformers	0 0
Cadmium	Electroplating	0.29	Zinc electroplating	0.06
Creosote treated wood	Piling		Concrete, steel	0.0 (reduced hazard)
Chlorofluorocarbons	Industrial solvents	70/81 = 0.9	Methyl chloroform; methylene chloride	0.9 (reduced hazard)
DDT	Pesticide	1.0+ manufacturing waste	Other chemical pesticides	(reduced hazard) 1.0+ manufacturing waste

* Quantity of hazardous waste generated per unit of product.

criticism has, however, been made of the variation in licensing efficiency and in the monitoring of licensed facilities in the UK.[30] Too many licences were considered not to specify adequately what a site might or might not accept and fencing was allowed to vary far more than was acceptable or safe.

The deposition of incompatible waste must be avoided to prevent undesirable reactions (page 495), e.g.[29]

- exothermic reactions resulting in fires or explosions
- generation of toxic gases, e.g. arsine, chlorine, hydrogen sulphide (refer to Table 14.17, page 847)
- generation of flammable gases, e.g. acetylene, hydrogen, methane
- dissolution of toxic compounds including heavy metals, e.g. complexing agents, chelates

Table 22.14 Technologies currently used for recovery of materials

Technology/description	Stage of development	Economics	Types of waste streams	Separation efficiency*	Industrial applications
Physical separation:					
Centrifugation: Spinning of liquids and centrifugal force causes separation by different densities	Practised commercially for small-scale systems	Competitive with filtration	Liquid/liquid or liquid/solid separation, i.e. oil/water; resins; pigments from lacquers	Fairly high (90%)	Paints
Filtration: Collection devices, e.g. screens, cloth, or other; liquid passes and solids are retained on porous media	Commonly used	Labour intensive; relatively inexpensive; energy required for pumping	Aqueous solutions with finely divided solids; gelatinous sludge	Good for relatively large particles	Tannery water
Flotation: Air bubbled through liquid to collect finely divided solids that rise to the surface with the bubbles	Commercial application	Relatively inexpensive	Aqueous solutions with finely divided solids	Good for finely divided solids	Refinery (oil/water mixtures); paper waste; mineral industry
Flocculation: Agent added to aggregate solids together which are easily settled	Commercial practice	Relatively inexpensive	Aqueous solutions with finely divided solids	Good for finely divided solids	Refinery; paper waste; mine industry
Gravity settling: Tanks, ponds provide hold-up time allowing solids to settle; grease skimmed to overflow to another vessel	Commonly used in wastewater treatment	Relatively inexpensive; dependent on particle size and settling rate	Slurries with separate phase solids, such as metal hydroxide	Limited to solids (large particles) that settle quickly (<2 hours)	Industrial wastewater treatment first step

Environmental pollution control

Process	Description	Status	Limitations	Application	Effectiveness	Examples
Component separation:						
Carbon/resin adsorption: Dissolved materials selectively adsorbed in carbon or resins. Adsorbents must be regenerated		Proven for thermal regeneration of carbon; less practical for recovery of adsorbate	Relatively costly thermal regeneration; energy intensive	Organics/inorganics from aqueous solutions with low concentrations, i.e. phenols	Good, overall effectiveness dependent on regeneration method	Phenolics
Distillation: Successfully boiling off of materials at different temperatures (based on different volatilities)		Commercial practice	Energy intensive	Organic liquids	Very high separations achievable (99+% concentrations) of several components	Solvent separations; chemical and petroleum industry
Electrolysis: Separation of positively/negatively charged materials by application of electric current		Commercial technology; not applied to recovery of hazardous materials	Dependent on concentrations	Heavy metals; ions from aqueous solutions; copper recovery	Good	Metal plating
Evaporation: Solvent recovery by boiling off the solvent		Commercial practice in many industries	Energy intensive	Organic/inorganic aqueous streams; slurries, sludges, i.e. caustic soda	Very high separations of single, evaporated component achievable	Rinse waters from metal-plating waste
Ion exchange: Waste stream passed through resin bed, ionic materials selectively removed by resins similar to resin adsorption. Ionic exchange materials must be regenerated		Not common for waste water	Relatively high costs	Heavy metals aqueous solutions; cyanide removed	Fairly high	Metal-plating solutions

Table 22.14 continued

Technology/description	Stage of development	Economics	Types of waste streams	Separation efficiency*	Industrial applications
Reverse osmosis: Separation of dissolved materials from liquid through a membrane	Not common: growing number of applications as secondary treatment process, e.g. metal-plating pharmaceuticals	Relatively high costs	Heavy metals; organics; inorganic aqueous solutions	Good for concentrations less than 300 ppm	Not used industrially
Solvent extraction: Solvent used to selectively dissolve solid or extract liquid from waste	Commonly used in industrial processing	Relatively high costs of solvent	Organic liquids, phenols, acids	Fairly high loss of solvent may contribute to hazardous waste problem	Recovery of dyes
Ultrafiltration: Separation of molecules by size using membrane	Some commercial application	Relatively high costs	Heavy metal aqueous solutions	Fairly high	Metal-coating applications
Chemical transformations: Chemical dechlorination: Reagents selectively attack carbon–chlorine bonds	Common	Moderately expensive	PCB-contaminated oils	High	Transformer fluids
Chlorinolysis: Pyrolysis in atmosphere of excess chlorine	Commercially used in western Germany	Insufficient US market for carbon tetrachloride	Chlorocarbon waste	Good	Carbon tetrachloride manufacturing
Electrodialysis: Separation based on differential rates of diffusion through membranes. Electrical current applied to enhance ionic movement	Commercial technology, not commercial for hazardous material recovery	Moderately expensive	Separation/ concentration of ions from aqueous streams; application to chromium recovery	Fairly high	Separation of acids and metallic solutions

Precipitation: Chemical reaction causes formation of solids which settle	Common	Relatively high costs	Lime slurries	Good	Metal-plating wastewater treatment
Reduction: Oxidative state of chemical changed through chemical reaction	Commercially applied to chromium; may need additional treatment	Inexpensive	Metals, mercury in dilute streams	Good	Chrome-plating solutions and tanning operations
Thermal oxidation: Thermal conversion of components	Extensively practised	Relatively high costs	Chlorinated organic liquids, silver	Fairy high	Recovery of sulphur, HCl

* Good implies 50 to 80% efficiency, fairly high implies 80%, and very high implies 90%.

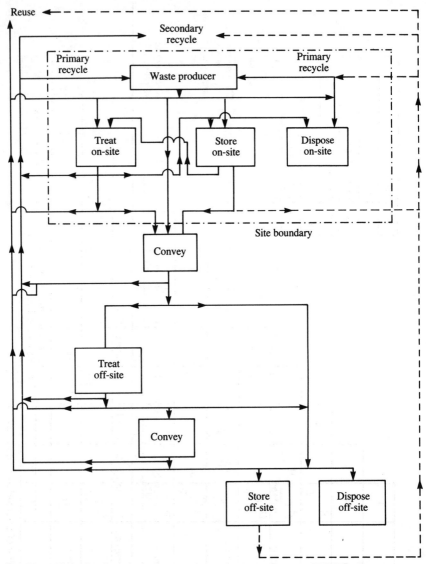

Fig. 22.4 Three basic techniques – treatment, storage, conveyance – to effect recycle or disposal of 'solid' waste either on- or off-site

Figure 22.5 provides an indication of the compatibility of selected hazardous wastes.[31]

Decomposition of household refuse inevitably generates landfill gas, typically with 60% methane and 40% carbon dioxide.[32] Possible hazards from this are

- Lateral migration underground resulting in accumulation in basements with the possibility of inadvertent ignition (relevant advice on the con-

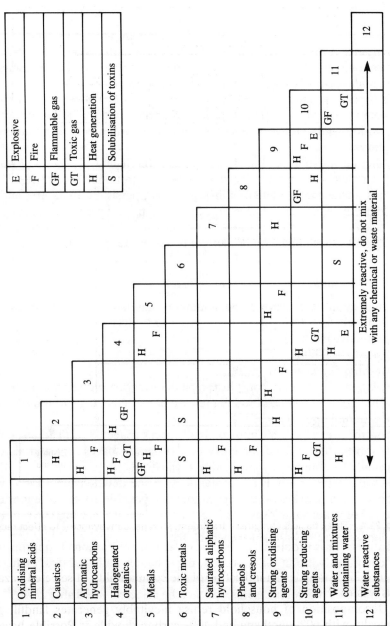

Fig. 22.5 Compatibility of selected hazardous wastes

struction of buildings adjacent to, or on, a site is given in refs. 31 and 33).
- Asphyxiation of workers in cable trenches or sewers.
- Tip fires.
- Accumulation and subsequent ignition beneath parked vehicles.

A nuisance problem may also arise from trace quantities of chemicals, e.g. mercaptans and hydrogen sulphide, present in the gas. Recommended design and operation are given in refs. 31 and 33.

It is now recognised that new landfill sites require location on impermeable strata, or be sealed with an appropriate membrane before waste deposition, and that leachate requires control in order to avoid pollution of acquifers.

Whether for solids or liquid/slurry wastes the technical aspects of waste disposal dump construction need careful consideration (Table 22.15). Hazardous wastes may be lethal, non-degradable, persistent in nature, and/or biologically magnified; hence the need for prior treatment chemically, physically or biologically using techniques listed in Table 22.16. Table 22.17 provides comparison of thermal treatment technologies.[27]

Precautions and control of pollution

During initial site surveys the minimum standard of safety protection should include[34]
- protective clothing
- basic first-aid box
- washing and changing facilities
- a flammable gas detector if historical data suggest old mine workings or other possible source of sudden release of methane, e.g. prior use of site for landfill operations
- control over any entry into confined spaces (page 981). (There should, however, be no need for entry into confined spaces as materials can usually be obtained by mechanical means.)

If soil samples do indicate contamination the detailed precautions will depend upon a risk assessment, to identify possible routes of exposure. The pH of water samples is one important indicator; very high or very low values would highlight a risk on skin contact. Dry ground conditions will exacerbate dust problems. A high standard of precautions is appropriate for chemicals with an 'Sk' notation, e.g. phenols and coal tars, or highly toxic by ingestion, e.g. lead or chromium compounds.

A range of control measures will then be selected: see Table 22.18.[34]

Gaseous waste

Airborne waste includes gases, dusts, mists and fumes. General treatment methods are listed in Table 22.19. Oxides of nitrogen are usually

Table 22.15 Engineered components of landfills: their function and potential causes of failure

Function	Potential causes of failure
Cover: To prevent infiltration of precipitation into landfill cells. The cover is constructed with low permeability synthetic and/or clay material and with graded slopes to enhance the diversion of water.	After maintenance ends, cap integrity can be threatened by desiccation, deep rooted vegetation, animals, and human activity. Wet/dry and freeze/thaw cycles, causing cracking and increased infiltration. Erosion; causing exposure of cover material to sunlight, which can cause polymeric liners to shrink, break, or become brittle. Differential settling of the cover, caused by shifting, settling, or release of the landfill contents over time. Settling can cause cracking or localised depressions in the cover, allowing ponding and increased infiltration.
Leachate collection and recovery system: To reduce hydrostatic pressure on the bottom liner, and reduce the potential for flow or leachate through the liner. Leachate is collected from the bottom of the landfill cells or trenches through a series of connected drainage pipes buried within a permeable drainage layer. The collection leachate is raised to the surface by a mechanical pump.	Clogging of drainage layers or collection pipes. Crushing of collection pipes due to weight of overlying waste. Pump failures.
Bottom liner: To reduce the rate of leachate migration to the subsoil.	Faulty installation, damage during or after installation. Deformation and creep of the liner on the sloping walls of the landfill. Differential settling, most likely where landfill is poorly sited or subgrade is faulty. Structural failure of the liner in response to hydrostatic pressure. Degradation of liner material resulting from high strength chemical leachate or microbial action. Swelling of polymeric liners, resulting in loss of strength and puncture resistance. Chemical extraction of plasticisers from polymer liners.

minimised in industrial gaseous emissions by combustion modifications and/or flue gas treatment technology.[35] (However control of emissions from vehicles, accounting for >40% of NO_x emissions in the US, requires some combination of engine modification and the use of removal from exhaust gas by catalytic converters.) Flue gas desulphurisation, e.g. using an alkaline solution/slurry is the main method of controlling sulphur oxide emissions. The primary control technologies for volatile organic com-

Table 22.16 Treatments for solid waste

Treatment	Examples/comments
Chemical	
Calcination	Gypsum
Chlorination	Tin removal
Cooking	Inedible offal
Froth flotation	Coal recovery
	Glass
Hydrolysis	Household refuse
	Vegetable waste
Incineration	Household refuse
Leaching	Gives an aqueous solution which may be treated as a liquid waste
Oxidation	Weathering
Pyrolysis	Household refuse
	Polystyrene
Sintering	Colliery spoil
	Millscale
Physical	
Adhesion	Household refuse
Agglomeration	Pulverised fuel ash
Ballistic separation	Household refuse
Baling	Cans
Centrifugation	Animal oil separation
	De-oiling swarf
Classification (air)	Household refuse
Classification (wet)	Plastic
Comminution	Mining wastes
	Cars
Compaction	Household refuse
Dewatering	Sewage sludge
Dissolution	Forms a liquid waste which may be treated
Drying	Filter cake
Electrostatic separation	Household refuse
Foaming	Slag
Freezing	Meat products
Granulation	Slag
Impalement	Household refuse
Jigging	Household refuse
Magnetic treatment	Iron removal from slag
	Household refuse
Melting	Selective non-ferrous metal recovery
Pelletisation	Iron and steel fines
Pulverisation	Household refuse
	Swarf
Quenching	Incinerator residues
Screening	Clinker
Settlement	China clay wastes
Shape separation	Household refuse
Sliding separation	Household refuse
Biological	
Anaerobic digestion	Farm waste
Bacterial leaching	Low-grade copper ore
Composting	Household refuse
Degradation	Plastic
Fermentation	Requires solution or dilute slurry – see liquid waste

Table 22.17 Comparison of thermal treatment technologies for hazard reduction

Advantages of design features	Disadvantages of design features
Incinerator designs	
Liquid injection incineration:	
Can be designed to burn a wide range of pumpable waste. Often used in conjunction with other incinerator systems as a secondary afterburner for combustion of volatilised constituents. Hot refractory minimises cool boundary layer at walls. HCl recovery possible.	Limited to destruction of pumpable waste (viscosity of less than 10 000 SSI). Usually designed to burn specific waste streams. Smaller units sometimes have problems with clogging of injection nozzle.
Rotary kilns:	
Can accommodate great variety of waste feeds: solids, sludges, liquids, some bulk waste contained in fibre drums. Rotation of combustion chamber enhances mixing of waste by exposing fresh surfaces for oxidation.	Rotary kilns are expensive. Economy of scale means regional locations, thus waste must be transported, increasing spill risks.
Cement kilns:	
Attractive for destruction of harder-to-burn waste, due to very high residence times, good mixing, and high temperatures. Alkaline environment neutralises chlorine.	Burning of chlorinated waste limited by operating requirements, and appears to increase particulate generation. Could require retrofitting of pollution control equipment and of instrumentation for monitoring to bring existing facilities to comparable level. Ash may be hazardous residue.
Boilers (usually a liquid injection design):	
Energy value recovery, fuel conservation. Availability on sites of waste generators reduces spill risks during hauling.	Cool gas layer at walls results from heat removal. This contains design to high efficiency combustion within the flame zone. Nozzle maintenance and waste feed stability can be critical. Where HCl is recovered, high temperatures must be avoided. (High temperatures are good for DRE.) Metal parts corrode where halogenated wastes are burned.
Applications	
Multiple hearth:	
Passage of waste onto progressively hotter hearths can provide for long residence times for sludges. Design provides good fuel efficiency. Able to handle wide variety of sludges.	Tiered hearths usually have some relatively cold spots which inhibit even and complete combustion. Opportunity for some gas to short-circuit and escape without adequate residence time. Not suitable for waste streams which produce fusible ash when combusted. Units have high maintenance requirements due to moving parts in high-temperature zone.
Fluidised-bed incinerators:	
Turbulence of bed enhances uniform heat transfer and combustion of waste. Mass of bed is large relative to the mass of injected waste.	Limited capacity in service. Large economy of scale.

Table 22.17 continued

Advantages of design features	Disadvantages of design features
At-sea incineration: shipboard (usually liquid injection incinerator):	
Minimum scrubbing of exhaust gases required by regulations on assumption that ocean water provides sufficient neutralisation and dilution. Could provide economic advantages over land-based incineration methods. Incineration occurs away from human populations. Shipboard incinerators have greater combustion rates, e.g. 10 tonnes/h.	Not suitable for wastes that are shock-sensitive, capable of spontaneous combustion, or chemically or thermally unstable, due to the extra handling and hazard of shipboard environment. Potential for accidental release of waste held in storage (capacities vary from 4000 to 8000 tonnes).
At-sea incineration: oil drilling platform based:	
Same as above, except relative stability of platform reduces some of the complexity in designing to accommodate rolling motion of the ship.	Requires development of storage facilities. Potential for accidental release of waste held in storage.
Pyrolysis:	
Air pollution control need minimum: air-starved combustion avoids volatilisation of any inorganic compounds. These and heavy metals go into insoluble solid char. Potentially high capacity.	Greater potential for PIC formation. For some waste produce a tar which is hard to dispose of. Potentially high fuel maintenance cost. Waste-specific designs only.

pounds include flares (both catalytic and thermal incineration), and carbon adsorption.[35] Particulate matter can be removed by filtration or using cyclones, etc.

Odour control

Within the UK the Environmental Protection Act includes odour emissions as a statutory nuisance. The introduction of EC odour control directives is planned for 1996.

Odours cannot be completely removed; what is required is for the impact of an odour to be reduced to a level at which it does not cause a problem. However, assessment of the acceptability of an odour is complex and involves both opinion and human judgement. The setting-up of testing panels, and the application of statistical techniques to the results, have been used to improve objectivity.[36]

One measure of odour is the Threshold Odour Number or TON. This is determined by a panel of volunteers who sniff two air streams, one odour-free and one the odorous air stream diluted with odour-free air; when 50% of the panel cannot detect any difference in smell between the streams, the dilution ratio of the odorous air is termed the TON. Further data on odour thresholds are provided in Chapter 21.

The volume of odorous air released is the second factor affecting accept-

Table 22.18 Precautions and control measures on contaminated land

Personal protection
- Protective clothing should include safety wellingtons, hand protection (impervious gloves with/without gauntlets depending on the work) and overalls. Eye protection or respiratory protection may be required on some sites.
- A good standard of hygiene should be achieved. The level of risk will determine the extent of the washing facilities required (page 319).
- A bootwash should be located immediately outside the entrance to the hygiene unit. This should include running water and fixed or hand brushes to facilitate removal of contaminated soils.
- A canteen or messroom should be provided on the clean side of the site, such that it can only be entered via the hygiene unit. No eating or smoking should be permitted in the dirty area.
- Air monitoring may be necessary to assess individual exposures. Health surveillance may also be advisable for some contaminants, as discussed on page 1625.

Organisation and equipment
- A 'dirty area' should be designated, based on a site survey. This should be clearly fenced off, e.g. with cleft chestnut pale fencing, and appropriate warning signs posted. Access to/ from this area should be solely via a hygiene facility for personnel or via a controlled gate with a wheel-wash for vehicles. The main site office should be situated in a clean area adjacent to the site.
- The complete site should be fenced. The fence and gate should be ⩾2 m high and difficult to climb over.
- Precautions, such as water sprays to dampen down dust, should be taken as appropriate. More stringent precautions will be essential if there is a risk of airborne asbestos contamination (refer to page 1682).
- Positive-pressure cabs should be considered to prevent entry of contaminants. Otherwise cabs of excavators, earthmovers and other vehicles used on site should be vacuumed clean at the end of each day to avoid the accumulation of contaminants.

Off-site movements
- A wheel-wash facility with a high-pressure hose should be provided. When a vehicle is to travel on public roads, vehicle washing should include the underbody and wheel arches. (If contamination is significant it will be necessary to collect water prior to disposal.)
- If contaminated materials are to be removed from site it will be necessary to provide sheeting for open lorries or skips. All waste must be disposed of to a suitable licensed facility. Suitable records must be kept as discussed on page 843.

---

ability to the public. Thus the likelihood of complaints has been related by the empirical formula

$$C = (2.2 \times E \times T)^{0.6}$$

where
C = complaint radius (m)
E = emission rate (m³/s)
T = Threshold Odour Number

Whilst only a guide, this identifies the two mains strategies for the reduction of odour nuisance, i.e. reducing the volume of air to be treated and reducing the strength of the odour.

A reduction in air volume can, in the simplest case, be achieved by covering the odour source and extracting the air from that volume only.

Table 22.19 Treatments for gaseous waste

General methods	Alternatives
Centrifugal techniques	Centrifuge
	Cyclone
Coalescence	
Condensation	
Destruction	Chemical reaction
	Incineration
Direct recycle	
Dispersion	
Electrostatic precipitation	
Filtration	Needle bonded fabric
	Reverse jet
	Reverse pressure
	Shaker type
Gravity settlement	
Total enclosure	
Wet scrubbing	Absorption tower
	Fluidised bed scrubber
	Impingement scrubber
	Irrigated target scrubber
	Pressure spray scrubber
	Rotary scrubber
	Self-induced spray scrubber
	Spray tower
	Venturi scrubber

Pollutant	Dry methods	Wet methods
Gas	Adsorption	Wet scrubbing (absorption)
	Destruction	
	Direct recycling	
	Dispersion	
	Total enclosure	
Liquid	Destruction	Centrifugal techniques
		Coalescence
		Destruction
		Gravity settlement
		Total enclosure
		Wet scrubbing
Solid	Centrifugal techniques	Electrostatic precipitation
	Destruction	Wet scrubbing
	Direct recycling	
	Electrostatic precipitation	
	Filtration	
	Gravity settlement	
	Total enclosure	

Alternatively, the air could be circulated in series, e.g. by taking air from the least odorous volumes and using this to sweep air from the most odourous volumes.

The strength of the odour can be reduced by treatment of the air stream.[36]

For an airstream with a TON of 10^5 and an air flowrate of 200 m³/h the complaint radius C is by calculation approximately 280 m. Treatment to reduce the strength of the odour would reduce this to

Per cent odour removal	Complaint radius C (m)
95	46
97	35
99	17

This strategy can be applied to bring the complaint radius within the factory site.

A combination of covering, to reduce the air volume needing treatment, and treatment to reduce odour strength is commonly used.

The two basic methods for odour treatment are adsorption, e.g. onto activated carbon or activated aluminium oxide, or chemical modification to inoffensive substances.

Odour control by activated carbon adsorption is used in a wide range of environments. It may serve as a passive vent filter, e.g. in which air exhausted from a vessel is passed through a carbon pack, or within an active system incorporating blower-fans. Humidities in excess of 85 per cent can limit the carbon's performance. A combination of types of alumina impregnated with reagents, e.g. potassium permanganate, uses both adsorption and oxidation to treat hydrogen sulphide-contaminated air. This system can cope with humidities in the range 10–95 per cent.[36]

A contaminant may be oxidised or neutralised to a less odorous substance by wet ozonation, wet chemical scrubbing or biological conversion.

Wet ozonation involves the use of ozone in a liquid phase to oxidise odours that are absorbed into it. A countercurrent tower is used in which the odorous gas passes up through a spray of an ozonated liquid. Oxygen and oxidation by-products result. This process is widely used in sewage treatment and has found some applications in fishmeal plants and foundries. Since ozone is a powerful oxidising agent, and highly toxic, careful precautions are obviously required to eliminate escapes from the unit.

Chemical scrubbing utilises oxidising or neutralising agents appropriate to the odorous materials. It is generally necessary to use a two-stage unit; an alkaline stage serves to remove hydrogen sulphide and organic acids and an acid stage removes compounds such as mercaptans and amines. Careful balancing of the reagents to the odorous material is required and the highly alkaline and highly acidic residues generated need treatment prior to discharge. This process is the most common for treating large volumes of air, e.g. at effluent treatment works.

Biological systems have yielded 95–99 per cent removal efficiencies in numerous applications. Naturally occurring sulphur-oxidising bacteria are used to remove hydrogen sulphide; different species are used for the removal of organic odour compounds. The equipment may comprise a bioscrubber, in which a microbiological film is immobilised on a gas/liquid transfer media, or a biofilter, which incorporates media with a very high surface area to support the bacteria but with low pressure drop characteristics. System choice is governed by the composition and strength of the odour, but applications have included pharmaceutical and fine chemical manufacturing units.[36]

Liquid waste

Aqueous waste is often discharged direct to sewers or rivers, but hazardous effluent requires prior treatment by non-biological or biological methods as determined by the biodegradability of the aqueous waste. General methods for treating liquid effluent are given in Table 22.20. However, as exemplified there, many still result in waste products for disposal or further treatment.

Non-biological processes

Since non-biological methods tend to be waste-specific, effluent must be characterised in terms of flow rate and composition prior to treatment. Methods for the removal of pollutants from aqueous effluents are summarised in Table 22.21. Treatment processes tend to rely on physical techniques such as filtration and adsorption (e.g. the general scheme for removal of soluble metals by precipitation/filtration is shown in Fig. 22.6, although in practice effluent may need adjustment for pH and metal content) or on reaction with a range of chemicals to convert toxic pollutants into less toxic substances, e.g.

- neutralisation (e.g. a typical scheme for a comprehensive pH adjustment/neutralisation system is shown in Fig. 22.7);
- reduction of pollutants, e.g. liquid waste containing chlorine can be air-stripped or reduced by a variety of reducing agents, thus

$$Cl_2 + SO_2 + 2H_2O \rightarrow 2HCl + H_2SO_4$$
$$Cl_2 + Na_2SO_3 + H_2O \rightarrow 2NaCl + H_2SO_4$$
$$Cl_2 + NaHSO_3 + H_2O \rightarrow NaCl + HCl + H_2SO_4$$
$$2Cl_2 + Na_2S_2O_5 + 3H_2O \rightarrow 2NaCl + 2HCl + 2H_2SO_4$$
$$Cl_2 + 3FeSO_4 \rightarrow FeCl_3 + Fe_2(SO_4)_3$$

- oxidation, e.g. for cyanide-containing effluent

$$NaCN + Cl_2 \rightarrow CNCl + NaCl \xrightarrow{2NaOH} NaCNO + 2NaCl + H_2O$$

Ozone and chlorine have widespread use as oxidants in aqueous pollution treatment processes, but with the formation of hazardous by-products, e.g. haloforms, epoxides and chlorates, their use has been questioned for certain treatment processes;[37]

Particular concern has arisen over the synthesis of organohalides upon

Table 22.20 Treatments for liquid effluent

Treatment	Example	Typical waste products
Chemical		
Cementation	Copper recovery	Chemical sludges for dewatering/disposal
Chlorination	Cyanide oxidation	Viscous oil tar reuse, special disposal/incineration
Coagulation – see Flocculation		
Demulsification	Soluble oil recovery	
Electrolytic processes	Metal recovery	
Flocculation	Sewage treatment	
Hydrolysis	Cellulose waste	
Incineration	Waste oils	Atmospheric emissions for control
Ion exchange	Metal recovery	
Leaching	Metal-bearing sludges	
Neutralisation	Waste acid	Chemical effluent but of controlled pH
Oxidation	Phenol removal	Chemical sludges
Ozonisation	Cyanide oxidation	
Precipitation	Metals	Chemical sludges for thickening/dewatering/disposal
Reduction	Hexavalent chromium	Chemical sludges
Thermal decomposition	Recycling hydrochloric acid	
Physical		
Adsorption	Removal of volatile organics	Contaminated carbon
Centrifugation		
Cooling	Water reuse	Humidified air
Crystallisation	Recovery of inorganic salts	
Desorption – see Absorption		
Dewatering – see Filtration		
Dialysis	Desalination	
Distillation	Solvent recovery	
Drying	Pig manure	High concentration salts
Electrodialysis	Desalination	
Evaporation		
Filtration	Sewage sludge	Wet solids for disposal
Flotation	Dairy wastes	Dewatered solids plus precoat Thickened sludge and supernatant stream

Table 22.20 continued

Treatment	Example	Typical waste products
Foam fractionation	Metal separation	
Fractionation – see Distillation		
Freezing	Desalination	
Heating	Demulsification	
Phase separation	Oily wastes	Viscous oil
Reverse osmosis	Desalination	
Screening	Sewage	Coarse wet solids may require dewatering
Sedimentation	Suspended solids removal	Organic sludge, 2–6% solids
Solvent extraction	Metal recovery	
Stripping	Ammonia removal	Contaminated air stream; may be adsorbed or secondary air to boilers
Ultrafiltration – see Dialysis, Reverse osmosis		
Biological		
Activated sludge	Sewage	Sludge
Anaerobic digestion	Food wastes	Biogas and high-strength sludge
Chemical production	Ethanol	
Disinfection	Sewage plant effluent	
High-rate filtration	Phenol removal	Sludge, odours
Oxidation – see Activated sludge, High-rate filtration, Trickling filter		
Reduction – see Anaerobic digestion		
Single cell protein production	Organic waste	
Trickling filter	Sewage	Sludge

Table 22.21 Methods available for the removal of pollutants from aqueous effluents

Methods	Aim	Application (examples)
A. Suspended or immiscible pollutants		
Centrifugation	Concentration of solids	In textile industry for recovery of wool greases and oil Removal of fat from bone-rendering operations Dewatering of sewage sludges and sludges in chemical industry
Coagulation or flocculation	Facilitation of the removal of colloidal particles (less than 10 μm in size)	Commonly used in removing metals from metal industrial wastewaters, and in the reduction of fibre in waste from the paper industry
Dewatering	Reduction of the water content of a liquid–solid system	Dewatering of sludges from chemical industry. Dewatering is a technique that reduces the water content of a liquid/solid system
Emulsion breaking	Separation of free oil from emulsions	Any aqueous waste containing immiscible organic material referred to as 'oily waste'
Filtration	Concentration of fine solids	Final polishing and sludge dewatering in the chemical and metal processing industries
Flotation	Removal of low specific gravity solids and liquids	Separation of oil, grease and solids in the chemical and food industries
Gravity settling/separation	Separation of solid particles from suspension	Solid waste recovery in non-ferrous metal effluents Oil in water
Polishing	Removal of suspended particles and/or dissolved material Final purification of water	Sometimes necessary to produce an effluent that is particularly free of suspended or dissolved material, for discharge directly to a watercourse or for water recycling
Screening	Removal of coarse solids	Vegetable canneries, paper mills, sewage treatment works
Sedimentation	Removal of settleable solids	Separation of inorganic solids, in ore extraction, coal and clay production
B. Dissolved pollutants		
Adsorption	Concentration and removal of trace impurities	Pesticide manufacture, dyestuffs removal
Aeration	Increase of dissolved oxygen content of effluent in biological treatment processes	Sewage treatment, food wastes

Table 22.21 continued

Methods	Aim	Application (examples)
Chemical reaction	Most effluent treatment involves chemical reaction at some stage	Reaction may be used to give a chemical precipitate, to destroy a toxic or noxious material, etc., but generalisations are not possible, and problems may need a specific solution
Dialysis and electrodialysis	Removal of a dissolved material	Recovery of nitric and hydrofluoric acid from pickling liquors Desalting of brackish waters (not sea water)
Distillation	Separation of a liquid mixture	Solvent reclamation
Drying	Final removal of water from a material	Sewage sludge
Electrolysis (electrowinning)	Recovery of certain materials from solution	Metal recovery from acid or alkali wastes
Evaporation	Concentration of aqueous solutions	Metal-processing wastes
Freezing (and thawing)	Concentration of liquids and sludges	Recovery of pickle liquor and non-ferrous metals
Heating	Reduction of viscosity of liquid under treatment	Emulsion breaking. Also has a range of other applications, e.g. accelerating specific chemical reactions, e.g. neutralisation
Incineration	Burning of waste, thermal breakdown of hazardous wastes	Applicable to effluents containing toxic or persistent materials (not asbestos or heavy metals)
Ion-exchange	Separation and concentration	Metal-processing wastes
Neutralisation	Adjusting the pH of waste	Acid wastes, alkali wastes
Oxidation/reduction	Removal of dissolved pollutants	Electroplating industry
pH adjustment	See Neutralisation	Acid wastes, alkali wastes
Precipitation (then as A processes above)	Separation of dissolved materials	Metal salts in acid or alkaline solutions See also Coagulation and Flocculation
Reverse osmosis	Separation of dissolved solids	Desalination of process and wash water
Solvent extraction	Recovery of valuable materials	Coal carbonising and plastics manufacture
C. Biological treatment		
Percolating filter	Use of micro-organisms to reduce the organic nature of wastes	High BOD wastes produced by the food industry, the treatment of sewage, etc.
Activated sludge	Use of micro-organisms to reduce the organic nature of wastes	As above
Anaerobic sludge digesters	As above	As above, for particularly strong/concentrated organic wastes

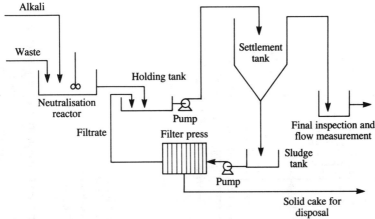

Fig. 22.6 Soluble metal precipitation and removal

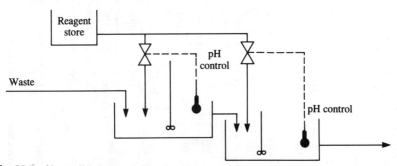

Fig. 22.7 Neutralisation and pH adjustment

chlorination of drinking water.[36b] Thus activated carbon filtration may be required in cities with > 100 ppb of chloroform and other trihalomethane compounds in drinking water.[36c]
- hydrolysis.

An overall effluent treatment scheme is depicted in Fig. 22.8.

Biological processes

Biological treatment processes which rely on micro-organisms to break down the pollutant include biological filters, activated sludge processes and anaerobic sludge digesters. The wastes from these are processed by sedimentation, dewatering by filtration, and centifugation.

Examples of pollution control practice

Many industries, e.g. the pulp paper industries,[37] have taken stock of their sources, nature and level of pollution and the following selected examples

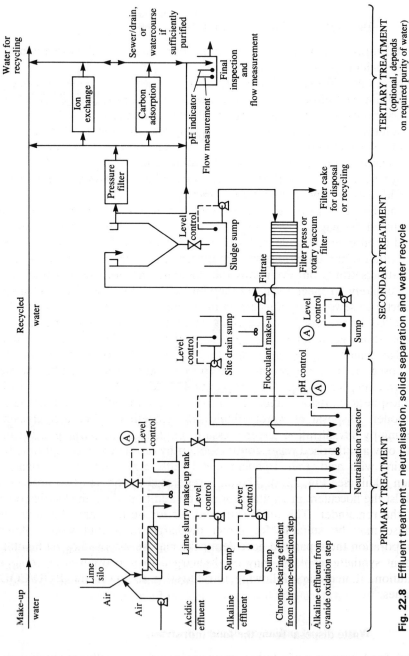

Fig. 22.8 Effluent treatment – neutralisation, solids separation and water recycle

illustrate how the general principles are applied to control pollution for a variety of situations and a range of chemicals.

Disposal of used metalworking fluids

Metalworking fluids are used for lubrication, cooling and swarf removal. They include
- Neat oils – These are based on refined mineral oils. Some heavy-duty variants contain extreme pressure additives comprising organosulphur, organophosphorus and/or organochloro compounds.
- Soluble oils – These are based on highly refined mineral oils and emulsifiers containing biocides, corrosion inhibitors and organosulphur, organophosphorus or organochloro compounds.
- Synthetic coolants – These are aqueous fluids based on polymers, with additives to prevent corrosion, bacterial growth, etc.

The health hazards and precautions associated with mineral oils are discussed on pages 1596 and 1597. During use all metalworking fluids tend to become contaminated with metallic swarf, hydraulic and gear oils, and other tramp impurities and debris. Their life can be extended by, e.g. filtration/settling, but eventually they become unfit for use. Because they are unlikely to meet local consent limits they must not be discharged directly to sewers without pretreatment, and water separators should be installed to prevent leakages from circulation systems entering surface water drainage systems (e.g. to prevent damage to bacteria in activated sludge plants). Spent neat oils tend to be sold for reclamation, or disposed of by incineration or deep-well landfill via authorised waste contractors, or by on-site incineration, taking care to comply with atmospheric emission standards, or by mixing with fuel oil and burning in a boiler. Disposal of water-soluble oils (sometimes after first concentrating by ultrafiltration under pressure) tends to be by burning either on site or via authorised waste contractors, or by on-site chemical splitting by mixing with acid, then trivalent metal salts, e.g. aluminium sulphate), then alkali (see Fig. 22.9). The oil is separated by the first two processes whilst the flocculent precipitate formed on addition of alkali absorbs some of the remainder. The oil is skimmed off for burning and the aqueous phase may be suitable for direct discharge to a foul sewer. After evaporation the separated flocculent cake may be disposed of via landfill. Spent synthetic coolants may be discharged to foul drain after copious dilution, if necessary following ultrafiltration to lower the BOD/COD values.[38]

Waste disposal from the food industries

The food processing industry produces a range of effluent discharges, including organic matter of high BOD. Some are discharged direct to

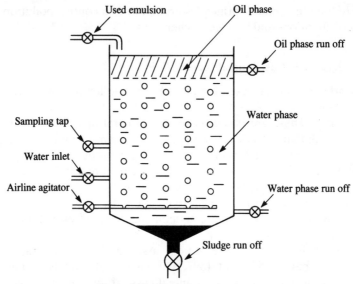

Fig. 22.9 Water-soluble oils disposal process

Table 22.22 Measures aimed at containing emissions

(No priority ranking is intended)

Equalisation of vapour spaces between road or rail tanker and bulk storage tanks.

Two-stage relief valve and bursting discs on vapour spaces, i.e. secondary containment.

Pressure control to minimise nitrogen consumption and associated losses from reactors.

Addition of reagents to reactors via sluice valves.

Low loss vacuum pumps, e.g. dry vacuum pumps, once through oil pumps, cryogenic solvent as pump seal liquid.

Covered basins in waste water treatment plant to contain VOC losses.

Vent collection and ducting from tank farms to central abatement systems.

Vent collection and ducting from reactors to central abatement systems.

Closed transfer systems from reactors to centrifuges to filters to dryers.

Double containment of underground tanks.

---

sewers but many require pretreatment to reduce the pollution load. Canning effluent can be screened to remove peelings, leaves and soil and the alkalinity can be reduced using acid. Trickling filters or honeycombed PVC and activated sludge are effective biological treatment operations which rely on micro-organisms acclimatised to high BOD wastes. Sedimentation is employed to remove solids, and sludges may need dewatering prior to disposal to landfill or incineration. Waste from, e.g. vegetable canning

Table 22.23 Measures for recovery and recycle

Separate organic and aqueous phase drains from process buildings.

Interceptor tanks at each process building.

On-site solvent recovery plants.

Off-site solvent recovery.

Water condensers on reactor overheads.

Refrigerated condensers on reactor overheads.

Cryogenic condensation on reactor overheads.

Carbon, organic liquid and polymer adsorption/desorption on vapour streams containing organics.

Aqueous scrubbing with solvent recovery.

---

includes large quantities of water containing sodium hydroxide (used in chemical peeling), water colorants and organic matter of high BOD values.

A cannery in the US was prohibited from discharging its effluent directly into a watercourse because when in full production no dissolved oxygen was found within five miles from the plant. In the UK the discharges from a milk processing factory during the year's production of Easter eggs in one week overloaded the primitive treatment facility. Resulting river pollution extended for 12 miles.[39]

Recycling water is common in the canning industry, but even so the waste becomes contaminated and requires pretreatment. Simple screening removes most of the coarse solids, e.g. leaves, soil, peelings, etc., which can be used for animal fodder or disposed of via landfill. Blending the watery waste or adjustment of pH neutralises the alkali. Sedimentation and dewatering of solids is usual prior to disposal to landfill or by incineration. In certain cases, e.g. sugarbeet, the high organic content of the water can be converted to usable methane by addition of anaerobic bacteria with final aerobic finishing before discharge.[39]

Whether a material is polluting depends in the main on, e.g. its 'toxicity' and the environmental loading. Even 'nutrients' (see page 1218) in excess pose problems. Thus during sugarbeet processing the waste water has a high organic content and chemical oxygen demand. Pretreatment techniques include aerobic processes (biological filters or activated sludge plants) with conversion of organic matter to usable methane by anaerobic fermentation.

Disposal of metal processing waste[39]

Metal processes include mining, ore processing, machining, degreasing, pickling, dipping, polishing, cleaning, plating, and anodising. Wastes are

Table 22.24 Sources of emission to air

From Synthesis Plants
- VOC losses during storage, filling and emptying of bulk solvent tanks and drums
- VOC losses from wet products/cake handling/transportation
- Vapour losses from in-process tanks holding mother liquors
- Stripping of VOCs and odorous compounds from open tanks in waste water treatment plants, resulting in releases to air and/or odour problems
- Off-gases from reactors including acids, e.g. mercaptans, amines and various solvents
- Vapour losses from open reactors
- Milling and packing of finished products
- Solvent vapours from drying operations
- Relief valve discharges, bursting disc discharges
- Incineration stack gases
- Building ventilation gases
- Final discharges from 'end of pipe' abatement systems

From fermentation plants
As for synthesis plants plus:
- Organisms in fermenter off-gases
- Stripped organic solvents from waste water treatment plant
- Dryer off-gases containing VOCs
- Odours from fermenter off-gases

From extraction plants
As for synthesis plants plus:
- Solvent losses in extracted materials
- Vapours from desolventiser exhaust

From formulation plants
- Solvent vapour losses from tablet coating
- Vapour losses from bulk solvent tanks
- Active ingredients in ventilation extracts

From cosmetic plants
- Alcohol losses and perfume filling
- LPG from aerosol filling, deodorants, hair sprays, etc.

VOC = volatile organic compound.
LPG = liquid petroleum gas.

varied: hence a range of treatments are required. As discussed in Chapter 21 discharge of metals into the aquatic environment is of prime concern because of the acute toxicity to aquatic life or chronic effects from accumulation in the food chain.

Particles can be removed by screening, and aqueous metal contaminants are usually removed by precipitation, usually at pH *ca.* 8–9, although in the case of Cr(VI) compounds these first need reduction to the less toxic Cr(III) species with, e.g. sodium bisulphite. Sodium carbonate rather than hydroxide is used to precipitate Pb. The liquid and solid phases are separated by settlement, flotation or filtration. Low concentrations of soluble metal compounds can be removed by ion-exchange processes. Sludges are dewatered by filtration or centrifugation. The resulting cake should preferably be subjected to metal recovery processes or disposed of on a tip

Table 22.25 Measures to reduce air emissions

Vapour incineration – thermal.

Vapour incineration – catalytic for non-chlorinated solvent streams.

Vapour incineration – regenerative. Option for non-chlorinated solvent streams.

Carbon adsorption as final treatment.

Biolfilters as final air treatment.

Selective chemical reaction scrubbers, e.g. hypochlorite for odour control of mercaptans, NaOH for acid removal.

Aqueous scrubbing of soluble VOCs for liquid phase biodegradation in waste water treatment plant.

Cyclones for removal of fermenter aerosol.

Filtration of fermenter exhausts.

Steam sterilisation of fermenter exhausts.

Liquid incineration.

Combined liquid/vapour incineration.

Individual reactor scrubbers for more selective treatment of each reactor exhaust.

HEPA filters – pharmaceutical dusts.

Bag filters – pharmaceutical dusts.

Incinerator flue gas quench system.

Incinerator flue gas scrubber for HCl and HF removal.

Dioxin removal filters (very new, not used in study sector).

Wet electrostatic precipitators.

Standby scrubbers as an incinerator backup for toxic or odorous releases.

Air or steam stripping of effluents for recovery or treatment.

Covering of trickling filters, bio-towers and waste water treatment basins for collection and treatment of off-gases.

Biofilters on WWTP off-gases.

---

which, in the case of toxic metals, must be sealed to prevent percolation to ground waters.

Wet scrubbing is commonly used to remove metal fumes in emissions from smelting operations.

Pollution control in rubber processes[40]

The rubber industry often involves the mixing and blending of natural or synthetic elastomer with a range of additives, e.g. pigments, sulphur, extender oils, accelerators, antioxidants and fillers. Selected processes and materials are the subject of offical guidance or regulation. In the UK BATNEEC should be applied to minimise atmospheric pollution. Emissions to air other than steam or water vapour should be free of

Table 22.26 Sources of emissions to water

From synthesis plants
- Seal losses from liquid ring vacuum pumps containing very volatile, non-biodegradable solvents
- Reactor wash water
- Scrubber liquors
- Aqueous phase from steam desorption of activated carbon
- Final discharge from 'end of pipe' abatement systems
- Fire water run-off
- Contaminated stormwater
- Solvent tank leaks

From fermentation plants
As for synthesis plants plus:
- BOD/COD in waste water from fermentation
- Organic solvent in waste water from extraction steps

From extraction plants
As for synthesis plants

From formulation plants
- Active ingredients in wash waters
- Fire water run-off
- Contaminated stormwater

From cosmetic plants
- Wash water
- Deliberate dilution of off-spec product
- Fire water run-off

---

persistent mist, fume, droplets, offensive odour or visible smoke. In any event the following concentrations in emissions should not be exceeded in the UK:

Emissions	Concentration
Total particulate carbon from operations involving carbon black	10 mg/m^3
Total particulates from other operations	50 mg/m^3
Isocyanates (as total NCO) excluding particulates	0.12 mg/m^3 (averaged over 2 h)
Volatile organics (total carbon excluding particulates)	50 mg/m^3
CO (from incinerators)	100 mg/m^3
NO$_x$ (as nitrogen dioxide)	100 mg/m^3

Emissions of organic solvent which meet the above criteria for volatile organics without being vented to arrestment equipment (fugitive emissions) should be minimised by, e.g. reduction of organic solvent in use, containment between process operations, substitution of organic solvent by water-based materials, containment of fugitive emissions and ventilation via systems for recovery or abatement. Compliance should be checked by monitoring. To illustrate, levels of particulate matter should be checked continuously where exhaust volume air flow exceeds 300 m^3/min for arrestment equipment, and where flow rates are less than 300 m^3/min

Table 22.27 Measures to reduce water emissions

Pre-treatment
- Steam stripping for removal of organohalogens prior to waste water treatment plant
- Heavy metals removal

Primary treatment
- Mechanical treatment
- pH correction
- Coagulation/flocculation

Secondary treatment
- Trickling filter – covered
- Trickling filter – uncovered
- Anaerobic digestion
- Low pressure wet air oxidation (not used in study sector)
- High pressure wet air oxidation (not used in study sector)
- Activated sludge
- Extended aeration
- Oxygen-enriched activated sludge (not used in study sector)
- PAC activated sludge (not used in study sector)
- Biotower reactors

Tertiary treatment
- Sand filters
- Ozonation/oxidation
- Activated carbon

Table 22.28 Sources of solid and viscous waste

From synthesis plants
- Fly ash and slag from incinerators
- Sludges from waste water treatment plants
- Contaminated packaging
- Still bottoms residue from solvent recovery plants
- Reject pharmaceutical materials

From fermentation plants
As for synthesis plants plus:
- Spent biomass in fermenter broths

From extraction plants
As for synthesis plants plus:
- Plant or animal residues from extraction processes

From formulation plants
- Active ingredients in dust collection systems
- Reject pharmaceutical substances
- Contaminated packaging and clothing (gowns, gloves, etc.)

From cosmetic plants
- Reject cosmetic products
- Packaging
- Waxes and other impurities removed in the blending process

Table 22.29 Measures to reduce or dispose of wastes

Waste water treatment plant sludge incineration
Solid waste encapsulation
Vitrification of incinerator ash
Engineered landfill of hazardous wastes
Off-site incineration of organic wastes
Still bottoms incineration – on or off-site
Incineration of pharmaceutical wastes and contaminated packaging – on or off-site
Landfilling WWTP sludge
Sanitary landfill of general wastes

Table 22.30 Efficiency of existing waste disposal and reduction measures

Technology	Reported efficiency %	Efficiency based on removal of
Liquid incineration	99.99	Individual organic compounds
Vapour incineration	99–99.99	Individual organic compounds
Carbon adsorption from vapours	90–96	Selected organic compounds
Absorption scrubbers	98–99	Selected compounds, organic and inorganic
Biofiltration of vapours	50–99	Depends on organic compound
Effluent treatment:		
Activated sludge processes	85–95	BOD
	70–80	COD
Trickling filter	60–70	BOD
Trickling filter + Extended aeration + Sand filter	98	BOD
Wet air oxidation	96	COD, including polishing stage
Biological treatment		

plant performance should be indicated by, e.g. pressure drop indication across absolute filters and pressure or flow indicators at the inlet duct and linked to visual and audible alarms activated on malfunction of arrestment plant. Care is essential during material handling, e.g. dusty materials should be stored in purpose-built silos, damaged bags should be placed in undamaged bags, empty bags of carbon black should immediately be placed in closed containers; dry sweepings of spilled material should be prohibited. Emissions from bulk storage vessels during off-loading should be vented to suitable arrestment plant or backvented to the delivery tanker. Gaskets and valves should be checked routinely. Processes should be arranged to minimise liberation of pollutant and transfer points may need to be ventilated via arrestment plant. Further guidance is given in ref. 40.

Pollution control in the paint industry[39]

In the paint industry the main sources of pollutants include paint sludges (containing, e.g. lead, cadmium, chromium and solvent) and waste solvents. Waste effluent containing solids and solvents is passed through settling tanks. Cleaning solvents can be recovered by, e.g. distillation. Scheduling compatible paint batches facilitates solvent recovery. Paint sludges are often sent for disposal via landfill or incinerated.

Pollution control for diisocyanate processes[41]

In the UK guidance based on BATNEEC has been drafted to control pollution from processes where, e.g. diphenyl methane diisocyanate is used in the production of rigid and flexible foams. Choice of blowing agents needs to be agreed with the local authority to obtain the least environmentally harmful variants. Emissions to air, other than steam and water vapour, should be colourless, free from persistent mist, fume, droplets and offensive odour. Emissions should be within the following standards:

Diisocyanates as total NCO	0.1 mg/m^3 (averaged over 2 h)
Total particulate	50 mg/m^3

A monitoring programme should be established and emissions of isocyanates checked at least six-monthly and particulate levels checked continuously, e.g. in exhausts from extracts over foam-sawing operations. Diisocyanates should ideally be stored in fixed tanks, and air displaced from bulk storage vessels during filling should be backvented to delivery tanker (page 281) or arrested on, e.g. carbon adsorption filters. Vents from storage tanks or mixing vessels should be equipped with adsorbents such as silica gel. Alarms should be fitted to prevent overfilling of bulk storage containers. Portable containers of isocyanates should be stored in ventilated conditions and their condition regularly monitored. Chimney height and vents should be no less than 3 m above roof and never <8 m above ground level. Vent design should anticipate efflux velocity of not less than 15 m/s at full load operation and vents should not be fitted with caps, cowls, etc. Additional advice is given in ref. 41.

Pollution control in the pharmaceutical industry[42]

The pharmaceutical industry converts hundreds of different raw materials and intermediates into an estimated 10 000 finished products within the European Union (EU). Other substances, e.g. solvents, are used as reaction media. Thus the principal inputs are fine organic chemical raw materials, organic solvents, bulk inorganic acids and alkalies, intermediate pharmaceuticals, absorbers, substrates for fermentation, plant and animal matter for extraction, etc. The principal processes used by the industry include chemical synthesis, fermentation, extraction and formulation. During these processes some substances are inevitably emitted to air or water or disposed of as hazardous waste. Strategies for control of pollution include containment, on-site recovery/recycle, and reduction of emissions. Typical measures used for containment are given in Table 22.22, whilst recovery and recycling techniques are given in Table 22.23. Clearly these measures are transferable to other industries.

Key sources of emissions to atmosphere are indicated in Table 22.24. The volatile organic chemical (VOC) emissions to atmosphere attributable to the pharmaceutical and cosmetics industries in the EU are estimated at 28 000 tonnes per year, which is <1% of the total VOC emissions from all sources. Techniques used for controlling atmospheric pollution are given in Table 22.25; again, many of these are adaptable to other industries.

The total volume of liquid effluent from the European drug industry is estimated at 480 000 m^3 per day, of which 150 000 m^3 may be discharged untreated to the aquatic environment. The main sources of discharges to water are listed in Table 22.26 and the approaches currently adopted for control are given in Table 22.27.

Hazardous waste comprises sludges from waste water treatment plants, spent activated charcoal, still bottoms, incinerator ash, vegetable matter, fermentation biomass, waste from fermentation plants, and waste from cosmetic plants. Sources of solid waste include those listed in Table 22.28. Measures to reduce or dispose of waste are indicated in Table 22.29.

There is no one BATNEEC solution to cover all pollutants and all plants. Some of the foregoing measures, the efficiency of which are indicated in Table 22.30, represent BATNEEC, whilst for others tighter controls are being proposed. These include: new technology; categorisation of plant according to risk; issue of consent conditions (for air emissions, ground level concentrations of odorous compounds, discharges to waterways, and minimum level of solvent recovery); licensing; environmental monitoring.

References

1. Bridgwater, A. V. & Mumford, C. J., *Waste Recycling and Pollution Control Handbook*, George Godwin Ltd., 1979.
2. Colon, F. J. & Hazelwinkel, *Environmental Technology, Assessment and Policy*, eds. R. K. Jain and A. Clark, Ellis Horwood, Chichester, 1989, page 198.
3. Wagner, T. P., *The Complete Guide to Hazardous Waste Regulations*, Van Nostrand Reinhold, New York, 2nd edn, 1991.
4. Anon., *The Chem. Eng.*, 1991, **11**, 496.
5. Department of the Environment, Environmental Protection Act 1990, *Waste Management – the duty of care*, Feb. 1992, HMSO.

6. McLeod, G. & Buxton, E., *Environmental Protection Bulletin*, 1993(025), 29.
7. Belshaw, C., Jones, A. P., Hosker, D. & Fox, A. G., *Inst. Chem. Eng. Symp. Series*, No. 117, 1990, 175.
8. Burnley, P., *Environmental Protection Bulletin*, 1990(004), 26.
    a) Cambridge Water Company versus Eastern Counties Leather, *Times Law Report*, 10 December 1993.
9. Control of Pollution Act, 1974, s.31 (now contained in the Water Act, 1989, s.107).
10. Alphacell Ltd versus Woodward (1972), 2 All ER, 475-88.
11. Impress (Worcester) Ltd versus Rees (1971), 2 All ER, 357.
12. EC Directive, *The assessment of the effects of certain public and private projects on the environment*, 85/337/EEC, adopted 27 June 1985.
13. Department of the Environment, *Environmental assessment*, DOE Circular 15/88 (Welsh Office Circular 23/88), 12 July 1988.
14. Department of the Environment, *Environmental assessment: a guide to the procedures*, 1989, HMSO.
15. Lee, N., *Environmental impact assessment: a training guide*, Occasional Paper 18, 2nd edn, Dept of Planning & Landscape, University of Manchester, 1989, 184 pp.
16. Anon., Environmental assessment, *Journal of Planning and Environment Law*, 1989 (Nov), 853-63.
17. Crossland, N., *Environmental Protection Bulletin*, 1990(009), 22.
18. Shillito, D. & Shepherd, P., *Environmental Protection Bulletin*, 1990(005), 16-20.
19. Anon., *Loss Prevention Bulletin*, 1990(094), 7.
20. The Hazardous Waste Inspectorate, 3rd report, June 1988, HMSO.
21. Spinks, P., *New Scientist*, 25 Nov. 1989, 36.
22. Perriman, R. J., *Environmental Protection Bulletin*, 1990(006), 15.
23. Freeman, H. M. & Hunt, G. E., *Environmental Protection Bulletin*, 1992(016), 17.
24. Her Majesty's Inspector of Pollution, 4th Annual Report (1990-91), HMSO, 1991.
25. Canter, L. W., in *Environmental Technology, Assessment and Policy*, ed. R. K. Jain and A. Clark, Ellis Horwood, Chichester, 1989, page 179.
26. Yaws, C., Yang, H.-C. & Pan, X., *Chem. Engineering*, 1991 (Nov.), 179.
27. Office of Hazard Technology Assessment, *Technologies and Management Strategies for Hazardous Waste Control*, OTA-M-196, US Congress, Washington, DC, 1983.
28. McMaster-Christie, A., *Environmental Protection Bulletin*, 1993(025), 29.
    a) Katin, R. A., *Chem. Eng. Prog.*, July 1991, 39.
29. O'Connor, D. G., *I. Chem. Eng. Symp. Series*, No. 117, 1990, 129.
30. Isaac, P., *Civil Protection*, 1990 (Spring), 14, 6.
31. Department of the Environment, *Landfilling Wastes*, Waste Management Paper No. 26, 1986.
32. Smith, A. J. & Parsons, P. J., *Environmental Protection Bulletin*, 1989(002), 16.
33. Department of the Environment, *The Control of Landfill Gas*, Waste Management Paper No. 27, 1991.
34. Collison J., *NEBOSH Diploma Course Notes*, ATM Safety Ltd., UK, 1992.

35. Stukel, J. J., in *Environmental Technology, Assessment and Policy*, eds. R. K. Jain and A. Clark, Ellis Horwood, Chichester, 1989, page 42.
36. a) Fiessinger, F., Mallevaille, J., Leprince, A. & Weisner, M., in *Environmental Technology, Assessment and Policy*, eds. R. K. Jain and A. Clark, Ellis Horwood, Chichester, 1989, page 77.
    b) Neutra, R. R. & Ostro, B., *Sci. Total Environ.*, 127, 1992, 91.
    c) Castleman, B. I., *Regulations Affecting Use of Carcinogens, in Cancer Causing Chemicals*, N. I. Sax, Van Nostrand Reinhold Co., New York, 1981, 4.
37. Kirkpatrick, N., *Environmental Issues in the Pulp and Paper Industries*, Pira International, 1991.
38. *The Disposal of Used Metalworking Fluids*, Burmah-Castrol plc, 1991.
39. Loss Prevention Council, *Pollutant Industries*, undated, Loss Prevention Council, London.
40. Department of the Environment, *Secretary of State's Guidance – Rubber processes*, PG6/28 (92), HMSO, Feb. 1992.
41. Department of the Environment, *Secretary of State's Guidance – Diisocyanate processes*, PG6/29 (92), HMSO, Feb. 1992.
42. *Technical and Economical Study on the Reduction (based on best available technology not entailing excessive costs) of Industrial Emissions from the Pharmaceutical and Cosmetics Industry*, Commission of the European Communities, Luxembourg, 1993.

CHAPTER 23

Storage, handling, utilisation and transport of liquefied petroleum gases

This chapter supplements the general guidance given in Chapter 4 for the safe handling of volatile, flammable materials, by specific reference to Liquefied Petroleum Gases (LPG). The hazardous properties are discussed together with their implications for appropriate precautions for handling these materials in bulk and on a small scale. Control measures are described for LPG storage and for use in the aerosol industry.

Hazards

Physico-chemical properties

As discussed in Chapters 3 and 19 an understanding of the chemistry and physico-chemical properties are essential for risk assessment of operations involving any substance. For example,
- The chemical structure will indicate the chemistry and dictate how the compound will behave in a fire.
- The boiling and melting points will indicate the physical state of the material under given conditions of temperature at atmospheric pressure.
- Vapour pressure provides a measure of the volatility. Where vapour and liquid coexist the vapour pressure is the saturated vapour pressure. It is independent of the amount of liquid present.

In an incident in Brazil in 1972 the relief valve on an LPG sphere failed to open and the pressure rose. Operators opened the drain valve in an attempt to reduce the pressure by draining-off water. The drain valve froze and when LPG was ignited some 37 fatalities resulted.

At the boiling point the vapour pressure is equal to the atmospheric pressure. The vapour pressure rises with an increase in temperature. Above ambient temperature the pressure required for liquefaction continues to rise with the temperature until the critical temperature is reached above which the substance can no longer exist in the liquid phase even if subjected to further increases in pressure. The vapour pressure at the critical temperature is the critical pressure. A knowledge

of the vapour pressure of a gas is crucial for specifying the design conditions for pressurised systems, e.g. cylinders, tanks and pipework, and for calculating gas offtake rates by natural vaporisation.
- The latent heat of a liquid is the amount of heat absorbed to enable vaporisation to take place. Such data are required for the design of vaporiser systems. When LPG is allowed to vaporise naturally the latent heat is taken from the liquid itself and from its immediate surroundings, causing a drop in temperature (auto-refrigeration).
- The density of the gas, in part, dictates how the vapour will disperse if allowed to escape (i.e. rise or slump). Liquid density and water solubility are important in predicting how the LPG will behave on mixing with, e.g. water.
- Coefficients of cubical expansion quantify the degree of expansion of liquid as the temperature increases, and are taken into account in specifying the maximum quantity of LPG permitted to be filled into a pressure vessel.
- The viscosity characteristics are important for the design of pumping and pipework systems.
- For a vapour system, the dew point indicates the temperature above which the system needs to be maintained to prevent condensation.
- Limits of flammability define the concentration range of LPG which will ignite in air when thoroughly mixed. In reality, gas escaping without ignition will tend to be over-rich near the leak source but form combustible mixtures on the fringes of the cloud which will then ignite on reaching an ignition source. (A 'visible' cloud due to refrigeration effects, noted below, does not necessarily show the limits of flammable gas; this may extend further on all sides.)

LPG is a mixture of propane and *n*- and iso-butanes, plus small amounts of the olefinic counterparts (I–III). Stenching agents are sometimes incorporated (see later). National standards, e.g. from the British Standards Institute in the UK, and the Gas Processors' Association and the ASTM in the USA,[1] define the requirements for commercial LPG. The major sources include natural gas wells, gas issuing from crude oil wells, and cracking of crude oil; the most important uses are as a fuel, as a feedstock in the chemical industry, and, increasingly, as an aerosol propellent.

The two liquefied petroleum gases in general use are commercial butane and commercial propane, supplied to product specifications, e.g. BS 4250. Commercial propane consists predominantly of propane (C_3H_8) and/or propylene.

Key physico-chemical properties of propane and butane are given in Table 23.1 (after refs 2–4). The compounds are gaseous at normal ambient temperatures and pressure but are easily liquefied by application of relatively low pressure. Any liquid leak vaporises and this chills the air in the vicinity; this causes condensation of water vapour and hence produces a visible cloud (the point of leakage may also frost over).

```
    H  H  H                H  H  H  H                 H
    |  |  |                |  |  |  |               H—C—H
 H—C—C—C—H            H—C—C—C—C—H              H   |   H
    |  |  |                |  |  |  |              |   |   |
    H  H  H                H  H  H  H          H—C—C—C—H
                                                   |   |   |
                                                   H  H  H
     Propane                 Butane                Isobutane
```

```
                                                            H
    H  H                 H  H  H                          C—H
    |  |       H       H    |  |  |                 H        H
 H—C—C=C           C=C—C—C—H              C=C
    |       H       H        |  |                 H        H
    H                       H  H                          C—H
                                                        H    H
   Propylene(I)          n-Butylene(II)         Isobutylene(III)
```

Liquefaction is accompanied by a considerable decrease in the volume occupied by LPG vapour, requiring much less storage space. If the pressure is subsequently released, the hydrocarbons again become gaseous. It is therefore stored and distributed in the liquid phase in pressurised cylinders and bulk containers at ambient temperature and allowed to revert to the vapour phase at or near the point of eventual utilisation.[5,6] Alternatively, LPG is stored in refrigerated vessels at a pressure close to atmospheric, e.g. for shipment by sea and for large-scale storage at marine terminals.

Butane and propane are colourless, odourless gases which are insoluble in water. The boiling point of propane is lower than for butane; it therefore exerts a greater vapour pressure under identical conditions. Since systems are often specifically designed for either butane or propane, a practical implication is that any butane system should be precluded from use with propane, although propane systems can be dual purpose. Thus it is important that propane or propane–butane mixtures are not mistakenly used for, e.g. portable heaters or blowtorches intended for butane.

The density of LPG in the liquid state is about half that of water whilst that of the gas is approaching twice the density of air. The latter tends to result in gas clouds 'slumping'.

> A gas release occurred when a line leaked allowing seven tonnes of liquid propane to escape in 40 minutes; fortunately it did not ignite. The gas cloud spread out rapidly, as shown in Fig. 23.1 although a compressor building and refrigerated ethylene storage tanks acted as partial barriers. Emergency action concentrated on preventing the cloud from reaching a furnace and steam-boiler areas using water jet guns and mobile water curtains. The cloud was then surrounded with mobile water curtains and water canons to force it out of the process area, a strategy which proved very effective.[7]
>
> The leak occurred from a 6 cm diameter start-up pipeline to a propane

Flammable and explosive hazards

Table 23.1 Physico-chemical properties

	Propane	n-Butane	Isobutane
Molecular weight	44.1	58.1	58.1
Vapour pressure at 21 °C, i.e. cylinder pressure (kg/cm² gauge)	7.7	1.15	21.6
Specific volume at 21 °C/1 atm (ml/g)	530.6	399.5	405.8
B.p. at 1 atm (°C)	−42.07	−0.5	−11.73
M.p. at 1 atm (°C)	−187.69	−138.3	−159.6
S.G., gas at 16 °C/1 atm (air = 1)	1.5503	2.076	2.01
Density, liquid at sat. pressure (g/ml)	0.5505 (20 °C)	0.5788 (20 °C)	0.563 (15 °C)
Density, gas at 0 °C/1 atm (kg/m³)	2.02	2.70	—
Critical temperature (°C)	96.8	152	135
Critical pressure (atm)	42	37.5	37.2
Critical density (g/ml)	0.220	0.225	0.221
Latent heat of vap. at b.p. (cal/g)	101.76	92.0	87.56
Latent heat of fusion at m.p. (cal/g)	19.10	19.17	18.67
Specific heat, liquid at 16 °C (cal/g °C)	—	0.5636	0.5695
Specific heat, gas at 16 °C:			
C_p (cal/g °C)	0.3885	0.3908	0.3872
C_v (cal/g °C)	0.3434	0.3566	0.3530
Specific heat ratio at 16 °C/1 atm, C_p/C_v	1.131	1.096	1.097
Gross heat of combustion at 16 °C/1 atm (cal/ml)	22.8	30.0	29.8
Viscosity, gas at 1 atm (centipoise)	0.00803 (16 °C)	0.0084 (15 °C)	0.00755 (23 °C)
Coefficient of cubical expansion at 15 °C (per °C)	0.0016	0.0011	—
Surface tension (dynes/cm)	16.49 (−50 °C)	16.02 (−10 °C)	15.28 (−20 °C)
Solubility in water at 1 atm (volumes/100 volumes water)	6.5 (18 °C)	—	1.7 (17 °C)
Flammable limits in air (% by volume)	2.2–9.5	1.9–8.5	1.8–8.4
Autoignition temp. (°C)	467.8	405	543
Max. explosion pressure (MPa)	0.86	0.86	—
Min. ignition energy (MJ)	0.25	0.25	—
Max. flame temperature (°C)	2155	2130	—
Max. burning velocity (m/s)	0.45	0.38	—
Necessary min. inert gas conc. for explosion prevention in case of emergent outflow of gas in closed volumes (% v/v):			
Nitrogen	45	41	—
Carbon dioxide	32	29	—

cracker feedline which was normally shut off. It was connected to the bottom of the feedline, as shown in Fig. 23.2, and hence collected water when not in operation. A 35 cm long crack developed in it due to the force exerted when this water froze. (Subsequently the line was modified so that it entered the top of the feedline and was isolated from it by a slip-plate immediately after the connecting valve.)

The ratios of gas volume to liquid volume at standard temperature and pressure are approximately 274 and 233 to 1 for propane and butane, respectively; the exact values for LPG vary with the blend and a value of

Storage, handling, utilisation and transport of LPG 1317

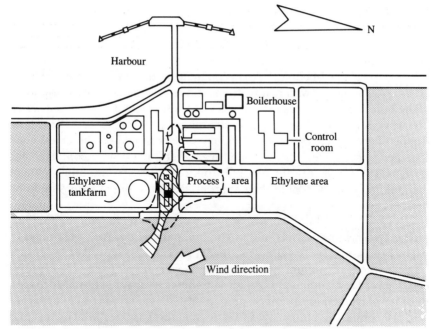

Fig. 23.1 Spread of propane gas-cloud in the incident at Rafnes described on page 1315.[7] (Dotted line indicates maximum gas-cloud spread; hatched area indicates gas-cloud spread when all water jet guns and water curtains were in operation)

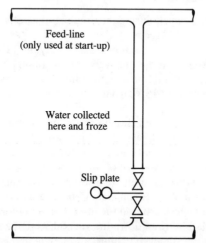

Fig. 23.2 Original piping arrangement allowing water accumulation

around 225–250 is usually taken (see page 88). Hence even small leakages of liquid LPG can result in a large volume of flammable vapour–air mixture, e.g. whereas one volume of vapour can form 10–50 volumes of flammable mixture, one volume of liquid can form 2500 to 12 500 volumes. Clearly, therefore, LPG must be securely contained and any liquid leak, resulting in a two-phase discharge, is considerably more hazardous than a similar-sized leak in the vapour phase.

The coefficients of cubical expansion indicate that LPG expands at least 10 times as much as water, a factor that needs to be accounted for when filling vessels in which free space has to be left (see page 99). Connecting pipework is equipped with hydrostatic relief valves wherever LPG can be trapped between closed valves. LPG containers should not be manifolded together since, under certain conditions, overfilling of one container can arise as a result of transfer of LPG from one vessel to the other. Where manifolding is unavoidable then appropriate non-return valves should be fitted.

LPG is in general non-corrosive so any common metals can be used for piping systems and vessels, although aluminium is normally restricted to vapour-phase operations unless stringent precautions are taken to ensure that all traces of sodium hydroxide carry-over in gas from refinery processes have been eliminated. As for any chemical plant, external corrosion must be prevented as described on page 101. Some non-metallic materials are susceptible to attack by LPG, e.g. natural rubber which becomes spongy, and certain plastics which become brittle or soft. It is advisable to consult with suppliers for advice on materials of construction for storage vessels, pipes, joints, valves, seals, gaskets, diaphragms, etc. All plant should be designed, installed and maintained as specified by competent engineers using appropriate Standards or Codes.

> In 1981 liquid butane escaped during the draining of caustic solution from a salt filter/drier vessel on an LPG recovery installation. The gas cloud ignited causing $1.5m damage to equipment but, suprisingly, no casualties. It was subsequently established that the plug valve on the drain line had been incorrectly assembled such that turning the plug clockwise would open the valve instead of closing it, and vice versa.[8]

> When the feed line to a light-ends fractionation unit ruptured a massive release of hydrocarbon followed which was ignited by static electricity. There were no personal injuries but the accident caused considerable damage to instrumentation and the unit was out of service for four weeks. The cause of the rupture was a thinning in the line as a result of external corrosion. This arose from extended exposure to moisture (from rainfall and mist from a nearby cooling tower) which entered the insulation in a break in the weatherproofing. The water ran down the vertical section and gathered at the horizontal section at the site of the failure.[5]

With LPG systems, as indeed was mentioned for pressure systems in general on pages 746 and 748, the pressure-testing must be comprehensive.

> A bad leak of LPG occurred during start-up involving the plant section shown in Fig. 23.3. Although the plant had been pressure-tested prior to start-up, the

Storage, handling, utilisation and transport of LPG

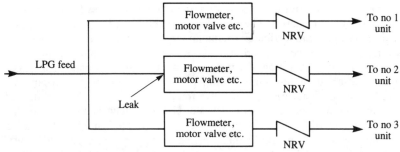

Fig. 23.3 Piping arrangement preventing pressure-testing of section of LPG unit

non-return valves in the pipelines rendered it impracticable to test the equipment to the left of them. (Subsequently the non-return valves were removed and replaced by a single valve in the common feedline.)[9]

Upon decompression LPG flows from pipes and vessels with a low negative temperature as a result of throttling (isoenthalpic pressure drop) or as a result of product being stored at low temperatures. Also as a result of the latent heats of vaporisation the very low temperatures reached by auto-refrigeration, particularly with butane, can pose a risk of severe cold burns for operators unless protective clothing is worn.

A liquefied gas line was protected by a small relief valve which discharged onto the ground as shown in Fig. 23.4. The ground was not levelled and, following

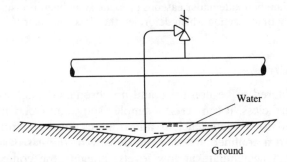

Fig. 23.4 Relief valve tail pipe with outlet submerged beneath a puddle of water which froze

heavy rain, the outlet became submerged beneath a puddle of rainwater. Freezing of this water resulted in overpressurisation in the line.[10]

All this also has a bearing on the choice of stop and control valves, materials of construction and safety measures. LPG can contain water in both the liquid and vapour states and, depending upon the temperature and pressure of the system, surplus water condenses and can lead to formation of hydrocarbon hydrates in pipelines and plugs of ice in, e.g. regulators.

The major LPG disaster in Feyzin is described briefly on pages 176 and 879, and more fully elsewhere.[11] The role of a frozen plug was identified although the primary cause was attributed to human error. Samples were taken from each LPG storage sphere on a routine basis for analysis. Since the refinery process led to the separation of caustic soda solution from propane on storage it was necessary to drain-off this solution prior to sampling the gas. Because of two previous incidents arising from valve freezing (due to propane hydrate and ice formation) a set procedure was devised. Unfortunately, on the day the plant operator operated the valves in the reverse sequence. He first opened the lower valve leading to atmosphere; he then opened the upper valve slightly to adjust the drawoff rate. A plug of ice, or propane hydrate, hampered the flow. When he opened the top valve fully a powerful jet of propane gushed out, causing frostbite to the operator and leakage of LPG which ignited killing 18, injuring 81 and causing $4.6m damage to the plant, a further $2m damage outside the refinery and necessitating the evacuation of 2000 local inhabitants.

LPG can undergo oxidation by atmospheric oxygen at temperatures well below the auto-ignition point. Hydrocarbons also undergo halogenation in daylight or upon UV radiation, some with explosive violence. The reaction in the liquid state can be initiated by heat or the presence of catalysts. Depending on the conditions, alkanes can react with inorganic materials such as nitric acid, sulphur dioxide, sulphuryl chloride, sulphur, carbon monoxide, etc., and with organic substances such as peroxides, acid chlorides, certain chlorocarbons, ketones, etc.

Heating barium peroxide under gaseous propane at ambient pressure caused a violent exothermic reaction which deformed the glass container.[12]

Toxicity

LPG is considered to be non-toxic and no chronic effects have been reported. Nevertheless, it can act as a simple asphyxiant (see Chapter 5) to those exposed to oxygen-deficient atmospheres resulting from, e.g. an LPG leak into a confined space. (The density of the gas is such that if released it will accumulate at low levels, hugging the contours of the ground, filling valleys, ditches, pits, pipe-trenches, sumps, etc.) LPGs possess mild anaesthetic properties and can cause depression of the central nervous system (see Chaper 5). Whilst concentrations as high as 100 000 ppm may be tolerated, butane vapour at a concentration of 10 000 ppm can lead to drowsiness after only a few minutes.[13] For the foregoing reasons coupled with its flammable properties (see below) an 8-hour time-weighted-average Threshold Limit Value (TLV)[14] and a Recommended Occupational Exposure Limit (see Chapter 5)[15] have been set at 1000 ppm for LPG with a 10-minute Short Term Exposure Limit of 1250 ppm in the UK.[15] However, the eight-hour time-weighted average TLV in the USA for 'Butane' is 800 ppm[16] whilst the corresponding UK Recommended

Occupational Exposure Limit for butane is 600 ppm with a short-term exposure limit of 750 ppm.[15]

Flammability

The major concern when handling LPG relates to its volatility and flammability. Because LPG is odourless, it is usual, during production, for a stenching agent, e.g. ethyl mercaptan or dimethyl sulphide, to be added at a ratio of about one pound of odorant per 10 000 gallons of LPG liquid. Thus under published standards the presence of the gas in air should be detectable at one-fifth of the lower limit of flammability (0.4 per cent of gas in air).

The effects can be nauseating.

> A works engineer and a fitter removed three of the four bolts securing an adaptor to the body of a drain valve of a 10 tonne LPG storage tank. About 5 tonnes of propane leaked over a period of several hours and a gusty wind helped to disperse the gas cloud. Around 2000 local residents and 100 students from an adjacent college were evacuated. A number of firemen were overcome by the smell and suffered from nausea, and a child similarly affected was detained in hospital overnight.[17]

However, the degree of risk of explosion when a leak occurs is difficult to assess from its odour due to the tendency of gas to 'slump' as described on page 95. For certain outlets such as food processing and the aerosol industry this odoriferous material is usually removed, e.g. by molecular sieves; clearly the sense of smell will then offer no warning of LPG leakages.

Although the flammable limits are rather narrow (approximately 2–10 per cent in air), the ratio of gas volume to liquid volume is such that, as described in Chapter 3, small amounts of liquid escape can produce an appreciably large volume of potentially flammable gas. For example, 4 litres of butane upon vaporisation produce about 0.9 m^3 of gas at atmospheric pressure and 15.6 °C which is equivalent to 18.12 m^3 of flammable mixture. This property, coupled with the propensity of leaking gas to accumulate at low levels in unventilated areas and travel considerable distances to sources of ignition, creates a considerable fire and explosion risk.[17]

> In the incident at Port Hudson (see page 873) 60 tonnes of LPG escaped to form a pancake-shaped cloud 3–6 m thick which rolled for 13 minutes before being ignited at a refrigeration plant 600 m away.[18]

At a typical ambient temperature (15 °C) loss of containment of liquid LPG under pressure contains sufficient air to vaporise nearly all the LPG released to atmosphere, resulting in a flammable vapour–air cloud.[19,20] This may of course, disperse harmlessly. Alternatively, it may ignite immediately, or drift until it reaches an ignition source of sufficient size.

Cloud dispersion is influenced by velocity and direction of discharge, temperatures of the air, vapour and ground, topography, wind conditions, density of vapour, and location in relation to buildings and plant.

Upon ignition the cloud may burn with or without explosion. Any cloud which is too rich to explode may commence to burn around the periphery and form a mobile fireball. The radiant heat from this may cause burns to personnel or ignite combustible materials depending upon their distance from the fireball. Convection currents in the wake of a travelling fireball may draw in and ignite debris.[19]

As described on page 873, an Unconfined Vapour Cloud Explosion – requiring a minimum of probably 5 tonnes of vapour – may result from the release of vapour and its admixture with air. When ignited this burns in free space with sufficient rapidity to generate pressure waves which propagate through the cloud and into the surrounding environment. The associated high overpressures result, as exemplified on page 876, in heavy damage.

A Boiling Liquid Expanding Vapour Explosion (BLEVE) is usually associated with failure of an LPG vessel exposed to fire. As the temperature of the liquid increases, bubble formation is followed by nucleate boiling with liquid swell. The pressure inside the vessel increases and eventually the relief valve lifts and releases vapour or a two-phase mixture, which may ignite as a jet flame. If heating is only on one side of the vessel recirculation will occur. If the liquid temperature continues to rise, a film of vapour (i.e. a transition to film-boiling) occurs at the vessel wall. In this condition, or when a flame impinges onto the vapour space, there is little internal cooling of this area of the vessel and the wall is continuously weakened under conditions of increasing internal pressure and is eventually likely to fail. If a pressure relief valve is fitted, failure may occur at pressures of around 1.5 MPa. Without relief, failure may occur at pressures up to the original burst pressure of the vessel, 3–8 MPa.[20]

Damage from a BLEVE usually arises, as summarised on page 877, from the blast wave due to the sudden release of internal pressure, thermal radiation and the projection of fragments over considerable distances.

At San Juan Ixhuatepec, Mexico City, a number of LPG storage vessels BLEVE'd over a 90-minute period. The horizontal bullets were of up to 140 tonnes capacity and the spherical vessels up to 1200 tonnes capacity. The majority of the fragments were within a radius of 400 m but the farthest distance reached was 1200 m. The pattern of damage is illustrated by Fig. 23.5.

Although the majority of BLEVEs have been initiated by fire impingement, similar phenomena have followed failure of a vessel due to mechanical damage, e.g. due to corrosion or by impact, or due to overfilling.

Upon release of LPG liquid, flash evaporation occurs. The fraction of liquid which will evaporate under specific conditions can be approximated as follows:

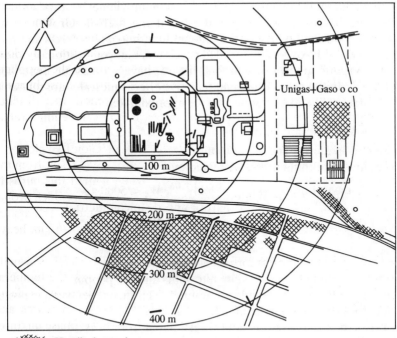

Heavily damaged area
— Bullets
○ Sphere fragments

Fig. 23.5 Damage area in the Mexico City disaster

Weight in vessel $= W_0$

Fraction flashed, $m = s\left(\dfrac{T_o - T_b}{L_b}\right)$, when $m \leq 1$

where s = liquid specific heat
L_b = latent heat at atmospheric boiling point T_b
T_o = temperature of vessel contents

However, entrainment of liquid droplets also occurs and for calculation purposes[22]

Fraction entrained, $e = m$ for $m \leq 0.5$, $e = 1 - m$ for $m > 0.5$

so,

Weight of cloud, $W_c = (e + m)W_0$ (flashing liquid)

If <0.35 is flashed then some fuel can be expected to burn in a pool fire at the point of release, rather than being consumed in the fireball.

With a fireball the strong buoyancy forces result in turbulent mixing and the formation of a mushroom cloud, often with a stem of flame emanating from the source of fuel. The fireball is generally described in terms of its duration, the time of burning, and the maximum diameter of the equivalent sphere based on the projected area of the fireball. Growth of the fireball is controlled by the initital momentum, buoyancy effects and radiative cooling.[21] Buoyancy-controlled processes, e.g. combustion of quiescent fuel vapour spheres and fireball lift-off, have a timescale dependent upon $W_c^{1/6}$. Initial momentum and radiative effect timescales are related to $W_c^{1/3}$.[21] Hence the flame radius may be approximated by,[22]

$$\text{Flame radius } r_F = 3.2 W_c^{1/3}$$

where r_F is in metres and W_c in tonnes.

The time of burning is approximated by

$$\text{Time of burning (s), } t_B = 1.1 W_c^{1/6}, \quad W_c < 5000 \text{ kg}$$
$$\text{(pre-mixed stoichiometric)}$$
$$t_B = 2.6 W_c^{1/6}, \quad W_c > 5000 \text{ kg}$$
$$\text{(diffusion controlled)}$$

The two regions of interest for risk assessment are
- Inside the fireball; which will ignite vents and kill people.
- Inside the flux where people are injured. This flux depends on exposure time to the inverse power of 2/3.

For calculation purposes the latter is estimated from[22]

$$\text{Distance to flux } I, \ x = r_F (I_o/I)^{1/2} \text{ metres}$$
$$\text{Distance to where } I < 47/t_B^{2/3} \text{ (safe dose)}$$
$$x = 10 W_c^{7/18}, \quad W_c < 5000 \text{ kg when}$$
$$I_o = 450 \text{ kW/m}^2$$
$$x = 12 W_c^{7/18}, \quad W_c > 5000 \text{ kg when}$$
$$I_o = 350 \text{ kW/m}^2$$

Ignition sources have included naked flames, incendive sparks, unprotected lights, non-flameproofed electrical equipment, static electricity, and hot surfaces. (Figure 23.6 shows the temperatures required for ignition of propane/air mixtures in the flammable range (after ref. 23).)

A driver was injured in an explosion whilst filling the liquid-gas tank of his stacker truck at an open-air filling station. There was also considerable material damage. The seal on the filler nozzle of the gas pressure tank, which was firmly connected to the stacker, was missing so that gas was able to escape at the connection point. Ignition of the flammable gas–air mixture produced was probably due to hot engine surfaces, i.e. exhaust pipes.[24]

Additional examples of accidents involving LPG are given elsewhere (e.g. pages 80, 95, 155, 176, 756, 760, 865 and 879) ranging from small-scale welding incidents to transport disasters and massive fires and explosions such as those at Feyzin and Mexico City. The worst-ever incident was the recent pipeline explosion in Russia.

LPG is delivered from the Western Siberian gas fields to chemical plants in the

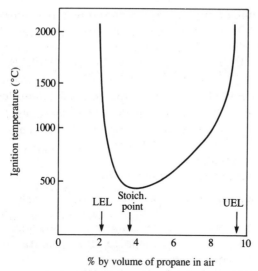

Fig. 23.6 Temperatures for ignition of propane–air mixtures within the flammable range

Urals by a cross-country pipeline hundreds of kilometres long. In June 1989 the pipeline fractured and gas leaked for several hours, possibly for two days, and accumulated in a railway cutting to produce a flammable cloud some two kilometres in diameter. This was ignited, probably by a spark from the wheels of a passing train on the track just over a kilometre from the pipeline, creating a huge fireball between 1.5 and 2 km wide which engulfed two crowded passenger trains hurling 28 railway carriages and their engines off the track and devastating extensive areas of forest. Early reports claimed a death toll in excess of 608. A total of 706 were hospitalised. Survivors recalled a strong acrid smell prior to the explosion. Blame was apportioned to negligence, irresponsibility, and lack of organisation. Indeed it would appear that instead of responding to the drop in pressure and complaints from local residents of a smell of gas by shutting-down the pipe, engineers apparently increased the pressure thereby exacerbating the accident.[6]

Despite this tragedy and other accidents involving LPG, this material can be handled safely on a vast scale in many operations with care and due attention to safety by properly trained and competent persons. This necessitates a full appreciation of the hazards and minimising the risk by segregating storage facilities, building plant to good engineering standards, and providing staff with adequate training. Previous incidents have resulted from a lack of attention to detail. For example,

> An unusual movement of LPG was to be made from one storage tank to another. The valve and piping arrangement was very complicated and two experienced men set up the wrong route; the LPG was pumped down a disused line and seven tonnes escaped from an open end in another plant before the leak was stopped. (The disused line was apparently last used some years

previously. It was believed to have been slip-plated at the receiving end but some time later the slip-plate was removed when changes were made in the receiving plant area.)

Water monitors were set up around the leak and no flammable gas concentrations were detectable outside the water curtain.[25]

In order to plan LPG storage and processing facilities, due consideration must be given not only to preventing leaks and subsequent ignition but to dealing with such mishaps should they occur. This demands an understanding of the ease with which LPG will ignite, and an appreciation of the heat evolved in such circumstances.

Bulk storage

The liquefaction of petroleum gases under certain conditions of temperature and pressure makes it possible to store and transport them conveniently in various quantities.

Pressure storage

Propane and butane are readily liquefied when subjected to pressures equivalent to their respective vapour pressures. This facilitates pressurised storage at pressures which are, of course, a function of temperature,[13,26,27] e.g. as shown in Table 23.2.

Table 23.2 Equilibrium vapour pressures at 15.6 °C and 37.8 °C

	15.6 °C (60 °F)	37.8 °C (100° F)
Butane	103 kPa (15 psig)	483 kPa (70 psig)
Propane	518 kPa (75 psig)	1400 kPa (210 psig)

Commercial grades of LPG are not pure, but are contaminated with small amounts of other low molecular weight hydrocarbons. The equilibrium vapour pressures of commercial propane and butane are hence somewhat higher than those of the pure materials, as shown in Fig. 23.7. Storage tanks must be designed to withstand the highest vapour pressure likely to be encountered in use, which will depend upon the highest temperature of exposure, usually by solar heating.

Static tanks are constructed in accordance with relevant Standards, e.g. BS 5500, ACTC Rules. The design pressure must not be <489 kPa for butane or <1470 kPa for propane. Tanks are fabricated from special steel and are inspected for soundness by a variety of sophisticated techniques for incipient cracking detection and weld integrity. They are usually inspected at five-yearly intervals whilst in service.

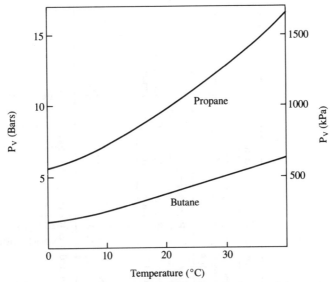

Fig. 23.7 Approximate maximum vapour pressures of commercial propane and butane

Tanks are sometimes submerged below ground to eliminate fire impingement and solar heating. This is a viable proposition in congested areas where spacing requirements cannot be met. Such tanks are provided with external protective coatings to resist soil corrosion; cathodic protection is also used. They must be securely fixed, otherwise ingress of water around even full tanks may produce severe flotation-induced stresses which could rupture associated piping.

Above-ground storage tanks are spaced to provide access for firefighting[28] and to avoid spread of a fire from tank to tank. The minimum recommended safety distances for tanks for industrial, commercial and bulk storage are summarised in Table 23.3.[29] Tanks are usually elevated above the ground, which is concreted and sloped away from the tank to prevent any accumulation of spilled LPG. Different recommendations apply to pressure storage at refineries and bulk plants.[29]

Installations are normally fitted with water and drench systems as discussed on page 176 and illustrated by Fig. 12.23. Their purpose is to discharge water onto a vessel at a rate sufficient to maintain an adequate flow of water over the entire vessel and supports if the vessel is threatened by radiant heat or direct flame impingement. The main objective is to keep the vessel sufficiently cool to prevent failure. However, cooling the wetted portion of a tank may also reduce the rate of vaporisation sufficiently for the vapour pressure to fall below the pressure relief valve setting. The valve will then close and prevent unnecessary discharge of vapour.

Application may be from fixed points, e.g. spray nozzles, drencher heads

Table 23.3 Location and spacing for pressurised bulk storage of LPG[29]

Maximum water capacity of any single tank in group		Maximum total water capacity of all tanks in group		Minimum separation distance (metres)					
				From building, boundary, property line (whether built on or not) or fixed source of ignition			Between tanks		
					Below ground				
Litres	Gallons	Litres	Gallons	Above ground	Buried portion	Exposed portion*	Above ground	Below ground
<450	<100	1350	300	—	3	3	—	1.5
>450 <2250	>100 <500	6750	1500	3	3	3	1	1.5
>2250 <9000	>500 <2000	27 000	6000	7.5	3	7.5	1	1.5
>9000 <135 000	>2000 <30 000	450 000	100 000	15	3	7.5	1.5	1.5

* Valve assembly and loading/unloading point above ground. (The isolating valves, filling valves and pressure relief valves located on the manhole cover of the underground tank.)

and monitors or mobile units; in any event 9.8 litres/m^2/min is required over the whole surface of the vessel.[30]

Since a BLEVE may occur within as little as 10 minutes after the commencement of fire exposure, water spray systems should be automatic.[18] To negate wind effects water jets should normally be ≤600 mm from the vessel surface. A remote safe operating point should also be provided.

Experience has shown that water discharge may need to be sustained for a prolonged period. Hence continuity of water supply must be ensured at the design stage.

All fittings on pressure storage vessels must be suitable for use at the temperatures and pressures likely to be encountered in service. The number of direct connections below the liquid level should be minimised; if possible only one branch, excluding drain lines, should be provided.

Each vessel must be fitted with at least

- one pressure relief valve connected directly to the vapour space
- one drain, or other means of removing the liquid contents
- one fixed maximum level indicator, and preferably also a contents gauge (the indicator should be independent of this gauge)
- a pressure gauge connected to the vapour space if the vessel capacity is ≥5000 litres
- a suitable earthing connection if the vessel capacity is ≥2250 litres[30,31]

Pressure relief valves should be fitted to protect the vessel under fire exposure, to vent at a pressure above the maximum working pressure at such a rate that further pressure build-up is prevented. No isolation valve is permissible between the vapour space of the vessel and the relief valve. If provision is made for multiple relief valves, interlocks are required so that if one is isolated for maintenance or testing the remaining valves can maintain the full venting capacity.[31,32]

For any vessel ≥5000 litres capacity the relief valve should have a vent pipe which ensures that in the event of ignition of discharging products flame does not impinge on the vessel or on any adjacent vessel, piping or equipment.

The vessel design code will specify the pressure at which relief valves begin to discharge and reach full flow. Each valve should be marked with this start-to-discharge pressure, certified capacity, and the manufacturer's identification.[18,31,32]

Drain connections to atmosphere should be <50 mm in diameter and have two drain valves in series. The length of piping between these valves should be ≥0.5 m, to reduce the risk of simultaneous blocking due to freezing of any water present. Both valves should have a means of actuation which is not readily removable, nor movable from the closed position except by intentional operation. No drain line should discharge within 6 m of a drainage system where this could create a hazard.

Adequate ullage is, as already noted, essential in storage vessels to avoid them becoming hydraulically full as a result of increased temperature.

Therefore instrumentation should be provided to alarm when the vessel approaches a full condition and to shut off delivery before overfilling occurs. Two independent level gauges should also be provided.

A high pressure alarm should be provided, set to below the pressure relief valve setting. A low pressure alarm should also be provided, set above the minimum safe operating pressure of the vessel.[30,32,33]

Pipelines and equipment which is normally full of liquid must be protected, by a combination of safety equipment and standard operating procedures, against hydraulic pressure resulting from expansion on temperature rise (page 99). Hydrostatic relief valves are therefore provided whenever liquid can be trapped, e.g. between closed shut-off valves; if located beneath a vessel these should not discharge any escaping liquid/vapour towards the vessel or adjacent accessways.[18,32] All liquid and vapour connections, with a few specific exceptions, should have shut-off valves located as close as practicable to the vessel.

The common storage tank configurations are shown in Fig. 4.11 for cylindrical tanks. The cylindrical pressure tanks range in size from 4.35 kg bottles to approximately 165 000 gallon 'blimps'; pressures range from 200 bar (for small sizes only) down to 2 bar. Spherical tanks are subject to the same limitations in size as horizontal tanks. Spheroidal tanks are generally used at pressures below 15 psig and can be made in very large sizes.

Small-scale storage of LPG is always in pressurised containers, ranging from small disposable cartridges, e.g. for cigarette lighters, to 100 kg 'bottles'. The cylinders are tested to the minimum pressures for commercial butane of 750 kPa and for commercial propane of 2200 kPa, and are provided with radically different coupling threads to prevent inadvertent misconnection. However, numerous incidents have occurred in which cylinders attached to portable appliances have been responsible for massive damage due either to slow leakage, accumulation and subsequent ignition of a flammable atmosphere or to catastrophic failure under fire conditions. Most cylinders are now being fitted with relief valves to counter the latter risk.

Refrigerated storage

Liquefaction of butane or propane by cooling below their respective boiling points facilitates storage as a liquid which exerts little pressure upon the container. The tank can be of relatively light construction and there are no problems with construction materials, comparable with the cryogenic shock associated with liquefied natural gas at approximately −160 °C. There is, however, a possible hazard from large refrigerated butane tanks.[34]

> 5000 and 10 000 tonne refrigerated storage tanks used at near atmospheric pressure were designed primarily to withstand the head of liquid. The mode of

failure of such a tank containing butane would differ from that of an identical tank containing oil. If a tank containing oil developed a split at a horizontal seam then the weight of the upper part of the tank would tend to keep the split closed. However, whilst relief valves for refrigerated storage are set to a pressure of only a few centimetres of water, with a similar tank containing butane this slight positive pressure would cause the top of the tank to lift and the split to gape open further. The outflow of liquid would thus increase and possibly engulf the surroundings.

It may be advantageous if all refrigerated storages of LPG utilise as refrigerant the same material as that stored.[26]

A large refrigerated tank was used to store pentane; the climate was warm so that refrigeration using low molecular weight hydrocarbon was applied to keep the pressure at a low value. The refrigerant line leaked and the presence of the low molecular weight hydrocarbon in the pentane increased the vapour pressure (as outlined on page 77) and, when the relief valve stuck, overstressed the tank; the top eventually blew off. The ensuing vapour cloud rose 45 m to 60 m and then settled and spread over the surrounding area; ignition occurred after about 30 s and a large firestorm resulted.

Tanks are usually heavily insulated with, e.g. expanded polystyrene, rock wool and perlite, the requirements including[26]
- ability to withstand the impingement of water jets
- resistance to vapour ingress
- maintaining effectiveness even if mechanically damaged
- resistance to fire, either intrinsically or behind protective cladding

The tanks are constructed with an outer cladding, e.g. aluminium sheeting, to protect and contain the insulation which may be 90 cm thick; the interspace may also contain inert gas. The insulation and refrigeration systems add considerably to the complexity of the installation. Therefore refrigeration is usually only applied to large static installations of 2000 tonnes upwards or for bulk carriage by sea.[25]

Tank layout

Bunding is not recommended around pressurised storage tanks or semi-refrigerated storage tanks, since this would create the potential for a large pocket of explosive vapour. One recommendation is that LPG vessels smaller than 136 m^3 should be surrounded by a layer of gravel only, and that vessels larger than this should stand on a concrete apron sloping down to a catchment area provided with gravel to assist vaporisation. A low protective wall may be provided[35] of 0.4 m height. The Health and Safety Executive in the UK recommend that there should be low walls around the vessels not exceeding 0.61 m high.[36] These walls should be at such a distance from the vessels as to allow free circulation of air, with a minimum distance of 1.5 m even for small installations. The walls are not designed to retain any specific quantity of liquid.

Bunds are generally provided for atmospheric storage tanks and for fully refrigerated storage tanks in order to contain spillages of refrigerated liquid. The object is to limit the rate of heat input to any spilled liquid and hence reduce the rate of vaporisation. Such containment of a large spill is also considered advantageous to facilitate 'mopping-up' operations when the liquid could be transferred into road or rail tankers capable of withstanding the vapour pressure on heating up to ambient temperature.

It is possible to envisage conditions under which bunding would be a disadvantage if a tank leaked. If the spill ignited, a contained – possibly sustained – fire would ensue and the light cladding used to protect the insulation would probably fail. The rate of heat input into the tank contents could then generate vapour in excess of the relief valve capacity and possibly lead to tank failure. The presence of a surrounding pool of cold butane would preclude the use of water to cool the tank wall since this water would drain into the bund. Under such circumstances water freezes to form an ice slurry which sinks to the bottom of the pool. In large quantities this would displace butane over the top of the bund, increasing the rate of evaporation and causing the incident to escalate in size and seriousness.[26]

The bund wall should be at a sufficient distance from the side of the tank to prevent any jet produced from a leak from a hole in the side of the tank escaping over it; alternatively the bund should be surrounded by an impervious surface sloped inward to the bund drain area. The corners of the bund are preferably rounded and it should be adequately drained with a valve on the drain situated outside the wall. Walls for full bunding should not be so high as to hinder firefighting; one Code recommends a maximum height of 2 m with an allowance of 0.15 m freeboard above the predicted liquid level to leave space for a foam blanket. There should be a minimum of two access points on opposite sides of the bund to allow for safe access in all wind directions.

Many modern installations incorporate all the safety measures summarised above; there are, however, still some which do not.

In 1978, an isobutane storage sphere at a US refinery being filled from an off-site pumping station via a pipeline ruptured and the leaking 'cloud' spread out before finding a source of ignition, possibly in a 'smoking shed'. There was a flash-back and gas burned beneath the sphere for 30–60 s before it exploded; three major pieces travelled in different directions, the farthest reaching 80 m. The fireball associated with the 800 m^3 of isobutane caused the death of seven operators and injured 10. The relief valve from the sphere was projected 120 m and caused a fire in a nearby refinery unit.[37]

Smaller explosions and fireballs ensued over the next 20 minutes as various vertical and horizontal tanks BLEVE'd. An adjacent sphere also BLEVE'd, the top section travelling 190 m and destroying a firewater tank and a fire pump.

The original sphere failed due to it being overfilled and overstressed when a single level monitoring device failed to indicate the correct level – remaining at 76% full for several hours – whilst LPG was pumped in. However, escalation of the incident was attributed to the following factors:
- the close proximity of the LPG facilities to the rest of the refinery;
- there were no fixed waterspray systems on the LPG vessels;
- the tank farm was not bunded and the ground under and around the vessels was flat;
- there were no gas detectors or alarms on the LPG vessels;
- fire protection of the vessels was considered inadequate.

Location and separation

To reduce the chance of interference, trespassing or sabotage all vessels, vaporisers and pumps should be enclosed within a security fence. This should be at least 1.8 m high and at least 1.5 m from any vessel or vaporiser. It should not interfere substantially with natural ventilation around the installation.

Appropriate separation distances are intended to
- minimise the potential hazards to, and from, the surrounding area
- protect the LPG facilities from effects of radiant heat from fires which may occur in other facilities
- minimise the risk of escaping LPG being ignited before it is dispersed or diluted

Vessels situated above ground should be in a well-ventilated position on an impervious base. This should preferably be sloped in order to prevent a spillage remaining beneath the tank resulting in the risk of engulfment in a fire. The slope should promote flow towards an area in which evaporation can occur safely, or to a catchment pit with a volume capable of containing the largest possible leakage. Obviously these areas should be sited at a sufficient distance from buildings, boundaries or potential ignition sources; the distance from the vessel itself should also be such that flame or significant thermal radiation from any fire in the pit will not affect the vessel.[30] In addition detectors which activate at 25 per cent of the LEL (Lower Explosive Limit, see Chapter 4) may be provided in the pit to provide an early warning of LPG leakage and accumulation.[33]

Storage and use of refillable cylinders

A BLEVE involving an 11 kg LPG steel container may result in a fireball 10 m wide and 15–20 m high.[38] Clearly, therefore, care is required in the storage and use of cylinders. However, in the 'domestic' use of LPG about 75 per cent of all accidents resulting in fires or explosions are due to leaks, e.g. from valve assemblies, pipework or appliances. Common causes are apparatus mishandling, vibration, perishing of flexible hoses, inadvertent extinction of burners by draughts or lack of ventilation, unlit burners being turned on and, with propane, blow-backs.[39] Incidents involving cylinders can be surprisingly costly in terms of fatalities and material damage.

At an ice show in 1963 at a fairground an LPG explosion occurred and of the 4500 in attendance 75 were killed and more than 300 injured. LPG radiant heaters were used to keep popcorn fresh. Just prior to the finale a propane cylinder fell over and emitted a greyish-white cloud. One employee tried to lift the cylinder but the explosion occurred uplifting stands, spectators, concrete and seats towards the central floor area. The cause was probably overheating of an over-filled cylinder by an adjacent heater and release of gas via the relief valve.[40]

Storage

Detailed recommendations have been published for the keeping of LPG in cylinders and similar containers[41] and for the special considerations on construction sites.[42] Reference should be made to these for specific advice but the general requirements for storage, other than rooftop, are summarised in Tables 23.4 to 23.6.

Separation distances are crucially important to protect against domino effects and also, in part, to mitigate against the risk from missiles.

At Mitcham in 1970 an 'empty' LPG cylinder was projected out of a fire and damaged factory premises 380 m away.[26]

Precautions in the use of LPG-fired appliances

Whenever LPG-fired equipment is used, careful consideration should be given to siting it in an area which is accessible but free from combustible material. Alternatively, a suitable barrier is required to separate it from combustible matter. Obviously the cylinder should be situated such that it is not heated by the equipment.

Adequate ventilation should be ensured of the space in which the burner is located to avoid incomplete combustion leading to an accumulation of carbon monoxide (see page 1717) and to dissipate any undetected minor leakage. Cylinders should be handled carefully, e.g.
- preferably on specially designed trolleys
- prohibiting throwing or dropping, and
- not using the cylinder valve for lifting or levering cylinders into position

All ignition sources must be eliminated before connecting up any LPG container to equipment. Cylinders should preferably be changed in the open air with prior inspection for any damage or faults. Cylinders should not be sited near to air intake vents since leaking gas could be drawn into

Table 23.4 General requirements for storage of LPG cylinders

- Store in the open in a well-ventilated area at ground level.
- Avoid other materials stacked near restricting natural ventilation. Flammable liquids and combustible, corrosive, oxidising or toxic substances and compressed gases should be kept separately.
- Avoid cylinders impeding or endangering means of escape from premises, or from adjoining premises.
- The store floor should be level. A load-bearing surface (concrete, paved or compacted) is required where cylinders are to be stacked.
- Security arrangements should be such as to prevent tampering or vandalism. Fencing should be of robust wire mesh which does not obstruct ventilation.
- The area should be at the separation distance from the property boundary, any building or fixed source of ignition quoted in Table 23.5.[39] This may only be reduced if suitable fire-resisting separation is provided.
- To give additional protection from thermal radiation from any cylinder stack fire, the minimum separation distance to any nearby building housing a vulnerable population should, for quantities >400 kg, be 8 m or as in Table 23.5 column 3, whichever is the greater. This may be reduced to those in Table 23.5 column 4 by the installation of fire-resisting separation or a fire wall.
- Prohibit all sources of ignition, including smoking, within a store or within the separation distance. (Exclude motor vehicles other than fork-lift trucks and those for delivery/ collection from open-air stores.)
- Avoid openings into buildings, cellars, or pits within 2 m or the separation distance, whichever is greater. (Any gulley or drain unavoidably within 2 m should have the opening securely covered or fitted with a water seal to prevent vapour ingress.)
- Electrical equipment suitable for a Zone 2 area (e.g. to BS 5345) and constructed to a recognised standard should be installed within the store and separation distance. Zone 2 areas are summarised in Table 23.6.[39]
- Clearly mark the area with notices indicating an LPG storage area, flammable contents, prohibition on ignition sources and procedures to follow in case of fire.
- Avoid accumulation of rubbish, small bushes, dry leaves, etc. within the separation distance. Remove weeds and long grass within the separation distance and up to 3 m from cylinders.
- Cylinders should be stored upright with the valves closed, with any protective cover, cap or plug in place. Cartridges *without* valves may be stored on their sides.
- Cylinders on a vehicle or trailer parked overnight rate as a single stack so that the separation distance in Table 23.5 column 3 is applicable.
- Cylinders received into store and taken out for delivery should be checked for damage or leakage. Stacks should be inspected daily for stability and that they contain no damaged/ leaking cylinders. (Any such cylinders should be dealt with as in Table 23.7.)
- Cylinders should be handled carefully to avoid personal injury or damage to them.

buildings and ignited. Manufacturers' instructions should be followed in connecting up, leak testing (by sound, smell or soapy water and never with a naked flame) and operation. A written procedure should be displayed by equipment since, e.g. the information in Table 23.7 is essential for safe operation.

Cutting and heating torches, roofing irons and other hand tools are commonly fed with LPG from portable containers. The container should always be used in an upright position and be adequately secured, e.g. on a trolley, to ensure that this cannot be knocked over. Only hoses specified for LPG gas service at the recommended operating pressure should be used to connect equipment to an approved regulator. Hoses should be

Table 23.5 Recommended minimum separation distances for total storage of LPG in cylinders or size of maximum stack. (The greater of the two distances is advised.)

(1) Total quantity LPG store Kilograms	(2) Size of largest stack Kilograms	(3) Minimum separation distance to boundary building or fixed ignition sources **from the nearest cylinder** (where no fire wall provided) Metres	(4) Minimum separation distance to boundary building or fixed ignition source **from fire wall** (where provided)*† Metres
From 15 to 400	Up to 1000	1‡	Nil
,, 400 ,, 1000		3	1
,, 1000 ,, 4000		4	1
,, 4000 ,, 6000	From 1000 to 3000	5	1.5
,, 6000 ,, 12 000		6	2
,, 12 000 ,, 20 000	,, 3000 ,, 5000	7	2.5
,, 20 000 ,, 30 000	,, 5000 ,, 7000	8	3
,, 30 000 ,, 50 000	,, 7000 ,, 9000	9	3.5
,, 50 000 ,, 60 000	,, 9000 ,, 10 000	10	4
,, 60 000 ,, 100 000		11	4.5
,, 100 000 ,, 150 000	,, 10 000 ,, 20 000	12	5
,, 150 000 ,, 250 000	,, 20 000 ,, 30 000	15	6
Above 250 000		20	7

* The distance from the nearest cylinder to a boundary, building, etc. should be not less than in column 3 when measured around the fire wall.
† Minimum distance from nearest cylinder to fire wall should normally be 1.5 m except as qualified.
‡ No separation distance is required for these quantities where boundary walls and buildings are of suitable construction.

Table 23.6 Areas classified as requiring Zone 2 electrical equipment

Location	Extent of classified area
Storage in open air	In the storage area up to a height of 1.5 m (5 ft) above the top of the stack, or beneath any roof over the storage place
	Outside the storage area or the space covered by any roof up to 1.5 m above ground level and within the distance set out for a fixed source of ignition in Table 23.5 column 3.
Storage within a specially designed building or in a specially designed storage area within a building	The entire space within the building or storage area and outside any doorway, low level ventilator or other opening into the store within the separation distance set out in Table 23.5 column 3 up to a height of 1.5 m above ground level

Table 23.7 Some factors in safe LPG-equipment operation

Leaking cylinder:
- If the leak cannot be stopped, remove cylinder to a well-ventilated open space remote from sources of ignition
- Leave it to leak with the valve uppermost
- Display notices prohibiting smoking or other naked lights
- Prevent access, e.g. by barriers
- Inform the supplier immediately

Gas escape from burner:
- Allow any gas to fully disperse before attempting to relight
- If the lighting-up procedure was wrong, interrupted or incomplete the gas must be turned off, unburned gas allowed to disperse and any fault rectified
- Equipment should be turned off if there is any odour of gas after ignition or during normal use

regularly inspected for wear or accidental damage. With torches, non-return valves should be provided in both the gas and oxygen supply lines; the use of flash-back arrestors is recommended.[42] Obviously the working area should be kept clear of combustible material, or if this is impracticable shielding should be provided between the operation and combustible matter. Control by a Hot Work Permit is advisable. Detailed guidance is available on welding and flame cutting precautions.[42]

LPG cylinders for bitumen boilers or cauldrons should be <3 m from the burner. General precautions in use include,
- locate boiler and cylinders where they are unlikely to be struck by site traffic.
- pay particular care to avoid the hose being damaged, e.g. by equipment or foot traffic, and that it is not a tripping hazard.
- avoid overfilling or boil-over of the boiler and select location with such an incident in mind.
- never leave the cauldron unattended with the burner alight.
- never transport or tow a cauldron with the burner alight.

A hired van was used to carry equipment including a free-standing bitumen boiler, three propane cylinders, and some petrol for use in cutting slots in a road surface and refilling with bitumen. Since obstruction of the road led to traffic congestion the van was moved with the boiler still lighted. It fell over and in the ensuing fire two of the propane cylinders BLEVE'd; they were projected 15 m across the road and flames were 6 m in diameter. One fireball enveloped a 12-year-old boy who was watching; he sustained 90% burns and subsequently died.[43] (The contractors were later convicted of offences under s.2 and 3 of the Health and Safety at Work, etc. Act.)

When LPG-fired equipment is used for drying operations measures are necessary to avoid local overheating which could result in scorching and fire. For example, heaters in drying rooms should be fitted with protective cages and care exercised to ensure that clothing does not restrict ventilation. In general adequate air supply and ventilation, and avoidance of hot-air currents impinging upon gas cylinders, are elementary precautions.

Precautions with LPG-fired equipment in small rooms or huts

Significant hazards are associated with the improper use of LPG-fired equipment, e.g flueless space heaters, gas rings and radiant heaters, in small rooms, huts, cabins and caravans.[40,43]

On construction sites in the UK over 50% of the incidents involving LPG occur in site huts. They concern either the accumulation of a flammable gas concentration from a leak, flame-failure or an inadequately turned-off valve or build-up of an unrespirable atmosphere through inadequate ventilation.

Precautions for use in site huts are summarised in Table 23.8.

Table 23.8 Precautions with LPG-fired equipment in site huts

- Locate all cylinders and regulators in a safe position outside. Protect cylinders against physical damage.
- Remove portable appliances (e.g. domestic flueless space heaters) incorporating cylinders to an outside store overnight.
- Use cylinders with valves uppermost and connect to appliances by rigid copper or steel piping, which is exposed to allow for leak checks and to avoid local gas accumulations from leaks. Short flexible hoses should only be used for final connections.
- Provide a stop valve at the point of entry of any fixed supply line into a hut if the appliance does not incorporate a control valve.
- Provide a good standard of general ventilation at both high and low levels as well as for individual appliances.
- Use only flueless space heaters which are fitted with atmosphere sensing devices.
- Regularly maintain appliances. Soot and smell are indicative of lack of air, dirty burners or incorrect adjustment.
- Display the correct operating procedure in each hut to ensure that, e.g. the fuel supply is turned off at the appliance and the cylinder after use, and that if a smell of gas is noticed no attempt is made to light any appliance and ignition sources are excluded.

Transport

Large quantities of propane and butane are transported safely and economically over long distances.

Disposable cartridges are used in large numbers, mainly in the leisure field, with content weights of up to approximately 0.45 kg. Their major disadvantage is that they cannot be removed from an appliance until exhausted; otherwise liquid may flash-vaporise causing cold burns and, if the vapour is ignited, a local fire.

Refillable cylinders are used mainly for domestic and light industrial applications. They range in weight from 5 kg to 46 kg capacity and, due to the significant difference in gauge pressure, are designed specifically for butane or propane (for 2 bar and 7 bar respectively). The fittings are not interchangeable. Cylinders require pressure-testing periodically, before and during use. The first retest is 10 years after the date of manufacture with subsequent retests at five-yearly intervals.[44]. Each cylinder should be marked with a serial number, the year of manufacture, the specification to which it was fabricated, the date of test, the test pressure and the minimum hydraulic capacity.

Road tankers are used for the transport of bulk loads between 1 and 20 tonnes and railcars between 10 and 90 tonnes. Tanks of more than 5 m^3 capacity are fitted with internal baffles to minimise internal surging of the contents. The maximum capacity of road tankers used in the UK is determined by the current Motor Vehicle (Construction and Use) Regulations and by the ullage required. Because of the volumetric expansion considerations referred to on page 99, it is necessary to ensure that the maximum amount of LPG in any road tanker should be such that the tank will not be >97 per cent full with the contents at the highest temperature likely in service. This Reference Temperature for Filling Ratios is in the UK 42.5 °C for tanks of 5 m^3 water capacity and below and 38 °C for tanks of larger size.[45] The maximum permitted LPG fill by weight is given by[45]

$$W_1 \text{ (kg)} = 0.97 \times V_1 \times \rho_{(ref)}$$

where V_1 = water capacity of vessel in litres

$\rho_{(ref)}$ = relative density of LPG at the reference temperature for filling ratio (RTFR)

and the tank should not in any case be filled so that a liquid-full condition could arise at 5 °C above the RTFR.

Since at low ambient temperatures the LPG vapour pressure may fall to less than atmospheric pressure, tankers must be designed to withstand this unaccustomed external stress, or alternatively be provided with vacuum relief. For this purpose a design temperature of −10 °C is assumed, at which the vapour pressure of n-butane is 35 kPa below atmospheric pressure.

LPG is transported by pipeline, usually submerged 1 m below ground.

Lines are currently wrapped in coal tar/fibre[35] and the problem of corrosion at points where wrapping was discontinuous has been largely overcome by the use of cathodic protection.

Large quantities of LPG are moved by pipeline from the gas-producing areas to coastal refineries. There it is processed to remove condensate and then loaded at terminals into marine gas carriers of up to 100 000 m^3 capacity for delivery to terminals in other countries. There the quantities are broken down to smaller volumes which are transported by small coastal or river gas carriers, road tankers and rail tankers to industries and to bottling plants.

The subject of pipeline and marine transport is, however, too extensive to be treated in full here.

Transport hazards

Transport hazards involving road, rail or ships include
- accidents en-route. (As a result of insulation and the fitting of head-shields and better couplers to prevent rupture, the numbers of LPG BLEVEs due to rail accidents have been reduced in the USA. Road containers may need to avoid high population centres.)
- stop-overs at parking places/marshalling yards/moorings
- during loading/unloading when transfer lines between vehicles and stationary facilities are frequently connected and disconnected. Here the main hazards include driving away whilst connected to the loading arm, failure of transfer connections, human error, and over-filling as illustrated by the isobutane case history on page 1332 and those below.

A rail tank car containing 80% butadiene and 20% butylene was overfilled. The car, which was not fitted with relief valves, overheated in the sunshine and burst releasing its contents which ignited within 20 s. The explosion caused major damage in a 50–100 m diameter area. A gas holder 200–250 m away was also damaged releasing acetylene. Losses amounted to $60m in 1986 and there were 60–80 fatalities.[40]

Workers connected liquid and vapour lines to a 128 m^3 (33 940 gal) rail tank car at an LPG distribution facility. The connector leaked, despite attempts to tighten it by striking it with an aluminium alloy pipe wrench. The leak ignited and two men were severely burned one of whom died. Torch fires impinged on the tank causing the relief valve to lift. Despite deluging the tank with water the tank ruptured and propelled itself 360 m along the track, and gave rise to a fireball *ca.* 50 m diameter. Storage and office facilities on the site were destroyed, 12 firefighters were killed and almost 100 other people injured.[40]

Control measures for loading/unloading vary between installations but key precautions include
- restricted access to the loading bay
- barriers to prevent driving away whilst still connected

- switching off LPG supply automatically if the connection is broken
- earthing of vehicles
- tare weighing and level instrumentation and alarms to prevent over-filling
- personal protection (e.g. to protect against physical hazards and exposure to leaking LPG and against frost-bite)
- LPG leak and fire detection/alarm systems
- emergency shut-down arrangements, including remote isolation of pumps
- water spray protection and fire extinguishment
- good communication between driver and site personnel
- operator training

A fire is more likely than an explosion following loss of containment of LPG from a rail tank car, barge, ship tank or pipeline. However, there have been both UVCEs and BLEVEs resulting from transport accidents.

> It has been stated that in the large majority of single transportable gas tanker incidents the relief valve will prevent a BLEVE.[46] However, such relief devices are mostly sized as for a static tank, i.e. assuming vapour only flow, and are put on top of the tanker, i.e. connected to the vapour space. If a tanker turns over so that the relief valve is below the liquid level, it would pass liquid and would be unlikely to prevent overpressure and failure of the tank barrel. (A device sized for liquid flow at a rate corresponding to the rate of evolution of vapour would need to be much larger, so that problems associated with leakage and spurious operation are increased.)[47]

With regard to the transport of cylinders, some 50 incidents per year are reported to the UK Health and Safety Executive which involved LPG cylinders on vehicles. However, information from the police suggests there is considerable under-reporting. The two main dangers are overheating and exploding, and gas leaks. Dangerous fragments may be thrown great distances.

> A fire and explosion in a closed van carrying 100 LPG cylinders resulted in a massive fireball and one cylinder was thrown 200 m.

Recommendations for transport include
- cylinders should be secured upright on *open* vehicles;
- there should be a fire extinguisher in the cab;
- drivers should have adequate training and instruction;
- drivers must have readily available information on the load and its hazards in case of emergencies.

In emergencies involving a leaking LPG vessel accompanied by a vapour cloud, or a vessel on – or exposed to – fire, the considerations are as follows.[48]
- Protection of people. All persons, except those required to deal with the emergency, should be evacuated >600 m from the cloud, wherever it

drifts. Firemen should approach from upwind of the leak or fire. Horizontal tanks should not be approached from the ends. If a cloud is present inside a building, firemen should only enter to complete a rescue since an explosion is very likely.
- Shutting off the gas. Valves at the container or remotely should be shut off (if this is not automatic). If valves cannot be located or used, every ignition source has to be shut off in the path of the vapour.
- Direction of vapour using water. As well as helping to protect firefighters closing valves, water fog can assist in dispersing a vapour cloud to a safer location. The spray should be directed across the vapour path with firefighters keeping low behind it and never entering, or approaching closely to, the cloud.

In December 1988, nine people were killed and >12 injured as a result of an incident in which a 10 000 US gallon semitrailer carrying propane overturned as it went around an inclined ramp on an interstate highway about two miles east of downtown Memphis and the Mississippi river. It struck part of a bridge and the front end of the shell ruptured; the liquid propane formed a vapour cloud which spread about 457 m up the expressway before it ignited forming a huge fireball. Several vehicles were caught in the vapour cloud. The tank shell rocketed approximately 120 m causing damage to houses and starting secondary fires.[49]

A major firefighting operation ensued with the Fire Department dealing with four different incidents at the site; two involved multiple building fires, one a single commercial building and the fourth the vehicles on the highway.

Road transport of LPG cylinders

Clearly LPG cylinders must be transported safely if they are not to pose a hazard to people and property. As already discussed, if they overheat and catch fire or explode fragments can be ejected considerable distances. The hazards can be reduced significantly by following simple procedures.[50]

Gas leaking from any container may be heard or smelled, or it may form frost. Soap solution can be used to trace the leak but obviously a naked flame must never be used. In any event smoking should not be permitted near to a cylinder.

Cylinders should be carried in open vehicles, kept upright and adequately secured, e.g. with a rope. Since serious accidents have resulted from tampering with unattended LPG cylinders, they should not be left in vehicles unsupervised. Bitumen boilers or cauldrons or similar equipment should never be transported with the burner alight. A fire extinguisher, e.g. 1 kg dry powder type, should be kept in the cab. (Its purpose is to keep a small fire under control until the emergency services arrive.)

In the UK, with certain exceptions, the driver of any vehicle carrying LPG cylinders, including 'empty' cylinders, must have received adequate training and instruction about the hazards of LPG, emergency procedures and driver duties.[50] Each driver must also have a Tremcard (page 813); this information must be kept readily available in the vehicle, e.g. on a clipboard in the cab. Relevant statutory duties[51] are summarised in Table 23.9.[50]

Table 23.9 Transport of LPG cylinders by road: duties of operators and drivers

Duties of the vehicle operator:
Check whether the regulations apply. There are four principal exceptions:
— where the cylinders have a capacity of less than five litres;
— when the cylinders are part of (i.e. connected to) equipment carried on the vehicle, e.g. tar boilers or burning gear;
— one spare cylinder for the equipment described immediately above is also exempt but only if the equipment uses only a single cylinder;
— LPG cylinders used in connection with the operation of the vehicle, e.g. for cooking in a mobile shop, water heating.
- Ensure the vehicle is suitable. Open vehicles should normally be used. (Closed vehicles should only be used for the carriage of a small number of cylinders; the load compartment should have adequate ventilation.)
- Ensure the driver is provided with adequate information in writing about the LPG so the nature of the dangers involved and emergency action to be taken are known (e.g. by a TREMCARD).
- Ensure the driver has received adequate instruction and training and keeps necessary records.
- Ensure loading, stowage and unloading are carried out safely. All cylinders should be packed, strapped, supported in frames or loaded so that they cannot be damaged as a result of relative movement and should be stowed with their valves uppermost.
- Ensure all precautions to prevent fire or explosion are taken including the provision of suitable fire extinguishers.
- Ensure the vehicle displays two orange plates if 500 kg of LPG is carried.
- Ensure the appropriate authority is informed of any fire which involves the consignment of any uncontrolled release or escape of the substance being carried.

Duties of the driver:
- Ensure information in writing about the substances supplied by the operator is always available during carriage (and that information about previous loads, or other substances not being carried, is destroyed, removed or locked-away).
- Ensure loading, stowage and unloading are carried out safely.
- Ensure all precautions to prevent fire or explosion are taken throughout the carriage.
- Ensure the orange plates are displayed when required by the regulations and are clean and free from obstruction.
- When 3 tonnes or more of LPG is carried, ensure the vehicle, when not being driven, is either parked in a safe place or is supervised by the driver or some other competent person over the age of 18.
- Provide appropriate information for a police officer or traffic examiner who wishes to inspect the vehicle and load.
- Ensure the operator is informed of any fire which involves the consignment or any uncontrolled release or escape of LPG.

LPG in aerosols

Background

The market for aerosols is diverse and includes insecticide, paint, and household, animal, industrial, automotive, food and personal products. In 1987 aerosol fillings world-wide approached 8 billion.[52]

1344 Flammable and explosive hazards

Since the early days of aerosol development chlorofluorocarbons (CFCs) were the favoured propellant. Alternative systems were then principally hydrocarbon blends which tended to be reserved for the cheaper end of the market. More recently CFCs which have also been used in refrigeration and air-conditioning units, as speciality solvents, and as a blowing agent for rigid polyurethane foams, have been linked via a free-radical mechanism to the depletion of the ozone layer in the stratosphere (see page 1243):

$$CFC \rightarrow Cl_\cdot$$
$$Cl_\cdot + O_3 \rightarrow ClO_\cdot + O_2$$
$$ClO_\cdot + O_3 \rightarrow Cl_\cdot + 2O_2$$

The USA therefore banned the use of CFCs in aerosols and the EEC introduced a voluntary reduction in the use of this class of propellant. One result has been the use of alternative propellants, e.g. dimethyl ether and carbon dioxide, plus an increased consumption of LPG as a more environmentally friendly propellant or co-propellant. However, the highly flammable properties of LPG introduce a hazard into what is otherwise a low risk operation. The industry has devised a series of codes as guidance on the safety requirements for filling with LPG, e.g. refs. 30 and 53–56.

Detailed requirements, specifically for handling LPG will be influenced by, amongst other factors, local legislation. This section, therefore, provides an overview of some of the main safety features to be considered when filling aerosols with this propellant as a general indication of the safety factors applicable to the handling of volatile flammable chemicals. It does not constitute comprehensive guidance for aerosol fillers who should consult more detailed sources.

LPG specification

As mentioned previously, there are national standards for LPG specifications. Stench gas is normally removed from LPG for aerosol usage. There are several grades of LPG available comprising different blends of propane and butane to afford a range of propellant pressure, as summarised in Table 23.10.

Table 23.10 Grades of LPG for aerosol use

Hydrocarbon propellant	Approximate composition (% by volume)			Vapour pressure at 25 °C	
	n-Butane	Isobutane	Propane	bar	psig
30	59	29	12	2.05	30
40	52	26	22	2.72	40
48	47	23	30	3.26	48

Aerosol filling plant

An aerosol plant, comprises a raw-material store (including LPG farm), concentrate production hall (for blending and filling all materials other than propellants, e.g. perfume, solvent, active ingredient), placement of valves onto containers, a gassing room, crimping of valve into position either before or after gassing, facilities for quality control of finished product (e.g. check weighing, pressure testing, leak testing by immersion of cans in warm water bath) and warehouse.

The general precautions for handling highly flammable liquids have been discussed in Chapter 4 and those for LPG in Table 12.20 on pages 732–3. This section deals more specifically with the use of LPG in gas filling operations: LPG storage is covered earlier. Selected key design features are described to highlight the practical approaches adopted for safe filling with LPG: more detailed advice is given elsewhere (e.g. refs. 53 and 57–9). The overall strategy for minimising risk is shown in Fig. 23.8.

General plant layout and design

Hydrocarbon filling lines are inherently prone to domino effects (see pages 869 and 892), i.e. a minor fire resulting in a minor or major Confined Vapour Cloud Explosion or Boiling Liquid Expanding Vapour Explosion (page 877) and/or a large fire.

> In Indiana, USA, on 12 August 1972, fire and explosion damaged a single-storey aerosol-filling plant just 10 minutes after workers had returned from lunch. The lightweight roof structure was blown off, relieving the main force of the explosion. However, windows were blown out on both sides of the factory and the brick walls were displaced 13 cm. Following the explosion flames were reported to have reached a height of 10 m. Twenty-two people were injured, eight seriously.[60]
>
> Aerosol products were being filled with a hydrocarbon propellant when a leak is believed to have developed due to a fault on the gassing head. Attempts were made to correct the fault without isolating the hydrocarbon supply. Continued leakage brought the propellant–air mixture within the flammable range and it was ignited by some unknown source, probably the shrink-wrapping machine. The flexible connection between the propellant main supply and the gassing machine is believed to have been damaged in the initial fire, resulting in additional leakage of hydrocarbon culminating in the explosion.

The siting and general design layout of LPG storage facilities, process equipment, control rooms, amenity buildings, access and emergency escape routes, the provision of adequate explosion vents and facilities for containing and fighting fire are crucial for safe operation and the avoidance of major domino effects in the event of mishap.

> A large contract aerosol-filling factory in Johannesburg was gutted by fire and two explosions on 2 March 1982. The first explosion occurred in the gassing room due to ignition of a hydrocarbon propellant leaking from a fractured

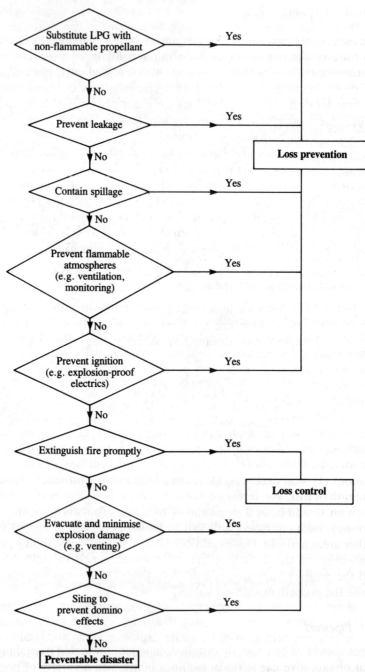

Fig. 23.8 Loss control strategy with aerosol plants

pipe. This triggered a second explosion in an acetone-paint mixing plant located above the gassing room. The bulk hydrocarbon storage was not involved, but the fire took 12 hours to bring under control. Eleven employees were killed and 100 injured. Damage was estimated at 5 million Rand. The cause of the initial explosion was ignition of propane or butane gas leaking from a fractured pipe to one of the gassing machines. This blew away the roof of the gassing room, which was constructed from 152 mm thick reinforced concrete and served as the floor of one of the mixing chambers directly overhead. This room was used for the preparation of acetone-based paint. The subsequent explosion and fire resulted in ignition of the acetone which was sucked back into the damaged gassing room. The gassing chamber was thoroughly mechanically-ventilated, equipped with flammable vapour detection systems, and generally designed to avoid the possibility of flammable mixtures developing. However, adequate explosion-venting was not apparently provided.[60]

LPG installations should be segregated from other high risk areas and should not be sited immediately uphill of areas where sources of ignition can be foreseen. Natural depressions, other than bunds to contain spillage, should be avoided. Access roads should be kept free of obstruction.

Gassing rooms should be at ground level and of single-storey construction. It is advisable to restrict their size to that required to house one filling machine, yet large enough to give easy access for maintenance. They should be dedicated to gas filling. Commonly, the gassing room is built with three 'hard' walls with the fourth composed of 'soft' material or provided with explosion vents of sufficient area. Similarly, the roof can be designed to withstand pressure from an explosion or equipped with preferential explosion venting. Roof vents should be protected from build-up of snow and all vents should be chained to prevent them becoming missiles in the event of an explosion. The areas of the soft wall or roof require careful calculation and their location should ensure relief is in a safe direction. Guidance is provided elsewhere;[56] as a rule-of-thumb the area of the soft wall or roof should be about 20 per cent of the total surface area of the gas filling room. Typical design pressures are 1400 kg/m^2 with relief pressures of one-tenth of this.

The room should have a minimum of two doors (outward-opening and self-closing), each equipped with sills to prevent seepage of spilled LPG into other areas and with hinges and catches capable of withstanding pressures in excess of that at which the vents are designed to yield. Doors around the primary enclosure should be interlocked to stop the machine and close the propellant shut-off valve.

Pipework

Pipelines should all be properly installed, labelled, protected from damage and not of excessive capacity. In facilities filling with a variety of propellants it is imperative to ensure the wrong gas supply cannot be connected to the filling machine. Flexible piping should only be used after consultation with the propellant supplier. Piping from drums or cylinders of LPG should be fitted with self-sealing quick-release couplings. Pressure relief mechanisms should be installed between closable valves. All pipework should be regularly inspected for damage and leaks (e.g. with soap solution or with explosimeters as discussed in Chapter 7) and should have good electrical continuity. As might be anticipated, the initial event in many fires and explosions involving LPG has been gas leakage from a pipe fracture, joint failure, or faulty hose.

A serious fire and explosion at a Nigerian aerosol-filling plant in June 1982 killed 10 people and injured 27, some with serious burns.[60] The incident followed severe flooding at the factory site after heavy rainfall. Part of the perimeter wall collapsed causing the hydrocarbon propellant supply pipework to the plant to fracture. A relatively small quantity of propane/butane escaped (i.e. 50 to 60 kg) before the automatic shut-off valve operated on the storage tank outlet. The liquefied gas was spread over the site by the flood water (see Table 23.1). The ignition source may have been a spark created on isolation of the electrical supply, due either to the flood or to the gas escape. A secondary fire occurred in the aerosol-filling plant.

The casualty rate was relatively high because a number of people had congregated to watch the flood water and the escaping LPG was distributed rapidly in the direction of the onlookers who were tragically engulfed in the ensuing fireball.

In February 1982 at 06.30 h just 15 minutes prior to the commencement of work, an explosion destroyed a West German factory which manufactured cosmetic products.[60] Material damage was estimated at approximately 14 million DM. Three people were killed and 20 were injured, some seriously. The casualty rate could have been much higher but because of the time of the incident few employees had arrived for work.

The blast from the explosion caused extensive damage to the roofs of houses up to 1 km away. Little was left of the factory but three bulk hydrocarbon propellant storage tanks survived intact. The explosion was heard 10 km away and one report claimed that 1500 buildings were damaged.

The Deputy Works Manager had entered the factory at about 06.15 h and switched on the lights, the compressor and the exhaust ventilation system. Possibly the initial vapour concentration was above the Upper Explosive Limit (UEL) and a flammable mixture was produced only on start-up of the ventilation system. The source of ignition was not identified but the propane/butane mixture is likely to have spread to other parts of the factory where an electrical spark, e.g. from a light switch, may have triggered the explosion. The leakage of an estimated 100 kg of hydrocarbon was suggested to have originated from bolted fastenings in the pipeline system in the gassing room which had been loosened through continuous vibration of the filling equipment.

Ventilation

A key precaution when handling LPG is to ensure leakages cannot occur. This is an ideal objective and, in reality, fugitive emissions will inevitably arise during normal use (e.g. during disengagement of cans from the filling head). Therefore proper ventilation is crucial to prevent flammable concentrations of LPG from developing and to prevent occupational health problems arising as a result of overexposure to LPG. Furthermore, by use of air flow or pressure sensors in the ventilation ducting, the system should be interlocked with equipment so that in the event of ventilation failure the gas supply is closed and plant is rendered inoperative. Ventilation failure is also usually signalled by audible alarms. (The use of sensors to detect fan motor failure is less reliable since the shaft can continue to rotate when fan fin failure has occurred, i.e. deterioration in ventilation performance would not be detected.)

Detailed ventilation design will differ for each operating company. It will be influenced by factors such as the speed and type of filling machine (which can be of a manual, semi-automatic or fast speed line design), the amount of LPG released under normal operation and in the event of foreseeable abnormalities, and the size of the room. Consideration of the detailed specification for ventilation systems is therefore beyond the scope of this text. In principle, however, there will be a need for both general and localised extraction and due cognisance must be taken of the density and flammable range of the gas. Thus the gas filling unit should be fitted with high and low level extraction with make-up air entering at high level. For example, to remove LPG at source fishtail local extract vents are often used around the filling head operating at or above 1 m/s duct-face-velocity. The maximum concentration of LPG in the duct should not exceed 50 per cent of the LEL. The system should have a facility for doubling the extraction rate in the event of hydrocarbon concentrations in the room reaching 10–20 per cent of the LEL. Evaporation of excess LPG from the valve cups may be encouraged by air jets directed towards the local extract ventilation nozzle and thereby preventing egress of LPG into the main production hall. To ensure air flow sweeps the entire floor area, leaving no dead spots, the positioning of bulky equipment needs careful planning and the air flow pattern should be checked with smoke tracers.

In the main production hall filled cans are tested for leaks by immersion in a water bath. This should be equipped with an extraction hood, the air flow of which should ensure capture of leaking LPG at any point within the bath. Similarly local extract ventilation should be provided at the button placer and waste can collection points. Like the gassing room the main hall should be fitted with a sufficient number of floor level extract vents to scavenge any gas escaping into the general milieu. All vents should exhaust externally to atmosphere via stacks of sufficient height above roof level and not sited near air intakes.

In calculating ventilation requirements it is recommended[61] that consideration be given to both (a) removal of the highest load of flammable gas that could escape during gassing operations, and (b) the need to provide a turnover of 1 to 1.5 times the gas house air volume per minute. Assuming:

1350   Flammable and explosive hazards

- gas house dimensions are 14 feet × 14 feet × 10 feet
- propellant leak rate is 3 ml per head
- gasser operates at maximum of 250 cans per minute (cpm)
- propellant present with lowest LEL is isobutane (LEL = 1.845 per cent)
- 1 gal of liquid isobutane produces 30.6 ft^3 of gas
- a safety factor of 20 per cent is applied to cope with contingencies, e.g. can rupture

then

- propellant release rate is 250 (cpm) × 3 (ml per can) = 750 ml/min.
- applying safety factor, 750 ml/min × 1.2 = 900 ml/min.
- converting to gallons, 900 (ml/min)/3785 (ml/gal) = 0.238 gal/min, or 0.238 gal/min × 30.6 (ft^3/gal) = 7.28 ft^3/min gas.
- the ventilation rate of the gas/air mixture is given by 7.28 ft^3/min × (100 per cent gas–air/0.184 per cent gas) = 3956.5 ft^3/min.
- the total ventilation requirements therefore are 3956 ft^3/min + 1 air change per min = 3956 + (14′ × 14′ × 10′) ft^3/min = 5916 ft^3/min, i.e. 6000 ft^3/min.

The emergency ventilation rate triggered when isobutane is detected at 20 per cent LEL would typically be two or three times the above rate.

General advice on design of extract ventilation is given in Chapter 12; more specific guidance for aerosol fillers is given elsewhere.[56,61]

Ignition sources

All unprotected sources of ignition should be excluded from areas where LPG operations are encountered. Obvious examples include cigarettes, lighters, matches, open flames, furnaces, internal combustion engines, hot pipes, etc. and electrical equipment needs careful consideration.

> In February 1980 a fire and a series of explosions occurred in a warehouse at Stoke-on-Trent. The entire contents of the warehouse including 49 tonnes of LPG and one tonne of petroleum mixtures were lost.[62]
>
> The gas source remains unidentified but probably resulted from leakage of pressurised LPG containers. The ignition source was probably the electrical system of a battery-operated fork-lift truck. It was concluded that the main warehouse should have been classified as a Zone 2 area and only suitably explosion-protected electrical apparatus should have been used in it.

All electrical equipment inside the gassing room should be to Zone 1 standard (i.e. an area where explosive atmospheres are *likely* to occur in normal operation) whilst that area external to the gassing room generally should be considered Zone 2 (an area where explosive atmospheres are *unlikely* to occur in normal operation) (see page 714). Portable non-explosion-proof electrics should not be allowed on the plant except for maintenance which must be controlled with a permit-to-work, to ensure the gassing line is not operational and that the area has been shown to be gas free.

On 22 October 1981, a plant in Massachusetts, USA, was seriously damaged by an explosion which ripped through the whole single-storey, steel-framed building. The blast destroyed the sprinkler system and blew out several of the plant's walls, leaving the steel framework and roof virtually intact. A large fire ensued. Five employees died as a result of the incident.

The cause was claimed to have been ignition of a hydrocarbon–air mixture which accumulated as a result of faulty seals on the isobutane filler passing propellant via a vacuum pump exhaust into a non-explosive-proofed section of the main building outside the gas-filling room. Frost built up on the vacuum pump exhaust and someone attempted to stop the process. A spark generated on opening the contact-breaker was the suggested source of ignition. The vacuum pump which was located outside the gas-filling enclosure apparently was not explosion-proof rated.

This incident appears to have been caused by a combination of an underrated vacuum pump located outside the high-hazard, gas-filling enclosure and the exhausting of the enclosure into the building as opposed to exhausting outside to atmosphere.[60]

Static electricity as a source of ignition is discussed on pages 144–7 and in Chapters 9 and 27. It can be safely dissipated by ensuring that all equipment is continuously bonded and connected to earth. Conveyor systems should be constructed of metal and the total electrical resistance from any point to earth should not be greater than 10 million ohms, but the value on commissioning should be much lower to allow for plant deterioration. High humidity also aids dissipation. The electrical potential of aerosol cans can reach unacceptably high values when product escapes uncontrollably from pin holes. The resulting vapour cloud becomes highly charged and can spark on contact with earthed objects. Such cans must thus be removed as soon as possible to a safe place.

Gas detection

Any LPG discharges which lead to significant accumulation of LPG as a result of unforeseen circumstances must be detected. Reliable LPG detection sensors must therefore be installed which are interlocked with an audible alarm system, with the extract ventilation and with the plant to shut off the gassing machine and propellant supply. Typically, concentrations of LPG in the ambient air at 20 per cent of the LEL would automatically trigger the ventilation booster whilst levels reaching *ca.* 40 per cent LEL would result in automatic and total shut-down of the plant (with the exception of the ventilation and emergency lighting). The 'warning' and 'alarm' states would be signalled by a different audible/visual system.

The main types of detectors used are pellistors, semi-conductors and infra-red analysers for which the operating principles are described in Chapter 8. These require regular calibration. Positioning of the detector head needs careful thought; common sites include close proximity to the line feed pump in the tank farm, floor level in the gas-filling room, the exit of the conveyor from the filling room and various locations in the plant area, generally at floor level. The main control modules, which should be self-checking with a system 'fail' indicator, are usually housed close to the production line in a non-explosion-proof area such as the gassing plant switch room.

Fire and explosion suppression

Installation of fire and explosion suppression systems offer a back-up to minimise loss. These are usually halocarbon systems either operated manually or linked to the fire alarm system to discharge gas upon activation to extinguish fire, e.g. in the propellant filling room. Guidance on the use of these systems is provided by the HSE.[63] In general water is not used in fighting LPG fires; it is normally reserved for preventing ignition from other sources and for cooling adjacent vessels. Pressurised dry powder extinguishers or BCF systems are employed. However, in large warehouses storing finished product water sprinkler systems are commonly encountered. To be effective these must be of sufficient capacity, properly installed and provided with a reliable water supply. The mix of stock in storage requires attention; indeed some authors have proposed classification of aerosol products themselves according to type as a means of specifying the appropriate sprinkler system.[64]

> A fire in a distribution warehouse resulted in a $114m loss. The 1.2 million sq ft facility contained extensive amounts of highly flammable materials including propane tanks, small arms ammunition, tyres, plastics, flammable liquids, and aerosols.[60] These products were stored on racks without in-rack sprinklers, under a 10 m high ceiling fitted with a sprinkler of only a 0.4 US gal/min/ft^2 design water density. This level of sprinkler protection was probably inadequate for the materials in storage. The aerosols were involved in 'rocketing-cans' phenomena and were therefore primarily implicated as the cause of this major fire loss.

> A massive fire at a Texan plant used to fill a range of aerosols with hydrocarbons and carbon dioxide propellants caused damage valued at $1.5 to 4.6 million.[65] It started out-of-hours on the second floor of the laboratory due to an unspecified accident and then spread to the warehouse. Because of threat to the installation's propane storage tank up to 1000 local residents were evacuated. Unfortunately the plant was not provided with a sprinkler system and the fire hydrants were some distance from the plant and somewhat inaccessible.

> On 24 January 1983, in Middelkerke, Belgium, a fire gutted a building used for

filling a wide range of aerosol and non-aerosol hair products, resulting in damage and loss of finished goods estimated at 60 million Belgian francs.[60]

The deputy manager checked the plant at 20.30 h to ensure that everything was in order and the power was turned off. At about 21.00 h he saw a flash of fire shoot through the roof at the back of the building. By the time the fire brigade arrived the entire building was alight and neighbouring premises and private dwellings were at risk. Seven fire brigades were called in and, because of the explosion risk, 20 private houses were evacuated.

Dealing with the fire was hampered by difficulties encountered with the water supply.

Management controls

Sound management systems are essential to back up the technical precautions for LPG handling. Accidents have happened despite the incorporation of safety devices, because they were poorly designed, poorly installed or maintained, or not fully appropriate. Attention to detail is pivotal to risk assessment and control, and formal techniques such as HAZOPS are useful aids in identifying potential problems, as described briefly on pages 741–4 and in more detail in Volume 4.

A can of aerosol deodorant exploded in a shrink-wrapping machine causing a series of aerosols to explode and a fire which damaged parts of the aerosol filling plant. The fire alarm operated within a few minutes but the rapidity with which the fire developed made it difficult to control. The shrink-wrapping machine and nearby conveyors were destroyed and there was extensive damage to plant and buildings.

The safety systems all operated correctly on the hydrocarbon propellant installation. The only propellant involved in the fire was therefore that in the filled cans which were destroyed, releasing an estimated 4 to 10 litres of alcohol and between 3 and 8 litres of hydrocarbon. The incombustible suspended ceiling formed a barrier to hot gases and combustion products which spread throughout the room and melted all plastic materials, including cable insulation, lamp holders, machine guards and air hoses above an elevation of about 2.5 m. The entire lighting, and much of the power wiring, was destroyed together with irreparable damage to parts of the cap-sorting machines.

The primary cause of the fire was a mechanical failure of the drive of the conveyor belt. All shrink-wrapping tunnels were provided with auxiliary power supplies to enable aerosol packs to be removed from the hot tunnels in the event of conveyor failure. (Subsequently motion detectors were installed on the conveyor belts, which switch off the heating and sound audible alarms in the event of conveyor failure.)[60]

Because of the sophisticated nature of many safety systems required for filling with LPG, their reliability is crucial to ensure they function properly when required and to provide credibility by a minimum of spurious alarms.

A well-engineered system was impaired by large volumes of stable shave cream foam, extracted from cans during the vacuum stage. These filled the tank of the pump located near to the gasser, and were discharged from the unit in such copious amounts that it impacted both the lower IR detector intake tube and much of the exhaust hood. The foam filled the lower portion of the system, partially blocking it. The upper detector tube supplied the IR unit with gas/air mixtures that read erratically up to 38% LEL and caused the emergency ventilation system to kick on and off in response. Upon investigation it transpired that the gas house had a history of such problems, with shave creams and other foam products. The problem was resolved by institution of a system of flushing and defoaming of these products.[61]

Once systems have been devised and instituted then management controls, e.g. training, control of entry into hazardous areas, provision of personal protection, permits to work, regular monitoring of work practices, maintenance, etc. are crucial in ensuring that technical safety systems and agreed safe working practices continue to operate, and that mechanical or human failures cannot lead to accidents such as those exemplified below.

A seal broke on a processing pump in the gassing room of an aerosol filling plant in Ohio. Three people were hospitalised with burns following ignition of the gas leaks.[66]

In Rhode Island, USA, an aerosol plant with five lines, with a capacity of 130 million units per annum on two-shift working, was badly damaged in January 1976 as a result of an explosion and fire which killed one employee and injured eighteen.

The incident occurred during a coffee break on a Sunday when only a small workforce was on hand to make a line change-over. A bleeder-valve was left open allowing isobutane to escape, and a faulty switch was in the 'off' position. Unknown to employees it activated a solenoid and the resulting spark initiated the explosion.[60]

A maintenance man unwittingly jeopardised a filling plant when he spray-painted the air intake louvres for the emergency ventilation system thereby preventing them from operating when needed. The potential hazard was discovered during a routine safety check.[61]

References

1. a) British Standard BS 4250, British Standards Institution, London.
   b) *Liquid Petroleum Gas Specification and Test Methods*, Gas Producers' Assoc., GPA Publication 2140–75, Tulsa, OK.
   c) ASTM Standard D1835–76, Pt.24, American Society for Testing and Materials, Philadelphia, PA, 1978.
2. *Matheson Gas Data Book*, eds. W. Braker & A. L. Mossman, Matheson Gas Products, New York, 5th edn, 1971.
3. Barker, D., LPG, The Hazards and Precautions, paper presented at Conference 5–6 March 1980, page 4, Fire Protection Association.
4. *Encyclopaedia of Occupational Health and Safety*, ed. L. Parmeggiani, 3rd edn, 1983, page 1244.

5. Anon., *Loss Prevention Bulletin*, 1989(087), 5.
6. a) *The Times*, 6 June 1989, page 8.
   b) *New Scientist*, 10 June 1989, page 25.
   c) *Sunday Times*, 11 June 1989, page A17a.
   d) *Financial Times*, 6 June 1989, page 2.
7. Institution of Chemical Engineers, *Preventing Emergencies in the Process Industries*, vol. 1, Hazard Workshop Module 006, 1987, Section 4.
8. Anon., *Loss Prevention Bulletin*, 1987(077), 27.
9. *Safety Newsletter*, No. 24, ICI Ltd, 1.
10. Kletz, T. A., *Learning from Accidents in Industry*, Butterworths, 1988, 140.
11. Anon., *Loss Prevention Bulletin*, 1987(077), 1.
12. Hoffman, A. B., *J. Chem. Educ.*, 1974, **52**, 419.
13. *LPG Installations*, Shell UK Oil, 1982.
14. Amer. Conf. Gov. Ind. Hygienists, *Documentation of the Threshold Limit Values and Biological Indicies*, 5th edn 1986, ACGIH, Cincinnati, OH.
15. Health and Safety Executive, Guidance Note EH40/93, HMSO, 1993.
16. *Threshold Limit Values and Biological Exposure Indices for 1988/89*, Amer. Conf. Governmental Hygienists, Cincinnati, OH.
17. 'Leakage of propane at Whitefriars Glass Ltd, Wealdstone, Middlesex, 20 November 1980', Report by the Health and Safety Executive, 1981.
18. Windebank, C. S., LPG – The Hazards and Precautions, paper presented at Conference 5–6 March 1980, Fire Protection Association.
19. Fire Protection Association, *LPG: The Hazards and Precautions*, 1980.
20. Considine, M., Grint, G. C., et al., *Bulk Storage of LPG – Factors Affecting Offsite Risk*, Institution of Chemical Engineers, 1982.
21. Pritchard, M. J., British Gas Lecture, Aston University, 1993.
22. Mecklenburgh, J. C. (ed.), *Process Plant Layout*, George Godwin, 1985.
23. Johnsen, M. A., *Aerosol Age*, 1987 (July), 25.
24. Anon., *Loss Prevention Bulletin*, Institution of Chemical Engineers, 1989(087), 7.
25. *Safety Newsletter*, No. 110, ICI Ltd, 4.
26. *The Bulk Storage of Liquefied Petroleum Gas at Factories*, Health and Safety at Work, No. 30, D.E.P., 1973, HMSO.
27. Jones, C. & Sands, R. L., *Great Balls of Fire*, 1981, published by the authors.
28. Code of Practice, *Installation and Maintenance of Bulk LPG Storage at Consumers' Premises*, LPGITA, 1974.
29. Health and Safety Executive, *The Storage of LPG at Fixed Installations*, Guidance Note CS5, HMSO, May 1981.
30. Health and Safety Executive, *The Storage of LPG at Fixed Installations*, HS(G) 34, HMSO, 1987.
31. Code of Practice, *Installation and Maintenance of Bulk LPG Storage at Consumers' Premises*, LPGITA, March 1978.
32. Institute of Petroleum Gas Engineers, *Liquefied Petroleum Gas; Vol. 1. Large Bulk Pressure Storage and Refrigerated LPG*, Wiley, Chichester, 1987.
33. Davenport, J. A., *Plant/Operations Progress*, 1987, **6**(4), 199.
34. Health and Safety Executive, *An Investigation of Potential Hazards from Operations in the Canvey Island/Thurrock Area*, First Report, 1978.
35. Hughes, J. R., *Storage and Handling of Petroleum Liquids – Practice and Law*, Griffin, 1967.

36. Lees, F. P., *Loss Prevention in the Process Industries*, Butterworths, 1980.
37. a) Anon., *Loss Prevention Bulletin*, Institution of Chemical Engineers, 1987(077), 11–15.
38. Health and Safety Executive, *Health and Safety Industry and Services*, 1975, HMSO.
39. *Manual of Firemanship, Practical Firemanship III*, Part 6c., 1984, 120, HMSO.
40. Davenport, J. A., *J. Haz. Mat.*, 1988, **20**, 3.
41. Health and Safety Executive, *The Keeping of LPG in Cylinders and Similar Containers*, Guidance Note CS4, HMSO, June 1986.
42. Health and Safety Executive, *Welding and Flame Cutting Using Compressed Gases*, HSW Booklet No. 50, HMSO.
43. Farmer, D., *Health and Safety at Work*, Feb. 1988, 19.
44. *Handbook Butane–Propane Gases*, 4th edn, Chilton Co., 1962.
45. Code of Practice 2, *Safe Handling and Transport of LPG in Bulk by Road*, LPGITA, 1974.
46. Ens, H., *Loss Prevention Bulletin*, Institution of Chemical Engineers, 1987(071), 34.
47. Hawkesley, J. L., *Loss Prevention Bulletin*, Institution of Chemical Engineers, 1990(091), 27.
48. Meidl, J. H., *Flammable Hazardous Materials*, 2nd edn, Glencoe Pub. Co., Westerville, OH, 1978, 128.
49. Adelman, R. M. & Adelman, B., *Loss Prevention Bulletin*, Institution of Chemical Engineers, 1990(094), 17.
50. Health and Safety Executive, *Transport of LPG Cylinders by Road*, Leaflet M30, 9/90.
51. The Road Traffic/Carriage of Dangerous Substances in Packages, etc. Regulations, 1986, and the amendments 1989.
52. Simpson, A., *Aerosol Age*, 1987 (Nov.), 25.
53. *An Introduction to Liquefied Petroleum Gas*, LPGITA, 1974.
54. *Recommendations for Prevention and Control of Fire Involving LPG*, LPGITA, 1977.
55. *LPG Piping – System Design and Installation*, LPGITA, 1990.
56. *A Guide to Safety in Aerosol Manufacture*, BAMA, 2nd edn, 1988.
57. Johnsen, M. A., *Aerosol Age*, 1987 (Dec), 42.
58. Johnsen, M. A., *Aerosol Age*, 1988 (Jan), 30.
59. Anon., *Aerosol Age*, 1988 (Dec), 30.
60. Private communication.
61. Johnsen, M. A., *Aerosol Age*, 1987 (Sept), 34.
62. Health and Safety Executive, *Report on the Fire and Explosion at Permaflex Ltd. Stoke-on-Trent, 11 February 1980*, HMSO.
63. Health and Safety Executive, *Gaseous Fire Extinguishing Systems: Precautions for Toxic and Asphyxiating Hazards*, Guidance Note GS16, HMSO, 1984.
64. Johnsen, M. A., *Aerosol Age*, 1989 (Feb) 26, (March) 42, (April) 36, (May) 26, (June) 34.
65. a) *Houston Post*, 28 Dec. 1983.
    b) *Hazardous Cargo Bulletin*, April 1984.
66. TV Reports, 10 April 1982 (WKYC-TV (NBC) CH3), Cleveland Action News.

CHAPTER 24

Spontaneous combustion and pyrophoric chemicals

Introduction

Fire is a chemical reaction emitting heat and usually light, although flames associated with some substances, e.g. hydrogen, are virtually invisible. The basic requirements are generally a fuel and air (or oxygen) within limits, as the reactants, and a means of ignition to provide the energy of activation. The interrelationship is often depicted as the 'fire triangle' (Fig. 4.1) or, to account for the free-radical combustion mechanism, the 'fire rectangle'.

Flammability limits and flash points (see Chapter 4) have been reported for hundreds of chemicals, as exemplified by Table 24.1, and flash point is conventionally used for the classification of liquids for legislative purposes. In general, the wider the flammable range and the lower the flash point of a chemical or mixture, the greater the hazards. (Materials with flash points below ambient temperature will ignite on exposure to a flame whereas those with higher values first require heating to the flash-point temperature.) However, in cases where an oxidising agent is present a supply of air may be unnecessary, or an exothermic reaction between chemicals within a mixture may result in ignition.

Also, a fire has started in the apparent absence of any identifiable ignition source. Sometimes combustible materials spontaneously inflame in the absence of any external means of ignition (Chapters 4 and 27) because they are at a temperature above their autoignition temperature (AIT) or as a result of oxidation with sufficient exothermicity to cause acute or chronic thermal runaways. Such accidents can be costly in terms of injury and material loss. This chapter, therefore, expands on the general principles introduced in Chapters 4, 9 and 11 for dealing with spontaneously flammable and pyrophoric chemicals by reference to selected materials. It deals with the overall hazards as opposed to just the fire dangers and guidance is given on the appropriate control measures. The supplier of any specific chemical and the literature need to be consulted for more detailed advice.

Chronic events

Spontaneous combustion is normally associated with materials commonly considered to be stable at ordinary temperatures, but which under certain

Table 24.1 Flammability data for selected substances

Chemical	Flammable range in air (%)	Flash point (°C)	Autoignition temperature (°C)
Acetaldehyde	4–57	−38	185
Acetic acid	4–16	43	426
Acetone	3–13	−18	538
Benzene	1.4–8	−11	562
n-Butane	1.9–8.5	−60	405
n-Butyl acetate	1.4–7.6	27	399
n-Butyl alcohol	1.4–11	29	365
n-Butyric acid	2–10	66	452
Carbon disulphide	1–44	−30	100
Cumene	0.9–6.5	44	424
Cyclohexane	1.3–8.4	−20	260
1,4-Dioxane	2–22.2	12	180
Decane	0.8–5.4	46	208
Diethyl ether	1.9–48	−45	180
Ethane	3.0–12.5		515
Ethyl alcohol	3.3–19	12	423
Ethyl chloride	3.6–15.4	−50	519
Heptane	1.2–6.7	−4	223
Hexane	1.1–7.5	−22	261
Hydrogen	4–75		585
Hydrogen sulphide	4.3–46		260
Isopropyl alcohol	2.3–12.7	12	399
Methane	5.3–14.0		537
Methyl alcohol	6–36.5	12	464
Nonane	0.8–2.9	31	220
Octane	1.0–3.2	13	220
Pentane	1.5–7.8	−40	309
Propane	2.2–9.5	−104	468
Pyridine	1.8–12.4	20	482
Styrene	1.1–6.1	31	490
Toluene	1.4–6.7	4.4	536
Vinyl chloride	4–22	−78	472
m-Xylene	1.1–7	29	528

conditions can undergo gradual self-heating within the chemical mass. In most fires involving spontaneous combustion, relatively large, enclosed or thermally insulated volumes of materials were present and spontaneous ignition occurred after prolonged periods of oxidative degradation during storage or transport. There are, however, exceptions, e.g. cross-linking of some plastics can result in spontaneous combustion.

Oxidation processes are exothermic and, with readily oxidisable compounds, the heat generated through reaction may be inadequately dissipated by, e.g. conduction to the surface, or by natural or forced convection. Thus the rate of heat gain is given by

$$dq/dt = kQV\phi_T - k'A(T - T_o)$$

where dq/dt is the rate of heat accumulation of the reacting mixture;
Q is the heat of reaction per unit mass of reactant consumed;

V is the volume of material reacting;
A is the surface area of the volume reacting;
T is the internal temperature;
T_o is the external temperature;
ϕT = reaction rate = mass of reactant consumed per unit volume, which is taken as a function of the internal temperature T;
k and k' are constants.

The first term on the right-hand side of the equation refers to rate of heat generation, and the second term describes the rate of heat loss. In most cases the rate of reaction, and hence the rate of heat generation, increases with T much faster than $(T - T_o)$. At $dq/dt = 0$ a steady state exists which is disturbed by increases in T_o or V/A and the temperature of the pile rises with possible ignition. The rate of an exothermic reaction, generally in this case oxidation by air, typically increases exponentially with temperature – curve H in Fig.4.6 (page 149).[1]

It is important to appreciate that the actual *initial temperature* of the pile is as important as T_o, e.g. if the material comes direct from a dryer.

The presence of certain inorganic impurities lowers the AIT. Several spontaneous fires have been attributed to the presence of catalytic quantities of rust (see above): sodium carbonate markedly lowers the AIT of coal and coke. Similarly, the presence of oxidising compounds increases the ease of spontaneous combustion.

Empty paper sacks which had contained sulphur and sodium nitrite were discarded along with general rubbish in a skip inside a building near to windows in direct sunlight. The resulting fire probably resulted from exothermic interaction of the nitrite and sulphur and damp paper with further heating aided by the insulating effects of the mound of rubbish in a draught-free skip warmed by direct sunshine. (Increased use of paper towels in laboratories poses a similar hazard if impregnated with oxidisers and the damp contaminated wipes are disposed of with other 'burnable rubbish'.)[2]

Aeration of non-ionic detergent slurry with air can lead to formation of potentially hazardous blue smoke in emissions from spray-drying towers with increased risk of spontaneous ignition of product. Use of nitrogen for aeration and incorporation of selected anti-oxidants in the slurry minimises the formation of hazardous oxidation products.

Reciprocating compressors with oil-lubricated cylinders may produce air/oil-mist mixtures in the air line capable of undergoing explosion and leaving cabonaceous deposits with potential for self-ignition.

Materials prone to gradual self-heating are listed in Table 24.2.[3] Most have a cellular, granular or fibrous structure with a low thermal conductivity. Thus from the onset of self-heating the core temperature is higher than the surface temperature. The rate of heating is mass-dependent, e.g.

Table 24.2 Materials liable to self-heating

Liquid materials susceptible to self-heating when dispersed on a solid

Bone oil	— moderate	Palm oil	— moderate
Castor oil	— very slight	Peanut oil	— moderate
Coconut oil	— very slight	Perilla oil	— high
Cod liver oil	— high	Pine oil	— moderate
Corn oil	— moderate	Rapeseed oil	— high
Cottonseed oil (refined)	— high	Rosin oil	— high
Fish oil	— high	Soyabean oil	— moderate
Lard	— high	Sperm oil	— moderate
Linseed oil (raw)	— very high	Tallow	— moderate
Menhaden oil	— high	Tallow oil	— moderate
Neatsfoot oil	— slight	Tung oil	— moderate
Oleic acid	— very slight	Turpentine	— slight
Oleo oil	— very slight	Whale oil	— high
Olive oil	— slight		

Solid materials susceptible to self-heating in air

Activated charcoal	Lagging contaminated with oils, etc.
Animal feedstuffs	Lamp-black
Beans	Leather scrap
Bone meal, bone black	Maize
Brewing grains, spent	Manure
Carbon	Milk products
Celluloid	Monomers for polymerisation
Colophony powder material	Oleic acid impregnated fibrous coal
Copper powder	Palm kernels
Copra	Paper waste
Cork	Pent
Cotton waste	Plastic, powdered
Cotton	Rags, impregnated
Cottonseed	Rapeseed
Distillers' dried grain	Rice bran
Fats	Rubber scrap
Fertilisers	Sawdust
Fishmeal	Seeds
Flax	Seedcake
Foam and plastic	Silage
Grains	Sisal
Gum rosin	Soap powder
Hay	Soya beans
Hemp	Straw
Hides	Sulphur
Iron filings/wood/borings	Varnished fabric
Iron pyrites	Wood chips
Ixtle	Wood fibreboard
Jaggery soap	Wood flour
Jute	Wool waste
	Zinc powder

Table 24.2 (continued)

Tendency for fibrous materials to self-heat when impregnated with the following vegetable/animal oils (tendency decreases down the list)

Cod liver oil
Linseed oil
Menhaden oil
Perilla oil
Corn oil
Cottonseed oil
Olive oil
Pine oil
Red oil
Soyabean oil
Tung oil
Whale oil
Castor oil
Lard oil
Black mustard oil
Oleo oil
Palm oil
Peanut oil

Other materials subject to self-heating (depending upon composition, method of drying, temperature, moisture content)

Desiccated leather	Garbage
Leather scraps	Leather meal
Dried blood	

Table 24.3 Ignition tests on mixed hardwood sawdust on hot surface

Depth of layer (mm)	Ignition temperature (°C)	Time to ignition (min)
5	355	9
10	320	24
20	290	50
25	280	100

the time taken for different-sized piles of hardwood sawdust placed on a thermostated hot surface to ignite is given in Table 24.3.[4]

General criteria for the possibility of spontaneous combustion of susceptible materials are as follows.
- Porous material soaked with a reactive liquid, e.g. catalyst support, absorbent or fibrous insulation.
- Accumulations, e.g. as layers, in heated process equipment (i.e. dryers), on hot surfaces (i.e. electric motors or hot bearings). Fires sometimes occur in dryers even when the highest temperatures appear to be well below the AIT of the substance in question. Self-heating may result in the dryer because of dust particles on ledges and heated sur-

faces, or in residues left in the dryer when it is shut down, or in dried material discharged to bins or hoppers, e.g. lumps of skim milk near the hot air inlet in a spray dryer.[5]

A tank was filled with 190 tonnes of oxidised bitumen and held at 240 °C for seven days. Over the following two days material was unloaded, leaving about 30 tonnes in the vessel. Ten hours later there was an explosion in the tank. The cause was attributed to either
— the extensive deposits of coke on the roof which had accumulated from three years' use without cleaning. This material, previously inerted by the steam purge, would have been at high temperature and exposed to a significant amount of air as hot material was withdrawn from the tank. Slow oxidation of the coke would commence accelerating due to retention of the heat, until the AIT of the flammable gas mixture was reached; or
— formation of pyrophoric iron (see later) below the coke due to the presence of sulphides in the bitumen. This would ignite on exposure to the air. Ignition would be delayed because of the protective layer of coke.
Both mechanisms could be avoided by inerting the tank and introducing a programme of regular cleaning.[6]

- Large piles of materials with restricted cooling.

Ricks of baled hay do not usually ignite as a result of spontaneous combustion because the air space between bales allows ventilation. (If there is doubt then temperature monitoring can be used, e.g. if the temperature exceeds 70 °C hay will probably ignite spontaneously when air reaches it on cutting-away.)[7]

- Over-dried material when moisture absorption may contribute to the rate of heat release.
- Finely divided material – exposing a large surface area, e.g. dust accumulations in extract ducts.

Aluminium dust arising from grinding and buffing operations has resulted in a number of fires and explosions.[8,9]

The role of water is complex. Thus, on the one hand, small quantities of water aid oxidation, whereas moisture also increases the thermal conductivity, thus assisting in the dissipation of heat.[1]

Water contamination may increase the risk of spontaneous combustion of steel turnings. Furthermore, the resulting fires cannot always be extinguished safely with water due to the ability of hot iron to react with water to generate hydrogen. Unless persevered with, even total immersion in water may prove inadequate in reducing the temperatures of the self-heating mass below critical levels.[10]

The precautionary measures in any specific case can hence be deduced, aided by small-scale tests[10,11] as illustrated by Table 24.3. For example, when processing powder prone to self-heating, if tests indicate a rise in temperature then cooling should be required prior to storage or packing. Automatic monitoring of temperatures, e.g. in silos, may be used to detect any rise in temperature above the norm.[12] Material, such as grain, may be recirculated to dissipate heat. Obviously, the hazard is reduced if the time for which a powder is at an elevated temperature is minimised.[12]

A particular hazard arises from oil contamination of fibrous lagging (insulation) materials operating at high temperatures, i.e. >100 °C. Ignition occurs because the oil-wetted lagging exposes a high surface area of oxidisable material to atmospheric oxidation under conditions where the heat of reaction cannot be dissipated at a sufficient rate. Prerequisites for such lagging fires may be liquid of sufficiently low volatility that it is not evaporated quickly and for it to deposit solid, to facilitate smouldering, e.g. due to cracking.[13]

> Asbestos-lagged, steam-traced pipework, containing molten petroleum jelly, ignited at 150 °C. The lagging had become soaked with jelly over some considerable time which had undergone gradual oxidation with eventual spontaneous combustion. The resulting fire was supported by additional jelly which had 'wicked' into the open cell structure of the lagging. Lagging is available with closed cell structure to avoid the phenomenon of wicking.[14]

Spillages of oils and fats frequently contaminate insulation because of leaks from pipe joints, pump glands, holed pipes or gearboxes. If this oil-soaked insulation is in contact with hot equipment (e.g. very hot process lines, still kettles or reboilers, or heat exchangers), then ideal conditions are created for spontaneous combustion, particularly for drying oils. If no action is taken then, based on experience, fire may occur in the lagging after about three hours following the spillage; the greatest hazard arises when the insulation is broken.

> A substantial run of lagging on pipework for mineral oil used for process heating was contaminated due to leakage from a defective pump. In the ensuing lagging fire smoke was first seen from lagging remote from the fire; the fluid then exceeded its set temperature of 270 °C by 50 °C and eventually flames enveloped the pump and a considerable length of adjacent piping.[15] (Apart from the obvious precaution of minimising oil leaks, measures may be needed to restrict oil and air ingress into lagging by use of metal cladding.)

It has been shown for several different oils impregnated in various insulating materials that the auto-ignition temperatures could be depressed by 100 °C to 200 °C below the values for the free oils.[16] The oil-soaked insulation samples were tested in an air stream, whereas in practice air available for combustion is limited, e.g. due to the equipment or piping surface and sheet metal cladding, so that the results represent worst-case conditions. However, they illustrate the large potential magnitude of AIT depression. Hence it is good practice to have fire-extinguishing equipment on stand-by when removing oil-soaked insulation, or the cladding, under 'hot process' conditions.[17]

Cotton waste contaminated with highly flammable liquids, e.g. polyurethane finishes or oil, are liable to ignite spontaneously, but petroleum oils are not as liable to self-heating as animal or vegetable oils. Rags soaked in petroleum oil and left for a period can, however, ignite spontaneously owing to the aerobic activity of bacteria. Best practice is to avoid placing oil-contaminated overalls, sacks, textile waste or rags near hot equipment or radiators. Waste rags, or other waste material liable to spontaneous combustion, should be placed in closed metal bins and removed regularly from the workplace to reduce the period available for self-heating.

No external heat source is required with unsaturated (i.e. natural) oils, nor indeed with a large pile of rags saturated with petroleum-based oils. Thus in the storage of turpentine at paper mills it is recommended that air should not be blown through transfer lines into storage tanks, and that monitoring of temperature, peroxide levels and pH should be carried out to avoid hazards from air oxidation.

Vegetable oils with 'iodine values' above 100 are more susceptible to self-heating and the higher the iodine value the greater the hazard. In descending order of hazard, commonly used oils include perilla, linseed (used for preserving leather and a constituent of some polishes), stillingia, tung, cic, hemp-seed, cod-liver, poppy-seed, soyabean, seal, whale (sperm), walrus, maize, olive, cottonseed, sesame, rape and castor.

> Cloths used for applying linseed oil to benches in a laboratory were disposed of into an open waste-bin. A fire started after a few hours destroying the laboratory.[2]

In process design the possibility of exposure of a large surface area in, e.g. mass transfer equipment may be a factor requiring exclusion of air.

> An unsaturated organic material with a known tendency to spontaneous oxidation was present on the knitted multifilament wire packing in a low-pressure distillation column. On shut-down temperature probes were removed from the cooled column allowing ingress of a continuous supply of air. Spontaneous heating resulted in cracking of the glass column and coating of the packing with oxides of iron. (Subsequently, in addition to the normal practice of cooling and introduction of nitrogen to bring the column up to atmospheric pressure, cleaning-out with solvent was recommended prior to engineering work involving the breaking of lines.)[18]

Spontaneous combustion without any external heating is a problem with some other materials with a high specific surface, e.g. soft grades of coal, porous charcoal. As coal is granulated, or pulverised, fresh reactive surfaces are created and hence the hazard is increased. Spontaneous heating is cumulative so that the coal is most hazardous 90–120 days after mining.[19] Alternate wetting and drying of coal cleans the surface and aids spontaneous heating. The liability to spontaneous heating is greater the softer and newer the coal, i.e. anthracite is least susceptible to it. The presence of significant proportions of coal fines increases the hazard; high stacks of powdered coal, where there is little heat escape, are most hazardous. The danger is enhanced by the presence of pyrites because of its liability to oxidise rapidly. Similarly, the tendency towards spontaneous ignition, and explosivity, of shale dust increases with decrease in particle size and with an increase in oil content.[20]

Charcoal is porous as freshly made or when surfaces are freshly exposed, e.g. by crushing or grinding; absorption of air generates heat which may induce slow oxidation and eventual ignition. It is often necessary to cover chemically active charcoals with a polythene liner as a protective oxygen-impervious layer. The risk is removed by moisture absorption but is re-created by drying out. Furthermore, wood itself may, if left in contact with a constant temperature source over a very long period of time, undergo chemical change resulting in the formation of charcoal which is capable of heating spontaneously. Thus 65 °C is generally considered the maximum temperature to which wood can be exposed in this way without risk of ignition. The phenomenon may arise with oil-impregnated wooden boards on scaffolding subjected to heat from process equipment. As with any other combustible solid, it is most vulnerable when in a finely divided state or in thin shapes or forms. Thus sawdust is sometimes prone to self-ignition, particularly if it is contaminated with resins or oils.

> The sawdust produced from green timber is liable to spontaneous combustion: the initial heating is probably microbiological action leading to the chemical oxidation stage. During a hot weekend rain penetrated a sawpit and the top surface of the sawdust caked-over. After 30 hours the sawdust was found to be glowing red with the wooden supports of the walls smouldering.[21]

Spontaneous heating of agricultural products can generally be prevented by control of the moisture content. Adequate curing and aeration will prevent heat build-up.[22] Oil seeds and oilseed cake and meals are occasionally susceptible to self-ignition, particuarly rape, or meal subjected to high mechanical action and then stored in bins or silos. The presence of natural anti-oxidants minimises the hazard.

Biological activity, particularly in moist dusts/mounds, may also raise the temperature above critical conditions resulting in spontaneous combustion. Whilst colonies of micro-organisms develop at an exponential rate the time to reach equilibrium may take 6–9 months.[10] Clearly, meaningful monitoring must extend over this time span. In general, temperatures above 80 °C would render the micro-organisms inactive. However, the temperature rise may be sufficient to initiate subsequent chemical oxidation and lead to spontaneous combustion. Thus dried spent grain from breweries/distilleries should be cooled to below 38 °C prior to storage.

Acute events

Liquids may ignite spontaneously when heated out of contact with air and then allowed to escape into the open atmosphere.

As described on page 147, a fire occurred, resulting in the deaths of three men and extensive damage to a plant, when hot oil above its autoignition temperature came out of a pump and ignited.[21]

Polymer had been made safely for years by the transesterification of dimethyl terephthalate with ethylene glycol. After increasing the heating capacity of the reactor several ignition incidents resulted. This was attributed to the formation of either acetaldehyde, by the thermal degradation of polymer, or dioxane from the glycol. Both have wide flammability ranges and low AITs. Ingress of air would then permit a vapour/air composition to develop within the flammable range and to ignite spontaneously.[2]

Under these circumstances the entire body of the material is heated to the AIT.

On expansion from high to low pressure at room temperature, most gases undergo cooling, i.e. the Joule–Thomson effect exploited in refrigeration, e.g. in liquefaction of gases. Hydrogen is an exception in that it exhibits a negative effect, i.e. it warms on expansion at room temperature; only below the inversion temperature does it cool upon expanding. Thus, high velocity escape of hydrogen, e.g. via the narrow orifice of gas cylinders/pressure regulators, may generate sufficient heat to raise the temperature of the ensuing gas beyond the AIT and cause the leak to ignite spontaneously. The greatest risk is when the gas temperature is already near the AIT, or in the presence of metal catalysts. Some incidents involving leaking hydrogen have also been attributed to static ignition:[2]

Release of hydrogen for filling balloons ignited when the aperture was rusted, a brush discharge being visible.

Release of hydrogen at 47.5 bar into a vented chromium-plated sphere caused explosive ignition.

Pyrophoric chemicals represent a sub-class of spontaneously combustible substances that ignite immediately on contact with air or water. Amongst the many pyrophoric chemicals identified in Volumes 1 and 2 (e.g. Chapter 4, pages 151–2) are water-sensitive materials such as alkali metals, hydrides, carbides and phosphides, and those which undergo such exothermic oxidation that they ignite almost immediately on contact with air, e.g. phosphorus and certain hydrides and metals. Some substances, e.g. certain organometallic compounds, are sensitive to both air and moisture.

Although pyrophoric materials are not widely encountered, they are of considerable value in specific industrial or laboratory applications.

Special measures are needed to control the risk from handling pyrophoric substances which is a function of the degree of phosphoricity, the quantity of material involved, and any surrounding circumstances (such as population density and neighbouring hazards). An awareness of the chemistry of the systems in terms of their physical form, solubility in a variety of solvents, thermal stability, chemical reactivity, toxicity, corrosivity and pyrophoricity is hence a prerequisite for determining their hazards and hence the appropriate safe handling arrangements.

Some chemicals are rendered pyrophoric as a result of vigorous exothermic reaction with a variety of other chemicals (e.g. chlorates upon mixing with sulphur or red phosphorus; glycerol or benzene on contact with solid potassium permanganate), or when subject to impact/friction, e.g. the common match. Consideration is restricted here to those chemicals which ignite spontaneously on contact with air (as a result of highly exothermic atmospheric oxidation) or following violent reaction with water or moisture. Some of these reactions liberate flammable gases. Examples are listed in Table 4.9. For convenience they can be divided into inorganic substances and those which are essentially organic in nature.

Inorganic substances

Carbides, hydrides and phosphides

Certain salt-like carbides and acetylides are potentially explosive whilst others may on contact with water liberate acetylene, methane, or allylene. These gases are not spontaneously flammable but possess wide flammable ranges and may be ignited by the heat of reaction. (Commercial acetylene produced from calcium carbide and water is scrubbed to remove impurities including pyrophoric phosphine.) Uranium carbide is pyrophoric in the finely divided state and the dicarbide emits brilliant sparks on impact, ignites on grinding in a mortar and ignites in steam.[2] Several phosphides, e.g. calcium and zinc phosphides, inflame on contact with water due to the formation of phosphine (see later). Night marine-location markers rely on the action of seawater on a mixture of metallic phosphides and carbides to liberate acetylene and phosphine which rise to the surface and burn with a brilliant flame. Many hydrides, such as diborane and sodium hydride, liberate hydrogen on contact with water and potassium hydride ignites spontaneously on contact with air.

> When an unprotected polythene bag containing sodium hydride was moved, some of the powder leaked through a hole, contacted moisture and immediately inflamed.[2]

> 0.5 kg of sodium hydride came into contact with water whilst being disposed of in a routine manner in a yard. The ensuing explosion caused a fire which spread to the factory and resulted in three men requiring hospital treatment for burns:[23]

$$NaH + H_2O \rightarrow NaOH + H_2$$

Even lithium, barium or calcium hydrides will inflame if in a finely divided state. Alkali metal tetrahydroborates are stable in dry air and potassium salts can be recrystallised from water. Aluminium tris(tetrahydroborate), however, is hydrolysed explosively and is violently pyrophoric in air. Some nitrides, e.g. calcium nitride, are pyrophoric in air whilst some transition metal nitrides are explosive.

Silicon hydrides exist in several forms and all tend to be spontaneously flammable in air: SiH_4 and Si_2H_6 are both gases at room temperature and pressure. Halosilanes and haloalkyl silanes ignite without heating or have very low auto-ignition temperatures (e.g. dichlorosilane: AIT = *ca*. 55 °C).

Silane from a cylinder at a pressure of 24 bar contained inside a ventilated cabinet inadvertently flowed back to the low-pressure side of a nitrogen cylinder regulator outside the cabinet. Upon failure of the regulator the leaking silane ignited and burned outside the cabinet, preventing access to the main valve. The leak re-ignited after the fire was extinguished:[2]

$$SiH_4 + 2O_2 \rightarrow SiO_2 + 2H_2O$$

Fires have also resulted when silane was produced *in situ*.

In an attempt to reduce a methyl ester to its alcohol using a mixture of triethoxysilane and titanium isopropoxide under nitrogen at 50 °C, an exothermic reaction started and the temperature rose to 90 °C. As a result of disproportionation the organosilane produced free silane which spontaneously ignited and exploded.[24]

Phosphorus forms two hydrides. PH_3 is a colourless poisonous gas with a b.p. of 88 °C. When dry it ignites spontaneously in air in the cold but when contaminated with traces of moisture it does not ignite in air below 150 °C. P_2H_4 is a colourless volatile liquid which emits an unstable vapour which spontaneously inflames in air.

$$2PH_3 + 4O_2 \rightarrow P_2O_5 + 3H_2O$$

The degree of reactivity of halophosphines towards air or water depends upon the degree of substitution of hydrogen by halogen, e.g. tetrachlorodiphosphane oxidises rapidly in air, often with ignition. Many of the lower alkyl non-metals and their hydrides are also pyrophoric in air, e.g. triethyl arsine and ethyldimethylphosphine.

Metals

Reactivity with air

Many of the metals in Table 24.4 are less electronegative than oxygen and hence undergo oxidation, albeit at varying rates. Thus at ambient temperature chromium, cobalt and nickel are resistant to attack by atmospheric oxygen, and iron forms oxides slowly. At the other end of the spectrum aluminium forms an oxide film as soon as it is exposed to air; this film is

Spontaneous combustion and pyrophoric chemicals

Table 24.4 Electrochemical series

Metal	Symbol	Electro-negativity	Occurrence	Reactivity with water
Lithium	Li	0.97		
Caesium	Cs	0.86		React with cold
Potassium	K	0.91	Never found	water to yield
Barium	Ba	0.97	uncombined	hydrogen
Strontium	Sr	0.99		
Calcium	Ca	1.04		
Sodium	Na	1.01		
Magnesium	Mg	1.23		
Aluminium	Al	1.47		Burning metals
Manganese	Mn	1.60		decompose water
Zinc	Zn	1.66		and hot metals
Chromium	Cr	1.56		decompose steam
Iron	Fe	1.64		
Cadmium	Cd	1.46	Rarely found uncombined	
Cobalt	Co	1.70		
Nickel	Ni	1.75		Very little reaction
Tin	Sn	1.72		unless at white heat
Lead	Pb	1.55		
Hydrogen	H	2.20		
Phosphorus	P	2.06		
Oxygen	O	3.50		
Bismuth	Bi	1.67		
Copper	Cu	1.75	Sometimes found	
Mercury	Hg	1.44	uncombined	
Silver	Ag	1.42		Inactive with water or steam
Platinum	Pt	1.44	Found uncombined	
Gold	Au	1.42	with other elements	

highly protective, rendering the metal passive to further attack under ambient conditions. Similarly the alkali metals, e.g. sodium and potassium, and the alkaline earth metals, e.g. magnesium, are quickly dulled in air as a result of oxide or carbonate formation. On heating to the respective melting point the oxide film is broken down and the metal reacts so vigorously as to burn. For example, magnesium melts at 651 °C and the ignition temperature of a solid lump is close to this melting point. However, fine shavings and loose scrap will ignite at <538 °C and powder at <482 °C; some magnesium alloys will ignite at <427 °C. These temperatures are exceeded by all common ignition sources. Heated sodium or potassium spontaneously ignites to emit choking fumes: depending upon circumstances the cold metals may inflame and explode on exposure to air.

> Some vessels about 1.2 m long by 1.2 m wide by 1.5 m high which had formed part of a magnesium dust ventilating plant were broken up in a scrapyard. They were coated inside with magnesium dust. Two were broken up without trouble

but the third burst into flames and the worker stood back to let the fire burn out. As soon as he applied his blow-torch to the fourth an explosion blew it some 10 m across the yard; he was knocked over by the explosion and his clothes caught fire.[24]

Lithium does not become oxidised much at temperatures <100 °C; it does, however, react with atmospheric nitrogen on standing to form the nitride, which can react further with water so vigorously to form ammonia as to burn. Calcium also forms a nitride under normal conditions although on burning in air it produces an oxide/nitride mixture. On prolonged exposure to air potassium forms potassium oxide which is itself oxidised to form an outer coating of yellow superoxide. If the latter is driven past the potassium oxide layer into contact with metallic potassium, e.g. during cutting of the dry metal or as a result of a hammer blow, a violent explosion ensues. The superoxide can also react explosively with organic matter.

When in finely divided form (powder, foil or small chips), even heavy metals, such as those considered above to be inert to air (e.g. lead, iron, cobalt) pose a real hazard and can spontaneously inflame because of the large surface area available for oxidation. For example, the addition of powdered 50/50 nickel aluminium alloy to aqueous caustic soda produces a solution of sodium aluminate together with a spongy form of nickel which has catalytic properties. Similarly, the finely divided form of metal produced by reduction of nickel oxide with hydrogen (Raney nickel) is a valuable catalyst, e.g. in the hydrogenation of oils and fats. However, both forms are highly pyrophoric and must be handled with care to avoid fires and explosion.

Two operators planned to feed Raney nickel catalyst suspended in methanol into a reactor filled with nitrogen. Prior to opening the feed inlet tube, they released pressure from the reactor, via a line leading to the connected acid wash. This line was under vacuum and air was sucked into the reactor. Sparks were observed flying around inside the reactor as the inlet tube was opened. This was immediately followed by an explosion and both operators suffered first- and second-degree burns. The remains of the pyrophoric catalyst had self-ignited in the air entering the reactor; this in turn ignited the methanol–air mixture. To avoid a repetition, a non-return valve was incorporated in the line leading to the acid wash, and the catalyst feed device was equipped with an airlock and inerted with nitrogen.[25]

An amine was produced by reduction with nickel catalyst, and the melt purified by filtration at 135 °C. On completion of filtration, the filter was blown through with compressed air to remove residual melt. During this process the contents of the collecting vessel ignited, the filter attaining temperatures of around 500 °C. The cause was spontaneous ignition of the finely divided nickel catalyst residues.[26]

Similarly, some metals even in the bulk state can become pyrophoric as amalgams, e.g. cerium, or in alloy form, e.g. thorium–silver. Thus, zirconium alloys have been used for more than 25 years for severe corrosive applications but misapplication of the alloys, or changes in operating conditions, can give rise to pyrophoric incidents. For example, surface reactions in certain corrosive media can produce a thin, highly combustible surface film and, although the base metal will rarely burn in this situation, any flammable vapours present can be ignited. The pyrophoricity is encouraged by elevated temperatures and oxygen-rich atmospheres.[27] For some situations, such as use in strong sulphuric acid environments, heat treatment of zirconium equipment is advocated as a means of improving weld corrosion resistance. Special care is required in this process.

> A shell and tube heat exchanger was rebuilt after eight years of service in sulphuric acid. The aim was to replace the steel shell and zirconium-clad tube sheets, whilst reinstalling the tube head. The tube-to-tube sheet steel welds were subjected to postweld heat treatment in a gas-fired furnace. The fuel gas was then turned off but not the combustion air supply. This resulted in oxygen enrichment within the furnace while the equipment was still subjected to elevated temperatures. The zirconium tubes suddenly ignited and continued to burn until they were essentially consumed.[27] (Precautions should have included rinsing the equipment with caustic soda solution, steam treatment for at least one hour, and avoidance of oxygen enrichment while the metal was red hot, e.g. by inerting with argon.)

> A hot-work permit had been issued for the flame-cutting of a heat exchanger on redundant chlorine plant. A violent fire broke out in the tube bundle. When water from a nearby hose (a requirement of the hot-work permit) was used on the burning tubes there was a loud explosion in another part of the tube. The heat exchanger comprised a carbon steel shell with thin-walled titanium tubes and tube plate facing. The titanium tubes were heated by the cutting flame to a point at which they burned in the presence of air. The water reacted with the titanium at 700 °C to evolve hydrogen which was ignited elsewhere in the heat exchanger and exploded.[28] Though not truly 'pyrophoric' this incident illustrates how metals can burn and can react with water to liberate hydrogen which can be ignited remotely.

The presence of impurities can increase the rate of oxidation of metals or metal salts to the point where an explosion occurs.

The presence of sulphur can increase the pyrophoricity of iron. Thus moist iron sulphide formed, e.g. when steel equipment is used with materials containing hydrogen sulphide or volatile sulphur compounds, is exothermically oxidised in air and may reach incandescence. This has been the cause of many fires and explosions when such plant, e.g. carbon steel equipment which has contained H_2S-bearing hydrocarbons (sour crude or sour naphtha), has been opened without effective purging.

> A 45 000 barrel spheroid collapsed completely due to an internal explosion during gas-freeing. The explosion occurred whilst draining water and filling and

overflowing to an established procedure. It was caused by a residual pocket of air reacting with iron sulphide.

Many other metal sulphides are also so readily oxidised as to be pyrophoric in air (Table 4.9).

A large dump of copper pyrites [copper–iron(II) sulphide] ore ignited after heavy rain. The thick layer and absence of ventilation were contributory factors to the accelerating aerobic oxidation which eventually led to ignition.[7]

Reactivity with water

Table 24.4 illustrates that many metals, especially the alkali and alkaline earth metals, are less electronegative than hydrogen and can therefore displace it from compounds such as water or alcohols. In practice, differing rates of reaction are again observed. If a pellet of sodium is placed on water it skims over the surface reacting exothermically to liberate hydrogen. The heat of reaction is, however, insufficient to ignite the gas unless the pellet is prevented from moving. (Thus the breakage of low pressure sodium lamps for disposal requires a controlled procedure to avoid the risk of fire. For example, breakage of a limited quantity in a dry atmosphere and within a large dry container, taking care to avoid flying glass, etc. may be followed by remote filling with water to produce weak sodium hydroxide solution.) Similarly, lithium reacts with water without ignition unless the metal is in a finely divided state. With potassium, however, the hydrogen invariably inflames spontaneously. In the bulk form ignition should be considered likely for all of these reactive metals. Indeed the hydrogen so produced could accumulate and be ignited by an external source of ignition, or in a closed system cause a rapid build-up of pressure. The difference in reactivity between sodium and potassium is incidentally exemplified by methods for their safe disposal in the laboratory.[29]

> Potassium reacts too vigorously with ethanol and the usual procedure is to cover the solid with glycerol.[29] The slow reaction gradually dissolves the potassium away; when little or no metal is present small quantities of ethanol are cautiously added. When addition of more ethanol yields no further reaction, after making the solution homogeneous, the flask contents may be added to water and the contents disposed of. Gram quantities of sodium may be disposed of by addition to ethanol in a beaker within a fume-cupboard. (Larger quantities should be cut into pea-size pieces and added to a cold flask containing ethanol and provided with a Liebig condenser.)

Calcium dissolves readily but not vigorously in water and magnesium forms an insoluble hydroxide film when exposed to moisture, rendering it passive to further attack. Those metals with electrode potentials close to that of hydrogen tend to react only slowly (e.g. rust formation from exposure of iron to water and air) or under forcing conditions, and the most electronegative elements are inactive.

Clearly, in addition to the risk of fire and explosion from incidents involving pyrophoric substances the materials can cause serious thermal and alkali burns if they come into contact with body tissue. There is potential for very severe injuries if the compound enters the body by ingestion or absorption, e.g. via a lesion.

> The jet of a sodium press became blocked during use, and the ram was tightened to free it. It suddenly cleared and a piece of sodium wire was extruded, piercing the operator's finger, which had to be amputated later. (Sodium in a blocked die should be dissolved out in dry alcohol.)[2]

Although not a true metal, it is convenient to consider phosphorus here. The element exists in two main allotropic forms, viz. red and white. The former is insoluble in water, and is only very slowly oxidised in air, and thus tends not to be pyrophoric. The reaction is, however, exothermic and stockpiles of commercial red phosphorus have been known to spontaneously ignite; the risk is increased by catalytic traces of metals such as iron and copper or by impact. Red phosphorus is relatively non-toxic. Conversely, white phosphorus is highly toxic, has a melting point of 44 °C, and inflames in air at around 35 °C. When white phosphorus burns it melts, forming a liquid which can flow and penetrate cracks and corners, where it burns rapidly.

As the electronegativity data in Table 24.4 would suggest phosphorus is not attacked by water. Therefore it is shipped under water for protection. The integrity of this water-cover is crucial so a periodic check is advisable of water levels/water leakage. The following case histories involving white phosphorus illustrate just how serious the risks from accidents involving pyrophoric chemicals may be.

> On 1 April 1978, a train of 85 loaded cars (one of which contained 65.6 cubic metres of white phosphorus under water and a carbon dioxide atmosphere), 14 empty cars and a caboose was derailed in open country near Brownson, Nebraska. The phosphorus car landed upside down with four other cars on top. A wheel burn cut through the outer sheet metal jacket of the phosphorus car and penetrated the glass fibre lagging and the steel inner shell.[30] The friction of the wheel burn melted the phosphorus which seeped out and immediately ignited. The resulting fire, in which flames reached a height of 15 m, heated the remainder of the tank car to about 540 °C. Since no water was available firefighters built a dyke around the tank and left the fire to burn itself out. Some seven hours later, when only 1 m^3 of phosphorus had burned, the phosphorus car BLEVE'd producing a fireball and a large dense white toxic cloud of phosphorus pentoxide and spraying the surrounding area with burning phosphorus which set fire to 30 acres of adjoining fields. Portions of the tank cars weighing up to three tonnes were blown 420 metres through the air. The fire continued to burn for three days. Six people were injured, two seriously and many people had to be evacuated. The inner tank was equipped with a relief valve but with the car inverted with other cars on top, this was unable to operate to limit build-up of internal carbon dioxide and steam pressure.

(This incident demonstrates that BLEVEs, normally observed with highly flammable materials such as liquid petroleum gas (LPG), can also result from other chemicals: the boiling liquid in the tank originated from the 5 cm water layer initially on top of the phosphorus.)

People in four seaside towns near Bristol were instructed to stay indoors with their windows and doors shut after drums of phosphorus caught fire at a warehouse in Portishead on 4 August 1990. Multiple explosions and fireballs occurred and dense fumes, probably consisting of a mixture of phosphorus pentoxide, phosphorus trioxide, phosphoric acid and highly toxic phosphine were blown from the warehouse towards the town. About 100 firemen brought the blaze under control in about two hours. This incident, like a smaller phosphorus fire at the plant 10 days earlier, was thought to have started as a result of evaporation of the protective water layer in the drums due to the hot weather. The 166 drums (each containing 200 kg phosphorus) were stored in a glazed building.[31] Possible reactions include:

$$P_4 + 3O_2 \rightarrow 2P_2O_3 \xrightarrow{3H_2O} 2H_3PO_4$$
$$P_4 + 5O_2 \rightarrow 2P_2O_5$$
$$2P_2O_3 + 6H_2O \rightarrow 3H_3PO_4 + PH_3$$
$$P_2O_5 + 3H_2O \rightarrow 2H_3PO_4$$
$$P_2O_3 + 3H_2O \rightarrow 2H_3PO_3$$
$$4H_3PO_3 \rightarrow 3H_3PO_4 + PH_3$$

Because white phosphorus is insoluble in water, will not burn beneath it, and is almost as twice as dense, drowning a white phosphorus fire with water – without allowing the hose streams to spread the liquid phosphorus – is the basis for extinguishment.[32] The phosphorus solidifies at <44 °C and can be covered with wet sand; premature removal of this cover should be avoided since re-ignition is likely.

Organic substances

Many organophosphorus compounds are not pyrophoric but others are, or possess low flash points (e.g. trimethyl phosphite is water-sensitive with a flash point at 27 °C).

An explosion and fire at a plant in Charleston resulted in six fatalities and 33 injuries, and extensive damage. Whilst the cause has not yet been revealed, the plant was processing trimethyl phosphite, dimethyl methylphosphonate and trimethyl phosphate, ironically to produce a flame retardant for the textile and polyurethane foam industries. Several years prior to this incident a chemical leak at the same site resulted in a fire and explosion causing 17 injuries.[33]

Organometallics

Organometallics can be equally, if not more, hazardous. Industrially important examples include Grignard reagents, the alkyllithiums and alkylaluminiums. However, all these materials can be handled safely if the chemistry and hazards are fully appreciated and adequate precautions adopted.

Organometallics are compounds that contain carbon–metal bonds. They have been known since 1849 when Franklin prepared diethyl zinc. Other notable developments include Grignard's synthesis and use of organomagnesium halides in 1900 and work by Ziegler and Natta in the 1950s on organoaluminiums and their application in the polymerisation of olefins. Later advances exploited the potential of organometallics in the synthesis of organic chemicals.

As a class organometallics exhibit a diversity of chemical and physical characteristics. They range from volatile gases, e.g. trimethyl boron, to infusible solids that cannot be distilled, e.g. di-n-butylmagnesium. The wide-ranging chemistry of organometallics is responsible for both their versatility in synthetic chemistry and their differing hazardous properties. Thus, many organometallics undergo spontaneous ignition when exposed to air whilst others do so, or explode, on contact with water. Evidently, therefore, when inert gas is used to blanket, or blow-over, such materials it should be pre-dried to a tightly specified moisture content; e.g. 10 ppm has been set for use with aluminium alkyls. In general, the organo derivatives of Groups I–III are the most highly reactive.

Organo derivatives of the Group IV metals[34]

Members of the Group IV metals constitute the greatest volume of industrially significant organometallics. Important applications include organosilicones as intermediates for silicones, organotin compounds as stabilisers for PVC plastics, and organolead compounds, e.g. lead tetraethyl, as anti-knock additives in petrol. Organometallics of Group IVA are characterised by their lower chemical reactivity when compared with their neighbours and the R4M compounds of Group IVA metals are non-reactive to water, air and in certain cases even dilute acid. The relative reactivities increase with increasing atomic weight of the metal.

Organo derivatives of Group III metals[34–39]

The organo chemistry of gallium, indium and thallium has not been industrially exploited. That of boron and aluminium is, however, more important, and key examples are given in Tables 24.5 and 24.6 respectively. Trialkylboranes are non-polar and insoluble in water but soluble in saturated aliphatic and aromatic hydrocarbons. The most commercially common aluminiumalkyls are triethylaluminium, triisobutylaluminium and diethylaluminium chloride which are mainly employed as catalysts in the polymer industry. Most low molecular-weight alkylaluminiums are clear colourless liquids at normal temperatures and pressures with high boiling

Table 24.5 Properties of selected organoboron compounds

Organoboron compound	Molecular formula	M.p. °C	B.p.$_{\text{pressure}}$, °C$_{\text{kPa}}$*	Density, g/cm^3	Refractive index	Heat of combustion, kJ/mol†	Heat of vaporsiation, kJ/mol†
dimethylborane	(CH$_3$)$_2$BH	−72.5	69–72$_{101}$				30.5
trimethylborane	(CH$_3$)$_3$B	−161.5	−21$_{101}$	0.63 (at −100 °C)		−2987	23.8
dimethylchloroborane	(CH$_3$)$_2$BCl		4.9$_{101}$				23.8
diethylborane	(C$_2$H$_5$)$_2$BH	−56.3	112$_{101}$				
triethylborane‡	(C$_2$H$_5$)$_3$B	−92.9	95$_{101}$	0.685 (at 20 °C)	1.397 (at 20 °C)	−4589	
diethylchloroborane	(C$_2$H$_5$)$_2$BCl	−84.6	78.5$_{101}$				
tri-n-propylborane	(C$_3$H$_7$)$_3$B	−65.5	156$_{101}$	0.725 (at 20 °C)	1.4135 (at 20 °C)		
di-n-butylborane	(C$_4$H$_9$)$_2$BH		40–41$_{0.03}$	0.765 (at 20 °C)	1.4375 (at 20 °C)		
tri-n-butylborane	(C$_4$H$_9$)$_3$B		209$_{101}$	0.756 (at 20 °C)	1.4260 (at 20 °C)		
di-n-butylchloroborane	(C$_4$H$_9$)$_2$BCl		173$_{101}$	0.879 (at 20 °C)			
triisobutylborane§	[(CH$_3$)$_2$CHCH$_2$]$_3$B	−88	188$_{101}$	0√35 (at 25 °C)	1.4203 (at 25 °C)	−8807	
diisobutylchloroborane	[(CH$_3$)$_2$CHCH$_2$]$_2$BCl		33$_{0.9}$	0.825 (at 25 °C)	1.4160 (at 25 °C)		
tri-sec-butylborane	(C$_2$H$_5$CHCH$_3$)$_3$B		60$_{0.3}$	0.766 (at 25 °C)	1.4349 (at 25 °C)	−8912	
tri-n-pentylborane	(C$_5$H$_{11}$)$_3$B		140$_{1.9}$	0.745 (at 20 °C)	1.4167 (at 20 °C)		
tri-n-hexylborane	(C$_6$H$_{13}$)$_3$B		101$_{0.01}$	0.761 (at 20 °C)	1.4200 (at 20 °C)		

* To convert kPa to mm Hg, multiply by 7.5.
† To convert kJ to kcal, divide by 4.184.
‡ Heat of formation (liquid at 25 °C) is reported as −197.5 ±15.5 kJ/mol, −215 kJ/mol, −190 kJ/mol.
§ Heat of formation (liquid at 25 °C) is reported as −383 kJ/mol, −343 kJ/mol.

points and low vapour pressures and low melting points.[32,40] They tend not to be particularly corrosive and many common materials of construction are usually suitable, but some metals and plastics should be avoided.

All alkylborons and alkylaluminiums undergo autoxidation, the degree of pyrophoricity tending to decrease in ascending an homologous series.

Limited air: $R_3Al + \frac{1}{2}O_2 \rightarrow R_2AlOR$
                $R_xAlCl_{(3-x)} + \frac{1}{2}O_2 \rightarrow R_{(x-1)}Al(OR)Cl_{(3-x)}$
Excess air: $R_3Al + O_2 \rightarrow Al_2O_3 + CO_2 + H_2O$
(fire conditions) $R_xAlCl_{(3-x)} + O_2 \rightarrow Cl_yAlO_x + CO_2 + H_2O + (3-x-y)HCl$

where $x = 1$ or 2, $y \leq 2$.

For higher molecular-weight homologues, e.g. C_{12} to C_{14}, the reaction progresses quietly, whereas the C_1 to C_4 trialkyboranes and trialkylaluminiums are spontaneously combustible in air at ambient temperature, irrespective of the quantity involved. The oxidation of triisobutylaluminium or diisobutylaluminium chloride to form alkoxides is an extremely exothermic process.[38]

Under controlled conditions oxidation can be carried out safely. Indeed when the same reaction occurs during partial oxidation, such as might be encountered in an improperly purged vessel, the main concern will be contamination of the trialkylaluminium. However, in the case of inadvertent or accidental oxidation the result could be disastrous since the heat of reaction is sufficient to degrade the organometallic into isobutene, leading to pressure build-up and hence an explosion in a closed vessel or to a fire hazard in an open system. The tributylboranes are borderline cases and the C_5 to C_8 alkylboranes may ignite at elevated temperatures or if spread over a large surface area. If contained over a small area they tend to just smoke heavily but the temperature can rise well above ambient temperature so the possibility of ignition cannot be ruled out. Tributylaluminiums in volumes greater than one litre may ignite spontaneously at elevated temperatures (Fig. 24.1 and Table 24.7) whereas their higher homologues smoke heavily in air but tend to burn only when ignited.

Although some alkylaluminiums are marketed in the neat undiluted form, in order to reduce their pyrophoric tendency, many are available as solutions in dry aromatic hydrocarbons such as toluene, or saturated aliphatics such as hexane or heptane. In general unsaturated aliphatic and alicyclic solvents and certain chlorocarbon solvents should be avoided. For solutions of alkyl homologues up to and including C_4, concentrations between 10 and 20 per cent are generally non-pyrophoric, and for C_5 and above the concentration can be increased to between 20 and 25 per cent. (However, even the diluted alkyls smoke in air, produce flammable vapours, and can cause skin burns.)

Increased branching on the α or β carbons tends to decrease the pyrophoricity of trialkylboranes. Alkyldifluoroboranes are less pyrophoric than

Table 24.6 General physical properties of aluminium alkyls and related compounds*

Chemical formula	Chemical name	Ethyl code name	Molecular weight	Normal boiling point, °F (°C)	Normal freezing point, °F (°C)	Density at 77 °F (25 °C) (lb/US gal)
$(CH_3)_3Al$	Trimethylaluminium	TMA	72.087	261.0 (127)	+59.5 (15.3)	6.24
$(C_2H_5)_3Al$	Triethylaluminium	TEA	114.17	365 (185)	−49.9 (−45.5)	6.95
$(n\text{-}C_3H_7)_3Al$	Trinormalpropylaluminium	TNPA	156.25	391 (199)	−76.0 (−60)	6.85
$(n\text{-}C_4H_9)_3Al$	Trinormalbutylaluminium	TNBA	198.33	464.1 (240)	−76.0 (−60)	6.87
$(i\text{-}C_4H_9)_3Al$	Triisobutylaluminium	TIBA	198.33	417.4 (213)	+33.8 (1)	6.53
$(n\text{-}C_6H_{13})_3Al$	Trinormalhexylaluminium	TNHA	282.49	587.3 (308)	<−76 (<60)	6.99
$(n\text{-}C_8H_{17})_3Al$	Trinormaloctylaluminium	TNOA	366.55	682.3 (361)	<−40.0 (<−40)	6.96
$(i\text{-}C_4H_9)_2AlH$	Diisobutylaluminium hydride	DIBAH	142.22	516.8 (269)	−112.0 (−80)	6.67
$(CH_3)_3Al_2Cl_3$	Methylaluminium sesquichloride	MASC	205.43	290.6 (143)	+73.0 (22.8)	9.70
$(C_2H_5)_2AlCl$	Diethylaluminium chloride	DEAC	120.56	417.4 (213)	−101.2 (−74)	8.10
$C_2H_5AlCl_2$	Ethylaluminium dichloride	EADC	126.95	397.5 (203)	+87.8 (31)	10.24
						Liq. at 95 °F (35 °C)
$(C_2H_5)_3Al_2Cl_3$	Ethylaluminium sequichloride	EASC	247.51	409.3 (209)	−6.3 (−21.3)	9.14
$(i\text{-}C_4H_9)_2AlCl$	Diisobutylaluminium chloride	DIBAC	176.67	554.2 (290)	−40.4 (−40)	7.61
$i\text{-}C_4H_9AlCl_2$	Isobutylaluminium dichloride	IBADC	155.00	467.6 (242)	−22.7 (−30.4)	9.39
$(C_2H_5)_2AlI$	Diethylaluminium iodide	DEAI	212.01	524.7 (274)	−46.1 (−43.4)	13.16
$(i\text{-}C_4H_9)_xAl_y(C_5H_{10})_z$	Isoprenylaluminium	IPRAL	230†		+63 (17)†	7.8†
$NaAl(C_2H_5)_2H_2$	Sodium aluminium diethyl dihydride	OMH-I	110.1 (Pure comp.)	—	<32 (<0)	7.35‡

Table 24.6 (continued)

Viscosity at 77 °F (25 °C) (cP)	Vapour pressure at 77 °F (25 °C) (mm Hg)	Critical conditions		Heat of: (Btu/lb)		Reaction with water at 77 °F (25 °C)	Liquid specific heat at 86 °F (30 °C) (Btu/lb/°F)
		Temperature, °F (°C)	Pressure (psia)	Vaporisation at NBP	Combustion at 77 °F (25 °C)		
1.03	12.09	581 (305)	550	119.4	17 640	2939	0.520
2.58	0.025	837 (447)	490	228.7	18 165	1811	0.502
5.32	9.6×10^{-3}	935 (502)	172	140.3	18 323	1296	0.516
11.5	6.3×10^{-4}	1008 (542)	146	109.0	18 402	994	0.523
1.90	0.133	722 (383)	269	94.3	18 420	1061	0.643
24.7	4.7×10^{-6}	1110 (599)	114	82.7	18 501	802	0.530
37.6	7.3×10^{-7}	1178 (637)	96	70.1	18 552	622	0.535
	at 104 °F (40 °C)						
13.5	3.2×10^{-4}	874 (468)	579	125.6	18 213	1285	0.494
8.5	5.507	598 (314)	347	76.7	6133	1664	0.328
1.5	0.167	722 (383)	259	76.6	11 387	1454	0.397
2.16	0.315	703 (373)	270	70.9	5425	1267	0.306
Liq. at 95 °F (35 °C)							
1.94	0.220	715 (379)	265	73.9	8325	1354	0.350
4.0	1.02×10^{-3}	848 (453)	208	62.8	13 681	956	0.431
3.8	0.029	770 (410)	245	—	—	—	—
3.4	3.5×10^{-3}	903 (484)	225	51.0	6471	864	0.232
6257†	—	—	—	—	—	1211	0.495
2.08‡	—	—	—	—	—	—	—

* Principal molecular forms exist as dimerised liquid at 77 °F except for TIBA as a monomer, DIBAH as a trimer and EADC as a solid.
† Dependent on C_5/C_4 ratio.
‡ 25% solution in toluene.
NBP = Normal Boiling Point.

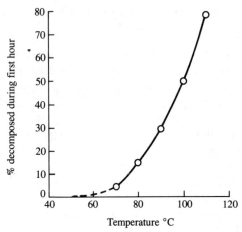

Fig. 24.1 Effect of temperature on decomposition of triisobutylaluminium (initial purity 95%)

Table 24.7 Temperatures for the onset of decomposition of alkylaluminiums

Compound	Temp. (°C) at which decomposition starts
Triethylaluminium	180–190
Tri-n-butylaluminium	170–175
Tri-n-hexylaluminium	165–170
Tri-n-octylaluminium	165–170
Triisobutylaluminium	50–60

their chloro counterparts. Alkylaluminium dihalides are also pyrophoric but less so than the corresponding dialkyaluminium halide.

The nature of the organoborates and organoaluminiums also dictates their reactivity with active hydrogen compounds. Thus, under normal conditions trialkylboranes fail to react with water or alcohols but some alkylaluminiums, such as trialkylaluminiums of chain length C_1 to C_9, react explosively (even in the absence of air such as when blanketed with nitrogen), liberating hydrocarbon and copious quantities of white smoke, apparently amorphous aluminium oxide (inhalation of which may induce the flu-like symptoms of metal-fume fever, a transient condition which usually lasts <36 hours).[38] The violence of the hydrolysis is determined by the nature of the alkyl moiety, the quantity of the aluminium alkyl, the quantity of water and the rate of mixing.

Limited water: $2R_3Al + H_2O \rightarrow (R_2Al)_2O + 2RH$
$2R_xAlCl_{(3-x)} + H_2O \rightarrow [R_{(x-1)}AlCl_{(3-x)}]_2O + 2RH$

Excess water: $R_3Al + 3H_2O \rightarrow Al(OH)_3 + 3RH$

$$R_xAlCl_{(3-x)} + 3H_2O \rightarrow xRH + (3-x)HCl + Al(OH)_3$$

where $x = 1$ or 2.

Water and steam are therefore generally unsuitable as heat transfer media and water is positively hazardous for fire-fighting purposes. Water and water-based foams can produce spectacular, violent reactions with low molecular-weight aluminium alkyls[39] although the use of water fog systems has been advocated as an effective means for dealing with non-flaming organoaluminium spillages.[44] The water fog delays the aluminium alkyl in solution and, although it increases the rate of generation of hydrocarbon vapour, it dilutes the vapour cloud with a steam blanket thereby hampering ignition. It also impedes any flame front and cools adjacent equipment.

Similarly, the vigour of the reaction with alcohols is determined by the organometallic, the nature of the alcohol and the quantities involved. For example, methyl and ethyl alcohols explode on contact with C_1-C_4 alkylaluminiums; isopropanol reacts explosively with undiluted triethylaluminium but *tert*-butanol merely reacts vigorously. Alkylaluminiums react with chlorine which becomes attached to the aluminium by displacement of an alkyl group.

$$R_2AlCl + Cl_2 \rightarrow RAlCl_2$$

Chlorocarbons such as carbon tetrachloride and chlorobromomethane also react violently so halocarbons are unsuitable as fire-extinguishing agents.

The 'open sextet' of electrons surrounding the aluminium atom are available for co-ordination with molecules possessing free electrons such as the lone pair of electrons on oxygen, nitrogen or sulphur atoms in ethers, tertiary amines, thioethers, etc. with which the organoaluminium reacts exothermically. In contrast, organoboranes tend not to form such strong donor–acceptor complexes.

The thermal stability of trialkylboranes facilitates their storage indefinitely under an inert blanket: no evidence of decomposition was detected after 18 months' storage in glass at room temperature. Trialkylaluminiums and alkylaluminium halides can be stored in glass, stainless steel or carbon steel containers without decomposition. At elevated temperatures breakdown of the alkyl compound to olefins and hydrogen leads to pressure build-up. Dialkylaluminium hydrides have greater thermal stability than trialkylaluminiums, the difference being particularly marked when β branching occurs in the alkyl moiety.

Organo derivatives of Group II metals[34,41]

Little is known of the organometallic chemistry of calcium and the heavier Group IIA metals. Organoberyllium derivatives are toxic, expensive, and of no importance industrially. Dialkylmagnesium compounds (R_2Mg) and

Grignard reagents (RMgX where X is usually chlorine) are used as intermediates in the preparation of many organic chemicals including pharmaceuticals, flavours and fragrances, and as catalysts in polymerisations, e.g. polyethylene production.

Grignard reagents are prepared by the reaction of organic halides and metallic magnesium in a dry organic solvent, generally a hydrocarbon. The C_1–C_4 straight-chain dialkylmagnesiums are white solids which are insoluble in hydrocarbons but freely soluble in donor solvents such as amines, ethers, etc., with which they form complexes. The solvent is often diethyl ether but, because of the latter's volatility and flammability, tetrahydrofuran (THF) or THF/toluene mixtures may be preferred. Heavily branched species such as di-*tert*-butylmagnesium are soluble in hydrocarbons. The formation of this intermediate and its subsequent reaction with a host of compounds is exothermic. Thus, Grignard reagents and dialkylmagnesiums can react violently with protic chemicals such as water to liberate the hydrocarbon which, if composed of fewer than five carbon atoms, will ignite spontaneously in air: pure diaryl magnesiums behave similarly.

$$RMgX + H_2O \rightarrow RH + Mg(OH)X$$

Grignard reagents also react vigorously with oxygen to produce an intermediate which readily hydrolyses to give the alcohol:

$$RMgX + O_2 \xrightarrow{RMgX} RO_2MgX \rightarrow ROMgX \xrightarrow{H_2O} ROH$$

Preparations are usually conducted in jacketed glass vessels, or on a plant scale in glass-lined steel reactors. Water is sometimes used as coolant but liquid or gaseous hydrocarbon refrigerants reduce the fire risk which could be associated with leakage of aqueous media.

Commercially available dialkyl magnesiums are stable indefinitely when stored at room temperature under dry nitrogen or argon but branched-chain compounds tend not to be as stable as their straight-chain counterparts. Grignard reagents can be supplied in 55-gallon stainless steel drums under an inert atmosphere. Spillages of organomagnesium compounds must be dealt with immediately using dry sand or earth; salvage is not possible because of contamination/hydrolysis. As with lithium alkyls, dry powder fire extinguishers must be used since Grignard reagents react vigorously with carbon dioxide, chlorinated hydrocarbons or water. Fires can generate toxic gases such as hydrogen chloride or hydrogen bromide, benzene, etc. and thus breathing apparatus is likely to be required during firefighting.

Of the Group IIB metals, organozincs have limited industrial use as organic intermediates (e.g. in the Reformatsky reaction) and organomercurials possess therapeutic and biocide properties and have found application in antiseptics, fungicides, etc. The decrease in reactivity of

Group IIB organometallics reflects the increase in electronegativity of the metal. Thus the lower dialkylzincs react explosively with water, whereas dialkylcadmiums hydrolyse only slowly and dialkylmercurials are stable towards hydrolysis. Dialkylzincs have linear structures and are monomeric compounds easily distilled: the boiling points of dimethylzinc, diethylzinc and di-n-propylzinc are 44, 117 and 139 °C, respectively. Chemically organozincs are similar to organomagnesiums but are less reactive. Nevertheless organozincs are pyrophoric. The lower dialkylmercurials are volatile liquids which tend to be stable when refrigerated. As mentioned earlier, as a class dialkylmercurials are not readily attacked by water.

Organo derivatives of Group I metals[34]

The organometallic chemistry of potassium, rubidium and caesium is not well developed, primarily because of their highly reactive properties and inherent instability. Conversely, industrial usage of organosodium compounds is significant, although somewhat in decline; the main use of alkylsodiums nowadays is in the synthesis of other organometallics. The structure of simple alkylsodiums is ionic and as a result the substances are involatile, white pyrophoric solids which are insoluble in organic solvents but which react explosively with water or other protic materials. Alkylsodiums tend not to be stable for prolonged periods, even at reduced temperatures.

Lithium derivatives[42,43] are in practice the most useful members of this class since they are easily prepared and are less reactive, and hence more controllable, than their sodium or potassium analogues. They also dissolve in organic solvents. Consequently, lithium alkyls are the most commercially important organoalkali compounds with butyllithium dominant. In the neat form, industrially available alkyllithiums, such as n-butyllithium and *sec*-butyllithium, exist as tetramers or hexamers and hence are viscous liquids. They are, however, usually supplied commercially as a solution in a hydrocarbon solvent such as hexane. The thermal stability of butyllithiums decreases in the order *tert*-butyl>n-butyl>*sec*-butyllithium, for given concentrations and temperatures. Thermal stability decreases with increased concentration of butyl compound[22] (Fig. 24.2) and reduces in the presence of alkoxides. Although soluble in ethers n-butyllithium is unstable in this class of solvent (Table 24.8). The decomposition reaction in diethyl ether is:

$$C_4H_9Li + C_2H_5OC_2H_5 \rightarrow C_2H_5OLi + C_4H_{10} + C_2H_4$$

Butyllithium also produces ethylene by cleavage of the ether linkage in furan. Interestingly, methyllithium, which is a crystalline material, being less reactive than the higher homologues, is stable for at least six months in diethyl ether at room temperature.

The degree of pyrophoricity is dictated by the structure of the organo-

Flammable and explosive hazards

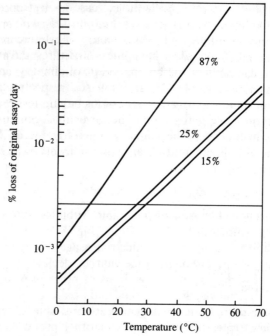

Fig. 24.2 Thermal decomposition of n-butyllithium in hexane: temperature versus percent loss at various concentrations

Table 24.8 Decomposition of n-butyllithium by ethers

Ether	$t_{1/2}$	Temperature, °C
Ethyl ether	6 d	25
	31 h	35
Isopropyl ether	18 d	25
Glycol dimethyl ether	10 min	25
	23.5 h	0
Tetrahydrofuran	5 d	−30

metallic species, its concentration, and the flash point of the chosen solvent. Thus, for given concentrations of lithium compound in a given solvent, there is an increase in pyrophoricity in the order n-butyl<sec-butyl<*tert*-butyl. At low concentrations, <15 per cent, the reaction of butyllithium with air may produce insufficient heat to ignite the solvent whereas exposure of higher concentrations (e.g. >25 per cent) to air will ignite the diluent spontaneously. Concentrations of 50–80 per cent butyllithium are considered very hazardous with pyrophoricity peaking at 60–75 per cent. Most commercial applications rely on 15 per cent solutions to minimise the pyrophoric risk. Because of the vigorous reaction between butyllithiums and water, high humidity and the presence of moisture enhance the likelihood of a spillage igniting. For aluminium alkyl homologues up to and including four carbon atoms, concentrations between 10 and 20 per cent alkyls are generally non-pyrophoric, and for C_5 and above concentrations can be increased to between 20 and 25 per cent.

Of those reactions given in Fig. 24.3, the exothermic oxidation of butyllithium by atmospheric oxygen and its violent reaction with water pose the greatest safety problems since oxidation can lead to spontaneous combustion whilst hydrolysis is accompanied by the evolution of heat and liberation of flammable gases which can ignite or result in pressure build-up with explosion potential.

One worker suffered burns to his neck and leg and another suffered from shock in a fire involving 300 kg of butyllithium. The fire started from a small leak which developed during the transfer between process vessels. The fire was left to burn itself out.[44]

A few instances have been reported in the literature in which use of organolithium reagents has led to violent chemical explosions.[45]

A pentane suspension (50 ml) of pentafluorophenyllithium was prepared under nitrogen at low temperature by the addition of butyllithium to bromopentafluorobenzene (ca. 2 g):

$$C_4H_9Li + \underset{\text{Bromopentafluorobenzene}}{C_6F_5Br} \longrightarrow \underset{\text{Pentafluorophenyllithium}}{C_6F_5Li}$$

The reaction vessel, a 100 ml two-necked round-bottomed flask provided with a magnetic stirrer, was almost completely immersed in an expanded polystyrene dish containing solid CO_2 and alcohol throughout the experiment. A stopper was removed from the flask and D_2O (0.2 g as a solution in 40 ml of ether) was slowly added to the stirred lithium reagent against a purging current of nitrogen. After about 50% of the D_2O had been added the contents of the flask suddenly became bright yellow and then, almost immediately, exploded violently. No flash or fire accompanied the explosion. The largest piece of apparatus traced subsequently was a glass sliver about 1¼ in × ½ in deeply embedded in the operator's neck; several hours of surgery failed to extract it.

The techniques described above had been used without previous mishap for some years and the operator in this case was an experienced post-doctoral fellow. Furthermore, ethereal solutions of polyfluoroaromatic lithium reagents had frequently been hydrolysed using normal distilled water and suspensions of solid C_6F_5Li in hydrocarbon solvents had been handled at temperatures up to about 10 °C. It is possible that the D_2O used in the above experiment contained small quantities of peroxide since solutions of Fe^{++}, when treated with the

1386 Flammable and explosive hazards

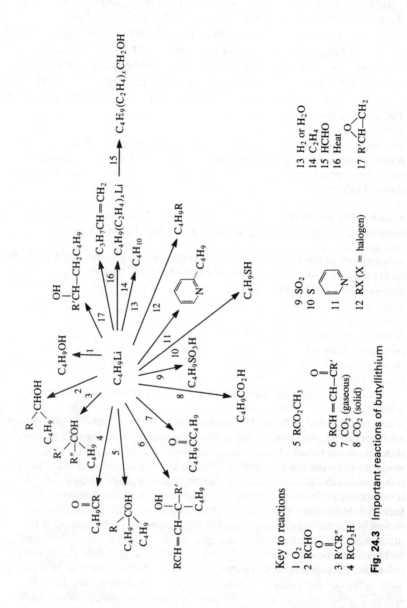

Fig. 24.3 Important reactions of butyllithium

Key to reactions
1 O_2
2 $RCHO$
3 $R'CR''=O$
4 RCO_2H
5 RCO_2CH_3
6 $RCH=CH-CR'=O$
7 CO_2 (gaseous)
8 CO_2 (solid)
9 SO_2
10 S
11 pyridine
12 RX (X = halogen)
13 H_2 or H_2O
14 C_2H_4
15 $HCHO$
16 Heat
17 $R'CH-CH_2$ (epoxide)

D$_2$O, gave positive tests for Fe^{+++}. The presence of a peroxide could have initiated the explosion of the lithium reagent.

A suspension of m-bromophenyllithium (0.01 mole) was prepared under nitrogen from m-dibromobenzene and n-butyllithium in hexane:

$$C_4H_9Li + \underset{m\text{-dibromobenzene}}{\underset{\displaystyle Br}{\bigcirc}\!\!Br} \longrightarrow \underset{m\text{-bromophenyllithium}}{\underset{\displaystyle Li}{\bigcirc}\!\!Br}$$

The reaction mixture had been largely freed by vacuum evaporation at ambient temperature from by-product butyl bromide, since this had been found to interfere in the subsequent reaction of the intermediate m-bromophenyllithium. The rate of nitrogen purging of the flask was increased and a stopper was removed, with the intention of adding ether to dilute the slurry. The contents of the flask immediately exploded violently, without flash or fire, the flask fragments causing some hand injuries and perforating a nearby window. Analysis of the sooty explosion residue showed the presence of an homologous series of polyphenyls and lithium bromide. This suggests that the explosion was caused by an intermolecular condensation reaction, possibly initiated by a particle of organolithium compound on the stopper becoming hot on exposure to atmosphere and falling-off into the flask contents, or by oxygen entering because of turbulence at the open neck. Once initiated in the viscous slurry the exothermic elimination reaction would rapidly accelerate and run away in the virtual absence of heat-absorbing diluent or mobile liquid phase. Suspensions of m-bromophenyllithium are very finely divided and correspondingly reactive in some circumstances, and p-fluorophenyllithium steadily eliminates lithium fluoride in ether solution at 0 °C, presumably in an intermolecular reaction.

Elimination of lithium halide would be expected to be facile in view of the relatively weak C–Li bond, with activation energy minimal for the m-isomer. It is also possible that the extent of solvation of the lithium atom, or of its relative adjacency to neighbouring halogen atoms in the crystal lattice, is involved in those cases where solid haloaryllithium compounds have exploded in the absence of solvent. Other such cases include p-fluoro-, m-chloro-, and m- and p-trifluoromethyl-phenyllithium. Instability was also noted with p-bromophenyllithium, which is also oxygen-sensitive. It was concluded that haloaryllithium compounds show inherent instability in the solid phase and should only, therefore, be used on a small scale with full, protective precautions. Explosions with m-bromophenyllithium in the presence of solvent had been observed previously. Isolated m- and p-dilithiobenzene and lithium chloroacetylide are also explosive.

The explosion hazards of pentafluorophenyl derivatives of aluminium and other metals have also been reported.[2,46]

In addition to their pyrophoric hazards butyllithiums are destructive to human tissue: on contact with the skin or eyes they react rapidly and exothermically with body moisture to produce caustic lithium hydroxide;

autoignition is also possible. The degree of damage will depend on the volume and concentration of butyllithium and the exposure time.

> In a minor pilot-plant accident, a solution of butyllithium caused a blister similar to an ordinary heat burn. Healing was normal and the operator suffered no unusual pain.[22] Conversely experience with isobutylaluminium derivatives on one plant indicated that all personnel burns were superficial but very slow-healing and extremely painful.[38]

Clearly, the organic diluents also pose health and fire hazards as described in Chapters 4 and 5. They will remove natural oils and fats from the skin, and the results of overexposure to solvent vapour include narcosis; relevant hygiene standards for the hydrocarbons have been set to protect employees. Hexane, for example, boils at 69 °C with a flammable range of 1.1–7.5 per cent. The flash point is below ambient temperature and the auto-ignition temperature is only 261 °C. Butyllithium itself has practically no vapour pressure so inhalation hazards do not exist, unless poor practices lead to formation of an aerosol which, if inhaled, could cause respirattory irritation. Systemic and chronic effects of butyllithium appear to be unknown.

Control measures

Controls for thermally unstable materials

Flammable materials heated above their auto-ignition temperature (AIT) can spontaneously inflame when exposed to air. Whenever possible, therefore, a material should not be heated above its AIT or if this cannot be avoided air should be excluded. Compounds known, or suspected, to be thermally unstable should not be subjected to uncontrolled heating. Some materials such as perchlorates and polynitro compounds can inflame or explode even in the absence of air. Thus, in the following case an exothermic runaway reaction, in which the decomposition rate after heating was too great for the heat liberated by the decomposition to escape from the mass of solid, resulted in exponential self-heating to a violent decomposition.[47a]

> Five employees were killed and several were injured following an explosion during maintenance work to remove two to three tonnes of solid waste from a still used for batch processing of nitrotoluenes. Two days after the fire some medical staff and firefighters were experiencing skin irritation and vomiting. [The vomiting has been attributed to a viral infection and not chemical poisoning.[47a]]
>
> The nitrotoluene residues left in the still were heat-sensitive and unstable at high temperature so instructions were given to heat to no more than 80 °C[47b] using a steam coil in the lower section in order to soften them. Unfortunately, the thermometer at the base of the still was not in contact with the waste

residue and gave a false reading of residue temperature – which it was later estimated must have reached >165 °C. The solid residues had poor properties of heat conductivity; the steam itself was also hotter than the operators believed. As a result of hot spots the residue ignited and a fierce jet of flame spewed out of the manhole opening, destroying the wooden control room and much in its path.[48,49]

Clearly such incidents are not precluded by the exclusion of air, i.e. inert gas blanketing.[47a]

In a similar, predictable incident, an explosion resulted when a tank containing iminobisacetonitrile (IBA) was heated with a steam coil to free a blockage. The heat of reaction caused overpressurisation and the IBA decomposed to HCN which then underwent exothermic polymerisation. Although personnel injuries were slight, plant damage and business interruption amounted to £10 million.[50]

Controls for inorganic compounds

Metals and their compounds

Some metals inflame spontaneously upon exposure to air or water whilst others, e.g. aluminium and magnesium, although not necessarily pyrophoric, can inflame under certain circumstances, e.g. in a finely divided state, especially in the presence of sparks. Precautions for the latter are therefore also briefly summarised.

Aluminium[51,52] – Aluminium and its alloys in the form of plate, sheet, extrusion wire or foil do not burn despite claims of their aluminium structure contributing to the burning of ships during the Falklands War. Indeed scrap metal is melted in open furnaces with flames playing directly onto the metal. Fire and explosion hazards may, however, arise during grinding and polishing operations of the metal and its alloys. Aluminium powder/dust may be ignited by sparks, e.g. from ferrous tools during cleaning/maintenance, from static electricity, or from electrical apparatus.

Dust should be extracted as close as practicable to the source, preferably via a wet scrubber. If dust is collected dry it should be protected from inadvertent water since under some circumstances hydrogen may be liberated. The collection system should ideally be located outside the main building and discharge should be either via a rotary valve or directly into a robust metal container firmly clamped to the discharge outlet. Dust from polishing operations (which also contains textile waste) should not be served by the same dust collection system as grinding, finishing, etc.

Neither water nor foam should be used to fight fires because of the risk from hydrogen generation. Halocarbons and certain dry powder extinguishers may intensify the fire. Proprietary powders or dry sand should be applied to smother the flames. After extinguishing the fire the mixture should not be disturbed.

Magnesium – The ease of ignition of magnesium is influenced by the shape and size of the material. In bulk it is difficult to ignite without raising its temperature to near the melting point (650 °C). As ribbon, shavings or chips with feather-like edges it can be ignited by sparks. In finely divided form it can inflame most easily and it reacts with water to liberate hydrogen. Once alight, magnesium burns with an intense white flame with a temperature of about 4000 °C and can be difficult to extinguish.

Bulk metal can be stored similarly to steel. Depending upon the quantities involved, metal powder should be stored in a detached non-combustible building or in a room isolated from other parts of the premises by fire-resistant walls and floors. The material should be protected from moisture and kept away from, e.g. halogens, acids, oxidisers, and ignition sources.[53]

Great care is needed with processes which generate metal or its alloys in finely divided form. Thus, whenever magnesium materials are machined, floors and equipment must be kept clean and dry. Tools should be kept sharp to provide clean heavy cuts so as to avoid fine turnings. Arrangements are required to prevent accumulation of turnings, which should be collected regularly (at least at the end of each working day) in specially designated containers. Tool-grinding operations which produce sparks should be avoided. Grinding should preferably be conducted in a detached single-storey building; the building and equipment should be equipped with explosion vents and plant should be maintained air-tight. Dust should be discharged into a wet collector or led away from machines in metal ducts and collected in metal bins via a cyclone. Rooms in which machining takes place should be used for no other purpose. Dust should be prevented from accumulating and floors, ledges, electric light fittings, shelves, etc. should be vacuumed regularly. Sources of ignition should be kept away from plant which should be totally enclosed and dust-tight.

In the UK the Magnesium (Grinding of Casting and Other Articles) Special Regulations of 1946 apply to magnesium alloys with a magnesium content of >20 per cent and require

- occupiers to provide and maintain plant, avoiding the use of equipment previously used for abrading ferrous materials, or tools capable of causing sparks
- dust to be intercepted as close as possible to point of escape and ventilated hoods and ducting to carry away dust safely and avoid accumulation, with access facilities for cleaning and equipped with, e.g. wet scrubber
- grinding/polishing facilities interlocked with ventilation; ventilation to run for at least five minutes prior to starting processing

- scrubbers to be emptied at least once per week; no ferrous tools to be used in the operation
- sludge removal from factory without delay, or burned in the open at a safe distance
- use of fire-resistant overalls and leather aprons with leather bibs
- prohibition of sources of ignition; no smoking; no ferrous tools
- all defects to be reported to management.

Sodium and potassium – As discussed, either sodium or potassium may ignite and explode on exposure to air depending upon conditions: potassium burns and splatters. They react with water to form hydrogen which may ignite spontaneously. Other hazards include violent reaction with other chemicals[2] and their caustic properties and ability to cause severe burns. The precautions listed in Table 24.9 (after ref. 54) therefore aim to maintain separation from air and water and avoid personal contact. Periodic inspections to check on the integrity of the protective media and of the container, e.g. in terms of corrosion, are essential: deteriorated product should be discarded.

> A sealed bottle containing sodium dispersed in xylene was stored in a tin for 16 years. A yellow-white solid layer exploded violently during scraping out, possibly attributable to peroxide formation.[55]

> A laboratory fire originated in a store cupboard possibly due to corrosive failure of an aluminium container used for potassium.[56] (Perforation of aluminium vessels has been attributed to the deliquescence of potassium carbonate, produced from atmospheric carbon dioxide, resulting in corrosive attack.[57])

In bulk, sodium is usually sold in bricks, in drums, or in tankcars. Because of its high reactivity it is packaged to minimise contact with air. The bricks are often packed under vacuum, in soldered tins or in gasketed pails. When transported in tankcars or drums, purging with dry nitrogen is important. The tankcars or drums are filled with molten sodium which is then allowed to solidify by cooling before shipment. It is then remelted for unloading at its destination. Sodium can also be dispersed in certain inert hydrocarbon media such as white oil or kerosene, to improve ease of handling. These dispersions can contain up to 50 per cent sodium by weight.

Fire-fighting requires breathing apparatus and full protection against contact with exposed parts of the skin. Water or any agent containing oxygen must never be used; dry sand, dry sodium chloride, dry soda ash or special dry powders (e.g. based on graphite) can be used to blanket the metal until it cools below its ignition temperature.

Titanium – Finely divided titanium in the form of a dust cloud or layer does

1392 *Flammable and explosive hazards*

Table 24.9 Summary of precautions for handling metallic sodium*

Uses	Production of tetraalykllead antiknock compounds for petroleum. Manufacture of refractory materials such as titanium, zirconium by reduction of their halides, of metals such as tantalum, silicon, magnesium and of many sodium compounds. Used in the manufacture of dyes, herbicides, pharmaceuticals, perfumes, etc., and as a general laboratory reagent and as a drying agent for certain solvents. Employed as catalyst in certain polymerisations.
Hazards	*Fire:* Spontaneously combustible on contact with air and water. *Explosion:* Reaction with water can be explosive. Hydrogen formation in enclosed space can result in pressure build-up. Hydrogen/air mixtures may reach explosive concentrations. *Health hazards:* Causes thermal and alkali burns. *Reactivity:* Can react violently with a range of substances, including water, carbon dioxide, carbon tetrachloride. See also ref. 54.
Firefighting	Use only dry powder inert extinguishers. DO NOT USE water, carbon dioxide, chlorocarbon or aqueous foams.
Storage	Facilities must be fire resistant and weatherproof, preferably segregated and dedicated for bulk storage. No sprinkler system, water or steam pipes should be present. Limit inventory. Provide adequate fire extinguishers. Label store appropriately. Store stocks off floor to avoid accidental contact with water. Because hydrogen is less dense than air (see Chapter 2) high level natural vents should be provided to help avoid accumulation of the gas.
Unloading bulk containers	Seek advice from supplier; special arrangements including provision of inert atmospheres will be needed for transfer of molten metal. Assume hydrogen is present in the headspace of drums and avoid sparks (e.g. hammering). Use non-sparking tools. Tools and surfaces must be dry; make sure no sodium is left on tools.
Handling	Avoid all contact with water. Avoid skin and eye contact and wear appropriate personal protection. Use spill trays beneath reactors to contain any spilled sodium which must be dealt with immediately. Reactors must not be heated with water or steam. Consult sodium supplier with regard to materials of construction and design of reactors/plant. During cleaning of vessels consider the possibility of accumulation of hydrogen and inerting gases. Special procedures will be required for large vessels (again consult with manufacturer). When charging solid sodium bricks to closed vessels a double gate with intermediate chamber capable of inerting with nitrogen should be considered. In the laboratory sodium is best handled in a glove box filled with dry nitrogen. When handled on the bench-top water must be excluded from the area. Cutting tools and presses must be clean and dry. Small quantities of waste sodium can be reacted carefully with alcohol.
Personal protective equipment	The level of protection required depends on the hazard in terms of the process and quantity of sodium in use. Eye protection ranges from safety spectacles to tight-fitting goggles and full safety visors. Skin protection is provided by rubber gloves and laboratory coats for bench-scale operations. Flameproof overalls and mitts provide production-scale protection. Hard hats and foot protection guard against general factory hazards including manhandling of drums, etc. Appropriate respiratory protection may be required when inspecting, repairing or cleaning plant or fighting fires.

Table 24.9 (continued)

First-aid	Speed is essential in removing sodium quickly from contact with skin or mucous membranes. If combustion occurs it should be extinguished as quickly as possible. *Skin contact:* Remove pieces of sodium carefully and quickly. Flush affected areas with copious volumes of water, preferably under a shower. Wash the area for 30 minutes. In all but the most minor mishaps seek medical attention. *Eye contact:* Immediately flush with plenty of water for 15 minutes and get medical attention.

* This is a brief résumé of some key precautions. Intending users should consult their supplier and the literature so that more detailed precautions can be devised as appropriate for the specific use in mind. Training of operatives in hazards, precautions, use/care of personal protective equipment, and in emergency procedures is also essential.

not ignite spontaneously. Spontaneous combustion of titanium has been reported, however, when fines, chips and swarf were coated with water-soluble oil.[58] Containers used for disposal should, therefore, be kept tightly closed and segregated.

Nickel – Particles of nickel powder are, if sufficiently fine, pyrophoric. Thus drums containing Raney nickel catalyst under water require careful handling, e.g. shaking to wet any catalyst at the top of the drum before being opened; displacement of air with inert gas before sealing is advised.[59]

> As the lid was taken off a drum of Raney nickel stored under water it was blown into the air by an explosion. Hydrogen in the head space was apparently ignited by dry, and hence pyrophoric, catalyst upon entry of air.[59]

Miscellaneous rarer metals – The pyrophoric risks with finely divided forms extend to some precious and rarer metals. For example, finely divided, i.e. catalytic, forms of platinum are hazardous when dry; explosions have occurred during air-drying and during manipulation on filter paper.[2] Powdered reactive uranium, from pyrolysis of the hydride, is also pyrophoric, and storage of foil in closed containers in the presence of air and water produces a reactive surface. Finely divided plutonium in the form of a dust cloud or layer may also ignite spontaneously.

Numerous fire and explosion incidents have involved pyrophoric, finely divided zirconium, moisture and friction plus possibly in some cases static discharges within plastic bags.[60] The precautions for the safe use of zirconium[61] tend, therefore, to exemplify the general requirements, i.e. exclusion of water, inerting to exclude air or oxygen, avoidance of moisture vapour, exclusion of oxidants, control over particle size and measures to limit exposure of personnel.

Hydrides – Calcium hydride is used as a reducing agent, a drying agent, a reagent for organic analyses, a portable source of hydrogen, and for cleaning blocked oil wells. It reacts with water to liberate hydrogen and it may incandesce or explode if ground with oxidising agents. The substance should be stored in sealed, properly labelled vessels, e.g. air-tight steel drums. Glass containers are unacceptable for all but the smallest quantities. There should be provision for venting hydrogen and regular checks for corrosion are essential. Depending upon quantities, stores should be ventilated, dedicated non-combustible single-storey buildings or be separated from other parts of the premises by fire-resistant walls and floors. In either case explosion vents should be considered. Water pipes, taps and all forms of moisture and sources of ignition should be excluded. Fires involving calcium hydride should be smothered with dry sand or proprietary dry-powder extinguishing agents; water, halogen or foams must not be used.

Lithium hydride and lithium aluminium hydride may similarly ignite upon contact with water or moist air.[58] The heat of reaction will probably ignite the hydrogen gas formed.

Sodamide is used for the preparation of certain nitrogenous organic compounds, in the manufacture of sodium cyanide, and as a laboratory reagent. It is flammable and may ignite spontaneously if rubbed or ground vigorously, or on contact with water or moisture (with which it reacts to liberate ammonia) as may be encountered in inadequately dried apparatus. It should be stored in air-tight steel drums, or in polythene bags in metal containers which should be checked periodically for damage or corrosion. Stores should be cool and dry, fire-resistant, ventilated and equipped with explosion vents. No combustible materials, acids, alkalis or oxidising agents should be stored in the same room, and water pipes, taps, etc. should be excluded. Dry sand or special powders should be used to extinguish fires, never water, Halons or foams.

Non-metals

Phosphine – Phosphine is a colourless gas at room temperature and pressure (b.p. −87.7 °C). It is odourless when pure but usually has a fishy or garlic-like odour due to impurities. It is shipped as a liquid in steel cylinders under its own vapour pressure of 592.7 psig at 21 °C: it is also often prepared *in situ*. Phosphine reacts explosively with oxygen and with chlorine.

Because of its toxic properties phosphine finds limited use for fumigation, usually via slow release from action of water on aluminium or magnesium phosphide. It is also employed in the production of semiconductors in the electronics industry and in the synthesis of organophosphines and organophosphonium compounds. Other sources of exposure are usually accidental, e.g. during production of acetylene, during conversion of yellow phosphorus into red phosphorus and its derivatives, from the action of water or damp conditions on calcium, magnesium, aluminium, and sodium phosphides, and also on ferro-silicon.[62]

Pure phosphine is not spontaneously flammable but the possibility of spontaneous ignition is increased where there are traces of the liquid hydride P_2H_6.[62] Hence, in addition to its toxic hazards, commercial phosphine is pyrophoric.

> Phosphine, generated from calcium phosphide, was dried by passage through towers packed with the same material. After the generator had been refilled and whilst it was being purged with argon an explosion occurred. Air displaced from the generator had reacted with dry phosphine in the towers; polyphosphine on the surface of the calcium may have catalysed the reaction.[63]

Key precautions, therefore, entail use of phosphine in sealed equipment and under well-ventilated conditions. All possible accidental sources should be identified and prevented (e.g. keep phosphides out of contact with water). Industrial plant is usually totally enclosed (including gas holders) and fitted with extract ventilation systems.

In the UK any plant using more than 100 kg would constitute a major hazard[64] and one tonne is a 'controlled quantity'.[65] On the smaller scale gas cylinders should be located out of doors or in a separate well-ventilated room and pipework should contain as few joints as possible. Stocks should not be stored with cylinders containing oxygen or other highly oxidising or flammable materials. All lines and equipment should be grounded and checked for leaks with nitrogen using soapy water. Gas monitoring systems are required to detect leaks and personal protection is required for certain operations (e.g. changing cylinders). Cylinders are not fitted with safety relief systems so should be stored away from heat sources. Defective systems may be rendered safe by bubbling the leaking phosphine through excess calcium hypochlorite solution. General precautions for handling gas cylinders are also appropriate (page 641).

Metal phosphides should be stored in sealed polythene-lined drums in well-ventilated locations. Formulated products should be stored in secure locked areas. Phosphide spillages should be kept dry. Respiratory protection is required to deal with emergencies. Leaking containers can, if small, be removed to deep pits and the phosphide allowed to hydrolyse slowly under supervision, the residue disposed of, and the land thoroughly washed.

White phosphorus – The main hazards with white phosphorus stem from its toxicity, pyrophoric properties and ability to inflict severe burns. Because phosphorus does not react with water, it is shipped and stored under water to prevent contact with atmospheric oxygen. The integrity of this water layer is crucial and periodic checks are advisable to ensure an adequate level is maintained (page 1373). To control corrosion of metal containers the pH of the water should be checked periodically and, if necessary, adjusted carefully with lime or soda ash. Alternatively phosphorus may be stored under an inert atmosphere in either metal drums or hermetically sealed cans. It should also be stored so as to prevent accidental contact with other compounds, particularly oxidising agents with which it may explode. Small leaks can be arrested temporarily by deluging with water to cause the phosphorus to solidify. Larger leaks should be covered with wet sand and water. Fire fighting measures are summarised on p. 1374.

Recommended practice is for returnable drums to be filled with water or to be washed thoroughly, inside and out, with hot water followed by soda ash solution. Non-returnable drums require similar washing, and may be heated in a furnace to destroy any residual phosphorus, prior to scrapping.

Silane – Silane is a colourless gas (b.p. −112 °C) with a vapour density of 1.1 and a repulsive odour. It is employed as a doping agent for solid-state devices and in the production of amorphous silicon and high-purity silicon oxides. The main hazards stem from its pyrophoricity and some violent reactions, e.g. with halogens.

Silane should be stored under positive pressure. Cylinders (of aluminium with red and yellow markings) should preferably be stored outdoors but under protection from rain and direct sunshine. The store should be securely fenced with cylinders located at least 2 m from the perimeter. The building should ideally be detached and constructed of non-combustible materials, with facilities for venting explosions, e.g. lightly constructed roof, and fitted with high- and low-level ventilation. Smoking and naked lights should be prohibited. Only silane cylinders should be housed in the store which should display prominent warning notices. When cylinders are not in use protective caps should be screwed down over the valves.

Regulators specially designed for silane should always be used for the proper control of outlet pressure. Valves and fittings should be kept clean and not be lubricated. Pipelines and equipment can be of iron, steel, copper, brass and other common materials. However, they should be checked regularly as silane decomposes to silica which can constrict or block lines and equipment.

After connecting up the cylinder and before opening the valve, the silane regulator and lines should be evacuated and purged with an inert gas, e.g. nitrogen (vented outdoors via a pipe small enough to ensure no back-diffusion), and checked for leaks with soapy water. Cylinder valves should be opened cautiously so that leakage at the valve can be detected before a serious escape can occur. Cylinders should be kept below 50 °C and not stored with oxygen or highly oxidisable or flammable materials. They should preferably be connected up in an isolated location, e.g. a ventilated metal locker attached to the outside of the building and the silane piped into the working area by the shortest possible route. Cylinder valves should always be turned off at the end of the working day.

Piping and equipment should be thoroughly pressure-tested above working pressure and be genuinely leak-free. Leaks may be indicated by flames which should be extinguished by turning off the supply. Silane should not be condensed ($-110\ °C$ or less) since serious accidents have occurred with liquid and solid silane. Diaphragm packless valves with resilient seats of materials such as PTFE are used, and back plates on rotameters and pressure gauge covers are removed. Extreme caution is needed when using silane in systems with halogenated compounds since even a trace of free halogen can react violently at room temperature. Mixtures containing 1 per cent silane in hydrogen or nitrogen have ignited spontaneously when mixed with air. All lines and equipment used with silane should be grounded.

In the event of fire, flames should be extinguished by closing the valve or plugging the leak if it is safe to do so. No attempt should be made to extinguish the fire in any other way but, if it is safe to do so, the cylinder and any adjacent ones should be cooled by spraying with water.

Controls for organometallics

The precautions for handling this class of material can be illustrated by reference to aluminium and lithium alkyls, e.g. butyllithium.[34,42,66-68]

World demand for butyllithium in 1984 was 600 tons per annum. A brief description of the manufacturing process provides an interesting example of chemical process and safety engineering. Production involves the direct reaction of butyl chloride with lithium metal, obtained from the electrolysis of lithium chloride, thus

(1) $\quad 2LiCl \xrightarrow{e} 2Li + Cl_2$
(2) $\quad 2Li + C_4H_9Cl \rightarrow C_4H_9Li + LiCl$

Stage 1 – A molten eutectic mixture of potassium chloride and lithium chloride is electrolysed at *ca.* 475 °C in carbon steel cells (suitably stress-relieved and metallised to withstand the operating temperatures) equipped with carbon anodes with the cell body acting as the cathode. Lithium metal is produced at the cathode whilst gaseous chlorine is evolved at the anode. Molten metal puddles on the salt bath from where it is periodically skimmed off and poured into moulds to cool and solidify.

Stage 2 – Lithium in finely divided form is prepared by rapid stirring of molten metal (m.p. 186 °C) in mineral oil dispersing agent under an inert atmosphere: argon is employed since lithium metal reacts with nitrogen to form the nitride. The metal is washed free of oil with an organic solvent, typically hexane, and butyl chloride is added in a carefully regulated manner to the lithium/hexane suspension. The exothermic reaction proceeds at ambient temperature and pressure and the lithium chloride which separates out is removed by filtration to yield a straw-coloured solution of butyllithium which is stored and handled under nitrogen.

Hazards

The main hazards of the process stem from the following.
- Exposure to molten lithium metal and to chlorine evolved during electrolysis. The corrosive and toxic properties of chlorine are discussed in Volumes 1 and 2 (e.g. Chapter 15) and in more detail in Chapter 30. The cell is, therefore, maintained under slight negative pressure induced by a fan located after the cell off-gas scrubbing system. The slight suction withdraws chlorine as it is liberated and also prevents its release from the cell during skimming. Since production of chlorine is regulated by the current flow, power to the ventilation system should be interlocked with the direct current feeding the electrolysis process such that failure of the fan supply automatically halts the reaction and hence generation of chlorine gas. Pressure sensors provide a useful back-up to detect failure in air flow, e.g. as a result of mechanical damage to the fans. Otherwise impeller breakage would result in poor suction despite continued rotation of the shaft. The off-gas is scrubbed with caustic soda solution to remove chlorine prior to discharge of effluent to atmosphere. Measurement of chlorine concentrations in the ambient atmosphere using grab samples and static background monitors (see Chapter 7) linked to an alarm system serves to provide an indication of the effectiveness of these controls, a check on employee exposure, and warnings of excursions beyond acceptable levels. Full personal protective equipment, including air-fed aluminised hoods and rayon overalls, is required during skimming as a back-up precaution against accidental exposure to chlorine, thermal radiation or splashes of molten metal.
- Accidental contact with caustic soda and sodium hypochlorite solution from scrubbing operations. Enclosure of the process, the use of protective clothing and the provision of eye-wash facilities and safety showers are essential requirements here, although it is to be noted that in order to prevent accidental contact with the molten metal, water must not be piped into the cell metal handling area itself. Chlorine detectors on the stack from the scrubber monitor chlorine levels in the effluent.
- Use of butyl chloride and hexane. Both materials are highly flammable and toxic. They boil at similar temperatures and have similar flammability profiles. Thus butyl chloride boils at 78 °C and has a flash point below ambient temperature. Fortunately, with a Lower Explosive Limit (LEL) of 1.9 per cent and an Upper Explosive Limit (UEL) of 10.1 per cent (see Chapter 4) its flammable range is not wide. The vapour has an autoignition temperature of 460 °C and a density of 3.2 (see Chapter 3). Hexane boils at 68 °C with an LEL and UEL of 1.1 per cent and 7.5 per cent respectively. The flash point is again below ambient temperature

and the autoignition temperature is only 261 °C. As with most organic solvents these materials can defat the skin. Skin contact and inhalation should be avoided.
- Reactivity of butyllithium. Physical properties of butyllithium are given in Table 24.10. The main hazards associated with use of butyllithiums

Table 24.10 Physical properties of butyllithium (100% product)

Property	Value	Reference
Molecular weight	64.05	69
Molar volume at 25 °C	84.3 cm^3/mole	
Density at 25 °C	0.765 g/cm^3	69
Viscosity at 25 °C	34.6 cP	69
Viscosity at 0.1 °C	119.7 cP	
Bond dissociation energy, D (CLi)	54 ± 9 kcal/mole	70
	55.5 kcal/mole	71
Dipole moment	0.97 Debye (indicating about 40% ionic character, i.e. similar to that of C–F bond)	72
Heat of formation	$\Delta H_1^0 = -31.4 \pm 0.7$ kcal/mole	70
	$= -32.0 \pm 1.7$ kcal/mole	71
Heat of reaction with water	$C_4H_9Li + H_2O \rightarrow LiOH + C_4H_{10}$	
	$\Delta H_R^{298} = -57.4$ kcal/mole	70
Vapour pressure at 60 °C	4.38×10^{-4} mm	71
Vapour pressure at 80 °C	44.4×10^{-4} mm	71
Vapour pressure at 95 °C	194×10^{-4} mm	71
	(The vapour pressure is due mainly to the evolution of butenes from the thermal decomposition of n-butyllithium)	
Solubility	n-Butyllithium is miscible in all proportions with normally liquid aliphatic and aromatic hydrocarbons, ethers and tertiary amines	
Thermal stability in hexane	Essentially stable indefinitely in sealed containers at room temperature in the absence of alkoxide	

include their pyrophoric properties and the danger of severe caustic and thermal burns resulting from personal contact. Kerosene rather than water is therefore used in the cooling jacket around the butyl chloride/lithium reactor and is refrigerated outside the main plant. Personal protection is used whenever exposure could arise.

Because of the flammability hazards, precautions include classification of the plant as a Zone 1 with provision of explosion-proof electrics (such as the agitator motor on the reactor). Argon is used as an inert atmosphere for molten lithium, and nitrogen as a means of providing a protective inert blanket for butyllithium and for product transfer, thereby avoiding use of pumps. Mineral oil is heated external to the main plant. All metal components are earthed to avoid static build-up. High- and low-level temperature and pressure alarms together with bursting discs vented in a safe remote area provide protection against runaway reactions. The plant is designated a 'No Smoking' area. A hydrocarbon detection/alarm system to monitor for hydrocarbon leaks is an additional potentially useful measure. Stocks of hexane and butyllithium solutions are stored in bunded external steel tanks coated with insulation to protect the tanks in the event of fire.

Handling precautions for lithium and aluminium alkyls

Despite the foregoing properties organometallics can be handled safely if the following technical and administrative precautions are adopted. These are used to illustrate the techniques appropriate for the handling of pyrophoric chemicals in the widest sense. Anyone intending to handle specific pyrophoric substances should seek more detailed advice from their supplier (e.g. see refs. 40, 42 and 73) and from the literature.

Whenever solvents are employed the pyrophoric nature of such organometallic preparations poses a real fire hazard, as two sides of the fire triangle (see Chapter 4) are present, viz. the organometallic as the ignition source and the solvent as the fuel. Solvents of low flash point and high volatility increase the chance of ignition, and even where spills do not ignite spontaneously the heat generated by the oxidation process accelerates the evaporation of solvent and flash-back fires can result. Contact with reactive or combustible materials or local overheating or presence of 'external' ignition sources also increases the probability of fire. The cornerstone of any strategy for the safe handling of pyrophoric organometallics is engineering control to ensure containment, segregation, exclusion of oxygen and water, and elimination of ignition sources. Reliance on use of personal protection and provision of firefighting facilities are secondary lines of defence, all supported by good housekeeping, regular maintenance, supervision and extensive training programmes.

Engineering control measures – Handling areas should be isolated and provided with ample fire extinguishant. Where danger of leaks and spills exists indoors it is highly likely that extract ventilation will be needed to supplement natural ventilation. In the laboratory all operations involving pyrophoric chemicals should be undertaken over spill-trays in a fume-cupboard to prevent spread of alkyl spillages. It has been suggested[74] that a fume-cupboard face velocity (see Chapter 11, page 549) of 0.75 m/s is required. Ventilation flow rates should be monitored regularly and checks for mechanical failure made prior to each operation. Operations necessitating opening-up of apparatus (e.g. certain transfers) should be carried out in a glove box flushed with dry nitrogen at 2.5–5 cm of water pressure. Laboratory operations such as reactions or distillations are best performed in clamped spherical ground-glass jointed apparatus. Stopcocks should be avoided where possible; they should be securely clamped if provided. Most greases are attacked by alkylaluminiums to varying degrees, which leads to freezing of taps, etc. and high vacuum silicone grease is preferred. PTFE taps should be considered. Tygon tubing is more resistant to attack by alkylaluminiums than ordinary rubber but both harden on exposure; polypropylene or polyethylene seems to be more acceptable for certain aluminium or alkyllithium preparations but not for solutions in aromatic solvents (see Table 24.11).

Table 24.11 Materials of construction for handling butyllithiums

Materials	Containers/piping	Gasketing	Valves
Dry glass	✓		
Steel	✓		✓
Stainless steel	✓		✓
Iron	✓		✓
Copper			✓
Brass			✓
Fibreglass	✓		
Viton® rubber		✓	
Teflon® rubber		✓	
Teflon® seated steel ball valves			✓

Viton and Teflon are trademarks of E. I. duPont de Nemours & Co. Inc.

Water-cooled condensers can be used for water-insensitive materials but for materials known to react vigorously with water hydrocarbon coolants should be employed (see page 492). Teflon, glass or stainless steel stirrer paddles and accessory equipment are adequate for reactions involving alkylaluminiums up to 200 °C. Copper and mercury are inert to alkylaluminiums and alkylborons and can be used in ancillary equipment, e.g. mercury traps, manometers, etc. Clearly, all glassware, syringes, tubing, sample bottles, etc. must be scrupulously oven-dried and flushed with nitrogen prior to use. When performing a reaction for the first time it is advisable to use the smallest scale feasible under constantly supervised conditions, behind a transparent shield in a vented area; 500 ml and 1000 ml are considered reasonably safe maximum volumes of material to distil for organoboranes and organoaluminiums, respectively.

Good plant design is the key to preventing accidents such as rupture or leakage. Totally enclosed systems are the preferred arrangement for handling pyrophoric organometallics and equipment and piping systems should be constructed of suitable materials as illustrated by Tables 24.11 and 24.12. Fixed pipe systems are preferred for transfer purposes to avoid unnecessary opening of equipment during operations. They should be

- routed through non-hazardous areas, avoiding normal working areas;
- placed as low as possible to minimise the hazard from overhead leaks;
- as short and direct as possible, welded with the minimum number of flanged connections;

Table 24.12(a) Recommendations for plastics and rubbers in aluminium alkyl service

	Neat aluminium alkyls		Aluminium alkyl blends with pentane, hexane, heptane, cyclohexane		Aluminium alkyl blends with benzene, toluene, xylene	
	Non-halide aluminium alkyls*	Aluminium alkyl halides†	Non-halide aluminium alkyls*	Aluminium alkyl halides†	Non-halide aluminium alkyls*	Aluminium alkyl halides†
Butadiene–acrylonitrile rubber (ABR)	NR	NR	NR	NR	NR	NR
Butadiene–styrene rubber (SBR)	NR	NR	NR	NR	NR	NR
Butyl rubber	NR	NR	NR	NR	NR	NR
Chlorinated polyether (Penton)	S	S	S	S	NR	NR
Chlorosulphonated polyethylene (Hypalon)	NR	NR	NR	NR	NR	NR
Ethylene–propylene rubber (EPR)	NR	NR	NR	NR	NR	NR
Fluorinated polymers						
(KEL-F)	R	R	R	R	R	R
(KYNAR)	S	S	S	S	S	S
(TEFLON)	R	R	R	R	R	R
(VITON)	NR	NR	NR	NR	NR	NR
(KALREZ)	R	R	R	R	R	R
Graphite–asbestos	R	R	R	R	R	R
Natural rubber	NR	NR	NR	NR	NR	NR
Neoprene	NR	NR	NR	NR	NR	NR
Nylon	S	S	S	S	S	S
Polyacrylate rubber (Thiokol)	NR	NR	NR	NR	NR	NR
Polyester resin	S	S	S	S	S	S
Polyethylene	S	S	S	S	S	S
Polypropylene	S	S	S	S	S	S
Polyurethane	NR	NR	NR	NR	NR	NR
Polyvinyl alcohol (Compar)	S	S	S	S	S	S
Polyvinyl chloride	S	S	S	S	NR	NR
Silicone rubber (Mosite 1402)	NR	NR	NR	NR	NR	NR
Tygon	NR	NR	NR	NR	NR	NR
Vinyl chloride–vinylidene chloride copolymer (PVC Type 2)	S	NR	S	NR	S	NR

* TMA, TEA, TNPA, TIBA, DIBAH, TNBA, TNHA, TNOA, IPRAL, OMH-1.
† MASC, DEAC, DEAI, EASC, EADC, DIBAC, IBADC. For abbreviations see Table 24.6.
R, Recommended; S, Suitable or satisfactory; NR, Not recommended.

Table 24.12(b) Recommendations for metals to be used with neat and blended aluminium alkyls*

Metals	Trialkylaluminiums, alkylaluminium hydrides and diethylzinc	Alkylaluminium halides
Aluminium	NR	NR
AMPCO aluminium bronze	NR	NR
Copper	S	S
Galvanised steel	S	S
Magnesium	NR	NR
Monel	S	S
Nickel	S	S
304 Stainless steel	R	NR
316 Stainless steel	R	NR
410 Stainless steel	R	NR
Steel	R	R
Yellow brass	R	R

* All compounds included in Table 24.6, neat or in solution.
R, Recommended; S, Suitable or satisfactory; NR, Not recommended.

- sloped to facilitate draining (traps preventing complete draining should be eliminated); and
- arranged such that individual sections can be isolated.

Flow rates are usually low so that pipe size and wall thickness are normally based on protection against mechanical damage. Valves for controlling flow should be operated manually from behind a protective wall or fitted with actuators. Valves and connections should be provided for flushing out residual butyllithium with hexane and for purging the system with inert gas. An oil seal pot containing ample oil should be inserted in the discharge of the purge line to prevent air from being sucked back into the system, and positioned so as to facilitate regular inspection of the oil level. All-welded joints are preferred to screw connections. Where the latter are needed PTFE tape seals are suitable for cold butyllithium but unsuitable for systems containing hot butyllithium solutions. Gaskets of appropriate material (e.g. Viton for butyllithiums and flexible graphite for aluminium alkyls) should be used in flanged joints protected where appropriate with metal shields to prevent spray from a leak. For organoaluminiums even small discharges would be detected since oxide fume would be noted from under the shield.

Valves should be of fire-safe design or insulated so as to prevent failure in the event of fire. Ball valves with PTFE seats are used for butyllithium service. Experiments with filtration materials showed that diisobutyl-aluminium chloride reacts violently with cloth filter cartridges, even under a nitrogen blanket, to leave a charred residue. This reaction was apparently due to the presence of binder in the filter cloth and not as a result of reaction with the cellulose since cheesecloth or industrial paper proved to be inert under similar conditions. Even so, Ethyl Corporation advise that wet cartridges, or cartridges containing cellulose or other hydroxyl-containing compounds, must not be used. The filter normally employed for filtering blends of alkylaluminiums comprises wound fibreglass or polypropylene elements for 5 μm particles. When filters are decontaminated with, e.g. isopropyl alcohol and washed it is important to ensure that all traces of these compounds are removed.

> An incident occurred when some decontaminating agent was left in a Micro Metallic filter. When the filter was placed in operation diisobutylaluminium chloride reacted with residual solvent and liberated isobutane causing rupture of the filter disc.[75]

Holding vessels are normally constructed of carbon steel with connections for pressurising with nitrogen. Overpressure relief should be provided to protect against inadvertent overheating. Because of the probability of ignition of pyrophoric and/or solvent occurring on release of pressure, relief lines should vent to a safe and isolated location and normally be equipped with flame arresters (see Chapter 12, page 701). However, one supplier of organoaluminiums does not recommend flame arresters on seal pot outlets as they are ineffective and prone to plugging.[73]

Butyllithium is often used to form an intermediate which is subsequently reacted with an electrophilic species and the butyl group is not incorporated into the molecule but is released as butane gas. Plant must therefore be designed to dispose of this gas safely: if the reaction is carried out at low temperatures then butane may remain dissolved in hexane until the final quench. Floors underneath vessels should be sloped to drain away spillage and drain trenches and sumps must be kept dry.

Since bottom connections can break and gaskets can leak, tanks and vessels to contain solutions of pyrophoric chemicals should ideally be designed with top fittings only.

> During the day shift a scaffold had been erected beneath a vessel to remove and repair a bottom run-off valve from a vessel which had contained residues of pyrophoric catalyst which had been swilled away. However, the faulty valve, which had been left on the scaffold for the fitter's mate to refit, had not been cleaned. When the mate arrived he found the scaffold boards smouldering and a hole had been burnt completely through the board as a result of ignition of catalyst which had accumulated in the valve and was glowing red hot. Fortunately, the incident was dealt with using a water hose and neutralisation with sodium nitrite in time to prevent a fire.[76]

Transfer of organolithiums and organoaluminiums is best achieved by pressurisation with nitrogen, using differential pressure to effect transfer via dip tubes extending through top connections. This eliminates pump maintenance and simplifies line layout. It is preferable to design plant such that reactions use a number of complete butyllithium containers. A pressure drop indicates when the container is empty and lines can easily be blown clear via the container and dip pipe prior to disconnection. This also avoids potential problems of valve blockage if part-full containers are stored. However, pumps of the correct type can be used. Thus butyllithium can be pumped from portable tanks, tankcars and rail tankers using magnetically coupled (seal-less) or self-priming centrifugal pumps. Diaphragm pumps can be employed for neat alkylaluminiums and centrifugal pumps with double mechanical seals, or canned pumps or magnetic drive pumps are available for solutions of aluminium alkyls. Thorough purging with dry nitrogen or argon is essential prior to start-up; but if the lines are short oxidation of residual traces of butyllithium resulting from improper purging is unlikely to generate sufficient heat to pose a fire danger because of the surface to volume ratio. Reliance on this, however, represents an unnecessary risk and when lines are emptied they should be flushed with hexane as opposed to nitrogen, since the latter will encourage evaporation of the hydrocarbon diluent, thereby increasing the concentration of the pyrophoric contaminant in the line and hence the probability of ignition when connections are broken. Metering is easily achieved by weighing the cylinder, although rotameters with stainless steel floats and PTFE gaskets are satisfactory for butyllithiums and turbine meters or metering pumps can be used for aluminium alkyls.

The basic approach for unloading butyllithium usefully illustrates the more general handling requirements. The unopened drum must be grounded and mechanically bonded prior to attachment to any transfer fitting to ensure a metal-to-metal contact with the ground cable. The gas inlet port of the container is connected to a controlled, pressurised supply of dry nitrogen; the liquid outlet port is attached to a dry nitrogen-purged pipe. Ingress of gas pressurises the upper section of the cylinder forcing butyllithium solution from the container via the dip tube as the main valve is opened. This is the last valve to be opened and the first to be closed after transfer. Clearly, provision must be made to relieve pressure as it builds up in the transfer process. Additional safety features include pressure regulation of the nitrogen supply monitored with a pressure gauge to ensure that the pressure does not exceed the rating of the cylinder or that of associated pipework and vessels. Filters are sometimes inserted in the product supply line. A similar procedure has been described for the transfer of Grignard reagents.[77]

The scheme given in Fig. 24.4 demonstrates the application of these principles in the dilution of concentrated butyllithium solutions.[42] The dilution vessel should be of sufficient capacity to receive the entire contents of the stock cylinder plus the required volume of diluent. Furthermore it should not be filled to more than 85 per cent capacity. The system should be purged with dry hexane/nitrogen. After charging the dilution vessel with the requisite volume of dry hexane (i.e. previously treated with e.g. molecular sieve absorbent), and with all the valves closed except for valve 1, nitrogen pressure should be applied to the stock container. Valves should

1406 Flammable and explosive hazards

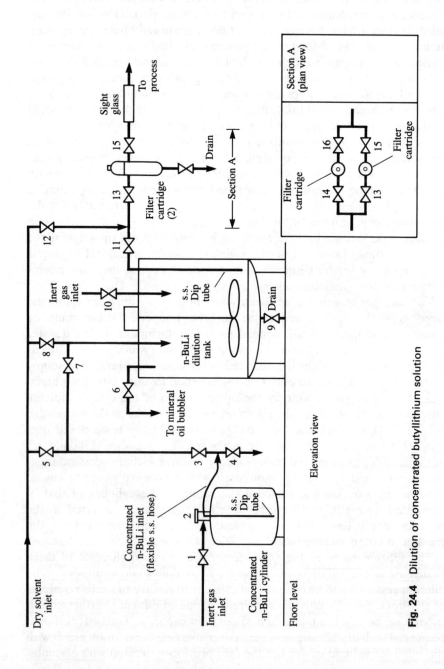

Fig. 24.4 Dilution of concentrated butyllithium solution

then be opened in the order 2, 3, 7 and 6. Upon completion of the transfer operation the valves should be closed in the order 6, 7, 3, 1, 2.

Guidance on blending facilities and sampling systems for aluminium alkyls delivered in a variety of containers is obtainable from the supplier: as an example the arrangements for material delivered in tankcars or trailers is given in Fig. 24.5 (see ref. 40 for further details). It is not

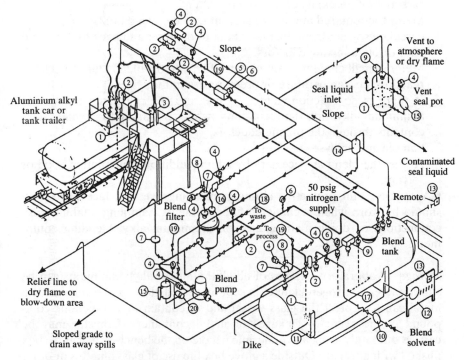

Notes:
Dike will be required around blend tank.
Flush connections for maintenance not shown.
Lines and vessels insulated and traced for aluminium alkyls which might freeze.

Fig. 24.5 Blending facility for aluminium alkyls from tankcars or trailers. 1. Dip pipe; 2. Ball valve with air operator. Should be connected to a specified emergency shutdown system; 3. Hand winch to aid hose handling; 4. Pressure indicator with chemical seal; 5. Integrating flow meter with remote readout. Weigh tank on scales can be substituted for flow meter. If a mass flow meter is used, it must be mounted to be free draining; 6. Temperature indicator; 7. Rupture disc; 8. Pressure relief valve; 9. High level alarm; 10. Positive displacement meter; 11. Load cells; 12. Control panel board with valve toggle switches, alarms, turbine meter and load cell readout. Located inside dike; 13. Emergency shutdown pushbutton. Locate one station at least 15 m from process area; 14. Rotameter for nitrogen purge of vent system; 15. Tank with level glass for circulating seal oil; 16. Sight glass (bullseye); 17. Mixing eductor nozzles; 18. Sample connection (blind flanged); 19. Flush connection (blind flanged); 20. Circulating pump for seal oil to blend pump mechanised seals.

advisable to connect hoses to delivery containers in the open during rain or snow. The blending procedure relating to Fig. 24.5 is as follows:
1. Check the solvent to be sure it is the intended material and contains no free water. Be sure the solvent is a hydrocarbon and not a reactive chemical. Check for moisture content and dry, if necessary, with molecular sieves or other type of dryer.
2. Be sure the blender is clean, dry, nitrogen purged and ready for blending. Check the oil level in the vent seal pot and ensure that a nitrogen purge is maintained. Check the level in the blend tank. Install new filter cartridges if necessary and purge the filter.
3. Meter the required amount of solvent into the blend tank.
4. Circulate solvent from the blend tank to the filter and back to the blend tank through mixing eductors.
5. Meter a small amount of aluminium alkyl into the blend tank. Check for reaction, i.e. temperature or pressure rise in the system.
6. If there is no reaction, meter or measure the calculated amount of aluminium alkyl into the blend tank, while continuing to circulate tank contents through the mixing eductors.
7. Sample and analyse.
8. Adjust the blend as necessary by adding additional aluminium alkyl or solvent.

CAUTION: Full protective clothing including eye and face protection should be worn when exchanging filter elements, sampling, making hose connections or doing maintenance on aluminium alkyl blending equipment.

Storage – Laboratory samples of alkyllithiums or alkylaluminiums can be stored under a nitrogen blanket in septum-sealed glass (preferably inside a metal can to protect the glass vessel and to contain organometallic compound in the event of leakage) or metal containers. Transfer may be carried out safely using a hypodermic needle on the bench or in a glove box flushed with nitrogen. Outside a glove box the use of glass pipettes or other fragile devices are not recommended because of the possibility of breakage and concomitant spills. An arrangement for sampling aluminium alkyls using this procedure is given in Fig. 24.6[73] but the supplier should be consulted for more detailed instructions. Alternative arrangements rely on the use of dip tubes.

Bulk storage facilities should be isolated from the main buildings in well-ventilated, sheltered areas equipped with dry powder extinguishers. Outdoor bulk storage tanks should be separated from each other and other buildings to prevent fire spread and bunded so as to contain the entire contents of a tank. The bund floor should be sloped to ensure material runs away from directly beneath the storage vessel. For aluminium alkyls susceptible to freezing (see Table 24.6), portable containers should be thawed in a warm room. Specially designed tank trailers and tankcars with insu-

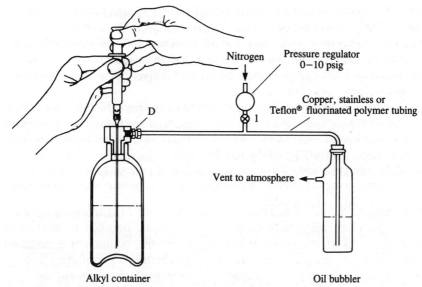

Fig. 24.6 Alkylaluminium unloading system without dip tube. Support and protective equipment for the glassware are not shown but must be used

lated jackets for thawing the contents are used for the large-scale transportation of these compounds.

Butyllithium should be stored under an inert gas (typically dry nitrogen) in dry, sealed containers of suitable design and free of scale or oxide films at temperatures not exceeding 20 °C. To reflect the range of needs, organoalkyls are shipped in a variety of approved containers ranging from laboratory-size cylinders to bulk trailers and tankcars (e.g. see Table 24.13),[68] all slightly pressurised with nitrogen. They are equipped with dip

Table 24.13 Commonly available cylinders used to supply butyllithium

Container	Total capacity (litres)	Fill capacity (litres)	Contained butyllithium (for given weight per cent of BuLi solution)		
			15	24	86
Dot 4BA 240 cylinders					
Type No. 11	11.8	10.6	1.08	1.76	4.93
Type No. 20	24.5	22	2.24	3.64	14.4
Type No. 100	108	97	9.89	16.16	63.4
Type No. 200	214	192	19.6	31.8	125.5
Type No. 420	450	405	41.3	67.1	264.7
Iso-containers					
7500	7500	6750	689	1118	4412
17 500	17 500	15 000	1530	2484	9804

tube and pressure relief devices ranging from relief valves on large tanks to removable safety plugs on cylinders that melt and release pressure in the event of the vessel overheating. Labelling should indicate the highly flammable, pyrophoric and corrosive nature of the contents. Shipments by road, rail and sea are permitted but postal deliveries and air freight are prohibited.

Upon receipt of cylinders in the store the container should be stood in an upright position and valves should be checked to ensure that they are closed and undamaged. Any seepage of butyllithium should be wiped up with a cloth saturated with mineral oil. If major defects are noted the cylinder should be isolated in a safe location and the supplier contacted.

Personal protection – Wherever possible pyrophoric organometallics such as butyllithium, alkylaluminiums, etc. should be handled in contained systems. Even so, fire-resistant personal protective equipment is essential where contact can arise as a result of deliberate or accidental loss of containment. For obvious reasons personal protection should be considered only as a second line of defence.

> The transfer of 25 per cent aluminium alkyl solution to a header tank was halted after 15 minutes because of erratic behaviour of a level indicator. An instrument fitter was requested to check the indicator and to blow it out with nitrogen. He isolated the supply to the tank, checked to his satisfaction that the pressure in it was zero, and began to unscrew the blank cap on the indicator. After it had been unscrewed a few threads the cap was blown off; alkyl solution was ejected and hit the ceiling 2 m above it. The fitter sustained soreness to his eyes, throat and chest because his face-shield was dislodged in the accident and dense fumes penetrated the upper floor and roof of the plant. (A valve in the vent line to a lute pot was subsequently found shut and the vent line was also restricted.)[78]

Pure cotton garments should not be worn since they absorb spilled material and become easily ignited. Indeed no apparel is 100 per cent impervious against this class of material but rubber aprons (for handling laboratory quantities) or PVC laminate suits (for plant operations) should be worn as appropriate for butyllithiums. Manufacturers recommend aluminised protective suits, hoods and gloves which are loose fitting (to aid removal in an emergency) composed of heat resistant, non-reactive material such as aluminised leather, preformed neoprene, or aluminised vinyl, plus foot protection when handling aluminium alkyls on the plant scale; goggles should be worn under the aluminised protective hood. For lithium alkyls eye protection should comprise, e.g. safety spectacles for small-scale use or face shields where larger volumes are involved.

On the plant rubber boots, loose-fitting rubber gloves and hard hats are additional essential requirements. Operatives should be trained in the use, and care, of their equipment which they should inspect for defects on a regular basis. In the case of accidental contact, contaminated clothing and shoes should be removed rapidly and a shower taken. Accidental spills on the skin should be copiously flushed with water and provision of a safety shower is essential. If the eyes have been affected they should be washed for 15 minutes with water. In all such incidents medical attention should be obtained.

Firefighting – Adherence to the foregoing procedures minimises fire risk but will never eliminate the possibility completely and adequate supplies of fire-extinguishing agent should be available. As mentioned earlier, pyrophoric compounds such as butyllithium, alkylaluminiums, etc. are a source of ignition and once alight the fires resemble petrol fires, with the added complication that harmful fumes of the metal hydroxide or oxide being generated in the combustion process. For fighting fires in enclosed spaces full protective clothing including breathing apparatus (see Chapter 6) should be worn. When fire breaks out the source should be eliminated where possible by closing the discharge valve and the fire tackled with a dry chemical extinguishant such as Ansul Purple K, Plus-Fifty B, or Monnex for butyllithium fires. Vermiculite, an expanded mica, is effective for small fires involving alkyl aluminiums, and *dry* sand, diatomaceous earth and other inert particulates are suitable for larger spills.

Water, water-based foams and chlorinated hydrocarbons or carbon dioxide must not be employed for fires involving any of these organometallics.

Once the fire has been extinguished any remaining organometallic can cause smouldering with re-ignition if the powder is disturbed prematurely before the complete oxidation has occurred. Allowing the fire to burn itself out should also be considered as an alternative strategy.

Maintenance – Maintenance programmes are crucial for continued safe operation. Because of the potential hazards involved in plant handling pyrophorics, maintenance and plant modifications should be the subject of a permit-to-work system (see Chapter 17, page 975). The plant should be checked regularly for leaks, particularly at valve bonnets, packing glands, gasket seals, etc. Prior to maintenance work lines should be drained completely and flushed with solvent or cleaned with nitrogen blow-down. Residues may be flushed from the line with high flash-point solvent, e.g. kerosene, into a slop tank for burning or disposal. Nitrogen for blow-down should be dry and contain less than 0.5 per cent oxygen. Lines should be broken cautiously in case pyrophoric material is trapped in the system.

Spillages and waste disposal – One method for dealing with small quantities of waste is controlled burning outside in a safe location, although this may not be permitted by some environmental regulatory authorities. For material which does not ignite spontaneously it can usually be initiated by sprinkling with water. These organometallics should not be discarded to the drain for obvious reasons. Alternatively, material can be heavily diluted with appropriate solvent and decomposed with water or alcohol (e.g. isopropyl alcohol) with efficient agitation under an inert atmosphere. For disposal of larger amounts the advice of the supplier should be obtained.

Small butyllithium spillages can be wiped up with rags impregnated with mineral oil and dumped in a suitable solvent for disposal as above. Spillages can be covered with dry fire extinguishant and left to oxidise under constant supervision. The residues can then be carefully shovelled up, working from the edges, by suitably protected, trained staff with consideration of the fact that fire may occur if oxidation is incomplete.

Poly- or oligo(dimethylsiloxane) (PDMS or ODMS), i.e. a silicone oil used for oil baths, has been claimed[81] to be a useful reagent for deactivating alkyllithiums since the quenching reaction is mild and careful addition is unnecessary. Furthermore no flammable gaseous by-product such as methane is produced but rather cleaved polymer with alkylsilyl groups and lithium silanolates. The basicity of the latter is weaker than that of the corresponding lithium alcoholate. A typical procedure is to add lithium alkyl to a 50 per cent solution of PDMS (e.g. Dow Corning 200 fluid) in tetrahydrofuran (100 g PDMS per mole of lithium compound) and to let stand for several hours. If need be, the resulting mixture can be neutralised by addition of acetic acid prior to discarding as common organic waste.

Training and supervision – The use of material safety data sheets is highly recommended in staff training and all tasks involving pyrophoric organometallics should be two-man operations and detailed in manuals of safe working practices. Management must enforce the agreed rules and all levels of staff should be trained to understand the risks involved, the reason for the rules, the tasks for which they are employed, the safe use, maintenance and inspection of personal protective equipment, dealing with spillages, first-aid, fire fighting and the emergency evacuation procedures (which should be rehearsed on a regular basis). Refresher training of existing staff is required in addition to induction training of new recruits (Chapter 17, page 1015).

References

1. Anon., *Loss Prevention Bulletin*, 1981(047), 15–19.
2. Bretherick, L., *Handbook of Reactive Chemical Hazards*, 4th edn, Butterworths, 1990.
3. Bond, J., *Sources of Ignition, Flammability Characteristics of Chemicals and Products*, Butterworth-Heinemann, 1991.
4. Bowes, P. C., *Self Heating: evaluation and controlling the hazards*, HMSO, 1984.
5. Buma, T. J. *et al.*, *Chem. Abs.*, 1977, **87**, 83300.

6. Anon., *Loss Prevention Bulletin*, 1986(071), 19.
7. Home Office (Fire Dept), *Manual of Firemanship*, Part 6B, 1973, 30.
8. May, D. C., *et al.*, *J. Haz. Mater.*, 1987 **17**(1), 61.
9. National Fire Protection Association, *Standard for the Manufacture of Aluminium and Magnesium Powder*, NFPA, Boston, MA, 1987, 651.
10. Gugan, K., *I. Chem. E. Symp. Series*, 1968, **25**, 8–16.
11. Spontaneous combustion in solid materials, *Fire Prevention*, No. 193, October 1985.
12. Cross, J. & Farrer, D., *Dust Explosions*, Plenum Press, New York, 1982.
13. Palkmer, K. N., *I. Chem. E. Symp. Series*, 1974, **39a**, 48.
14. Anon., *Loss Prevention Bulletin*, 1983(051), 13.
15. Gugan, K., *I. Chem. E. Symp. Series*, 1974, **39a**, 28–43.
16. MacDermott, P. E., *Chem. Britain*, February 1976, 69.
17. Craig, T. O., *Loss Prevention Bulletin*, 1993(110), 27.
18. Chemical Safety Summary, *C.I.A.*, 1979, **50**, 8.
19. Meidl, J. H., *Flammable Hazardous Materials*, 2nd edn, Glencoe Publishing Co., 1978, p. 166.
20. Attwood, M. T., *et al.*, *12 Oil Shale Symp. Proc.*, 1979, 299.
21. Bradford, W. J., in *Fire Protection Handbook*, Ch. 6, National Fire Protection Association, Boston, MA, 16th edn, 1986.
22. Underdown, G. W., *Practical Fire Precautions*, Gower Press, 1971, 87.
23. Anon., *RoSPA Bulletin*, 1985 (March), 2.
24. Health & Safety Executive, *Repair of Drums and Small Tanks: Explosion and Fire Risk*, HSW No. 32, HMSO, 1970.
25. Anon., *Loss Prevention Bulletin*, 1990(095), 16.
26. Anon., *Loss Prevention Bulletin*, 1983(051), 16.
27. Condliff, A. F., *Materials Performance*, 1988, **27**(10), 70.
28. Anon., *Loss Prevention Bulletin*, 1987(075), 22.
29. Gaston, P. J., *The Care, Handling and Disposal of Dangerous Chemicals*, The Institute of Science and Technology, Northern Publ. Aberdeen Ltd, 1970.
30. a) Anon., *Hazardous Cargo Bulletin*, 1984, **5**(1), 24.
    b) Hymes, I., *Loss Prevention Bulletin*, 1985(061), 17.
31. a) Private Communication.
    b) Cox, J. *Loss Prevention Bulletin*, 1994.
32. Meidl, J. H., *Flammable Hazardous Materials*, 2nd edn, Glencoe Publishing Co., 1978.
33. Thayer, A., *Chem. Eng. News*, 1991 (24 June), 4.
34. Malpas, D. R., *et al.*, Organometallics, in *Kirk-Othmer Encyclopedia of Chemical Technology*, ed. Grayson, M., Interscience, New York, 3rd edn, Vol. 16, 1984, 555.
35. Mirviss, S. B., *et al.*, *Ind. Eng. Chem.*, 1961, **53**(1), 53A.
36. Heck, W. B. & Johnson, R. I., *Ind. Eng. Chem.*, 1962, **54**(12), 35.
37. Zeigler, K., *et al.*, *Angew. Chem.*, 1955, **67**, 424.
38. Knap, J. R., *et al.*, *Ind. Eng. Chem.*, 1957, **49**(5), 874.
39. Heck, W. B. & Johnson, R. I., paper presented at *Ind. Eng. Chem. Division, Amer. Chem. Soc. (Washington, DC), 140th Meeting*, Chicago, 3–8 Sept. 1961.
40. *Handling Procedures for Aluminium Alkyl Compounds and Other Organometallics*, Ethyl Corporation, Louisiana, USA, 1989.
41. Rakita, P. E., *et al.*, *Chem. Engineering*, 1990 (March), 110.

42. *Butyllithium: Guidelines for the Safe Handling*, FMC (Lithium Division) brochure, UK, 1988.
43. Wakefield, B. J., *The Chemistry of Organolithium Compounds*, Pergamon Press, 1974.
44. Anon., *The Chem. Eng.*, 1992 (12 March), 7.
45. Kinsella, E., *Chem. & Ind.*, 1971, 1017.
46. Chambers, R. D., *J. Chem. Soc.* (C), 1967, 2185. Chambers, R. D., Tetrahedron Lett., 1965, 2389.
47. a) Cook, J. S., *Loss Prevention Bulletin*, 1993(111), 27.
47. b) Accident Technical Report, Team Brief Special Edition, Hickson and Welch, 26 Nov. 1992.
48. Anon., *Chem. Ind.*, 1992 (5 Oct), 715.
49. Anon., *The Chem. Eng.*, 1992 (10 Dec), 7.
50. Anon., *Loss Prevention Bulletin*, 1993(109), 7.
51. Aluminium Federation fact sheet (undated), *Aluminium and Fire*, Birmingham, UK.
52. Aluminium Federation, *Fire and explosion hazards in the grinding and polishing of aluminium and its alloys*, 3rd edn, 1991.
53. *Standard for the Storage, Handling and Processing of Magnesium*, NFPA, Boston, MA, 48.
54. *Sodium, Metallic*, Chemical Safety Data Sheet SD-47, Manufacturing Chemists Assoc., Washington, DC, 1974.
55. Anon., *Sichere Chemicarb.*, 1989, **41**(5), 58.
56. Anon., *Fire Precautions*, 1988, **213**, 46.
57. Taylor, D. A. H., *Chem. Britain*, 1974, **10**, 101.
58. *Fire Protection Handbook*, NFPA, Washington, DC, 16th edn, 1986, 6–48.
59. Anon., *Jahresber.*, 1978, 72.
60. *Zirconium Fire and Explosion Incidents*, TID-5365, USAEC, Washington, DC, 1956.
61. *Standard for the Production, Processing, Handling and Storage of Zirconium*, NFPA, Boston, MA, 1987.
62. a) *Fumigation using phosphine*, Guidance Note CS10, Health and Safety Executive, HMSO, 1991.
b) *Phosphine – health and safety precautions*, Guidance Note EH20, 1979, Health and Safety Executive, HMSO, 1978.
c) Environmental health Criteria 73, *Phosphine and selected metal phosphides*, WHO, Geneva, 1988.
63. McKay, H. A. C., *Chem. Ind.*, 1964, 1978.
64. The Control of Industrial Major Accident Hazard (CIMAH) Regulations 1984 as amended 1988, 1990, UK.
65. The Planning (Hazardous Substances) Regulations, HMSO, 1992.
66. Demelyk, E. W., *Ind. Eng. Chem.*, 1961, **53**(6), 56A.
67. *n-Butyl Lithium in Hydrocarbon Solvents*, Chem. Safety Data Sheet SD-91, Manufacturing Chemists Assoc., Washington, DC, 1966.
68. Anderson, R., *Chem. Ind.*, 1984 (19 March), 205.
69. Lewis, D. H., *et al.*, *Chimia*, 1964, **18**, 134.
70. Fowell, P. A. & Mortimer, C. T., *J. Chem. Soc.*, 1961, 3793.
71. Lebeder, Yu. A., *et al.*, *Dokl. Akad. Nauk. SSSR.*, 1962, **145**, 1288.
72. Rogers, M. T. & Young, A., *J. Am. Chem. Soc.*, 1946, **68**, 2748.

73. n-Butyl Lithium Product Bulletin 3/89 FMC, Lithium Division, 1989.
74. Mirviss, S. B., et al., Ind. Eng. Chem., 1961, **53**(1), 53A.
75. Knap, J. E., et al., Ind. Eng. Chem., 1957, **49**(5) 874.
76. Anon., Chem. Safety Summary, 1985, **56**(222), 31.
77. Rakita, P. E., et al., Chem. Engineering, 1990 (March), 110.
78. Anon., Chem. Safety Summary, 1980, **51**, 10.
79. Susuki, T., Chem. Eng. News, 1991 (2 Feb.), 11.

CHAPTER 25

Dust explosions

The phenomenon of a dust explosion is discussed in Chapter 4 (page 135) and Chapter 11 (page 607) and some precautionary measures are referred to on page 719. The present chapter serves to expand upon these and, in particular, to deal with the prevention and mitigation of the effects of such explosions which can be extraordinarily energetic. Spectacular destruction may result if the initial explosion is completely confined to the plant and the plant is too weak to withstand the full pressure of the explosion. Additional hazards include fires, burns and production of combustion gases which can be deficient in oxygen or toxic. These can pose a risk to operators, particularly if they are momentarily rendered unconscious because of the blast, or to the emergency services.

Numerous industries manufacture or handle explosible dusts, and sources of ignition are ubiquitous.

Virtually any combustible solid can, if in a finely divided state, form an explosive dust cloud in certain conditions. Therefore, unless there is positive knowledge to the contrary, it should be assumed that any organic or carbonaceous material may produce a dangerous dust.[1] As exemplified by Tables 4.5 and 11.24, this includes many naturally occurring products of animal and vegetable origin, e.g. fish meal, grain, seeds, cork, coal, malt, starch, wood, sugar and resins. Many products of organic chemical syntheses are also included, e.g. synthetic resins and plastics, dyes and intermediates, fine chemicals and pharmaceuticals. Indeed the very first reported dust explosion which occurred in Italy in 1785 involved flour dust. Many other easily oxidisable powders, e.g. metal powders such as magnesium and aluminium and non-metals such as sulphur, are also vulnerable. Sources of ignition include flames and smouldering particles (e.g. from dryers, grinders, furnaces, kilns, ovens), friction, static discharges, spontaneous combustion, and sparks from electrical switchgear.

The phenomenon of a secondary dust explosion following dispersion of dust layers by the pressure wave of a primary explosion, and their ignition by the following flame-front, was described on page 135.

A series of explosions in a flour mill at Bremen in 1979 resulted in 14 fatalities, injuries to 17 others and mass destruction of the site.[2] A dust explosion on the bridge connecting the store to the main buildings was followed by a chain of secondary explosions until the final major explosion occurred. Large pieces of masonry were projected >100 m, windows were broken at 1.5 km and the resulting fire took 3 weeks to finally extinguish.

Tertiary explosions may also occur if burning material is transported along conveying or dust collection equipment to the next flammable dust–air mixture.[3]

Comprehensive textbooks are available on the principles of dust explosions, their causes, effects and protective measures, as illustrated by the selected examples given in ref. 4.

Explosion prevention relies, as noted on page 159, on the avoidance of
- a dust concentration within the flammable range (generally below the lower limit), and/or
- a source of ignition of the required energy for the particular dust cloud.

Limiting the spread and effects of an explosion require precautions such that
- the quantity of material involved is minimised;
- spread of an explosion within plant is prevented by the use of individual plant units, or by the sub-division of continuous plant with chokes; and
- the flame and hot gases are contained within the plant, suppressed, or relieved to a safe place in the open air.

The latter may necessitate adequately designed ducting to vent the explosion outside.[5]

An explosion occurred in a fluidised bed dryer due to ignition by a static discharge. The dryer was provided with an explosion vent but the relief ducting leading to outside the building was too weak; it ruptured allowing dispersion of burning particles into the building. Building damage and a fire resulted.[3]

Limitation of the effects may also require emergency shut-down of transfer equipment to avoid normal spreading of burning material.

A small dust explosion in an elevator ruptured the relief panels at the top and bottom. However, the elevator and downstream conveyor continued to operate and transferred smouldering material into a receiving silo. A dust cloud ignited in the silo which was destroyed.[3]

Principles of dust explosions

When a powder is in a layer or heap, or in a dense dust cloud, insufficient oxygen may be present to allow the exothermic oxidation reaction necessary to proceed so rapidly that an explosion ensues. In order for a dust cloud to explode, for most dusts the particles in air should be close enough for the heat of one particle to initiate reaction in the next but far enough apart to allow free access for oxygen; hence there is an optimum particle density. This explains the concept of a maximum and minimum explosive concentration.

The minimum or lower explosive (or flammable) limit is one parameter by which a combustible dust is classified as illustrated in Table 25.1.[6] The maximum explosive limit is not, however, a well-defined parameter and is seldom measured but is assumed to be several kg/m^3. In any event it is unsafe to assume that a dust above a specific particle density is outside the hazardous range, because dust layers may burn or smoulder slowly; they may also self-heat to ignition. A dust layer may thus create an explosive cloud if disturbed accidentally.

The severity of a dust explosion is related to the maximum explosion pressure (P_{max}) and rate of pressure rise as exemplified by Tables 4.5 and 25.1. Thus the most hazardous metal powders, e.g. magnesium, aluminium and their alloys, can produce a maximum rate of pressure rise >10,000 psi/second. Although they may produce similar maximum pressures, the rate of pressure rise with plastics or agricultural products tends to be lower. Hence dusts are classified according to the associated rates of pressure rise as shown in Table 25.2.[7]

Particle size and shape are important parameters. Thus, as discussed on page 118, smaller particles will remain in suspension longer and expose a larger surface area per unit mass, and hence be more prone to oxidation. Values of rate of pressure rise, the minimum ignition energy, etc. are influenced by particle size: the smaller the particle the greater the rate of pressure rise and the lower is the minimum ignition energy. The typical effect of average particle diameter on maximum explosion pressure is illustrated in Fig. 25.1.[1] Hence care should be exercised when using published data for design purposes. The finest dust in a plant, probably from bag filters at the end of an extraction system, is the fraction to be considered for the design of explosion vents (page 1455).

A thin flat particle will also possess a larger specific surface area than a spherical particle and hence be more easily ignited.

Dust clouds may comprise a mixture of materials and particle sizes. Coarse material may be rendered hazardous by the presence of a few percent of fines.

Incombustible material, such as inert solid particles, tends to reduce the flammability of a dust cloud; the action is by chemical inhibition and by the cooling effect of the particles. Specimen results are shown in Fig. 25.2.[1]

Non-flammable volatile components act similarly, as does moisture for which specimen results are shown in Fig. 25.3.[1]

Dusts with moisture contents in excess of 15 per cent would generally exhibit a significant reduction in explosibility, and dusts with moisture contents >30 per cent are unlikely to be the initiators of serious dust explosions.[8]

Of course, particle agglomeration associated with high moisture levels will also reduce the tendency for a cloud to reach the lower explosive concentration.

Whilst the presence of moisture in the dust particle raises the minimum temperature required to ignite the dust cloud the presence of moisture in the gas surrounding the dust has little effect.

Chemical dusts may contain flammable solvents; this will increase the risk.

> A number of explosions occurred in a bag filter dust-collection system associated with a water-cooled sanding drum in which sheets of styrene–butadiene rubber were sanded at a temperature below 60 °C. Ignition was probably due to very hot dust particles, some of which accumulated in the base of the sander and some were swept up by the extraction system. Because the sanded material decomposed slightly and yielded styrene vapour on heating, the dust was probably more explosive than samples tested subsequently.[9]

The minimum electrical ignition energy (MIE) of dusts and powders is some 100 times greater than that for flammable vapours. However, fine dust is much easier to ignite than coarse dust, as illustrated in Fig. 25.4 (after ref. 10). Care must be taken in the interpretation of such data, e.g. in the context that plant ignition sources may have sufficient energy to ignite dust with an MIE >500 mJ, since[9,11]

- dust suspensions wet with flammable solvent, or in a solvent–air mixture as mentioned previously, can be much easier to ignite than dry dust; indeed, dusts which are not flammable in air can become flammable with the addition of even small amounts of flammable gas or vapour. This increases the explosion risk and considerably reduces the minimum ignition energy to initiate an explosion;
- plastic liners in bags can result in sparks of static electricity sufficient to ignite flammable vapour when powder is emptied into a vessel containing flammable liquid; and
- classification by gravity (page 118) can produce an ignitable cloud from fine dust in a mixture which possesses a relatively high overall average particle size and MIE.

Dust explosibility testing

With rigorous enforcement of explosion prevention methods, of the type discussed later in this chapter, the frequency of explosions can be minimised. However, they cannot be ruled out completely, because, for example, of human error and the occasional unpredictable behaviour of equipment.

Indeed, prevention against a dust explosion may be inherently more difficult than against a gas explosion because it is often impossible to prevent the build-up of dust concentrations within a plant.

Table 25.1 Dust explosion characteristics of combustible solids

Dust	Minimum ignition temperature (°C) Cloud	Minimum ignition temperature (°C) Layer	Minimum explosible concentration (g/litre)	Minimum ignition energy (mJ)	Maximum explosion pressure (lb/in²)	Maximum rate of pressure rise (lb/in² s)	Maximum oxygen concentration to prevent ignition (% by volume)	Notes
Acetamide	560	—	—	—	—	—	—	Group B dust
Acetoacetanilide	560	—	0.030	20	90	4,800	—	
Acetoacet-p-phenetedide	560	—	0.030	10	87	>10,000	—	
Acetoacet-o-toluidine	710	—	—	—	—	—	—	
2-Acetylamino-5-nitrothiazole	450	450	0.160	40	137	9,000	—	
Acetyl-p-nitro-o-toluidine	450	—	—	—	—	—	—	
Adipic acid	550	—	0.035	60	95	4,000	—	
Alfalfa	460	200	0.100	320	88	1,100	—	
Almond shell	440	200	0.065	80	101	1,400	—	
Aluminium, atomised	650	760	0.045	50	84	>20,000	—	
Aluminium, flake	610	320	0.045	10	127	>20,000	—	
Aluminium–cobalt alloy	950	570	0.180	100	92	11,000	—	
Aluminium–copper alloy	—	830	0.100	100	95	4,000	—	
Aluminium–iron alloy	550	450	—	—	36	300	—	
Aluminium–lithium alloy	470	400	<0.1	140	96	6,000	—	
Aluminium–magnesium alloy	430	480	0.020	80	86	10,000	—	
Aluminium–nickel alloy	950	540	0.190	80	96	10,000	—	
Aluminium–silicon alloy	670	—	0.040	60	85	7,000	—	
Aluminium acetate	560	640	—	—	59	950	—	
Aluminium octoate	460	—	—	—	—	—	—	
Aluminium stearate	400	380	0.015	10	86	>10,000	—	Guncotton ignition source in pressure test
2-Amino-5-nitrothiazole	460	460	0.075	30	110	5,600	—	
Anthracene	505	Melts	—	—	68	700	—	

Dust explosions

Substance							Notes
Anthranilic acid	580	—	—	—	84	6,500	—
Anthraquinone	670	—	—	—	—	—	—
Antimony	420	330	0.030	35	—	300	—
Antipyrin	405	Melts	0.420	—	28	—	—
Asphalt	510	500	—	1920	53	—	—
Aspirin	550	Melts	0.025	25	94	4,800	—
Azelaic acid	610	—	0.015	16	87	7,700	—
α,α'-Azo isobutyronitrile	430	350	0.025	25	76	4,700	—
Barley	370	—	0.015	25	134	8,000	—
Benzethonium chloride	380	410	0.020	60	91	6,700	—
Benzoic acid	600	Melts	0.011	12	95	10,300	—
Benzotriazole	440	—	0.030	30	103	9,200	—
Benzoyl peroxide	—	—	—	21	Did not ignite	—	—
Beryllium	910	540	0.080	100	87	2,200	15 Contained 8% oxide Inert gas carbon dioxide
Beryllium acetate, basic	620	—	—	—	—	—	—
Bis (2-hydroxy-5-chlorophenyl)-methane	570	—	0.040	60	70	2,000	13 Inert gas carbon dioxide
Bis (2-hydroxy-3,5,6-trichlorophenyl)-methane	Did not ignite	450	—	—	—	—	—
Bone meal	490	230	—	—	11	100	— Guncotton ignition source in pressure test
Boron	730	390	Did not ignite	—	41	200	— Guncotton ignition source in pressure test
Bread	450	—	—	—	—	—	—
Brunswick green	360	—	—	—	—	—	—
p-t-butyl benzoic acid	560	—	0.020	25	88	6,500	—
Cadmium	570	250	—	4,000	7	100	—
Cadmium yellow	390	—	—	—	—	—	—
Calcium carbide	555	325	—	—	13	—	—
Calcium citrate	470	—	—	—	—	—	—
Calcium gluconate	550	—	—	—	—	—	— Group B dust
Calcium-DL-pantothenate	520	—	0.050	80	105	4,600	— Group B dust

Table 25.1 (continued)

Dust	Minimum ignition temperature (°C)		Minimum explosible concentration (g/litre)	Minimum ignition energy (mJ)	Maximum explosion pressure (lb/in²)	Maximum rate of pressure rise (lb/in² s)	Maximum oxygen concentration to prevent ignition (% by volume)	Notes
	Cloud	Layer						
Calcium propionate	530	—	—	—	90	1,900	—	
Calcium silicide	540	540	0.060	150	86	20,000	—	
Calcium stearate	400	—	0.025	15	97	>10,000	8	
Caprolactam	430	—	0.07	60	79	1,700	—	
Carbon, activated	660	270	0.100	—	92	1,700	—	Guncotton ignition source in min. expl. conc. and max. expl. pressure tests
Carbon, black	510	—	—	—	—	—	—	
Carboxy methyl cellulose	460	310	0.060	140	130	5,000	—	
Carboxy methyl hydroxy ethyl cellulose	380	—	0.200	960	83	800	—	
Carboxy polymethylene	520	—	0.115	640	76	1,200	—	
Casein	460	—	—	—	89	1,200	—	
Cellulose	410	300	0.045	40	117	8,000	5	Inert gas nitrogen
Cellulose acetate	340	—	0.035	20	114	6,500	7	
Cellulose acetate butyrate	370	—	0.025	30	81	2,700	—	
Cellulose propionate	460	—	0.025	60	105	4,700	—	
Cellulose triacetate	390	—	0.035	30	107	4,300	—	
Cellulose tripropionate	460	—	0.025	45	88	4,000	—	
Charcoal	530	180	0.140	20	100	1,800	—	Guncotton ignition source in pressure test
Chloramine-T	540	150	—	—	7	150	—	
o-Chlorobenzmalononitrile	—	—	0.025	—	90	>10,000	—	

Dust explosions

Material							Notes
o-Chloroacetanilide	640	—	0.035	30	94	3,900	—
p-Chloroacetanilide	650	—	0.035	20	85	5,500	—
Chloroamino toluene sulphonic acid	650	—	—	—	—	—	—
4-Chloro-2-nitroaniline	590	120	<0.750	140	123	3,500	—
p-Chloro-o-toluidine hydrochloride	650	—	—	—	—	—	—
Chocolate crumb	340	—	—	—	—	—	—
Chromium	580	400	0.230	140	56	5,000	—
Cinnamon	440	230	0.060	30	121	3,900	—
Citrus peel	500	330	0.060	100	51	1,200	—
Coal, brown	485	230	—	—	—	—	See also Lignite
Coal, 8% volatiles	730	—	—	—	—	—	
Coal, 12% volatiles	670	240	—	—	—	—	
Coal, 25% volatiles	605	210	0.120	120	62	400	Standard Pittsburgh coal
Coal, 37% volatiles	610	170	0.055	60	90	2,300	
Coal, 43% volatiles	575	180	0.050	50	92	2,000	
Cobalt	760	370	—	—	—	—	
Cocoa	500	200	0.065	120	69	1,200	
Coconut	450	280	—	—	—	—	
Coconut shell	470	220	0.035	60	115	4,200	
Coffee	360	270	0.085	160	38	150	Inert gas carbon dioxide 10
Coffee, extract	600	—	—	Did not ignite	47	—	
Coffee, instant	410	350	0.280		68	500	
Coke	>750	430	—	—	—	—	Guncotton ignition source in min. expl. conc. and max. expl. pressure tests
Coke, petroleum, 13% volatiles	670	—	1.00	—	36	200	
Colophony	325	Melts	—	—	—	—	—
Copal	330	Melts	—	—	68	—	See also Gum manila

Table 25.1 (continued)

Dust	Minimum ignition temperature (°C) Cloud	Minimum ignition temperature (°C) Layer	Minimum explosible concentration (g/litre)	Minimum ignition energy (mJ)	Maximum explosion pressure (lb/in²)	Maximum rate of pressure rise (lb/in² s)	Maximum oxygen concentration to prevent ignition (% by volume)	Notes
Copper	700	—	—	Did not ignite	Did not ignite	Did not ignite	—	
Copper—zinc, gold bronze	370	190	1.00	—	44	1,300	—	
Cork	460	210	0.035	35	96	7,500	—	
Corn cob	450	240	0.045	45	127	3,700	—	
Corn dextrin	410	390	0.040	40	124	7,000	—	
Cornflour	390	—	—	—	—	—	—	
Cornstarch	390	—	0.040	30	145	9,500	—	
Cotton flock	470	—	0.050	25	94	6,000	—	
Cotton linters	520	—	0.50	1,920	73	400	5	
Cottonseed meal	530	200	0.055	80	89	2,200	—	
Coumarone-indene resin	550	—	0.015	10	93	11,000	11	
Crystal violet	475	Melts	—	—	—	—	—	
Cyclohexanone peroxide	—	—	—	21	84	5,600	—	
Dehydroacetic acid	430	—	0.030	15	87	8,000	—	
Dextrin	410	440	0.050	40	99	9,000	—	
Dextrose monohydrate	350	—	—	—	—	—	—	
Diallyl phthalate	480	—	0.030	20	90	8,500	—	
Diaminostilbene disulphonic acid	550	—	—	—	—	—	—	Groub B dust
Diazo aminobenzene	550	—	0.015	20	114	>10,000	—	
Di-t-butyl-p-cresol	420	—	0.015	15	79	13,000	9	
Dibutyltin maleate	600	—	—	—	—	—	—	
Dibutyltin oxide	530	—	—	—	—	—	—	

Dust explosions 1425

Substance								
Dichlorophene	770	—	—	—	72	3,000	—	
2,4-Dichlorophenoxyethyl benzoate	540	—	0.045	60	84	2,200	—	
Dicyclopentadiene dioxide	420	—	0.015	30	89	9,500	—	
Dihydrostreptomycin sulphate	600	230	0.520	—	42	200	7	
3,3'-Dimethoxy-4,4'-diamino diphenyl	—	—	0.030	—	82	>10,000	—	
Dimethylacridan	540	—	—	—	—	—	—	
Dimethyldiphenylurea	490	—	—	—	—	—	—	
Dimethyl isophthalate	580	—	0.025	15	84	8,000	—	
Dimethyl terephthalate	570	—	0.030	20	105	12,000	6	
S-S'-Dimethylxanthogene thylene bisdithiocarbamate	400	—	0.300	3,200	84	1,500	—	
Dinitroaniline	470	—	—	—	—	—	—	
3,5-Dinitrobenzamide	500	Melts	0.040	45	163	6,500	—	
3,5-Dinitrobenzoic acid	460	—	0.050	45	139	4,300	—	
Dinitrobenzoyl chloride	380	—	—	—	—	—	—	
Dinitrocresol	340	Melts	0.030	—	—	—	—	
4,4'-Dinitro-sym-diphenylurea	550	—	0.095	60	102	2,500	—	
Dinitrostilbene disulphonic acid	450	—	—	—	—	—	—	
Dinitrotoluamide	500	—	0.050	15	153	>10,000	—	
Diphenyl	630	—	0.015	20	82	3,700	—	
4,4'-Diphenyl disulphonylazide	590	140	0.065	30	143	5,500	—	
Diphenylol propane (Bisphenol-A)	570	—	0.012	11	81	11,800	5	Inert gas nitrogen
Egg white	610	—	0.14	640	58	500	—	
Epoxy resin	490	—	0.015	9	94	8,500	—	
Esparto grass	—	—	—	—	94	7,300	—	
Ethyl cellulose	340	330	0.025	15	112	7,000	—	
Ethylenediaminetetra-acetic acid	450	—	0.075	50	106	3,000	—	
Ethylhydroxyethyl cellulose	390	—	0.020	30	94	2,200	—	
Ferric ammonium ferrocyanide	390	210	1.500	—	17	100	—	
Ferric dimethyl dithiocarbamate	280	150	0.055	25	86	6,300	—	
Ferric ferrocyanide	370	—	—	—	82	1,000	—	

Table 25.1 (continued)

Dust	Minimum ignition temperature (°C) Cloud	Minimum ignition temperature (°C) Layer	Minimum explosible concentration (g/litre)	Minimum ignition energy (mJ)	Maximum explosion pressure (lb/in²)	Maximum rate of pressure rise (lb/in² s)	Maximum oxygen concentration to prevent ignition (% by volume)	Notes
Ferrochromium	790	670	2.00	—	—	—	—	
Ferromanganese	450	290	0.130	80	62	5,000	—	
Ferrosilicon (45% Si)	640	—	—	—	—	—	—	
Ferrosilicon (90% Si)	Did not ignite	980	0.240	1,280	113	3,500	—	
Ferrotitanium	370	400	0.140	80	55	9,500	—	
Ferrous ferrocyanide	380	190	0.400	—	—	—	—	
Ferrovanadium	440	400	1.300	400	—	—	—	
Fish meal	485	—	—	—	—	—	—	
Fumaric acid	520	—	0.085	35	103	3,000	—	
Garlic	360	—	0.10	240	57	1,300	—	
Gelatin, dried	620	480	<0.5	—	78	1,200	—	
Gilsonite	580	500	0.020	25	78	4,500	—	
Graphite	730	580	—	—	—	—	—	
Grass	—	—	—	—	56	400	—	
Gum arabic	500	260	0.060	100	117	3,000	—	
Gum Karaya	520	240	0.100	180	116	2,500	—	
Gum manila (copal)	360	390	0.030	30	89	6,000	—	
Gum tragacanth	490	260	0.040	45	123	5,000	—	
Hexamethylenetetramine	410	—	0.015	10	98	11,000	11	
Horseradish	—	—	<0.100	—	96	1,600	—	
Hydrazine acid tartrate	570	—	0.175	460	30	200	—	
p-Hydroxybenzoic acid	620	—	0.040	—	37	—	—	

Dust explosions

Material							Flame ignition source in pressure test
Hydroxyethyl cellulose	410	—	—	40	106	2,600	—
Hydroxyethylmethyl cellulose	410	—	—	—	—	—	—
Hydroxypropyl cellulose	400	—	0.025	30	96	2,900	—
Iron	430	240	0.020	—	—	—	—
Iron, carbonyl	420	230	0.105	100	47	8,000	—
Iron pyrites	380	280	1.00	8,200	5	100	—
Isatoic anhydride	700	—	0.035	25	80	4,900	—
Isinglass	520	—	—	—	Nil	Nil	—
Isophthalic acid	700	—	0.035	25	78	3,100	—
Kelp	570	220	Did not ignite	—	19	200	—
Lactalbumin	570	240	0.040	50	97	3,500	—
Lampblack	730	—	—	—	—	—	—
Lauryl peroxide	—	—	—	12	90	6,400	7
Lead	790	290	Did not ignite	—	3	100	9
Leather	390	—	—	—	—	—	—
Lignin	450	—	0.040	20	102	5,000	—
Lignite	450	200	0.030	30	94	8,000	—
Lycopodium	480	310	0.025	40	75	3,100	—
Magnesium	560	430	0.030	40	116	15,000	—
Maize husk	430	—	—	—	75	700	—
Maize starch	410	—	—	—	—	—	—
Maleic anhydride	500	Melts	—	—	—	—	—
Malt barley	400	250	0.055	35	95	4,400	—
Manganese	460	240	0.125	305	53	4,900	7
Manganese ethylene bisdithiocarbamate	270	—	0.07	35	—	—	—
Manioc	430	—	—	—	—	—	—
Mannitol	460	—	0.065	40	97	2,800	—
Melamine formaldehyde resin	410	—	0.02	50	93	1,800	—
DL-Methionine	370	360	0.025	35	119	5,700	7
1-Methylaminoanthraquinone	830	Melts	0.055	50	71	3,300	—
Methyl cellulose	360	340	0.030	20	133	6,000	—
2,2-Methylene bis-4-ethyl-6-t-butyl phenol	310	—	—	—	76	7,300	—

Table 25.1 (continued)

Dust	Minimum ignition temperature (°C) Cloud	Minimum ignition temperature (°C) Layer	Minimum explosible concentration (g/litre)	Minimum ignition energy (mJ)	Maximum explosion pressure (lb/in²)	Maximum rate of pressure rise (lb/in² s)	Maximum oxygen concentration to prevent ignition (% by volume)	Notes
Milk	440	—	—	—	—	—	—	
Milk, skimmed	490	200	0.050	50	95	2,300	—	
Milk sugar	450	Melts	—	—	31	—	—	
Molybdenum	720	360	—	—	—	—	—	
Molybdenum disulphide	570	290	—	—	—	—	—	
Monochloracetic acid	620	—	—	—	—	—	—	
Monosodium salt of trichloro-ethyl phosphate	540	—	—	—	—	—	—	
Moss, Irish	530	230	Did not ignite		21	—	—	
Naphthalene	575	Melts	—	—	87	300	—	
β-Naphthalene-azodimethyl aniline	510	Melts	0.020	50	70	2,300	—	
β-Naphthol	670	—	—	—	—	—	—	
Naphthol yellow	415	395	—	—	—	—	—	Group B dust
Nigrosine hydrochloride	630	—	—	—	—	—	—	
p-Nitro-o-anisidene	400	—	—	—	—	—	—	
p-Nitrobenzene arsonic acid	360	280	0.195	480	77	900	—	
Nitrocellulose	—	—	—	30	>256	>20,900	—	
Nitrodiphenylamine	480	—	—	—	—	—	—	
Nitrofurfural semicarbazone	240	—	—	—	>143	8,600	—	
Nitropyridone	430	Melts	0.045	35	111	>10,000	—	
p-Nitro-o-toluidine	470	—	—	—	—	—	—	
m-Nitro-p-toluidine	470	—	0.030	—	—	—	—	
Nylon	500	430	0.030	20	95	4,000	6	

Dust explosions 1429

Oilcake meal	470	285	—	—	—	—	—
Onion, dehydrated	410	—	0.130	Did not ignite	35	500	—
Paper	440	270	0.055	60	96	3,600	—
Para formaldehyde	410	—	0.040	20	133	13,000	—
Peanut hull	460	210	0.045	50	116	8,000	—
Peat	420	295	—	—	—	—	—
Peat, sphagnum	460	240	0.045	50	104	2,200	—
Pectin	410	200	0.075	35	132	8,000	—
Penicillin, N-ethyl piperidine salt of	310	—	—	—	—	—	—
Pentaerythritol	450	—	0.030	10	90	9,500	7
Phenol formaldehyde	450	—	0.015	10	107	6,500	—
Phenol furfural resin	530	—	0.025	10	88	8,500	—
Phenothiazine	540	—	0.030	—	56	3,000	—
p-Phenylenediamine	620	—	0.025	30	94	11,000	—
Phosphorus, red	360	305	—	—	—	—	—
Phosphorus pentasulphide	280	270	0.050	15	64	>10,000	—
Phthalic acid	650	Melts	—	—	62	—	—
Phthalic anhydride	605	Melts	0.015	15	72	4,200	11
Phthalimide	630	—	0.030	50	89	4,800	—
Phthalodinitrile	>700	Melts	—	—	43	—	—
Phytosterol	330	Melts	0.025	10	76	>10,000	—
Piperazine	480	—	—	—	72	1,400	—
Pitch	710	—	0.035	20	88	6,000	—
Polyacetal	440	—	0.035	20	113	4,100	—
Polyacrylamide	410	240	0.040	30	85	2,500	—
Polyacrylonitrile	500	460	0.025	20	89	11,000	—
Polycarbonate	710	—	0.025	25	96	4,700	—
Polyethylene	390	—	0.020	10	80	7,500	—
Polyethylene oxide	350	—	0.030	30	106	2,100	5
Polyethylene terephthalate	500	—	0.040	35	98	5,500	—
Polyisobutyl methacrylate	500	280	0.020	40	74	2,800	—
Polymethacrylic acid	450	290	0.045	100	97	1,800	—
Polymethyl methacrylate	440	—	0.020	15	101	1,800	7

Table 25.1 (continued)

Dust	Minimum ignition temperature (°C) Cloud	Minimum ignition temperature (°C) Layer	Minimum explosible concentration (g/litre)	Minimum ignition energy (mJ)	Maximum explosion pressure (lb/in^2)	Maximum rate of pressure rise (lb/in^2 s)	Maximum oxygen concentration to prevent ignition (% by volume)	Notes
Polymonochlorotrifluoroethylene	600	720	—	Did not ignite	—	—	—	
Polypropylene	420	—	0.020	30	76	5,500	—	
Polystyrene	500	500	0.020	15	100	7,000	—	
Polytetrafluoroethylene	670	570	—	Did not ignite	—	—	—	
Polyurethane foam	510	440	0.030	20	87	3,700	—	
Polyurethane foam, fire retardant	550	390	0.025	15	96	3,700	—	
Polyvinylacetate	450	—	0.040	160	69	1,000	11	Inert gas carbon dioxide
Polyvinyl alcohol	450	Melts	—	—	78	—	—	
Polyvinyl butyral	390	—	0.020	10	84	2,000	5	
Polyvinyl chloride	670	—	Did not ignite	—	—	—	—	Flame ignition source Group B dust
Polyvinylidene chloride	670	—	—	—	38	500	—	
Polyvinyl pyrrolidone	465	Melts	—	—	15	—	—	
Potassium hydrogen tartrate	520	—	—	—	79	9,500	—	
Potassium sorbate	380	180	0.120	60	97	1,000	—	
Potato, dried	450	—	—	—	—	—	—	
Potato starch	430	—	—	—	93	1,400	—	
Provender	370	—	—	—	95	1,500	—	
Pyrethrum	460	210	0.100	80	—	—	—	
Quillaia bark	450	—	—	—	—	—	—	
Rapeseed meal	465	—	—	—	—	—	—	
Rayon, viscose	420	—	—	—	—	—	—	

Material							
Rayon, flock	—	—	0.03	—	—	—	—
Rice	440	240	0.050	50	105	2,700	—
Rosin	390	—	0.015	10	87	12,000	—
Rubber	380	—	—	—	—	—	13
Rubber, crude, hard	350	—	0.025	50	80	3,800	—
Rubber, crumb	440	—	—	—	84	3,300	—
Rubber, vulcanised	360	—	—	—	40	—	—
Rye flour	415	325	—	—	35	—	—
Saccharin	690	Melts	—	—	—	—	—
Salicylanilide	610	—	0.040	20	73	4,800	—
Salicylic acid	590	—	0.025	—	84	6,800	—
Sawdust	430	—	—	—	97	2,000	—
Sebacic acid	440	—	—	—	74	400	—
Senna	400	—	0.010	105	49	300	9
Shellac	Did not ignite	—	0.020	10	73	3,600	—
Silicon	—	760	<0.10	80	94	13,000	—
Soap	430	600	0.085	100	77	2,800	—
Sodium acetate	590	—	0.030	35	90	4,600	—
Sodium amatol	580	Melts	0.140	—	65	800	—
Sodium benzoate	560	680	0.050	80	91	3,700	5
Sodium carboxymethyl cellulose	320	—	1.10	440	49	400	—
Sodium 2-chloro-5-nitrobenzene sulphonate	550	440	—	—	—	—	—
Sodium 2,2-dichloropropionate	500	—	0.260	220	68	500	—
Sodium dihydroxynaphthalene disulphonate	510	—	—	—	—	—	—
Sodium glucaspaldrate	600	—	—	—	—	—	—
Sodium glucoheptonate	600	—	—	—	—	—	—
Sodium monochloracetate	550	—	—	—	—	—	—
Sodium m-nitrobenzene sulphonate	—	—	—	—	92	400	—
Sodium m-nitrobenzoate	—	—	—	—	87	2,900	—

Group B dust

1432 Flammable and explosive hazards

Table 25.1 (continued)

Dust	Minimum ignition temperature (°C) Cloud	Minimum ignition temperature (°C) Layer	Minimum explosible concentration (g/litre)	Minimum ignition energy (mJ)	Maximum explosion pressure (lb/in²)	Maximum rate of pressure rise (lb/in² s)	Maximum oxygen concentration to prevent ignition (% by volume)	Notes
Sodium pentachlorophenate	Did not ignite	360	—	—	Did not ignite		—	
Sodium propionate	479	—	—	—	70	700	—	
Sodium secobarbital	520	—	0.100	960	76	800	—	
Sodium sorbate	400	140	0.050	30	87	6,500	—	
Sodium thiosulphate	510	330	—	—	11	<100	—	
Sodium toluene sulphonate	530	—	—	—	—	—	—	
Sodium xylene sulphonate	490	—	—	—	—	—	—	
Soot	>690	535	—	—	Did not ignite		—	
Sorbic acid	440	460	0.020	15	106	>10,000	5	Guncotton ignition source in pressure test
L-Sorbose	370	—	0.065	80	76	4,700	9	
Soya flour	550	340	0.060	100	94	800	9	
Soya protein	540	—	0.050	60	98	6,500	—	
Starch	470	—	—	—	—	—	—	
Starch, cold water	490	—	—	—	—	—	—	
Stearic acid	290	—	—	25	80	8,500	—	
Steel	450	—	—	—	—	—	—	Inert gas nitrogen
Streptomycin sulphate	700	—	—	—	—	—	—	
Sucrose	420	Melts	0.045	40	86	5,500	—	
Sugar	370	400	0.045	30	109	5,000	—	
Sulphur	190	220	0.035	15	78	4,700	—	
Tantalum	630	300	<0.20	120	55	4,400	—	

Dust explosions   1433

Material							Ignites in carbon dioxide
Tartaric acid	350	—	—	—	—	—	—
Tea	500	—	—	—	93	1,700	—
Tea, instant	580	340	Did not ignite	—	48	400	—
Tellurium	550	340	—	—	—	—	—
Terephthalic acid	680	—	0.050	20	84	8,000	—
Tetranitro carbazole	395	Melts	—	—	—	—	—
Thiourea	420	Melts	—	—	29	100	—
Thorium	270	280	0.075	5	79	5,500	—
Thorium hydride	260	20	0.080	3	81	12,000	—
Tin	630	430	0.190	80	48	1,700	—
Titanium	375	290	0.045	15	85	11,000	3
Titanium hydride	480	540	0.070	60	121	12,000	—
Tobacco	485	290	—	—	—	—	—
Tobacco, dried	320	—	Did not ignite	—	85	1,000	—
Tobacco, stem	420	230	—	—	53	400	—
Tribromosalicyl anilide	880	Melts	—	—	—	—	—
Trinitrotoluene	—	—	0.070	75	63	2,100	—
s-Trioxane	480	—	0.143	—	85	600	—
α,α'-Trithiobis (N,N-dimethyl-thioformamide)	280	230	0.060	35	96	6,000	—
Tung	540	240	0.070	240	74	1,900	—
Tungsten	730	470	—	—	Did not ignite		—
Uranium	20	100	0.060	45	69	5,000	—
Uranium hydride	20	20	0.060	5	74	9,000	—
Urea	900	—	—	Did not ignite			—
Urea formaldehyde moulding powder	460	—	0.085	80	89	3,600	9
Urea formaldehyde resin	430	—	0.02	34	110	1,600	10
Vanadium	500	490	0.220	60	57	1,000	—
Vitamin B1 mononitrate	380	190	0.035	35	120	9,000	—
Vitamin C	460	280	0.070	60	88	4,800	—
Walnut shell	420	210	0.035	60	121	5,500	—
Wax, accra	260	—	—	—	—	—	—

Group B dust

Table 25.1 (continued)

Dust	Minimum ignition temperature (°C)		Minimum explosible concentration (g/litre)	Minimum ignition energy (mJ)	Maximum explosion pressure (lb/in²)	Maximum rate of pressure rise (lb/in² s)	Maximum oxygen concentration to prevent ignition (% by volume)	Notes
	Cloud	Layer						
Wax, carnauba	340	—	—	—	—	—	—	
Wax, paraffin	340	—	—	—	—	—	—	
Wheat flour	380	360	0.050	50	109	3,700	—	
Wheat grain dust	420	290	—	—	43	—	—	
Wheat starch	430	—	0.045	25	100	—	—	
Wood	360	—	—	—	90	6,500	5	
Wood, bark	450	250	0.020	60	103	5,700	—	
Wood, flour	430	—	0.050	20	94	7,500	7	
Wood, hard	420	315	—	—	66	8,500	—	
Wood, soft	440	325	—	—	63	—	—	
Yeast	520	260	0.050	50	123	3,500	—	
Zinc	680	460	0.500	960	70	1,800	—	
Zinc ethylene dithiocarbamate	480	180	—	—	45	300	—	
Zinc stearate	315	Melts	0.020	10	80	10,000	—	
Zirconium	20	220	0.045	5	75	11,000		Ignites in carbon dioxide
Zirconium hydride	350	270	0.085	60	90	9,500	3	

Table 25.2 Dust explosion classification (based on strong ignition source; 10 kJ chemical igniter and 1 m³ test apparatus) (after ref. 7)

Dust explosion group	K_{st} (bar.m.s^{-1})	Explosion characteristics
St 0	0	no explosion
St 1	$>0 \leq 200$	weak/moderate
St 2	$>200 \leq 300$	strong
St 3	>300	very strong

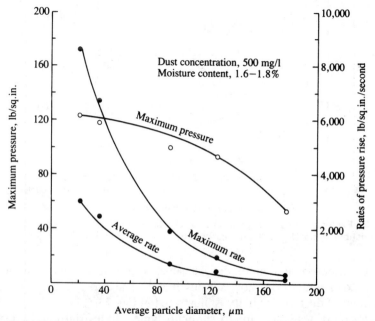

Fig. 25.1 Explosion pressures generated by clouds of starch dust: effect of particle size[1]

Explosion protection methods are therefore essential, and within the UK are required under s. 31 of the Factories Act,[12] i.e.

> Where there is present in any plant used in any such process as aforesaid dust of such a character and such an extent as to be liable to explode on ignition, then, unless the plant is so constructed as to withstand the pressure likely to be produced by any such explosion, all practicable steps shall be taken to restrict the spread and effects of such an explosion by the provision, in connection with the plant, of chokes, baffles and vents, or other equally effective appliances.

The selection and design of explosion prevention and protection methods requires assessment of the particle size and of the explosibility of

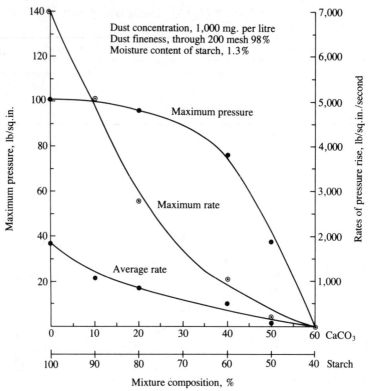

Fig. 25.2 Explosion pressures generated by clouds of starch dust: effect of incombustible material, i.e. calcium carbonate[1]

dusts by experimental methods. These experiments resolve whether a hazard exists and, if so, how serious it is.[13] The procedure for overall assessment of dust explosibility is illustrated in Fig. 25.5.[14]

Explosibility tests attempt to simulate typical industrial conditions in terms of dust concentrations, turbulence and probable sources of ignition. The main aim is to recreate the worst probable conditions, so that the most serious hazards are identified.

Even so, as mentioned on page 1418, caution may be required with regard to interpretation of published data. For example the minimum ignition temperature and minimum ignition energy are determined by established methods under laboratory conditions, whereas due to other factors such as geometry, particle size, presence of impurities, changes of material with time, etc., the actual values in plant conditions may be much lower. Similarly, area classification methods and suitability of electrical equipment are less well defined for dusts than for gases.[15] The assessments currently available are given in Table 25.3.

Techniques ranging from 20-litre laboratory spheres to large-scale experimental mines have recently been reviewed.[16]

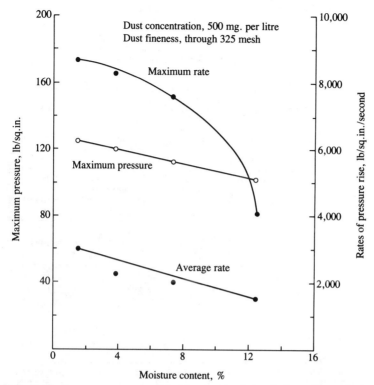

Fig. 25.3 Explosion pressures generated by clouds of starch dust: effect of moisture[1]

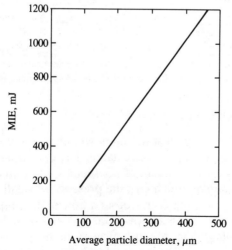

Fig. 25.4 Effect of average particle diameter on the MIE of cellulose acetate moulding powder[9]

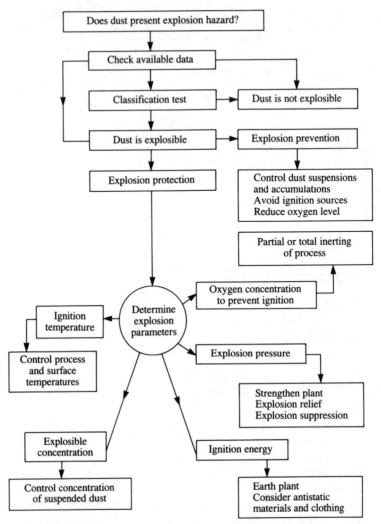

Fig. 25.5 Procedure for overall assessment of dust explosibility[13]

Explosibility classification

If the explosibility of a dust is unknown, it is essential to determine whether it will ignite and propagate flame. The classification test therefore provides a qualitative assessment of whether a suspended dust is capable of initiating and sustaining a flame in the presence of small ignition sources. Classification is based on visual observation of flame propagation in a vertical tube apparatus with an open top. An explosible dust, ('Group A'), is one which enables flames to move away from the ignition source. A non-explosible dust, ('Group B'), is one which does not propagate flame away from the ignition source.

Table 25.3 Selected useful dust explosion assessments

Dust layers	Dust clouds
Flammability	Particle size distribution
Burning behaviour	Explosibility classification (to determine
combustibility tests	whether dust will explode (Class A) or not
burning rate	(Class B))
Deflagration	Explosible limits
screening test	Explosion pressure/explosion violence (e.g.
laboratory test	20-litre sphere)
Smoulder tests	Minimum ignition energy
smoulder temperature	Ignition temperature
Autoignition temperature	
Exothermic decomposition	
open vessel	
differential thermal analysis	
Explosibility	
impact sensitivity	
friction sensitivity	
thermal sensitivity	

Additional quantitative data are normally required for a Group A dust so that appropriate safety precautions can be specified.

Minimum ignition temperature

For a dust suspension the minimum ignition temperature is determined in a vertical vitreosil furnace which can attain temperatures up to 1000 °C. The test should always be performed with the finest dust produced, or likely to be encountered, within relevant equipment or at the factory. The minimum ignition temperature is the lowest temperature at which a dust suspension in air will ignite, and represents a maximum surface temperature which should never be exceeded during plant operation. It is of particular relevance to plant with relatively large heated areas, e.g. a furnace wall or a dryer surface. Operation at <70 per cent of the minimum ignition temperature is advised.[13]

For dust layers, e.g. where the hazard relates to spontaneous ignition of deposits on hot surfaces, ignition tests are performed using layers of material on a thermostatically controlled hotplate.[13] The ignition temperatures for various layer thicknesses can then be matched to the industrial situation without the need for extrapolation.

Minimum explosible concentration

The minimum explosible concentration is the smallest amount of dust suspended in a measured volume that is capable of being ignited and sustaining flame propagation. As shown in Table 25.1, values as low as 0.01 g/litre have been determined.

Minimum ignition energy

The minimum ignition energy is determined in a vertical tube apparatus with the specific dust concentration being varied so as to cover the optimum conditions for an explosion. Energy from a high-voltage source is reduced until flame propagation is not observable in the dust dispersion. As shown in Table 25.1, values range from <5 mJ to >8 J. They are particularly relevant to hazards from ignition by static discharges, i.e. the majority of incidents have been associated with dusts with an ignition energy <25 mJ.[13]

Limiting oxygen concentration

The limiting oxygen concentration is determined in an apparatus in which dust is dispersed into a measured reduced-oxygen atmosphere, e.g. containing nitrogen or carbon dioxide. The data, summarised in Tables 4.6 and 25.1, are of use in explosion prevention by inerting as discussed on page 138.

Maximum explosion pressure and rate of pressure rise

Dust explosion pressures and rates of pressure rise may be determined by either of two common techniques, i.e. using a Hartmann bomb apparatus or a 20-litre sphere. The data are not interchangeable and applications for it differ.[13] The maximum rate of pressure rise decreases according to a cubic law thus

$$(dp/dt) V^{\frac{1}{3}} = K_{st}$$

where V is the volume and K_{st} is a dimensional constant characteristic of the specific dusts. As mentioned on page 287 these can be used to classify explosions. K_{st} values for selected materials are given in Table 25.4.

The maximum rate of pressure rise in the Hartmann apparatus is used in the vent ratio method to calculate the requisite size of an explosion relief vent.[17,18]

The 20-litre sphere is now preferred for explosion pressure data measurement using a standardised procedure.[13] The data can be utilised in nomographs to estimate vent areas having covers of known opening pressures for vessels of varying volume and strength provided that the vessels can withstand >0.2 bar and have a length:diameter ratio of <5.[13,17,18]

Table 25.4 Explosion characteristics of selected powders

Dust	P_{max} (bar.g)	K_{st} (bar.m.s^{-1})	Classification
PVC	8.5	50	St 1
Sugar	8.0	80	St 1
Coal	7.7	85	St 1
Polyethylene (coarse)	9.2	115	St 1
Metallic soap	8.4	140	St 1
Polyethylene (fine)	9.2	150	St 1
Starch	9.4	150	St 1
Cellulose	10.0	160	St 1
Maize	9.7	195	St 1
Dextrin	8.7	200	St 2
Organic pigment	10.0	286	St 2
Aluminium	11.5	555	St 3
Red phosphorus	~6.0	~570	St 3

Control measures

Techniques for minimising risk from dust explosions include measures both to prevent explosions occurring and to mitigate the consequences in the event of an explosion taking place. Precautions to guard against the former include substitution by less hazardous materials, minimising the opportunity for dust formation, elimination of ignition sources, inerting, modification of the process, and careful attention to plant design. The effects of an explosion can be limited by containment, venting, use of thermal insulators, suppression, and good housekeeping. Selected examples are discussed in more detail.

Dust generation and dispersion

The manufacture, size reduction, collection, transfer and use of dry powders inevitably results in a dusty environment. Moreover, as explained on page 118, the settling velocities of particles <10 μm, termed 'float dust', are extremely low.

Metal dusts are generated in numerous manufacturing operations, e.g. grinding with an abrasive wheel, pulverising, finishing or scurfing with an abrasive sheet, scratch brushing, and mechanical polishing or buffing. The potential for dispersion to result in a flammable dust–air mixture is therefore present in all these operations and indeed whenever metals are crushed or surface-finished; or dust is conveyed, collected, dried, screened, graded, blended, weighed or packed. The hazard is also present in special manufacturing operations, e.g. the production of zinc powder by distillation; the production of aluminium particles by the Blown Powder, Stamped Powder and Milled Powder methods; and in electrostatic paint spraying.

Explosive dust concentrations are, in general, most likely to arise close to a dust source or within an enclosed space. Thus common industrial health precautions, which enclose and confine dust sources, may well concentrate the dust and result in a potential explosion problem.

Design considerations

Design of equipment and process systems should attempt to minimise the generation of fine particulate matter, e.g. it is unlikely that particles of size >500 μm will initiate a dust explosion although coarse particles may rub together and generate fine particles. Design should in addition aim at reducing dust levels. (These objectives are of course interrelated since, as discussed on page 118, fine particles remain in suspension for longer periods.)

Process selection

Formation of a dust cloud may be prevented by using a liquid instead of air as the dispersing medium. This technique is, for example, applied in magnesium grinding.[19] It is advantageous also if the final product is a paste or a slurry, e.g. aluminium is ground in ball mills with white spirit in paint manufacture.

Prevention of ignition

Clearly all potential sources of ignition (page 1512 *et seq.*) should be identified and effective exclusion aimed for. Given the wide variety of possible sources, e.g. with metal particles even collisions in a turbulent airstream may be a factor,[6] a combination of measures is often required for control. Examples applicable to bucket elevators and storage bins (after ref. 7) are as follows.

Bucket elevators

- Dust-tight casings should be fitted and should not be readily removable. Maintenance procedures should emphasise the importance of replacing casings.
- Elevator bands should be checked regularly for slipping.
- All buckets should be securely fixed. Damaged, or missing, buckets should be replaced.
- Belt speed meters should be incorporated to detect belt slip. Belt slip detection should result in shut-down.
- Anti-runback devices should be installed.
- Bearings should be inspected regularly and lubricated. Bearing temperature sensors may be an advantage.

- The drive should be external to the casing.
- If a material is sensitive to ignition by static electricity, antistatic belting should be used.
- Feed rate should be controlled within the design limits. Overloading should not be permitted.
- Electrical equipment should conform to BS 6467 (1985) or equivalent.
- An overload trip should be installed on the drive.
- Bucket elevators should, where possible, be installed outside buildings.

Storage bins

- All-metal storage bins are preferred if the material is sensitive to ignition by static electricity. Resistance to earth should be <10 ohms and should be checked regularly. Special precautions may be necessary for plastic containers for high resistivity powders.
- Electrical wiring should not be strapped to the outside of bins.
- Metal items, e.g. chain measures or metal tapes, should not be lowered into the bin. Level indicators should be of approved dust-tight design. They must be adequately earthed. Electricity supply cables should not run inside the bin.
- Feed to the bin should be shut-off if it contains excessively hot material. Infra-red sensors may be used to detect hot material.
- Cutting and welding operations on the bin should only be carried out in accordance with a strict permit-to-work procedure.
- Bins which store materials liable to spontaneous heating should have provision for discharge to a safe place in isolation from the rest of the plant.
- Carbon monoxide monitors may be used to detect spontaneous combustion.
- Electrical equipment should conform to BS 6467 (1985) or equivalent.

Spontaneous ignition

The temperature of any combustible dust should be kept at a safe margin below its ignition temperature (Table 25.1). If processing causes a rise in temperature the dust should be allowed to cool completely before storage or packing. Silo and bin temperatures should be monitored.

Equipment and building design should avoid ledges, corners, crevices, etc., which allow dust accumulation.

Certain dusts manufactured under an inert atmosphere may ignite spontaneously on contact with air (page 1447). To prevent this a small amount of oxygen may be added under controlled conditions during the latter stages of manufacture.

1444   Flammable and explosive hazards

Frictional ignition

Frictional ignition sources arise from either rubbing friction or frictional sparks. The major sources are listed in Table 25.5.

Table 25.5   Major sources of friction heating or impact sparks

- Overloading of mechanical equipment resulting in material becoming clogged and localised heating.
- Presence of foreign objects in process material, or falling into it from equipment components, which spark on impact with the plant.
- Overheated bearings.
- Misalignment of plant, c.g. fans or mill rolls, resulting in moving parts striking a surface.
- Inappropriate use of power tools, hammers, etc.
- Frictional rubbing and belt slip.
- Contact of highly reactive products with iron or steel tools.

Solids handling equipment inevitably contains many moving parts so that poorly designed/installed or inadequately maintained components can act as 'rubbing friction' ignition sources, e.g. slipping belt drives or overheated bearings.

An explosion occurred about 1.5 to 2 minutes after a sugar bucket elevator was switched on following a nine-day shut-down. Sugar had accumulated in the elevator pit (via a flap valve from another part of the plant) and dredging through this put abnormal strain on the elevator, causing a misadjusted tensioning device to fail. The out-of-alignment elevator buckets rubbed hard against the bucket guides and were struck heavily by the sprocket wheels. Sufficient heat was generated by friction to ignite a sugar dust cloud.[9]

An explosion occurred in a finished product dust collector/hopper connected to an air-flow bar type hammer mill used to grind a rubber additive. It was initiated by an assembly fault in the mill which caused overheating. Fortunately the explosion was vented by one of two explosion-relief doors on the collector and an ensuing fire was extinguished within a few minutes using a fixed monitor.[20]

The main preventive measures to avoid overheated bearings involve a combination of sound design and regular maintenance. Bearings should, wherever possible, be located outside equipment. The frequency of inspection of inaccessible bearings may otherwise be irregular and they may also be vulnerable to dust accumulation over long periods.

Frictional sparks may occur due to metallic objects, e.g. tramp metal, or hard non-metallic objects entering mechanical grinding or milling equipment. This may cause a 'hot spot' in the dust resulting in ignition.[21] The origins of tramp metal vary from parts of equipment to 'lost' hand tools or other metal components.

This problem may be controlled by the addition at feed points of magnetic separators to remove ferrous metals. Magnetic separators are no good when stainless steel nuts and bolts are around. Velocity or momentum traps are extremely effective provided they are cleared out periodically. Other material may also become trapped which could be harmful. They rely on an enlargement of the pipework or a section to trap material as it travels around a curve (Fig. 25.6). In some situations pneumatic separators serve to remove non-ferrous metals, stones, etc.

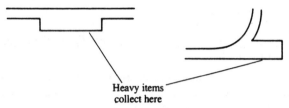

Fig. 25.6 Momentum/velocity traps for tramp metal

It is considered advisable to install tramp metal detectors at appropriate points in grain processing and storage facilities or animal feed mills, e.g. where sparking could occur in contact with duct or conveyor sides. However, dust explosions involving grain, flour or animal feed are unlikely to be initiated by single accidental impacts between tramp metal and anvils of metal, corroded metal, stone or concrete at net impact energies <20 J.[22]

Many of the USA silo explosions were caused by the metal tag used to seal the bottom doors of the railway wagons after weighing, being dropped into pneumatic conveying systems or mechanical conveyors along with the grain when the doors were open. This produced friction ignition.

Disintegration of equipment running at high speed can obviously result in friction sparks.

Ignition of powdered milk caused an explosion in a spray drier. Mechanical failure of an atomiser rotating at 7500 rpm caused frictional sparks.[23]

Hence, extractor fans should have anti-spark copper linings, in case a fan blade loosens and touches the housing. Ironically even relief provisions can act as a source of ignition.

A spark from an explosion relief panel initiated a serious dust explosion in the air filter unit associated with a zinc grinding mill.[24]

The use of hand tools may cause friction resulting in intense local heating. Therefore they should only be used after equipment shut-down, removal of all local dust, and issue of a permit-to-work.

Electrical sources

Sparks generated by electrical equipment, e.g. during the operation of switchgear, or when fuses blow or cables or equipment are damaged, may readily provide a source of ignition.

> A valve malfunction in a corn starch pneumatic conveying system resulted in overfilling of a feed hopper. The hopper was displaced and 30 kg of starch was ejected under pressure. This was ignited by a short-circuit from an electrical cable due also to the hopper displacement.[25]

Accumulations of dust on equipment with high surface temperatures may also be prone to ignition.

> A dust explosion within a grain silo complex was initiated by an inspection lamp left on inside it. Components were blown up to 100 m and a 0.5 m thick reinforced concrete wall was blown 12 m.[26]

> Dust, known to be capable of a giving rise to a dust explosion (minimum ignition energy of 152 mJ and minimum ignition temperature of 560 °C), had accumulated on the walls of a feed hopper in a polymer compounding plant. With the aid of an inspection lamp, an operator began to 'wet-down' the hopper but because the polymer was difficult to wet the first spray of water created a dust cloud which exploded. The operator, who was exposed to the flame front, suffered burns to the face, hands, arms and torso from which he eventually died. The lamp had cracked, probably due to water splashes or as a result of being dropped, resulting in ignition from the arcing filament.[27]

Moreover, dust may penetrate most types of conventional electrical equipment and this can be ignited, e.g. during the operation of switchgear.

When practicable electrical equipment should therefore be located where it is unlikely to be exposed to combustible dust. For example, switches, contact breakers, commutator motors, fuses, etc., may be situated in a clean air area remote from dust. Hence switchgear may be in a control room and lighting may be installed behind dustproof windows.

Silo or bin lights may be installed in dustproof casings, the requirement being that the maximum surface temperature must be well below that capable of igniting the specific dust.

Intrinsically safe electrical instruments can be installed; otherwise all electrical equipment should be totally enclosed and dust-tight and constructed to an appropriate standard.[28]

Spontaneous combustion

The phenomenon of spontaneous heating of dusts is discussed at length on page 1364 *et seq*. Unless the heat is dissipated the dust may reach its ignition temperature. The normal consequence is a fire but if the dust is subsequently dispersed as a cloud an explosion may ensue.

When processing of powder prone to self-heating results in a rise in temperature it should be cooled prior to storage or packaging. Automatic monitoring of temperatures, e.g. in silos, may be provided to detect any rise in temperature above the norm. Material may be recirculated to dissipate heat; the hazard is reduced if the time for which a powder remains at an elevated temperature is minimised.

Self-heating may be the major potential ignition source in some spray-drying operations.[29] Thus oxidation of components of milk, e.g. fats, proteins and sugars, occurs 10^3 times faster at 100 °C than at 20 °C. In a thin deposit heat is able to dissipate and the temperature will not rise to the point of incandescence. Heat loss is so slow with thicker deposits that the temperature can rise to 700 °C and initiate a fire or explosion. At 100 °C the critical thickness for self-heating is approximately 1 cm.[30]

> In 1993 an explosion occurred at a milk powder factory damaging vibrofluidisers, situated downstream of a spray drier, and fines return cyclones. A fire started some three to five minutes later in the vicinity of the exhaust fan from these cyclones.
>
> The proposed ignition source was an incandescent lump of milk powder which may have fallen-off the spray nozzle support lances and into the fluid bed where it continued to heat and char. It may have broken-open on passage through a rotary valve and glowed or ignited within the fluid bed. [Subsequent recommendations included inclusion of the exhaust fans from all baghouses in the emergency shutdown mechanism, relocation of butterfly valves to improve containment, consideration of monitoring the temperature probe in the vibrofluidiser exhaust – to obtain early warning of any smouldering fire – and of incandescent particles in the dryer and/or vibrofluidisers, e.g. using an infrared detector.]

Clearly, control of dryer temperature, minimisation of residence time and regular cleaning of surfaces are important considerations. In other types of equipment, however, surfaces may become overheated if a fault increases heat production or prevents adequate cooling.

Electrostatic discharges

Movement of air, e.g. in dust collector units, can create static electricity. The static discharge from dusts can reach energy levels of up to 50 mJ. If the minimum ignition energy for a dust cloud is below this value then static electricity may be all that is required for ignition to take place. Some dusts can be ignited by only a few millijoules. Charge generation for powders in a series of unit operations are given in Table 25.6.[31]

Pneumatic conveying of combustible powders can readily result in static accumulation followed by discharge. The electrostatic energy available in isolated parts of such systems is exemplified in Table 25.7.[32] Methods

1448 Flammable and explosive hazards

Table 25.6 Charge generation on medium resistivity powders[31]

Operation	Mass charge density µC/kg
Sieving	10^{-3} to 10^{-5}
Pouring	10^{-1} to 10^{-3}
Scroll feed transfer	1 to 10^{-2}
Grinding	1 to 10^{-1}
Micronizing	10^2 to 10^{-1}
Pneumatic conveying	10^3 to 10^{-1}

Table 25.7 Electrostatic energy available in isolated conducting parts of pneumatic conveying systems (based on voltage field E = 2,200 kV/m in the receivers and E = 311 kV/m in the pipe)[32]

Item	Gap thickness (in.)	Sizes and length (ft)	Width (ft) or diameter (in.)	Energy (mJ)
Isolated pipe	0.10	20.0	4 dia.	2.0
	0.20	20.0	4 dia.	4.0
Coupling in pipe	0.20	20.0	4 dia.	0.2
Coupling in receiver	0.03	1.0	4 dia.	1.8
	0.06	1.0	4 dia.	3.2
	0.12	1.0	4 dia.	6.3
	0.20	1.0	4 dia.	10.6
Manway door at receiver (1.0 inch wide gasket)	0.02	2.0	2	4.1
	0.06	2.0	2	12.1
	0.10	2.0	2	20.1
	0.20	2.0	2	40.1
Baghouse door	0.06	1.5	3.3	13.5
	0.10	1.5	3.3	22.5
	0.20	1.5	3.3	45.0
Receiver isolated from pipe and ground	0.03	5.0	4	95.0
	0.06	5.0	4	380.0

available to minimise electrostatic discharge (ESD) ignition are summarised in Table 25.8.[32]

Static charges are likely to develop on equipment constructed of non-conducting materials, e.g. plastic sacks, bag filters, conveyor belts and rubber sleeves. The problem is exacerbated with any dust of high electrical resistivity, e.g. $>10^8$–10^9 Ω/cm^3, since it will retain a charge for a considerable period.

> Phenacetin powder was sieved from a high level sieve on metallic staging via a terylene sleeve into a metal receiving bowl on nylon wheels at ground level. The sleeve was affixed to both sieve and bowl. When about 100 kg of powder

Table 25.8 Methods available to minimise ESD ignition of dusts in pneumatic transport systems[32]

Reduce ES charging by:	Reduce voltage fields in systems by:

• using large particles (>60 mesh)
• reducing velocity
• using round particles
   (low surface/volume ratio)
• reducing volatiles
• increasing moisture content
• changing particle-size distribution
 	
Increase charge drainoff by:
• adding moisture
• adding conductive particles
• adding antistatic material
   (water getter)
• mixing with more conductive material in process
	A. Sparking
	• grounding all conductive materials/ components
	• reducing voltage fields
	• using large particles
	
	B. Propagating brush discharge
	• reducing size of vessels
	• using smaller particles
	• reducing voltage fields
	
	C. Reduce dust explosion potential:
	• reducing/removing fines (-60 mesh)
	• preventing dust explosions
	• reducing ESD potential

had been transferred an incendive spark was generated, because the bowl was not earthed, and ignited the dust cloud within the sleeve.[33]

Where practicable anti-static conducting rubbers etc. should be substituted for non-conducting materials.

> An explosion occurred within a flaking machine used to produce solid organic biphenyl products and a flash of flame vented through an open viewing port in the machine enclosure. In operation a cooled drum rotated slowly in a trough of molten material and picked up a thin layer of solid. This was continually scraped off with a knife and pieces fell into a pack-out hopper. It was concluded that a recently installed knife constructed from phenol formaldehyde resin resulted in dust accumulation and that when some of this fell off it was ignited by a low energy electrostatic discharge in the pack-out hopper because of its very low ignition sensitivity.[34]

All metal parts should be electrically bonded together and to earth. All ducts should have a minimum of bends and all flanged joints should be earth-linked.

The electrostatic hazard associated with filter bags can be reduced by lowering the resistivity of the materials used, e.g. by incorporating stainless steel filaments or carbon fibres into the weave. These are termed 'epitropic' filters. They must be connected to an effective grounding point on the outside of the dust collector. Fabric resistance is less than 10^{-4} ohms. (Conversely the hazard can be increased if the earthing of the filaments fails or if the filaments break before the fibres wear out. So the resistance to earth of such bags, and their wear, should be checked regularly.)

Inerting

As discussed on page 137, inerting reduces the oxygen concentration in equipment to a level that cannot support an explosion. Carbon dioxide and nitrogen are effective inerting agents for carbonaceous materials but some metal powders can ignite, and burn, in carbon dioxide and even in certain circumstances in nitrogen.

The effect of inerting on pressure measurements is illustrated by Table 25.9.[35] Except for metal powders, reduction of oxygen concentration to 11

Table 25.9 Explosion pressure measurements in oxygen/nitrogen mixtures[35]

Dust	Oxygen concentration (%)	Maximum explosion pressure (bar)	Maximum rate of pressure rise (bar/s)
Calcium stearate	21 (air)	8.6	1,040
Zinc stearate	21 (air)	7.8	1,500
Calcium stearate	16	6.7	110
Zinc stearate	16	5.9	170
Calcium stearate	12	1.9	8
Zinc stearate	12	0.7	2
Calcium stearate	10	—	—
Zinc stearate	10	—	—

Explosion vessel closed, 1.2-L volume; —, no flame propagation; 1 bar = 10^5 N/m^2.

per cent v/v using carbon dioxide as inert gas or to 8 per cent v/v using nitrogen will prevent flame propagation of substantially all combustible dust clouds from electric sparks.[11] Less oxygen is necessary for effective inerting of metal powders; for example for atomised aluminium any oxygen concentration >3 per cent would allow combustion.

In practice safety margins should be applied giving a lower permissible oxygen concentration. For example about 5 per cent v/v oxygen is required to prevent ignition of dust clouds by a surface at 850 °C; 3–4 per cent v/v oxygen may be necessary for dusts containing carbon.

> Sulphur powder which is both relatively easy to ignite, as shown in Table 4.5 (indeed a minimum ignition energy of 3 mJ has been reported),[36] and prone to static electrification has been ground in a unit in which cooled flue gas has reduced the atmospheric oxygen concentration to <10%.

It is necessary for oxygen concentration in the inert gas blanket to be monitored continuously, since levels which would support combustion can arise accidentally.

Inerting in open-circuit dust-collection systems (i.e. which bring in and exhaust air) may be uneconomic because of the constant loss of expensive inerting gas.

Addition of thermal inhibitors

In the coal mining industry dust explosions are prevented by a combination of strategies including elimination of ignition sources, ventilation, and 'rock dusting'. The latter entails use of inert solid such as limestone such that the dust created by an explosion pressure wave produces an inert atmosphere, so arresting flame propagation by preventing coal dust reaching explosive limits.

A practice has been reported of inerting coal in a mill, by the addition of finely ground limestone placed on top of it, to prevent an explosion during start-up after any involuntary shut-down.[37] The addition of a diluent, e.g. calcium fluoride has also been suggested to avoid a magnesium dust explosion when vacuuming airborne dust from a graphite reactor in a magnesium reduction plant.[38]

The technique is analogous to inerting with a gas, e.g. nitrogen.

Thermal inhibitors may also function by chemical inhibition and cooling.

Application of solid thermal inhibitors to other industries or processes depends on the nature of the dust in question and whether it is an intermediate, an end-product or a by-product of little value. Rock dust may provide useful protection of plant during non-routine operations such as welding.[39]

Limiting the effects of an explosion

Limitation of the effects of an explosion relies upon design features such as:
- designing plant to withstand explosions (even after years in service) or to isolate them;
- explosion relief on plant, so that blast and flames are vented safely via vents, ducts or bursting panels;
- explosion suppression systems, e.g. where the initial pressure increase triggers release of a chemical which suppresses the explosion (Fig. 12.20);
- plant location, so that the effects of an explosion in one area do not result in personal injury or spread to other areas.

Each case must be assessed separately but it has been suggested that special precautions would not normally be required for dust-handling plants with a volume <0.5 m^3 for dusts with a maximum rate of pressure rise <100 bar/ms and with sufficient vent areas to atmosphere through inlets and outlets.[40] Thus, as a rule-of-thumb, the following volumes of items of plant with small dusty air volumes may be regarded as not requiring explosion relief vents:

Plant	*Volumetric limit (m^3)*
General dust collectors	0.57
Dust collectors, flour industry	2.27
Flour or sugar bins	10
Sugar-refining industry	20

Explosion-resistant construction

In limited situations it is possible to design vessels, equipment and piping to withstand the maximum pressure and rate of pressure rise that can be generated by an explosion (Table 25.1). However, practical difficulties tend to restrict this approach to equipment of limited size and simple geometry; connected fittings must clearly be similarly pressure-rated. For example, commercially available dust collectors are often designed to withstand only 7–14 kN/m^2 (1–2 psi) which is insufficient to contain an explosion. Pipework is *usually* sufficiently strong to withstand an explosion but not ducting.

Explosion-resistant construction covers all equipment which is explosion pressure-resistant and/or explosion pressure shock-resistant. Explosion pressure-resistant equipment can withstand the expected explosion pressure several times without suffering permanent deformation. Explosion shock-resistant equipment is constructed to withstand the expected explosion overpressure without rupturing but permanent deformations may result.

Even if no deformation is visible after any explosion it is necessary to check that the remaining strength continues to satisfy the design requirements.

Isolation

To prevent the spread of an explosion, isolation of one section of plant from another can be achieved using chokes. A power-driven rotary valve or double-acting plate valve which will prevent the passage of a blast wave may be used for this purpose. Transfer of burning dust from one section to another on a conveyor can be reduced by providing for cutting off the drive power in the event of an explosion. With a worm conveyor, the dust is itself used to form a plug as in Fig. 25.7(a). Alternatively with an inclined

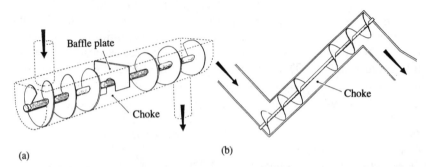

Fig. 25.7 Use of screw conveyors to prevent propagation of an explosion between different parts of a plant. (a) Removal of one flight and insertion of baffle plate in horizontal conveyor. (b) Removal of one flight of worm on inclined conveyor

conveyor if one flight is omitted the worm will not empty to a level below that of the missing flight so that the foot of the elevator remains sealed, as illustrated in Fig. 25.7(b).

If the physical properties of the dust make it impracticable to use such chokes, the use of isolated reservoirs provides a compromise between small isolated plant units and continuous plant fitted with chokes.

Suppression

Any explosion requires a finite time for the development of maximum pressure (Table 25.1). For example, explosion pressure takes about 10–50 milliseconds to build up before it exceeds the safe level for a particular plant. Therefore it is practicable to 'quench' a developing explosion, i.e. to arrest the growth of pressure before it reaches a dangerous level, by the introduction of a suppressant or inerting agent to displace the oxygen and impede combustion. Halon, water or dry powder (ammonium phosphate) may be used.

The principle of suppression is illustrated in Fig. 25.8. The detector may

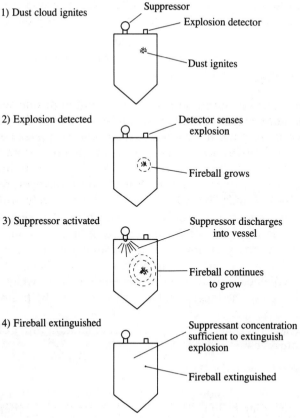

Fig. 25.8 Principle of suppression

operate by detecting either thermal radiation or the initial pressure rise. The controller responds by shutting down the equipment in a controlled manner, to avoid fuelling the fire/incipient explosion, and to protect auxiliary plant; it then triggers suppressant release. The suppressant acts by quenching the flame or by inhibiting the chemical chain reaction.

Unlike vents, suppression relies upon active systems. Therefore the design, installation and testing procedures will affect overall system reliability and experts should always be consulted.

Clearly both detection of an incipient explosion and the introduction of suppressant must take place within the vessel where the explosion is initiated. The suppressant is generally ejected from hemispherical suppressors, or high rate discharge bottles. In addition to extinguishing flame by chemical action and cooling, it inerts any unburned flammable mixture. An automatic plant shut-down facility is often included, to provide a warning to operators that an incident has occurred and to facilitate a safety inspection of the plant.

For an explosion initiated at atmospheric pressure, the pressure rise is limited by suppression to normally ≤ 2 psi. Therefore in most cases a capability of equipment to withstand a pressure of at least 3 psi is satisfactory.

Such systems have been demonstrated to provide reliable protection. They can, however, fail if the design is flawed. Such failures have included explosion detected too late, flow of suppressant too low, or insufficient total quantity of suppressant.

Suppression systems are sometimes operated in conjunction with rapid-acting isolation valves located in both the inlet and outlet headers of dust collectors, particularly if toxic dusts are involved. These valves prevent an explosion from discharging dust into the ducting into, and out from, the collector.

Advance inerting

Advance inerting relies upon the introduction of a suppressant into a section of the plant away from the centre of an explosion. The aim is to prevent a secondary explosion or fire resulting from the propagation of flame or the passage of burning material. The volume to be inerted is generally connected by ducting to the equipment in which an explosion may be initiated; the length of ducting determines the times available for inerting.

This strategy is generally employed in combination with isolation and the other protective methods.

Explosion relief

The function of explosion relief is to open coincident with a set increase in pressure in order to limit the internal pressure to a predetermined safe value.

The principle is illustrated in the typical pressure–time curves shown in Fig. 25.9.[41] Hence properly sized and located reliefs can prevent, or con-

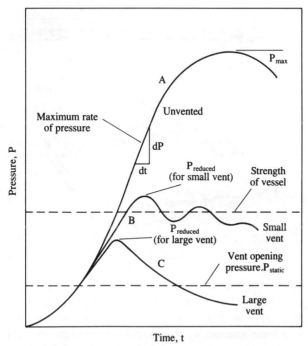

Fig. 25.9 Typical pressure–time histories of vented and unvented explosions (after ref. 45)

siderably reduce, damage to plant which is insufficiently strong to withstand the pressure likely to be developed by an explosion (Tables 25.1 and 25.4).

If ducting is used the peak pressure in equipment during a vented explosion is increased. The following combination of measures will maintain the peak pressure to <10 kN/m²:
- restrict duct length to 3 m.
- size the duct cross-sectional area as, at least, equal to the vent area.
- allow no more than one very shallow bend in the duct.
- raise any hat over the duct outlet by at least one duct diameter. (Clearly it should be sufficiently strong to withstand the pressure wave).

Since the increase in peak pressure is proportional to the square of the duct length, any ducts >6 m in length require additional vents in their walls.

The relief facility must be dust-tight, have sufficient mechanical strength to resist wear, but operate very rapidly when required. It usually takes the form of a series of bursting panels, explosion doors or displacement panels.

A bursting panel consists of a sheet of thin material (e.g. waterproofed paper, varnished cloth, metal foil) placed across the vent opening. In an

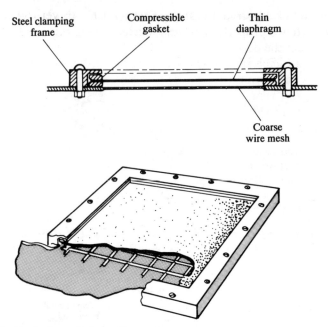

Fig. 25.10 Explosion relief panel[1]

explosion the membrane simply bursts and vents the combustion products. Alternatively the membrane may be displaced from the panel, e.g. as illustrated in Fig. 25.10. The bursting pressure depends upon, e.g. material, thickness, vent area, method of fixing, and the rate at which pressure builds up. During normal equipment operation the panel may require support on the inside, e.g. by wide-gauge wire mesh, to avoid damage. Whilst bursting panels provide good dust-tight seals, they suffer from the disadvantages that following an explosion any fans may blow burning dust out of the open aperture, or air may enter the equipment precipitating a secondary explosion. The membrane should be weatherproof, able to withstand the normal fluctuations in pressure, capable of withstanding abrasion and chemical attack from the dust being processed, and conductive to avoid accumulation of static electricity.

An explosion door comprises a lightweight rigid cover over a vent opening. Pressure generated within the plant simply displaces this cover and is relieved. Light hinged doors have been found to be nearly as effective as open vents for relieving the pressure from any explosion which does not exhibit a very high rate of pressure rise. The doors must be as light as practicable, in order not to retard operation, and a rim of oiled felt may be used to provide a dust-tight seal around the edges of the door.

Considerable increase in force may be added by corrosion or freezing between the door, the casing and the restraint interfaces. Often doors are held in place by non-corrosive magnetic strips and in a vertical position so as to eliminate water traps when used outside. Wire ropes are sometimes used to prevent the door from flying off during an explosion.[42]

Clearly if an explosion vent is called upon to operate then a pressure wave, flames and burning dust particles will be discharged. Therefore areas in front of relief vents must have restricted access (the area of restriction depending upon the type and size of the relief device) and workers should not be allowed into them whilst the plant is in operation.

> A dust explosion within a spray drier and filter used for the manufacture of a sulphur product was relieved by adequately dimensioned, self-closing pressure vents. Although notices prohibited access to the drier platforms a worker on the steps, which were provided with diverter plates, suffered serious burns to the hands and face.[43]

Since flame and burning dust from relief vents can be ejected over considerable distances this should be taken into account when locating buildings and equipment.

Housekeeping and maintenance

> Vibration raised a cloud of polymethylmethacrylate (PMMA) dust from a grating platform. As the dust settled it contacted an open propane flame from a manual shrink-wrapping device causing the PMMA dust cloud to explode. Lessons were to resite the shrink-wrapper and to clean dust from the platform at the end of each shift.[44]

The avoidance of accumulations of dust on floors, walls, ledges and beams by regular, thorough cleaning is always an important consideration to avoid – or minimise the scale of – a secondary explosion.

> In 1983 a dust explosion occurred at a plant which manufactured fine aluminium powder for paints resulting in damage costing >£1 million.[45]

> The process involved atomisation of molten metal in two production lines located in the open air. Each line was fed by a holding furnace adjacent to the atomising head in a building at one end of the site; the lower walls were of brick and the upper walls and roof of lightweight steel cladding. Droplets of atomised metal were rapidly cooled by a flow of air and powder was collected in multi-tubular collectors which were at an elevation of 20 m. Powder was discharged from these via rotary valves into bins which were taken on trolleys to a screen house, a lightly clad steel-framed building in which four screens were situated in concrete rooms. After screening, powder was transferred to the bagging room and then to a store, both of lightweight steel cladding construction.

> Safety features included the following:
>
> - Maintaining the average powder concentration in a line below the minimum

explosible concentration (an interlock between the induced draught fan and the atomiser served to automatically stop production in the event of air failure).
- Provision of vacuum cleaning facilities in the screen and bagging rooms.
- Control of ignition sources where powder was likely to be present, e.g. the all-metal plant was bonded and connected to earth, electrical equipment was dust-tight and naked flames were controlled. (However, complete elimination of ignition sources was not possible, e.g. molten metal at the atomiser or sparks due to metal-to-metal contact upon failure of moving equipment parts.)
- Powder collectors had explosion-relief panels in the sides facing each other.
- Buildings were of lightweight construction with access to screen rooms prohibited during operation. Production lines were inside a fenced enclosure.

At the time of the explosion only one line was in operation and the initial explosion probably occurred in the collector, probably initiated by a spark resulting from displacement of a component part. The primary explosion resulted in the ejection of burning and unburned powder and disturbed powder in the collectors of the second line and in the screen house and was hence followed by secondary explosions.

The largest missile, a 500 kg section of exhaust trunking, was projected 50 m and fell through the store roof. With the exception of some small missiles of lightweight material, most were restricted to within about 50 m of the perimeter fence but burning aluminium caused patch scorching of the ground and started brush fires up to 150 m away.

In this incident the explosion venting performed well but damage would have been reduced if the panels had released quicker and if an increased area had been provided on the collectors on the second line.

To avoid important ignition sources such as hot overheated bearings, stuck elevators, and piles of smouldering dust, proper maintenance of equipment, lubrication of bearings, adjustment of transfer belts, clearing of stuck elevators, good housekeeping and training are crucial.

A checklist for operation with combustible dusts is given in Table 25.10.[46] Clearly for technical and administrative controls to offer protection they must be properly installed, maintained and monitored.

An explosion took place in a bag filter after a rotary dryer. It was followed by a fire which was extinguished after 20 minutes. There were no injuries, but the bag filter itself and surrounding insulation and electrical wiring and equipment were damaged. The dust which exploded was polymeric and the source of ignition was probably self-heating. The plant was designed to minimise damage in the event of dust explosion. The vent panels relieved properly and most of the damage that did occur was due to the subsequent fire. The explosion suppression was effective in preventing the spread of fire but failed to protect the filter itself. The drying circuit was interlocked to shut down when an explosion occurred but the interlock on the main blowers to the dryers did not operate. The dust collector had an automatic water deluge fire protection system to extinguish fire but this failed to operate due to plugging and other

Table 25.10 Checklist for operation with combustible dusts (after ref. 43)

- Be aware and report unusual noise and leakage of dusty materials from the equipment.
- Monitor the rotary speed of rotating blades and augers in mixers, blenders, screw feeders, etc. and interlock the sensors to automatically shut down the equipment when the speed drops below a safe threshold value.
- Check rotating equipment (mixing blades rotating arms, etc.) on a regular basis. Look for signs of eccentricity in the bearings.
- Vessels with rotary valve discharge spouts should not be completely emptied, to provide an explosion/flame barrier.
- Provide interlocks to prevent conveying or processing equipment from operating when dust collectors are not operating or are operating inefficiently.
- Use dust ignition-proof vacuum cleaners.
- Electrically bond and earth dust collector bag holders.
- Use lift trucks approved for dust environments in process areas where an abnormal condition could result in combustible dust becoming airborne.
- Electrically bond and earth metal platform grating inside dust collectors or other equipment.
- Provide $\lessdot 60\%$ relative humidity air entering pulverisers and in pulveriser room enclosures to minimise electrostatic charge accumulation.
- Evaluate air flow capacity of dust collectors to prevent excessive air flow and the resultant removal of more dust than is necessary. Air flow velocity at pick-up is critical; it must be adequate but not excessive.
- Instrument particle size reduction equipment to detect excessive vibration. Provide audible alarm to alert the operator.
- Bond and earth screens in screeners and sifters.
- Check instrumentation inside storage silos and hoppers to assure electrical earth when their power lines are disconnected.
- Install blown vent sensors on explosion relief vent panels. Wire these sensors to shut down the process when a relief device is activated.
- Design a start-up sequence of process to delay the backpulsing cycle of a dust collector until the system is at a steady state condition.
- Eliminate common ducts to prevent propagation from one vessel to another.
- Remove air hoses from areas where they can cause combustible dust clouds to be generated.
- Restrict access to authorised associates in rooms that are explosion vented. Associate exposure to operating equipment must be determined to be safe.
- Prohibit maintenance operations within the confines of the process areas where hazardous dust conditions may exist while systems are operational.
- Install low level material detectors interlocked to shut off rotary valves at a predetermined value that allows these valves to function as an explosion barrier.
- Inspect pneumatic conveying line couplings on a per shift basis.
- Provide cleaning procedures of clean air outlet lines that are used on blower system in conjunction with pneumatic transfer.
- Prohibit filling of silos during lightning storms.

Table 25.10 (continued)

- Regularly clean all areas with special attention to overhead surfaces in rooms handling dusty materials.
- Clean magnetic separators once per shift.
- Provide supervision during delivery of dry, raw materials by outside contractors to monitor transfer/filling operation.
- Provide natural bristle brooms for sweeping. Avoid use of spark-producing or static-generating tools.
- Electrically earth and bond pneumatic conveying lines. Inspect these systems for electrical continuity on a quarterly basis and after any maintenance is performed.
- Require quality controllers to look for smouldering fires inside delivery vehicles during sampling of incoming dry, raw materials.
- Dust leaks resulting in material free-falling through the air should be corrected immediately.
- Provide a double check procedure system to check tolerance settings when working on pulverisers, grinders, and air classification mills.
- Prohibit use of plastic pipe in product transfer systems.
- Provide dust control at emergency back-up dump stations.
- Any use of a torch or other open flame device in a dry material handling area requires a welding permit. Other ignition sources should be controlled by an appropriate hot-work permit.

---

faults. Thus whilst the risks had been recognised and adequate safety systems had been provided, some secondary damage that occurred would have been minimal if all the systems had been in working order.[34]

Design and operating practices

Explosion prevention and explosion protection are often used in combination to control a perceived risk. Thus, for example, following the explosion in the flaking machine summarised on page 1448[34]:

Explosion prevention included improved control of the feed temperature (assumed to enhance ignition sensitivity). The roll was altered and a steel knife used. Grounding improvements were introduced and nitrogen purging was adopted in the packout equipment and in the vicinity of the knife.
Explosion protection involved modification and strengthening of the whole system to enable an explosion to be vented safely. A new rapid closing shut-off valve prevented transfer of pressure or flame from the filter to the packout extract system. Pressure-resistant glass viewing ports were provided.

Explosion protection encompasses precautions which ensure that the effects of an explosion are minimised. This involves some combination of limiting the quantity of material involved, preventing material spreading throughout a plant, and containing flames and hot gases within the plant or relieving them to a safe place.[47]

One major precaution is to locate sensitive items of plant in the open air, or in a special chamber with a light roof. Alternatively location may be near an outside wall so that explosion relief from the equipment can vent to outside.

Important design criteria for combustible dusts are summarised by Table 25.11.[46]

Typical dust emission rates for a variety of unit operations are given in Table 6.2. Clearly however, dust levels inside plant during manufacture, conveying, separation, drying, screening, and storage/disposal will be much higher and often within the explosible range. Interestingly, Table 25.12 shows the equipment/plant involved in dust explosions from a study of over 300 incidents.[48] Typical dust explosion control techniques for a range of chemical process equipment are illustrated by Table 25.13 (after ref. 49), although ultimate solutions must be the result of more detailed analysis.

Bagging operations

Dust-producing operations, e.g. bagging-up, should be extracted at source.

In certain sectors of industry the use of bulk bags has become popular. However, hazards including electrostatic ignition may arise during filling, storing and discharging the contents. Spark discharges can be prevented by bonding, i.e. grounding the conductive inner lining of the bulk bag. If flammable vapours or solvents are present brush discharges must be eliminated.[50]

The rapid rotating blades of bag slitting machines which cut-open bags, sieve out solids and eject packaging, are potential sources of ignition for combustible dusts. Explosion suppression systems in the entrance hood will quench a primary explosion in this critical part of the machine and also prevent propagation into the workroom and so avoid a more serious secondary explosion. Relief devices on the hood ensure excessive pressure does not build up and serve to vent combustion gases and extinguishing powder. The activation pressure of the suppression system must be below the static activation pressure of the relief device. The main machine casing should also be fitted with relief.

Dust conveyance equipment

The main forms of conveying are manual, mechanical, and pneumatic. Clearly, wherever possible, it is preferable to select systems which do not generate dust clouds. Whereas there is limited free volume for dust to form in a fully-loaded screw conveyor, friction can lead to ignition of a sensitive dust and overload trips should be installed on the motor driving the screw. Usually a conveyor needs to be enclosed for a dusty material. Free volumes

1462 Flammable and explosive hazards

Table 25.11 Design criteria for reducing frequency, and limiting effects, of explosions (after ref. 46)

- **General**

Equipment location
— Layout to minimise exposure to people, other important processes, and utilities.
— Use separate building or isolate hazardous operation.
— Locate outdoors or on outside wall.
— Avoid exposure to high traffic, busy areas.
— Maintain clear space to property line and public ways.

Dust control
— Design systems for dust containment.
— Use negative pressure systems.
— Provide dust collector pickups at dust generating/transfer points.
— Select equipment whose inside surfaces will collect and retain minimal quantities of combustible dust.

Explosion relief venting
— Flow area of vent should be based on the combustion characteristics of the dust, the yield pressure of the vessel, the volume and the geometry of the vessel, and the vent line through which gases are exhausted. The maximum pressure reached during venting should be $\not> 90\%$ of the yield pressure of the weakest element in the vented vessel. Permanent deformation of the vessel is thus prevented.
— Weight of the vent panel should not exceed $12.2 \, kg/m^2$ ($2.5 \, lb/ft^2$).
— Vents must discharge to a safe outdoor area with minimum of 15 m radius marked 'WARNING'.
— Venting into associate work areas is prohibited.
— Frangible type vents are preferred. Hinged or latched types are acceptable but require preventive maintenance.
— Non-frangible type vents require a retaining device to prohibit damage from airborne debris.
— Install blown vent sensor on explosion relief vent panels, wire sensor to shut down the process when vent is activated.
— Ducts are best avoided but when connected to vents: duct length should not exceed 3 m except for very weak explosions; bends should not exceed 22° with minimum 2 duct diameters between bends; strength of ducting system should be $\not<$ yield pressure of equipment being vented.

Tramp material separation – magnetic/electronic
— Install at bulk receiving stations, strategic points throughout system and upstream of pulverisers, grinders, and air classification mills.
— Prohibit use of non-conductive components in systems that contain moving material.
— Design systems so that the loss of a single component bond does not result in the loss of a total system earthing.
— Bond and earth across all material discharge points and their respective portable or stationary receiving bins.

Enclosures for hazardous operations
— Design vent panels for 1.2 kPa (25 psf) resistance and remaining walls, floor and ceiling areas for 4.8 kPa (100 psf) resistance.

Table 25.11 (continued)

— Arrange doors to swing into the room provided there is no conflict with local code requirements.

— Interior doors should be designed for 4.8 kPa (100 psf) resistance.

— Adjust vent panel design pressure upwards to account for local maximum wind loadings.

Lightning protection

— Locate equipment such that it does not act as a lightning rod.

— Tall silos present unique problems and require special consideration.

Flame permit

— A permit system is required before using cutting and welding equipment or other open flame device in a dry material handling area.

- **Equipment**

Silos

— Locate outdoors.

— Explosion vent entire top by allowing it to lift off or provide equivalency.

— Provide restraining chains to keep vent from flying away indiscriminately.

— Arrange equipment and walkways to prohibit restraining relief of vent.

— Avoid interconnecting silo dust collection systems.

— Polyester silos represent unique problems and require special consideration, including for any modifications.

Hoppers, bins, receivers, etc.

— Provide explosion relief venting (exception: manual bag dump stations). Vessels smaller than $1\,m^3$ in volume may be excluded in some applications; this requires specific evaluation.

— Bin vent socks violate system integrity and cannot be used.

Pulverisers, grinders, air classification mills, etc.

— Locate on outside wall, isolate and vent room.

— For existing interior locations, vent through roof and strengthen wall enclosure.

— Feed inlets and discharge outlets require an explosion barrier such as a screw conveyor.

— For pneumatic inlet and discharge ducting systems, design for 690 kPa (100 psig).

— Isolate air inlet ducts either by using explosion barrier valves or locating intake outdoors.

— Bin vent socks violate system integrity and cannot be used.

— Design flexible connections to contain 103 kPa (15 psig).

Ovens, dryers and roasters

— Locate on outside wall.

— Identify ignition characteristics of material (dry) and operate equipment at 50°C below ignition temperature of dust layer.

— Design to minimise product accumulation, provide self-cleaning feature, avoid surfaces and crevices that collect fines.

— Provide excess temperature thermocouple to monitor discharge air and equip with alarm and shutdown features.

— Identify opportunities for condition monitoring; for example, temperature increase over time required to compensate for material build-up in banking oven which ultimately suggests a need to clean oven or ignition threshold of deposits will be reached.

Table 25.11 (continued)

— Vacuum ovens require specific consultation.

Pneumatic conveying lines

— Locate and isolate away from ignition sources.

— Use of static generating non-conductive material such as plastics is prohibited.

— Provide adequate support to prevent vibration and swaying.

— Positive pressure systems are undesirable. If required they must be securely anchored, equipped with loss of pressure alarm, and designed to zero leakage.

— Ensure all system components are properly earthed.

Mechanical conveying systems (screws, belts and buckets)

— Install non-contact type motion detectors on mechanical conveyors that handle dry, combustible materials. Set alarms to signal operators when conveyors stop or slow down due to drag.

— Install motion detector sensors at locations on screw conveyors where the failure of couplings will not result in false readings.

— Use vertical, inclined (one flight removed), or horizontal (one flight plus baffle plate) screw conveyors as explosion barriers wherever possible.

— Install alarmed whisker-type electrical switches between all pulleys and inside housings of bucket elevators.

Dust collectors

— Collector to be designed for a minimum 1 bar yield pressure.

— Provide individual collection systems to avoid the dust collection system interconnecting multiple operations.

— Locate outdoors or near outside wall to allow for ease of venting.

— Earth bag holders and other internal parts.

— Duct clean air exhaust outdoors.

— Strength of clean air exhaust duct should not be less than yield pressure of the collector.

— Inlet ducts require special consideration when selecting design strength based on length, diameter, location and combustion characteristics of dust; in most cases, 690 kPa (100 psig) is adequate.

— Sprinkler protection within the collector may be important.

Flexible connections

— Design strength should be no less than 103 kPa (15 psig) using non-plastic material.

— Provide adequate clamps to maintain design strength.

Screener and sifters

— Provide earthing for all system components.

— Do not use materials that have volume resistivities greater than 10^{11} ohm-cm.

— Provide dust control systems.

Emergency bag dump stations

— Locate and isolate away from ignition sources.

Table 25.12 Different types of equipment/plant involved in the 303 incidents reported to HSE (1979–1988)

Types of equipment/plant	Number of incidents	% of total
Mill/grinder	51	17
Filter	47	16
Dryer	43	14
Silo/hopper	19	6
Duct	15	5
Conveyor	13	4
Cyclone	8	3
Mixer/blender	7	2
Elevator	5	2
Other	95	31
Total	303	100

are less with drag link systems, which transfer dust in a solid mass with virtual elimination of dust clouds, than belt conveyors. Attention must be given to eliminating static and overheated bearings.

Wherever the system comprises rapidly moving mechanical components (e.g. bucket elevators) then explosion-resistant construction is recommended. Bucket elevators can be constructed with cylindrical, pressure-resistant legs, preferably mounted along the outside wall and vented directly to atmosphere. Chokes should be provided at the feed and delivery points where practicable. Strong, dust-tight casings of fire-resistant construction are required. Explosion reliefs should be provided at the top and bottom and with long elevators, at intervals. Design can help eliminate sources of ignition and the use of exhaust ventilation and explosion suppression will further reduce the risk. (A summary of ignition prevention measures was given on page 141 *et seq.*).

Pneumatic conveying systems normally comprise a pick-up unit, transfer ducting, a powder–air separator (e.g. cyclone, cartridge filter, or a baghouse) and a receiver or storage silo. Generally the system is vacuum or pressure driven. Electrostatic discharge can occur where high enough voltage fields will cause breakdown across isolated material and components. Values can be calculated based on size of gap and component which yields capacitance.[33]

Control measures for pneumatic systems depend upon design. They include

- selecting a solid:air ratio well below the lower flammable limit and ensuring powder is fed at optimum rates (which may not be the case on start-up and shut-down)
- installing air relief valves and overload trips at the feed point of positive-pressure systems
- elimination of static electricity, e.g. bonding to earth, use of electrically non-conducting flexible links in the system

Table 25.13 Appropriate means for preventing and mitigating dust explosions in chemical process plant (X = most appropriate, (X) = possible but not often used)

	Dust concentr. < min expl. concentr.	Inerting by adding inert gas	Intrinsic inerting	Evacuation of process equipment	Addition of inert solids	Elimination of ignition source	Explosion resistant equipment	Explosion venting	Automatic explosion suppression	Explosion isolation
Powder mixers										
With mixing tools:										
High-speed		(X)	X		(X)			X	(X)	(X)
Low-speed		(X)	(X)		(X)		X	(X)	(X)	(X)
Without mixing tools:										
Drum mixers			(X)		(X)		X	(X)		
Tumbling mixers			(X)		(X)		X	(X)		
Double cone mixers			(X)		(X)		X	(X)		
Air flow mixers:										
Fluidised bed mixers							X	(X)	(X)	
Air mixers							X	(X)	(X)	

Dust explosions 1467

						(X)					

Powder/dust conveyors and dust removal equipment

Equipment	1	2	3	4	5	6	7	8	9	10	11
Screw conveyors		(X)				(X)			(X)(X)		
Chain conveyors		(X)				(X)			(X)(X)		
Bucket elevators		x							x		
Conveyor belts											
Shaker loaders		(X)(X)				x x x x		x x	(X)(X)		
Rotary locks											
Pneumatic transport equipment		(X)				(X)		x	(X)	(X)	
Dust filters and cyclones		x						(X) x			
Industrial vacuum cleaning installations		x									

Crushing and milling equipment

Equipment											
Ball mills		x (X)		(X)			x x x	x x x	(X)(X)	(X)(X)	
Vibratory mills				(X)(X)				x x x	(X)(X)	(X)(X)	
Crushers	x x						(X)(X)	(X) x x x x			
Roll mills	x x						x x				
Screen mills		(X)(X)					(X)(X) x x		(X)(X)	(X)(X)(X)	(X)
Air jet mills		(X)(X)		(X)(X)							
Pin mills		(X)(X)		(X)							
Impact mills	(X)										
Rotary knife cutters			x x								
Hammer mills		(X)	(X)								

Powder dryers

Equipment											
Spray dryers (nozzle)	x x	(X)(X)				x		x	(X)(X)	(X)(X)	(X)
Spray dryers (disc)	x x	(X)(X)						(X)(X)	(X)(X)	(X)(X)	(X)
Fluidised bed dryers	(X)	(X)						x	(X)(X)	(X)(X)	
Stream dryers											
Spin-flash dryers	x x						x x	(X)(X)			
Belt dryers	x x			(X)			(X)(X)	x			
Plate dryers	x x						(X)				
Paddle dryers	x			(X)			(X)	(X)	(X)	(X)	

Dryers

Indirectly-fired dryers are preferable to direct-fired dryers (page 474). If the latter are used,

- flame failure protection is obligatory
- fuel supply to the burners should be impossible unless air flow to the dryer is above a predetermined minimum

In either case dust should be prevented from entering any combustion chamber or heat exchanger unless the temperature is so low that decomposition/ignition of the dust will not occur. Any combustion chamber needs to be located outside a hazardous area or completely isolated from it by a fire-resistant enclosure.

Fluid bed dryers pose an explosion hazard particularly if combustible dusts are contaminated with organic solvent vapours. Design and mechanical strength of the equipment must be assured by testing. Observation windows represent a particular weakness, and if blown out of rubber profiles they can lead to a secondary explosion. As a generalisation, because of their greater mechanical strength, round rather than square equipment is preferred if explosions are a possibility. Explosion relief, preferably on the clean side of the filter, must be ducted to vent external to the work area. Explosion suppression can be provided for St 2 dusts (see page 1435) and for dust/solvent mixtures. Depending upon circumstances, use of barriers and flame arresters may prevent propagation of an explosion into the air inlet side of the equipment. The generation of static electricity must be avoided, e.g. by earthing of all metal parts.

Spray drying relies on rapid vaporisation of liquid from atomised solutions or slurries injected at the top of the tower into a stream of hot air. Spray drying towers normally operate at dust concentrations significantly below the lower explosible limit. However, dust deposits on walls, auto-oxidation, etc. increase the risk of an explosion. Explosion control measures include

- constructing towers to have sufficiently-high mechanical strength, e.g. minimum pressure resistance of 1 bar
- inerting (using recycled oxygen-deficient combustion gases from the burner, or using nitrogen)
- explosion venting in the side of the tower via panels, the sizing of which can be gauged from nomograms unless the height of the tower is more than 5 times the diameter when the entire roof of the tower must be used for venting
- detector activated rapid action valves or extinguisher barriers to prevent explosion propagation via pipelines into other parts of the installation
- explosion suppression

Dust collection equipment

Dust collectors, e.g. bag filter units and cyclones usually at the end of the system, e.g. after product has been deposited into a silo, pose an inherent hazard when handling dry dust since at the product stream inlet the dust cloud is likely to be within the flammable limits, and when shaken or 'blown back' bag filters release a dust cloud which is invariably within the limits. Thus a minimum of transfer operations is desirable.

> It was concluded from a survey of grain bagging plants that dust once collected should not be returned to grain in the sacks. At the first point of reception material should be enveloped in large bundles, e.g. in synthetic fibre sacks, and be left undisturbed until, e.g. grain cleaning.[51]

Flammable dust clouds may also be generated during the intermittent shaking of bag filters in order to displace dust into the base for removal.

If the dust constitutes waste, e.g. dust from ground or polished articles, this hazard can be eliminated by the use of a wet scrubber.

Whilst specialised dust collectors have been designed to withstand the maximum pressure generated during an explosion, commercially-available systems will not.

Wet-type dust collectors are recommended when the nature of the process permits their use. For particularly high risks, e.g. magnesium grinding, their use is obligatory.

Dry-dust collectors (see Fig. 25.11) should be situated in the open air or outside occupied workrooms, i.e. in locations such that no injuries to personnel would result from an internal explosion. Small unit collectors are preferable to large dust-collection systems. No dust collector should serve a mixture of processes or operations with differing explosion risks. Adequate explosion reliefs should be installed and chokes should be fitted at delivery points. Relief vents should be sited on the 'dirty' (unfiltered) side of filters, as the filter media itself acts as a barrier to the expanding gases.

Grinders

Grinders and milling equipment are potential ignition sources of explosive dusts, because of the energy used in the comminution process and as a result of friction or impact ignition in the presence of tramp material. These, and any associated equipment, must always be fitted with protective measures. Inerting is most common for batch mills whereas for other types the equipment is often made sufficiently strong to withstand an internal explosion. When tested for pressure shock resistance, all inlet and outlet openings should be tightly closed and the pressure resistance (pressure-shock resistance) of the grinder should be at least equal to that of the entire grinding installation. If an explosion could flash back into the grinder from a connecting vessel the pressure resistance of the grinder must be higher. Ferrous contaminants in the feed can be removed by magnets, but the metal collected on the magnets must be removed regularly.

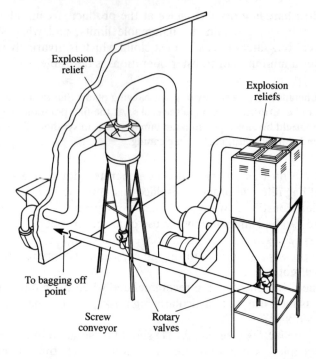

Fig. 25.11 Dry collectors for combustible dust. Explosion reliefs and chokes provided on plant. Open-air location[1]

Powder coating

In the spray application of powder coatings, pneumatic pressure and electrostatic charging serve to direct polymeric powders onto glass or metallic targets from a spray gun. In addition to the precautionary practices summarised earlier it is recommended that[52]

- Airflow must be sufficient to keep the powder concentration below the minimum explosible concentration. A level of 50 per cent of the minimum is advisable; if the design exhaust-dust concentration exceeds this, explosion suppression equipment is required.
- For fixed equipment, approved flame detection apparatus should be installed capable of reacting to the presence of flame within 0.5 seconds to shut down all energy supplies, close segregation dampers and actuate an alarm.
- Design and operating procedures should minimise powder hold-up, e.g. on horizontal surfaces or in ductwork.
- All metal parts of the spray booth, powder collection system and workpieces must be grounded with a resistance <1 mΩ.
- Explosion venting ducts should be as short as possible and lead to the outside.

References

1. Department of Employment and Productivity, *Dust Explosions in Factories*, HSW 22, HMSO, 1970, UK.
2. Weinsheimer, L. & Richter, W., et al., *Rheinische Post*, 8 Feb., 1979.
3. Cartwright, P., *Chemical Engineer*, 1990 (Sept.), **27**, 22.
4. a) Eckhoff, R. K., *Dust Explosions in the Process Industries*, Butterworth-Heinemann, Oxford, 1991.
   b) Bartknecht, W., *Dust Explosions*, Springer, Berlin, 1989.
   c) *Dust Explosion Hazards*, Video Training Package, Institution of Chemical Engineers, Rugby, 1993.
5. VDI Richtlinie 3673, *Druckenblastung von Staubexplosionen*, 1979.
6. Grossel, S., *J. Loss Prev. Ind.*, 1988 (April), 64.
7. Schofield, C. & Abbott, J. A., *Guide to Dust Explosion Prevention and Protection, Part 2 – Ignition, Prevention, Containment, Inerting, Suppression and Isolation*, Institution of Chemical Engineers, Rugby, 1988.
8. Field, P., *Explosibility assessment of industrial powders and dusts*, Department of the Environment, Building Research Establishment, HMSO, 1983, UK.
9. Cross, J. & Farrer, D., *Dust Explosions*, Plenum Press, New York, 1982.
10. Hartmann, I. & Nagy, J., *US Bur. Mines Rept. Invest.* 1944, **2751**.
11. Anon, *Fire Prevention*, 1986, **194**, 45.
12. Factories Act 1961, Part II, s. 31, UK.
13. Field, P., *Loss Prevention Bulletin*, Institution of Chemical Engineers, 1985 (066), 1.
14. Field, P., in reference 6.
15. British Standards Institution, *Electrical apparatus and associated equipment for use in explosive atmospheres of gas or vapour other than mining applications*, BSCP 1003.
16. Bodurtha, F. T., *Industrial Explosion Prevention and Protection*, McGraw Hill Book Co., New York, 1980.
17. Gibson, N. & Harris, G. F. P., *Chem. Eng. Progress*, 1976, **72**(11), 62.
18. Lunn, G., *Guide to Dust Explosion Prevention and Protection, Part 1 – Venting*, 2nd edn, Institution of Chemical Engineers, Rugby, 1992.
19. *Magnesium (Grinding of Castings and Other Articles) Special Regulations 1946*, and National Fire Protection Association, *Dust Explosion Prevention: Magnesium Powder*, NFPA Standard Code 651, Washington, DC, 1987.
20. May, D. C., et al., *J. Hazardous Materials*, 1987, **17**, 1.
21. Bartknecht, W., *Use of Protective Measures Against the Threat of Dust Explosions in Process Equipment*, in *The Hazards of Industrial Explosion from Dusts*, Conference Papers, Scientific and Technical Studies, New Orleans, 1981.
22. Pedersen, G. M. & Eckoff, R. K., *Fire Safety J.*, 1987, **12**, 153.
23. Sweis, F. K., *J. Hazardous Materials*, 1987, **17**, 241.
24. Anon, *Jahresber.* 1978, 72.
25. Health & Safety Executive, *Corn Starch Dust Explosion at General Foods Ltd., Banbury, Oxfordshire, 18 Nov. 1981*, HMSO, 1983, UK.
26. Field, P., *Dust Explosions*, in *Handbook of Powder Technology*, Elsevier Scientific, Amsterdam, 1982, Vol. 4.
27. Anon, *Loss Prevention Bulletin*, Institution of Chemical Engineers, 1990(095), 1.

28. British Standards Institution, *Electrical Apparatus with Protection by Enclosure for Use in the Presence of Combustible Dusts. Specification for Apparatus.*, BS 6467, 1985 Part 1 and *Electrical Apparatus with Protection by Enclosure for Use in the Presence of Combustible Dusts. Guide to Selection, Installation and Maintenance*, BS 6467 1986 Part 2, BSI, UK.
29. Pineau, J. P., *Protection Against Fire and Explosion in Milk Powder Plants*, Explosion Research in Practice, 1st Int. Symp., Antwerp 1984 (April), Part 2.
30. Beever, J. F., *Journal of Food Technology*, 1985, **20**, 637.
31. Cartwright, P., *Chem. Eng. Progress*, 1992, **88**(9), 76.
32. Dahn, C. J., *Chem. Eng. Progress*, 1993 (May), 17.
33. Pay, F. J., *Loss Prevention Bulletin*; Institution of Chemical Engineers, 1987(078), 1.
34. Skinner, S. J., *Plant/Operations Progress*, 1989, **84**, 211.
35. Palmer, K. N., *Chem. Eng. Progress*, 1990, **86**(3), 24.
36. Enstad, G., paper presented at Euro. Fed. Chem. Engrs. Working Party Mtg., 16 March 1975.
37. Anon, *Loss Prevention Bulletin*; Institution of Chemical Engineers, 1994(115), 9.
38. Anon, *Loss Prevention Bulletin*; Institution of Chemical Engineers, 1990(095), 13.
39. Amyotter, P. R. & Pegg, M. J., *Plant/Operations Progress*, 1992, **11** (3), 166.
40. Lunn, G. A., *Venting Gas and Dust Explosions – A Review*, Institution of Chemical Engineers, Rugby, 1984.
41. Field, P. & Abrahamsen, A. R., *Dust Explosion – The respective roles of the Hartmann bomb and 20-litre sphere in prescribing the size of explosion relief vents: a preliminary study*, Building Research Establishment Note N81/81, 1981.
42. Palmer, K. N., *Dust Explosions and Fires*, Chapman & Hall, London, 1973.
43. Anon, *Sichere Chemiearbeit*, 1985 (Jan.).
44. Anon, *Loss Prevention Bulletin*, Institution of Chemical Engineers, 1989(090), 27.
45. *Lessons of Aluminium Dust Explosion*, International Environment & Safety News.
46. Anon, *Loss Prevention Bulletin*, Institution of Chemical Engineers, 1990(095), 7.
47. Scholl, E. W., *Technical Safety Systems for Dust Explosion Protection*, in *Explosion Protection in Practice, Part 2 – Dust Explosion Protection*, European Information Centre for Explosion Protection, Antwerp, 1984.
48. Porter, B., in *Industrial Incidents*, paper presented at *Dust Explosions: Assessment, Prevention and Protection*, 24 Nov. 1989, IBC Technical Services Ltd, London.
49. Noha, K., *Auswahlkriterien fur Explosionsschutzmassnahmen*, VDI–Berichte 701, 681, VDI–Verlag GmbH, Düsseldorf, 1989.
50. Bunderer, R. E., *Chem. Eng. Progress*, 1993 (May), 28.
51. Beck, H., *Explosions in Grain and Feedstuff Bagging Plants and in Mills and Measures for their Prevention*, BAU Occupational Safety Series, 1979, **24**, 31.
52. National Fire Protection Association, *Spray Application Using Flammable and Combustible Materials*, NFPA 33, Boston, MA, 1985.

CHAPTER 26

Chemicals that may explode

Materials liable to spontaneous combustion are discussed in Chapter 24, and the hazards and precautions for unstable and reactive substances were briefly introduced in Chapters 10 and 11. The present Chapter expands on the dangers and control measures for those unstable, energetic chemicals that are capable of undergoing rapid exothermic decomposition or polymerisation which can lead to explosion. For those intending to work with these chemicals, expert technical and legal advice should be sought from, for example, the Health and Safety Executive in the UK, and dedicated texts such as in references 1–3 should be consulted.

Explosions are a rapid transformation of matter giving rise to large emissions of energy. If these are formed in the open air in a very short time, or are produced in a confined space, they will be subjected to varying degrees of pressure, which when released produces varying degrees of mechanical effects and noise. Explosions arise due to:
- nuclear reactions
- physical forces causing failure of vessels (e.g. hydraulic or pneumatic pressure within a confined space)
- chemical reactions

This Chapter deals with chemical reactions.

The hazards of reactive chemicals are dictated by molecular structure, temperature/pressure variations, temperature/time variations, inhibitor depletion, catalytic effects of impurities or of materials of construction, and increased surface contact. All chemical explosions release large quantities of gas and heat. They can be deflagrations or detonations. In the former the chemical reacts relatively slowly, whereas in a detonation the flame front travels as a shock wave followed closely by a combustion wave which releases energy to sustain the shock wave. Detonation shock waves travel at, or more rapidly than, the speed of sound.

To identify explosive chemicals it is important to study both the thermodynamic properties (the potential for a reaction under specified conditions) and the kinetic characteristics (to indicate whether a reaction will actually occur and with what speed). Chemical explosions can result from:

1 ignition of flammable vapour or gas clouds in either confined or unconfined conditions;
2 unwanted or accidental vigorously self-reacting chemicals with thermodynamically unstable structures; or chemicals which react in an uncontrollable manner with common contaminants, e.g. water, oxygen, sunlight, or pyrophoric substances;
3 explosives designed and used to produce explosions.

Type 1 explosions were discussed in Chapter 15. Explosibility assessment and testing were introduced in Chapter 20 and basic precautions were mentioned on page 508. The present chapter concentrates on the hazards and precautions associated with selected examples of type 2 and type 3 chemicals (with pyrophoric substances being covered more fully in Chapter 24). However it is crucial for legal and practical reasons to differentiate between the two classes of chemical. Thus, whether a chemical is termed an explosive rather than being classified as unstable or reactive is often a matter of definition and intended use, rather than degree of hazard.[4] Chemicals that may 'explode' as a result of the energy release during violent self-reaction/decomposition are exemplified by substances with the atomic groupings listed in Tables 10.4, 10.5 and 20.2 as typified by relatively high degrees of unsaturation; significant nitrogen content; nitrogen to halogen bonds; oxygen to oxygen bonds; compounds of oxygen and the halogens; ammonium nitrate; salts of aromatic nitro compounds, azides, fulminates, acetylides, etc. As a generalization, bonds between highly electronegative atoms (e.g. N—O and Cl—O) tend to be less stable than those between atoms of differing electronegativity. Molecular orbital calculations predict the instability of ions such as azides, N_3^- and fulminates, CNO^-. In structures such as the nitrate ion, NO_3^- and perchlorate ion, ClO_4^-, there is a highly electronegative atom with a large, positive oxidation number (+7 for Cl in ClO_4^-) which indicates electron deficiency and explains their reactivity as electron acceptors (oxidizers). Some of these groupings feature in true explosives whilst others are more associated with chemicals that may explode accidentally.

As discussed in Chapter 10, and exemplified in detail in reference 4, many unstable chemicals can pose a serious explosion hazard, even though they are not formally classified as 'explosive'.

Type 2 chemicals: accidental or unwanted explosions involving reactive chemicals

These chemicals are not normally classed as 'explosives', but rather chemicals capable of exploding. Chemicals liable to vigorous reaction with incompatible substances under uncontrolled conditions were discussed in Chapter 10. Other type 2 chemicals are typified by, but are not restricted to, those with positive enthalpy of formation, organic peroxides, and monomers which can polymerize with the release of energy if not properly inhibited and if the temperature is inadequately controlled.

Chemicals of high positive enthalpy of formation

The enthalpy of formation, ΔH_f, is defined as the heat absorbed or evolved during the formation of a substance from its elements at 25 °C and 1 bar. If the heat of formation is positive the compounds are endothermic; this represents stored energy which may be released on decomposition or in reactions. Examples of some relatively simple chemicals with a high positive ΔH_f are shown in Table 26.1.

Table 26.1 Selected compounds with high positive heat of formation (ΔH_f)

Substance	Structure	ΔH_f (kJ/mol)
Cyanogen	$N{\equiv}C{-}C{\equiv}N$	+308
Benzotriazole	(benzotriazole structure)	+250
Nitrogen trichloride	NCl_3	+230
Acetylene	$HC{\equiv}CH$	+227
Iodoform	CHI_3	+211
Allene	$H_2C{=}C{=}CH_2$	+192
Diazomethane	$CH_2{=}\overset{+}{N}{=}\bar{N}$	+192
Cyanogen chloride	$ClC{\equiv}N$	+138
Hydrogen cyanide	$HC{\equiv}N$	+130
Ethyleneimine	$\underset{CH_2{-}CH_2}{\overset{H{-}N}{\diagup\diagdown}}$	+127
1,3-Butadiene	$H_2C{=}CH{-}CH{=}CH_2$	+112

As part of a process for the pilot-scale synthesis of diiodotyrosine, a 10-gallon digestor was left overnight containing ammonia to neutralise the iodine liquors. When the operator went to remove the lid it exploded violently; no part of the lid was found. The cause was the formation of nitrogen triiodide an extremely explosive black powder which, although insensitive when damp, is shock-sensitive when dry.[6] $NI_3.NH_3$ complexes readily form in systems containing ammonia and iodine. This unstable detonator can be initiated by minimum energy in the form of light, heat, sound, nuclear energy or mechanical vibration. The heat of initiation on contact with many chemicals is sufficient to cause explosion. Desiccated crystals explode spontaneously. In all cases large volumes of gas are produced.

$$8NH_3.NI_3 \rightarrow 5N_2 + 6NH_4I + 9I_2$$

Exothermically formed compounds can explode only if the sum of the heats of formation of the products of the decomposition is greater than the heat of formation of the compound itself. Heats of formation can be determined by experimentation or empirically: the energy that is released by a reaction (including decomposition) can be calculated if the reactants, the products and their enthalpies of formation are known.

Organic peroxides

In the present context peroxides are taken to embrace all those chemicals with —O—O— linkages as illustrated by the examples in Tables 10.4 and 10.5. Of the many known organic peroxides, relatively few are of industrial importance because of cost, application or hazard. Commercial peroxides may be of technical grade or be supplied as preparations which incorporate inert diluents such as solvents, or for solids, calcium carbonate. Organic peroxides find use as polymer initiators, hardeners for resins, and bleaching agents.

The dangers from inadvertent formation of peroxides were discussed on pages 600–602. Organic peroxides are hazardous because of their flammable, toxic, corrosive and reactive properties. When involved in a fire the active oxygen content increases the intensity of combustion. In addition, some peroxides may decompose if exposed to only slight heat, friction, mechanical shock or contamination, to yield combustible gases and vapours which mix with air to form explosive mixtures. Indeed, unlike many of the type 2 chemicals, it is the very instability of peroxides to produce free radicals, that forms the basis of their exploitation and that can lead to explosions due to the gaseous products. Whilst peroxides are low-power explosives they have a relatively high sensitivity to mechanical shock and to heat: some are more sensitive to impact than true 'explosives', e.g. mercury fulminate, $Hg(ONC)_2$. Also, since brisance is only partially dependent upon power, the shattering properties are often underestimated. Thus, organic peroxides are rated as having specific blast energies in the range 0.1–0.4 that of TNT.[7]

> When a chemist removed 5 g of crystalline acetyl peroxide from a refrigerator it exploded spontaneously causing the scientist to lose both hands.[8]

There is, however, a wide spectrum of reactivity and thermal stability. Thus, trifluoromethyl peroxide is stable up to about 220 °C, whereas methyl peroxide is highly dangerous: pure peracetic acid will explode when subjected to friction even at −20 °C. Table 26.2 illustrates the varying thermal stabilities for a selection of peroxides (after reference 2) and Table 26.3 classifies selected peroxides according to their fire and explosion hazard.[9] The decomposition threshold can be lowered significantly by the presence of catalysts.[10] Especially sensitive are the ketone peroxides, hydroperoxides, peroxyketals and some acyl peroxides. The most common contaminants which cause problems include redox agents such as cobalt

Table 26.2 Self-accelerating decomposition temperatures (SADTs) for selected commercial organic peroxides (after ref. 2)

Compound	Structure	Decomposition temp., SADT (°C)	Control temp. must not exceed (°C)
3,3,6,6,9,9-Hexamethyl-1,2,4,5-tetraoxacyclononane		140	(a)
Tertiary butyl peroxide	$(CH_3)_3C-O-O-C(CH_3)_3$	>110	(a)
Benzoyl peroxide	$Ph-\overset{O}{\underset{\|}{C}}-O-O-\overset{O}{\underset{\|}{C}}-Ph$	88	(a)
Lauroyl peroxide	$CH_3(CH_2)_{10}\overset{O}{\underset{\|}{C}}-O-O-\overset{O}{\underset{\|}{C}}(CH_2)_{10}CH_3$	55	(a)
Isononanoyl peroxide	$(CH_3)_3C-CH_2-\underset{\underset{CH_3}{\|}}{CH}-CH_2-\overset{O}{\underset{\|}{C}}-O-O-\overset{O}{\underset{\|}{C}}-CH_2-\underset{\underset{CH_3}{\|}}{CH}-CH_2-C(CH_3)_3$	20	0
Octanoyl peroxide	$CH_3(CH_2)_6\overset{O}{\underset{\|}{C}}-O-O-\overset{O}{\underset{\|}{C}}(CH_2)_6CH_3$	25	+10
Acetyl cyclohexane sulphonyl peroxide	$CH_3-\overset{O}{\underset{\|}{C}}-O-O-SO_2-Ph$	15 to 20	−5
Tertiary butyl perbenzoate	$(CH_3)_3C-O-O-\overset{O}{\underset{\|}{C}}-Ph$	60	(a)

Table 26.2 (continued)

Compound	Structure	Decomposition temp.; SADT (°C)	Control temp. must not exceed (°C)
Tertiary butyl perisononanoate	(CH$_3$)$_3$C−O−O−C−CH$_2$−CH−CH$_2$−C(CH$_3$)$_3$ ∥ | O CH$_3$	60	(a)
Tertiary butyl peroctoate	(CH$_3$)$_3$C−O−O−C−C−(CH$_2$)$_3$−C(CH$_3$)$_3$ ∥ | O C$_2$H$_5$	35	20
Tertiary butyl perpivalate	(CH$_3$)$_3$C−C−O−O−C(CH$_3$)$_3$ ∥ O	25	−5
Tertiary butyl hydroperoxide	(CH$_3$)$_3$C−O−OH	88	(a)

Cumene hydroperoxide	CH$_3$
	     |
	Ph−C−O−OH
	     |
	     CH$_3$
		>80	(a)

(a) Ordinary storage temperatures will cause no serious decomposition.

Chemicals that may explode 1479

Table 26.3 Some typical organic peroxide formulations classified according to their fire and explosion hazard

Class	Definition	Peroxide	Concentration, wt %	Diluted with	Maximum individual container size
Class I	A serious explosion hazard. Class I organic peroxides can undergo easily initiated, rapid and explosive decomposition. Class I includes formulations that are relatively safe only under closely controlled temperatures such as refrigerated storage, or when diluted	Acetyl cyclohexane sulphonyl peroxide	60–65	Water	1 lb R
		Benzoyl peroxide	98+	—	1 lb
		t-Butyl hydroperoxide	90	t-BuOH	5 gal
		t-Butyl peroxyacetate	75	OMS	5 gal
			98	—	1 gal R
		Di-n-propyl peroxydicarbonate	85	OMS	1 gal R
Class II	Severe fire hazards and intermediate explosion hazards. The explosive decomposition is not as rapid, violent or complete as those of Class I formulations. Class II includes some formulations that are relatively safe under controlled temperatures including refrigeration, or when diluted	Acetyl peroxide	25	DMP	5 gal
		t-Butyl hydroperoxide	70	DTBP and t-BuOH	55 gal
		t-Butyl peroxy-2-ethyl-hexanoate	97	—	55 gal R
		Peroxyacetic acid	43	Water, acetic acid and hydrogen peroxide	30 gal
Class III	Severe fire hazards and moderate explosion hazards. They burn rapidly and give off intense heat. Many Class III formulations are diluted or require refrigerated storage	Benzoyl peroxide	78	Water	25 lb
		Benzoyl peroxide paste	50–55	BBP	380 lb
		Cumene hydroperoxide	86	Cumene	55 gal
		Decanoyl peroxide	98.5	—	100 lb R
		Di-t-Butyl peroxide	99	—	55 gal
		Diisopropyl peroxydicarbonate	30	Toluene	5 gal R
		MEK peroxide	9% active oxygen	DMP	5 gal R

Table 26.3 (continued)

Class	Definition	Peroxide	Concentration, wt %	Diluted with	Maximum individual container size
Class IV	Moderate fire hazards that are easily controlled. Many are diluted and some require refrigerated storage	Benzoyl peroxide	70	Water	25 lb
		Benzoyl peroxide paste	50	BBP and water	380 lb
		Benzoyl peroxide slurry	40	Water and plasticiser	380 lb
		Benzoyl peroxide powder	35	Starch	100 lb
		t-Butyl hydroperoxide	70	Water	55 gal
		Lauroyl peroxide	98	—	110 lb
		MEK peroxide	5.5% active oxygen	DMP	5 gal
		MEK peroxide	9% active oxygen	Water and glycols	5 gal
Class V	A low or negligible fire hazard. Their combustible packaging materials may pose more of a hazard	Benzoyl peroxide	35	Dicalcium phosphate dihydrate or calcium sulphate	100 lb
		1,1-Di-t-butyl peroxy-3,5,5-trimethyl cyclohexane	40	Calcium carbonate	100 lb
		2,5-Dimethyl-2,5,-di-(5-butylperoxy) hexane	47	Inert solid	100 lb
		2,4-Pentanedione peroxide	4% active oxygen	Water and solvent	5 gal

Abbreviations: R – Refrigeration required; DTBP – Ditertiary butyl phthalate; *t*-BuOH – tertiary Butanol; BBP – Butyl benzyl phthalate; OMS – Odourless mineral spirits; gal – 1 US gallon = 3.785 litres; DMP – Dimethyl phthalate; MEK – Methyl ethyl ketone.

salts and other accelerators and promoters, as well as various metal ions and ionising acids, e.g. sulphuric acid. Thus, unless carefully controlled, accidental explosions can occur whenever peroxides are manufactured, stored, transported or used, as illustrated by the following case histories.[2]

> A fire started during the unloading of 15 tons of methyl ethyl ketone peroxide. An explosion soon followed which killed four firemen and shattered windows up to 2 km away. The cause was attributed to friction from a parcel during the unloading.

> Four people were killed and extensive damage caused when 220 kg of methyl ethyl ketone (MEK) peroxide product was left in a filter apparatus for 6 hours where it corroded the metal gauze and resulted in catalytic decomposition of product by nickel and iron salt impurities.

> During the manufacture of peracetic acid in hastily improvised apparatus there was a temperature excursion in the reaction mixture. The resulting explosion killed the operator and metal fragments were found more than 100 m away.

> About a ton of benzoyl peroxide exploded during storage in a depot. The damage was comparable to the effects of a 300-kg gunpowder explosion.

As a result selected precautions include the following.

Manufacture

- follow written procedures and only deviate with permission: tests may first need to be conducted
- provide fast-emptying facility for drenching compounds in the event of unacceptable temperature excursions
- consider the need for production facilities akin to those for sensitive explosives, e.g. strong walls with loose-fitting skylight, remote controls, tight-fitting heavy doors on cubicle
- ensure proper materials of construction, e.g. polytetrafluoroethylene (PTFE) as packing material, stainless steel, or if this is not sufficiently corrosion-resistant then titanium or tantalum can be used: avoid certain metals, e.g. copper, brass, lead and other substances likely to rapidly decompose peroxide, e.g. wood and some steels, aluminium alloys, zinc and galvanised metal
- ensure equipment is scrupulously clean
- for dangerous peroxides, when practical, work with dilute solutions

> Following the incident described on page 507 the manufacturer reduced the concentration of MEK peroxide from 60% to 50%.[2]

- continuous production (e.g. as used for lauroyl peroxide) is a safer means of providing material consistently within specification than batch processes

Use

- seek advice from suppliers and other experts
- enforce good housekeeping of the workplace and the apparatus
- store and handle under well-ventilated conditions
- containers of stainless steel, pure aluminium with polished surfaces, or impermeable plastic are often used and easily cleaned
- wash apparatus before use, e.g. with water and detergent and then with solvent, and then passivate with, e.g. strong acid followed by water rinse
- segregate peroxides from flammable materials and salts of heavy metals, especially activators, and consider, e.g. separate scales for weighing peroxides
- prohibit smoking, since it poses an ignition source and the ash contains traces of manganese salt which could have a catalytic effect
- containers of peroxides should have loose-fitting lids and solvents/diluents should be prevented from evaporating
- restrict material in use to, e.g. sufficient for 0.5 day use
- deal with spillages immediately: solids should be wetted or mixed with inert powder, whilst liquids should be absorbed on perlite or vermiculite and swept up
- use clean cloths only for small drips and then discard to special bins
- never pour unused peroxide back into original stock container
- always add peroxide gradually to reactives, never the other way around
- ensure accelerators have been well dispersed with resin before adding peroxide catalyst
- follow written procedures and wear appropriate protective equipment, e.g. goggles, face mask, gloves, etc.
- provide adequate supplies of water to wash spillages and fight fire
- eliminate sources of ignition; use explosion-proof electrics
- assume novel peroxides are dangerous and treat as 'explosives'

Storage

- store in accordance with legislation in properly labelled containers under the correct environmental conditions, preferably in a cool place so as to minimise active oxygen content
- do not store on wooden pallets
- large stocks (e.g. >150 kg) should be stored in specially designed buildings separated from other occupied buildings, roads, etc. as per regulations: in the absence of official guidance a separation distance of 15–25 m is considered good practice depending upon the peroxide in question
- consult the manufacturer's literature for properties of specific compounds. Methyl ethyl ketone peroxide gradually decomposes to liberate flammable gases and needs to be stored in vented containers; it also explodes on contact with acetone and should therefore be kept segregated from it and never be diluted with this solvent

- segregate from incompatible materials
- where practical store in several small containers rather than one large one
- end user should keep material in container as supplied, often polythene in protective cardboard boxes
- do not store containers on top of one another or too close together

Disposal

There are no all-encompassing rules but the following guide is helpful provided that only competent staff undertake the operation:
- acyl peroxides and peresters can be gradually fed into stirred methanolic solution of soda
- peroxydicarbonates can be destroyed by adding gradually to 5% aqueous soda
- peroxides or peroxydicarbonates that are unstable at room temperature may be spread thinly on the ground outside at a safe location, away from ignition sources, and left to decompose
- solid peroxides can be spread in a thin layer on a bed of sawdust at a safe location and set alight
- liquid peroxides can be absorbed on vermiculite or perlite and spread out as a thin layer in a safe location and ignited
- small quantities of soluble peroxyacids can be diluted with at least twenty times their weight of water and carefully neutralised, and then flushed to drain

Monomers

Monomers are the building blocks for polymers. They comprise compounds with reactive groupings such as vinyl, acid, alcohol, or strained rings such as in ethylene oxide or ethylenimine (see Table 26.1). These permit linking in chains either to themselves or to other reactive monomers to form macromolecules.

$$>C=C< \quad >C=C< \quad >C=C< \;\rightarrow\; -C-C-C-C-C-C-$$

The monomers are not explosives but they can result in explosion because of
- their flammability
- the heat released from the polymerisation reaction leading to 'runaways'
- the formation of unstable intermediates or by-products
- decomposition

as exemplified by hydrogen cyanide, ethylene, ethylenimine and ethylene oxide below.

Hydrogen cyanide

In addition to the toxic hazards referred to in Chapter 31 hydrogen cyanide possesses other hazards including its flammability (flash point of -18 °C; autoignition temperature of 538 °C; flammable range 6–41%), its reactivity with a host of chemicals, its cryogenic properties on evaporation, and its exothermic homopolymerisation. For example, between pH 5 and 11, usually after an incubation period, the onset of reaction to produce black polymers or tetramers is rapid and if confined can prove to be violently explosive.

$$\left(\begin{array}{c} C-CH-NH \\ \| | \\ NH CN \end{array} \right)_n \qquad \begin{array}{c} H_2N CN \\ \diagdown \diagup \\ C \\ \| \\ C \\ \diagup \diagdown \\ H_2N CN \end{array}$$

Homopolymer   Tetramer

A hydrogen cyanide tank was ruptured by a violent explosion resulting in damage equivalent to 4.5 kg to 7 kg of TNT. The vapour released ignited but the 6 m to 15 m high flames were extinguished by water from a ruptured line. It was subsequently concluded that a rapid polymerisation with a temperature and pressure rise had occurred in the tank and the vent was unable to cope. Conditions which favoured polymerisation were a residence time of 12 days, dilution to 92%, the temperature of some added material being 30 °C to 40 °C instead of the normal 4 °C to 5 °C, inadequate stabilisation, and the possible presence of mercury as a promoter.[5] (Previous examples of the rapid polymerisation of HCN have been cited.[5])

Since polymerisation is encouraged by water and heat, stored hydrogen cyanide should contain less than 1 per cent water, be kept cool and should be inhibited by the presence of acetic, phosphoric or sulphuric acid and the storage life should be restricted to say 90 days. Anything which depletes the concentration of the stabiliser (e.g. the presence of contaminants, heating the cylinder to increase vaporisation which would also cause the acid inhibitor to react with the cylinder walls) can result in explosive polymerisation. Thus in the event of formation of a yellow-brown colour and liberation of heat the material should be promptly reacidified and cooled.

Ethylene

As with any flammable gas, clouds of ethylene within the flammable range can explode if ignited.

Failure of a 10 mm compression fitting on a 14 MPa ethylene line in a trench resulted in spillage of 180 kg of ethylene. The gas ignited in the partial confinement of a courtyard between buildings and the resulting explosion equalled 100–270 kg TNT and caused multimillion dollar damage.[11]

When heated, ethylene undergoes decomposition to produce carbon, methane and hydrogen, or aromatics depending upon conditions. At elevated temperatures and pressures ethylene undergoes polymerisation to form polyethylene. This can be catalysed by, e.g. anhydrous aluminium chloride, to such an extent as to trigger explosions. Explosive gums can form between nitrogen oxides and ethylene compounds. Thus precautions include careful control of reaction conditions, prevention of flammable atmospheres, removal of sources of ignition, and ensuring the absence of impurities.

Many other vinyl derivatives have been involved in explosions. The importance of stabilisers such as hydroquinone monomethyl ether to inhibit polymerisation of, e.g. acrylic esters, in storage was stressed on page 527. In many cases, the presence of oxygen is essential for these additives to function. This, however, is not universal and it is claimed that in the case of vinyl acetate monomer the presence of oxygen may destabilise it at elevated temperatures.[12] There is an optimum oxygen concentration for the stabilisation of methacrylic acid.[13]

Ethylenimine

Ethylenimine is used as an intermediate for the manufacture of selected organic compounds and to make cationic polymers for enhancing binding effects such as adhesion, strength, and flocculation of cellulosic materials, e.g. paper. Ethylenimine is toxic (see page 333) and highly irritant. Physical properties are given in Table 26.4. It is a very reactive chemical and

Table 26.4 Properties of ethylenimine

Property	Value
boiling point	55–56 °C
flash point, °C	−11.1
autoignition temperature, °C	323
flammable range	3.6–46%
heat of vaporisation, kJ/mol	34.11
heat of combustion of liq at 25 °C, MJ/mol	1.59
heat of formation at 25 °C, kJ/mol	
gas	127.2
liq	93.3
heat capacity of liq, J/(g·K)	
at 0 °C	2.39
at 25 °C	2.51
at 42 °C	2.59
heat of mixing to produce 20% by wt soln in H_2O, kJ/mol	13.8
dielectric constant at 25 °C	18.3
dipole moment, C·m	$(7.74–7.84) \times 10^{-30}$
specific conductance $(\Omega \cdot m)^{-1}$	8×10^{-6}

subject to aqueous auto-catalysed exothermic polymerisation. If not controlled, reactions may be violent. Thus undiluted ethylenimine tends to polymerise exothermically and with explosive violence in the presence of acids and oxidising agents, and frequently requires cooling. The substance is normally stored over solid caustic alkali to minimise carbon dioxide catalysed polymerisation, and storage tanks should be provided with nitrogen blankets, water sprays, high-temperature alarms, and relief valves or rupture discs vented via an acid scrubber. Mild steel, stainless steel or glass are suitable materials of construction, whereas copper and its alloys, and silver must be avoided. Plastics are also often unsuitable because of swelling but PTFE or ethylene-propylene elastomers are acceptable for gaskets. Cylinders should be stored in well-ventilated locations segregated from acids, ignition sources, and highly flammable materials.

Ethylene oxide

Similarly, ethylene oxide,[2,14] which is used for fumigation, sterilisation, and as a precursor to ethylene glycol, polyethylene glycols, nonionic surfactants and a host of other organic chemicals, is also a highly reactive compound.[14]

It is a flammable gas with extremely wide flammability limits, low flash point, and low autoignition temperature, as indicated by Table 26.5.

It undergoes a very exothermic homopolymerisation at about 100 °C to produce polyethylene glycol and releasing more than 2000 kJ/kg. In the presence of catalysts the polymerisation can occur even at sub-zero temperatures and with tremendous violence, causing explosions in closed vessels.

$$\overset{O}{\overset{/\,\backslash}{CH_2-CH_2}} + \overset{O}{\overset{/\,\backslash}{CH_2-CH_2}} \rightarrow -(CH_2-CH_2-O-)$$

It isomerises to acetaldehyde, liberating 113 kJ/mole. This normally occurs gradually at 300–400 °C but at lower temperatures in the presence of catalysts.

$$\overset{O}{\overset{/\,\backslash}{CH_2-CH_2}} \rightarrow CH_3.CHO$$

The resulting ethylene oxide/acetaldehyde mixture has a lower autoignition temperature than pure ethylene oxide.

It disproportionates, requiring higher activation energies than polymerisation but liberating the same amount of heat. Since reaction rates are much faster at elevated temperatures, the heat release rates are proportionally higher.

$$4C_2H_4O \rightarrow 3C_2H_4 + 2CO_2 + 2H_2$$

Table 26.5 Fire and explosion data for ethylene oxide

Property	Value
Liquid properties	
electrical conductivity (-25 °C), pS/m	4×10^6
reaction threshold temperature (ARC, 0.02 °C/min, $\phi=1$), °C	~200
flash point, °C	<18
Vapour properties	
standard heat of formation, kJ/mol	-52.63
heat of decomposition (max), kJ/mol	-134.3
minimum decomposition pressure, kPa	40
decomposition flame temperature (P constant), K	1277
decomposition pressure ratio (V constant)	~11.0
autodecomposition temperature, K	~773
fundamental burning velocity, m/s	2.7×10^{-2}
minimum ignition energy, J	~1
heat capacity, C_p, J/(mol·K)	48.28
thermal conductivity, W/(mol·K)	0.124
Properties of vapour–air mixtures	
lower flammable limit, mol %	3.0
upper flammable limit, mol %	100
deflagration K_G index J, (MPa·m)/s	18.8→24.6
deflagration P_{max} index, kPa	968→975
minimum deflagration pressure (~5–60 mol %), kPa	1.3–2.6
stoichiometric composition, mol %	7.72
detonable range, mol %	5–30
recommended TNT efficiency, %	10
low heat of combustion, kJ/mol	1217
high heat of combustion, kJ/mol	1305
theoretical flame temperature (P constant), K	2402
theoretical deflagration pressure ratio (V constant)	9.91
autoignition temperature, K	718
fundamental burning velocity, m/s	1.08
minimum ignition energy (10.4 mol %), J	6×10^{-5}

$$5C_2H_4O \rightarrow 4C_2H_4 + 2CO_2 + H_2 + H_2O$$
$$6C_2H_4O \rightarrow 5C_2H_4 + 2CO_2 + 2H_2O$$

It undergoes exothermic vapour phase concentration-, pressure- and temperature-sensitive autodecomposition via free radicals to form carbon, hydrogen, carbon dioxide, ethylene and acetaldehyde with the evolution of some 32 kcal/mole. The reaction can be catalysed by high surface area metal oxides such as γ iron oxides which may produce rapid local heating and vapour ignition under near-adiabatic conditions.

The most rapid rate of self-heating occurs when conditions promote disproportionation rather than isomerisation or polymerisation. The handling of ethylene oxide requires care to avoid explosion since the manufacture and use has resulted in a number of explosions such as those described on pages 525, 704, 706 and 881, and the case history below.

In March 1991 an ethylene oxide reduction plant in Texas exploded. The force of the blast and the ensuing fire caused one fatality and extensive damage to the plant. A previously unknown iron oxide-catalysed reaction raised the temperature to produce a hot-spot in excess of the 500 °C auto-decomposition temperature which then generated gaseous products such as carbon monoxide and methane. No free oxygen was required for the reaction to occur. The resulting flame, which progressed into the column base, accelerated upward and caused the explosion within a second. The energy released exceeded by four times the maximum allowable working pressure of the still.[15]

Implications for safe handling include:
- use carbon steel and stainless steel for all equipment in ethylene oxide service
- store in containers with extremely clean walls; all-welded construction is recommended. Inert with nitrogen
- during transfer avoid sudden pressure shocks which could heat the vapour by adiabatic compression
- earth tanks and metal equipment to prevent static buildup
- locate plant in a well-ventilated environment and use explosion-proof electrics
- ensure catalytic impurities are absent during storage and reaction
- use non-return valves in lines carrying fluids which may react vigorously with ethylene oxide.

Type 3 chemicals: characteristics of purpose-designed explosives

The development of explosives in Europe started with the use of black powder in the 13th century, but the main impetus came from advances in nitration chemistry in the 19th century.

In the Silvertown munitions factory explosion in 1917 sixty-nine people, mostly within 250 m of the source, were killed. The incident involved some 53 tonnes of TNT. Damage to houses was slight beyond 650 m.[16]

Explosive chemicals find use in military and space research, mining and chemical synthesis. They can be used safely if their properties are understood and the requisite precautions devised and enforced. However, since they require specialist knowledge and facilities, chemicals which are 'explosive' *per se* will only be considered briefly.

Basically an explosive comprises a mixture of solids, or of solids and liquids, which upon rapid and violent decomposition produces large volumes of gas, commonly hydrogen, nitrogen and its oxides, oxides of carbon, and water, e.g.

$$2C_3H_5(NO_3)_3 \rightarrow 3CO+2CO_2+6NO+4H_2O+H_2CO$$
(nitroglycerin)

The heat and gases produced exert sudden pressure on the surroundings; thus the greater the heat and gas produced per unit volume the more powerful the explosive. As a generalisation explosives yield *ca.* 1000 cm^3 of gas and 4.2 kJ per g.

> Over 2000 tonnes of TNT exploded at a UK munitions store in 1944. A nearby farm was completely destroyed and seven people killed. Within a radius of 10 km there was extensive damage to property and a covering of a layer of debris; there were a further 21 fatalities within this area, involving people in or near buildings.[17]

The power output of explosives depends on the rate at which energy is liberated and may vary by several orders of magnitude. 'Explosives' therefore are categorised as either 'low' or 'high' explosives depending upon the reaction rate.

- 'Low' explosives, e.g. nitrocellulose or black powder, are highly energetic and tend to burn intensely but do not normally detonate, i.e. they deflagrate at subsonic velocity up to 100 m/s. Thus although the burning rate of 'smokeless powder' made from nitrocellulose is extremely rapid, the gases generated as the explosive is consumed can serve to propel a projectile, e.g. from a gun barrel.
- Detonation (high) explosives can also deflagrate, but when properly stimulated a pressurised detonation wave is produced. The velocity of the detonation wave determines the brisance, i.e. the effect on the surroundings.

There is a wide variation in the susceptibility of high explosives to detonate. Thus those termed 'primary explosives' or 'initiating explosives' (e.g. lead azide or mercury fulminate), which are used in initiations or detonators, explode instantly upon ignition. They are also sensitive to other forms of initiation, e.g. mercury fulminate when dry may be initiated by flame, heat, impact, friction or intense radiation.[18]

Main charge explosives, e.g. TNT or dynamite (e.g. nitroglycerin adsorbed by diatomaceous earth), require either a strong mechanical shock or an initiator explosion in order to explode.

Specific definitions of 'Explosives' appear in legislation, e.g. in the UK under the Explosives Act 1875 as amended which covers:

- high explosives which detonate to produce shock waves of supersonic velocity (1–10 km/s). Materials which are easily detonated by mechanical or electrical stimuli are termed 'primary explosives'. The properties of selected primary explosives are given in Table 26.6 (after ref. 19). Those requiring an impinging shock wave to initiate them are termed 'secondary explosives' and are exemplified by aliphatic nitrate esters, nitramines and nitroaromatics.
- pyrotechnics which burn to produce heat, smoke, light and/or noise. Pyrotechnic substances evolve large quantities of heat but less gas than

Table 26.6 Properties of selected primary explosives

Property	Mercury fulminate	Lead azide	Silver azide	Normal lead styphnate	Diazodinitro-phenol	Tetrazene
Molecular weight	285	291	150	468	210	188
Colour	grey	white	white	tan	yellow	light yellow
Crystal density, g/cm^3	4.43	4.93	5.1	3.10	1.63	1.7
Crystal form	orthorhombic	orthorhombic monoclinic		cubic	tabular	
Melting point, °C	160, explodes		252	explodes	157	140–160, explodes
Hygroscopicity	0.1					0.8
Solubility in H$_2$O at 20 °C, g/100 g	−0.925					
Heat of formation, kJ/g	3.93	−1.45	−2.07	17.9	4.00	1.13
Heat of combustion, kJ/g	1.79	2.64	4.34	5.24	13.58	2.75
Heat of detonation, kJ/g	316	1.54	1.90	1.91	3.43	
Gas volume, cm^3/g at STP	29.8	308		368	876	
Activation energy, kJ/mol	10.8	172	146	259	230	
Collision constant, log$_{10}$/s	5.4	14.0		22		
Detonation rate, km/s at density, g/cm^3	4.2	6.1	6.8	5.2	6.9	
	0.50	4.8	5.1	2.9	1.60	
Specific heat, J/(g·K)		0.46	0.50	0.67		
Compressive strength, MPa	1837	1.4–21				
Thermal conductivity, W/(m·K)×10^{-4}	explodes	1256	837			
Vacuum stability at 100 °C, ml gas per g per 40 h at STP	<1	<1	<1	<1	<1	>5
Weight loss at 100 °C, %	<1	<1	<1	<1	<5	<5
Explosion temperature at 5 s, °C	190–260	345	290	265–280	195	160
Effect of prolonged storage	detonates at 80 °C	stable	stable	stable	stable	stable
Relative impact test value, % TNT	5	11	18	8	15	5

Friction pendulum	reacts	reacts	reacts		
Static discharge max. energy for non-ignition, J	0.07	0.01	0.007	0.001	0.25	0.036
Relative energy, output, % TNT						
lead block	50	40	45	40	110	50
ballistic mortar					95	
sand test	45	40		25	105	50
plate dent test		60				

propellants or explosives. The properties of selected pyrotechnics and explosives are compared in Table 26.7 (after ref. 19).

Table 26.7 Comparison between properties of pyrotechnic compositions and some explosives

Composition	%	Heat of reaction, kJ/g	Gas volume, cm^3/g	Relative brisance, % TNT	Ignition temperature, °C	Impact test, % TNT
Pyrotechnic						
Delay:						
barium chromate	90					
boron	10	2.010	13	0	450	12
Delay:						
barium chromate	60					
zirconium–nickel alloy	26					
potassium perchlorate	14	2.081	12	0	485	23
Flare:						
sodium nitrate	38					
magnesium	50					
laminac	5	6.134	74	17	640	19
Smoke:						
zinc	69					
potassium perchlorate	19					
hexachlorobenzene	12	2.579	62	17	475	15
Photoflash:						
barium nitrate	30					
aluminium	40					
potassium perchlorate	30	8.989	15	15	700	26
High explosive						
TNT		4.560	710	100	310	100
RDX		5.694	908	140	260	35

- propellants which burn to produce heat and gas as a means of pressurising pistons, starting engines, propelling projectiles and rockets.

In the UK explosives have been classified thus:

    Class 1 Gunpowder
         2 Nitrate mixture
         3 Nitro compounds
         4 Chlorate mixture
         5 Fulminate
         6 Ammunition
         7 Fireworks

For the purpose of safety distances in connection with the issue of licences for factories and magazines, explosives have been categorised thus:

Category	Explosives having
X	fire or slight explosion risks or both, with only local effect
Y	mass fire risk or moderate explosion risk but not mass explosion risk
Z	mass explosion risk with serious missile effect
ZZ	mass explosion risk with minor missile effect

Characteristics associated with propellant burning, explosive detonation and conventional fuels are compared in Table 26.8. Whether a compound

Table 26.8 Characteristics of burning and detonation

Characteristics	Burning		Explosive detonation
	Fuel	Propellant	
Typical material	coal–air	propellants	explosives
Linear reaction rate, m/s	10^{-6}	10^{-2}	$(2-9) \times 10^3$
Type of reactions	oxidation–reduction	oxidation–reduction	oxidation–reduction
Time for reaction completion, s	10^{-1}	10^{-3}	10^{-6}
Factor controlling reaction rate	heat transfer	heat transfer	shock transfer
Energy output, J/g	10^4	10^3	10^3
Power output, W/cm^2	10	10^3	10^9
Most common initiation mode	heat	hot particles and gases	high temperature–high pressure shock waves
Pressures developed, MPa	0.07–0.7	$(0.7-7) \times 10^2$	$7 \times 10^3 - 7 \times 10^4$
Uses	source of heat and electricity	controlled gas pressure, guns, and rockets	brisance, blast munitions, civil engineering

burns or detonates is influenced not only by chemical structure but also by the type and intensity of initiation, the degree of confinement and the physical and geometric characteristics of the substance. Thus many explosives that normally detonate may burn under controlled conditions with gentle ignition to avoid shockwave formation.

In the USA the Department of Transport classifies explosives in order of decreasing sensitivity as follows.

Class A	Materials which possess detonating or otherwise maximum hazard, e.g. dynamite, black powder, mercury fulminate, lead azide, desensitised nitroglycerin plus blasting caps, detonators, detonating primers and certain smokeless propellants.
Class B	Materials which have a high flammable hazard and which function by rapid combustion rather than detonation,

	e.g. special fireworks, photographic flash powders and most smokeless powder propellants.
Class C	Explosives, including manufactured articles, containing limited quantities of Class A or Class B explosive, e.g. fireworks, detonating cord or explosive rivets. Such materials will normally not detonate under fire conditions.
Blasting agents	Which are not cap sensitive so require a strong primer. Generally considered safer than Class A, B or C explosives.

Low explosives

Black powder

Black powder, comprising a mixture of approximately six parts of potassium nitrate, one part of charcoal and one part of sulphur, was the first gunpowder. It has been superseded as a propellant explosive but still finds commercial use in blasting fuses, sporting cartridges, fireworks and pyrotechnic preparations. Unfortunately its manufacture and use, both commercially and by amateurs, has resulted in numerous serious accidents since the first mentioned on page 1541.

> Two students in a chemistry class on the mixing of solids mixed, under supervision, potassium nitrate, sulphur and carbon. An explosion ensued whilst one student was holding the mortar, causing damage to his hand, burns to his face, and perforation of the ear-drums.[20]

In fact, despite having one of the lowest detonation rates of all explosives, and being much less sensitive to shock than most high explosives, as summarised in Table 26.9, black powder has proved extremely dangerous

Table 26.9 Properties of black powder

Property	Description
Sensitivity	Relatively insensitive to shock
	Relatively insensitive to friction
	Relatively insensitive to static electricity
Ignition	Ignites from spark or heat that reaches ignition temperature 282 °C from any source (spark, fire)
	Blasting accessories (blasting caps, detonating cord)
Velocity	Open burning, 5 seconds per foot
	In steel pipe (coarse granulations) 168 m/s
	In steel pipe (fine granulations) 620 m/s
Fumes produced	Carbon monoxide and hydrogen sulphide

to manufacture, store and use.[4] This results from its extremely low ignition temperature and rapid speed of burning. It is very sensitive to ignition from sparks, heat or friction and burns violently even when 'loose'; it generates volumes of black smoke on exploding.

The properties of black powder tend to be unpredictable since they vary with proportion and exact type of constituents (e.g. sodium nitrate may replace potassium nitrate and powdered coal may replace charcoal) and with particle size.

Smokeless powder

The majority of smokeless powders are now double-based propellants relying on nitroglycerin and nitrocellulose with added stabilisers. Because they do not normally detonate when exposed to heat, in a fire involving small-arms ammunition the shells explode individually. Moreover, since with an unconfined shell the gases are able to escape readily, any bullet will not be ejected at a velocity anywhere approaching that from a firearm. However, there is an additional hazard from missiles from fracturing cartridge cases. Thus protection from exploding, unconfined ammunition is provided by barriers, or for fire-fighters in protective clothing, by a minimum distance with a normal quantity of ammunition of 6 m.[4]

High explosives

Nitro-explosives

Common nitro-explosives and their explosive properties are summarised in Table 26.10. All of these have sufficiently high energy contents to produce disastrous explosions. However, the content, i.e. the potential explosive power, between compounds in the group is proportional to the percentage of NO_2 present in the molecules. This procedure is applicable to estimation of the relative explosive power of substances for which data are unavailable. The more important factors in selection of explosives may, however, be the tendency to detonate, i.e. the brisance already referred to, and the thermal and shock sensitivities. Thus, industrially the ability to handle an explosive safely tends to be more important than maximum explosive potential. Hence the long-established preference for TNT.

Nitric ester explosives

Common nitric ester explosives and their explosive properties are summarised in Table 26.11. The explosive powers of compounds in the group can be compared by reference to the percentage of ONO_2 present in the molecules. Moreover, comparison of the equivalent percentage of NO_2 present also facilitates comparison with the data in Table 26.11.

Clearly the nitric ester explosives have in general a higher energy content than the nitro-explosives. Thus, in terms of the TNT equivalent,

Table 26.10 Explosive properties of nitro-explosives

Name	Energy content, %NO$_2$	Brisance	Sensitivity
Nitromethane	75.2	moderate	low
Dinitromethane	89.2	high	high
Trinitromethane	91.6	high	high
Tetranitromethane	93.0	high	high
Nitroethane	61.5	moderate	low
Nitrobenzene	37.4	nil	very low
Dinitrobenzene	54.6	low	low
Trinitrobenzene	65.2	moderate	moderate
Dinitrotoluene	50.6	low	low
Trinitrotoluene	60.8	moderate	moderate
Trinitrophenol	60.6	high	high
Trinitroresorcinol	56.8	high	high
Trinitroaniline	61.0	high	high
Dinitrochlorobenzene	45.5	low	low
Trinitrochlorobenzene	56.3	high	high
Trinitrobenzoic acid	54.1	high	high
Trinitrocresol	57.2	high	high
Tetranitronaphthalene	60.0	low	low

Table 26.11 Explosive properties of nitric ester explosives

Name	Equivalent %NO$_2$	Brisance	Sensitivity
Methyl nitrate	108	high	high
Ethyl nitrate	91.5	high	high
Ethyleneglycol dinitrate (EGDN)	110	high	high
Nitroglycerin	111	high	high
Nitrocellulose	85.2	high	moderate
Ammonium nitrate	98.5	high	high

nitroglycerin	1.83
ethyleneglycol dinitrate (EGDN)	1.82
ammonium nitrate	1.62
nitrocellulose	1.40
trinitrotoluene (TNT)	1.00

It is clear from this why ammonium nitrate either in isolated storage or in an industrial process is, if present above a threshold quantity, regulated by the CIMAH Regulations (page 896) with the proviso for ammonium nitrate-based fertilisers of the presence of 28% weight N. The assessment of hazards leading to a detailed Safety Case for such materials is dealt with in ref. 21.

Nitroglycerin is manufactured by adding glycerol in a thin stream to a cold mixture of concentrated nitric and sulphuric acids as shown on page 523. It is a poisonous, oily liquid insoluble in water. It usually burns quietly when ignited but explodes violently when heated rapidly or struck or detonated. It is difficult to use in unmixed form and is often combined with other explosives to form the basis of dynamites and blasting gelatines.

Nitrocelluloses are made by immersing cellulose in nitrosulphuric acid mixture to provide varying degrees of nitration. They are divided into two groups according to their nitrogen content:
- those with nitrogen content >12.6%, e.g. highly nitrated gun cotton, used mainly in military applications.
- those with nitrogen content <12.6% which find use in paints, varnishes, lacquers.

Early forms of nitrocellulose were unstable and its rapid decomposition resulted in several plant explosions which led to abandonment of industrial manufacture until a more stable product was invented by washing the impure form to remove unstable substances. In the dry state, fibrous nitrocelluloses are highly sensitive explosives to both impact and initiation. They can detonate. Wetted versions are rendered much less sensitive. Gelatinised versions, even when dry, are less sensitive than their fibrous counterparts.

Fulminates, azides and styphnates are initiating explosives which are highly sensitive and dangerous. The duration of explosion is short and violent and they are used with other substances as initiators in the form of caps of detonators.

Nitrates provide the oxygen in the explosives mixtures, of which ammonium nitrate is the most common. Whilst not explosive in the pure state, they are explosive when mixed with selected chemicals and are then used for civil and military applications.

Chlorates are akin to nitrates in their ability to supply oxygen, and it is usually the sodium or potassium salts that are used. They are not used as blasting explosives but find limited application in cap and detonator formulations and are primarily used in fireworks.

About 90 per cent of blasting operations in the USA use non-nitroglycerin materials; the commonest agent is ANFO based upon a mixture of ammonium nitrate prills with fuel oil. Powdered aluminium is sometimes added to increase the general strength of the material.[22]

Effects of high explosives

Explosives and blasting agents are liable to produce a disastrous explosion when subjected to fire as exemplified on pages 880 and 881. Elimination of sources of fire is hence the only effective method for fire protection.[23] Considerations therefore follow those listed in Table 17.13.

As noted earlier the most devastating effect from a given quantity of

1498 Flammable and explosive hazards

high explosive is caused by detonation from which a shock wave is generated in the surrounding environment. The effect of the air blast can be roughly scaled by the cube root of the mass of the explosive charge.

The hazards with high explosives are exemplified by Table 26.12 which shows the explosive effects of small quantities of high explosives in a 6 m × 6 m single-storey building.[23]

Figure 26.1 relates the size of fireball to quantity of burning pyrotechnic, high explosive or propellant.[23] With pyrotechnics the hazard is related to

Table 26.12 Explosive effects of small quantities of high explosives in 6 m × 6 m room

1 g of explosive:
- any person holding the explosive could receive serious injury.

10 g of explosive:
- any person close to this quantity of explosive at the time of initiation would receive very serious injuries. 1% of persons at a distance of 1.5 m away are also liable to ear-drum rupture.

100 g of explosive:
- 50% of windows in room likely to be blown out.
- 1% ear-drum rupture at distance of 3.5 m.
- 50% ear-drum rupture at distance of 1.5 m.
- persons in very close proximity to explosion (e.g. holding the explosive) almost certainly killed.

500 g of explosive:
- complete structural collapse of brick-built building is most likely.
- steel or concrete framed building would probably survive.
- persons very close to blast almost certainly killed.
- persons close to blast will be seriously injured by lung and hearing damage, fragmentation effects, and from being thrown bodily.
- almost all persons within the room will sustain perforated ear-drums.

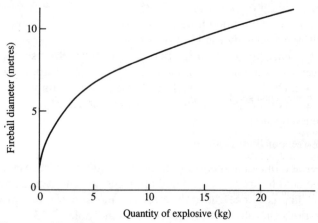

Fig. 26.1 Diameter of fireball versus quantity of explosive

the violence with which the chemical burns. One scheme used to classify pyrotechnics is given in Table 26.13. This is used to restrict quantities in use/storage and for selection of the appropriate safety precautions. Propellant hazards are akin to those for pyrotechnics except that confinement can lead to detonation.

Table 26.13 Classification of pyrotechnics

GROUP 1
Compositions (burn very violently):
    Chlorate and metal perchlorate report or whistling compositions.
    Dry non-gelatinised cellulose nitrates.
    Barium peroxide/zirconium compositions.

Articles (mass explosion risk):
    Flash shells (maroons).
    Casings containing flash compositions.
    Sealed hail preventing rockets.

GROUP 2
Compositions (burn violently):
    Nitrate/metal/sulphur compositions.
    Compositions with >65% chlorate.
    Black powder.
    Nitrate/boron compositions.

Articles (accelerating single-item explosions):
    Large firework shells.
    Fuse unprotected signal flares.
    Non-pressed report bullets (bird scarers).
    Report cartridges, unpacked.
    Black matches, uncovered.

GROUP 3
Compositions (burn fast):
    Nitrate/Metal compositions without sulphur.
    Compositions with up to 35–65% chlorate.
    Compositions with black powder.
    Lead oxide/silicon with >60% lead oxides.
    Perchlorate/metal compositions other than report.

Articles (burn very violently with single-item explosions):
    Large firework shells.
    Fuse protected signal flares.
    Pressed report cartridges in primary packagings.
    Quickmatches in transport packagings.
    Waterfalls; Silver wheels; Volcanoes.
    Black powder delays.

GROUP 4
Compositions (low/medium speed burning):
    Coloured smoke compositions.
    White smoke compositions (except those in Group 5)
    Compositions with <35% chlorate.
    Thermite compositions.
    Aluminium/phosphorous pesticide compositions.

Articles (single-item ignitions/explosions):
    Large firework shells without flash compositions in transport packagings.
    Signal ammunition without flash compositions, up to 40 g of composition.
    Small fireworks, fuse protected (except volcanoes and silver wheels).

GROUP 5
Compositions (burn slowly):
    Slow burning heating compositions.
    White smoke compositions based on hexachloroethane with zinc, zinc oxide and <5% of aluminium, or <10% of calcium silicon.

Articles (slow single-item ignitions/explosions):
    Small fireworks in primary packagings.
    Signal ammunition in transport packaging.
    Delays without black powder.
    Coloured smoke devices.
    Sealed table bombs.
    White smoke devices unpacked (see Group 5 composition).

1500 Flammable and explosive hazards

If a fire does occur involving explosives, once the fire has reached them fire-fighting should be abandoned and the area evacuated, e.g. to 600 m. Remote control equipment can of course be left in position.

In the ammonium nitrate explosion referred to on pages 881 and 1554 fragments of the ship were blown 1500 to 2000 m in the air and some landed 2 km away. Oilwell drillings rods weighing some 0.5 tonnes were projected about 3 km and embedded in the ground on impact.[24]

Precautions with explosives

The sensitivity of some explosives, and the devastating effects which may result from the explosion of relatively small quantities, necessitates that expert advice is sought *before* they are brought onto that site. In developed countries extensive legislation regulates the purchase, storage and use of explosives, e.g. the licensing of premises, import/export, labelling, storage and handling, and disposal. Extensive advice relevant to laboratory, pilot plant and production operations is provided in ref. 25; manufacture, transport, storage and use are covered by ref. 26. Selected precautions for work on a laboratory scale are described on page 607 *et seq*. General considerations are given in Table 26.14 and exemplified by the following.

Nitrocellulose

During the manufacture of nitrocellulose the explosion risk is low. However, explosive decomposition is possible when the product is spun dry. Care is needed to ensure that the nitrocellulose still contains a high level of acid liquor and in controlling the centrifugal force of the dryers used. It is important when using material in flock form that the substance always contains the same percentage of water or alcohol as when it left the manufacturer, i.e. prevent drying out on storage. Dust can form and it is advisable for operatives to wear headgear to prevent dust entering their hair; it is also advised that floors are kept wet. Unwanted/soiled material should be stored in water-filled, air-tight containers awaiting disposal by, e.g. burning off as a thin layer on wood chips, or by denitrating with warm ammonium sulphide solution.

Precautions for mixing test quantities of pyrotechnics include:
- use only small quantities (1–2 g) of novel mixtures
- eye protection is mandatory
- pre-grind the components individually to fine particle size; never grind oxidiser and fuel together
- mix no more than 2 g in a mortar with a soft brush
- use a fire-proof board on which to conduct burning tests
- ensure flammable materials are kept away from the working area

Table 26.14 General considerations for work with explosive chemicals

- Consult with experts on the hazards and on technical, administrative, and legal requirements.
- The chance of accidental initiation is related to the energy imparted to the substance and the sensitivity of the compound. Hence the sensitivity of compounds should be established (e.g. pages 621–2) prior to devising appropriate control measures. (Many sensitive explosives have ignition energies of 1–45 mJ whilst some very sensitive materials have ignition energies <1 mJ.)
- Consider the advantage of continuous vs batch operations.
- Consider intraline distances to minimise explosion propagation.
- Depending on scale, specially designed facilities may be required remote from other buildings, accessways or populated areas. Remote handling procedures may be required possibly with closed-circuit TV monitoring utilising concrete outbuildings or bunkers. Access to hazardous areas must be rigidly controlled.
- Consider fire protection, detection and suppression requirement, and means of escape, alarms, etc.
- Minimise stocks and segregate from other chemicals and work areas. Where appropriate keep samples dilute, or damp, and avoid formation of large crystals when practicable. Add stabilisers if possible, e.g. to vinyl monomers. Store in specially designed, well-labelled containers in 'no smoking' areas, preferably in several small containers rather than one large container. Where relevant store in dark and under chilled conditions, unless this causes pure material to separate from stabiliser (page 508).
- Do not decant in store.
- Consider need for high/low temperature alarms for refrigerated storage; these should be inspected and tested regularly.
- Consider need for mitigatory measures (fire, blast, fragment-resistant barricades/screens), electrical and electrostatic safeguards, personal protection, disposal, etc.
- Stores and work areas should be designated 'No Smoking' areas and access controlled.
- Depending upon scale, explosion-proof electrics may be required and static eliminated. Grounded equipment is essential with black powder. Non-sparking rods must be used to open containers. Operators must wear anti-static boots.[2]
- Ensure cleanliness throughout, remembering that some raw materials or products are also toxic.
- Deal with spillages immediately and make provision for first-aid.
- All staff should be fully trained and written procedures provided.

- provide supervision by a competent and experienced person
- use remote working arrangements for certain mixtures

Lead azide

This material, which is used as a primary military explosive, is prepared by reaction between solutions of lead nitrate and insensitive, but toxic, sodium azide:

$$Pb(NO_3)_2 + 2NaN_3 \rightarrow 2NaNO_3 + Pb(N_3)_2$$

It is stable at ambient temperatures but crystalline material is sensitive to impact, particularly if contaminated with small quantities of foreign matter. Detonation velocity builds up rapidly on ignition and even single crystals will detonate. The substance hydrolyses at high humidities to form hydrazoic acid which reacts with copper, lead, cadmium, silver, mercury or their alloys to produce very sensitive metal azide.

Precautions therefore include preparation of small batches (e.g. 5 kg) by remote control in double-walled stainless-steel vessels using distilled or deionised water. Nucleating agents are used to give free-flowing crystals and to prevent spontaneous explosion during precipitation. Attention to providing scrupulously clean conditions and avoidance of dusting are essential during detonator loading.

Nitroglycerin

Nitroglycerin (glyceryl trinitrate) is used mainly as an explosive in dynamites and as a plasticiser for nitrocellulose. It is very susceptible to shock, friction and impact. Because of the hazards of handling bulk quantities, batch processing, for which careful control of temperature is essential (page 524), has been superseded by a continuous, totally automated process using emulsion technology in highly polished stainless steel plant equipped with detonation traps and fail-safe features, controls and alarms and monitored by TV from a remote bunker.

> In June 1988 an explosion in an explosives plant killed two operators in the building and damaged other buildings. Fires were started by debris; one led to a smaller secondary explosion about 15 minutes later which destroyed another building.[27]
>
> The primary explosion was where nitroglycerin was mixed batchwise with other ingredients to make a final explosive for cartridges. The most likely cause was contamination of the explosive in the operating mixers by grit or similar hard materials.
>
> The two operators were actually in the area of the mixing components, and both mixers were in operation, when the explosion occurred. This was contrary to an explicit operating instruction; the operators should have been in the control room blockhouse or similar safe location. Detailed changes were made subsequently. New interlocks and other equipment were fitted to render it much more difficult to operate the plant with any person present in the mixing area. The supervisor's duties were strengthened and clarified, and his responsibility to ensure safe working practices were made clear to all.

Attention to effluent quality is important since nitroglycerin is soluble in water (*ca.* 0.2 g/100 g H_2O at 20 °C) and pH needs control. The product, if desensitised with, e.g. dibutyl phthalate or absorbent solids, can be transported safely.

Detonator loading plants

Because of the sensitivity of some primary explosives to electrostatic energy, precautions during detonator loading include preferably loading at site of manufacture (or transported wetted with water or alcohol suspended in rubberised cloth bags cushioned with sawdust or other inert media), atmospheric humidification, grounding of equipment, and operators' use of antistatic clothing.

Disposal

Disposal of explosive waste and the repair or dismantling of contaminated plant needs extreme care. Table 26.15 provides guidance on techniques for disposal of the more commonly encountered explosives by experts at proper disposal sites.[29] Collection of the waste should be in well-labelled,

Table 26.15 Disposal techniques for explosives

Explosive	Disposal method				
	Burn	Detone	Dissolve	Chemical	Drowning
NG-based blasting explosives	/	/	x	x	/
ANFO	/	/	/	x	/
Pyrotechnic compositions	/	x	/	x	/
Blackpowder	/	x	/	x	/
Detonators	/	/	x	x	x
Detonating cord	/	/	x	x	x
Nitroglycerin	/	x	x	/	x
Slurry explosives	/	/	/	x	/
Contaminated paper waste	/	x	x	x	x
Fireworks (finished)	/	x	x	x	x
Initiatory explosives	x	/	x	/	x
Propellants	/	x	x	x	x
Non NG-based blasting explosives	/	/	x	x	/
Shotgun cartridges	/	x	x	x	x
Sporting/small arms ammunition	/	x	x	x	x

NG = nitroglycerin.

distinctive, specially designed containers. Cleaning and decontamination of plant comprises removal of gross contamination under wet or solvent conditions using tools made of soft material, final cleaning with solvent or chemical reagent, and finally 'proving' of the equipment by heating to temperatures exceeding those for decomposition of contaminant. Repair work should be the subject of a permit-to-work system (page 975); it should be assumed that explosives may have penetrated threads, joints and other crevices, bolts, flanges, etc. These should be thoroughly decontaminated prior to dismantling. Operatives should be suitably protected.

Transport

Within the UK the standards for the packaging of civilian explosives for transportation are specified in the Classification and Labelling of Explosives Regulations 1983; essentially similar standards for military explosives are controlled by an Explosives Storage and Transport Committee.

> A van carrying 800 kg of nitroglycerin-based commercial explosives, together with a number of detonators and fuseheads, caused a massive explosion in a busy industrial estate in Peterborough, UK.[29] A fire brigade officer was killed and 80 other people injured; 20 buildings and 100 vehicles were damaged. The blast left a crater 0.6 m deep and 3.1 m in diameter. The vehicle was not labelled to indicate to the emergency services what was being carried and the initial '999' emergency call is said to have warned the firemen only that there was a 'tanker alight'.
>
> Subsequent investigation revealed that fusehead combs in a box were ignited by the jolting of the vehicle when it passed over a speed ramp; the fire spread to the rest of the load which detonated 12 minutes later. It was found that
> - the fuseheads had been carried in unauthorised and unsafe packages;
> - the method of packaging was unsuitable – leading to a build-up of loose explosive material in the rusty metal boxes thereby increasing the risk of ignition;
> - the quantity in each package was excessive.[30]

Subsequent to this incident the UK Health and Safety Executive made the following recommendations regarding the transport of explosives.
- Operators should develop systems to ensure that all activities involved in transporting explosives are safe and that they meet the requirements of the relevant regulations. These should include proper classification and labelling and correct packing of explosives.
- Safety and quality systems should apply not only to new or modified products but should also include a periodic review of established ones.
- Transport crews should be kept informed about changes in circumstances such as new products among their load and their training should include measures needed when carrying specific products and mixed loads.
- All transport of explosives should be pre-planned, including measures to ensure that explosives in different compatibility groups are prevented from coming into contact with each other if their packaging is degraded, e.g. through fire or minor explosion.[31]

The Road Traffic (Carriage of Explosives) Regulations 1989 cover constructional and operational matters concerned with road transport of explosives, limit the quantities of certain groups of explosives that may be transported, and clarify how vehicles must be labelled.

Within the UK the security of explosives is regulated by the Control of Explosives Regulations 1991. The explosives covered include blasting explosives, detonators, fuses, ammunition, propellants, pyrotechnics and fireworks. Some controls are also placed on 'restricted substances', *viz.* materials which are not generally known as explosives (for the purposes of storage, transport, etc.), but which could be used as such, e.g. explosives that have been diluted or multi-part explosives.

Under the provisions of the Dangerous Substances in Harbour Areas Regulations 1987, persons wishing to carry, handle, load or otherwise bring explosives into harbour areas require an explosives licence from the HSE. The terms of such a licence will include the type and quantity of explosives that may be handled, the areas in which they may be handled, and in some cases the times at which loading/unloading is permissible.

References

1. Conkling, J. A., *Chemistry of Pyrotechnics*, Marcel Dekker, NY, 1985.
2. Meddard, L. A., *Accidental Explosions*, Ellis Horwood, Chichester, 1989.
3. Home Office, *Technical Bulletin 1/1992: Explosives Guide*. HMSO, 1992.
4. Meidl, J. H., *Explosive and Toxic Hazardous Materials*, Glencoe Press, 1970.
5. Bond, J., *Loss Prevention Bulletin*, Institution of Chemical Engineers, October 1991(101), 3.
6. Gaston, P. J., *The Care, Handling and Disposal of Dangerous Chemicals*, The Institute of Science and Technology, Northern Publishers (Aberdeen) Ltd., 1964.
7. Marshall, V. C., *Major Chemical Hazards*, Ellis Horwood, Chichester, 1987.
8. Kuhn, L. P., *Chem. Eng. News*, 1948, **26**, 3197.
9. NFPA Code 43B, *Code for Storage of Organic Peroxide Formulations*, National Fire Protection Association, Boston, MA, 1986.
10. McCloskey, C. M., *Plant/Operations Progress*, 1989, **8**(4), 185.
11. Lenoir, E. M. & Davenport, J. A., *Process Safety Progress*, 1993, **12**(1), 12.
12. Levy, L. B., *Process Safety Progress*, 1993, **12**(1), 47.
13. Nicholson, A., *Plant/Operations Progress*, 1991, **10**(3), 171.
14. June, R. K. & Dye, R. F., *Plant Operations Progress*, 1990, **9**(2), 67. Britton, L. G., *ibid.*, 75. Curtis, J. S., *ibid.*, 91. Brockwell, J. L., *ibid.*, 96.
15. Viera, G. A., Simpson, L. L. & Ream, B. C., *Chem. Eng. Progress*, 1993 (August), 66.
16. Walker, J., Studio Vista, 1973.
17. Withers, J., *Industrial Major Hazards, Their Appraisal and Control*, Gower Technical Press, 1988, 24.
18. Bretherick, L., *Handbook of Reactive Chemical Hazards*, 4th edn, Butterworth, 1990.
19. Linder, V., in *Kirk-Othmer Encyclopaedia of Chemical Technology*, ed. M. Grayson, 3rd edn, 1980, **9**, 561, John Wiley, New York.
20. Anon., *Chemistry in Britain*, July 1993, 559.
21. Fertiliser Manufacturers Association, Industrial guidance on the safety case: 2, ammonium nitrate, in *Safety Cases*, eds. F. P. Lees and M. L. Ang, Butterworth, 1989, 155.
22. Porter, S. J., Explosives and blasting agents, in *Fire Protection Handbook*, 16th edn, NFPA, Boston, MA, 1986, 7.
23. HSE, Specialist Inspector Report No. 31, *Safe Handling Requirements during Explosive, Propellant and Pyrotechnic Manufacture*, HSE, 1991.
24. Baker, W. E., Cox, P. A., Westire, P. S., Kuesz, J. J. & Strehlow, N. R. A., *Explosion Hazards and Evaluation, Fundamental Studies in Engineering*, Elsevier, Amsterdam, 1983.

25. *Explosives, Propellants and Pyrotechnic Safety Covering Laboratory, Pilot Plants and Production Operations*, Manual AD–272–424, US Naval Ordinance Laboratory, 1962, Washington, DC.
26. NFPA Code 495, *Code for the Manufacture, Transportation, Storage and Use of Explosive Materials*, National Fire Protection Association, Boston, MA.
27. Anon., *Loss Prevention Bulletin*, Institution of Chemical Engineers, April 1991(098), 18.
28. HSE, HSG 36, *Disposal of Explosive Waste and the Decontamination of Explosives Plant*, HMSO, 1987.
29. Anon., *Hazardous Cargo Bulletin*, May 1989, 71.
30. Anon., *Health & Safety at Work*, December 1990, 5.
31. Health & Safety Executive, *The Peterborough Explosion*, 1990.

CHAPTER 27

Fire protection practice

A brief summary of aspects of fire protection practice is provided in Chapter 4. Examples have been quoted throughout the text of fires which escalated, resulting in substantial losses, because of failures to follow elementary principles. Moreover in major fires secondary or domino effects may be serious.

> In January 1980 Essex Fire Brigade was called to a fire at the premises of a chemical company. The fire threatened a building containing 2 tonnes of sodium chlorate and 1 tonne of sodium cyanide. Because of the possible explosion and toxic hazards, which in the event did not ensue, some 3000 local residents were evacuated for one night.

Precautions and mitigatory measures are considered in greater depth in this chapter; in particular detailed references are provided to assist the solution of specific practical problems.

Fire prevention

The first principle of fire protection is fire prevention. This can be achieved by the control of combustibles, atmospheres and of ignition sources (page 159).

Control of combustibles

As mentioned on page 1, a fire broke out in wickerware stored with ammonium nitrate and compound fertilisers and potassium nitrate in a warehouse. The fire spread rapidly through the ground floor section of the building fanned by a north-west wind and fuelled by the stored combustibles, e.g. wickerware and charcoal, and enhanced by the powerful oxidising agents. Within an hour a series of deflagrations occurred showering the neighbourhood with debris and causing firemen to withdraw; these explosions probably resulted from a reaction between the potassium nitrate and nearby stacks of charcoal. The fire was not brought under control until 6 hours after it started.

> Between 750 and 1000 people were evacuated from an area south-east of the

warehouse because dense clouds of smoke were blown in that direction; the fear was that there might be a significant concentration of oxides of nitrogen from the burning ammonium nitrate fertilisers. (In the event measured levels were below those which might be expected to create a public danger.)[1]

Control of flammable atmospheres

Atmosphere control involves maintaining conditions such that a gas–air mixture within the normal flammable limits is avoided. The principles are described on page 131 *et seq*.

If control depends upon maintaining the temperature of a liquid reliably below its flash point then the requirements are those listed on page 123. One important consideration is the tendency for mist or spray formation, which may be ignited at temperatures below the flash point. Such liquid–air dispersions may, for example, be produced from leaks at flanges, by rapid, turbulent filling, or jetting, or agitation. Any source of ignition may vaporise the mist droplets in its vicinity and ignite that vapour; the flame can then propagate by successive vaporisation and burning of droplets through the mass of mist.

> The flash point of kerosene (page 124) is about 38 °C but a kerosene mist can be ignited at below −17.8 °C.

In such a situation, the heat generated above the liquid surface by combustion of the droplets may so increase the rate of vaporisation as to bring its vapour within the flammable range.

If control over a flammable atmosphere in a vessel, equipment or pipework relies upon 'inerting' then essential requirements are as follows:
- continuity of supply (possibly requiring a standby supply or reservoir);
- monitoring the quality of the inert gas supply, particularly if it is from an inert gas generator but also to avoid any confusion if it is from cylinders;
- monitoring the pressure within the protected enclosure together with some form of low-pressure trip;
- monitoring the atmosphere within the protected enclosure with, depending on the circumstances, appropriate alarm and automatic shutdown provisions.

> A flare system, comprising a knockout vessel, a water seal and a flare stack, was decommissioned (to insert a blind in the flare header) by first isolating the header and purging with nitrogen until the flame was extinguished. The nitrogen supply was then cut off, steam introduced, and the water seal vessel drained. Since it became apparent that hydrocarbon residues remained in the vessels, nitrogen was reintroduced. As the fitters separated the flanges of the header there was a rumble in the pipework and soot was blown out of the gap between the flanges. As they tried to bolt up the blind a second, more severe, explosion occurred which killed one fitter.
>
> Some 40 tons/day of nitrogen was actually required to replace steam of 13

tons/day that had been introduced for 16 hours. Only 10 tons/day of nitrogen was available and, as a consequence, air had entered the system as the flanges were broken to create a flammable hydrocarbon/air mixture which was ignited by the pilot burner at the flare tip.[2]

A control unit containing sparking electrical equipment was made safe by nitrogen pressurisation and locating in a Zone 2 area. The arrangement is shown in Fig. 27.1[3] If the nitrogen pressure fell the pressure switch isolated the electrical supply.

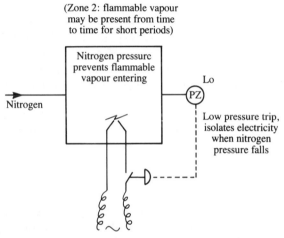

Fig. 27.1 System for explosion prevention with sparking electrical equipment in a Zone 2 area

An explosion occurred because the trip was bypassed, so that when the nitrogen was contaminated with flammable vapour a fall in pressure allowed ingress of air; the flammable mixture was ignited when the supply was switched on. (In this situation use of compressed air was subsequently recommended since the supply was more reliable and it was less likely to be contaminated.)

An explosimeter or combustible gas detector is normally used to monitor for the presence of flammable vapour (although for specialised applications other methods, e.g. oxygen analysis, are also used). An appreciation of the correct method of use, and the limitations of, this instrument is essential to avoid false readings or misinterpretation of the results.[4,5] In summary, the problems are as follows (after ref. 6).

- *False calibration*: Explosimeters are often calibrated using a known mixture of gases, e.g. methane and air. Use with other vapour–air mixtures may require correction factors.
- *Insufficient oxygen*: Explosimeters generally operate by oxidation of the combustible gas on a catalyst surface; the resultant heat output is then related to the flammable gas concentration. Therefore with high gas levels the instrument may give a low reading because of the absence of

sufficient oxygen; dilution with air may subsequently bring the atmosphere into the flammable range. (Similar problems arise with the failure of explosimeters when inert gas is used, so specialised measuring apparatus may be required.)
- *Poisoning*: The catalytic element can be poisoned, e.g. by silicones, chlorinated hydrocarbons and organic lead compounds. The instrument then 'fails to danger'. Thus calibration of the meter before and after use is the requisite safeguard.
- *High temperature*: If calibrated within temperature limits close to ambient, then use outside these limits, e.g. in steamy atmospheres, may cause false readings. This may also occur if a long probe is used inside a hot tank because condensation of vapour may take place before the explosimeter.
- *Stratification*: The atmosphere in large volumes may become stratified (page 95). Careful monitoring at several points, using extended probes, may be necessary.

Despite accurately indicating that flammable atmospheres are absent, subsequent heating of 'empty' vessels can result in creation of flammable atmospheres by thermal breakdown of involatile deposits.

Clearly, therefore, thorough training is required for personnel using an explosimeter.

Prevention of leakage of flammable liquids, vapours or gases into the open air is obviously another facet of atmosphere control. Equipment design features, and measures to maintain its integrity, are discussed in Chapters 12 and 24.

> The first single-compartment LNG storage tank failed in Cleveland in 1944, just one year after it was built. There was little provision to contain liquid spills, so LNG flowed into nearby streets, buildings and sewers; it eventually ignited some distance away causing extensive damage.[7] (The construction material was a 3.5 per cent nickel alloy which is not, by today's standards, acceptable for LNG temperatures.)

The need for secondary containment is established with nuclear reactors and other situations where leakage can have particularly serious consequences. For example, for the fully-refrigerated storage of LNG it has become normal to use a double containment tank of the type illustrated in Fig. 27.2. The primary liquid containment is surrounded with a secondary shell, separated from it by a gap of up to 6 m, which is capable of holding liquid but is not designed to contain vapour released by product leaking from the inner shell. (There are also full containment tanks in which the outer shell is designed to hold vapour as well as liquid; in this case the gap between the two shells is 1 to 2 m.) The material of the inner tank is selected for the product and the relevant design code; the outer shell is generally of steel, prestressed concrete or reinforced concrete with an

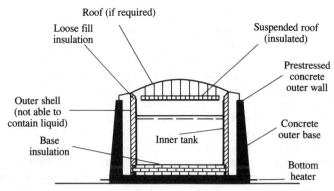

Fig. 27.2 Double-containment refrigerated tank

earthen embankment to protect it against the liquid impact which would follow a failure of the inner tank.[7]

Irrespective of the quality of the design, however, as the examples on pages 100, 147 and 155 illustrate, well-maintained systems of work are essential for fire prevention.

> An explosion at a high pressure polyethylene plant in Texas in October 1989 resulted in 23 fatalities and injuries to a further 130 people.[8] It was caused when nearly 40 tonnes of gaseous ethylene leaked from an open pipe and, after less than 2 minutes, exploded. Two isobutane tanks exploded 15 minutes later.
>
> The pipe had been opened up to clear a choke but, unfortunately, a compressed air-operated isolation valve was left open and there was no back-up protection, i.e. a double valve or blind flange insert (Table 17.7, page 978).

Hence the importance of the various systems summarised on page 967 *et seq*.

> In 1988, 167 men died when the *Piper Alpha* oil production platform was engulfed by a series of fires and explosions. A massive explosion had been triggered by a leak of light oil following steps taken by night-shift personnel to restart a pump which had been shut down for maintenance. Unknown to them, a pressure relief valve had been removed from the relief line of this pump and a blank flange which had been fitted was not leak-tight.
>
> Lack of awareness concerning removal of the valve stemmed from a breakdown in communications and a failure in the permit-to-work systems.[9]

The need for full procedures has been proved on numerous occasions.

> The incident summarised on pages 1 and 1507 was probably caused by conducted heat or a spark from welding, igniting paper wrapping on cane chairs. This welding was being performed on metal window frames immediately outside the building. The welders questioned the fact that stacks of cardboard boxes could be seen just inside the windows but were apparently led to understand that these could not be affected by heat from the welding. No other instructions were given to them about fire precautions or fire-fighting and no special comment was made regarding the presence of nitrate fertilisers.[1]

A fire which eventually involved ten out of twelve tanks in a tank farm burned for seven days despite strenuous efforts at fire-fighting using large quantities of foam. The fire was started by flame-cutting operations between two of the tanks; flames flashed across fuel spills and grass, passed through an uncovered channel in a bund wall and reached the bund containing two tanks; this bund also apparently contained spilled fuel. Fanned by the wind and assisted by a tank 'boil-over' the fire then spread inexorably. (Apart from inadequate control of flame-cutting work, it was subsequently concluded that the tank spacing was inadequate, the quantities of fire-extinguishing agents at hand were insufficient, fire engines encountered a lack of access routes to get near the source of the fire and the local fire brigade was insufficiently experienced or organised to combat a fire of this magnitude.[10] Lessons from the mechanisms of fire spread have also been reviewed.)[11]

Control of ignition sources

The wide range of ignition sources arising in industrial situations is discussed on pages 141–57 and summarised in Table 27.1. They are the subject of a recent in-depth review by Bond.[12] In addition to the obvious flames, hot surfaces, electrical sparks, etc. they include the less common sources, e.g. electrostatic discharges, electrical system earth faults and stray earth currents, surge phenomena and lightning, thermite reactions and friction sparking. Procedures and practices for their control are discussed here in more detail.

General

Table 27.2[13] indicates the temperatures associated with selected ignition sources. To inflame, the temperature of a flammable gas/vapour–air mixture must be raised above the auto-ignition temperature (AIT), examples of which are included in Table 24.1 with additional data for petroleum fractions in Tables 4.1 and 4.2. However, such values are not absolute but depend upon the temperature of surfaces in contact with the mixture, the contact time, whether the surface is active or inert, whether contaminants enhance or inhibit combustion, etc.[14] Furthermore AITs vary according to the method of determination and can be scale dependent. Experience has shown that minimum ignition temperatures in industrial process vessels are likely to be significantly lower than values quoted from small-scale laboratory determinations. Also, ignition delay times will increase with increasing vessel size.[15] The interrelationship between flammable limits, flash point and AIT is illustrated by Fig. 27.3.[16]

In general AIT values are of the order of several hundreds of degrees C, although a few chemicals possess low values (e.g. AIT for carbon disulphide is 100 °C (reflecting its very low MIE) and with a vapour pressure of 360 mm Hg at room temperature and a boiling point of only 46 °C vapours can easily reach the flammable range of 1.3–44 per cent in air and be

Table 27.1 Sources of ignition

Mechanical sources	
Friction	Metal to metal
	Metal to stone
	Rotary impact
	Abrasive wheel
	Buffing disc
	Tools, drill
	Boot studs
	Bearings
	Misaligned machine parts
	Broken machine parts
	Choking or jamming of material
	Poor adjustment of power drives
	Poor adjustment of conveyors
Missiles	Hot missiles
	Missile friction
Metal fracture	Cracking of metal
Thermal sources	
Hot surface	Hot spot
	Catalyst hot spot
	Sparks from incinerators, flarestacks, chimneys
	Car exhaust
	Steam pipes
	Refractory lining
	Foreign metal in crushing and grinding equipment
	Electrical heater
	Smoking
	Drying equipment
	Molten metal or glass
	Heat transfer salt
	Hot oil/salt transfer lines
	Boiler ducts or flues
	Electric lamps
	Hot process equipment
	Welding metal
Self-heating	Oxidation
	Reaction
	Activated carbon
Flames	Pilot light
	Cutting, welding
	Burners, arson
Compression	Pressure change
	Piston
Engines	Exhaust
	Engine overrun
	Hydraulic spray into engine air intake
Diffusion	High pressure change
Chemical sources	
Peroxides	Oxygen release
	Unstable
	Decomposition
Polymerisation	Exothermic reaction
	Catalyst
	Lack of inhibitor
	Crystallisation

1514 *Flammable and explosive hazards*

Table 27.1 (continued)

Spontaneous	Pyrophoric metals
	Deposits
	Water reaction
	Sulphides
	Oily rags
	Heat transfer
Reactions with other substances:	
Thermite reaction	Rust
	Aluminium
Unstable substances	Acetylides
Decomposition	Initiator
	Temperature
	Catalyst
Electrical sources	
Electrical current	Switch gear
	Cable break
	Vehicle starter
	Broken light
	Electric motor
Electrostatic	Liquid velocity
	Surface charge
	Personal charge
	Rubbing of plastic
	Liquid spray
	Mist
	Water jetting
	Powder flow on plastic
	Water settling
Lightning	Direct strike
	Hot spot
	Induced voltage
Stray currents	Railway lines
	Cable break
Radio frequency	Aerial connection
	Intermittent contact

ignited by hot steam-pipes). On heating some materials explode without the application of a flame (Table 27.3).

Flame cutting, welding, etc.

Dangerous sparks, i.e. globules of molten, burning metal or slag, are generated in welding and cutting operations. Sparks from cutting, especially oxy-fuel gas cutting, tend to be more hazardous because there are more of them and they travel greater distances. The example quoted on page 142 demonstrates that it is necessary to isolate, or protect, combustible materials in the work area. For example, sparks may fall through cracks or other openings in floors and partitions or be carried beyond platforms due to their trajectory, or bouncing.

Following air quality tests inside, and near to, a pipeline which gave negative

Table 27.2 Approximate temperatures of common ignition sources

Source	°C
Flames or sparks	
Candles	640–940
Matches	870
Manufactured gas	900–1340
Propane	2000
Light bulb element	2483
Methane	3042
Electrical short circuit, or arc	3870
Non-flame sources	
Steam pipes (normal pressure)	100
Steam pipes at 10 lb/in^2 (69 kN/m^2)	115
Light bulb, normal	120
Steam pipes at 15 lb/in^2 (103 kN/m^2)	121
30 lb/in^2 (206 kN/m^2)	135
50 lb/in^2 (345 kN/m^2)	148
75 lb/in^2 (517 kN/m^2)	160
100 lb/in^2 (690 kN/m^2)	170
150 lb/in^2 (1035 kN/m^2)	185
200 lb/in^2 (1380 kN/m^2)	198
300 lb/in^2 (2070 kN/m^2)	217
500 lb/in^2 (3450 kN/m^2)	243
1000 lb/in^2 (6900 kN/m^2)	285
Cigarette, normal	299
Soldering, iron	315–432
Cigarette, insulated	510
Light bulb, insulated	515

results a work permit was issued for welding. The line was approximately 6 m above the ground and a hot piece of slag bounced and fell onto a sump 2.5 m to the side; the sump cover was loose and oil inside caught fire.[17]

The minimum requirements for combustibles control are:[18,19]
- move all combustibles a safe distance of at least 10.6 m horizontally away and ensure there are no openings in walls or floors within this distance; *or*
- move the work to a safe location; *or*
- if neither of these is possible, protect the exposed combustibles with suitable fire-resistant guards (e.g. flameproofed covers, or metal guards, or curtains) and arrange for a trained fire-watcher to be readily available with extinguishing equipment.

However, properly trained and instructed workers are essential since such procedures are of no avail if the operation being attempted, or the equipment, or its manner of use is unsafe.

Repair work or demolition of tanks or vessels, of various sizes, which have held flammable materials is, as exemplified on pages 87 and 142, an activity requiring careful planning and supervision. The recommended

Table 27.3 Approximate temperatures at which selected substances will explode, without the application of a flame[2]

Substance	Temperature (°C)*	
	Lower	Higher
Solids		
Gun cotton (loose)	137	139
Cellulose dynamite	169	230
Blasting gelatine (with camphor)	174	
Mercury fulminate	175	
Gun cotton (compressed)	186	201
Dynamite	197	200
Blasting gelatine	203	209
Nitroglycerin	257	
Gunpowder	270	300
Gases		
Propylene	497	511
Acetylene	500	515
Propane	545	548
Hydrogen	555	
Ethylene	577	599
Ethane	605	622
Carbon monoxide	636	814
Manufactured gas	647	649
Methane	656	678

* The higher temperatures are applicable when the heat rise is very rapid, i.e. if the rate of rise is slow then the explosion will occur at the lower temperatures. The figures quoted are the temperatures at which the substance itself explodes, not the temperatures at which its container ruptures with the possible subsequent ignition of the contents.

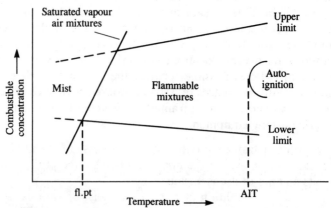

Fig. 27.3 Effect of temperature on limits of flammability of a combustible vapour in air at a constant initial pressure

procedure is exemplified in Fig. 27.4. In general 'hot work' should be controlled by a permit-to-work (page 986) the contents of which may include the following.
- Location, date and period of validity.
- Work to be done.
- Equipment permitted/not permitted.
- Precautions* taken, e.g. isolation, purging, clearance of surroundings.
- Precautions* to be taken, e.g. presence of firewatch and appliances, provision of non-combustible sheeting.
- Testing requirements, e.g. initial, repeated at intervals.
- Clear instructions that work is restricted to that covered by the permit.
- Cancellation of the permit following a defined 'emergency'.
- Authorising signatures for start-of-work, completion and hand-back.
- Copies issued to, listed.

  * Between them these must cover, if appropriate, the following.
  - Removal of combustibles or, if this is impracticable, their protection.
  - Notification of fire authority, arrangements for attendance, laying of hose lines, selection and provision of extinguishers.
  - Covering to nearby sewers.
  - Provision of natural or forced ventilation.
  - Cleaning, flushing, purging of equipment. Isolation.
  - Wetting of the area.
  - Personal protection required.

Numerous other hazards are associated with cutting and welding and detailed procedures and safety checks are advisable each time.

During oxy-acetylene cutting, the oxygen cylinder was nearly empty and the diaphragm within the regulator was cracked. The acetylene cylinder was lying on its side and therefore (page 639) a mixture of liquid acetone and acetylene was being fed to the burner. When the oxygen was used up excess pressure forced an acetone–acetylene mixture back up the oxygen line and, via the cracked diaphragm, into the cylinder. The plant was destroyed in the ensuing explosion.[20]

Incidentally extreme caution is required following any fire at which acetylene cylinders are present, since local overheating can trigger self-propagating decomposition of acetylene causing a cylinder to explode. A detailed procedure has been proposed for dealing with such cylinders.[21]

Vehicles

The ability of vehicles to ignite flammable atmospheres is dramatically illustrated by numerous examples including the major disaster at Gothenburg as described on page 134 and by the following case-history.

A road tanker containing 26 200 litres of motor spirit overturned and an emergency team was set up to recover the hydrocarbon. After about four hours

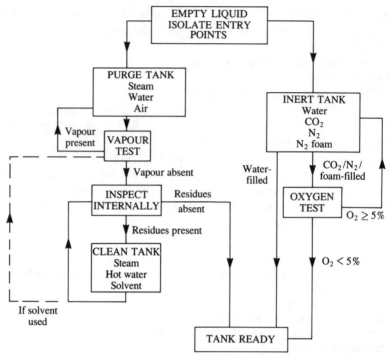

Fig. 27.4 Procedure for cleaning and gas freeing of tanks containing flammable residues for repair work. (Serious accidents are not infrequent through inadequate preparation; the diagram is a summary only and detailed procedures, e.g. ref. 6 should be consulted)

of successful transfer, ignition occurred resulting in superficial burns to two people and shock to the entire team. One person broke both heels when he jumped from the top of the vehicle. Ignition was attributed to the portable uplift pump driven by a diesel motor.

The use of a permit-to-work system to control vehicle access to certain areas was referred to on page 985.

Devices are available to protect diesel engines for operation in areas where there may be leaks of flammable vapour or gases; they serve to shut off the air and fuel supply in an emergency.

Hot hydrocarbon vapour, which escaped from plant during maintenance, was sucked into the air inlet of a diesel engine. Isolation of the fuel supply failed to stop the engine racing, simply because fuel was being fed in via the air inlet. Two men were killed when the hydrocarbon was ignited by a flash-back.[22]

In addition spark arrestors and flame arresters should be fitted to the exhaust, the exhaust pipe surface temperature must be less than the auto-ignition temperature of the vapour or gases, electrical equipment requires protection and any decompression control should be disconnected.

Detailed recommendations for the use of diesel engines are as follows.[23]

Engines which operate for >1000 h per year in Zone (Division) 1 or 2 areas should be fully protected with a device to enable them to be stopped if fuel is sucked in through the air inlet, spark and flame arresters on the exhaust, and suitable electrical equipment. Compressed air or spring starters should be used instead of electric starters. The exhaust temperature should be kept below the auto-ignition temperature of materials which might leak and the decompression control should be disconnected. Whenever possible fixed diesel engines should be located in safe areas, or air supplies should be drawn from safe areas and exhausts should discharge to safe areas. Vehicles for occasional use, e.g. for maintenance operations, can be protected to a lower standard provided they are never left unattended; however, they should be fitted with a device which enables them to be shut down if fuel is being sucked in through the air inlet. The exhaust temperature should be below the auto-ignition temperature of materials which might leak, and the decompression control should be disconnected. If there are no combustible gas detector alarms in the area then portable ones should be installed. Vehicles just 'passing through' need not be protected at all but should only enter with permission; this should only be given if conditions are steady and there are no leaks.

They also discouraged the use of cranes, welding sets, etc., in Zone (Division) 1 and 2 areas whilst plants were on-line but fell short of stopping it. The use of hydraulic cranes was encouraged as a replacement for diesel-electric cranes.

As summarised on page 16, only properly trained, authorised drivers should be permitted to handle vehicles within the factory.

A propane cylinder was punctured when it was hit square-on by a fork-lift truck. The escaping vapour cloud was ignited, probably by an electric discharge, and several persons were injured in the ensuing explosion and fire. The truck was found to have defective brakes and was operating on a slippery surface in a restricted area; however, the main reason for the accident was loss of control of the vehicle by an inexperienced, untrained driver who had only been working full-time for the firm for three days. (The elements of training for fork-lift-truck drivers are summarised on pages 17–22 and 1023.)

Space heaters

Hazards may arise from space heaters due to
- Improvised heating by portable appliances.
- Misuse of an appliance. For example deposition or storage of materials so as to reduce the heat dissipation from an appliance to a degree at which a dangerous high temperature can develop locally.
- Inherent hazards, e.g. from fuel stocks, flames and sparks, overheating or overpressure due to faults.
- Incorrect installation, e.g. inadequate clearance between flue pipe and combustible elements of a structure.

- Inadequate maintenance, e.g. reaction of leaking heat transfer medium with process materials.

Good practice is described in ref. 24.

Pyrophoric chemicals

Pyrophoric chemicals are discussed in Chapter 24 and listed elsewhere, e.g. page 151.

Static electricity

Static electricity occurs commonly in industry and in everyday life. It can be a source of danger as well as of discomfort and inconvenience. Static electricity can cause operational problems during manufacturing and handling processes, e.g. causing articles to adhere to each other or to attract dust. It can also ignite flammable gas or dust atmospheres as described on pages 144 and 1448. It is generated in many operations, including the flow of liquids, powders and saturated vapours, the contact and separation of solids, and in gases carrying minute particles of dust, rust or water droplets (ice particles). The chemical industry is particularly susceptible to electrostatic sources of ignition. Between 4 and 9 per cent of fires and explosions in the industry have been attributed to electrostatics,[25] and one company concluded that 30 per cent of its incidents over a 20-year period were due to static.

> During the sulphonation of toluene with oleum at *ca.* 120 °C vapour escaped from the glass-lined vessel, ignited and exploded. Fortunately the operators managed to escape injury and damage was restricted to the site.
> The prime cause of the incident was the absence of restraints which were essential in order to prevent excessive movement of glassware and a bellows connecting the reactor to a pair of condensers. The ignition source was electrostatic discharge to earth from an insulated metal clamp on the glassware.[26]

> During the batch crystallisation of chemical from hexane in a 1000 gal glass-lined reactor under nitrogen, contamination was detected in quality control checks on samples. This was traced to significant damage to the glass-lining. After reglassing the problem reappeared. Investigation revealed multiple pinholes in a ring formation at or above the liquid surface level, multiple spalling on the agitator blades, and several large areas of exposed steel adjacent to a tantalum earthing tip. Throughout one run, bright flashes of static electricity sparks were seen emanating from the agitated liquid as electrostatic potential which had built-up in this liquid, due to the insulating properties of the glass, caused breakdown of the glass lining. This caused electrical breakdown of the glass-lining: the rate of charge loss through the glass was less

than the charge generation rate. Presumably fire and explosion was avoided by use of the inert atmosphere.[25]

Charge formation and retention

There are three basic ways in which electrostatic charging can arise:
(i) contact electrification
(ii) charging by induction
(iii) charge transfer.

Contact electrification: This can occur at solid/solid, liquid/liquid, or liquid/solid interfaces.

Bringing two surfaces of solids into close contact will cause electron transfer between the materials, due to uneven distribution of electrons at the boundary. On separation charged particles may remain on the two separated surfaces; thus resultant charging can occur. If the two materials are good electrical or heat conductors this charge is usually rapidly recombined before the surfaces are obviously separated. If one material is an insulator it may be left with considerable charge which may result in a very high potential reaching many kilovolts, depending on the degree of mechanical action, despite the small amount of charge involved.

In liquids electrification depends on the presence of ions or submicroscopic charged particles. These ions of one polarity are absorbed at the interface and attract ions of opposite polarity to form a diffuse layer of charge close to the surface. If the liquid is then moved with respect to the interface, the diffuse layer is carried away and the separation mechanism occurs, leaving very high voltages as a result of the work done in separating the charged layer from the interface (see Fig. 27.5).

Charging by induction: There is an electric field around or between any charged object. A conducting material brought into the influence of this field causes field distortion and causes a separation of opposing charges in the conductor to take place. If this conductor is otherwise insulated from earth it will take up a potential, dependent on its position in the field; it is then 'charged by induction'. This potential, coupled with the separated charges that it carries, makes it capable of an electrostatic discharge. If one point of the conductor, whilst it is still in the field is momentarily earthed it potential is reduced to zero, but there is a net charge imbalance left on the conductor. If now the conductor is removed from the vicinity of the field it is left with an availability to cause a spark. The moment it comes within reach of either a similar object with opposite sign net charge or an earthed object, charge equalisation will occur causing a spark. This mechanism is shown progressively in Fig. 27.6.

Charge transfer: This occurs whenever a charged object makes contact with

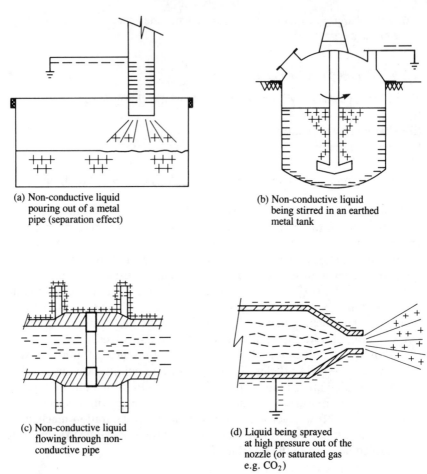

Fig. 27.5 Some examples of electrostatic charge generation

(a) Non-conductive liquid pouring out of a metal pipe (separation effect)

(b) Non-conductive liquid being stirred in an earthed metal tank

(c) Non-conductive liquid flowing through non-conductive pipe

(d) Liquid being sprayed at high pressure out of the nozzle (or saturated gas e.g. CO_2)

one that is uncharged; the charge is shared between them to the extent that their conductivity will allow. The most common form of this involves a charged spray, mist, or dust impinging on a solid object.

Ignition capability

In order to ignite explosive mixtures a minimum concentration of energy is required locally. The parameter for ignition is the minimum ignition energy (MIE), defined as the amount of energy which will just ignite the most easily ignited gas/air mixture.

This value varies from about 0.25 mJ for middle hydrocarbon distillates to as low as 0.019 mJ for hydrogen, and 0.009 mJ for carbon disulphide. Comparatively, ignition energies for dust–air mixtures (excluding explosives and other reactive substances as summarised in Table 25.1) lie between 10 and 100 mJ, and thus are considerably higher than gas/air mixture

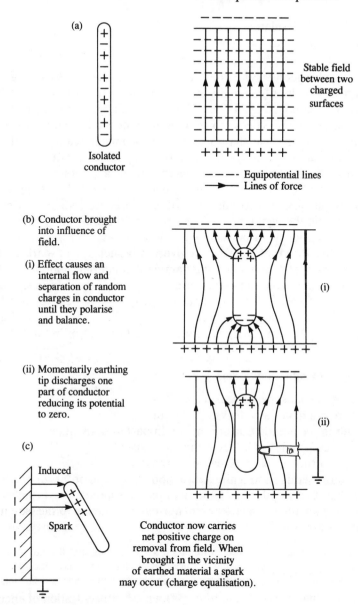

Fig. 27.6 Electrostatic charging by induction

energies for ignition. The lower end of this range is less than the energy content of a typical electrostatic spark.

Mechanisms for capacitive sparks of ignition capability

Electrostatics can produce ignition in any of three ways:
(i) discharge of isolated conductors

(ii) discharge of insulators
(iii) human discharge

Sometimes the geometry of the plant and processes are such that spark discharges of ignition capability will occur in certain process conditions between two conducting bodies. Very large charge accumulations may create spark discharges between charged insulators and conductors. Cylindrical and hemispherical shapes of >5 mm diameter readily develop the right type of discharge to ignite gas and/or vapour air mixtures, and such shapes are typical of pipes, elbows, screws, rivets and tools, etc.

As shown in Fig. 27.6 the conductor becomes a hazard as it approaches earthed metalwork. This condition is often caused by people walking into an electric field, being momentarily earthed by touching an earthed object and thus being left with net charge, walking away, and this charge then equalising in a different location. Inadvertently such a final spark could be at an open port of a vessel holding flammable material, igniting the flammable mixture. Persons in this condition complain of receiving a shock, but the cause is often not the object they touch, as they believe; often it is somewhere quite different (and can be quite difficult to locate), that has caused them to become charged as the isolated conductor in Fig. 27.6.

Electrostatics and solids

In the case of solids, examples include[20]
- flake from a flaker as it leaves the drum
- powder in pneumatic transfer line on impact with the pipe
- micronised powder as it leaves the microniser
- during pouring, sieving, screw-feed transfer, grinding, etc.

The energy required to ignite a dust cloud is several orders of magnitude greater than that required to ignite flammable vapours. Nevertheless sufficient energy can be generated in normal production processes unless precautions are taken.[27]

> Phenacetin powder was sieved from a high level sieve on metal staging down into a metal bowl on nylon wheels at floor level. The sieved powder fell down a terylene sleeve affixed between the sieve and bowl, which was not earthed. After collection of some 100 kg of powder an incendive spark ignited the dust cloud within the sieve.[27]

Numerous incidents have been recorded involving ignition due to an electrostatic spark produced when a powder was being charged from a plastic bag into a vessel already containing flammable liquid.[28] Such powders may be added as polymer, inhibitors, viscosity adjusters, fire retardants, fillers, additives or antioxidants. Alternatively they may be reactants. Suitable precautions are particularly necessary when flammable materials are involved.

A polyester resin was added to a styrene mixture, at a temperature above its flash-point, in an agitated vessel. This was followed by a powdered filler which was tipped from bags through a manhole; the bags were of woven polypropylene with a polyethylene liner. As the eleventh bag was withdrawn empty via the manhole, a static discharge, probably generated by flow of powder over the bag surface, ignited styrene vapour; the ensuing fire caused burns to the face and arms of the operator.

Subsequently the procedure was altered by lowering the styrene temperature during addition, the use of nitrogen in the vessel, and the use of an earthed chute.[28]

In grinding and micronising the charge per unit mass exceeds that from sieving and pouring because of the smaller particle size. Pneumatic transfer produces a greater charge per unit mass than most other operations because of the greater amount of work involved and the effective large contact area. The charge is also affected by the chemical composition and the surface properties. The speed with which the powder is able to dissipate its electrostatic charge is determined by its bulk resistivity. It is important to recognise that even if powders with large resistivities are placed in metal plant, charge delay can be days or even weeks.

Most common plastics and glass have high resistivites of about 10^{17} and 10^{13} ohm-m, respectively and hence act as electrical insulators. Thus, insulators are widely encountered in chemical process plant, e.g. nylon wheels on hoppers and buggies, PTFE-lined dryers and centrifuges, glass-lined reactors, polystyrene filter cloths, plastic scoops and sampling cups, epoxy-coated floors, synthetic protective clothing and footwear, pipe couplings, etc.

With solids the selection of anti-static packaging materials may be a desirable precaution to avoid electrostatic sparks.

Cellulose acetate butyrate in powder form was being poured manually from a 25 kg polyethylene sack into a solvent vessel during paint manufacture. A static discharge ignited the solvent vapour–air mixture in the stirred vessel and a jet of flame leapt out causing burns to approximately 25 per cent of the operator's body surface.[29] (Only anti-static packing materials had been in use for several years but old stock was being processed.)

A 1000 gal glass-lined reactor was washed with acetone and left to drain 24 hrs. 1000 kg of fine tray-dried powder was added from a series of 50 kg polyethylene-lined fibreboard drums. As the sixth drum was charged an explosion occurred which resulted in transmission of a flame front and pressure wave from the reactor's manway which left the two operators badly burned. The ignition source was an electrostatic spark from the fibreboard drum to the grounded reactor. The drum had become electrically charged as a result of induction charging and contact charging between the powder and the plastic drum liner. The operators gloves were PVC and their footwear synthetic, both preventing flow of the accumulated charge on the drum to the ground.[25]

Electrostatics and liquids

Charges with ignition potential come from liquids which on expanding into a larger vessel cause charge separation at the pipe outlet (see Fig. 27.5). The resultant charge then stays in the liquid in the tank for a time until it discharges either naturally or violently, as it comes across a deformation in the internal field, e.g. a dipstick being introduced into the tank. Filling a non-inerted vessel with liquid above its flash point will result in formation of a flammable atmosphere within the confined ullage space. In the case of drums, ignition of the headspace atmosphere, e.g. by brush discharges or sparks from static electricity, will, in most cases, cause drum failure. If the upper chime fails the outcome may be limited to a minor flash fire with some splashing. If the drum is not held in place, e.g. by a fixed filling lance, upper chime failure may cause the drum to tip over. The expanding gases from lower chime failure can propel the drum into the air, dispersing the contents and causing more serious fire.[30]

Different liquids have different charge retention characteristics, and the way in which equipotential bonding techniques are carried out is dependent on the dielectric constants of the materials being handled, e.g. acetone is significantly different to hexane. In a grounded, unlined drum, only liquids of low conductivity (usually less than 50 pS/m) can accumulate significant static. Compounds which *generally* tend not to accumulate static include most alcohols, aldehydes, ketones, acids, epoxides and nitriles. Classes of low conductivity liquids, on the other hand, are exemplified by ethers, hydrocarbons, carbon disulphide and certain silicone-based liquids.[30] Furthermore, the problems are multiplied if a second liquid, or particular phase, is involved (e.g. water in toluene, or crystals in a solvent in a chemical reactor). Agitator systems in vessels may also produce liquid layer separations, and certain centrifuge operations can also produce similar effects (see Fig. 27.5b).

Use of bonding and earthing, and increasing the conductivity of liquids by additives, in order to facilitate the removal of charge before sparking potentials are reached is noted on page 146.

> An operator was draining toluene from a kettle into a drum when vapour in the drum exploded. Pressure vented through the bung with such velocity that a fireball reached a mezzanine level above the work area. The drum was severely bulged but did not rupture. Toluene continued to drain onto the floor but an operator shut the drain valve so the spill did not reach the chemical sewer.
>
> A static electrical spark was the likely source of ignition, the significant features being:
> - The solvent was dehydrated toluene which is very prone to develop static charge because of its low electrical conductivity and dielectric constant.
> - A full-flow filter was installed in the drain line just before the nozzle (a position most likely to result in a highly charged liquid). The filter contained a polyethylene cartridge which would tend to accumulate charge.
> - The drum was grounded to structural steelwork by a spring-loaded clamp

and grounding cable. This may not have been effective at the time, since the teeth on the clamp were worn and may not have penetrated paint on the drum.
- Contrary to the in-house standard, there was no ground from the discharge spout to the building ground nor jumper cable between the spout and the drum.

Improvements were subsequently introduced in training and procedures, including relocation of the filter as far away from the spout as possible, and replacement of the short drain spout by a filling lance extending to the bottom of the drum and equipped with a permanently attached grounding cable. Resistance tests on grounding systems were also incorporated into scheduled maintenance.[31,32]

Earthing and grounding is desirable irrespective of container size.

As an operator was pouring a liquid inhibitor from a plastic bucket into a mixing vessel he noticed that sparks were being generated by static electricity. He promptly reported this and investigation revealed that this was the first time a plastic bucket had been used in preference to a metal bucket. Metal buckets were reintroduced and a possible ignition source hence eliminated.[33]

An ungrounded metal drum was being splash-filled with vinyl acetate from a height of 0.6 m via a 5 cm rubber hose when, after a few minutes, a violent explosion blew out the ends of the drum. One man sustained burns which subsequently proved fatal.[17]

A tank truck was splash-filled with a gas oil with a flash-point of 60 °C. The mist was ignited by a static charge resulting in a fire, the flames of which ceased when all the mist had burned.[17]

(This phenomenon was referred to on page 146 and two other incidents have been cited of static ignition when filling ungrounded metal containers as distinct from large tanks.[17]) Care must be taken to provide special earthing provisions on, for example, equipment mounted on rubber anti-vibration mountings.

The maintenance of earthing and bonding provisions is obviously crucial to safety. Therefore if equipment is dismantled, moved, insulated or painted, a recheck is subsequently necessary to ensure that earthing has not been impaired.

Ground clamps must be designed to penetrate any paint layers on the drum and commercial ground indicators might be considered. Whilst drum lining cannot impede the grounding of any bonded metallic component, it can impede the dissipation of charge from liquid inside the drum under certain conditions. Thus bonding and grounding may not always be enough since static discharges may originate from charged liquid in the drum, unaffected by grounding.

A drum exploded with failure of the upper chime. Fortunately, no operators

were in the vicinity and the outcome was minor. Investigation revealed a number of adverse factors, viz.
- the liquid being handled was a toluene/heptane mixture which is a highly flammable, low conductivity solvent;
- the drum was lined with a thin coating;
- a polyester filter was located immediately upstream of the lance;
- the lance was partially inserted and had a rounded tip.

The most striking observation, however, was that the bonding and grounding were still intact and ignition must have been inside the drum. Following the incident two similar accidents occurred, one of which resulted in a fatality.[30]

Some general conclusions from one study[30,34] are:
- smooth bore chemical hoses generate less static than certain hoses with an internal grounding spiral;
- dip pipes should have the largest practical diameter and have pointed terminations at the liquid ejection end;
- thin internal drum coatings can be neglected;
- pointed dip pipes should be used in poly-steel drum loading operations, particularly if the drum liner thickness exceeds about 2mm.

Additional techniques applied industrially are as follows.[35]
- *Use of static collectors and neutralisers*: Static collectors, e.g. tinsel bars, needle-pointed copper combs or flexible metallic brushes, may be used to prevent the accumulation of static on moving paper, belts or other materials. Collectors should be located very close to the charged surface, and be earthed via uninsulated flexible copper pipe, to minimise the opportunity for incendiary sparks to develop.
- *Humidity control*: Raising the relative humidity of the surrounding atmosphere will produce a very thin layer of moisture on objects and so allow charges to leak away to earth. An optimum of 60–75 per cent relative humidity is recommended. The use of this technique is limited to where:

  — working conditions for personnel remain comfortable and healthy (page 60);

  — process materials will not be affected, i.e. changed dimensionally or damaged;

  — materials have low moisture absorption properties;

  — no fire or explosion hazard could be introduced, e.g. from chemical reactions (page 491).

  Generally humidity control is applicable to small and enclosed buildings.
- *Ionisation*: A conducting path through which charge may dissipate may be provided by ionisation of the surrounding atmosphere. High voltage eliminators may be used, in an atmosphere free of flammable vapour, in order to eliminate or completely neutralise static charges. The voltage is carefully regulated to prevent addition of a charge to the neutralised

surface after neutralisation of the static charge. Electronic static bars may be used where high voltage eliminators would be unsafe; they avoid overcharging of the statically charged area. The voltages are closely balanced so that sparking potentials do not occur and shock hazards are negligible. Radioactive static charge eliminators ionise the air in contact with the charged body. The radioactive source, which should be controlled in accordance with the appropriate legislation,[36] is commonly an alpha particle emitter with a range of 10 cm. (Other types of radiation may be used to achieve a greater range.) Radiation shielding may be required with some sources (page 457).

Use may also be made of
- antistatic footwear. In many situations it may be necessary to prevent static discharge from people or vehicles. Use may be made of anti-static flooring and footwear. The resistance of such footwear is sufficiently high to protect the wearer from shocks from normal electricity supplies, e.g. due to insulation failure on equipment. Alternatively, conductive flooring and footwear, of much lower resistance, provide a much higher degree of protection against static accumulation, e.g. where explosives are handled, but do not, of course, maintain protection against electrocution. Vehicles may be protected by the use of conducting rubber tyres.
- inerting (page 132)
- reduced pumping/mixing speeds.

It may be that in some cases the problem of electrostatic charging can be eliminated by process modifications.

The manufacture of a cosmetic product required vigorous dispersion of a finely divided silica in a volatile silicone fluid followed by heating of the suspension to 80 °C and addition of a second powder. During the first production trial crackles were heard when the silica was being dispersed; on heating vapour above the suspension ignited. Subsequently reversal of the powder addition procedure so that the second, ionically dissociable, powder was added first eliminated the electrostatic charging hazard.[37]

Electrostatics and gases/mists

Static electricity generated by foam jets is believed to have provided the source of ignition for a fire in a 90 000 m^3 naphtha tank in February 1991 and for two earlier tank fires. Possible charging mechanisms with low conductivity flammable liquids include the direct impact of foam jets on the liquid surface and water droplets sinking through the fuel as water drains from the foam. (A changed procedure for storage tank incidents is summarised on page 1558, however since the charging mechanism requires a minimum liquid depth of 2 to 3 m, foam can safely be applied to shallow fuel spills.)[38]

Leaking steam mains can also convey charge to unearthed metalwork on which the leak impinges (see Fig. 27.7). A special case of ignition by electrostatic sparking occurs when carbon dioxide gas is discharged as a fire-fighting medium. Major secondary explosions have been caused by CO_2 fire protection systems, but it is often overlooked as a potential ignition risk. Electrostatic voltages up to values of 10 kV are quite easily generated if the CO_2 system has not been properly designed to take account of the risk.

Arising from the Joule Thompson effects as the gas is ejected at sonic velocity through the nozzle, ice particles are ejected in it. The result is a charged cloud and a charged nozzle. Discharge of this on surrounding metalwork can cause an ignition to occur if enough energy can be brought to the point of discharge, and in the presence of volatile vapours. Electrostatic ignition of escaping hydrogen was discussed on page 1366.

Mechanical sources

Following an explosion aboard a marine tanker a 1.5 m by 2.1 m piece from the hull structure was blown 180 m and struck the conical roof of a storage tank containing ethanol. It penetrated the roof and ignited the ethanol.

Mechanical sources may cause ignition due to heat generated by friction, impact of missiles, or by metal fracture.[39]

Frictional ignition was discussed briefly on page 153. It may be grouped into three energy levels.

(i) Low energy: including ignitions that involve approximately 10 J, such as might be produced in manpowered operations with hand tools (equivalent to 500 g falling 2 m).
(ii) Medium energy: including ignitions that involve aound 1 kJ of energy (equivalent to 25 kg falling 4 m) which might be produced by a powered hand tool or moving parts of machinery.
(iii) High energy: including ignitions that involve energy in excess of 1 MJ such as might be produced by a crashing road tanker or ships in collision.

The range of events that have been recorded are extensive, e.g. some 110 such incidents were listed between 1958 and 1978. The three key ways of generating friction sparking are:
(i) impact
(ii) rubbing
(iii) grinding and cutting.

Impact accounts for by far the largest proportion of frictional ignitions, representing some 68% of ignitions for gases and vapours, and some 65% of ignitions of powders and dusts.

Rubbing, for example, overheated bearings, belt friction on pulleys, or between buckets and bucket guides on elevators have all been responsible for ignitions.

Grinding and cutting account for the least number of ignitions but have nevertheless been recorded.

Whilst bending of metal usually does not produce enough heat to ignite most chemicals, friction sparks which result when heat is generated by friction or impact initially heats a particle sufficiently to ignite flammable gaseous mixtures.

> On the 8 January 1979, the oil tanker *Betelgeuse* exploded with the loss of 50 lives at Bantry Bay, Ireland. The circumstances are summarised on page 777. Ignition of the gas was attributed to the grinding of fractured metal plates (friction heat).[39,40]

Depending upon the ease of oxidation and the heat of combustion of the metal particle, the freshly exposed surface may oxidise, increasing the temperature until the particle is incandescent. Examples include
- shoe nails on concrete floors
- foreign metal in crushing, grinding, mixing equipment

> Polystyrene was expanded with pentane and a 14-day period was normally allowed for pentane to be lost, by diffusion, from the plastic. When new material was fed to a waste chopping machine impact of a rotor blade on a piece of metal generated a friction spark and ignited a pocket of pentane–air gas.

- steel tools impacting upon concrete.

Mechanical sparks generally cool quickly. However, they may start fires if they fall into loose fibrous combustible solids, into combustible dust piles/ layers or onto explosive materials. The avoidance of mechanical sparks in areas where flammable vapour may be present is discussed on page 153; this precaution should, of course, extend to the avoidance of grinding wheels.

In thermite reactions a distinction is made between general frictional sparking which is dealt with above and reactions associated with light metals (see page 153). Such metals, e.g. aluminium, magnesium and titanium, are characterised by their ability when finely divided to react exothermically with atmospheric oxygen and, as a result, to ignite a flammable atmosphere. The term 'light alloy' refers to an alloy containing at least 50% of a light metal by atomic proportions. Aluminium smearing on rusty steel provides an ideal base for a thermite reaction if the smear is struck by a glancing blow from a portable tool. The iron oxide renders the oxygen readily available for a chemical reaction which releases 15.5 kJ per gramme of aluminium and produces a spark of ignition capability.

$$2Al + Fe_2O_3 \rightarrow Al_2O_3 + 2Fe$$

Attention should, therefore, be given to the need to use alloys in the plant design and where it is unavoidable advice should be sought on their suitability for the particular application.

Aluminium and magnesium often form part of the components used in electrical systems, and their use should be approved by a professional electrical engineer. Cognisance should be taken of the use of aluminium paints to protect plant. Similar sparks are produced when pyrites deposits (FeS_2) are struck, because the sparks from this sulphide burn very easily and have a high incendiary capacity with explosive gas mixtures.

Smoking

As noted on page 142 smoking and the use of matches and automatic lighters are potentially serious causes of fire; the problem is exacerbated by the movement of smokers about a factory or construction site.

Obviously a system of work is required to ensure that no person ever smokes in a place in which flammable gas or vapour, or highly flammable liquid, or combustible solid is present and the circumstances are such that smoking would introduce a risk of fire or explosion. Notices should be clearly displayed at all access points to these areas. Alternatively a clear, bold notice should be displayed at every factory entrance indicating that smoking is prohibited throughout the factory except at places where there are notices displayed to indicate the contrary. Areas set aside for smoking should be clearly identified.

> A fire occurred in a crude oil storage tank whilst sludge was being cleaned out by contractors. Vapour from the sludge was ignited by illegal smoking by cleaners who had actually removed their breathing apparatus in order to see better and to smoke.[41] On the previous day the measured concentration of vapour in the tank was 25 per cent of the Lower Flammable Limit – in fact a limit at which consideration should have been given to stopping work, but such concentrations can vary, particularly when sludge is disturbed. (Vapour concentration is not evenly distributed, depending on ambient temperature, ventilation and the extent to which sludge is disturbed.) One man, a non-smoker, who had been driving a hydraulically powered tracked vehicle fitted with a rubber-edged wooden scraper was killed – possibly because his airline got entangled with support pillars and hindered escape, or he may have slipped or tripped over hoses or other equipment.

Smoking presents a particular problem if oxygen enrichment occurs.

> Leakage of oxygen from a subterranean pipe conduit into a basement workshop and permeation of oxygen into an oxygen regulating platform on a large oxygen plant have been reported.[42] Fatalities and ignition due to smoking occurred in both cases.

Spontaneous combustion

The phenomenon of spontaneous combustion is described on page 148. Materials liable to self-heating are exemplified by Table 24.2 and detailed descriptions and precautions are summarised in Chapter 24.

Compression

A liquid may gain ignition energy due to compression, e.g. from pressure surges due to quick-acting valves or in pumps. A compression hot spot in the liquid may arise from compression of any small vapour bubble. An adiabatic compression apparatus may be used to test liquids for sensitivity but many parameters are involved.

Open flames

Open flames pose a potential fire hazard in many situations unless proper control is exercised over their use. Therefore, in any factory or warehouse where an open flame is permitted care must be taken to avoid the unexpected presence of flammable vapour or dust.

> As a fitter walked across a 6 m grating platform vibrations dispersed a cloud of polymethylmethacrylate dust. This entered the open propane gas flame of a manual shrink-wrapping device resulting in a dust explosion.[43]

On construction sites any heating appliance incorporating naked flames, e.g. space heating equipment, bitumen boilers or mastic asphalt cookers, must be sited remote from any source of flammable vapour or gas; alternatively they must be incapable of causing ignition.[44]

> A 0.5 tonne propane cylinder was being moved by a fork-lift truck when it rolled off the forks and damaged a valve connection. Leaking propane ignited on a nearby stove and the resulting explosion caused extensive damage to buildings.[45]

Safe positions may generally be those at least 4 m from which highly flammable liquids are stored, used or manipulated.

> An industrial painter sustained fatal burns while painting the interior of a 1.37 m pipe on a hillside slope with a gradient of 1 in 4. The site owners had dewatered the pipe and issued a permit to work. The contractors specified that low voltage flameproof lighting would be used for illumination inside the pipe but, in the event, an LPG gas powered lamp was used. It was accidentally knocked over and rolled down to set alight thixotropic gelled bitumen paint in a tray (flash point slightly >32 °C). The painter attempted to escape past the flames but unfortunately his paint-impregnated overalls were ignited.

Painting is often attended by risks which are difficult to control, e.g. blow lamps have caused numerous fires during burning-off operations when flame has reached concealed timber or other combustibles which as a result smouldered for hours before bursting into flame. The problem is exacerbated since paint thinners and solvents, rags used for cleaning, and materials used for sheeting-up may contribute to the rapid development and spread of fire. A summary of fire precautions applicable to painting operations is given in Table 27.4 (after ref. 35 and in refs. 46 and 47).

Table 27.4 Fire precautions for painting operations. (Additional precautions apply to spray painting, e.g. to avoid static, capture overspray and provide local exhaust ventilation)

- **Burning-off old paint**

  The use of blowlamps should wherever possible be avoided in fire-hazardous areas and safer paint stripping methods used. Specific procedures should be laid down for blowlamp refilling or recharging procedures. LPG-fuelled equipment should be regularly inspected for leaks.

  Combustible and flammable materials should, where practicable, be removed before burning-off operations. Approved screening or sheeting is an alternative.

  On completion the area should be inspected; if the area is to be left unattended there should be a further check to ensure that no potential fires are left. Such inspections should include adjacent materials and materials at lower levels in open plant and on perforated floors.

- **Flammable liquids (e.g. thinners and cleaning solvents)**

  Means of ignition should be prohibited. There should be adequate ventilation. Provision is required for safe storage and disposal of rags or other waste contaminated with liquids or paint.

  Stocks of flammable paints or solvents in the area under decoration should be limited to that required for a day's work.

- **Sheeting-up**

  Materials for protecting stock or equipment should ideally be non-combustible or, as a minimum, be treated with a flame-retardant compound or be of material which will not contribute to a rapid spread of fire.

- **Fire protection**

  Suitable fire extinguishers should be provided for the duration of the redecorating work. Operators should be competent to use the fire extinguishers. Fire-fighting personnel should be present, as appropriate, in hazardous areas.

- **Change of paints**

  The correct paint system should be selected for the conditions of service.[46,47] Paints pigmented with materials that may result in incendiary sparking if struck under dry conditions, e.g. aluminium (page 153), should be excluded from locations where a flammable atmosphere may arise.

Whilst naked flames such as in furnaces may be sited remote from areas where flammable materials are normally handled or processed, leaks may result in flammable gases or vapours travelling to 'find' an ignition source: if ignition sources are difficult to eliminate completely precautions should include ensuring discharges are vented to a safe place.

> One man was killed and a two-storey building demolished by a violent factory explosion. Nearby residential property was damaged by the blast and heat radiation; a 200 litre drum landed on one roof and more drums landed outside

the site boundary. The cause was ignition of hexane vapour, discharged after loss of cooling water to the condenser of the pot still in which 6000 litres of hexane were being distilled in a flame-proof room. The source of ignition was attributed to the flame of an oil-fired steam boiler situated in a nearby room. The fire spread rapidly to the remainder of the site where drums of solvent were stored.[48]

In laboratories, although their use is decreasing, naked flames, e.g. bunsen burners, gas rings and other sources of ignition, need to be confined to specific areas where measures ensure the absence of a dangerous concentration of flammable vapours.

Rubbish burning, e.g. on construction sites, should be strictly controlled. If necessary it should be carried out on open ground not less than 10 m downwind of any hazardous area.[44] Highly flammable liquids should never be used as a booster for brightening such fires.

Electrical sources

As noted on page 143, electrical equipment may be a cause of ignition due to sparks (e.g. during switching on *or* off) or overheating, either during normal operation or from faults in the equipment.

During the loading of low flash-point hydrocarbon into a road tanker some spillage occurred onto the concrete. On completion of the loading operation the driver started the engine of the tanker and vapours from the spillage were ignited by the main electrical isolation switch of the tanker, the terminals of which were corroded which led to arcing.

Clearly, therefore, potential electrical sources of ignition should be avoided in dangerous atmospheres or in proximity to combustible material.

The categories of risk for the zoning of electrical equipment, and the types to be chosen for each area, are summarised on page 714. In addition, main switchgear should be sectionalised in groups, labelled and readily accessible for safe operation in case of fire.[35]

There are occasions when areas containing flammable chemicals do not justify being equipped with explosion-proof electrics if other ignition sources are present. Thus a welding shop contains both acetylene (highly flammable gas) and oxygen. However, since a flame burns continuously throughout the work process, installation of expensive explosion-proof electrics would be pointless. Similarly, it may be inappropriate to install special electrical systems in areas containing pyrophorics since these chemicals require no ignition source, only exposure to the air. The presence of highly flammable diluents may conversely render use of explosion-proof electrics sensible so as to avoid ignition of solvent vapour.[49]

In chemical plant, feeder cables should be arranged to pass through areas of least fire risk. They should be protected or dispersed such that a relatively minor fire will not cause extensive damage to the power supply. Holes should be filled in where cables pass through walls.

> LNG leaked from an inadequately tightened pump seal at a reception facility. Gas passed through 70 m of underground electrical conduit and entered a sub-station where it ignited, causing one fatality and $3m damage including demolition of the sub-station.[50]

Electrical conductors which carry alternating current should be located so as not to result in varying magnetic fields of sufficient intensity to cause local overheating of adjacent steelwork.

Electrical checks should be recorded periodically on
- earth and continuity resistance
- insulation resistance
- rating of fuses and setting of overload devices

Apparatus should not be opened until 'dead' and the temperature of the internal unit is at a safe value. Intrinsically safe instruments should be used for testing in danger areas.

Whilst the above precautions apply equally to laboratories, some laboratory instruments, ovens and other electrical equipment are not always obtainable to fully protected standards. This, however, does not lessen the need for precautions.

In these circumstances (after ref. 51):
- the best available standard of equipment should be obtained;
- the equipment should be located and used in an area where all steps have been taken to ensure that a dangerous concentration of vapour may not reasonably be expected to be present;
- adequate ventilation should be provided to prevent the accumulation of a dangerous concentration of flammable vapour;
- the quantity of highly flammable liquid present in the area should be restricted to that needed for immediate use.

In all industrial situations the use of temporary electrical equipment should be minimised, e.g. by the use of alternative energy such as compressed air where practicable. Any temporary installation should be disconnected and dismantled immediately after it has served its purpose. It should be provided with means for a speedy disconnection from the supply on all poles and be suitably protected from the weather and mechanical damage. Supplies to portable lighting and equipment should not exceed 25 V a.c. or 125 V d.c. above earth potential. All cables should be supported clear of walkways, stairs and working areas and so as to avoid mechanical damage.

Stray currents may arise unintentionally. Thus when using electric arc welding equipment the cables should be carried over any pipelines on insulated bridging. Earths should be connected directly to the welding equipment and positioned close to the job. Installed equipment must not be used for earthing.

Earth faults and stray earth currents arise from malfunction of the electrical installation. Any malfunction must be detected and the cause eliminated as soon as possible since they constitute a serious ignition risk.

Earth faults

Large fault currents arising from the sudden failure of the system insulation are controlled by protective devices. These devices operate in milliseconds to cut off the fault current before too much thermal energy is released into the environment. Systems are usually designed so that the probability of this type of fault in the presence of a flammable mixture is avoided.

However, low values of earth fault current, which may not be detected by normal protective devices, can persist for long periods of time. Such currents may well generate sparks of ignition capability.

High resistance joints in the path of such low currents can also give rise to local hot spots as the energy equivalent of the current is dissipated through the joint.

Stray earth currents

Electrical systems and components are well-insulated, in order to function properly, and to prevent personal contact with dangerous voltages. However, even insulation conducts a very small leakage current. Insulation will deteriorate during its normal life and the leakage current increase. Deterioration is accelerated by internal and external effects, such as overheating due to excessive load currents and/or ambient temperature, dielectric stress, chemical effects (solvents, vapours) and physical effects (e.g. solar radiation, mechanical impact), etc.

Leakage currents which follow random paths via metal enclosures of electrical systems to earth are termed 'stray currents'. In general they are led away via the metal enclosure to earth, since normally the path to earth will be permanently available.

When, however, this path is not permanently available high impedance gaps are formed, e.g. in case of corrosion build-up, vibration, loose contacts, and the stray current may bridge these gaps creating sparks which depend on gap distance, circuit inductance and the magnitude of the current.

Surge phenomena and lightning

Surges in the electricity supply system

Internal surge effects on electrical systems are usually controllable by electrical engineering design. Since they may cause maloperation of the factory system and cause a process plant to operate in an unsafe manner, they are usually taken care of within the system design.

Lightning strikes

The more obvious 'electrical disturbance' is that of lightning during a thunderstorm and the need to provide earthing for lightning discharge was mentioned briefly on page 155.

Lightning strikes may ignite flammable materials because of[39]
- direct entry into a vapour space;
- sparking, e.g. at a flange joint, of lightning conducted through piping;
- local heating of a metal plate by a strike, so that vapour on the opposite side exceeds the autoignition temperature;
- fast changes in magnetic fields, as a result of the fast current pulses of up to 2 000 000 A, inducing voltages and currents in circuits and support structures (if followed by spark-over of sufficient energy).

> A flash of lightning struck a cone roof gasoline storage tank resulting in an explosion. The most likely cause of ignition was spark discharges at an internal temperature measuring assembly.[53]

There are two strokes associated with a lightning discharge to earth, the leader stroke (from cloud to ground) which travels at around 2×10^5 m/s with a charge deposit along the length of about 1 to 10 coulombs. When this reaches the ground intense luminosity occurs back along the path and this is known as the return stroke. In effect this constitutes an electrical short circuit. The leader stroke has a narrow central core, surrounded by a much larger corona envelope. The current flowing in the return stroke is most concentrated in the core of the leader channel. The diameter of this core is about two centimetres and its maximum temperature about 30 000 K.

This leads to the basic range of effects from a lightning flash:
- thermal effects
- mechanical effects
- electrical effects

Thermal effects arise from the very high core temperature, but these temperatures are only maintained for several microseconds. Where the discharge takes place across a high resistance joint, e.g. two good conductors in poor contact, heavy sparking occurs because of the heat generated at that point. This is a particular problem with overlapping corrugated metal roof panels, etc. There is also the problem of the possible puncturing of these metal skins at the point of strike. For example an aluminium sheet of 3 mm thickness will be punctured in 0.05 seconds with the application of only 40 coulombs of charge.

The average lightning flash can convey up to 400 coulombs of charge. Thus when lightning strikes an insulating material or a poor conductor the local temperature at the point of contact will be very high and a clean hole may be punctured right through the material. If the insulant contains a trace of moisture, or some other conducting material, the discharge current will flow preferentially along the path of best conductivity. This moisture can be converted to steam and the resulting pressure can cause explosive fractures.

> In an incident in South Africa a discharge furrowed 250 m along some rock interstices dislodging 70 tonnes of material, and the forces were calculated to be the equivalent of 250 kg of TNT.

The thermal effect can ignite combustible materials, e.g. timber and thatch, as well as flammable dusts, solvents, vapours, etc.

Mechanical effects arise in two different distinct ways, firstly the shock wave produced by the return stroke and secondly the bending forces exerted on conductors through which the discharge current flows.

Due to the very high temperature of the core produced in a matter of microseconds the air surrounding the stroke core has to expand at an extremely high rate, thus producing a pressure wave which is initially supersonic. The result of this is evidenced by tiles having lifted on the roofs of buildings that have been struck.

Two parallel conductors sharing in the discharge of a lightning current are subjected to a force of attraction which is proportional to the square of the current and the inverse of the distance between them. This can squash electrical conduits and other lightly formed tubular conductors, etc. Sharp right angle bends in conductors are also subjected to a force which tends to straighten them out. This latter force is proportional to the square of the amplitude of the current and is usually not greater than a few thousand kilograms.

Electrical effects: When the electrical field between two separately conducting systems that are not electrically interconnected at a low enough value of resistance becomes higher than the electrical insulating breakdown value of the air separating them, a flashover will occur. This aspect of lightning surge phenomena is termed 'side-flash'. Basically as the lightning strike contacts the building the point of contact may be raised to values of 1 megavolt or higher due to both the resistance between this point and the main mass of earth, and the inductive impedance of the path to the equalisation point. If the electrical breakdown strength between this path and another separately connected conductor is high enough, then a sideways discharge will occur with reduced thermal/mechanical effects, but nevertheless potentially disastrous results. A reasonable absolute separation distance between the main conductor and unconnected conductors or conducting material is about 0.4 metres. This makes presumptions about electrical breakdown strengths associated with impulse waves rather than steady state values.

In general above-ground tanks containing flammable or combustible

1540 Flammable and explosive hazards

liquids or flammable gases at atmospheric pressure are considered to be reasonably well protected if constructed entirely of steel and[54]
- all joints between steel plates are riveted, bolted or welded;
- pipes entering the tank are metallically connected to the tank at the entry points;
- all vapour openings are closed or provided with flame arresters, although the dangers from blockage from dust and polymer must be considered. Styrene tanks are usually left open. Pressure vacuum valves are commonly used on tanks and vessels;
- the tank and roof are of a minimum 4.8 mm thickness (so holes will not be burned through);
- the roof is continuously welded, bolted or riveted and caulked to the shell to provide a vapour-tight seam and electrical continuity.

If additional protection is required the internal support members of the roof may be bonded to the roof at 3.05 m intervals; alternatively an overhead ground wire system or mast protection may be installed.[55]

Steel tanks in direct contact with the ground or connected to extensive, properly grounded, metal piping are considered as inherently grounded.[54] Floating roof tanks with hangers located within a vapour space may be protected by bonding the roof to the shoes of the seal at 3.05 m intervals around the circumference and by providing insulated bonds around each pinned joint of the hanger mechanism.

Above-ground storage tanks containing liquids or liquid petroleum gas (LPG) under pressure are considered safe because the vapour is within the tank and the vapour-air mixture is too rich.[54]

Detailed advice has been published for lightning protection.[55,56]

Unusual ignition sources

An unusual potential source of ignition is illustrated in Fig. 27.7. Here portions

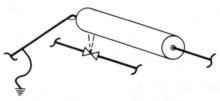

Fig. 27.7 Arcing as a result of steam impingement on an unearthed metal jacket on insulation

of an overhead pipe were left uninsulated following an equipment rearrangement. A neighbouring steam line developed a small, almost invisible, leak which impinged onto an uninsulated section of pipe. Arcing occurred between the metal jacket of the insulation and an uninsulated portion of the same line.

Aluminium rotors in refrigeration compressors can lead to a fire if
- the aluminium rotor is run at high speed;
- the rotor moves into contact with a stationary metal part, not necessarily aluminium, e.g. due to a bearing failure or a rotor moving along its shaft;
- the refrigerant is a chlorofluorocarbon.

Such conditions can expose a fresh, unoxidised hot surface of aluminium to the refrigerant; the aluminium can react exothermically creating aluminium chloride, aluminium fluorides and carbon. These aluminium rotors are not, therefore, recommended in centrifugal refrigeration compressors.[57]

A bursting tyre on a road tanker has constituted an unusual source of ignition.

> Following a crash involving a 20-ton ethylene tanker liquid spilled from a sheared connection at the front and from damaged connections at the rear. Nearly two hours after the incident a frozen tyre failed due to embrittlement and ignited the spill. Tyres which explode emit showers of sparks due to the steel wires incorporated in their construction.[58]

Energy generated from a radio-frequency electromagnetic field can only serve as an ignition source if there is an electromagnetic radiation of sufficient intensity, and some structure to act as a receiving aerial to generate an induced current, and an intermittent contact so that the received energy is discharged as a spark.

> A ship owner reported sparking between a cargo hook suspended from a wire and the cargo deck while the radiotelegraph transmitter was in operation. It burnt the deck paint.[39]

For electromagnetic fields to function as ignition sources the radio-frequency range of concern is above 15 kHz. A guide is now available on the prevention of ignition.[59]

Exothermic reactions can also act as a source of ignition for fuels as illustrated in Chapter 24.

> In 1788 Berthollet, having discovered potassium chlorate, considered using it as a replacement for potassium nitrate in black powder production. During a demonstration of the preparation of the new powder in a crushing mill a violent explosion resulted in two fatalities.

Exothermic reactions are not generally so violent but their propensity to runaway is discussed on pages 509–14. 'Self-heating', which is a particular case, has often resulted in fires.

> A new drier for an animal feedstuff, a by-product of distillery operations, was cylindrical with a rotating central shaft fitted with paddles. It was heated by a steam jacket at 174 °C. A fire occurred in the drier within a few weeks of start-up.[52] Analysis of self-heating test results indicated that the critical layer

thickness at 174 °C was 22 mm whereas deposits up to 50 mm thick could form near the drier outlet. Hence no drier malfunction was necessary to result in a fire. (Fairings were subsequently introduced at the outlet to prevent hot, dry material collecting in corners.)

Fire protection

Protection against fires in factories generally involves some combination of
- inventory control
- limitation by space separation and compartmentation;
- use of automatic fire detection systems and watch systems;
- automatic fire extinguishment and control systems;
- use of appropriate hand extinguishers and hose reels;
- reliance on industrial fire brigades and the fire services.

Minimising stocks limits the impact of fires and segregating oxidising agents and reducing agents ('fuels' and 'oxygen' sources) is crucial in avoiding fires (page 157).

Two incompatible chemicals AZDN (azo-diisobutyronitrile), which is a thermal initiator, and sodium persulphate were kept in the same store. The AZDN was stored in cardboard kegs on a shelf close to a steam pipe which had not been correctly disconnected. On decomposition at >50 °C it expanded and released nitrogen; the kegs ruptured and it spilled onto bags of sodium persulphate one of which is believed to have been pierced by a keg lid. Upon ignition by an unidentified source (possibly a static discharge or a falling metal lid) a fire ensued followed by a substantial explosion.[60]

Space separation is the distance between buildings, or between external tanks or equipment, to prevent radiation from a large fire in a building starting a fire in a neighbouring building or plant. The distance is calculable from estimates of the exposure and the size of the flames. Compartmentation is introduced by building structural items, e.g. walls or floors, through which fire will not pass in a specified time and which will continue to bear their design load. They are selected on the basis of fire-resistance tests, the thermal diffusivity of the materials being an important factor in imparting fire resistance to the element concerned.[56]

Automatic fire detection/watch systems and the effectiveness of efforts by the fire services are closely related since the success of any fire-fighting action depends on the rapidity of detection.

Fires started in the small hours of the night are about five times more likely to become large fires than fires started during the day because of delays in detection.[61]

Hand extinguishers are generally only useful before a fire is greater than 1 to 2 m^2 in area. Anyone who has not been specially trained would not normally be expected to approach a larger fire. However, an ordinary factory fire which has not extended more than a few tens of square metres can be quickly brought under control by the skilled application of about 100 litres of water, i.e. a fire in an 18 m² room may be controlled with about 45 litres of water. This is well within the scope of a fire appliance carrying water but as a fire becomes substantially bigger the quantity of water needed increases much more than proportionately.

> For a large fire one large jet delivering 400 to 700 litres/min is used for every 10 m of fire perimeter. Thus it is important for the fire brigade to arrive on the scene before a fire has grown >10 to 100 m².[62]

Building design

The desirable features of a building design are that it should withstand fire and prevent it spreading.

Some important aspects for the design or selection of buildings for the storage of combustible or flammable materials are summarised in Table 27.5 (after ref. 35). Clearly, however, building size, duty, location, occupancy and the value of the contents determine – in conjunction with local legislation and economics – the features actually incorporated.

For the purpose of assessing risk, it is usual to take into account the features summarised in Table 27.6.

The intensity of a fire depends upon room or building size, the contents and the degree of ventilation, i.e. air supply. Hence high stocks of combustible solids, e.g. paper, or flammable liquids or gases are a particular concern. Problems may also be created by materials which may flow in a fire, e.g. plastics, tar, rubbers or waxes. Limitation of the spread of a fire within a building obviously requires consideration of the control of inventory, segregation and spacing of combustible and flammable materials. Other factors to consider are as follows.

- The transfer of heat to combustibles by radiation (page 868). The flux received decreases in proportion to the square of the distance so that spacing is one strategy.
- The potential for burning 'brands' to blow across open floorways or over open ground.
- The potential for fire spread at roof level due to combustible roofs and linings (see Fig. 4.8, page 160).
- The contribution which combustible floors and walls, or internal finishes can make to fire spread including effects on collapse.

> A fire at Kings Cross Underground Station in 1987 resulted in the deaths of 31 people and injured many more. A carelessly discarded match found its way beneath the escalator and ignited a flammable mixture of grease and dust (detritus) which had accumulated over the years. This fire served to pre-heat the plywood sides and steps of the escalator, rendering them more susceptible to ignition and spread of fire. After 10 minutes the development of the fire became

Table 27.5 Features of buildings for the storage of combustible or flammable materials

- Single-story buildings with no basement are preferable.
- Adequate protection is required by distance, or other means, where the building or its contents could be exposed to external fire.
- Buildings should be of non-combustible construction if used for fire-hazardous materials (except for small units which are relatively safe because of their position in relation to the site boundary and adjacent risks). Buildings for highly flammable or explosive materials should comply with appropriate legislation.
- Openings in external walls depend on the relationship between the building and the site boundary and appropriate legislation. Fire doors are required in openings in dividing walls from other activities.
- All exterior openings should be weatherproof. Buildings should be reasonably secure against unauthorised access.
- Floors in multi-storey buildings should be designed to carry stored material and loads imposed by fire-fighting water, which may be absorbed by the material. Any watertight floors should be provided with drains or other suitable means to carry away fire-fighting water to a secure place and not to drain to river. Stairways and other vertical shafts should be enclosed or sealed off at each floor level with construction to the same fire-resistance rating as the building.
- Fire-resisting walls, fire doors and floors should be maintained in good repair. Fire venting of the building may be applied as appropriate.
- With dusty materials, ledges where flammable residues may accumulate should be avoided.
- High-risk areas should be segregated into small units with dividing walls of appropriate strength and fire resistance. Explosion venting is required for dust explosion risks.
- Measures should be taken to prevent heat build-up from the sun's rays where heat-sensitive materials are stored, i.e. window and roof lights should be avoided, good ventilation provided and the roof area should be suitably insulated.
- Floors may be of special construction, e.g. anti-static or non-sparking.
- Consider appropriateness of sprinkler systems.

extremely rapid leading to a flashover, which caused the bulk of the damage and injuries. A trench-effect formed by the escalator sides and steps served as a medium for flashover causing the flames to lie down in the trench instead of rising to the ceiling. Air-flow changes caused by train movements accelerated the process.[63]

- The potential for fire spread rapidly through openings and undivided areas, as exemplified by the K-mart Warehouse fire (page 1352) and the Braehead Warehouse fire (page 162).
- The propensity of hot combustion gases to rise through staircases, shafts and ducts and hence to spread fire to other floors.

A fire started in a fifth-floor university chemistry laboratory in a six-storey building, originating in an area where electrical equipment had been in continuous use the previous day. Soon after discovery an explosion occurred causing considerable spreading. Considerable quantities of flammable liquids were present in the laboratory and an 18-litre can of solvent burst. Lateral and

Table 27.6 Risk assessment of buildings[78]

High Risk: If there is present in the building (or part) any material, dust, vapour, gas or liquid in such quantity or disposition, or of such a nature, as will be likely when ignited to cause fire, or smoke or fumes to spread rapidly. Other factors are as follows.

Undesirable structural features, e.g.
- stairways which cannot be separated from workrooms, unenclosed vertical ducts
- wooden floors supported by wooden joists, particularly if oil-soaked
- large areas of flammable surfaces (either as stacked stock or materials, or walls/ceilings).

Any unusual circumstances relating to occupants, e.g.
- large numbers relative to building size
- occupants mainly/predominantly disabled persons
- individuals/small groups working in isolated parts of building.

Normal Risk: A high percentage of factories (or parts). Any outbreak of fire is likely to remain localised or to spread only slowly, and there is little risk of any part of the structure of the building taking fire readily. Offices normally come into this category.

Low Risk: Factories (or parts) where there are few flammable and no explosive materials, little risk of fire breaking out, and no likelihood that fire or smoke or fumes will spread rapidly. Examples include buildings (or parts) used for heavy engineering, where only 'wet' processes are carried on or only non-combustibles are present.

---

vertical spread of the fire was permitted via ventilation ducting constructed partly of plastic and partly of metal. Several branches of this ducting within the laboratory joined a common duct over the fifth-floor corridor which passed through the concrete floor above into a sixth-floor plant room. When the plastic ducting collapsed completely it left openings through which the fire spread to the plant room causing severe damage.[64]

Thus where ducting passes through ceilings, floors or walls, fire dampers operating on fusible links should preferably be installed.

- The potential for collapse of unsupported roof supports, columns or trusses in the heat of a fire

Thus basic principles of use in building design are as follows.

- Sub-divide the building into the smallest practicable compartments of fire-resistant construction (i.e. with fire-resistant floors, walls, ceiling and self-closing doors).

A fire started in an inorganic chemistry laboratory located on the top storey of a four-storey building. (The cause was either overheating of an oil bath due to failure of a thermostat or ignition of flammable vapour by wiring contacts in a domestic refrigerator.) The fire had broken through the roof before it was noticed, late at night, by a passer-by. Whilst the fire inside the laboratory was intense, fed by flammable liquids from exploding Winchester bottles, it was confined to a section of the top storey only because of a fire-door.[64]

- Close all vertical and horizontal openings with tight-fitting fire-resisting doors and hatches.

In 1980 a massive warehouse fire which resulted in damage of £72.5 million was caused by ignition of an unprotected parts store by an exploding power transformer. This transformer was located inside the warehouse and isolated from the storage area only by wooden doors which were quite ineffective.[65]

- Ensure each compartmentalised area can contain the fire and ensure building stability.

In the incident described on page 1545 when some of the metal alloy hangers holding up part of the suspended ceiling failed the fire entered an undivided space beneath a flat roof. It spread through the void and destroyed most of the roof.[64] Fire-resisting doors and partitions were not continued through the false ceiling and hence could not prevent fire spread.

Sub-division is hence preferable to full building height, and indeed above if the roof is not fire-resistant.
- Use fire-resisting construction for walls, beams, etc. and for partitions which form protected enclosures (i.e. lift, hoist and stair enclosures).
- Protect structural steelwork by encasing it in brick, concrete or other insulation. In chemical process plants fire protection may be applied to load-bearing members of steel structures in accordance with Table 27.7[66] and as illustrated in Fig. 27.8. The top of any such fire protection is

Table 27.7 Typical fire protection requirements for structural steel in chemical process plants

---

Equipment-supporting structural elements

Fire protection is applied if:
- Structures are in areas where 'flammable liquids' are processed/handled so that the possibility of a sustained fire arises. If 'non-flammable liquids' are processed and are directly adjacent to the above area, fire protection may be omitted if the possibility of a sustained fire is considered remote, e.g. due to layout, etc.
- Structures supporting the equipment are >1.5 m high.
- Equipment contains >500 litres of flamamble liquid.
- Structure supports two or more pieces of equipment containing together >500 litres of flammable liquid and <6 m apart.
- Structure carries a piece of equipment containing non-flammable material and a total weight of ⩾2,500 kg.
- Structure supports two or more pieces of equipment containing an aggregate weight of ⩾2,500 kg and located <6 m apart.

Fire protection for these structures is applied from grade level to a height of 4.5 m above the hazardous level, and includes all stanchions, equipment–support beams, etc. Examples are shown in Fig. 27.5.

(An alternative for load-bearing elements is to install water spray protection.)

Pipe-support structures

Fire protection is applied to steel structure for overhead pipe tracks with a height of ⩾1.5 m and located with 5 m from a source of fire hazard, and is applied to the stanchions only. (A source of fire hazard is any location at either grade or elevaated level, where an accidental release of hydrocarbon may cause a fire due to spontaneous ignition or ignition from contact with hot equipment, etc.) If pipe-support structures are combined with structures supporting fin-fan coolers, the stanchions, cooler supporting beams and all members which reduce the effective buckling length of the stanchions are protected including the supports of the tube bundles. Examples are shown in Fig. 27.5a.

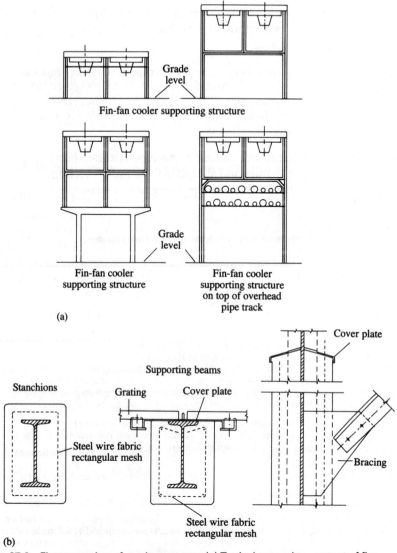

Fig. 27.8 Fire protection of steel structures. (a) Typical examples – extent of fire protection. (b) Typical details – concrete fire protection

protected from rainwater ingress between the steel structure and the applied fire protection by cover plates continuously welded to the steel structure, e.g. as in Fig. 27.8(b) and by water-proofing. The purpose of such fire-proofing is to prevent collapse since with equipment containing, or conveying, flammable material this could add significantly to a fire.
- Design floors to convey the weight of water from hose reels or sprinklers and arrange to collect and free-drain this maximum water loading.

- Fire-stop ducts, cable trays, etc. In the past the inadequacy of fire-stops in the openings through which cables penetrated walls, floors or floor–ceiling assemblies has sometimes permitted cable fires to propagate from one fire area to another.

Sparks from a bark-burning boiler ignited an accumulation of sawdust and bark in a 60 cm wide cable tray in a pulp and paper mill. This then ignited 191 signal and control cables to power and recovery boilers. Poor housekeeping, lack of approved fire-stops and the absence of fire protection for the tray resulted in damage requiring a seven-day shut-down.[67]

Attention is therefore recommended for cable-penetration fire-stops, use of highly flame-retardant cables and adequate fixed fire protection.[67]

- Use appropriate surface finishes depending upon location, as exemplified in Table 27.8.

Table 27.8 Broad classification of materials groups for surface finishes, or walls and ceilings, in buildings[78]

A: Inorganic group
Brickwork, blockwork, concrete, plasterboard, ceramic tiles, plaster finishes (including rendering on wood or metal laths), asbestos boards.

B: Cellulosic group
(Not flame retardant treated)
Timber, hardboard, particleboard (chipboard), blockboard.
Acceptable in small rooms of floor area not exceeding 30 m^2.
Acceptable in small areas of other rooms, these areas not exceeding half the floor area of the room or 60 m^2, whichever is the lesser.
Not acceptable on escape routes, i.e. staircases, corridors, entrance halls.

C: Cellulosic group
(Flame retardant treated to achieve Class 1 surface spread of flame rating)
Acceptable in all rooms, provided evidence of suitable treatment is available.
Not acceptable on escape routes, i.e. staircases, corridors, entrance halls.

D: Woodwool slab
Acceptable in all locations.

E: Plastics – thermosetting
(Decorative laminates)
Acceptable as for Group B unless shown to be of flame retardant grade (to Class 1 surface spread of flame rating evidence to be provided) in which case acceptability will be as for Group C.

F: Plastics – thermoplastics
(Expanded polystyrene wall and ceiling linings)
Acceptable on inorganic surfaces in thicknesses not exceeding 5 mm on walls, 12 mm on ceilings, provided not finished with gloss paint, in same situations as Group B. (Expanded polystyrene tiles or other expanded polystyrene surfaces which have been painted with gloss paint should be removed.)

Within the UK the main requirements relating to fire resistance and surfaces of factory building structure (and means of escape, means for giving warning in case of fire, fire-fighting equipment, and fire instruction and drills) are encompassed by the Fire Precautions Act 1971.

With regard to fire resistance and surfaces of a building structure, standards recommended for the fire resistance of the elements of a building structure (e.g. floors, walls, etc.) – where this exists or is provided because it is essential – are given in Table 27.9.

Table 27.9 Recommended standards for the fire resistance of elements of a factory building

	Walls (minutes)	Fire-resisting doors (minutes)	Floors (minutes)
Floor immediately over a basement	—	—	60*
All separating floors	—	—	30*
Separating a stairway	30	30†	—
Separating a protected lobby	30	30	—
Separating a lift well	30‡	30‡	—
Separating a lift motor room	30	30	—
Separating a protected route	30	30†	—
Separating compartments	60	60	—
In a corridor to sub-divide it	—	30	—
In a stariway from ground floor to basement	—	2 × 30 or 1 × 60	—

* Excluding, e.g. a gallery floor.
† Except a door to a WC containing no fire risk.
‡ Except a lift well contained within a stairway enclosure.

Apart from small areas the surface finish of walls and ceilings should correspond to a standard not less than

    Class 0 (means as prescribed by Building Regulations)   in escape routes
    Class 1 (BS 476 Part 7: 1971)   in rooms
    Class 3 (BS 476 Part 7: 1971)   in small rooms

Fire detection and extinguishment are considered later.

Access

Free access and means of escape (page 63) need to be considered at the design stage. The need for adequate access for emergency services was noted on page 671. A height of >9 m will facilitate access for larger appliances, e.g. turntable ladders or hydraulic platforms. As an example, a building >85 000 m^3 and 9 m high would require a suitable access along all perimeter walls to cater for the larger type of fire appliance.[68]

Access requirements are crucial in the layout of process plants storing substantial quantities of flammable materials.

In the major oil tank fire summarised on page 160, at one period the fire-fighting operations involved over 150 firemen, 45 pumps, 7 major foam cannons (mobilised from other refineries) plus other appliances and road tankers carrying foam compounds. Some difficulties were encountered because of the congestion around three elevations of the tank and on roads not designed for such a large number of foam tankers and appliances.

However, in addition to supply considerations the ability to fight less devastating fires from a reasonable distance is also a factor.

A rim seal fire was being fought with foam from a fire-truck boom. The fire was almost extinguished when the floating roof exploded. Product is believed to have leaked into one pontoon compartment and created a flammable vapour–air mixture for the explosion.

Automatic fire detection

Fire losses are often aggravated by fires being discovered at a late stage in their development, and this may be accompanied by a delay in reporting the fire through lack of training or poor communications.[35] Such delays, which are particularly likely to arise outside normal working hours, can be avoided by automatic detection installations. This should often be in conjunction with automatic transmission of the alarm to the fire-fighting service.

Such systems can generally detect fires which are of the order of $0.1 \, m^2$ in size, with smoke detectors being more sensitive than heat detectors.[61] On the assumption that within most urban areas a fire brigade can be in attendance in a matter of minutes, an automatic fire detection system with a direct line to the brigade is a powerful form of defence.

However,
- any automatic detection and transmission system must be kept in perfect working order. Hence automatic or regular testing, and correct and effective maintenance procedures, are essential.
- any system is of no value unless it initiates rapid action to extinguish the fire.
- automatic systems providing blanketing, which could result in asphyxiation (e.g. carbon dioxide in switch rooms), need to be disarmed if a person enters the room.

So even with a well-designed, reliable automatic detection system it is necessary to make an accurate assessment of the likely rate of fire development and for this to be related to the time taken for fire-fighting to commence. If it is apparent that a fire may reach major proportions before fire-fighters will arrive then an appropriate type of automatic extinguishing system is required. Water sprinklers are the most common type; agents such as bulk carbon dioxide, dry powder, foam or high expansion foam, or vaporising liquids are used in special situations.

The 'early warning' provided by an automatic fire detection system will generally serve for some combination of the purposes listed in Table 27.10 (after ref. 35).

Table 27.10 Functions of automatic detection systems

- Early detection of fire in remote, unobserved/unattended or enclosed plants.
- Operation of audible alarms to warn personnel.
- Signal an emergency to management.
- Summon the Fire Brigade via a direct line.
- Actuation of fixed fire protection installations.

A short-circuit in a mains-operated calculator ignited the plastic body and started a slowly developing fire in a first-floor study adjoining a research laboratory. Since it started in the early hours of the morning, the fire was not discovered until it spread into the laboratory and activated heat detectors which summoned the fire brigade through an auto-dialler system. (This demonstrates that a detector system should preferably be installed to cover all areas, otherwise a fire can reach considerable proportions before the alarm is given.)[64]

Gas detectors

In some situations in which flammable gas or vapour may be present due to leakage, particularly within rooms or modules but also in open-air locations, warning that a pre-fire condition is being created can be provided by appropriately located flammable gas detectors. The main types are:
- catalytic pellistor detectors, which work on the principle that when a flammable gas passes over a hot surface containing a catalyst combustion occurs on the surface and the temperature rise can be detected.
- infra-red (IR) absorption band detectors, the principles of one type being as shown in Fig. 27.9.

Pellistor detectors are widely used but most are easily poisoned and are consequently unreliable for certain environments. Therefore, for land-based application, where there is a high risk of catalyst poisoning, e.g. as on many chemical plants, IR band detectors are recommended.

Fire detection

The types of fire detection systems available, and their likely areas of use, are as follows.[69]

Smoke detectors

Smoke detectors are available as either single-point ionisation or optical detectors. Ionisation chamber detectors are based on the principle that smoke entering a small volume of ionised air across which there is an

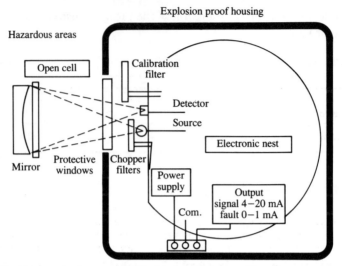

Fig. 27.9 Schematic diagram of typical infra-red detector[63]

electrical potential will cause a small, but detectable, change in the current flow. Optical detectors usually use the scattering effect of smoke on light; scattered light from a light-emitting diode inside the detector head is picked up by a photo-sensor.

Optical detectors are generally more reliable for smouldering fires and ionisation chamber detectors for fast-burning flaming fires.

Heat detectors

Heat detectors include line, sprinkler head, and point detectors, operating on either rate of rise (e.g. 5 °C per minute) or fixed temperature (e.g. 70 °C).

Flame detectors

The two main types of flame detectors are the infra-red (IR) and ultra-violet (UV), depending on the wavelength of light to which they are sensitive. IR detectors are more suited to indoor applications. UV detectors are suited to outdoor use but are prone to unwanted alarms from ionising radiation, quartz-iodine lighting, welding or lightning.

Flame detectors are suitable for flammable liquid and gas risks but need to be located fairly close to the source, e.g. ≤15 m.

Unwanted or false alarms from automatic fire detection systems are a real problem, e.g. amounting to twenty unwanted alarms for every one real alarm. Some trends in the number of unwanted alarms per 1000 detections per annum, which will of course vary with detector manufacture and situation, are given in Table 27.11. Typical sub-categories of unwanted alarms are given in Table 27.12.

Table 27.11 Number of unwanted alarms per 1000 detections per year for different detector types[69]

Detector type	Alarm rate (events per 1000 detections per annum)
Smoke – ionisation	20
Smoke – optical	70
Heat	10
Flame	10

Table 27.12 Sub-categories of unwanted alarms per 100 unwanted alarms[69]

Unwanted alarm sub-category	Events rate per 100 unwanted alarms
Equipment fault	30
Human error	21
Detected non-fire disturbance	26
Other	12
Unknown	11

Conversely an automatic fire detection system may not function when it is required because
- the detector was unsuitable for the type of fire involved; or
- the detector was badly positioned, and therefore did not 'see' the smoke, heat or flame; or
- the system was not working due to an unannounced fault.

Effective inspection and maintenance is therefore crucial.

Reliability considerations

The best way to deal with a fire is often to rely on passive means, e.g. fire walls, separation and redundancy of safety systems. Water spray systems will be needed in many situations to reduce the potential consequences of any fire to an acceptable level. However, the reliability of water spray systems has a significant impact on overall safety.[69]

The probability of a sprinkler system failing to operate in a fire, because of an equipment fault, may be 1×10^{-2} compared with, e.g. 5×10^{-3} for a traditional automatic fire detection system with three-monthly test intervals, or 1×10^{-4} for a microprocessor-based analogue system. Hence well-engineered fire detection systems together with effective manual fire-fighting may conceivably be cheaper and more reliable than traditional approaches. However, a water spray system may be more appropriate on occasions when, despite rapid automatic detection, the delay in initiating manual fire-fighting will be too long for effective fire control.

Caution is, however, required in using reliability data. Some examples of the variation between equipment types are given in Table 27.13.[69]

Table 27.13 Variations between types of fire detection equipment

Detector type	Faults per year (some of which cause unwanted alarms)
Smoke detectors	0.026–0.35
Heat detectors	0.0088–0.39
UV flame detectors:	
Industrial use	0.1
Offshore use	0.6

Moreover, reliability data in data banks often relate to standards for particular environments and periods of maintenance. Alternatively, if an assumed three-monthly maintenance/test cycle is not followed there will be a reduction in reliability. Fire detectors located in inaccessible positions may tend to go unsuspected. Measures are also necessary to prevent the same heads being tested at the specified intervals, i.e. all the heads require equal attention.

Fire extinguishment

In the initial stages of the incident described on page 881 the fire was discovered in the ammonium nitrate fertiliser packed in paper bags on the *SS Grandcamp*. Two pails of drinking water and the contents of a soda acid extinguisher were expended on it. A hose line was called for but an order was then given not to use it in case the cargo was damaged or the ammonium nitrate swelled. (Incidentally the cause of this fire was possibly a carelessly discarded cigarette end.)

The majority of fires commence slowly initially, so if suitable fire-fighting equipment is readily available – and is used promptly and properly – they can be extinguished before they escalate. Damage and risks to occupants of a building or site are hence minimised.

Common fire-extinguishing agents are described on page 167 *et seq.*; the special provisions with organometallics are summarised on page 1411 and for LPGs on page 1352.

The normal range of first attack equipment, described on page 174, is needed to enable employees to tackle fires in their early stages.

Appropriate means should be provided having regard to the fire classification given there. In summary provision of fire-fighting equipment for Class A fires (see page 174) is appropriate in most industrial situations. Hence water-type extinguishers or hydraulic hose reels are required. The approved standard for hose reel installations is that with ≯45 m of hose no part of the area to be protected is >6 m from the nozzle of a reel when the hose is fully extended. Hose reels must be permanently connected to an adequate mains water supply and be located where they are both conspicuous and accessible. Portable extinguishers should be provided in sufficient numbers to give adequate cover to those parts of a site not covered by hose reels. The equivalent of one 9 litre capacity water-type extinguisher is normally sufficient for ≤210 m^2 of area, provided there are not <2 extinguishers per floor.

For class B fires it is usually appropriate to provide portable extinguishers of the powder, CO_2, foam or Halon types. Replacements for Halon 1301 include a total flooding gaseous mixture of nitrogen, argon and carbon dioxide which extinguishes fire within a room by dilution of the oxygen level to between 12 and 14%. An alternative mixture comprises 50% nitrogen and 50% argon. For machinery spaces, e.g. engine, generator and boiler rooms, a suppression system has been developed which utilises atomised water. Guidance on the minimum scale of provision is given in Table 27.14. The only effective action against Class C fires is to stop the flow of gas. Hence no special provision can be made for dealing with such fires. Special powders should be provided for dealing with Class D fires.

Some additional practical considerations in fire-fighting are summarised here.

Foams

The common range of foam concentrates is summarised in Table 27.15. As discussed in Chapter 4, foam is applicable to
- extinguishment of burning liquids of lower density than water;
- blanketing of spills to prevent ignition;
- insulating and protecting exposures from radiant heat;
- extinguishment of surface fires involving ordinary combustibles.

However, covering a spillage with foam will not always reduce the rate of evaporation since with a cold liquid, such as LNG or chlorine, water draining from the foam may increase the rate of evaporation. An estimate of the evaporation produced may be based on the assumption that 25 per cent of the water in the foam will drain in the first 15 minutes.[70] The higher the expansion of the foam the less the drainage but, as discussed on page 171, high expansion foams may be blown away by the wind. Furthermore, cases have been reported in which application of foam to some fuels could cause ignition due to formation of static electricity generated at the foam nozzle.[71]

Foam is not suitable for
- fires involving gases;
- liquid jet fires;
- water-reactive material fires;

Table 27.14 Minimum scale of provision of extinguishing media to deal with a fire involving exposed surface of contained liquid

1	2				3
	Quantity of extinguishing media				
Largest undivided* exposed area in the room	Powder (kg)	Carbon dioxide (kg)	Foam (litres)	Halon (kg)	Maximum aggregate area per unit in column 2 (m^2)
Less than 500 cm^2	1	2	4.5	0.75	0.2
500 cm^2 to 0.1 m^2	1	2	4.5	0.75	0.4
0.1 m^2 to 0.25 m^2	2	3.5	4.5	2	0.8
0.25 m^2 to 0.5 m^2	3	4.5	9	3	1.6
0.5 m^2 to 1 m^2	3	7 (with trolley)	See below†	3	2.4
1 m^2 to 2 m^2	8	—		8	4.8
2 m^2 to 3 m^2	14 (with trolley)	—		—	6.4

* Areas less than 1.5 m apart should be considered to be undivided.
† In the case of foam the quantity of extinguishing media needed may exceed the capacity of the largest available unit, e.g. two 9-litre foam extinguishers can cater for an undivided area of 0.5 to 1 m^2 or four 9-litre foam extinguishers for 1 m^2 to 2 m^2 etc.

Note: Some figures in column 2 may not equate with extinguisher sizes commercially available. The extinguisher size above the recommended figure should normally be chosen.

Table 27.15 A selection of foam concentrates

Type	Trade names	Use on fires	Vapour sealant (flammable/toxic)	Expansion ratio	Recommended storage construction	Base injection use	Grades (%) available
Protein	Nicerol	Hydrocarbon	Hydrocarbon Liquid chlorine	8–1	Mild steel/GRP	No	3%, 6%
	Regular/Aero-foam	Hydrocarbon	H/carbon & L/chlorine	8–1	Mild steel/GRP	No	3%, 6%
	Pro Foam 803/806	Hydrocarbon	H/carbon & L/chlorine	8–1	Mild steel/GRP	No	3%, 6%
Fluoro protein	FP70/FP570	Hydrocarbon	H/carbon & L/chlorine	8–1	Mild steel/GRP	Yes	3%, 6%
	Plus F/XL3/XL6	Hydrocarbon	H/carbon & L/chlorine	8–1	Mild steel/GRP	Yes	3%, 6%
	Fluoro Foam 806/803	Hydrocarbon	H/carbon	8–1	Mild steel/GRP	Yes	3%, 6%
Synthetic	Expandol	Combustible material in enclosed units	LPG	From 25–1 to 1000–1	Mild steel/GRP	No	2%–6%
	HEX/SYNDET		LNG		Mild steel/GRP	No	2%–6%
	SYNDET		Liquid chlorine		Mild steel/GRP	No	2%–6%
	HI-FOAM		Liquid Ammonia		GRP stainless	No	2%–6%
Alcohol type	Alcoseal F.F.F.P	Polar solvents	Polar solvents	8–1	Stainless/GRP	No	4%–6%
	Master Foam*	Alcohols	Alcohols	8–1	Stainless/GRP	No	4%–6%
	Universal	Water soluble	Water soluble	8–1	Stainless/GRP	No	4%–6%
	A.P. Foam	Fl. liquids	Flammable liquids	8–1*	Stainless/GRP	No	6%, 9%
	Light Water ATC						
Fluoro chemical (aqueous film forming)	Tridol	Hydrocarbon	Hydrocarbon } Some can be used at 50–1	8–1	Stainless/GRP	Yes	3%, 6%
	Fluorofilm	Hydrocarbon	Hydrocarbon	8–1	Stainless/GRP	Yes	3%, 6%
	Aero-Water	Hydrocarbon	Hydrocarbon	8–1	Stainless/GRP	Yes	3%, 6%
	Light Water	Hydrocarbon	Hydrocarbon expansion + in non-aspirated form	8–1	Stainless/GRP	Yes	1%, 3%, 6%
Fluoro aqueous film forming	Petroseal	Hydrocarbon	Hydrocarbon	8–1	Stainless or mild steel with epoxy protection/GRP	Yes	3%, 6%

* In non-aspirated form.

- live electrical equipment;

and judgement is necessary in applying it to heated liquids above 100 °C.

A critical rate of foam application is necessary to counter the breakdown of foam through vaporisation, due to heat transfer from the fire, and drainage caused by heating and the action of the flammable liquid.

As a result of the possibility of static electricity generated by foam jets, referred to on page 1529, one company has changed its procedures for dealing with storage tank incidents and recommends that foam should not be used unless a tank is actually on fire or there is an imminent and immediate risk of ignition which cannot be removed. This is contrary to the standard firefighting practice of blanketing exposed flammable liquid with foam before it has the chance to catch fire. If foam is used it should be directed at the tank walls rather than the liquid surface and care should be taken that foam, not water, is flowing from the nozzle. Foam should preferably be applied from fixed 'pourers' on the tank, avoiding monitor jets or hand-held hoses.[38]

Sprinklers

Sprinklers are widely used for protection of buildings incorporating Class A fire risks (pages 166 and 168).

Sprinkler systems require an unobstructed flow of water in order to control a fire. Thus, if an obstruction arises in the sprinkler piping (e.g. due to scale formation, poor installation practice or material introduced during repair or fire-pump maintenance) it may prevent adequate water reaching the first sprinklers that open in a fire. The fire will not be controlled, an excessive number of sprinklers will open (decreasing further the water supply to each sprinkler) and the fire can spread. Routine inspection, and cleaning if necessary, are recommended.[71]

For the majority of warehousing a sprinkler system is generally the most suitable fixed fire protection. For high rack storages an 'in-rack', rapid response, sprinkler system is recommended.[68] The two main requirements are

- to detect and control a fire and prevent its horizontal and rapid vertical spread. This will enable fire-fighters to deal with the outbreak as a 'small fire';
- to cool the steel racking system and so prevent it collapsing during the early stages of fire-fighting.

In the case of any warehouse with narrow aisles and containing high bay racking fire-fighters would encounter problems of entering and penetrating into the building and in locating the seat of a fire, e.g. several metres above ground level. The use of ladders and fire-fighting jets would also be severely restricted. Therefore automatic fire ventilation connected to an automatic smoke detector system is advisable in order to keep the layer of smoke and hot gases above storage height. This will improve visibility at lower levels and help to restrict lateral fire spread. As the vents are in the roof of the building they are mainly used in single-storey structures. They may also require the sub-division of the roof space by vertical, non-combustible, curtains to prevent unrestricted flow of smoke across the underside of the roof.[72] Vents are unlikely to give adequate protection to high risk storage if used without sprinklers.

In order to minimise water damage the following precautions are advisable.

- Sprinkler stop valves should be inspected and maintained regularly.
- Operators should be familiar with the positions of sprinkler system sectional control valves, pit valves and water valves, in case the sprinkler riser valve is inoperative or inaccessible.
- If sprinkler valves are locked, an adequate number of keys and a bolt cutter should be provided on a leather strap (or fibre rope used to secure the hand-wheel).

> Whilst floor tiles were being replaced using a naked flame a fire started, believed to be due to the ignition of rags in the area. One sprinkler head was set off. Although the fire itself caused very little damage, considerable damage did result from the inability to stop the water flow from the fused sprinkler head after the fire was out. The sprinkler stop valve was faulty and plant personnel could not shut it. A sprinkler stopper – which could have been inserted between the sprinkler's deflector and orifice to stop the flow – was available. Water ran for about 45 minutes resulting in an overflow of about 2500 gallons of water into a room below, until the Fire Department closed the valve using a crowbar.

Water supplies

The provision of ground hydrants depends mainly on the size of fire which it is anticipated may have to be tackled within a building. It is dependent on the fire loading within an uncompartmented building, e.g. as in Fig. 27.10. A storage building may be expected to have a higher fire loading than other buildings as indicated.

Hydrants should be installed generally on a minimum 10 cm service pipeline. Their number is determined by the availability of other water supplies, accounting for the maximum water supply required for fire-fighting in any reasonable eventuality.

Drainage of fire-water

The potential pollution problems associated with contaminated fire-water drainage are highlighted on pages 1225 and 1542.

> The Sandos fire referred to on page 911 could not be extinguished by foam alone and enormous quantities of water were required to prevent the fire from escalating to involve neighbouring warehouses and production buildings. It was

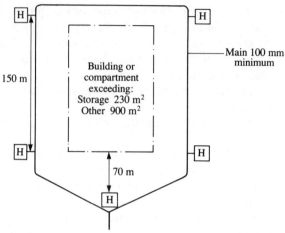

Fig. 27.10 Provision of ground hydrants. Hydrants are not more than 70 m from any entry to the building and not more than 750 m apart. Open water supplies can be taken into account; town main hydrants can also be taken into account

drainage of some of this water which caused severe ecological damage. Subsequently, two catch basins were constructed to provide greater retention capacity for fire-water, one of 15 000 m^3 and the other 2500 m^3 capacity.[73]

Following the incident described on page 1542 the fire reached major proportions with flames reaching a height of over 90 m (Fig. 27.11). Extinguishing/cooling fire-water was applied at about 3500 m^3/h. Approximately 90 m^3 of foam concentrate were used.[74]

The surface drainage system could not cope with these quantities and extensive surface flooding occurred. Hydrocarbons released from the unit floated on the water and, although a foam blanket was applied, breaks in the foam allowed flash fires to occur intermittently.

Improvements were subsequently made to the fire-water drainage system.

Where it is not feasible to increase existing drainage systems, if they are possibly inadequate, consideration should be given to directing the flow of excess water to less hazardous areas and to the provision of mobile pumps to remove excess water.[74]

Impairment of fire protection systems

Even when there is a regular maintenance procedure fire protection systems may be impaired occasionally, i.e. reduced to less than normal design effectiveness for some period of time. This may occur because of a mistake, a breakdown or a deliberate decision, e.g. for maintenance. So far as sprinkler systems are concerned the causes may be:[75]

- closed sprinkler valves; this is the most common fault (e.g. for new

Fire protection practice 1561

Fig. 27.11 Fire at a chemical plant which resulted in river pollution (courtesy West Yorkshire Fire Service)

equipment installation, relocation or building alterations, replacement of damaged sprinklers);
- freezing of piping in unheated areas, and subsequent cracking and discharge;
- inoperative fire pumps;
- failure of electric power;
- failure of public water supply;
- failure of pipework, e.g. due to poor installation or building settlement.

Many losses have escalated because impairments were not handled effectively. General guidelines for shutting a sprinkler control valve are summarised in Table 27.16;[75] the generality of such principles to other protection systems (e.g. carbon dioxide system, explosion suppression system, Halon system, dry chemical system) is obvious.

Consequent upon the fire described on page 1560 guidelines on the options for dealing with fire-water run-off are to be published by the Health and Safety Executive.[76]

Warning in the event of fire

Within the UK means for giving persons in the building warning in case of fire are a specific requirement of the Fire Precautions Act 1971.

In most buildings of 'high risk' an electrical alarm system should be provided. A less sophisticated type of warning system may, however, be acceptable in single-storey premises.

In 'normal risk' buildings either an electrical fire alarm, or an internal telephone system associated with a distinctive alarm, e.g. a siren or hooter

Table 27.16 Some guidelines for closing a sprinkler system control valve

- Keep protection in service wherever possible.
  Explore the possibility of not shutting down any protection.
  Leave the system in service if new project work can be conducted around the charged system without shutting the control valve (especially if cutting and welding is involved).
- Restore protection promptly.
  Plan so that as much work as can be done without shutting down protection is completed first; then shut down and complete the project as quickly as practicable, i.e. working continuously if possible.
  Restore protection at the earliest possible stage, not after the end of the project or cleaning up.
- Limit the size of the unprotected area.
  If only one system needs to be impaired avoid shutting down protection to the whole plant.
  Cap the ends of appropriate branch-lines to the affected area, leaving the majority of the system in service.
  For work on, or near, the sprinkler riser or underground feed main the system can be fed from nearby systems or hydrants and the affected area can be isolated. (Care must be taken not to cross-connect potable and non-potable sources.)
- Ensure the workforce involved remain alert so that protection is restored quickly.
  Alert contractors and construction personnel to the fact that fire protection is out-of-service.
  Alert plant personnel also, so that fire safety is given the highest priority.
- Eliminate ignition sources, including where possible shutting down hazardous operations.
  Plan if possible for impairments to be in idle plant hours when fewer ignition sources are generally present.
  Prohibit smoking for the duration of the impairment.
  Relocate combustible storage to fully protected areas, if possible.
- Ensure any fire can be quickly detected, e.g. by establishing a continuous fire watch.
- Alert the in-house Emergency Organisation and the Local Fire Brigade.
  Alert emergency response team members.
- Have the sprinkler valve closed by an operator who, if possible, remains stationed at it for the period of impairment ready to open it in an emergency.
- Connect up a temporary fire hose ready to use in the affected area.

sounded from a position which is manned continuously during working hours, will be suitable. One of the methods noted below for 'low risk' buildings may be acceptable in single-storey buildings or in two-storey buildings of limited floor area.

In multi-storey 'low risk' buildings a warning system similar to that for a 'normal risk' may be advisable. Otherwise dependence may be placed upon simple hand-operated devices, e.g. gongs, triangles, klaxons or rope-operated bells. Whistles or hand bells may be acceptable provided it can be ensured that they will always be in place.

All electrical alarm systems should comply with an appropriate standard, e.g. British Standard Code of Practice CP 1019: 1972. Call points should preferably be of the release button type and should, like points for any other system, be situated such that a person can reach one within 30 m to raise the alarm.

The audible alarm, and any visual warning signals, should be distinctive. They must be sited such that the warning is perceptible throughout all parts of the building or relevant site. An indicator panel is required if the building or site is of such an extent that a delay could occur in ascertaining the location of any fire; this panel should be sited with due regard for the layout of, and the fire alarm procedure for, the site.

> For warehouses a manual alarm system, incorporating automatic fire detection equipment, is recommended to ensure a rapid response by employees. This can also provide an automatic call to the local fire brigade. The automatic fire detection equipment may be coupled to an automatic fire ventilation system, fire-doors, shutters, etc. This reinforces other provisions for restricting fire spread.

Fire instruction and drills

All persons should be instructed and trained to understand the fire precautions and the action to be taken in the event of fire. This should include the distinction between stand-by and evacuation signals and should be based on written instructions. For example, a typical Personal Fire Instruction Card, for each individual to carry with them, is reproduced as Fig. 27.11.

Instruction should be given by a competent person at least once every year and provide for the details summarised in Table 27.17. Certain staff (e.g. engineering and maintenance staff, chemists, security staff, telephonists) will require instruction and training relevant to their special responsibilities at the time of a fire.

Fire Instruction Notices stating clearly action necessary upon discovering a fire and on hearing the fire alarm should be displayed at conspicuous positions in all parts of a site.

A practice fire drill should be carried out at least once a year. This should simulate conditions in which one or more of the escape routes from a building is obstructed, although this is not, of course, an acceptable situation. Arrangements must be made, including the provision of assistance, to ensure that those handicapped permanently of temporarily (e.g. leg in plaster) can escape. Lifts should not be used.

> In the Pepcon disaster two people died, the manager who tried to put out the fire and a disabled person left in the office.

(Within the UK there is now a wider duty imposed upon an employer under the Management of Health and Safety at Work Regulations 1992 to establish and where necessary give effect to appropriate procedures to be followed in the event of 'serious and imminent danger' to persons at work in his undertaking. There is also a requirement to nominate a sufficient number of competent persons to implement evacuation procedures.)

Means of escape[77,78]

The means of escape in case of fire should be of adequate size, not involve an excessive travel distance and lead to outside at ground level. Escape should

If you discover a fire

1. Break the glass in the nearest fire alarm.
2. Attack the fire with the fire extinguishers provided. *Take no risks.*
   (If in doubt whether it is a fire do not investigate for yourself but call the Fire Brigade)

If you hear the evacuation signal

3. Leave the building immediately by the nearest exit
4. Report to your assembly point and make sure your name is entered on the roll call
5. Do not run
6. Do not go to the cloakroom
7. Do not stop, or return, to collect personal belongings

The evacuation signals will never be sounded for surprise tests

If danger is not imminent

8. Stop machines
9. Shut off gas and electric power, but *not* lighting
10. Close doors and windows

(Reverse)

You should know:

1. The position of the nearest fire alarm point
2. The position of the nearest fire extinguisher and how to operate it. (Read instructions printed on it before a fire occurs)
3. The nearest exit and the route to it
4. Your assembly point
5. Report to your next senior if:

You do not hear the regular test of the evacuation warning
Any exit door is locked, or any exit route is obstructed by loose material
Any fire extinguisher is missing, obstructed, damaged or apparently out of order
Any fire alarm point is obstructed

Your Assembly Point is ....

Fig. 27.12 Personal Fire Instruction Card (after ref. 72)

Fire protection practice 1565

Table 27.17 Instructions and training requirements for all personnel

- Action to be taken upon discovering a fire.
- Action to be taken upon hearing the fire alarm.
- Raising the alarm, including the location of alarm call points, internal fire alarm telephones and alarm indicator panels.
- Correct method of calling the fire brigade.
- The location and use of fire-fighting equipment.
- Knowledge of the escape routes.
- The importance of fire doors and the need to close all doors at the time of a fire and on hearing the fire alarm.
- Stopping machines and processes and isolating power supplies where appropriate.
- Evacuation of the building.

not involve going towards a fire and 'protected routes' should not be penetrated by smoke or fire. Essentially the objective is that any person can escape:

- by their own efforts rapidly, e.g. in not normally >2.5 minutes, without using a lift;
- via clearly marked routes into which doors and windows are normally closed;
- to an assembly area where a formal check on escapees can be made, e.g. against a roster. (Alternatively fire wardens may perform a check on building occupancy.)

The main factors taken into account when assessing the requirements for means of escape in case of fire are:

- building construction, which will, to some extent, determine how it is likely to behave under fire conditions;
- occupancy, e.g. the number of persons, their density, distribution and physical condition;
- building use, e.g. the nature of the contents, type and quantity of any ancillary operations such as assembly and packing.

The evacuation times, travel distances and number of exits are assessed from these factors. Increased travel distances to a place of safety are generally accepted in certain warehouses compared with other buildings.

Depending on building use and design, emergency lighting is normally required to ensure that the means of escape can be used safely and effectively at all times.

General guidelines are available for distances of travel.[77] They relate to the distance between any point in a building and the nearest fire exit, or door to a stairway which is a protected route, or a door for means of escape in a compartment wall.

Recommended maximum distances of travel are given in Table 27.18.

Where a room is an inner room (i.e. a room accessible only via an access room) the distance to the exit from the access should be a maximum of 6 m if

Table 27.18 Recommended maximum escape distances

Category of risk	Distance of travel within room, work room or enclosure (metres)	Total distance of travel (metres)
(a) Escape in more than one direction		
High	12	25
Normal	25	45
Low	35	60
(b) Escape in one direction only		
High	6	12
Normal	12	25
Low	25	45

the inner room is of 'high risk', 12 m if the access room is of 'normal risk', or 25 m if the access room is of 'low risk'.

With regard to travel within rooms, more than one exit is required if a room is occupied by >60 persons or if the distance of travel to the only exit exceeds that in Table 27.18(a).

The width of any exit should not normally be <75 cm. If it is necessary to have more than one exit, the aggregate width of all exits less any one should not normally be <75 cm for up to 100 persons or <1.1 m for up to 200 persons, with an extra 7.5 cm for each additional 15 persons (or part).

Exits will be satisfactorily sited if the angle between routes from any point to the exits is $\geqslant 45°$; or, if the angle is <45°, the travel distance to the nearest exit is less than that recommended in Table 27.18(b).

The contents of any room occupied by employees should be arranged or disposed so that there is free passageway to a means of escape.

With regard to travel from rooms to a stairway or final exit, escape in more than one direction is available from any point from where there are different routes leading to
- separate stairways which are protected routes;
- separate final exits;
- a combination of these;
- a combination of one of these with a door for means of escape in a compartment wall; or
- separate doors for means of escape in compartment walls, provided a final exit/door to a protected stairway can be reached in both the adjoining compartments.

In a building (or part) of high risk where escape from any point is in one direction only, and any part of the route is a corridor, the corridor walls and partitions should be of fire-resisting construction and doors should be fire-resisting. A main corridor should not normally be <1 m wide. Corridors >30 m in length should be sub-divided by fire-resisting doors to prevent the free travel of smoke.

Generally two or more stairways are advisable except in a building (or part) of low risk with $\ngtr 2$ floors above the ground floor, or in a building (or part) of normal risk with $\ngtr 1$ floor above the ground floor.

Stairways should generally be $\geqslant 80$ cm wide and the aggregate width of the stairways should be sufficient for the likely number of escapees. Consideration should be given to the risk of one stairway being inaccessible. Stairway separation should be such that a person need not pass through the enclosure to reach an alternative route. If this is impossible an alternative route may bypass the stairway by means of balconies or intercommunicating doors between adjacent doors.

In any building with a floor >18 m above ground level, the only doors into a stairway should be to WCs, to protected lobbies, to lift wells contained within a stairway enclosure or to final exits.

All stairways in buildings (or part) of high risk, or in buildings (or part) of normal risk, except if there is $\leqslant 1$ floor above ground level and >1 escape route, should be separated from the remainder of the building by fire-resisting construction and fire-resisting doors. Any stairway in a low-risk building should be similarly protected it there are >2 floors above the ground floor.

It should be ensured that any external stairway can be used at the time of a fire (i.e. with regard to weather protection, and smoke and flame issuing from openings in the external building walls). Spiral stairways would only be suitable for $\ngtr 50$ able-bodied adults. Portable ladders and throw-out type ladders are not considered suitable. Automatic lowering lines and other manipulative emergency devices for self-rescue are only for use in exceptional circumstances. Any ramps should not be steeper than 1 in 10. Lifts and hoists are unacceptable as means of escape.

It should not normally be necessary to rely on wall hatches or floor hatches for means of escape. Balconies, bridges, walkways, etc. can often provide suitable escape routes. Persons should not normally have to ascend to a higher level in order to escape.

Escape doors should open in the direction of escape, or sometimes slide. All fire-resisting doors should generally be fitted with positive-action self-closing devices. Restrictions apply to the use of wicket doors in a larger door or in a shutter.

Loading doors, goods doors, shutters, up-and-over doors, etc. may not provide satisfactory exits. Similarly windows are not generally acceptable as exits.

Wherever possible escape doors should be kept unlocked and unfastened when a building is occupied. In any case they must be easily and immediately openable from inside. Appropriate notices are required on doors, e.g. 'Fire-door – keep shut', 'Push bar to open', 'Keep shut when not in use', 'Slide to open' or 'Fire exit – keep clear'.

1568 Flammable and explosive hazards

Any exit, other than a normal route, should have a 'Fire exit' notice. A similar notice with a directional arrow should be placed at suitable points along an escape route, where there is any doubt as to location.

References

1. Health & Safety Executive, *The fire and explosions at Cory's Warehouse, Toller Road, Ipswich, 14 October 1982*, HMSO, UK.
2. Anon., *Loss Prevention Bulletin*, 1992(107), 23.
3. Kletz, T. A., *Cheaper Safer Plants or Wealth and Safety at Work*, Institution of Chemical Engineers, Rugby, 1984.
4. Health & Safety Executive, *Hot Work*, Booklet HS(G) 5, HMSO, UK.
5. Health & Safety Executive, *Industrial use of flammable gas detectors*, Guidance Note CS1, HMSO, UK.
6. Health & Safety Executive, *The cleaning and freeing of tanks containing flammable residues*, Guidance Note CS15, HMSO, UK.
7. Dharmadhikari, S. & Heck, G., *The Chemical Engineer*, 499, 27 June 1991, 17.
8. *The Phillips 66 Company Houston chemical complex explosion and fire*, OSHA, 1991, USA.
9. *The Public Inquiry into the Piper Alpha Disaster (Cullen Report)*, Department of Energy, 1990, HMSO, UK.
10. Anon., *Loss Prevention Bulletin*, Institution of Chemical Engineers, August, 1988(082), 19.
11. Browning, B. & Searson, A. H., *Loss Prevention Bulletin*, Institution of Chemical Engineers, August 1990(094), 19.
12. Bond, J., *Sources of Ignition, Flammability Characteristics of Chemicals and Products*, Butterworth-Heinemann, Oxford, 1991.
13. Dennett, M. F., *Fire Investigation*, Pergamon Press, Oxford, 1980.
14. Ashmore, F. S. & Blumson, D., *Loss Prevention Bulletin*, June 1987(075), 1–10.
15. Snee, T. J., *Loss Prevention Bulletin*, 1988(081), 25.
16. Zabetakis, M. G., in *Handbook of Laboratory Safety*, ed. N. V. Steere, The Chemical Rubber Co., Ohio, 2nd edn, 1971.
17. Kletz, T. A., *What Went Wrong: Case Histories of Process Plant Disasters*, 2nd edn., Gulf Publishing Co., Houston, TX, 1988.
18. *Fire Protection Handbook*, 16th edn., National Fire Protection Association, Boston, MA, 1986, 10–68.
19. National Fire Protection Association, *Standards for Fire Protection in Use of Cutting and Welding Processes*, NFPA 51B, Boston, MA.
20. Anon., *CISHC Chemical Safety Summary*, 1976, **47**, 33.
21. Ashmore, F. S. & McNally, M., *Fire Prevention*, November 1987, 37.
22. Anon., *Chemical Age*, 12 December 1969, 40.
23. Kletz, T. A., *Learning from Accidents in Industry*, Butterworth, 1988, 43.
24. National Fire Protection Association, *National Fire Code Vol. 4*, NFPA, Boston, MA.
25. Cartwright, P., *Chem. Ind.*, 1991, 18 February, 115.
26. Health & Safety Executive, *The Fire and Explosion at Manro Products Ltd., Stalybridge, 11 December 1982*, HMSO 1983, UK.

27. Pay, F. J., *Loss Prevention Bulletin*, Institution of Chemical Engineers, December 1987(078), 1.
28. Bond, J., *Loss Prevention Bulletin*, Institution of Chemical Engineers, August 1989(088), 21.
29. Anon., *Loss Prevention Bulletin*, Institution of Chemical Engineers, October 1990(095), 15.
30. Britton, L. G. & Smith, J. A., *Plant/Operations Progress*, 1988, **7**(1), 33.
31. Anon., *Loss Prevention Bulletin*, Institution of Chemical Engineers, February 1991(097), 23.
32. Institution of Chemical Engineers, *Controlling Electrostatic Hazards*, Video Safety Training Package, Rugby.
33. Anon., *BP Shield*, 1988, 3.
34. Britton, L. G. & Smith, J. A., *Plant/Operations Progress*, 1988 **7**(1), 63.
35. British Standards Institution, *Code of Practice for Fire Precautions in Chemical Plant*, CP 3013, 1974, BSI.
36. *The Ionising Radiations (Sealed Sources) Regulations 1969*, HMSO, UK.
37. Anon., *Loss Prevention Bulletin*, Institution of Chemical Engineers, December 1987(078), 9.
38. Coates, R., reported in *The Chemical Engineer*, 25 March 1993, 6.
39. Bond, J., *Sources of Ignition*, Butterworth-Heinemann, Oxford, 1991.
40. *Report on the disaster at Whiddy Island, Bantry, Co. Cork*, 8 January 1979, Stationery Office, Dublin.
41. Health & Safety Executive, *The fires and explosion at BP Oil (Grangemouth) Refinery Ltd. – A report by the Health & Safety Executive into fires and explosion at Grangemouth and Dalmeny, Scotland, 13 March, 22 March and 11 June 1987*, HMSO, 1989, UK.
42. Göller, O., *Proc. 1st. Int. Loss Prevention Symposium*, Hague, 28–30 May 1974, Elsevier, Amsterdam.
43. Anon., *Loss Prevention Bulletin*, Institution of Chemical Engineers, December, 1989(090), 27.
44. Health & Safety Executive, *Highly Flammable Materials on Construction Sites*, HMSO, 1978.
45. *Hazardous Cargo Bulletin*, March 1984, 40.
46. British Standards Institution, *Code of Practice for Painting of Buildings*, BSI, CP 231, UK.
47. British Standards Institution, *Code of Practice for the protection of iron and steel structures from corrosion*, BSI, CP 2008, UK.
48. Health & Safety Executive, *The explosion and fire at Chemstar Ltd., 6th September 1981*, 1982, HMSO, UK.
49. Ahern, A., *Chem. Eng. Progress*, July 1991, 65.
50. *Hazardous Cargo Bulletin*, July 1980, 14.
51. Health & Safety Executive, *Highly flammable liquids in the paint industry*, 1978, HMSO, UK.
52. Beever, P. F., *Hazards X*, Institution of Chemical Engineers Symposium Series, **115**, 1989, 223.
53. Erdöl und Koble-Erdgas Petrochemie, June 1966, 19.
54. *Fire Protection Handbook*, 16th edn., National Fire Protection Association, Boston, MA, 1986, 12–92.

55. National Fire Protection Association, *Lightning Protection Code*, NFPA 78, Boston, MA.
56. Anon., *Loss Prevention Bulletin*, Institution of Chemical Engineers. February 1987(073), 1.
57. *The ICI Engineer*, Issue 7, Spring 1988, 10.
58. Anon., *Loss Prevention Bulletin*, Institution of Chemical Engineers, August 1990(094), 7.
59. British Standards Institution, *British Standards guide to prevention of inadvertent ignition of flammable atmospheres by radio-frequency radiation*, BS6656: 1986, UK.
60. Anon., *Civil Protection*, Winter 1992, **25**, 12; Anon., *The Chemical Engineer*, 11 February 1993, 4.
61. Rasbash, D. J., *The Chemical Engineer*, November 1970, CE 385.
62. Anon., *Loss Prevention Bulletin*, Institution of Chemical Engineers, October 1986(071), 13.
63. Fennel, D., *Investigation into the King's Cross Underground Fire*, 1988, HMSO, UK.
64. Anon., *Laboratory Practice*, October 1985, 13.
65. Heels, R., *Loss Prevention Bulletin*, Institution of Chemical Engineers, December 1988(084), 7.
66. Anon., *Loss Prevention Bulletin*, Institution of Chemical Engineers, 1981(043), 5.
67. Anon., *Loss Prevention Bulletin*, Institution of Chemical Engineers, 1981(041), 5.
68. Mackintosh, A., *Loss Prevention Bulletin*, Institution of Chemical Engineers, 1988(084), 1.
69. Finucane, M., *Loss Prevention Bulletin*, Institution of Chemical Engineers, 1984(059), 1.
70. Harris, N. C., paper presented at International Symposium on Preventing Major Chemical Accidents, 3–5 February 1987, American Institute of Chemical Engineers, Washington, DC.
71. Howells, P., *Loss Prevention Bulletin*, Institution of Chemical Engineers, 1994(114), 1.
72. Palmer, K. N., *Loss Prevention Bulletin*, Institution of Chemical Engineers, 1989(086), 30.
73. Anon., *Loss Prevention Bulletin*, Institution of Chemical Engineers, June 1987(075), 11.
74. Anon., *Loss Prevention Bulletin*, Institution of Chemical Engineers, 1989(089), 13.
75. Anon., *Loss Prevention Bulletin*, Institution of Chemical Engineers, April 1987(074), 17.
76. Anon., *The Chemical Engineer*, 27 January 1994, 4.
77. Underdown, G. W., *Practical Fire Precautions*, Gower Press Ltd., 1971, 87.
78. *Guide to the Fire Precautions Act 1971, 2 Factories*, 1985, HMSO, UK.

CHAPTER 28

Control of substances hazardous to health

Introduction

A wide variety of chemicals which are hazardous to health are encountered at work. Whilst most can be tolerated at the levels of exposure involved, others may provoke harmful toxic effects, either locally or systemically as discussed in detail in Chapter 5. About one third of the 100,000 substances on the market in the UK are considered as being probably hazardous to health. Substances can range from simple pure chemicals to complex impure mixtures or formulations. Exposures occur not only to operatives on the plant where the chemicals are produced but may also affect every workplace, and hence occupations as diverse as office worker, cleaner, garage mechanic, chemical-tanker driver, machinist using cutting oils, gardener applying pesticides, production engineer, laboratory scientist, paint-sprayer, lone tradesperson, sewer worker, drycleaner, car park attendant, etc.

A secretary suffered severe scarring which lasted several years after spilling typewriter cleaning fluid on her forearm.

A number of cleaners required medical treatment when toilet cleaner and bleach were used simultaneously (see page 495.)

A man tiling a toilet compartment using a solvent-based adhesive died after inhaling the vapours.

In two separate incidents plumbers, who wore no protective devices whilst repairing plastic pipes in a confined space using glue containing tetrahydrofuran (THF), developed nausea, blurred vision, dizziness, chest pain and cough.[1] They had signs of irritation of the mucous membranes and mild effects on the central nervous system. THF is a colourless, volatile liquid (b.p. 66 °C and vapour pressure 176 mm Hg at 25 °C) with a slight ethereal odour. It is readily absorbed across the alveolar membrane and from the digestive tract, and can cross the skin in toxic amounts. It is considered to be one of the least toxic cyclic ethers.

The control of health risks in the manufacture, transport and use of chemicals depends upon identification of the inherent hazards – and, if possible, quantification of the risks.

The hazard of any substance is its potential to cause harm. Possible routes of entry into the body are discussed in Chapter 5. (In the laboratory injection is also conceivable, e.g. when dosing experimental animals, but in the factory this route is rare. Acute accidental episodes can, however, be associated with high pressure systems.) However, the risk to health which any substance poses can only be assessed by looking at the extent and the likelihood of it causing harm in the actual circumstances in which it is used. Thus control is of risk, not of the inherent hazard.

> A chemical compounder died from lung cancer in 1989 following some seven and a half years' employment up to 1978 on work which took him near to a chloromethylation process. The management perceived in 1978 that bis-chloromethyl ether (BCME) could conceivably exist very temporarily as an intermediate during this reaction. However, tests did not at that time detect BCME in the work environment. This recognition of a 'theoretical hazard' coincided with the death from lung cancer of a thirty-nine year old process worker and the process was stopped. A report in 1985 identified a potential risk in the process and there had indeed been further lung cancer deaths all involving men of similar ages. A second report demonstrated that there was a substantially higher likelihood of lung cancer developing amongst those who worked near the process, however transitory their contact. It was concluded that BCME is a highly potent carcinogen and that the process should be severely curtailed, or restricted to only the most rigorously controlled environments.[2]

The first step in the control of any specific risk is the recognition that it exists, and the requirement for assessment is discussed later in this chapter. However, it is, unfortunately, still common experience that many occupational health problems arise through ignorance.

> A worker engaged on the repair of blood-pressure measuring equipment which contained mercury exhibited symptoms akin to a perpetual state of drunkenness plus lethargy, indigestion, tingling of the fingers and loss of manual dexterity attributable to exposure to mercury. The company, and employee, were quite unaware that mercury vapour could cause poisoning; no mechanical ventilation was provided to extract fumes, there were no lipped trays to contain spillage nor proper storage for the mercury itself.[3]

> In a separate incident involving mercury, an inspector visiting a factory where mercury was recovered by vacuum distillation, discovered atmospheric levels of the vapour seven times the current Short-Term Exposure Limit of 0.15 mg m^{-3}. Mercury levels in workers' blood and urine were three times the upper levels of acceptability. Most of the factory was found to be contaminated with mercury, including the still room, a covered storage area, and even the carpet in the main office. Prohibition and Improvement Notices were served on the company. After decontamination, laying a new floor, improved general ventilation and provision of local exhaust ventilation, environmental and

biological monitoring indicated exposures had fallen to within acceptable levels.[4]

(The hazards of mercury and precautions required with it are described in ref. 5 and on page 627.)

> During the demolition of two power stations inspectors found asbestos lagging being stripped at the top of a 39 m building was allowed to drop like snowflakes onto workers below. Some were dressed in jeans and sweaters without protective clothing.[6]

General strategies for control of substances hazardous to health are described in Chapter 6. The present chapter addresses important, recent, related legislation and the special considerations associated with hazards and precautions associated with allergens (sensitisers) and carcinogens, in dealing with mixed toxic hazards, and in the understanding of sick building syndrome. Chapters 29–32 provide more detailed guidance for the handling of toxic gases, liquids, solids and pathogenic micro-organisms by reference to specific examples.

Hazard evaluation

Control measures depend upon the degree of risk ascribed to a substance, or substances, in use. Hazard categories in use in the UK are
- extreme hazard
- high hazard
- medium hazard
- low hazard

The general guidelines given in Table 20.6 which make use of relevant statutory legislation[7] are applicable to this categorisation. Substances of unknown toxicity should be regarded as being at least in the high hazard category.

All routes of entry into the body should be considered. For example, in some cases oral or percutaneous absorption may be a critical factor.

> Hydrogen cyanide gas can penetrate the skin to such an extent that, without inhalation exposure, the body can take up enough cyanide from the atmosphere to cause severe or lethal intoxication.[8]

Estimation of risk may be a difficult task since a substance with a high hazard potential may present an acceptable risk if the exposure potential is low, or a substance with a relatively low hazard potential may present an unacceptable risk if there are high exposures. Exposure potential for the substance is assessed taking into account many factors including (after ref. 9):
- quantity stored/in use
- physical form and properties
- volatility and/or dustiness

1574   Toxic hazards

- concentration (if in solution)

However, deviations during handling and processing need to be taken into account and all the physico-chemical principles outlined in Chapters 4 and 19 are of relevance. For example,

> The volatility of a chemical cannot be predicted from its physical form. Methyl mercury which is a solid is 5.7 times more volatile than liquid metallic mercury.[8]

Factors included for the activity are:
- intrinsic potential for exposure, e.g. use of open containers, likelihood of aerosol/mist/vapour or dust generation
- frequency of exposure
- duration of exposure
- likely exposure route, i.e. inhalation, ingestion, skin/mucous membrane absorption, inoculation
- personal decontamination procedures
- air sampling/biological monitoring data
- potential for interaction with other activities

One technique to facilitate ranking of exposure potential as High, Medium or Low is given in Table 28.1. This evaluation is based only on estimates of exposure potential from the activity, ignoring additional control measures, e.g. secondary containment.

Risk assessment and control

Combination of the estimates of hazard and exposure leads to an overall classification of an activity as High, Medium or Low Risk; each of these is assigned an appropriate containment regime and, where necessary, supplemented by additional measures, e.g. personal protective equipment (PPE) and defined operating procedures.

How this applies to laboratory operations is illustrated in the matrix reproduced as Table 28.2.[9]

If the hazard is unknown it may be possible to make a judgement based, e.g. on comparison with chemically related substances. This must, however, be done cautiously (the difficulty is exemplified by the differing precautions required between toluene and benzene, or between carbon tetrachloride and 1,1,1-trichloroethane).

Therefore it may be more convenient to conduct activities involving unknown hazards in a high level of containment with appropriate back-up provisions as exemplified on page 333.

Loss of containment, or other mishap, or other foreseeable phenomena should be taken into account in the assessment. It is, for example, foreseeable that, because chemical reactions used in many manufacturing processes do not reach completion in the finite time available, traces of reac-

Table 28.1 Basis for estimating potential for exposure to toxic chemicals

Score	1	10	100
(A) Quantity of substance	<1 g	1–100 g	>100 g
(B) Physical characteristics of substance	Dense solids, non-volatile liquids, no skin absorption	Dusty solids, lyophilised solids, volatile liquids, low skin absorption	Gases, highly volatile liquids, aerosols, solutions that promote skin absorption
(C) Characteristics of operation activity	Predominantly enclosed system, low chance of mishap	Partially open system, low chance of mishap	No physical barrier, any operation where chance of mishap is medium or high

Exposure potential is estimated by multiplying individual scores $A \times B \times C$.
Here scores: $A \times B \times C < 1,000$ would achieve a 'Low' exposure potential.
$A \times B \times C$ = between 1,000 and 10,000 would rank as 'Medium' exposure potential.
$A \times B \times C > 10,000$ would rank as 'High' exposure potential.

Time factors, e.g. frequency and duration of the activity, should also be considered. Short duration tasks, involving a few seconds' exposure at infrequent intervals, should not affect the initial estimate, whereas continuous operations on a daily basis would probably raise the estimate to the next higher category.

Table 28.2 Matrix for a general evaluation of risk assessment and control determination (Risk = Hazard × Exposure potential)

Hazard category (see Table 28.1)	Exposure potential (see Table 28.1)		
	Low	Medium	High
Extreme	Risks presented by the handling of such substances are unsuited for this procedure and must be addressed on an individual basis		
High	2	2/3	2/3
Medium	1	2	2
Low	1	1	1

Containment regime determined from the above:
1. Open bench
2. Fume cupboard (or other specially vented area)
3. Special facility

In addition to the above containment regimes, it may be necessary to specify personal protective equipment (PPE) or other control measures, particularly where there may be exposure by the dermal route.

tants may remain in the finished products. This is particularly relevant in polymer production.

> Polyvinylchloride (PVC) powder will, when freshly made, contain a little unreacted vinyl chloride. A proportion of this will be lost before the polymer is used, e.g. during processing into PVC compounds or dry blends, but some ventilation is recommended in areas where the latter are stored in bags to keep the vinyl chloride concentration in the store as low as possible. Release during processing operations will depend upon the quantity of unreacted monomer present initially, the temperature to which the polymer is raised and the time these conditions exist for, and the exposed surface area. Recommendations have therefore been published as to ventilation requirements on PVC compounding and fabrication processes.[10]

With some substances used for certain purposes the risks are so great – for reasons discussed in Chapters 6 and 29–31 and in the current section on carcinogens – that a general prohibition is considered the only practicable control strategy. For example in the UK the substances listed in column 1 of Table 28.3 are prohibited to the extent listed in column 2.[11]

Mixtures of chemicals

Industrial exposures to chemicals frequently involve a mixture of components rather than single compounds, or substances containing a common entity, e.g. 'isocyanates', for which the majority of Occupational Exposure Limits have been set. (There are exceptions, e.g. 'rubber fume', 'welding fume', 'rubber process dust'.)[12] Such mixed exposures – to dusts, mists, gases – can arise

- from the use of, or from work upon, materials which contain a mixture of substances:

Table 28.3 Substances hazardous to health prohibited for certain processes in the UK

Item no.	Description of substance	Purpose for which the substance is prohibited
1	2-naphthylamine; benzidine; 4-aminodiphenyl; 4-nitrodiphenyl; their salts and any substance containing any of these compounds, in any other substance in a total concentration exceeding 0.1 per cent.	Manufacture and use for all purposes including any manufacturing process in which a substance described in column 1 of this item is formed.
2	Sand or other substance containing free silica.	Use as an abrasive for blasting articles in any blasting apparatus (see note 1).
3	A substance – (a) containing compounds of silicon calculated as silica to the extent of more than 3% by weight of dry material; or (b) composed of or containing dust or other matter deposited from a fettling or blasting process.	Use as a parting material in connection with the making of metal castings (see notes 2 and 3).
4	Carbon disulphide.	Use in the cold-cure process of vulcanising in the proofing of cloth with rubber.
5	Oils other than white oil, or oil of entirely animal or vegetable origin or entirely of mixed animal and vegetable origin (see note 4).	Use for oiling the spindles of self-acting mules.
6	Ground or powdered flint or quartz other than natural sand.	Use in relation to the manufacture or decoration of pottery for the following purposes: (a) the placing of ware for the biscuit fire; (b) the polishing of ware; (c) as the ingredient of a wash for saggars, trucks, bats, cranks, or other articles used in supporting ware during firing; and (d) as dusting or supporting powder in potters' shops.

7	Ground or powdered flint or quartz other than –
	(a) natural sand; or
	(b) ground or powdered flint or quartz which forms parts of a slop or paste.
		Use in relation to the manufacture or decoration of pottery for any purpose except –
		(a) use in a separate room or building for –
		  (i) the manufacture of powdered flint or quartz, or
		  (ii) the making of frits or glazes or the making of colours or coloured slips for the decoration of pottery;
		(b) use for the incorporation of the substance into the body of ware in an enclosure in which no person is employed and which is constructed and ventilated to prevent the escape of dust.

Table 28.3 (continued)

8	Dust or powder of a refractory material containing not less than 80 per cent of silica other than natural sand.	Use for sprinkling the moulds of silica bricks, namely bricks or other articles composed of refractory material and containing not less than 80 per cent of silica.
9	White phosphorus.	Use in the manufacture of matches.
10	Hydrogen cyanide.	Use in fumigation except when – (a) released from an inert material in which hydrogen cyanide is absorbed; (b) generated from a gassing powder (see note 5); or (c) applied from a cylinder through suitable piping and applicators other than for fumigations in the open air to control or kill mammal pests.

Notes
1. 'Blasting apparatus' means apparatus for cleaning, smoothing, roughening or removing of part of the surface of any article by the use as an abrasive of a jet of sand, metal shot or grit or other material propelled by a blast of compressed air or steam or by a wheel.
2. This prohibition shall not prevent the use as a parting material of the following substances: natural sand, zirconium silicate (zircon), calcined china clay, calcined aluminous fireclay, sillimanite, calcined or fused alumina, olivine.
3. 'Use as a parting material' means the application of the material to the surface or parts of the surface of a pattern or of a mould so as to facilitate the separation of the pattern from the mould or the separation of parts of the mould.
4. 'White oil' means a refined mineral oil conforming to a specification approved by the Health and Safety Executive and certified by its manufacturer as so conforming.
5. 'Gassing powder' means a chemical compound in powder form which reacts with atmospheric moisture to generate hydrogen cyanide.

- from work upon materials which generates a mixture of substances. Electric arc or gas-flame cutting/welding, for example, will produce a range of pollutants even in the simplest case involving a 'pure' metal. Metal oxide fume, NO_x, CO and in the case of electric welding some ozone, are the primary pollutants. (The situation is particularly complex if alloys, plastic-coated, metal-coated, painted, or oil-covered surfaces are involved, and Table 28.4 summarises some of the possibilities.) Combustion processes generally produce a mixture of pollutants.

In order to dismantle a 2.4 m square by 1.6 m deep galvanised metal tank a welder climbed inside and cut it into sections using oxyacetylene equipment. He worked alone through a full seven-hour shift within this confined space which was itself inside a boiler room. He subsequently suffered delayed pulmonary oedema from inhaling nitrous fumes (page 1726); symptoms of metal fume fever, attributable to the inhalation of zinc oxide fume, were recognisable as a complication.[13]

The main risk to a person who enters a slurry storage system on a farm is from gases generated by bacterial decomposition of slurry, including carbon dioxide, ammonia, methane and hydrogen sulphide, and the possibility of oxygen deficiency.[14]

- from the use of several different substances simultaneously:

    A solution of 1 to 2 per cent copper sulphate neutralised with hydrated lime, termed Bordeaux mixture, is sprayed either manually or mechanically onto vines to control mildew. Lung disease has apparently occurred amongst some vineyard workers after prolonged exposure. An excess of carcinoma of the lung requires confirmation.[15]

- from the use of several different substances within a workshift. For example, in an electroplating factory there may be a vapour degreasing facility (involving a chlorinated solvent), acid pickling, nickel solutions, chromic acid solutions, and cyanide solutions.

Examples of the miscellany of contaminants likely to be encountered in any particular industry are summarised in Table 5.1, and the common situation of environmental pollution involving mixtures is discussed in Chapter 21.

In all cases scientific assessment of risk and monitoring are rendered more difficult if a mixture is present. Such mixed exposures, which may even arise from a worker performing different jobs in different areas of a factory, or in different factories, over a period may also make diagnosis of disease difficult for medical practitioners.

Pure siderosis due to the deposition of iron oxide particles in the lungs is considered a benign condition not productive of pulmonary fibrosis. But fume produced from covered welding rods is a complicated mixture since – as indicated earlier – in addition to iron oxide it may contain other metal oxides, silicates and fluorides and alloy constituents;[16] NO_x and ozone may also be encountered. Therefore 'Welders Siderosis' may be accompanied by symptoms which reflect the combined effects of exposure.[17]

Multiple chemical sensitivity

A more contentious effect possibly associated with exposures to very low levels of mixtures of chemicals is termed Multiple Chemical Sensitivity (MCS). The symptoms vary widely, e.g. headaches, rashes, watery eyes, asthma, nasal congestion, fatigue and depression.[18]

A middle aged man was referred to clinic because of delayed recovery from an episode of pneumonia that had resulted from a chemical spillage. As his X-ray cleared, rather than improving, his condition deteriorated. Exposure to chemical odours caused breathing difficulty and chest pain. Upon returning to work he collapsed on several occasions after the slightest exposure to fumes and even exposure to household products or environmental contaminants caused respiratory distress. Clinical tests failed to define his condition pathophysiologically.[18]

This is an inadequately defined, and poorly understood, syndrome. One basis is as follows:

Table 28.4 Significant pollutants generated by hot work[16]

	Particulates													Gases			
	Nuisance particulates	Barium, Ba	Beryllium, Be	Cadmium, Cd	Chromium, Cr	Cobalt, Co	Copper, Cu	Fluoride, F	Lead, Pb	Manganese, Mn	Nickel, Ni	Silver, Ag	Thorium, Th	Carbon monoxide, CO	Oxides of nitrogen, NO$_x$	Oxygen enrichment, O$_2$	Ozone, O$_3$
Welding																	
Oxygas																	
Mild steel	1														1	1	1
MMA																	
Mild steel	2	(2)						1						1	1		1
Manganese steel	2	(2)						1		2				1	1		1
Stainless steel	2	(2)			2									1	1		2
Metal-arc gas-shielded																	
Mild steel: CO$_2$ gas	2						1△							2			1
Mild steel: Ar gas	2						1△							1			1
Stainless steel	2				2						2			1			2
Manganese steel	2									2				1			1
Aluminium	2																2
Copper	2		(3)				3										1
Mild steel: flux-cored wire	2	(2)						1		1				1			1
Tungsten-arc gas-shielded																	
Mild steel	1				1						1		x		2		2
Stainless steel	1												x		2		2
Aluminium	1		(3)												2		2
Copper	2						2				2		x		2		2
Nickel	2												x		2		2

Control of substances hazardous to health

Process							
Submerged-arc steels	1						
Cutting and gouging							
Oxygas Mild steel	2		2*		1	2	2
Air-arc Mild steel	2	2	2*		1	1	2
Oxyarc Mild steel	2		2*		1	1	2
Plasma Mild steel	2		2*		1	1	2
Aluminium	2					1	2
Brazing							
All processes	2	(2)		2			
Soldering							
All processes	2		1				
Preheating							
Flame					2	1	

Key: ( ) occasional occurrence only
× possible significant radiation hazard when grinding electrodes only
△ from electrode wire antirust plating
* lead-painted or lead-coated steel
1 significant amounts of fume, but not usually exceeding TLV
2 precautions recommended, such as local ventilation
3 potential danger from quantity and toxicity
Constituents with TLVs higher than the nuisance particulate level (10 mg/m^3) at the time of compilation (1980), such as aluminium and iron, have been omitted. The ratings are based on continuous use of the process in an open shop.

Patients must have symptoms or signs related to chemical exposures at levels tolerated by the population at large. (Reactions to such well-recognised allergens as moulds, dusts and pollen are not included.) The symptoms must wax and wane with exposures and may be expressed in one or more organ systems. A chemical exposure associated with the onset of the condition does not have to be identified, and preexistent or concurrent conditions – such as asthma, arthritis or depression – should not exclude patients.[19]

Furthermore the syndrome is acquired in relation to documentable environmental exposure(s), insult(s), or illness(es). Symptoms involve more than one organ system and recur and abate in response to predictable stimuli. They are elicited by exposure to chemicals or diverse structural classes and toxicologic modes of action and by exposures that are demonstrable (albeit of low level). Exposures that elicit symptoms must be very low (several standard deviations below levels known to cause adverse human responses). No single widely available test of organ system function can explain symptoms.[19]

A more restrictive definition may be favoured[20] but in any event the mechanisms involved are speculative and psychological mechanisms may play some part. However, one categorisation for MCS patients is reproduced as Table 28.5;[21] those cases attributed to 'tight buildings' are dis-

Table 28.5 Categorisation of Multiple Chemical Sensitivity patients[21]

Group	Nature of exposure	Demographics
Industrial workers	Acute and chronic exposure to industrial chemicals	Primarily male; blue-collar; 20 to 65 years old
Tight-building occupants	Off-gassing from construction materials; office equipment or supplies; tobacco smoke; inadequate ventilation	Female more than male; white-collar office workers and professionals; 20 to 65 years old; schoolchildren
Contaminated communities	Toxic waste sites; aerial pesticide spraying; groundwater contamination; air contamination by nearby industry and other community exposures	Male and female, all ages; children or infants may be affected first or most; foetuses could possibly be affected; middle to lower class
Individuals	Heterogeneous – indoor air (domestic); consumer products; drugs; pesticides	70–80% female; 50% 30 to 50 years old; white, middle to upper class and professionals

cussed further under Sick Building Syndrome on page 1613. Whilst existence of the disease is not universally accepted it is increasingly recognised in US government regulations and in the courts. It has been recommended that[19] treatment should be individualised but may include the following:

- non-judgmental, supportive therapy
- enhance patient's sense of control

- reduce psychosocial stress and/or patient's response to stress
- biofeedback, relaxation response
- treatment of coexisting psychiatric illness
- behavioural desensitisation to *low-level* chemical exposures
- pharmacologic treatments to control symptoms
- increase in physical and social activity
- treatment of other coexisting medical illnesses

Assessment

The problem of assessing exposures by inhalation when mixtures of chemicals may be present was briefly noted on pages 404–6.

Firstly the assessment should be based upon the concentrations of each of the constituents 'in air' to which a worker is exposed. The relative concentrations of each component in air may differ considerably from those in liquid or solid source material, e.g. because of vapour pressure differences (page 76) or particle size differences (page 117), depending upon the nature of each component and the conditions of use. Hence the composition of a bulk mixture cannot be relied upon for assessment unless there is good evidence to support this approach.

Close examination of toxicological data is necessary to determine which, if any, of the main types of interaction, viz. synergistic, additive or nil (i.e. the substances act independently), is likely with a particular mixture. These are discussed below in descending order of consideration.

Synergistic substances

With synergistic substances, the overall effect of exposure to a mixture is considerably greater than the sum of the individual effects. This may be due to mutual enhancement of the effects of the constituents, or because one substance acts to 'potentiate' another so that it acts in a different way. Such substances are the most problematic in effect and require strictest control.

Whilst known cases of synergism are relatively infrequent they are the most difficult to assess.

The phenomena of synergism depends upon the effect of one compound on the metabolism of another. If these metabolic processes are stimulated the result of exposure to a chemical can be increased or reduced.

Phenobarbitane potentiates many metabolic processes on chemicals, e.g. carbon tetrachloride, carbon disulphide, chloroform or nitrosamine become more toxic; alfatoxin becomes less toxic.[8]

The National Institute for Occupational Safety and Health (NIOSH) studies indicate that workers exposed to ortho-toluidine and aniline at one plant have greatly increased chances of developing bladder cancer, e.g. 27 times more than the general population. Companies have been advised to reduce workers' exposures to the lowest feasible concentration.

No cases were observed among workers employed in the exposure area for less than five years and NIOSH were unable to separate the effects of ortho-toluidine and aniline from each other epidemiologically. (The carcinogenic effect of ortho-toluidine has, however, been demonstrated at lower doses than aniline.)[22]

Specialist advice may be advisable whenever there is reason to suspect interactions.

Disparate results as to the effect of recurrent exposure to low concentrations of welding gases, e.g. ozone and nitrogen dioxide, have been attributed mainly to lack of control populations, variable smoking habits, unmatched controls and exposure to other respiratory hazards, e.g. asbestos or free silica.[23]

Promoting effects, such as carcinogenicity where another substance which may in itself be harmless appears to promote the action of a carcinogen, may be one factor of particular importance when assessing discharge of a chemical into the environment. For example, animal tests have shown that very much lower concentrations of benzo[a]pyrene were required to produce skin tumours when it was administered with a straight-chain C^{12} hydrocarbon; ethanol enhanced the activity of a nitrosamine in the oesophagus.[24]

Additive substances

For some additive substances there is reason to believe that the effects of exposure to a mixture are 'additive', e.g. the substances act on the same organs, or by similar mechanisms, so that the effects reinforce each other.

A refinery employee collapsed whilst assisting with the removal of valves from a molecular sieve dryer in a gas recovery unit. He was pronounced dead on arrival at hospital. The dryer had been purged with nitrogen for seven days and tests performed for both H_2S and hydrocarbons; a slight nitrogen purge was left on. H_2S gas trapped in a section of piping was displaced by the purge and released via valve flanges as the bolting was loosened. The victim's face was very close to the flanges. Whilst hydrogen sulphide poisoning was determined as the cause of death, reduced oxygen intake due to nitrogen inhalation may have been a contributing factor. (This incident also highlights a possible hazard with adsorption dryers with an ability to remove H_2S even after purging since desorption may continue.)[25]

Thus exposure to two solvents may result in a more advanced narcotic effect, or a more severe liver damage, than to one of them. Direct methaemoglobin-forming agents exert an additive effect on the formation of methaemoglobin, an inactive form of oxygen-carrying pigment in the blood.[8]

The exposure limits are based upon the same health effect. The mixed exposure may then be assessed by means of the formula on page 405.

However, this is strictly only applicable where the additive substances have been assigned Occupational Exposure Standards (OESs) and EL_1, EL_2 etc. relate to the same reference period in the approved list.

Independent substances

With independent substances, no synergistic or additive effects are likely within a mixture. They act on different body organs, or by different toxicological mechanisms.

The substances are then considered to act independently. It is generally considered desirable to treat all non-synergistic systems as though they were additive, particularly where the toxicity data are scarce or difficult to assess.[12] This avoids the requirement to distinguish between additive and independent systems. Application of the additive formula as on pages 404–6 then enables exposure to be assessed. However, if one of the constituents is assigned an Maximum Exposure Limit (MEL) the 'additive' effect also needs to be taken into account in deciding the extent to which it is reasonably practicable to reduce exposure further, as noted on page 261.

There have been several salutory lessons over the years which show, in hindsight, the need to err on the side of caution when assessing health hazards from chemicals. The case of vinyl chloride is noted on page 1598 but possibly the most damaging example of underestimation of health hazard potential is that of asbestos.

In following such an approach it is of course necessary to take a broad view, so as not to overlook the problems which can arise with so-called 'inert' materials. For example, the phenomenon of oxygen deficiency summarised on page 242, and exemplified on pages 982 and 1703, is a serious occupational risk. Indeed it has been suggested that nitrogen may have resulted in more fatalities than any other single substance, e.g. due to its accidental use instead of air, or from its unexpected presence in a confined space.[26]

Carcinogens

Introduction

Some chemicals possess the ability to cause long-term health effects which are very difficult, or impossible, to cure and the effects can be irreversible. Examples include chemicals with the potential for interfering with the reproduction processes, e.g. causing the prevention of pregnancy, or resulting in damage to the developing embryo or foetus, or in inducing miscarriages. Both male and female exposures are of potential importance depending upon the specific chemical. Mutational damage or interference with sperm production in man can induce sterility or affect the development of the child.

In 1977 the pesticide dibromochloropropane (DBCP) was identified as a potential testicular toxin. Workers at a manufacturing plant noticed that since the men had started work with DBCP none of their wives had borne any children. Sperm tests indicated that the overwhelming majority had zero, or severely reduced, sperm counts. There were no other noticeable effects.

An offspring's development can be affected by its mother's environment before and at conception, and during the early period of pregnancy when teratogenesis can be initiated.[27] In laboratory tests a substance is considered a teratogen if there is irreversible interference with the development of an embryo or foetus at a dose which has no detectable effects on the health of the mother. There are few known human teratogenic chemicals and most are therapeutic substances (Tables 5.6 and 5.7).

In addition to these mutagenic and teratogenic substances, hazards are also associated with chemicals which possess some potential to cause cancer.

> There is a disease peculiar to a certain set of people which has not, at least to my knowledge, been publicly noticed; I mean the chimney sweeper's cancer. The fate of these people seems singularly hard; in their early infancy they are most frequently treated with great brutality, and almost starved with cold and hunger; they are thrust up narrow and sometimes hot chimneys, where they are bruised, burned and almost suffocated, and when they get to puberty become peculiarly liable to a most noisome, painful and fatal disease.[28]

This observation in 1775 was the first linking skin cancer with occupation, i.e. 'sooty cancer' – cancer of the scrotum – due to exposure to pyrolysis products such as 3,4-benzopyrene (benzo[a]pyrene; page 209), from coal. There was an increased incidence amongst English sweeps in comparison with those in either Germany or Scotland. This difference derived, as do many others in occupational health experience (e.g. as on page 1588), from variations in the system of work, i.e. in England coal replaced wood at an earlier date, there was earlier employment of young children inside flues rather than pulling up the debris from outside, and there were significant differences in personal hygiene. Cancer of the scrotum was next associated with cotton mule spinners, the fronts of whose trousers became soaked with mineral oil as the men bent over the mules on which it accumulated as it was thrown off the spindles. Oil shale workers were also affected and, more recently, e.g. toolsetters using automatic machine tools.[29]

The wide range of chemicals associated with cancer in humans is discussed more fully in Chapter 5 and in ref. 30. Hazards can arise from occupational, para-occupational (e.g. of mesothelioma due to painters or fitters being exposed to asbestos), and non-occupational exposures (e.g. Love Canal described on page 1205). Moreover, problems may arise from the formation of potential carcinogens due to unintended chemical reactions. For example, nitrosamines, which can occur in food, drink and tobacco smoke, have been shown to be carcinogenic in animal experiments, even in single small doses. Fortunately they are little used in industry but some synthetic, metal-grinding and metal-cutting fluids may contain traces from reactions between a nitrite salt and amines such as diethanolamine; specific guidance has therefore been issued to eliminate or control this hazard:[31]

$$R_2NH + NaNO_2 + HCl \rightarrow R_2N-N=O + NaCl + H_2O$$
(secondary amine such as diethanolamine where $R = (CH_2CH_2OH)_2$)

The formation of bis(chloromethyl) ether, a potent carcinogen from formaldehyde and hydrogen chloride, was mentioned on page 498. The substantially increased probability of carcinoma of the lung developing amongst those with even transitory contact with a chloromethylation process was subsequently confirmed as described on page 1572.[2]

Carcinogens merit special attention as substances hazardous to health for a variety of reasons which include the following.

- The disease does not necessarily arise at the site of entry of the chemical into the body and it is often difficult to prove a causal link between a particular chemical and cancer in humans. Epidemiological data, which would provide a link, are often sparse due to the sheer numbers of subjects required to be of sufficient significance, or because of the delay between exposure and effect. Often factors other than industrial exposure, e.g. smoking, snuff-taking, alcohol abuse, diet, may overwhelm industrial causes. (Clinical and epidemiological aspects have recently been reviewed in detail.[32])
- Most cancers carry a high risk of premature death and prevention may be the only control if no cure exists.
- There is commonly a long delay between exposure and occurrence of cancer, sometimes decades.
- The mechanisms by which carcinogens exert their effects are not fully understood. In the absence of such information, justifying thresholds may be difficult when setting control limits. Exposure below these levels still carries a residual risk, i.e. exposure relates more to the probability of cancers occurring in the exposed population rather than the severity of the outcome in individual cases, although in most cases the proportion of cancers that may be caused or exacerbated by occupational exposure to carcinogens is unknown. Estimates vary according to source, but even a 1 per cent contribution would result in twice as many deaths as those resulting in reported work accidents.[33]
- Emotion and fear associated with cancer.

Scope

An introduction to the toxicology of carcinogens is given in Chapter 5 along with tables of key examples. The routes of entry into the body are as for toxic chemicals generally discussed in Chapter 5. Although ingestion is possibly of less significance it cannot be discounted.

> Girls employed in painting luminous dials on clocks and watches using a luminous paint containing radium-226 evidently swallowed small amounts because of the practice in the USA of pointing the brushes between their lips. Many experienced ill-effects including severe anaemia, necrosis of the jaw, spontaneous fracture and sarcoma (malignant tumours) of the bone.[34] (Elsewhere painting was done with a stylet which was not sucked and no untoward effects were observed.)[35]

Tumours can be benign or malignant. Cancer can arise from exposure to certain chemicals either as a direct consequence of the toxic properties of the substance or as a result of formation of toxic metabolites in the body. Some substances do not cause cancer directly but may promote, or initiate, it on exposure to additional chemicals. In many circumstances where an industrial carcinogenic hazard may be clearly demonstrated the nature of the (presumptive) chemical carcinogen(s) remains totally obscure. This is illustrated by the adeno-carcinoma of the ethmoid sinuses in woodworkers in the furniture industry which was established from clinical observations.[36] Subsequent epidemiological surveys confirmed that the risk is linked with inhalation of hardwood dust by workers in this industry but not among carpenters.[37] Both the site of this tumour and its histology are rare and such neoplasms – like hepatic angiosarcomas and mesotheliomas – provide valuable marker tumours in epidemiology. For woodworkers, prolonged exposure to hardwood dust of aerodynamic diameter around 5 μm, which becomes trapped in the nasal passages, seems to be an important factor. However, the causative agent has not been identified, but since some of the thousands of components present (which include polysaccharides, cellulose and hemicellulose, more complex phenyl propane polymers, lignin extractable compounds such as phenolics (e.g. tannins, flavonoids, lignins), terpenes, aliphatic acids (mainly as esters), alcohols as esters or glycosides) have been shown to possess carcinogenic potential, the dust *per se* must therefore be considered to have carcinogenic potential.[38] Dust exposure varies with process (see page 1657). Less than 25 per cent is respirable. In addition wood may contain fungi, wood treatment chemicals and chemicals formed as wood chemistry changes during processing.

In other cases good fortune, or a timely scientific intervention, has limited the extent of occupational exposures.

> 4-Aminodiphenyl, which has been shown to be a bladder carcinogen, was used by American chemical manufacturers in the production of certain rubber antioxidants, and some of these were imported and used by British rubber manufacturers prior to 1949. The compound was, however, never made or marketed in Britain because of timely warnings given in 1952 by Walpole *et al*.[39]

The traditional methods of indicating whether a substance may be carcinogenic to humans are as follows.

- By chemical analogy, i.e. by comparison of the chemical structure with known materials of proven or suspected carcinogenicity. This method is of limited value;[40] the difficulties are exemplified by the differences established between benzene and toluene.
- From knowledge of the substance's metabolism.
- From studies on experimental animals. Substances are administered by feeding, inhalation, skin painting, implantation, etc., to test animals under controlled conditions to determine whether tumours are produced. Whilst this does not prove conclusively that a substance is a definite carcinogen to humans, it does indicate a risk, particularly if more than one animal species is involved.
- From short-term tests using bacterial and mammalian cell culture assays to detect damage to, or mutation of, DNA induced by chemicals. Although mutation may be a precursor to cancer, and most proven human carcinogens are mutagenic in laboratory test systems, the reverse has not been unequivocally established and the most severe limitation of these tests is the lack of quantitative relationships between their results and cancer incidence in experimental animals.[41] Thus demonstration of carcinogenicity in experimental animals is seen as a possible indicator of carcinogenicity in man; substances found to be mutagens in laboratory tests are considered suspect carcinogens.
- From epidemiology, which involves studying mortality and cancer registration patterns in a population. This may lead to identification of a group of workers who, because they exhibit a raised incidence of deaths from a particular cancer, appear to be at a greater risk than a control group. For example, data as in Table 28.6 compare the numbers of cancer registrations with what might be expected on a population-wide basis, and hence provide a means for estimating the numbers of cancer cases in different classes of workers due directly to individual exposure.[42] Problems in relying upon epidemiological studies are that persons may be being exposed, and be contracting cancer, during what is inevitably an extended time period; and that the latency period with some cancers may be 20 years or more. However, epidemiology is particularly valuable in identifying cancer risk, since:[41]

    Many substances have never been tested for carcinogenicity, and most that have were tested as single substances. Humans encounter bewildering mixtures of carcinogens that no laboratory testing can mimic. So one important aspect of epidemiology is that it can detect human risks that have not been predicted from animal studies and probably cannot be.

In the UK carcinogens which are embraced by the COSHH Regulations are given in the associated Approved Code of Practice. They are:

1) Substances assigned the R45 risk phrase 'May cause cancer' under the

Table 28.6 Occupational cancer in men, 1966–69

Cancer site and occupation	Cancer registrations in men aged 15–74 1966–69	Expected registrations	Excess registrations due to occupational exposure	Excess cancer risk factor
Lip				
Agricultural labourers	68	15	53	5.0
Farmers	56	22	34	2.5
Construction workers	88	43	45	2.0
Labourers	203	116	87	1.8
Salivary gland				
Armed forces	34	13	21	2.6
Oesophagus				
Agricultural labourers	69	42	27	1.6
Textile workers	53	35	18	1.5
Stomach				
Miners, quarrymen	1,014	709	305	1.4
Glass and ceramic workers	90	65	25	1.4
General metal workers and jewellers	192	136	56	1.4
Boiler firemen	138	104	34	1.3
Rectum				
Miners and quarrymen	462	362	100	1.3
Liver				
Carpenters and joiners	31	20	11	1.3
Pancreas				
Coal miners	33	13	20	2.5
Tailors	23	12	11	2.0
Nose				
Woodworkers	29	15	14	2.0
Larynx				
Machine tool operators	80	58	22	1.4
Lung, trachea, bronchus				
Ceramic formers	71	48	23	1.4
Foundry moulders, coremakers	306	214	92	1.4
Fettlers, metal dressers	114	81	33	1.4
Metal plate workers, riveters	372	281	99	1.3
Plumbers, lead burners	491	387	104	1.3
Bricklayers, tile setters	845	668	177	1.3
Plasterers	238	184	52	1.3
Charmen, window cleaners, chimney sweeps	260	198	62	1.3
Skin				
Farmworkers, foresters, fishermen	1,148	890	258	1.3
Prostate				
Farmworkers, foresters, fishermen	667	496	171	1.3
Bladder				
Clothing workers	103	83	20	1.2
Rubber workers	42	26	16	1.5
Brain				
Farmers, market gardeners	114	78	36	1.5
Hodgkin's disease				
Machine tool operators	66	50	16	1.3

Classification, Packaging and Labelling of Dangerous Substances regulations, 1984 (Chapter 13).
2) The following substances and/or processes.
Aflatoxins
Arsenic and its inorganic compounds
Beryllium and its compounds
Bichromate manufacture involving the roasting of chromite ore with the addition of calcium compounds to the kiln feed mix
Mustard gas
Calcining, sintering or smelting of nickel copper matte or acid leaching or electro-refining of roasted matte
Coal soots, coal tar, pitch and coal tar fumes
The following mineral oils:
 – unrefined and mildly refined vacuum distillates
 – catalytically cracked petroleum oils with final b.p. >320 °C
 – used engine oils
Auramine manufacture
Leather dust in boot and shoe manufacture, arising during preparation and finishing
Hardwood dusts
Ipa manufacture (strong acid process)
Rubber manufacture and processing giving rise to rubber process dust and rubber fume
Magenta manufacture
4-Nitrobiphenyl
o-Toluidine

Other carcinogenic materials are covered by different legislation, e.g. asbestos and vinyl chloride, and alternative lists are included in Chapter 5. The UK COSHH Regulations specifically revoke the Carcinogenic Substances Regulations, 1967.

Control measures

In general, the following technical and administrative controls, which are aimed at preventing exposure to the compound or at least ensuring exposures are extremely low so as to ensure minimum risk, are applicable for working with teratogens, mutagens and carcinogens. Whilst there is no sex-related difference in the hazards faced from working with carcinogens and mutagens, the question arises as to whether pregnant women should be prevented from work with teratogens. The policy varies from company to company. Difficulties for the company (and the employee) may be in knowing when the individual is indeed pregnant. Furthermore some chemicals pose risks predominantly for males. Such philosophical debate is

Toxic hazards

outside the scope of this book but clearly implications of sex discrimination in industry need careful consideration.

> In 1982 eight female employees at a battery manufacturing company became pregnant while maintaining blood lead levels in excess of 30 µg per decilitre. As a result the company introduced a foetal protection policy such that 'Women who are pregnant or who are capable of bearing children will not be placed into jobs involving lead exposure or which could expose them to lead...'. The company defined women capable of bearing children as 'all women except those whose inability to bear children has been medically documented'.
>
> In 1991 the US Supreme Court unanimously ruled that women cannot be barred from jobs that expose them to hazardous chemicals on the grounds that such exposure might harm potential foetuses since it was a clear violation of the Civil Rights Act. The company policy was considered biased since fertile men but not fertile women were given the choice as to whether they wished to risk their reproductive health for a particular job.[43]

Prohibition

Where a substance is both a potent carcinogen and of limited industrial value the legislative policy is often to institute a ban on it. If the compound is of natural origin or has many uses then the strategy tends to be a ban on specified processes. Asbestos is intermediate; blue asbestos is no longer used but in the UK, because benefits are considered to outweigh the disadvantages and some material still exists in older establishments, the approach has been to set a control limit. Employment in the manufacture and use, including any process resulting in the formation of 2-naphthylamine, benzidine, 4-aminodiphenyl or 4-nitrodiphenyl and their salts and any substance containing any of these compounds in a total concentration greater than 0.1 per cent, is banned as is the importation of the four substances above.

At one time the permitted concentration was 1 per cent and it is possible that some workplaces such as laboratories may still unwittingly hold old samples which are outside the present specification. Where such substances are in existence they should be adequately disposed of, or in the unlikely event that continued usage is essential, application should be submitted to the HSE for an exemption from this aspect of the Regulations and a formal risk assessment carried out, the results of which should be discussed with employees.

Handling procedures

Because of the serious, and often irreversible, toxic effects arising from exposure to carcinogens, prevention of exposure is paramount.

In most situations a combination of control methods must be planned to provide adequate protection, with no single measure being relied upon exclusively.[40] Critical tactics include the following.

- Making every effort to replace the material by non-carcinogenic, or less harmful, substances or to use processes which do not form or release carcinogenic products, intermediates, by-products or waste into the workplace. Caution must, however, be exercised since substitution by a substance not adequately tested could be as harmful as, if not worse than, the original hazard.[40]
- Informing competent authorities: in certain cases it is a legal requirement to request authority to handle material.
- Appropriate design of building and plant. For example, the principles of prime importance with bladder carcinogens of lesser potency than those for which manufacture and use has now ceased are summarised in Table 28.7.[44]

Table 28.7 Principles of building and plant design for control of bladder carcinogens[44]

- Buildings should be large and airy with high ceilings.
- Walls should be of material which cannot become impregnated with chemicals and which can be easily cleaned; floors should have an adequate slope and be constructed of materials which will withstand physical and chemical damage. No wood should be used in the construction of stairs, platforms or handrails. Floor grids must be easily removed, and drains should have sufficient gradient to ensure quick flow.
- Forced ventilation should be installed with general airflow from the operator towards the plant units. No plant items should be vented within the working space, and scrubbers should be provided.
- The object in all plant design must be to contain the material within an enclosed system, and to automate all operations as far as possible. All plant should be designed to minimise chances of blockages and frequency of maintenance, and should be capable of easy and complete decontamination before any maintenance is carried out. Lagging should be enclosed in an impervious casing.
- Automated enclosed filters should be employed and the final product, if possible, should be delivered to a user plant as a slurry or solution. If drying is essential, an enclosed self-discharging drier should be employed, with adequate ventilation at the discharging point. Grinding should be avoided altogether.
- All plant must be decontaminated completely, preferably by chemical methods, before any maintenance is carried out.
- Containers must be designed for easy filling and discharge, must be completely sealed to avoid spillage, and must be capable of complete decontamination. Damaged containers must be discarded.
- Special attention must be paid to the sampling of toxic substances or mixtures or solutions containing them. The protection of laboratory personnel is of equal importance to that of the operatives.

- Instituting control measures to minimise risk; explaining these to, and discussing with, employees. Examples include
  - total containment, e.g. mechanically or by operating under negative pressure with adequate filtration;
  - minimise emissions and control exposure to as low a level as is reason-

ably practicable, and in any event below hygiene standards, by engineering means (including ventilation);
  – reduce the number of persons exposed to a minimum, and exclude non-essential personnel.
- Using personal protective equipment only where essential, e.g. to regain adequate control of the process, and its use, application and maintenance must be in compliance with the legal requirements.

  > In geothermal energy production the steam may contain a very low concentration of arsenic. Since the carcinogenicity of inhaled or ingested inorganic arsenic has been demonstrated under certain circumstances special precautions must therefore be taken in the cleaning of turbine blades encrusted with arsenic salts. This operation is performed infrequently by few persons who are provided with adequate protection by means of respiratory equipment and protective clothing.[45]

- Where carcinogens are transported/stored all appropriate measures should be taken to avoid spillage or leakage or contamination.
- Because of the seriousness of overexposure, environmental monitoring is normally essential for risk assessment and to maintain a check on the adequacy of containment measures. Data can be compared with approved occupational exposure limits or, in their absence, in-house standards. Surfaces should also be swabbed and checked for contamination.
- Although biological monitoring techniques are not available for many substances, where they do exist they offer valuable warnings of overexposure (e.g. arsenic levels in the urine of arsenic workers; chromosome damage in the white blood cells of vinyl chloride workers, exfoliate cytology of the urine as an indicator of early signs of bladder cancer in workers exposed to certain aromatic amines or nitro compounds).
- Health surveillance may be essential unless exposure is insignificant. Some requirements are listed in Schedule 5 to the COSHH Regulations. For skin carcinogens (e.g. arsenic, coal soots, coal tar, non-solvent refined mineral oils, contaminated used mineral oils) health surveillance should include regular inspection of the operative's skin by a suitably qualified person, or enquiries directed to a responsible person following self-inspection by employees. Health surveillance is generally of limited value in identifying susceptible individuals and in the early stages when treatment is likely to be most effective. Medical surveillance by an employment medical advisor or appointed doctor is required in those cases included in Schedule 5. Skin cancer is, however, an example of where the disease can be detected at an early stage when it can be cured.
- Providing information since there is a latent period between exposure and onset of adverse health effects. An example of advice is the leaflet for pitch and tar workers shown in Fig. 6.11.
- Instruction and training are very important since there is no apparent

and immediate effect and because of the need to keep exposure to as close to zero as reasonably practicable; there is a continuous need to keep employees aware of risk.
- Records of employees' names, the chemicals they handle, monitoring data, equipment (e.g. ventilation), personal protective devices, medical condition, etc., are particularly relevant here under the COSHH Regulations.
- A high standard of general hygiene should naturally be maintained (page 318). This would encompass the application of 'good housekeeping' techniques, e.g. the use of vacuum cleaners with adequate filtration; prohibition of eating and drinking in any workplace where substances are used; and provision of good washing facilities to encourage a high standard of personal hygiene.

Obviously control measures inside the workplace should not be applied in ways which may accentuate the hazard to the outside. Considerations include:[40]
- emission to atmosphere both from the process itself and from the exhaust ventilation controls applied to it. Hence filters, scrubbers, incinerators, etc., may be required as described on pages 328, 854 and 1292.
- effluent discharge, and waste (solid and liquid) disposal must comply with statutory requirements.
- contamination by transfer, on protective clothing (e.g. footwear and overalls), on wheels of vehicles (including private cars), on uncovered loads, etc.
- inadvertent discharge, e.g. if reactions get out of control, or vessel contents escape from pressure relief systems or other parts of the plant.

Further guidance on requirements is given in ref. 46. However, it is important to note that, in addition to the use of rigorous control measures for proven or suspected carcinogens, it is essential to adopt a general precautionary policy for controlling the use of all substances until fully evaluated. Otherwise a serious problem can arise if a substance not known to be toxic or carcinogenic is used indiscriminately or without particular care and is subsequently found to be carcinogenic.

> Vinyl chloride (see later) provides an example of a substance which was not considered to be acutely toxic (indeed it was used as an anaesthetic) and which was used industrially for many years before it was found to be a human carcinogen. Eventually research showed it to be definitely carcinogenic, with the main risk being from inhalation of vapour; steps were immediately taken to control exposure to vapour levels as near to zero as possible, to restrict entry into the danger area, and to monitor control by measuring vapour levels, etc. Before the discovery of evidence of carcinogenicity, vinyl chloride monomer had been used in a relatively careless manner throughout the world.[40]

The application of these general principles in industry can be exemplified by reference to use of mineral oils and vinyl chloride, whilst the procedures for working with asbestos described in Chapter 29 provides further illustrative practical requirements.

Mineral oils

Hazards

Mineral oils are used in a wide range of engineering processes, e.g. as lubricants, coolants in metal cutting and as quenching agents in the tempering of metals. When used in engineering machine shops as coolants, because of the high speeds and tempos of work used, copious flows of coolant are required. Some of this may be sprayed into the air as fine mist. Thus there is a risk of cancer of the respiratory tract and digestive system through inhalation.[47] Moreover, toolsetters often have to adjust the tool while the machine is in motion. In spite of splash guards, unless impervious protective clothing, e.g. aprons and elasticated armlets, are worn, gross contamination of clothing may occur and the skin on thighs and genital regions as well as on arms can become affected by a variety of skin complaints including cancer. Thus certain mineral oils are known or suspected of causing cancer and are subject to the COSHH Regulations.

Distillates from crude oil are termed mineral oils. The carcinogenicity is normally, but not always, associated with the presence of polycyclic aromatic hydrocarbons (PAH). The PAH content depends on the raw material, and its refining and finishing process. Unrefined and mildly refined oils are used in printing inks and the rubber industry. These products contain substantial quantities of PAH and have been shown to cause cancer in experimental animals. There is no evidence of a carcinogenic risk from vacuum-distilled, highly refined oils which are used as machinery lubricants. When unrefined vacuum distillates are finished by solvent extraction the extracts are rich in PAH. Some have been assigned the R45 risk phrase. Catalytically cracked oils with b.p. >320°C contain high levels of PAH, again capable of inducing cancer in animals, but of uncertain potency. Whilst engine oil is usually highly refined and low in PAH when supplied, combustion products (including PAH) concentrate in the crank case oil particularly in petrol as opposed to diesel engines.

Although the acute toxicity of mineral oils is low and occasional skin exposure poses no real problems, repeated exposure can result in defatting and skin problems. Inhalation of oil mist can result in respiratory problems and theoretically lung cancer, although epidemiological studies have not confirmed this to date. Thus in risk assessment it is important to establish the nature of the oil in order to classify its carcinogenicity. A full discussion is provided in refs. 48 and 49.

Precautions

Oils classified as carcinogenic should be replaced by highly refined oils to minimise risk. Where this is not feasible, employers should take adequate measures to avoid exposure by engineering control. Where skin contact is possible then personal protective equipment should be provided. Those handling used engine oil, e.g. fleet mechanics and waste oil recoverers, will be at greatest risk. Good personal hygiene and housekeeping are important, e.g. washing thoroughly prior to eating, smoking, leaving work, and giving prompt attention to cuts, etc., and cleaning of contaminated surfaces. Oil-soaked clothing should be removed and overalls should be changed regularly. Disposable wipes are preferred to rags. Mist formation should be prevented where possible and exposure kept below the OES for highly refined non-carcinogenic mineral oils. The presence of additives as biocides and corrosion inhibitors also needs to be taken into account.

Skin condition should be monitored regularly and medical staff informed as soon as any skin abnormalities are detected.

Waste oils should not be poured on the ground or down sewers. Proper routes include recycling, disposal via licensed contractors, or incineration as discussed on page 854. Employees should receive regular reminders of the nature of their work and the importance of any control features, use and maintenance of personal protective equipment, etc.

Vinyl chloride

Hazards

The main industrial route to vinyl chloride is controlled chlorination of ethylene followed by pyrolytic dehydrochlorination of the resulting ethylene dichloride. The hydrogen chloride by-product is also reacted with the ethylene feedstock in the presence of oxygen, viz.

$$Cl_2 + CH_2=CH_2 \rightarrow ClCH_2-CH_2Cl \xrightarrow{\Delta} CH_2=CHCl + HCl$$
$$\text{(ethylene)} \qquad \text{(ethylene dichloride)} \qquad \text{(vinyl chloride)}$$
$$2HCl + \tfrac{1}{2}O_2 + CH_2=CH_2 \rightarrow ClCH_2-CH_2Cl + H_2O$$

Key physical properties are given in Table 28.8.[50] Although handled industrially as a liquid (b.p. $-13.4\,°C$), under normal temperature and pressures vinyl chloride is a colourless, flammable gas (large fires of vinyl chloride are difficult to extinguish and the vapours pose an explosion hazard). The hazards of chlorination processes are discussed in Chapter 10.

Vinyl chloride is one of the largest commodity chemicals and finds extensive use as an organic intermediate and, most importantly, as a precursor to PVC and a range of copolymers; at one time it was a component of aerosol propellants.

Potential human exposure to vinyl chloride is mainly from inhalation and, less frequently, through skin absorption. Important effects of acute exposures are listed in Table 28.9.

Effects of long-term exposure include fatigue and abdominal pain,

Table 28.8 Physical properties of vinyl chloride

Property	Value
Molecular weight	62.499
Melting point, °C	−153.8
Boiling point, °C	−13.4
Specific heat, J/(kg·K)[a]	
vapour at 20°C	858
liquid at 20°C	1352
Critical temperature, °C	156.6
Critical pressure, MPa[b]	5.60
Critical volume, cm³/mol	169
Compressibility factor	0.265
Pitzer's acentric factor	0.122
Dipole moment, C·m[c]	5.0×10^{-30}
Latent heat of fusion, J/g[a]	75.9
Latent heat of evaporation, J/g[a]	330
Standard enthalpy of formation, kJ/mol[a]	35.18
Standard Gibbs energy of formation, kJ/mol[a]	51.5
Vapour pressure, kPa[b]	
−30°C	50.7
−20°C	78.0
−10°C	115
0°C	164
Viscosity, mPa·s (= cP)	
−40°C	0.3388
−30°C	0.3028
−20°C	0.2730
−10°C	0.2481
Explosive limits in air, vol.%	4–22
Self-ignition temperature, °C	472
Flash point (open-cup), °C	−77.75
Liquid density (at −14.2°C), g/cm³	0.969

[a] To convert J to cal, divide by 4.184.
[b] To convert MPa to psi, multiply by 145.
[c] To convert C·m to D, divide by 3.336×10^{-30}.

Table 28.9 Effects of acute exposure to vinyl chloride (after ref. 51)

Exposure concentration (ppm)	Effects
2,000–5,000	Odour threshold
8,000–10,000	Vertigo
16,000	Impaired hearing and vision
70,000	Narcosis and loss of consciousness
120,000	Death

neurotoxic symptoms and angioneurotoxic disorders (Raynaud's phenomenon), changes in the skin and skeleton, modifications to the liver and spleen, and angiosarcoma of the liver with the possibility of tumours at other sites. Evidence of carcinogenicity in man stems from groups occupationally exposed to high concentrations for many years; the American Conference on Governmental Industrial Hygienists (ACGIH) concluded that if the average exposure does not exceed 5 ppm there will be no increase in the incidence of cancer, specifically of angiosarcoma of the liver.[52] The need for tight hygiene standards is clear.

Workers at greatest risk of exposure are those engaged in vinyl chloride synthesis and especially operatives employed in PVC production plants. One estimate[53] suggests that in the USA more than 3.5 million workers may be exposed to the chemical with a further 4.6 million living within 5 miles of industrial sites from which environmental emissions occur; 3–12 ppb vinyl chloride was detected in such emissions.

PVC is produced by bulk, emulsion and predominantly suspension polymerisation techniques. The process is batch-operated in large autoclaves as exemplified by Fig. 28.1.[54] Since the conversion to PVC after a 5–8-hour

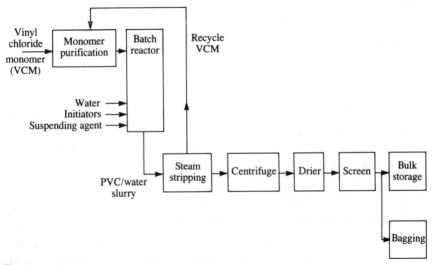

Fig. 28.1 Polyvinyl chloride process flow

reaction cycle is less than 100 per cent, the autoclave is gas freed, the unchanged monomer being recovered, before the reactor is emptied. Polymer is stripped and dried. Polymer crusts that accumulate on the reactor walls are removed by high-pressure water jets, but it is still necessary for operatives to enter the reactor periodically to carry out checks or to clear stubborn deposits. This task poses the greatest cause for concern with levels of 10 000 ppm VCM being reported, although other sources of potential exposure include leakages from glands and pumps, and liberation of residual monomer during polymer finishing/blending and also during subsequent processing such as injection moulding. Slow release of trace VCM residues within PVC prevents no serious emission hazards in the workplace during normal storage, etc., although higher levels may be encountered under bulk transport/storage conditions.

Precautions

In the USA vinyl chloride is an OSHA-regulated chemical. The EPA and FDA have banned its use as an aerosol propellant (whereas in some countries up to the mid-1970s some aerosol products, including some hairsprays, contained >50% by weight as propellant), and the EPA have addressed levels of vinyl chloride in water, whilst the Clean Air Act addresses vinyl chloride emission from production and manufacturing facilities.

In the UK relevant legislation includes the COSHH Regulations 1988 and an Approved Code of Practice has been published.[55] The latter, which supplements the Approved Codes of Practice for the (a) Control of Substances Hazardous to Health and the (b) Control of Carcinogenic Substances, applies to persons who

1) are employed in works in which vinyl chloride is produced, reclaimed, stored, discharged into containers, transported or used in any way whatsoever or in which vinyl chloride monomer is converted into vinyl chloride polymers, and
2) are liable to be exposed to vinyl chloride monomer in a working area.

In the UK an 8-hour time-weight maximum exposure limit of 7 ppm and an annual exposure limit of 3 ppm have been set with recommendations for action levels set at

15 ppm when measured over 1-hour period
20 ppm when measured over 20-minute period
30 ppm when measured over 2-minute period

To ensure safe use of vinyl chloride (VC), plants should be designed, built, operated, maintained and controlled to avoid or minimise VC leaks into the work environment. Hence plants should be located in the open air, or be provided with good general ventilation supplemented with local exhaust ventilation where emissions of VC are likely (e.g. when purging injection moulding machines[56] or for entry into confined spaces containing VC). The need for maintenance or breaching of, or entry into plant containing VC, should be minimised.

The number of manual operations can be reduced by automation of the production cycle, automatic cleaning, etc. With improved autoclave technology, closed introduction of additives, and the use of methods to reduce fouling, it has become possible to operate the reactor for an increasing number of cycles without the need for opening.[51] Emission rates have been minimised by use of closed systems and superior valve and pump seals, by collecting samples in bombs from closed loops and by a reduction in the level of residual monomer in the polymer at the various conversion stages. Storage and process vessels should be vented safely; residual emissions can be absorbed or incinerated, and all effluent should be adequately treated.

Even with such controls leaks are bound to occur and entry into confined spaces containing VC will be needed on occasions. The level of containment and operative exposure should be checked by environmental monitoring using background multi-point samplers equipped with automatic alarms as described on page 283 and in ref. 26, coupled with use of personal dosimeters as appropriate. Potential leakage points should be regularly and thoroughly examined by competent persons. Any anomalies, breakdowns and operating difficulties should be reported and investigated immediately.

There should be procedures for dealing with emergencies (leaks, spillages, etc.) and entry into confined spaces. Prior to entry into a reactor it should first be evacuated and the use of local exhaust ventilation and adequate personal protection (e.g. including special clothing, compressed airline or self-contained breathing apparatus) is likely to be required, and the operation controlled by a permit-to-work system. When taking samples of powder from tankers and bulk storage containers, exposure can be minimised by use of long-handled scoops. If entry into storage containers is essential then full precautions, including use of air-supplied respirators, should be taken. When entry into 'empty' PVC containers is needed, e.g. for cleaning operations, the level of VCM should be reduced below the Occupational Exposure Limit (OEL) (as determined by headspace analysis), e.g. by blowing air through the container, and staff provided with relevant personal protection including protective clothing and an appropriate dust mask.

Adequate washing and showering facilities should be provided and good standards of personal hygiene and general housekeeping should be encouraged. Eating, drinking and smoking should be prohibited in VC/PVC areas.

Within the UK health surveillance upon recruitment and at least at 12-month intervals is required for VC workers as defined in the Approved Code of Practice. This includes a full medical and occupational history, clinical assessments of, e.g. the abdomen, skin and extremities, and possibly radiology, and laboratory tests which may include urine analysis. (VC is metabolised to chloroacetaldehyde which conjugates (page 201) with glutathione or cysteine, or oxidises to monochloroacetic acid. The main urinary metabolites are hydroxyethyl cysteine, carboxyethyl cysteine or their N-acetylated derivatives, monochloroacetic acid and thiodiglycolic acid.)[55]

Keeping records of environmental/operator exposure data, health surveillance, exhaust ventilation performance, training, etc., which are an integral component of any control strategy for work with substances hazardous to health, is of particular importance for carcinogens.

Conclusions regarding carcinogens

Many occupational cancer risks have been reduced during the past century and one estimate is that only about 4 per cent of cancer deaths can be attributed to occupational causes.[57] For example, exposure to ionising radiations has been reduced by many orders of magnitude in the last few decades, and exposure to asbestos has been, and continues to be, substantially reduced[58] although it was recently estimated that because of the long latency period, deaths due to mesothelioma associated with asbestos exposure will triple in the next decade. Substitution has been widely introduced, e.g. in the replacement of neat mineral cutting oils by synthetic coolants, and in the phasing-out of polychlorinated biphenyls (PCBs) as dielectric fluids (although there are difficulties in the interpretation of animal studies in which they were demonstrated to be liver carcinogens).[45] Many potent carcinogens are no longer in use, e.g. in dyes, antioxidants or pesticides. By contrast,

> A survey finally reported in 1954 of the incidence of bladder cancer amongst workers exposed to certain dyestuff intermediates revealed rates as high as 50% to 100%, the latter amongst workers involved in the distillation of β-naphthylamine.[59,60]

Carcinogenic metals such as beryllium, and certain processes involving nickel, cadmium and chromium are now recognised as such and handled accordingly.[58] The risks from chemicals used in bulk, e.g. chlorinated hydrocarbons, vinyl chloride, nitro- and amino-aromatic hydrocarbons, benzene and others have been substantially reduced by combinations of the measures summarised earlier. The residual risks therefore appear (after ref. 58) to be:

(a) from chemicals whose carcinogenicity is not well understood or acknowledged. In this context a rapid management response is clearly essential once a hazard is recognised;

> In January 1974 it was reported that three workers involved in PVC production had died of angiosarcoma of the liver. Within months the US hygiene limit, which had just recently been lowered from 500 ppm to 200 ppm, was reduced to 50 ppm. Almost immediately this was reduced again to 'no detectable level' and finally replaced by a limit of 1 ppm.[61] (The current standard in the UK is an MEL, long-term exposure limit, of 7 ppm, subject to an overriding annual maximum exposure limit of 3 ppm.)

(b) from accidental exposures, e.g. arising from system of work failures, neglect of personal hygiene and improper waste disposal practices.

As to (b), improved performance clearly requires proper education, training and instruction, particularly since prolonged exposure to very low levels can be serious. Such exposures may arise in poorly-controlled routine manual operations, e.g. on demolition or waste-disposal sites.

Benzo[a]pyrene is a powerful experimental carcinogen, and occupational exposures may occur to roofers, asphalters and hot-pitch waterproofers. Higher death rates from lung cancer have been suspected amongst such workers exposed for >20 years.

Table 28.10 identifies the contribution of cancer to occupational respiratory disease in one recent UK study.[62] The mineral dust diseases (lung cancer in the presence of pulmonary fibrosis, malignant mesothelioma, and pneumoconiosis) accounted for 34 per cent of the cases, and the suspected agents are listed in Table 28.11(a).[62] Table 28.11(b) identifies the high-risk

Table 28.10 Cases of work-related respiratory diseases diagnosed in the UK in 1989

Diagnostic category	No.	(%)	Crude rate/ million/y*
Inhalation accidents	72	(3.4)	3
Acute pulmonary oedema	15		
Other	57		
Allergic alveolitis	133	(6.3)	5
Asthma	554	(26.4)	22
Building related illness	217	(10.3)	9
Byssinosis	23	(1.1)	1
Infectious diseases	100	(4.8)	4
Lung cancer (with pulmonary fibrosis)	51	(2.4)	2
Malignant mesothelioma	340	(16.2)	13
Pneumonoconiosis	322	(15.3)	13
Benign pleural disease	221	(10.5)	9
Other diagnoses	68	(3.2)	3
Bronchitis/emphysema	25		
Lung cancer	12		
Other	26		
Unspecified	5		
Total	2101	(100%)	84

*Based on the 1988 Labour Force Survey estimate of the UK working population.

occupations associated with asbestos-related disease.[62] However, the well-established risk to laggers, who in the past applied and stripped asbestos-based thermal insulation, often in semi-confined spaces, is concealed by their inclusion with other building and construction workers.

Sensitisers

Introduction

As discussed in Chapter 5 exposure to certain agents may result in an individual becoming 'sensitised'. Such materials pose special challenges in avoiding occupational ill-health since sensitised workers may over-react upon subsequent exposure to levels of the substance well below that needed to induce initial sensitisation, or to that required to bring about apparent untoward effects in unsensitised workers.

Sensitisation can result from ingestion (e.g. food allergies) but in the industrial context skin and respiratory sensitisation are the most common. Such reactions are not necessarily restricted to occupational exposures.

> A six-year-old developed a cough, itchy eyes, a watery nose and asthmatic wheezing usually coinciding to days when an odour of methylamines emanated from a chemical plant situated 1.5 miles from her home. A strong reaction was produced in a skin test with a solution of methylamine.[63]

Some substances may be both skin and respiratory sensitisers. Thus, colophony (or rosin), a natural pine resin, has caused contact dermatitis in workers handling the material or even handling nails coated in the rosin to keep them embedded in wood. Colophony (as the residue from turpentine distillation) is also used as a flux in solder. Some workers exposed to solder fume have developed occupational asthma.

> A television engineer developed wheezing, shortness of breath and chest infection. Because symptoms occurred when away from work in the evenings it proved difficult to trace the cause. Upon investigation, however, occupational asthma was diagnosed and he had developed an allergy to colophony used in soldering. He was unable to return to the same job as symptoms would recur. This is a well-recognised and preventable disease and less costly to avoid than replacing a skilled worker.

Mechanisms

The immune system is one of the body's natural defence mechanisms which enables foreign chemicals entering the system to be recognised and dealt with. The immune system comprises antigens (materials which stimulate formation of antibodies or alter the reactivity of certain cells), lymphocytes (white blood cells) and immunoglobins (Ig)/antibodies (proteins of which there are five types, viz. IgM, IgA, IgD, IgE and IgG). Certain natural synthetic chemicals are capable of stimulation of the immune system. These can be classified as:
- those substances of sufficient molecular size to function as complete antigens, e.g. allergens of pollens, house-dust mites, animal urine;
- those of small molecular size which behave as haptens, i.e. they must conjugate with protein before they can initiate an immune response.

Once sensitised by an initial exposure, the immune response may be triggered upon subsequent exposures to trace quantities of the allergen, or to a related material as a result of 'cross-sensitisation'. This may occur days, weeks, months or even years after the first exposure. The effects are dose-dependent, but such minute exposures may be required to elicit responses in sensitised individuals that hygiene standards are extremely low. For potent sensitisers even these may be of little practical significance in preventing untoward health effects; at best they may avoid workers becoming sensitised but offer no guidance in protecting those already affected. For example, the documentation data for *p*-phenylene diamine states:

> The TLV for this chemical is believed to be sufficiently low to minimise the number of persons who become sensitised but it is recognised that the limit is not low enough to prevent exacerbation of asthma in those already sensitised to para phenylene diamine.

Basically four types of immune response are recognised, viz.

- Type I reactions are characterised by antigen–antibody reaction on the surface of mast cells and basophils, with liberation of histamine. This increases vascular permeability. Also involuntary muscles such as those associated with conducting airways may contract. Hay fever and asthma are examples. IgE antibodies are believed to be involved.
- Type II reactions involve antibody combining with an antigen which is on, or part of, a cell. Metals or low-molecular-weight proteins can be absorbed onto cellular membranes or tissue surfaces, and subsequently take part in Type II reactions.
- Type III allergies involve antigens and IgG antibodies where the molecules of both are freely dispersed throughout the tissues and tissue fluids.
- Type IV reactions differ from the others in that immunoglobulins are not involved and the reaction takes place between the antigens and lymphocytes which appear to possess antigen-specific receptors.

The differing features of allergic sensitivity as between atopic and nonatopic subjects are summarised in Table 28.12.[64]

Contact allergic dermatitis

The key signs of contact dermatitis are skin cracking, flaking, redness, swelling, and blistering. Visually there is no distinction between allergic contact dermatitis and irritant dermatitis but the underlying mechanisms are quite different. This has implications for control strategies.

Skin sensitisation results when the chemical, generally of low molecular weight, penetrates the protective barrier layer of skin and then provokes cell-mediated Type IV immune responses. This induction period of sensitisation typically requires 7 days to complete. Thereafter further skin contact with that chemical elicits allergic contact dermatitis usually within 6–48 hours after contact. Individual susceptibility varies widely and sensitisation can persist for life.

Skin sensitisers are not necessarily skin irritants or respiratory sensitisers. Thousands of skin sensitisers have been reported; selected examples of industrial importance are given in Table 5.19. Since some chemicals are ubiquitous in industry it is crucial to take cognisance of all sources of exposure; for example Table 28.13 identifies key industrial sources of

1606 Toxic hazards

Table 28.11(a) Suspected agents for the mineral dust diseases

	Disease, No. (%)			
Agent	Lung cancer	Mesothelioma	Pneumoconiosis	Benign pleural disease
Asbestos	33 (65)	297 (87)	186 (58)	216 (98)
Coal	0	0	66 (20)	0
Silica	0	0	43 (13)	0
Other	5 (10)	2 (<1)	9 (3)	0
Not known	0	9* (3)	0	0
Not stated	13 (25)	32 (9)	18 (6)	5 (2)
Total	51 (100)	340 (100)	322 (100)	221 (100)

*No known exposure to asbestos.

Table 28.11(b) Asbestos-related disease* in men in high risk occupations

Occupational group	No. of cases	Population	Rate/million/y	(95% CL)
Electricians and power plant operators	40	354,579	113	(81–154)
Building and constructions workers (including laggers)	100	728,202	137	(122–167)
Plumbers and heating engineers	33	184,872	179	(123–251)
Boiler operators	14	9,862	1,420	(776–2,412)
Shipyard and dock workers	77	47,792	1,611	(1,271–2,014)

* Includes cases of asbestosis, lung cancer with pulmonary fibrosis and malignant mesothelioma.

chromates. Furthermore domestic sources of exposure can aggravate the risk, e.g. sensitisation resulting from industrial exposure to nickel can give rise to contact allergic dermatitis in individuals wearing nickel-plated jewellery, i.e. 'nickel itch'.

Allergic contact dermatitis from wood arises mainly amongst woodworkers, e.g. in the furniture industry; the principal cause is contact with airborne dust. The main sensitising woods are summarised in Table 28.14. Mucosal inflammatory irritation of the respiratory tract and allergic respiratory disease may accompany the dermatitis.[65]

Respiratory sensitisation

Sensitisation of the respiratory system may occur following inhalation of certain compounds, or as a result of systemic sensitisation following entry into the body via another route. Parts affected include the nasal passage, major airways and the fine structure of the lung.[66] Symptoms range from mild, but unpleasant, rhinitis and conjunctivitis, to more severe and potentially fatal alveolitis and asthma (attacks of wheezing, chest tightness and breathlessness resulting from variable constriction of the airways). As

Table 28.12 Comparison of the features of allergic sensitivity in atopic and non-atopic subjects

Type of allergy Mainly	'Atopic' subjects Immediate, Type I	'Non-atopic' subjects Arthus, Type III
Sensitisation	Constitutional predisposition to be readily sensitised by ordinary exposure to daily life.	More extensive exposure required.
Antibody	Reaginic antibody, IgE.	Precipitating antibody IgG and IgM.
Mechanisms of reaction	Combination of allergen with reaginic antibody attached to surface of mast cells, causes rapid release of histamine and other mediators of tissue reactions.	Immune complexes, of allergen and antibody formed, which fix and activate components of complement enzymatically.
Tissue response	Oedema, eosinophilia, hypersecretion, reversible and not tissue damaging	Perivascular inflammatory response, mononuclear cells mainly, granuloma formation, fibrosis, tissue-damaging reaction.
Speed and duration of reaction typically	Rapid 'immediate' onset in minutes, lasts 1½ to 2 h, as seen on skin and inhalation tests.	Slow, 'late' onset in about 3 to 4 h, lasts 24 to 36 h, as seen in skin and inhalation tests.
Upper respiratory tract and eyes	Rhinitis, conjunctivitis, itching, sneezing, rhinorrhoea, nasal obstruction, lacrymation.	Not a well-defined feature.
Bronchi	Asthma – rapid onset. Immediate, Type I, reaction. Wheezing and dyspnoea.	(a) Asthma – slow 'late' onset, Type III reaction. Wheezing and dyspnoea.
Peripheral gas-exchanging, 'alvelor' tissues	Not affected	(b) Extrinsic allergic alveolitis (Farmer's Lung type of disease), Type III reaction. Dyspnoea, cough, no wheezing, crepitant rales present. (a) and (b) may occur together in some subjects.
Systemic reactions	Eosinophilia of blood tissues and secretions. Nothing else characteristic.	Eosinophilia not a feature. Fever, malaise, myalgia and loss of weight, which may be considerable.
Pulmonary function	Ventilatory obstruction – expiration decreased and prolonged.	Restrictive ventilatory effect – decrease in gas-transfer and elasticity of the lungs.
Radiography of lungs	Nil characteristic.	Extrinsic allergic alveolitis – miliary nodular infiltration, fibrosis and cystic changes.
Pathology	Oedema and eosinophilia of bronchial mucosa.	Infiltration of alveolar walls with lymphoid histiocytic and plasma cells. Epithelioid cell granulomata, fibrosis and cystic changes.

Table 28.13 Industrial sources of chromates

Anti-corrosives in water systems, e.g. inhibitors	Leather tanning
Anti-rust coatings	Magnetic tapes
Ashes	Milk testers
Cement	Oils
Chrome polishing	Paints
Colour television sets	Paper industry
Electroplating	Photographic chemicals
Food laboratories	Primer paints
Foundry sand	Printing
Galvanised sheets	Solvents
Glass polishes, stains and glazing	Welding fume
Glues	Wood preservatives

noted on page 221, asthma is a prescribed disease in the UK for industrial causes arising from exposure to specific agents. However, many other substances are capable of causing respiratory sensitisation, including nickel and chromium in stainless-steel manual arc-welding fumes; acrylic polymer dust; alkaline persulphates (e.g. used in the manufacture of hydrogen peroxide); perfumes, preservatives, dyes, amines, hardwoods and other vegetable matter. Table 28.15 identifies selected materials claimed to have caused occupational asthma.

Table 28.14 Main sensitising woods

African mahogany	Oregon pine
Balsa wood	Rosewoods
Coco-bolo	South African boxwood
East African camphorwood	Teak
East Indian satinwood	Western red cedar
False rosewood	White perobe
Mansonia	

Persistent allergic symptoms were complained of by several upholstery workers in a furniture factory. They handled several different materials including glue, silicone spray, upholstery fabrics and felt. A detailed medical investigation revealed that the cause was in fact castor bean allergens in the felt, which was manufactured from sacks, some of which had been used to store castor beans.[67] (Castor-bean allegy is a well-established phenomenon which has been observed in workers handling the bean and in people residing near castor bean processing plants.)[68]

Thus in one study of 2101 cases of work-related respiratory disease reported in the UK for 1989, asthma accounted for 25 per cent as shown in Table 28.16.[69] Only in about 50 per cent of the cases were suspected agents on the prescribed list. Occupations at greatest risk from asthma are given

Table 28.15 Agents *claimed* to have been implicated in occupational asthma

Abietic acid
Acacia
Acacia gum
Acaridae
*Acid anhydride hardening agents
Acrylic fibres
Acrylic precursors
Actinomyces
Alicyclic amines
Aliphatic aldehydes
Aliphatic amines
Alkyl phosphates
*Amine hardening agents
Ampicillin
*Animals including insects and other arthropods or their larval forms
Anthraquinone dyestuffs
*Antibiotics
*Azodicarbonamide
Bacillus subtilis
Barley (see 'Dusts')
Benzalkonium chloride
Benzylpenicillin
β-lactamines
Biocides
Bromelain
Candida tropicalis (proteins)
Carbamates
*Castor bean dust
Castor oil
Cats
Chlorthion
Chlorine
Chromium
*Cimetidine
Cobalt
Cockroaches
Colophony (*fumes from the use as a soldering flux)
Colorado beetles
Cotton
Cows
*Crustaceans
DDVP
Diazinon
Diazomethane
Diethanolamine
Diethylene diamine
Diethylene triamine
Dogs
*Dusts arising from barley, oats, rye, wheat or maize, or meal or flour made from such grain
Epoxy resins and hardeners
Ethylhexylamine
Exotic woods
Flax

Flour or meals
Formaldehyde
*Fumes or dusts arising from the manufacture, transport or use of hardening agents (including epoxy resin curing agents) based on phthalic anhydride, tetrachlorophthalic anhydride, trimellitic anhydride, or triethylene tetramine)
*Fumes from stainless steel welding
*Glutaraldehyde
Graminaceous pollens
*Green coffee bean dust
Ground nuts
Guinea-pigs
Gum arabic
Hair, horns, feathers, etc. of animals
Hamsters
Hemp
*Henna
Hexamethylenetetramine
Industrial perfumes
Insecticides
*Ipecacuanha
*Isocyanates
*Ispaghula powder
Jute
Karaya gum
Laboratory animals
Lead
Liquorice
Locusts
Mercury diphenyl
Mercury (organic compounds)
Metampicillin
Mice
Mites
Moulds
Nickel
Nitric oxide
Oats
Oil cake
Oleandomycin
Organic isocyanates (see Isocyanates)
Organophosphorus compounds
Panonychus ulmi
Papain
p-Dichlorobenzene
p-Formaldehyde
p-Phenyldiamine
Penicillins
*Persulphate salts
Pesticides
Phenyl-formaldehyde resins
Phenylglycine
Phenylhydrazine
Phenylmercuric nitropropionate

Table 28.15 (continued)

Phosphoramines	Spiromycin
Phthalic acid (see 'Fumes')	Tea dust
Piperazine	Textiles, natural
*Platinum salts	Textiles, synthetic
Polyamides	Thrombin
Polyesters	Triethylene diamine
*Proteolytic enzymes	Triethylene tetramine
Pyrethrum	Trimellitic anhydrides (see 'Fumes')
Quinine	Trypsin
Rabbits	Urea-formaldehyde resins
Rats	Vanadium
*Reactive dyes	Vanillin
Red spiders	Vegetables (pharmacological action)
Rice	Viscose
Rye	Water fleas
Sericin	Welding fumes
Silk	*Wood dust
*Soya bean	Wool

*Agents to which occupational exposure is prescribed for the purposes of occupational asthma as Prescribed Disease D7 in the UK.

in Table 28.17 and include spray painters and workers in plastics manufacture and processing. The incidence of occupational asthma varies widely and for some substances only isolated cases have been reported. Also the mechanisms by which they operate vary with the substance. Thus, data suggest a Type I mechanism for platinum salts, which are extremely potent respiratory sensitisers also capable of causing dermatitis, whereas formaldehyde elicits Type IV responses, isocyanates Type I or Type IV responses, or both, and complex immunological patterns in workers exposed to hardwoods show evidence for Type I, Type III and Type IV responses. Lists of materials capable of inducing asthma also neglect respiratory sensitisers which may result in milder symptoms of distress.

Control measures

The key general precautions for working with sensitising agents are as follows.
- Establish the nature and potency of the effect from the supplier, from literature, or by research.
- Substitute where possible with safer material.

> Three workers in the bonding section at a bicycle factory developed occupational dermatitis after using epoxy resins. Their job entailed immersing bicycle components in industrial methylated spirits to remove dust and then application of an epoxy resin adhesive using a small spatula. In some cases, to make their jobs easier, workers had removed protective gloves and worn only cotton liners. The company are now investigating the possibility of using safer adhesives on an automatic process. They should also introduce health surveil-

Table 28.16 Suspected agents in occupational asthma

Prescribed causes	Compensated cases* (no.)	SWORD‡ cases (no. (%))	Non-prescribed causes	SWORD‡ cases (no. (%))
Flour/grain dusts	40	42 (8)	Other plant dusts, fungi and animal antigens	38 (7)
Wood dusts	28	31 (6)	Formaldehyde	7 (1)
Other prescribed plant dusts	1	0 (0)	Ethylene diamine	5 (1)
Laboratory animals/insects	9	21 (4)	Chlorine	5 (1)
Isocyanates	64	120 (22)	Ammonia	3 (<1)
Azodicarbonamide	5	6 (1)	Methylene chloride	2 (<1)
Antibiotics	6	5 (1)	Glutaraldehyde	2 (<1)
Proteolytic enzymes	2	5 (1)	Other pharmaceuticals	3 (<1)
Platinum salts	12	5 (1)	Other chemicals	36 (7)
Solder flux	24	32 (6)	Aluminium potroom emissions	11 (2)
Hardening agents	31	15 (3)	Cobalt	8 (1)
Total	222	282 (51)†	Welding fume	4 (<1)
			Chrome	3 (<1)
			Paints	15 (3)
			Glues	12 (2)
			Hair/cleaning products and disinfectants	11 (2)
			Inks and dyes	7 (1)
			Cutting oil	7 (1)
			Insecticides/fungicides	3 (<1)
			Fibreglass	3 (<1)
			Other specified agents	29 (5)
			Total	214 (39)†

*New cases qualifying for disablement benefit in Great Britain in 1988 by causative agent (Department of Social Security).
†In a further 58 cases (10%) the suspected agent was unknown or not specified.
‡Surveillance of Work-related and Occupational Respiratory Disease.

Table 28.17 Incidence of occupational asthma in high risk occupations

Occupational group	Cases	Population	Rate/million/y	(95% CL)
Welders/ solderers/electronic assemblers	35	220,068	159	(111–221)
Laboratory technicians and assistants	26	127,478	204	(133–299)
Metal making and treating	14	56,270	249	(136–417)
Plastics making and processing	27	66,005	409	(270–595)
Bakers	29	70,839	409	(274–588)
Chemical processors	31	73,189	424	(288–601)
Coach and spray painters	35	54,737	639	(445–889)
Other painters	21	201,225	104	(65–160)

lance for exposed workers and provide training to ensure workers appreciate the proper use of control measures.

- Use material in a safer form, e.g. replace dusty powder by marumes, prills, noodles, granules, etc.
- Provide high levels of containment and where necessary exhaust ventilation and personal protection, and operate to very stringent hygiene standards. The personal protection may include a suitable respirator (Chapter 6) so as to provide adequate protection, even against peak exposures, and clothing which is suitably impervious and close-fitting at the neck and wrists. The washing facilities should be adequate and conveniently situated, and include showers where appropriate.
- Undertake pre-employment medicals and screen out, where relevant, people who may be unsuitable to handle the substances on health grounds. For example, for work with isocyanates workers with hay fever, recurrent acute bronchitis, pulmonary tuberculosis, asthma, chronic bronchitis, interstitial pulmonary fibrosis, atopic eczema, occupational lung disease or impaired lung function should not be exposed to further hazard.[70]
- Where practicable segregate the work with sensitisers from other activities, either physically or by performing it at a time when other employees are not present.
- Undertake regular environmental and/or biological monitoring.
- Restrict access to areas handling sensitisers to authorised personnel.
- For sensitised individuals prevent exposure to any concentration of allergen, especially potent sensitisers. For example with isocyanates, sensitisation usually means that the individuals can no longer work in situations where isocyanates are in use and need to be transferred to other duties where no exposure can occur.[71] Similarly, clinical tests are used to determine whether individuals should work on the manufacture of enzymatic fabric washing products (page 1651); those with, e.g. established chest disease, poor lung function or symptoms of atopy are usually excluded. However, when their economic situation has necessitated it, construction workers sensitised to chromates in cement, and

machinists sensitised to coolant constituents, have continued at work using impervious gloves.
- Provide adequate information, supervision, instruction, training, maintenance, etc.

These are amplified in Chapter 26 by reference to the detailed arrangements adopted for processing with detergent enzymes and by reference to recommended precautions for working with other selected sensitisers.

Sick building syndrome

Occupational health problems in offices, etc., with clearly identifiable causes include Legionnaires' disease and Pontiac fever (Chapter 32); these can be serious and possibly even fatal.

A growing incidence of illness with far less specific aetiology, termed 'sick building syndrome', is apparently linked to physical stressers such as lighting, visual display units (VDUs), air ion levels, humidity, temperature, noise, draughts and indoor air pollution at very low levels in environments where chemicals are not handled routinely in quantity.[72] Emissions may arise from carpets, wall coverings, wood treatments, building insulation, paints, etc. Major sources are listed in Table 28.18 (after ref. 73).

Table 28.18 Major sources of indoor air pollutants

Source	Pollutants
Internal sources	
Tobacco	NO_2, CO, particulates, polyaromatic hydrocarbons, formaldehyde, VOCs
Combustion appliances	NO_2, SO_2, particulates, polyaromatic hydrocarbons, formaldehyde, VOCs
Consumer/commercial products (paints, cleaners, correction fluids, personal care products)	VOCs
Building materials	Asbestos, VOCs, formaldehyde, radon
Occupants, pets, plants	Biological contaminants, pesticides
HVAC systems, water damaged materials	Biological contaminants
Photocopiers	Ozone
Humidifiers	Biological contaminants, particulates
Attached garages	CO, NO_2, VOCs, particulates
Office furnishings	Formaldehyde, micro-organisms, dust (e.g. detergent after shampooing)
External sources	
Soil/rock	Radon, pesticides
Stret traffic, loading docks	CO, NO_2, VOCs
Ambient air	Biological contaminants, VOCs, CO, NO_2, particulates
Water supplies	Radon, VOCs

VOC = volatile organic compound.

However, a problem amongst a given group may be due to some feature of the workers themselves, the material or technology they work with (e.g. VDUs), the jobs they perform (e.g. stress) or some feature of the building in which they work.[74]

Symptoms commonly reported are[75]
- eye, nose and throat irritation
- sensation of dry mucous membranes and skin
- erythema (skin rash)
- mental fatigue
- headaches, high frequency of airway infections and cough
- hoarseness, wheezing, itching and unspecified hypersensitivity
- nausea, dizziness

So far as volatile organic chemicals (VOCs) are concerned, the populations most likely to be affected by low-level exposure include
- industrial workers
- residents of light buildings
- residents in communities with air and water contamination
- anyone with unusual exposure to certain types of chemicals

Medical controversy has arisen, however, because of the lack of any pattern among persons exposed to the same chemicals and apparent wide differences in sensitivity.

In summary the problem occurs when – due to some combination of building design, structural components, ventilation systems and management practices – air quality inside a building has deteriorated to the point at which occupants become ill. Generally[73]
- the symptoms are non-specific, e.g. headache, nausea, throat irritation, and not unique to specific classes of contaminants;
- the problem can seldom be traced to one source or one contaminant;
- low-level mixtures of many contaminants are present, each often well below known adverse health thresholds;
- microbial contamination is often a very important factor;
- the design, operation and maintenance of the ventilation system is an important consideration.

The syndrome is described as a relatively new phenomenon associated with energy-efficient structures, e.g. where systems meet physiological needs and provide comfort but peak load conditions are handled with minimum ventilation rates.

In the 1960s the US General Services Administration ventilation standard was 30 ft^3/min per occupant. By the late 1970s it was 5 ft^3/min and many buildings were sealed to prevent air infiltration and well insulated to save energy.[73] (The ASHRAE standard 62 is under review and there may be an increase to 15 ft^3/min.)

Some indication of the low-level chemical exposures in houses in the USA is given by Table 28.19.[76] (Additional pollutants, e.g. CO_2, CO,

Table 28.19 Indoor pollutants found in houses in five different geographical areas in the USA[76]

• Chemicals usually found: Benzene m- and p-Dichlorobenzene Ethylbenzene Tetrachloroethylene 1,1,1-Trichloroethane o-, m-, and p-Xylene • Chemicals often found: Chloroform Styrene Trichloroethylene	• Chemicals occasionally found: Bromodichloromethane Bromoform Carbon tetrachloride Chlorobenzene Dibromochloromethane Dibromochloropropane o-Dichlorobenzene 1,2-Dichloroethane Vinylidene chloride

NO_x, SO_x and HCHO, are not included in this list.) Significantly higher levels of the chemicals in this list were found inside houses than in the ambient air outside regardless of geographic location. Generally, however, levels of exposure were very low, e.g. the benzene was attributed to smoking. Sources include various furnishing materials, e.g. adhesives, wallpaper adhesive, furniture glues, carpet backing; construction materials, e.g. fibre-board, insulation, flooring tiles; consumer products, e.g. deodorants; and, in offices, machinery, e.g. copying machines.

Formaldehyde, e.g. from urea-formaldehyde insulation and certain types of board, is an irritant and may cause some symptoms similar to 'sick building syndrome'. However, some surveys have discounted a link with it.[75]

The problems of exposure to airborne micro-organisms are discussed separately on pages 1786 to 1798.

Prevention

Research on emission patterns and emission rates of harmful chemicals from consumer products and various furnishing materials should lead to more enlightened construction and furnishing of buildings.[73]

Various air volume systems of building ventilation, which supply air at constant temperature and modulate the amount to obtain comfortable conditions, may result in a variation in the quantity of air reaching any spot in a building. Hence the quantity may be insufficient. It has been claimed that 70–80 per cent of complaints caused by pollution can be minimised by bringing in the proper amount of air and distributing it properly.

In a study of sick building syndrome amongst office workers in the UK the prevalence of work-related symptoms was extremely high with 80 per cent of workers reporting at least one.[74] Overall, buildings with ventilation from local or central induction fan coil units had more work-related symptoms per worker than buildings with 'all air' systems which, in turn, had more than naturally or mechanically ventilated buildings. There was a substantial increase in symptoms once the air supply was chilled or humidified. The same contamination with bacteria, fungi, protozoa, etc., which may occur in humidifiers can arise in the chiller which, because of its location in ceiling spaces and walls rather than in a central plant room, is often difficult to service.

There is some evidence that common landscaping plants assist in the provision of pollution-free homes and workplaces. (For example it has been demonstrated that indoor plants and their soil/root zone area can remove formaldehyde, benzene and trichloroethylene from sealed experimental chambers.)

There is some conflicting evidence as to the precise causes of, and appropriate mitigatory measures for, sick building syndrome. However, in summary (after ref. 75):

- *Noise.* Whilst noise levels in offices are unlikely to impair hearing they can cause distraction, loss of concentration and annoyance.
- *Ventilation rates.* There is insufficient evidence to stipulate a minimum safe rate but in a number of cases symptoms have been reduced by increasing the fresh air input.
- *Temperature and air movement.* Some evidence suggests that high, uniform temperatures and lack of air movement result in more symptoms. Comfort rates are shown in Fig. 28.2.[65]

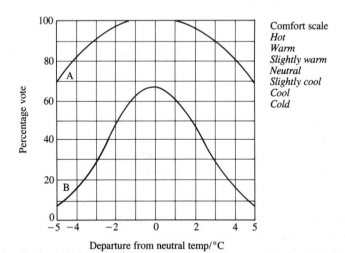

Fig. 28.2 Percentage comfort rate around the 'neutral' temperature: **A** = per cent vote for those citing the three central descriptions; **B** = per cent vote for neutral/comfortable only

- *Humidity.* Low relative humidity may sometimes result in erythema and eye irritation. A combination of humidity and warm temperatures

produces air perceived to be 'dry'. Static can also cause discomfort if relative humidity falls below 50 per cent.
- *Lighting standards.* Symptoms have been more prevalent in buildings with certain types of lighting, e.g. where the lighting and decor are dull and uniform and there is little daylight.
- *Airborne pollution.* Pollution may have been responsible in some cases, although pollution levels have been very low.
- *Airborne organic matter* from the air-conditioning system, particularly humidifiers, may be responsible for some chest symptoms.
- *Low morale and general dissatisfaction.* Whilst it is considered unlikely to be a psychogenic illness, sick building syndrome appears to be partially caused or exacerbated by general dissatisfaction with work and/or the workplace environment.

UK developments in control

Introduction

The predominant legislation dictating control measures in the UK, and representing a radical change since the Health & Safety at Work etc. Act 1974 in the application of statutory restrictions, is the Control of Substances Hazardous to Health Regulations 1988.[77] These are previewed on page 935 and came into force on 1 October 1989; in any particular situation the full Regulations should be consulted.

The Regulations cover virtually all substances hazardous to health. (Asbestos, lead, and materials producing ionising radiations and substances below ground in mines, which have their own legislation, are excluded.) Substances 'hazardous to health' include substances labelled as dangerous (i.e. very toxic, toxic, harmful, irritant or corrosive) under any other statutory requirements, agricultural pesticides and other chemicals used on farms, and substances with exposure limits. They include harmful micro-organisms and substantial quantities of dust. Indeed any material, mixture or compound used at work, or arising from work activities which can harm people's health, is apparently covered.

The substances actually designated in Regn. 2[78] as being hazardous to health are given in Table 28.20.

In determining whether a substance is hazardous to health, it is advised that the following additional factors be taken into account:[78]
- different forms of the same substance may present different hazards, e.g. a solid may present negligible hazard but, when ground into dust of a respirable size, may be very hazardous;
- many substances contain impurities which could present a greater hazard than the substance they contaminate, e.g. crystalline silica is often present in minerals which would otherwise present little or no hazard;

Table 28.20 Substances listed as hazardous to health in COSHH Regulation 2[78]

Substances listed in Part 1A1 of the *Approved List* as **Dangerous for supply** within the meaning of the Classification Packaging & Labelling Regulations 1984[79]	Defined as very toxic toxic harmful corrosive irritant
Substances with an **MEL** (maximum exposure limit) or **OES** (occupational exposure standard)	All MELs listed in Schedule 1 of The Regulations; all OESs approved by HSE and given in HSE Guidance Note EH 40/– *Occupational Exposure Limits*.[80]
Micro-organisms which create a hazard to health of any person	Where the risk arises out of or in connection with work that is under the control of an employer. In practice all group 2, 3 and 4 pathogens (as listed by The Advisory Committee on Dangerous Pathogens[81]) will be covered by the COSHH Regulations. Some 'non-pathogenic' micro-organisms or their products may cause an allergic response in some people, or simply be toxic, despite an inability to infect. These properties of the micro-organisms are still covered by the COSHH Regulations.
Dust of any kind when present in 'substantial quantities'	Substantial quantities are: 10 mg/m^3 total inhalable dust (8-hour time-weighted average (TWA)); 5 mg/m^3 respirable dust (8-hour TWA) except where hazardous properties indicate a lower exposure value is required, or one is specified in Guidance Note EH 40.[80]
Any other substances	Which create a comparable hazard to health to any of the above.
The Regulations are not, however, applicable to	
Lead	Already covered by the Control of Lead at Work Regulations 1980.
Asbestos	Already covered by the Control of Asbestos at Work Regulations 1987.
Substances dangerous solely by virtue of physical properties	E.g. radioactive, explosive, flammable, high or low temperature, high pressure.
Substances administered for human medical treatment	The Regulations do not apply to patients given substances for medical reasons (including clinical trials), but staff involved in the administration (pharmacists, nurses, doctors, etc.) are covered by the Regulatons. Veterinary surgeons and researchers who administer substances to animals (for treatment or clinical trials) are also covered by the COSHH Regulations.

- some substances have a fibrous form which may present a potentially serious hazard to health, if the fibres are of a certain size or shape;

- some substances may be known to cause ill health but the causative agent may not have been identified, e.g. certain textile dusts causing byssinosis;
- combined or sequential exposures to various substances may have additive or synergistic effects;
- a 'substantial' concentration of dust should be taken as a concentration of 10 mg/m^3, 8-hour time-weighted average, of total inhalable dust or of 5 mg/m^3, 8-hour time-weighted average, of respirable dust, where there is no indication of the need for a lower value (the current edition of ref. 4 provides an explanation of 'inhalable' and 'respirable' dust);
- epidemiological data which indicate that a micro-organism or its products is the cause of a hazard to health at work.

The Regulations set out essential measures that employers (and sometimes employees) have to take. They are as follows:
- Assess the risk to health arising from work and what precautions are needed.
- Introduce appropriate measures to prevent or control the risk.
- Ensure that control measures are used and that equipment is properly maintained and procedures observed.
- Where necessary, monitor the exposure of the workers and carry out an appropriate form of surveillance of their health.
- Inform, instruct and train employees about the risks and the precautions to be taken.

The *assessment* requirement is particularly relevant and involves a systematic review[15,82] by 'competent persons'. As a guide Table 5.1 identifies contaminants for a range of processes and industries but all the substances involved or arising in individual process-operations must be identified in each case. Table 6.1 indicates emission rates from process equipment, and Tables 6.3 and 6.4 highlight sources of chemical exposure for selected operations and processes, respectively. These, together with Table 28.21[83] which ranks exposure sources, provide indicators for risk assessment. Clearly, however, the health risk is influenced by other factors such as process conditions (temperature, pressure), properties of the substances in question (toxicity, physico-chemical characteristics) and duration of exposure, etc., associated with the operations subject to assessment.

The main relevant factors to be considered are summarised in Table 28.22 (after ref. 84) and amplified below.

(1) What substances are present? In what form? This requires a check on
- substances brought into the workplace (raw materials, solvents, catalysts, formulated products)
- substances given off during any process or work activity including intermediates, by-products
- substances produced at the end of any process or work activity including all wastes

Table 28.21 Exposure sources in the chemical process industry:[83] M = maintenance, P = personal protection, W = work practices, E = engineering

	Intermittent or continuous	Episodic	Worker activity	Importance	Control
Fugitive emissions					
Pump seal leaks	Either	No	No	High	M
Flange leaks	Cont.	No	No	Low	M
Agitator seal leaks	Either	No	No	Med	M
Valve stem leaks	Cont.	No	No	High	M
Process operations					
Sampling	Int.	Yes	Yes	Med	W,E
Filter change	Int.	Yes	Yes	Low	W,P
Gauging	Int.	Yes	Maybe	Low	E,W
Venting and flaring	Either	No	No	Med	E
Extruding	Either	No	Yes	Med	E
Material handling					
Solid addition	Int.	Yes	Yes	Med	
Liquid transfer	Either	No	No	High	E
Bagging	Cont.	No	Yes	High	E
Drumming	Cont.	No	Yes	High	E,W
Bag dumping	Int.	No	Yes	High	E,W
Screening	Cont.	No	No	Med	E
Open mixing	Int.	No	No	Med	E,P
Banbury mixing	Either	No	Yes	High	E,P
Milling	Either	No	No	Med	E,P
Maintenance					
Equipment opening	Int.	Yes	Yes	High	W,P
Instrument line draining	Int.	Yes	Yes	Med	W,P
Welding	Int.	Yes	Yes	High	E,W,P
Painting	Int.	Yes	Yes	Med	W,P
Sandblasting	Int.	Yes	Yes	High	E,P
Insulating	Int.	Yes	Yes	High	W,P[a]
Insulation removal	Int.	Yes	Yes	High	W,P
Chemical cleaning	Int.	Yes	Yes	Med	W,P
Degreasing	Int.	Yes	Yes	Low	E
Cutting and burning	Int.	Yes	Yes	Med	W,P
Catalyst handling	Yes	Yes	Yes	High	W,P
Waste handling					
Bag house cleaning	Int.	Yes	Yes	High	P
Drain and sewer venting	Either	No	No	High	E
Spill clean up	Int.	Yes	Yes	Med	P
Sweeping	Int.	Yes	Yes	Low	W
Incineration	Either		Maybe	Med	E
Waste water treatment	Cont.	No	No	Med	E
Sludge handling	Int.	Yes	Yes	Med	W,P

[a] Substitution of less toxic materials for asbestos is the most common control.

(service activities are included). Substances 'hazardous to health' can be identified by

- for brought-in substances, checking safety information from labels and that legally obtainable from the suppliers

Table 28.22 Relevant factors in a COSHH Assessment

PROJECT DETAILS/WORK PROCEDURE
1. Identify substances hazardous to health, quantities, grouping, mixtures
2. Establish how they could enter the body and potential effects
3. Consider
   - who is exposed (remember general public, other groups of employees, contractors, etc.)
   - under what circumstances (include breakages and spills, emissions to atmosphere)
   - how much they would be exposed to and for how long
4. Prevention of exposure – is it possible?
   - elimination
   - substitution
   - enclosure of equipment/apparatus
   - exclusion of people from work area
5. If prevention is not possible, consider control measures, e.g.
   - engineering controls
   - safe systems of work
   - personal hygiene needs
   Keep test records
6. If PPE* or RPE* necessary, information on types required
   Keep RPE test records
7. Emergency procedures following spillage including first-aid
8. Routine exposure monitoring requirements
   Keep records
9. Health surveillance requirements
   Keep records
10. Personnel training needs
11. Storage arrangements for raw materials, disposal arrangements for products
12. Any further action needed to comply with the regulations
13. Review date for assessment

* PPE = Personal Protective Equipment; RPE = Respiratory Protective Equipment.

- use of existing knowledge, e.g. past experience, knowledge of the process, understanding of relevant current best industrial practice, information on related industrial health problems
- seeking advice from a trade association, others in a similar business, consultants
- checking whether a substance is mentioned in any COSHH Regulations or Schedules, or listed in Guidance Note EH 40[80]
- examination of published literature or documentation, trade data, HSE guidance information
- checking Part 1A of the Approved List under the Classification, Packaging and Labelling of Dangerous Substances Regulations 1984. (Anything listed as very toxic, toxic, corrosive, harmful or irritant is covered by COSHH.)

(2) What harmful effects are possible (toxicity of substances, review of health surveys)?
(3) Where and how are the substances actually used or handled? For example,

- where and in what circumstances are the substances handled, used, generated, released, disposed of, etc.?
- what happens to them in use (e.g. do they change form, such as from bulk solid to dust by machining, or from bulk liquid to aerosol by spraying, etc.?)
- identify storage and use areas
- identify modes of transport (bulk, drums, cylinders, etc.)

(4) What harmful substances are given off, etc.?
(5) Who could be affected, to what extent and for how long?
- Identify employees, including cleaners, security staff, contractors, and members of the public who could be affected.

(6) Under what circumstances?
- Is some of the substance likely to be inhaled?
- Is it likely to be ingested following contamination of fingers, clothing, etc.?
- Is it likely to cause skin contamination or be absorbed through the skin? (Some materials have a definite 'Sk' notation in EH 40)
- Is it reasonably foreseeable that an accidental leakage, spill or discharge could occur (e.g. following an operating error or breakdown of equipment, or failure of a control measure)?

Consider
- how are people *normally* involved with the substance?
- how *might* they be involved (e.g. through misuse, spillage)?

(7) How likely is it that exposure will happen?
- Check control measures currently in use. Check on their effectiveness and whether they are conscientiously/continuously applied.

(8) What precautions need to be taken to comply with the rest of the COSHH Regulations? Having regard to
- who could be exposed,
- under what circumstances,
- the possible length of time,
- how likely exposure is,

together with knowledge about the hazards of the substance (i.e. its potential to cause harm), conclusions are reached about personal exposure.

An assessment should be reviewed wherever there is evidence to suspect that it is no longer valid. This may, for example, arise because of[77]
- the results of periodic thorough examinations and tests of engineering controls;
- the results of monitoring exposure at the workplace;

- the results of health surveillance, or a confirmed case of occupationally induced disease;
- new information on health risks;

or where there has been a significant change in the relevant work, e.g.
- in the substances used or their source;
- plant modification, including engineering controls;
- in the process or methods of work;
- in the volume, or rate, of production.

The employer's duty is to ensure that the exposure of employees to a hazardous substance is prevented, or if this is not reasonably practicable, adequately controlled.

Thus conclusions from assessments should indicate whether existing control measures are adequate, or if unsatisfactory what further action is required to provide the necessary degree of control. In all but the simplest cases risk assessments should be recorded to indicate the substances involved, how the materials are used, how the assessment was performed including sources of information, the control measures, occupational health and hygiene data and any recommendations with regard to changes in controls, environmental monitoring, health surveillance, provision of information, and training, together with a review date.

Control

As exemplified in Chapters 6 and 29–31, control may mean preventing exposure by
- removing the hazardous substance by a change in the process;

    Ultraviolet radiation can be used for water treatment, eliminating the use of chlorine or sodium hypochlorite.

- substitution of the hazardous substance by a safer one, or using it in a safer form;

or, if the above are not reasonably practicable, by
- total enclosure of the process
- using partial enclosure and extraction equipment
- general (i.e. dilution) ventilation
- use of safe systems of work and handling procedures and exclusion of personnel and reduction in periods of exposure
- training, instruction and provision of information
- supervision

However, in choosing method(s) to control exposure COSHH limits the use of personal protective equipment (e.g. respirators, dust masks, protective clothing), as the means of protection to those situations *only* where other measures cannot adequately control exposure.

Again control measures should be documented, for example on the Risk

Assessment forms, in codes of safe working practice, etc. This is helpful in demonstrating compliance with the COSHH Regulations and as proof of discharging the duty under Common Law.

A health authority cleaner in a hospital developed dermatitis. The health authority were unable to provide evidence that all domestic staff had been issued with gloves and told to wear them, nor that they had been given a handbook on gloves. Nor did the court accept that a notice had been posted about the availability of gloves. It was irrelevant that disposable gloves were available on the wards; the domestic staff worked on the corridors and toilets. There was evidence that gloves had been ordered but none that they had been supplied to the supervisors; nor was there evidence that the supervisors had been instructed to ensure that the cleaners wore the gloves. The health board were held liable for the plaintiff's dermatitis (£5,900 damages were awarded) because they had failed to supply her with rubber gloves for protection.[85]

In a similar case a cleaner was awarded £5,750 damages for dermatitis as the Health Authority had failed to inform her of the risks of working without gloves.[86]

Exposure limits

Exposures require control such that nearly all people would not suffer any adverse health effects even if exposed to a specific substance (or mixture of substances) day after day. For certain substances there are set Occupational Exposure Limits.
- For substances assigned Maximum Exposure Limits (MELs) the level of exposure should be reduced so far as is reasonably practicable and, in any event, it should not exceed the MEL.
- For substances with Occupational Exposure Standards (OESs) it is sufficient to ensure the level of exposure is reduced to that standard.

Guidance is available for setting in-house hygiene standards, from bodies such as the UK Chemical Industries Association.

Maintenance, examination and testing of control measures

An employer has a specific obligation to ensure all control measures are kept in an efficient working order and good repair. Engineering controls should be examined and tested at suitable intervals, e.g. local exhaust ventilation equipment must be tested at least once every fourteen months. Tables 28.23 and 28.24 indicate the types of records to be kept on local exhaust ventilation for judging its continuing efficiency and for test data respectively. Records should be kept for at least five years. Respirators and breathing apparatus must also be examined frequently.

Monitoring

COSHH requires that the exposure of workers should be monitored in certain cases.[88] Examples are

Table 28.23 Type of record on local exhaust ventilation for judging continuing efficiency

- Enclosures/hoods:
  Maximum number to be in use at any one time
  Location or position
  Static pressure behind each hood or extraction point
  Face velocity
- Ducting:
  Dimensions
  Transport velocity
  Volume flow
- Filter/collector:
  Specification
  Volume flow
  Static pressures at inlet, outlet and across filter
- Fan or air mover:
  Specification
  Volume flow
  Static pressure at inlet
  Direction of rotation
- Systems which return exhaust air to the workplace:
  Filter efficiency
- Concentration of contaminant in returned air

Table 28.24 Type of record on local exhaust ventilation (LEV) test data

- Name and address of employer responsible for the plant
- Identification and location of the LEV plant, process and hazardous substance concerned
- Date of last thorough examination and test
- Conditions at time of test: normal production or special conditions (e.g. maximum use, stood down)
- Informaiton about the LEV plant which shows:
  (a) its intended operating performance for controlling the hazardous substance for the purpose of Regulation 7
  (b) whether the plant now still achieves the same performance
  (c) if not, the repairs required to achieve that performance
- Methods used to make judgement on (b) and (c) above (e.g. visual, pressure or air flow measurements, dust lamp, air sampling, filter integrity tests)
- Date of examination and test
- Name, designation and employer of person carrying out examination and test
- Signature of person carrying out examination and test
- Details of repairs carried out (to be completed by the employer responsible for the plant. The effectiveness of the repairs should be proved by a re-test)

- where it is not certain that particular control measures are working properly
- where it is not possible to be sure that exposure limits are not being exceeded
- where there could be serious risks to health if control measures were to fail or deteriorate.

Specifically, monitoring should take place
- for substances/processes listed in Schedule 4 of the Regulations

- usually where carcinogens are handled, because exposure can result in serious health effects (a separate code applies to carcinogens[78])
- where vinyl chloride monomer is used; where an electrolytic chromium process is carried out from which vapour/spray is given off
- usually during, and on completion of, a fumigation operation, except where exposure can be prevented.

The monitoring strategy, introduced in Chapter 7, should cover the following.

- Timing of monitoring. If exposure is intermittent, monitoring should coincide with specific operations so that a realistic peak and average exposure estimate is obtained.
- Location of monitoring. Everyone who may be exposed should be covered, e.g. cleaners, maintenance workers, security staff.
- Sampling and measuring methods used. Personal monitors are preferred to static or background monitoring systems to provide a more accurate estimate of individual exposures. Advice is provided for measurement methods for some specific substances or groups;[87] validated methods should otherwise be used. Accurate measurements may not be feasible for all substances.
- Location of measurements. Persons doing similar jobs may be grouped and measurements taken of representative members of the group. (This involves the assumption that all persons perform the work, and any anciliary jobs, in an identical manner; this should obviously be checked.)
- Frequency of monitoring. Where necessary, monitoring should be performed at least once in every twelve months (more often in the case of work with vinyl chloride monomer or an electrolytic chromium process or where previous exercises have indicated exposures close to the hygiene standard, or where the process has been changed).
- Biological monitoring is relevant for substances which can be absorbed through the skin, where there is dependence on wearing personal protection, to reflect differences in rates of absorption by different individuals, etc. Examples include analysis of blood, exhaled breath and urine.
- Interpretation of records. Causes of high or increasing exposure – in comparison with an MEL, OES or self-imposed standard – should be taken into account and further surveys performed to verify improvements.
- Keeping of records. It should be ensured that monitoring results can be compared with any health records. Personal monitoring results should be kept for at least thirty years and other results for five years.[78]

Once control measures are introduced management have a duty to set up systems to monitor that they are used. Environmental analysis for levels of chemicals is only one indicator of control. Other systems requiring monitoring include performance of ventilation systems (e.g. face velocity measurements) and visual inspections of exhaust ventilation and the con-

Control of substances hazardous to health

dition of personal protective devices. Inspection of standards of hygiene and general cleanliness of the work area will also offer opportunities to detect falling standards.

Health surveillance

Health surveillance is necessary (Regn. 11)
- where carcinogens are handled (unless exposure is not significant)
- where people are involved in any of the processes and with any of the substances listed in Schedule 5 to the Regulations (health surveillance here must be under supervision of an employment medical advisor or appointed doctor)
- where people are exposed to any substance that can produce an identifiable disease or adverse health effect, detectable by valid techniques, and for which there is a reasonable likelihood of the disease or effect arising as a result of the particular conditions at work.

Surveillance should not detract resources from achieving control of exposure; it is not a substitute for obtaining and sustaining adequate control.

The procedures chosen should be appropriate to detect a disease or adverse health effect arising from exposure to specific substances, or groups of similar substances.

Health surveillance always includes the keeping of an individual health record, and usually also requires additional procedures, e.g.
- biological monitoring, e.g. for mercury, cadmium;
- enquiries about symptoms, e.g. about respiratory symptoms;
- inspection, e.g. of skin for dermatitis;
- medical surveillance, e.g. clinical examinations and measurements by a suitably qualified doctor;
- review of records.

These and similar procedures must lead to some action which will be of direct benefit, either in the short or in the long term, to the health of people concerned.

Following each round of health surveillance the results should be considered carefully. If they indicate that control may be inadequate for work with a particular substance, or group of substances, urgent action should be taken to regain control. The assessment should then be reviewed.

For all people subject to health surveillance the health records should be kept for thirty years. The particulars required have been listed.[78] Each person should be informed of his personal health surveillance result. Employers should also provide information on collective, anonymised health surveillance results to staff and their safety representatives.

Information, instruction and training

There is a requirement (Regn. 12-(1)) for employees to be provided with sufficient information, instruction and training to ensure they are aware of any health risks associated with exposure to particular substances and the precautions which should be taken.

Information should cover
- nature and degree of risk involved, the consequences of exposure, and any factors which might increase the risk, e.g. smoking, consumption of alcohol;
- the control methods adopted, the reasons for them and their proper use;
- why, and where, personal protective equipment is necessary;
- monitoring procedures, including arrangements for access to results and special notification if an MEL is exceeded;
- the role of any health surveillance, the need to attend for health surveillance procedures, access to individual health records and collective results of health surveillance.

Instruction should cover
- what to do when dealing with substances hazardous to health, the precautions to take, and when;
- what cleaning, storage and disposal procedures are required, why they are required and when;
- what emergency procedures to follow. (In the UK, establishment of procedures to follow in the event of 'serious and imminent danger' is also required by the Management of Health and Safety at Work Regulations 1992, Regn. 7 (a).)

Training should be provided on
- methods of control;
- personal protective equipment;
- emergency procedures.

Records should be kept to indicate the names of those who were trained and the trainers, the date of training and a synopsis of the topics covered.

Legislative developments within the European Community

Introduction

Selected EC Directives relating to safety were mentioned on page 946. There have been several important developments within the European Community with implications for health and safety, notably the Single European Act implemented in January 1992, the prime aim of which is removal of barriers to trade and free movement of goods between Member States. Obstacles identified include varying standards that exist between Member countries in the construction and safety of different products, and the varying standards of safety and health in the workplace.

Standards

The development of harmonised standards is an integral component of any programme targeted at promulgation of common levels of safety through

Control of substances hazardous to health 1629

ganisation has therefore been established for setting
e Comité Européen de Normalisation (CEN) deals
rds, and the Comité Européen de Normalisation
ELEC) deals with electrical standards. Examples
andards include EN 136, 142, 145 and 148 which
tive devices; eye protection is embraced by EN
rganisation considers whether existing standards
are acceptable as European Standards. For
parts 1 and 2 have been recognised as a basis for
Electro-sensitive Safety Systems and BS 5305:
principles for safeguarding machines, has been
a harmonised standard. To prevent duplication
also liaise with the International Organisation
which develops standards with world-wide

hin the EC include embryonic proposals plus
ctives (e.g. in the 1982 Major Hazards Direc-
Chapter 15 – quantity limits for a number of chemicals have been
revised. Thus limits have been reduced for phosgene, chlorine, methyl
isocyanate, and cobalt and nickel oxides, carbonates and sulphides. Ammonium nitrate *per se* is now differentiated from fertilisers and different limits set accordingly. Sulphur dioxide has been added to the list of 'isolated storage' and oxygen and sulphur dioxide included in the list for industrial installations.) All of these measures are designed to influence safety at work and the control of substances hazardous to health.

Some Directives are 'enabling' in style. Others are more specific, supporting daughter Directives including legislation on specific substances. Illustrative examples are discussed below as an indication of the impact that output from the European Community has on local legislation in relation to substances hazardous to health, and the direction, and scope, of likely future legislation within Member States of the European Communities. Some of the Directives are directly relevant to toxic substances, e.g. Carcinogens and Personal Protective Equipment Directives.

Framework Directive on the Protection of Workers from Risks Related to Exposure to Chemical, Physical and Biological Agents (80/11707/EEC)

This enabling Directive, adopted in 1980, requires member states to introduce legislation for subsequent implementation of daughter directives listed as an appendix. It requires member states to ensure worker exposures are kept as low as is reasonably practicable and to introduce health surveillance for workers exposed to asbestos and lead, and to allow workers access to information on the hazards of asbestos, arsenic, cadmium, lead and mercury. Limit values should be set and data from environmental and biological monitoring should be available to the workforce. When introducing national legislation on specific substances Member States are to consider various protection measures specified in the Directive.

This Framework Directive has been implemented in the UK by the Control of Substances Hazardous to Health Regulations, 1988 discussed earlier.

The Introduction of Measures to Encourage Improvements in the Safety and Health of Workers at Work (89/391/EEC)

One of the 1992 initiatives on the social front is to enhance levels of safety and health of workers and develop more uniform standards of safety and health in the workplace of Member States. Thus a Framework Directive has been drafted with supporting daughter Directives. In broad terms the Directive equates with the UK Health and Safety at Work Act and the requirement for risk assessments is mirrored in UK regulations, such as in The Control of Substances Hazardous to Health Regulations and the Management of Health and Safety at Work Regulations 1992. However, the Directive embraces two important philosophical points with implications for the UK, viz.

- Apparent rejection of the concept of 'reasonably practicable' in favour of a *'force majeure'* effectively making the requirements of the Directive absolute.[14]
- The proposal for workers to be given the right to 'balanced participation', an undefined phrase of uncertain meaning based on German systems.

The scope of the Directive covers both the public and private sectors and includes industry, agriculture, commerce, administration, services, culture and leisure activities.

A key element is the requirement for employers to integrate 'safety' into normal activities with obligations to

- avoid risk or evaluate level of unavoidable risk
- control risk at source using most up-to-date techniques
- adapt work to worker taking cognisance of the individual's limitations
- seek substitutes for hazardous materials (page 278)
- issue a safety policy (e.g. page 923)
- provide proper instruction and information. Thus if danger exists in the workplace the workers must be told and instructed in the measures required to avoid injury. Results of risk assessments, accident reports, and data from official and regulatory bodies must all be made available to employees. Training must be provided on recruitment, on transfer to

new jobs, on introduction of new plant/technology or equipment, and on a refresher basis. Contractors' employees must also be trained
- consider safety and health when planning new plant or introducing new chemicals
- provide first-aid, fire-fighting and evacuation measures
- introduce health surveillance appropriate to the risks.

To assist with implementation of these requirements employers must designate competent personnel to take charge of health and safety activities, or use competent outside services. Employers have to assess risk and to inform, train and consult with employees on the risks and precautionary measures.

Duties of employees are to
- use correctly machinery, equipment, personal protection etc.
- refrain from interference with safety devices
- notify employer of observed unsafe conditions
- co-operate with employer to ensure that the workplace is safe

This Directive was adopted to be implemented by Member States by 31 December 1992.

Directive Concerning the Minimum Safety and Health Requirements for the Workplace (89/654/EEC)

This 'Workplace Directive' applies to most fixed permanent workplaces and addresses employers' duties with regard to
- provision of clear escape routes and emergency exits
- cleaning and maintenance of workplaces
- keeping checks on safety equipment.

Employers must comply with one of two Annexes of minimum safety and health requirements. One covers workplaces used for the first time after 31 December 1992 as well as modifications, extensions and conversions of existing workplaces; the other is aimed at existing workplaces. Common requirements include provisions on structural stability, electrical installations, emergency escapes, fire precautions, ventilation, temperature, lighting, doors and gates, dangerous areas, rest rooms, washing facilities and first-aid facilities.

The Directive was adopted to be implemented by Member States by 31 December 1992 and is mirrored in the UK Workplace (Health, Safety and Welfare) Regulations 1992.

Exposure Limits Directive (88/642/EEC)

Contentious concepts such as adopting as standards one-fifth of the 8-hour value and setting short-term excursion limit values at five times the 8-hour value led to the delay in progress on early drafts of this Directive. Revised proposals are for establishment of scientifically based 'indicative limit values'. The provisions apply to a list of 27 indicative limit values for substances or groups of substances specified in an Annex. Priority is given first to 25 compounds with the same or similar hygiene standards already within Member States and then to another group of chemicals for which scientific data are available but which have different standards in various Member States. Technical definitions and requirements for measurement procedures are annexed to the proposals. The draft has been agreed by the Committee of Representatives of Member States under Article 9 of the old Framework Directive 88/642/EEC as amended by Directive 88/642/EEC in May 1991, but publication in the *Official Journal* is still awaited.

The Minimum Health and Safety Requirements for the Use of Personal Protective Equipment at the Workplace (89/656/EEC)

There are two important Directives on PPE, one relating to the manufacture and this one, which is targeted at the safe use of PPE. It is the third individual Directive under 89/391/EEC. Where health risks are known to exist the philosophy is elimination of the risk by alternative strategies, e.g. substitution by less hazardous materials, use of extract ventilation, etc. Only where this is not feasible should reliance be placed on PPE and employers will then have obligations to

- carry out a risk assessment
- define the characteristics of the PPE necessary and compare with the specifications of available equipment
- ensure compliance with appropriate EC standards and appropriateness to the risks involved, and ensure that no additional risks are created by its use
- consider the ergonomics such that workers can cope and their awareness of warnings is in no way hampered. Pre-existing medical and mental conditions should be considered and account must be taken of individual limitations
- ensure that any combination of PPE used to deal with a range of risks occurring simultaneously are compatible
- provide PPE on a personal basis where possible
- inform, instruct and train workers in terms of the hazards and in the use of PPE.

Annex II provides a list of items of personal protective equipment whilst Annex III identifies activities and sectors of activity which may require the provision of personal protective equipment. The relevant UK legislation is the Personal Protective Equipment at Work Regulations 1992.

The Approximation of the Laws of the Member States Relating to Personal Protective Equipment (89/686/EEC)

This Directive aims to harmonise laws relating to the manufacture, quality, and materials of construction for PPE and to ensure that PPE is marketed only if it preserves the health of users. Manufacturers are obliged to ensure that while protecting the user the device itself poses no risk to others. PPE must also satisfy certain identified basic safety and health criteria. Where PPE is in compliance with the requirements and is stamped with the EC logo there must be no restriction on their sale or marketing. Proof of compliance will be by a prescribed certificate of conformity. The mechanism for issue will be dictated by the level of risk which the equipment is intended to protect against.

Protection of Workers from the Risks Related to Benzene at Work

This draft document, the 'Benzene Directive', contained proposals for an 8-hour limit value of 5 ppm and an action level of 1 ppm with specific requirements for air monitoring. Lack of agreement by the Council of Ministers led to the dropping of the proposals. Because benzene is classified as an 'R45' substance and carries the risk phrase 'may cause cancer' (page 783), this compound has now been included, more sensibly, in a generic draft Directive on Carcinogens (see later).

Protection of the Health of Workers Exposed to Vinyl Chloride Monomer (78/610/EEC)

VCM was the first workplace chemical to be the subject of an EC Directive. Protective measures include technical controls, establishment of limit values, environmental monitoring, provision of personal protective equipment where necessary, informing employees of the risks and precautions, maintaining a register of workers and their exposures, and for medical surveillance. It requires worker exposure to VCM to be reduced to the lowest possible levels and contains the novel concept of a 'technical long-term limit value' (3 ppm) measured over one year. In the UK the Directive has been implemented by the publication of the Vinyl Chloride Approved Code of Practice under the COSHH Regulations.

Protection of Workers from the Risks Related to Exposure to Asbestos at Work (83/477/EEC)

This Directive was adopted in 1983 as the second daughter Directive under 80/1107/EEC. It requires an assessment of workers exposed to asbestos and where levels can exceed 0.25 fibres/ml or 1500 fibre-days/ml over 3 months stipulated actions need to be instituted. Quantities of asbestos must be limited to the minimum reasonably practicable: the number of workers exposed should be the lowest possible. Practices must be drawn up for work involving asbestos, buildings involved should be capable of regular cleaning, storage and transport of asbestos should be in suitable containers, and waste should be collected in sealed, labelled containers as soon as possible. The need to cordon off areas and post warning labels, e.g. during demolition work, is covered and where it is technically not feasible to contain asbestos such that exposure will be below the limit value the need for respiratory protection is emphasised. Importance is given to informing and consulting with employees and their representatives, and for undertaking environmental monitoring. Practical advice for clinical assessment is given in an Annex. This Directive was implemented in the UK by the Control of Asbestos at Work Regulations 1987 as amended by the Control of Asbestos (Amendment) Regulations 1992 which are not embraced by COSHH.

The Directive precedes the Single European Act and an amendment was adopted by the Council of Ministers in June 1991 which must be implemented by Member States by 1 January 1993; the provisions of this Directive must be reviewed again by 31 December 1995. The amending Directive provides for

- dual action levels for two groups of asbestos to reflect their differing toxicities, viz. 96 fibre-hours per ml of air for chrysotile, and 48 fibre-hours per ml of air for all other asbestos forms. A cumulative exposure to asbestos over a 12-week period is taken as an action level to trigger other control measures identified in the Directive.
- revised limit values for the two groups of asbestos, viz. 0.6 fibres per ml of air for chrysotile over an 8-hour period, and 0.3 fibres per ml of air for other forms of asbestos. (A Limit Value is the maximum concentration of asbestos in air to which workers can be exposed without the use of respiratory protection).
- an extension to the ban on asbestos spraying by prohibiting the use of friable asbestos products (i.e. with density less than 1 g/cm^3) as used, e.g. for insulation or sound proofing.
- specified features that must be included in plans of work drafted before work commences on stripping asbestos from buildings and which must be communicated to the competent authorities.

Council Directive 87/217/EEC addresses environmental pollution by asbestos.

Protection of Workers from the Risks Related to Exposure to Metallic Lead and its Ionic Compounds at Work (82/605/EEC)

This was the first daughter Directive to the Framework Directive. The main features include risk assessment wherever lead absorption could occur (with examples provided in Annex 1), environmental and biological monitoring, medical surveillance, maintenance of records and provision of information and monitoring data to the workforce. It has been implemented in the UK by the revision of the Approved Code of Practice governing the 1980 Control of Lead at Work Regulations and is outside the scope of COSHH.

Directive on the Protection of Workers by the Banning of Certain Specified Agents and/or Certain Work Activities (88/364/EEC)

This Proscription Directive was the fourth Directive and prohibits production of 4-nitrobiphenyl and 2-naphthylamine, 4-aminobiphenyl and benzidine and their salts. These are highly potent carcinogens which in the UK were prohibited under the Carcinogenic Substances Regulations of 1967, subsequently revoked under the provisions incorporated into COSHH. The Council can extend the list of substances included in the Directive.

The Protection of Workers from the Risks Related to Exposure to Carcinogens at Work (90/394/EEC)

This sixth individual Directive within 89/391/EEC defines a carcinogen as a substance which in Annex I qualifies for the R45 risk phrase 'may cause cancer' (page 783), a preparation which must be labelled with the R45 phrase, or a substance, preparation or process involved or which is released by processes referred to in Annex I, as follows:

Annex I. List of substances, preparations and processes

1 Manufacture of auramine
2 Work involving exposure to aromatic polycyclic hydrocarbons present in coal soots, tar, pitch, fumes or dust
3 Work involving exposure to dusts, fumes and sprays produced during the roasting and electro-refining of cupro-nickel mattes
4 Strong acid processes in the manufacture of isopropyl alcohol

Key requirements include
- regular assessments of exposure risk
- substitution wherever technically possible by less dangerous substances
- provide contained systems or ensure exposures are reduced to as low a level as technically possible
- limit quanitities used and released
- minimise the number of workers likely to be exposed
- use local extract ventilation, environmental monitoring
- provide personal protection where necessary, ensure high levels of hygiene
- segregate work area and post warnings and restrict access to authorised people

- draft emergency plans
- provide safe systems for storage, handling, transportation and disposal
- label all containers clearly
- where the outcome of risk assessments reveal a risk to workers, provide competent authorities with details of the material, process, quantities, workforce, control measures, degree of exposure and cases of replacement
- inform and train employees and consult regarding unforeseen and foreseeable exposures
- establish relevance of health surveillance and keep medical records for at least 40 years.

The Protection of Workers from the Risks Related to Exposure to Biological Agents at Work (90/679/EEC)

This 'Biological Agents' Directive considers micro-organisms in the broadest sense but distinguishes between exposure from deliberate work with them, e.g. in microbiological laboratories, as distinct from exposure which may arise incidentally, e.g. during health care and farming. The provisions of this Directive are akin to those for the handling of any substance hazardous to health, i.e. imposing responsibilities on employers to carry out risk assessments, prevent or minimise exposure, provide information, instruction and training for workers, retain records of exposed workers, provide health surveillance, notify competent authorities, etc. There are also requirements for special measures for health care, industrial processes, laboratories and animal rooms. The classification scheme is the subject of a separate Directive which closely resembles UK and OECD guidance.

The Directive was adopted for implementation by Member States by 26 November 1993.

The Minimum Safety and Health Requirements for Use of Work Equipment by Workers at Work (89/655/EEC)

This second daughter Directive under the Framework Directive describes the proposed minimum safety and health requirements for the use of work equipment and addresses employers' obligations to provide and maintain the correct equipment for use at work and to limit its use to trained people. The main text sets out the objectives in a series of articles covering
- the relationship between this Directive and the Framework Directive
- the definition of 'work equipment'
- employers' obligation to ensure that equipment is suitable for the work
- the condition of equipment – 'new' items have to comply with relevant Directives

- the duty of employers to ensure that only authorised workers operate equipment
- employers to provide adequate information and instruction on use of equipment covering normal and abnormal operation
- employers to provide suitable training covering any risks involved
- information to be given to workers and their representatives and the need for consultation
- changes to the Annex to reflect technical developments
- Member States to enact local laws to comply with the Directive.

More detailed requirements are given in a supporting Annex, examples of which include:

Control switches
- properly labelled
- clearly visible
- located outside danger zones
- designed so as to avoid the creation of hazards upon failure or faulty/inadvertent operation
- designed such that a deliberate action is necessary to activate machine after stoppage
- start controls must be subservient to stop controls
- each piece of equipment should be equipped with an isolator clearly identified with the machine it controls

Physical protection
- provided where falling or projecting components pose a threat
- adequate guards fitted to overcome risk of contact with moving parts and be so designed so as to be robust, difficult to defeat and positioned away from danger source
- to protect against exposure to high or low temperatures

Maintenance
- machine must be shut down or special precautions introduced

Chemical emissions
- containment or ventilation should be provided where escape of gases, vapours or liquids poses a threat.

The Directive must be implemented by Member States by 31 December 1992 and is encompassed in the UK by the Provision and Use of Work Equipment Regulations 1992.

In addition to the above there is also a Machines Directive (85/392/EEC) to harmonise the level of machine safety within the Community by ensuring that it is built in at the design and construction stages, by ensuring that health and safety considerations extend beyond those operating the machine, and by identifying 'Essential Safety Requirements' (ESRS), which represent the minimum safety measures necessary to enable machines to be operated with safety.

References

1. Garnier, R., Rosenberg, N., Puissant, J. M., Chauvet, J. P. & Eithylmiou, M. L., *Brit J. Ind. Med.*, 1989, **46**, 677.
2. Burgess, M. J., *Loss Prevention Bulletin*, Institution of Chemical Engineers, 1990(091), 28.
3. Anon., *Health & Safety at Work*, May 1986.
4. Report by HM Chief Inspector of Factories, 1986–87, Health & Safety Executive, 1987.
5. Health & Safety Executive, *Mercury*, Guidance Note MS12, HMSO, 1979.
6. Anon., *Health & Safety at Work*, April 1987.
7. The Classification, Packaging and Labelling Regulations 1984, SI 1244. (To be replaced by Chemicals (Hazard Information and Packaging) Regulations.
8. Muir, G. D., *Hazards in the Chemical Laboratory*, 2nd edn, The Chemical Society, 1977.
9. The Royal Society of Chemistry, *COSHH in Laboratories*, The Royal Society of Chemistry, Cambridge, 1989.
10. British Chemical Industry Safety Council, *Vinyl Chloride Monomer, Advisory Note for the Guidance of PVC Processes*, Chemical Industries Association, 1974.
11. Health & Safety Commission, *Control of carcinogenic substances*, Approved Code of Practice, 1988.
12. Health & Safety Executive, *Occupational Exposure Limits*, Guidance Note EH40, HMSO, 1990.
13. Atherley, G. R. C., *Occupational Health and Safety Concepts, Chemical and Processing Hazards*, Applied Science Publishers, 1978, 337.
14. HM Agricultural Inspectorate, *Slurry Storage Systems, Guidance on Promoting Safe Working Conditions*, ALC 1986/155, Health & Safety Executive.
15. a) Health & Safety Executive,*COSHH Assessments: A Step by Step Guide to Assessment and the Skills Needed for It*, HMSO, 1988.
    b) Pimentel, J. C. & Margues, F., *Thorax*, 1965, **24**, 678–88.
16. The Welding Institute, *Health and Safety in Welding and Allied Processes*, 3rd edn, ed. N. C. Balchin, The Welding Institute, UK, 1983.
17. Health & Safety Executive, *Welding*, Guidance Note MS15, HMSO, 1978.
18. Hilema, B., *Chem. Eng. News*, 1 April 1991, 4; 22 July 1991, 26.
19. Sharp, P. J., *et al.*, *J. Occup. Med.*, 1994, **36**(7), 718 and 731.
20. Cullen, M. R. (ed.), *Workers with Multiple Chemical Sensitivities*, Hanley & Belfus, Philadelphia, 1987.
21. Ashford, N. A. & Miller, C. S., *Chemical Exposures: Low Levels and High Stakes*, Van Nostrand Reinhold, New York, 1991.
22. Anon., *The Chemical Engineer*, 1991, **496**, 11.
23. a) Doig, A. T. & Challen, P. J. R., *Ann. Occup. Hyg.*, 1964, **7**, 223–9.
    b) Challen, P. J. R., *J. Soc. Occup. Med.*, 1974, **24**, 38–47.
24. Walker, E. A., *The Disposal of Hazardous Wastes from Laboratories*, The Royal Society of Chemistry, Cambridge, Spring Meeting 1983, 6.

25. Anon., *Loss Prevention Bulletin*, Institution of Chemical Engineers, 1991(057), 7.
26. Kletz, T. A., *Proc. 3rd Int. Symp. Loss Prevention & Safety Promotion in the Process Industries*, Basle, SSCI, 1980, 1515–27.
27. Barlow, S. M., Dayan, A. D. & Powell, C. J., 'Reproductive Hazards at Work' in *Hunter's Diseases of Occupations*, 8th edn, 1994, p. 728.
28. Pott, P., *Chirurgical Observations Relative to the Cateract, the Polypus of the Nose, the Cancer of the Scrotum, the Different Kinds of Ruptures, and the Mortification of the Toes and Feet*, Hawes, Clark and Collins, 1775.
29. Kipling, M. D., Oil and the Skin, in Annual Report of H.M. Chief Inspector of Factories, 1967, 105.
30. Fletcher, A. C., *Reproductive Hazards of Work*, Equal Opportunities Commission, 1985.
31. Health & Safety Executive, *Nitrosamines in Synthetic Metal Cutting and Grinding Fluids*, Guidance Note EH49, HMSO, 1987.
32. Harrington, J. M. & Saracci, R., *Occupational cancer. Clinical and epidemiological aspects*, in *Hunter's Diseases of Occupations*, 8th edition, Edward Arnold, 1994, 654.
33. Dunster, H. J., *Tech Report No. 1*, Health & Safety Executive, 1981.
34. Martland, H. S., *J. Amer. Med. Assoc.*, 1931, **92**, 466.
35. Hunter, D., *The Diseases of Occupations*, 5th edn, The English Universities Press Ltd., London, 1975, 879.
36. MacBeth, R., *J. Laryngol.*, 1965, **79**, 592.
37. Acheson, E. D., Cowdell, R. H. & Rang, E. H., *Brit. J. Ind. Med.*, 1981, **38**, 218.
38. Health & Safety Executive, *Carcinogenic Hazard of Wood Dusts and Carcinogenicity of Crystalline Silica*, Toxicity Review 15, HMSO, 1986.
39. H.M. Factory Inspectorate, *Occupational Cancer of the Renal Tract*, Technical Data Note No. 3 (Rev.), Dept of Employment, August 1971.
40. Health & Safety Executive, *Toxic Substances: a Precautionary Policy*, Guidance Note EH18, HMSO, Dec. 1977.
41. Neal, R. A. & Gibson, J. E., in *Reducing the Carcinogenic Risks in Industry*, ed. P. F. Diesler, Jr, Marcel Dekker, New York, 1984, 39.
42. ASTMS, *The Prevention of Occupational Cancer*, Association of Scientific, Technical and Managerial Staffs, February 1980, 18.
43. Long, J., *Chem. Eng. News*, March 25, 1991, 6.
44. CIA, *Tumours of the Bladder in the Chemical Industry. A Study of Their Cause and Prevention*, Chemical Industries Association, February 1976.
45. Sagan, L. A. & Whipple, C. G., in *Reducing the Carcinogenic Risks in Industry*, ed. P. F. Diesler, Jr, Marcel Dekker, New York, 1984, 159.
46. CIA Carcinogens Task Force, *Guidance Document on Carcinogens in the Workplace*, Chemical Industries Association, 1987.
47. Curler, R. L. & Symington, T., *Health and Safety Trends*, 1975, **7**, 47.
48. World Health Organisation, *Selected Petroleum Products*, Environmental Health Criteria 20, WHO, Geneva, 1982.
49. Health & Safety Executive, *The Carcinogenicity of Mineral Oils*, Guidance Note EH58, HMSO, 1990.
50. Cowfer, J. A. & Magistro, A. J., in *Encyclopaedia of Chemical Technology*, 3rd edn, ed. R. E. Kirk and D. F. Othmer, John Wiley, New York, 1978.

51. Viola, P. L., in *Encyclopaedia of Occupational Health and Safety*, ed. L. Parmeggiani, 3rd edn, I.L.O., 1983, 2256.
52. ACGIH, *Documentation of Threshold Limit Values*, American Conference on Governmental Industrial Hygienists, Cincinnati, OH, 1986.
53. *Fourth Annual Report on Carcinogens (Summary)*, US Dept of Health and Human Services, 1985, 200.
54. Lynch, J., in *Industrial Hygiene Aspects of Plant Operations*, ed. L. V. Cralley and L. J. Cralley, MacMillan Publishing Co. Inc., Vol. 1, 1982, 440.
55. Health & Safety Commission, *Control of Vinyl Chloride at Work, Approved Code of Practice*, Health & Safety Executive, 1989.
56. Mutchler, J. E. & Proskie, K. G., in *Industrial Hygiene Aspects of Plant Operations*, ed. L. V. Cralley and L. J. Cralley, MacMillan Publishing Co. Inc., London, Vol. 2, 1984, 261.
57. Doll, R. & Peto, R., *The Cause of Cancer*, Oxford University Press, 1981.
58. a) Higginson, J., in *Reducing the Carcinogenic Risks in Industry*, ed. P. F. Diesler, Jr, Marcel Dekker, New York, 1984, 1.
    b) Pierce, A., *The Times*, 7 January, 1994.
59. Case, R. A. M., Hosker, N. E., McDonald, D. B. & Pearson, J. T., *Brit. J. Ind. Med.*, 1954, **11**, 75.
60. Case, R. A. M. & Pearson, J. T., *Brit. J. Ind. Med.*, 1954, **11**, 213.
61. Atherley, G. R. C., *Occupational Health and Safety Concepts*, Applied Science Publishers, 1978, 23.
62. Meredith, S. K., Taylor, V. M. & McDonald, J. C., *Brit. J. Ind. Med.*, 1991, **48**, 292.
63. Waldbott, G. L., *Health Effects of Environmental Pollutants*, 2nd edn, C. V. Mosby Co., 1978, 201.
64. Lacey, J., Pepys, J. & Cross, T., Actinomycete and fungus spores in air as respiratory allergens, in *Safety in Microbiology*, ed. D. A. Shapton and R. G. Board, Academic Press, New York, 1972.
65. Morris, L., *The Safety Practitioner*, March 1987, 4.
66. Parkes, W. R., *Occupational Lung Disorders*, 2nd edn, Butterworth, 1982.
67. Topping, M. D., Tyrer, E. H. & Lowing, R. K., *Brit. J. Ind. Med.*, 1981, **38**, 293-6.
68. Mendes, E., Asthma provoked by castor bean dust, in *Occupational Asthma*, ed. C. A. Frazier, Ch. 17, Van Nostrand Reinhold, New York, 1980.
69. Meredith, S. K., Taylor, V. M. & McDonald J. C., *Brit. J. Ind. Med.*, 1991, **48**, 292.
70. Health & Safety Executive, *Isocyanates – Medical Surveillance*, Guidance Note M58, HMSO, June 1977.
71. Hardy, H. L. & Devine, J. M., *Ann. Occup. Hyg.*, 1975, **22**, 421.
72. Stolwijk, J. A. J., *Environmental Health Perspectives*, 1991, **95**, 99.
73. Bisio, A., *The Chem. Eng.*, 24 May 1990, 18-21.
74. Burga, S., Hedge, A., Wilson, S., Bass, J. H. & Robertson, A., *Ann. Occup. Hyg.*, 1987, **32**, 4A, 493.
75. Sykes, J. M., *Sick Building Syndrome: A Review*, Special Inspector Report No. 10, Health & Safety Executive, 1988.
76. Environmental Protection Agency, *Total Exposure Assessment Methodology (TEAM)*.

77. SI 1988/1657, The Control of Substances Hazardous to Health Regulations 1988, HMSO, ISBN 0 11 087657 1.
78. Control of Substances Hazardous to Health Regulations 1988, Approved Code of Practice Control of Substances Hazardous to Health and Approved Code of Practice Control of Carcinogenic Substances, HMSO.
79. Information approved for the Classification Packaging and Labelling of Dangerous Substances for Supply and Conveyance by Road (Authorised and Approved List), 2nd edn, HMSO, 1988.
80. Health & Safety Executive, *Occupational Exposure Limits*, Guidance Note EH 40/90 (and subsequent editions), HMSO, ISBN 0 11 885420 8.
81. Pathogens (see COSHH).
82. a) Health & Safety Executive, *COSHH Introducing Assessment: a simplified guide for employers*, IND(G)64(L), 1988.
b) The COSHH Regulations: A Practical Guide, D. Simpson and W. G. Simpson (eds), The Royal Society of Chemistry, Cambridge, 1991.
83. Lipton, S. & Lynch, J., *Health Hazards Control in the Chemical Process Industry*, John Wiley, New York, 1987.
84. Executive Services Advisory Committee, *COSHH: Guidance for Universities, Polytechnics and Colleges of Further and Higher Education*, HMSO, 1990.
85. Ralson-v-Grecter Glasgow Health Board (1987), *Scots Law Times* 386.
86. Campbell-v-Lothian Health Board (1987), *Scots Law Times* 665.
87. Health & Safety Executive, *Methods for the Determination of Hazardous Substances*, Series Nos 1–68, HMSO.
88. Health & Safety Executive, *Monitoring Strategies for Toxic Substances*, Guidance Note EH42, HMSO, 1989.

CHAPTER 29

Toxic solids

The precautions for handling toxic solids were introduced in Chapter 6.

> In the early days chrysotile asbestos was – apart from its extensive use in thermal insulation – used in small quantities to generate 'snow' in the film industry, and for Christmas decorations, and to manufacture Santa Claus whiskers.[1]

This exemplifies how the manner in which a toxic solid is handled and processed depends upon knowledge of the possible health hazards. Thus in the UK 12 fibres (>5 µm in length) per ml was the standard for control of airborne concentrations of asbestos in 1968, reducing to a hygiene standard of 2 fibres/ml in 1970 and further reducing to, e.g., 0.5 fibres/ml for chrysotile and 0.2 fibres/ml for amosite or crocidolite in 1984; since 1988, as noted on page 1683, all work with asbestos has been even more strictly regulated.

The dispersion of toxic solids may, as described in Chapter 21, result in nuisance or adverse health effects to populations outside the site of operation.

> Clouds of fine dust residue of castor beans issuing from the chimneys of a linseed and castor oil factory resulted in an epidemic of asthma involving 85 patients.[2] As already noted on page 1610 castor bean dust is a strong sensitising agent. (An epidemic of asthma has also been attributed to soya bean dust dispersed by an inshore breeze during the unloading of a ship in a port.)

The risks from exposure to solids in finely divided forms include ingestion, inhalation, and absorption by contact with the skin or mucuous membranes.

> A laboratory technician cut his little finger on a broken arsenic bottle. He cleaned up the wound and finished work normally but became ill at home 3 hours later. He subsequently died in hospital from suspected arsenic poisoning.

The potential hazards upon application of heat to produce vapours or fumes were referred to on, e.g. pages 217 and 221.

> Workers dismantling a missile site in New Mexico suffered from lead

intoxication following the use of acetylene torches to cut through structural steel coated with red lead paint.

Polymer fume fever (page 217) may be caused by exposure to thermal degradation products when polytetrafluoroethylene (PTFE) is heated to >300 °C.

> In a textile plant using a PTFE compound at temperatures limited to 165 °C all workers who exhibited the classic symptoms (i.e. cough, aching or weakness, fever and chills, and shortness of breath) were found to be cigarette smokers. The temperature within the combustion zone of a cigarette reaches 875 °C and the symptoms were deduced to have resulted from conversion of PTFE contamination of the cigarette into hazardous fumes. Elimination of smoking solved the problem.[3]

Similar examples are given on pages 217–18.

Preventive measures in areas where fluorine-containing polymers are produced or machined, etc., therefore include the provision of special ventilation if temperatures may exceed 200–250 °C, instruction of workers regarding hygiene, and the incineration of waste in special furnaces.[4]

In other situations a hazard can arise, in the absence of heating, due to the evolution of adsorbed or unreacted vapours at ambient temperature.

> A shift electrician, who did not normally work with isocyanates, was required to perform an emergency repair to an exhaust fan. Whilst removing the fan he developed a sudden, typical asthma attack. Subsequent analysis of the type of dust found in the fan showed that it might contain up to 14% unreacted isocyanate. (This incident occurred over 25 years ago; the more stringent practices now applicable with volatile isocyanates are exemplified on page 1655.)

Manual handling of solid chemicals, as outlined in Table 2.4 and page 326 involve other obvious risks. Thus it has been concluded that the major occupational health problems relating to fertilisers in agricultural work arise more from handling the bags than from their chemical contents; they are often densely packed in plastic bags which become very slippery to handle.[5]

The present chapter illustrates how the general precautions applicable to handling toxic solids, especially to avoid excessive exposure to particulates, are implemented in practice by reference to the hazards and industrial control measures for handling sodium hydroxide, detergent enzymes, asbestos, therapeutic substances, and lead compounds. These are materials with differing toxic properties but all with potentially serious occupational health risks from inhalation: the inclusion of discussion of toxic solids in the rubber industry highlights the problems and solutions associated with complex mixtures of toxic solids. More detailed advice on handling any of these specific materials should be sought from the literature (including the references provided here and in Chapter 18), from the industry (suppliers, relevant trade associations), and regulatory authorities. Furthermore, in the UK any arrangements must comply with the Control of Substances Hazardous to Health Regulations, 1988 and associated Approved Codes of Practice as discussed further in Chapter 28.

Detergent enzymes

The use of enzymes, such as trypsin, to aid detergency by digesting certain stubborn biological soils such as fat and protein was first patented in 1913 by Otto Rohm, co-founder of Rohm and Haas. Subsequent developments in enzyme technology led to various presoak formulations. It was, however, the availability of Alcalase, an enzyme stable and active under the alkaline conditions of fabric washing processes, that provided the impetus for world-wide exploitation of proteolytic enzymes by the detergent industry in the 1960s. Advances in biotechnology and genetic engineering have resulted in other enzymes appearing on the market, e.g. Esperase and Savinase (truly alkaline proteases), amylases (to catalyse the hydrolysis of starch), cellulases (to prevent fabric such as cotton from becoming harsh as a result of repeated washing), and lipases (to catalyse the hydrolysis of triglyceride components of fatty stains).

> In the late 1960s workers at a detergent factory complained of chest symptoms, e.g. bronchitis, bronchial spasm, asthma or influenza. Investigation confirmed that certain individuals handling proteolytic enzymes derived from *Bacillus subtilis* suffered from breathless attacks lasting from several hours to days, and in a few cases several months, chest pain, general weakness and malaise. Many had been exposed to non-enzyme detergent dust for several years without apparent ill-effects. A few patients became ill after an alleged single, high exposure (e.g. from working alongside unenclosed enzyme dosing points) but most suffered delayed responses. In the majority of cases previous exposure was known to have occurred without symptoms. Non-pulmonary effects were confined to slight primary irritation to exposed areas of high perspiration such as the back of the neck, or where sleeves chafed at the wrists of those handling the neat enzyme powder.[6] (This irritant effect could have resulted from exposure to the detergent *per se*.)

Extensive studies have subsequently been undertaken on the adverse health effects of overexposure to proteolytic enzymes in detergent workers.[7-12] Inhalation of proteolytic enzyme particulates is now recognised as a possible cause of respiratory allergy in the form of hay-fever or asthma-type reactions in certain workers. Within the UK occupational enzyme asthma is now a Prescribed Disease.[13] Those most susceptible are smokers and atopics (showing immediate positive skin reaction to common allergens such as house dust, pollen); the non-atopic population do not become sensitised so easily. Skin and eye irritation can also occur. For example, occupational dermatitis has been reported from handling proteolytic enzymes but this is of an irritant nature, due to the inherent proteolytic activity of the enzyme, and there is no evidence of allergic contact dermatitis.[14]

Adverse health effects are avoided if levels of airborne enzyme are kept as low as is reasonably practicable and operator exposure maintained below recommended Occupational Exposure Standards. There is no consumer risk of respiratory or skin sensitisation to biological detergents arising from domestic use of these products.[15]

Toxic mechanism

A discussion of the mechanism by which allergens operate is provided in Chapter 28. Proteolytic enzymes result predominantly in Type I responses although some evidence has been found of a Type III reaction.[6] Thus persons sensitised to natural allergens such as grass pollens and house-dust mites (which may be as high as one-third of the general population) are more likely to react adversely upon exposure to detergent enzymes. In contrast, the development of asthmatic symptoms from exposure to toluene diisocyanate (see page 220) is not necessarily related to an atopic state; this is considered later. More detailed discussion of sensitisation is given in ref. 16.

Handling precautions

The crucial factor in preventing occupational ill-health when handling detergent enzymes is avoidance of exposure to particulate enzymes. (In some operations, however, enzymes are handled as liquids so that control of mists or sprays is a factor; for completeness, these control measures are also summarised briefly in this chapter.)

The incidence of sensitisation depends on the amounts of enzyme antigen in the work rooms, duration of exposure and to some extent the susceptibility of the workers. When dust levels are high all those exposed, atopics and non-atopics, are at risk although not all become sensitised.[17] At low dust levels atopics are at most risk. If only traces of the dust occur few, if any, individuals become sensitised. Generally, if overall factory dust levels are low then atmospheric enzyme levels are also low as demonstrated by Fig. 29.1.[11] Enzymes can be, and are, handled safely in the detergent industry by a combination of modification to enzyme raw-material specification, engineering controls, health screening of employees, monitoring of dust and enzyme levels in the workplace atmosphere, and operator training. (In most liquid detergent manufacturing plants little dust is produced but aerosols can be generated, e.g. around the pack filling head. Hence the need for control features and environmental monitoring of enzyme levels.)

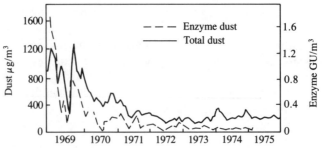

Fig. 29.1 Total dust and enzyme dust concentrations in packing department

Raw-material specification

Originally, the enzyme concentrates used in the detergent manufacturing process were dusty powders. Subsequent developments led to the introduction of non-dusty, coarse >150 μm granules in which the enzyme was protected by fixing onto, e.g. granular sodium tripolyphosphate and rendering the resulting particles essentially dust-free by spraying on tacky non-ionic material such as ethoxylated linear aliphatic alcohols.[17,18] Other systems include mixing the enzyme with, e.g. fibrous cellulose and titanium dioxide and coating the particles with a waxy material such as polyvinylpyrrolidone.[19] Other forms of non-dusty enzyme particles, such as marumes or prills, became available. Prills, for example, were produced by spray-cooling a molten mixture of the enzyme and non-ionic organic material to form regularly shaped non-friable beads.

Such advances have enabled tight specifications to be set for the dustiness of the encapsulate, as determined in standard elutriation tests designed to simulate attrition of encapsulates.[18] Encapsulated enzymes are supplied in containers ranging from 40 kg kegs with separate liners, to returnable 1000 kg 'big bags', and in some parts of the world in 50 or 200 kg drums. (Enzymes are also provided commercially as solutions or slurries in plastic-lined metal containers for use in heavy-duty fabric washing liquid formulations.)

Engineering control

Containment is the key objective of plant design. Even so, scope for some enzyme escape is unavoidable. In normal operation the greatest risk of exposure to enzyme dust is from handling the enzyme encapsulates during discharge of stock containers. A typical arrangement is shown in Fig. 29.2. This operation should be undertaken in a dedicated area segregated from the main plant. The area should be equipped with efficient general ventilation (e.g. 10–15 air changes per hour) coupled with local extract ventilation with a recommended face velocity of 1 m/s (Chapter 12).

Design of extract ventilation systems for bag handling facilities should

Toxic solids 1647

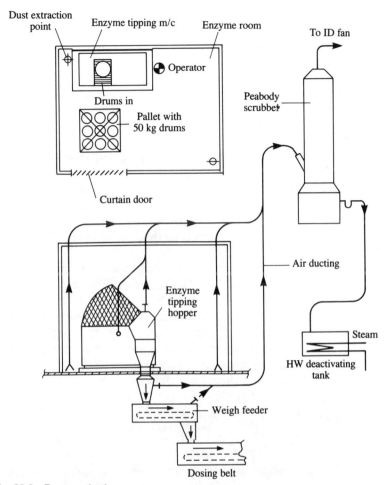

Fig. 29.2 Enzyme tipping room

prevent excessive dust formation during the folding of the bags. The extract ventilation system should vent via high efficiency microfiltering units, or a wet scrubber designed and located so as not to pose noise problems. Detection of pressure drops across the filter, or water supply, linked to visible or audible alarms provides an indication of the efficiency of the ventilation system which is particularly important in the enzyme handling area.

(Liquid enzymes are delivered in tote bins and the contents transferred to stock tanks using a Bredel hose pump connected via self-sealing, reinforced clip-on hose and the tote-bin cap removed to prevent implosion. Open decanting should not be permitted since enzyme aerosols can form. Pump and pipe flanges on plant containing liquid enzymes should be contained to avoid aerosol releases; this is conveniently achieved by wrapping them in knitted wire-mesh, of the type used in demisters, and cladding in removable sheet metal covers. Floors should be constructed so as to facilitate wet mopping and be fitted with a covered water sump so that spillages of enzyme slurry can be washed to drain: high-pressure hoses should be avoided since these generate aerosol.)

Gross spillages of encapsulate on floors or in machines should be vacuumed; mobile systems should be fitted with absolute filter units on the exhaust. Brushing should be prohibited since dust can result.

Weigh-belt dosing systems should be enclosed and under negative pressure with an air velocity of 1 m/s at any opening and vented via highly efficient dust filters or wet scrubbers (ensuring that detergent dust does not vent into the scrubber, thereby causing foaming and reduced efficiency). Powder transfer points should be enclosed and ventilated; ventilation of conveyors needs optimisation since excessive draughts promote dust formation. (In-line dosing is one preferred method for incorporating enzyme liquids.)

In the powder packing room the filling heads tend to be enclosed, and general ventilation is supplemented by local extract ventilation around the machine packing head and at the point where the cartons exit from the packing machine, since these are potential sources of dust. (Additional considerations for packing of enzyme-containing heavy-duty liquids include the potential for aerosol formation around the filling machine nozzles, vacuum pump exhaust vents, pressure equalisation valves, etc. Therefore, each filling nozzle should be equipped with local extract ventilation of the type illustrated in Fig. 29.3, providing an air velocity of 1 m/s

Fig. 29.3 Local extract ventilation provisions around filling head on filling plant for heavy-duty liquid detergents containing enzymes

at the nozzle. Feed pipes to the stock tank of formulated heavy-duty liquid should extend to the bottom of a tank equipped with extract ventilation so that the rate of extraction exceeds the air displacement. Extracted air should be vented via a coalescing filter.)

Environmental monitoring and hygiene standards

A general discussion on hygiene standards is provided in Chapter 5 and advice on techniques and strategies for air sampling, analysis and data interpretation in Chapter 7. Enzyme activity is usually expressed as Glycine Units (GU), measured by release of amino acid from a standard protein and estimated colorimetrically. Other units sometimes encountered are the Anson and Delft (1 Anson Unit = 623 000 GU = 243 000 Delft Units; 1 µg of 1.5 Anson Units of protease activity as received from the enzyme supplier is approximately equivalent to 1 GU/mg enzyme concentrate).

Before hygiene standards can be set the antigenic potential must be established for all new enzymes of interest and it is important that the enzyme samples tested are typical of commercial concentrates used in the preparation of encapsulates, i.e. that they contain all the antigens present in the factory environment.

The specific activity of purified Alcalase (i.e. Subtilisin A) was formerly used to convert the target level for the environment and the encapsulate specification expressed as proteolytic activity into an amount of protein (0.5 GU/m^3 = 15 µg (0.015 mg) protein/m^3 and 165 GU/60 g enzyme encapsulate = 5.2 mg protein/60 g enzyme encapsulate). However, like other commercial detergent enzymes, Alcalase is not a purified preparation and contains more than one protein component. (Alcalase contains three major proteins: Subtilisin A, Alcalase C and a smaller amount of amylase. All these protein components are antigenic and allergenic. Lipolase contains three major proteins; the lipase enzyme, an amylase and a relatively minor unidentified component.) To convert the target level for the factory environment and the encapsulate specification into an amount of enzyme activity, it is more correct to use the specific activity of the enzyme concentrate and not that of the purified enzyme. Since enzymes do not behave synergistically the hygiene standards are additive when dealing with mixtures of enzymes.

Historical data indicate that some employees could develop enzyme asthma if regularly exposed to more than 3 GU/m^3. The current UK Occupational Exposure Standard[20] is 0.00006 mg/m^3 of pure crystalline subtilisin, equivalent to 1.92 Glycine Units (GU)/m^3. It is common practice to set an in-house action level of 0.5 GU/m^3 enzyme: for excursions between 0.5 and 1.0 GU/m^3 investigation of the cause is advisable. More stringent actions, such as increased frequency of monitoring and reliance on respiratory protection, may be required for levels in excess of 1.0 GU/m^3 and eventual shut-down if continued high exposures are confirmed. Because of the relationship between dust and enzyme levels it is prudent to set standards for general dust levels such as <500 μg/m^3 for packing and encapsulate tipping, and e.g. <1000 μg/m^3 elsewhere on the plant.

Airborne levels of enzyme are determined by collecting dust on filter discs in high-volume background 'Galley' samplers, operating at about 600 1/min for several hours with subsequent enzyme analysis on the weighed material. (The basis of the analysis is degradation of acetylated casein by the enzyme and a colorimetric reaction of the amino bodies so formed with trinitrobenzene sulphonic acid, the intensity of which is measured in the 414–420 nm range.) The amount of enzyme in the ambient air is then calculated in GU/m^3.

Galleys are placed at strategic points close to potential sources of leaks, e.g. in the enzyme handling and dosing areas, packing machine heads, and in special areas dedicated to recovery of powder from damaged containers or from packing machines. Sampling frequency is dictated by circumstances but two-hour sampling is considered appropriate when enzymes are first handled on the plant with sampling periods extended to four hours once commissioned plant is running within the Occupational Exposure Standard. (Since non-enzymatic material can interfere with the analysis to give false positives, 'background' levels should be established prior to introducing enzymes into plants.) To date no personal dosimetry techniques are suitable for enzyme workers. Galley samplers operating in facilities handling liquid enzymes should be run at 300 l/min to avoid liquid aerosol passing the filter.

More detailed advice on sampling and analytical procedures for enzymes is given in refs. 18 and 21.

Personal protection

Enzymes are skin irritants so operatives handling the encapsulates, e.g. in the tipping room, should wear gloves in addition to their normal protective overalls and shoes. Provision should be made for overalls to be laundered at least weekly, and staff should be encouraged to adopt a high standard of personal hygiene. Skin barrier creams are not recommended.

Adoption of the foregoing precautions eliminates the need for respiratory protection during normal operations. That the measures have, in general, been successful is indicated by the currently low rates of sensitisation.

> A few workers have become so sensitive that the most minute traces of enzymes are sufficient to cause a sensation of tightness of the chest. Association with employees with contaminated work clothes may apparently be sufficient to

trigger an attack in some. A few may even develop symptoms in the detergent aisle of a supermarket.[22]

However, approved respirators (e.g. BS 2091B or BS 6016 type 2) with a Nominal Protection Factor of 10 (Chapter 6) are required when exposures can arise, e.g. during replacement of exhausted bags, pumping enzyme concentrate from supply to storage tanks, dealing with spillages, cleaning or maintaining plant, especially filter units and ducting, etc. (Eye protection is also required when handling neat enzyme solution, e.g. when emptying tote bins.)

Health screening and monitoring

Pre-employment health checks have included
- medical history, particularly with respect to atopicity (e.g asthma, hay fever, eczema)

One study amongst workers producing enzyme washing powders showed that an atopic individual is about twice as likely to become sensitised as the non-atopic.[8]

- Previous work experience and smoking habits
- Physical examination
- Establishment of lung function using spirometry (see pages 262 and 264)
- Skin prick testing to common allergens

Chest X-rays are only relevant subject to clinical findings. The suitability of individuals for employment in enzyme areas will then be based on the level of enzyme exposure associated with the job and the clinical assessment. Subjects with a significant history and/or findings such as chest disease, and/or poor lung function, are likely to be considered unsuitable for employment in the manufacture of biological washing products.

For employees engaged in work with enzymes the record of respiratory sickness absence should be kept under regular review, along with periodic checks on smoking habits and vitalograph measurements. Individual findings influence advice by the medical officer on the timing of the next health check and of the worker's suitability for continued work in the enzyme facility. Skin prick tests may be of value in certain circumstances.

Further guidance on health screening and monitoring is given in ref. 23.

Management responsibilities

The legal responsibilities of management to provide and maintain safe plant and systems of work were identified in detail in Chapters 16 and 17. In the present context these should minimise risk to employees and visitors from inhalation of enzyme dust or aerosol. Arrangements must be instituted for pre-employment health checks of new recruits and for their training in the hazards, and precautions for the safe handling of enzymes and other detergent ingredients. The workforce should be adequately supervised and conversant with emergency procedures such as enzyme spillage, fire, etc. Management must then prohibit access to the 'enzyme areas' by unauthorised personnel. Procedures should be introduced to prevent risk to Third Parties, e.g. contractors working on the plant, disposal firms involved in incineration of empty 'big bags', companies laundering protective overalls, etc.

In the UK the COSHH Regulations (Chapter 28) require management to perform formal risk assessments of their processes. These have a particular bearing on enzyme handling, so atmospheric monitoring of enzyme levels forms an integral component of the assessment. Permit-to-work systems are appropriate for engineering/maintenance work on enzyme plant. Results from environmental analyses must be kept for 30 years and be available for inspection by, amongst others, employees. Similarly, health screening records must conform to COSHH requirements and be retained for 30 years; the collective results should be available to employees, albeit in a form to prevent data being linked to any particular individual. There is also a need for regular, recorded checks on the performance of safety features such as respirators and ventilation plant. Thus under the UK COSHH Approved Code of Practice (page 1617) all ventilation equipment should be visually inspected at least once per week; the local ventilation plant exhausting enzyme dust should be examined at least at 14-monthly intervals, and possibly every 3 months for ventilating equipment containing the neat encapsulate. Additional guidance on ventilation is given in Chapter 12 and in ref. 18.

Conclusion

Whilst detergent enzymes are respiratory sensitisers, they can, and indeed are, handled safely in the industry by a combination of containment, ventilation, personal protection, health screening and environmental and biological monitoring. It is important to appreciate that the same measures may be equally as relevant to avoidance of respiratory distress, in e.g. laboratory workers, nurses, and others handling even small quantities of proteases or other highly purified enzyme preparations such as lipase, cellulase, etc., for example in diagnostic procedures.[16]

> A 29-year-old atopic plant pathologist undertook laboratory cloning experiments, part of which entailed study of a mould as a cause of potato blight. He developed shortness of breath with exertion, cough, generalised aching, and malaise. His symptoms were worse at, or shortly after, work.
>
> A few months later he opened a box of powdered enzymes used to digest plant cells. The work was performed in a laminar flow bench (page 1665) where the direction of air-flow was across the bench towards him. Within a few minutes he experienced a sensation of severe swelling in his throat and

shortness of breath which caused him to seek emergency treatment. After hospitalisation and steroid treatment he was able to return to work but discontinued the practice of mixing enzyme solutions.

The enzymes he had worked with were Cellulase and Macerozyme. Investigations with the patient and other laboratory workers showed that the enzymes were capable of eliciting Type I hypersensitivity. Skin prick tests with Cellulase produced a 15 mm wheal which was also associated with symptoms of rhinitis and itchy eyes. The patient experienced a massive late onset response beginning six hours after the skin test and lasting more than 24 hours.[24]

A 29-year-old atopic research chemist regularly handled crystalline chymotrypsin in analytical quantities. Within a year he noted increasingly frequent and distressing lachrymation and nasal congestion coincident with handling even trace amounts of purified enzyme. Symptoms became so severe that it was necessary for him to cease work with the material. Months later he developed hay-fever-like symptoms when using trypsin, a substance hitherto only encountered intermittently. These materials had potential to sensitise via the same mechanism underlying development of pollen allergy.[25]

A 53-year-old paediatric nurse worked in the same unit for several years before developing nasal discharge, shortness of breath, wheeze and chest tightness. Symptoms developed about half an hour after mixing powdered pancreatic extract with infant feeds, subsiding 2 to 3 days after avoiding exposure. She stopped this task but remained on the ward; her symptoms persisted. She had long suffered from hay-fever. She moved to another ward in which the extract was not handled and 2 months later her symptoms were improved, although she needed the aid of an inhaler. The pancreatic extract contained free protease, lipase, and amylase, and it is advisable to avoid exposure to pancreatic extracts in powder form and where possible to use the substance as granules or capsules.[26]

Other substances capable of sensitising

Many compounds are capable of causing respiratory sensitisation (Table 28.14) and, as indicated on page 255, predictive tests exist for new substances. The foregoing general considerations are equally applicable to avoiding adverse health effects when handling any material with similar toxicological properties to enzymes. This is illustrated by the following case histories and the additional examples cited in the section on therapeutic substances and microbiological hazards.

Peracid precursors

Sodium isononanoyl oxybenzene sulphonate (SINOBS) has recently been exploited as a surfactant and bleach activator in fabric washing products. Such compounds contain facile leaving groups which increase the rate of perhydrolysis and thereby enhance bleaching in low-temperature washing processes.[27]

$$\underset{\text{(SINOBS)}}{H_3C-\underset{\underset{CH_3}{|}}{\overset{\overset{CH_3}{|}}{C}}-CH_2-\overset{\overset{CH_3}{|}}{CH}-CH_2-\overset{\overset{O}{\|}}{C}-O-\underset{}{\langle\bigcirc\rangle}-SO_3Na} \xrightarrow{HO_2^- \text{ (i.e. peroxide species)}}$$

$$H_3C-\underset{\underset{CH_3}{|}}{\overset{\overset{CH_3}{|}}{C}}-CH_2-\overset{\overset{CH_3}{|}}{CH}-CH_2-\overset{\overset{O}{\|}}{C}-OOH + HO-\langle\bigcirc\rangle-SO_3Na$$

Rashes, rhinitis, conjuctivitis and wheezing were noted in several detergent workers involved in a 12–18 month development programme on SINOBS. Delayed contact hypersensitivity to SINOBS was probably responsible for some of the rashes. One employee then developed asthma from an estimated exposure of between 0.02 and 1.0 µg of SINOBS. This patient did not smoke and had no pets or other relevant domestic exposures but skin prick tests gave positive immediate reactions to a range of common allergens. Upon return to work after three weeks' absence there was a slight recurrence of his symptoms and he again ceased work for a fortnight. Inhalation challenge tests with nebulised SINOBS solutions gave late asthmatic reactions which were dose related. The mechanism by which sensitisation is induced differs from that of Aspirin, a compound of some structural similarity to SINOBS.[28–30] Interestingly, when SINOBS was eventually registered on ELINCS under regulations governing marketing of new substances it was classified as a skin sensitiser, a danger to health by prolonged exposure through inhalation, toxic if swallowed, and a skin and eye irritant.

Platinum salts

Platinum, a corrosion-resistant metal capable of forming soluble complex salts, finds wide use in the laboratory, electronics industry, in jewellery, and as a catalyst in petroleum refining and environmental pollution control. Inhalation of soluble platinum salts can give rise to adverse health effects including allergic rhinitis and respiratory distress (termed platinosis) with latency periods ranging from a few months to years. Allergic skin disorders have also been reported. Clearly a high degree of containment is required to ensure exposures are below the low hygiene standard of 0.002 mg/m^3 (as platinum). This value was set to avoid development of skin or respiratory sensitisation. It is recommended[31] that atopics are screened out from work with these compounds and periodic health checks are introduced to include symptomatic enquiries, skin tests for sensitisation, clinical chest examinations and respiratory function tests. Again, environmental monitoring will indicate levels of compliance with hygiene standards and engineering targets.

Isocyanates

Isocyanates find extensive use in industry and those at risk not only include operators manufacturing the compounds but could include aircraft builders, lacquer workers, organic chemists, painters using polyurethane paints, plasticiser workers, polyurethane foam makers, rubber workers, ship builders, etc. Isocyanates may be volatile liquids such as toluene diisocyanate (TDI) or solids such as dianisidine diisocyanate or diphenylmethane diisocyanate. Exposure can arise by skin contact or inhalation of vapours from volatile compounds or when vapourised on heating during processing, or by generation of aerosol or dust.

TDI
2:4 isomer
2:6 isomer

Diphenylmethane diisocyanate

Inhalation may produce primary irritation of the respiratory tract and respiratory sensitisation with asthmatic attacks[32] with the immediate or delayed onset of bronchiospasm.

> A 41-year-old night-shift worker with a rubber company suffered attacks of breathlessness which became so severe that he could not walk more than 150 yards on the flat without panting. He had been employed on a machine that sprayed an isocyanate-containing glue onto extruded rubber. Prior to this episode of ill-health he had been fit and enjoyed sport and had never suffered with asthma or other lung disorders. He was diagnosed as suffering from asthma associated with exposure to isocyanates.
>
> Investigation revealed that spraying took place in a fully automated spray-booth equipped with extract ventilation. Whilst air flow across the booth opening was adequate, leaks occurred around the fan casing. The leaks were sealed and joints repaired, and a maintenance programme instituted. The affected worker was transferred to another part of the plant but his health remained poor and he suffered from chronic asthma and he could no longer take part in sport.[33]

Again, because of the low hygiene standards (8-hour time weighted average Control Limit of 0.02 mg/m^3 and 0.07 mg/m^3 short-term value), high-level containment and exhaust ventilation backed up where necessary by use of personal protective equipment are essential. Pre-employment medicals are required with periodic follow-up health checks. Screening out atopics by skin prick tests is not considered appropriate. Entrance to isocyanate areas should be restricted to authorised personnel. Once sensitised, individuals should not be further exposed to any level of the compound since they may suffer asthmatic attacks upon exposure to concentrations well below the Control Limit. Monitoring the ambient air by dosimeters, or static background monitors, equipped with direct readout and visual/audible alarms (see Chapter 7 and ref. 34) are an essential check on levels of containment and employee exposures. In the event of spillage the area should be immediately evacuated and the spillage cleaned up by suitably equipped personnel using absorbent such as sawdust and 5 per cent aqueous ammonia decontamination mixture. Isocyanates should not be washed to drain since solid polymers can form and plug the sewer system. Isocyanates can react violently with acids and alkalis; the presence of water liberates carbon dioxide which can lead to a dangerous build-up of pressure in drums and plant, as illustrated in Chapter 19.

Organic acid anhydrides

Anhydrides find wide industrial use. For example, phthalic anhydride is used in organic synthesis for the manufacture of resins, dyes, drugs, insecticides, polyesters, plasticisers, saccharin, etc. Trimellitic anhydride and maleic anhydride are also versatile raw materials, e.g. the latter is used in the production of polyester and alkyd coating resins and in the preparation of tartaric and maleic hydrazide. As a class these substances are highly toxic and potent irritants of the skin, eye and respiratory tract. In addition exposure to the dust or fumes from molten material can cause pulmonary sensitisation by Type I reaction.

> After some time working at a plastics processing company mixing batches of granules for subsequent processing, a 30-year-old man developed nose and eye discomfort. This was associated with chest tightness and wheezing, often delayed for several hours. His level of physical fitness reduced. The cause proved to be exposure to sacks of anhydride cross-linking agent. The managers were aware of the hazards of this class of substance but did not have a material safety data sheet and had not provided the workers with any information, instruction or training. The man was diagnosed as suffering from occupational asthma and was advised to avoid further exposure. After four years away from exposure he remained unemployed and physically unfit.[33]

The current Occupational Exposure Limit (OEL) for phthalic anhydride is only 1 ppm (eight-hour time weighted average) and 4 ppm (10 minute short-term value) whilst the corresponding eight-hour value for maleic anhydride is just 0.25 ppm with no short-term standard set. Clearly their handling requires careful technical and administrative controls to prevent ill-health.

Wood dust

Overexposure to wood dusts can cause skin problems, or obstruct the nose. Long-term exposure to hardwoods may, rarely, cause a rare cancer of the nasal cavity and sinuses. Furthermore, as Chapter 28 indicated, certain wood dusts can cause a type of asthma. The immunological pattern is complex with evidence for Type I, Type III and Type IV responses. Asthma was reported to be most frequently associated with iroko, red cedar, abirucana, mahogany, ramin and oak.[16] Conjunctivitis, rhinitis and asthma are well-established symptoms of exposure to western red cedar.[35] An eight-hour Control Limit has been set at 5 mg/m^3 for hardwood dust and an eight-hour Recommended Exposure Limit of 5 mg/m^3 for softwood dust.

Wood dust is created by a variety of processes. As an indication of risk, Table 29.1 summarises results from personal monitoring in several furniture factories.[35] Data indicate that properly designed and maintained exhaust ventilation is needed to ensure dust concentrations remain below the Control Limit. This should also alleviate exposure to wood additives

Table 29.1 Dust concentrations from various processes in furniture factories (after ref. 36)

Process	Types of machine	Dust concentration (mg/m^3)	Range (mg/m^3)	Percentage of results above 5 mg/m^3
Sawing	Circular, straight line, dimension, band saws	3.4	1–30	6
Cutting (excluding sawing)	Planers, thicknessers, moulders, shapers, mortisers, tenoners, copy lathes, routers, spindle moulders, borers	2.3	0.3–9	7
Sanding (machines)	Narrow belt including horizontal and vertical belt sanders	8.5	1–20	56
	Drum, bobbin and brush	4.4	1–51	12
Sanding (hand)	Paper and block, portable hand machine (no ventilaton)	8.1	1–27	48
	Paper and block sanding with exhaust ventilation	2.2	1–5	0

such as fungicidal and insecticidal wood preservatives of the type shown in Table 29.2. Where such technology is not currently available, e.g. hand

Table 29.2 Wood preservatives

Use	Compound
Diluting agents:	Benzene
	Ketones
	Fuel oil
	Anthracene oil
	Aliphatic hydrocarbons
	Aromatic hydrocarbons
	Halogenated hydrocarbons
	Polyglycols
Fixative agents:	Dehydroabietylamine
	Linseed oil
Water repellent agents:	Paraffins
	Abietic resins
	Chlorinated resins
	Coumarin resins
	Epoxy resins
	Glycerophthalic or alkyd resins
Dispersing agents	
Wetting agents	
Insecticides:	Aldrin
	Alkaline chromates and dichromates
	Alkali pentachlorophenates
	Alkali phosphates
	Arsenicals
	Borates
	Chlorobenzenes
	Copper sulphate
	DDT
	Dieldrin
	Fluorides
	Hexachlorocyclohexane
	Mercury dichloride
	Phenol
	Zinc chloride

sanding, then respiratory protection is required.[36,37] Table 29.3 provides data from a separate study[37] with static samplers placed close to machines to collect respirable dust under worst-case conditions, e.g. when the machine was used by several operators. Results were interpreted as offering reassurance since most levels were below the Control Limit. It is noted that the hygiene standard could be exceeded as a result of orbital and belt sanding. Since dust levels will vary according to type of wood, specification of the ventilation system, and task, the risk needs to be assessed in individual circumstances.

Additional general guidance on handling allergens in the workplace is given in ref. 38.

Table 29.3 Typical airborne wood dust levels[37]

Machine	Local exhaust ventilation fitted	Dust concentration (mg/m^3)	Type of wood
Wide-belt sander	Yes	1.4	Meranti
Double-ended tenon	Yes	1.0–3.6	Meranti
Hand sanding (belt)	No	4.5–8.3	Various
Computer router	Yes	>7.5	Various
6-Cutter moulder	Yes	0.4–0.8	Meranti
Joinery (manual)	No	0.6–4.0	Various

Therapeutic substances

Introduction

The pharmaceutical industry develops and manufactures therapeutic substances for the treatment of human and animal diseases. Obviously the products, and some intermediates and wastes, are of necessity biochemically active substances likely to produce effects in humans at relatively low dose-levels. Therefore, only extremely low levels of occupational exposure tend to be permissible; in terms of occupational safety in the pharmaceutical industry, therefore, 'risk to health' embraces 'unacceptable pharmacological effects'.[39]

The pharmaceutical sector is extremely diverse in terms of process range, size and product type. The principle production processes are chemical synthesis, fermentation, extraction and formulation. Fine chemicals and basic ingredients from primary production include synthetic organic and inorganic compounds, e.g. analgesics, aluminium hydroxide, magnesium oxide, sulphonamides, and vitamins; antibiotics from fermentation processes, e.g. ampicillin, cephalosporins, penicillin, streptomycin, and tetracyclines; biological extracts and their derivatives, e.g. insulin, quinine and morphine; and biological materials such as vaccines, antibodies, and antigens. The industry also produces wadding, gauzes, bandages, sterile surgical catgut, etc.

The processes are normally operated batchwise and a wide variety of chemical raw materials, finished products and chemical operations are involved. Secondary manufacture entails physical operations such as milling, compounding, compression, dispersion, drying, granulation, and packaging.

Hazards

Few studies have been published on the risk of occupational ill-health to workers engaged in the manufacture and formulation of pharmaceuticals, and those available are confusing or inconclusive.[40,41] Reports of increased incidences of cancer, suicides and spontaneous abortions, etc., within the industry were considered to be either not work-related or of no statistical significance.

Disturbance of powders with consequential release of dust is inherent in certain manufacturing processes, particularly small-scale batch operations which have to rely upon manual handling and transfer of materials.[42] In addition to the solid preparations exemplified by Table 29.4, many of the

Table 29.4 Characteristics of solid pharmaceuticals

Dosage form	Constituents, properties	Uses
Bulk powder	Comminuted or blended, dissolved in or mixed with water	External, internal
Effervescent powder	CO_2-releasing base ingredients	Oral
Dusting powder	Contain also absorbents	Skin treatment
Insufflations	Insufflator propels medicated powder into body cavity	Body cavities
Lyophilised powders	Reconstitution by pharmacist of unstable products	Various uses, including parenteral and oral
Capsules	Small-dose bulk powder enclosed in gelatin shell; active ingredient plus diluent	Internal
Massed or moulded solid pills	Adhesive or binding agents facilitate compounding; prepared by massing and piping	External
Troches, lozenges, pastilles	Prepared by piping and cutting or disc candy technology; compounded with glycero-gelatin	Slow dissolution in mouth
Tablet triturates	Small moulded tablets intended for quick complete dissolution (e.g. nitroglycerin)	Oral
Granules	Particle size larger than powder	Oral
Compressed tablets	Dissolved or mixed with water; great variety of shapes and formulations	Oral and external
Pellets	For prolonged action	Implantation
Coated tablets	Coating protective; slow release	Oral

manufacturing processes for making liquids or creams also involve solids handling. Aerosol can be generated from liquid handling. Airborne particulates are therefore an inevitable by-product of pharmaceutical operations. As in most industries uncontrolled dust may create fire and explosion hazards. It may also result in impaired product quality by cross-contamination. Furthermore, therapeutically active components of pharmaceutical formulations are often of high potency. They therefore pose potentially damaging health risks for workers overexposed to the material as a result of inhalation, ingestion or skin contact, particularly when it is in a pure uncompounded, or undiluted, state.

Toxic solids

At the manufacturing stage, the biological properties of these materials are likely to have been extensively explored to support intended use, but the focused routes of administration are likely to differ from those experienced accidentally by workers during the processing operations. Furthermore, whilst biological effects and side reactions may be acceptable in patients suffering serious illness, these may be unacceptable hazards for healthy operators. There may also be a dearth of knowledge at the bench or pilot-plant stage of product development.

From the scope of products involved, potential hazards include skin and respiratory irritation, dermatitis, occupational asthma, liver disorders, adrenocortical suppression, feminisation, cancer, blood disorders, etc., as the following case histories illustrate.

> Accidental exposure to chloramphenicol resulted in pronounced leukopenia, thrombocytopenia and cytogenetic changes in lymphocytes, similar to those observed in cases where treatment was prescribed. Disease has also resulted from occupational exposure to neuroactive materials.[43]

> Fourteen employees of a pharmaceutical company exposed to different organic dusts over a period of 1–18 years developed airways obstruction and bronchial hyperactivity.[44] The symptom intensity abated over weekends and vacations but did not subside when they were moved to other departments. The dust which the workers assumed to be the cause of the complaint, consisted mainly of pancreatin, pepsin, bile acids and filler, e.g. starch. Pancreatin is a heterogeneous pancreas extract, containing principally trypsin, amylase, lipase and all pancreatic proteins. It is produced from porcine pancreas by vacuum drying, defatting and milling. Levels of dust exposure varied throughout the plant, but were highest where pancreatin powder was produced when dust density often inhibited visibility. Eczematous skin lesions were common in the workers, particularly in humid body regions. Tablet production and packing areas produced less dust.
>
> Skin tests and bronchial provocation tests confirmed hyperactivity to pancreatic extract, with enzymes such as lipase and trypsin and especially α-amylase proving to be the causative allergen for Type I allergy symptoms. Lung X-rays of patients with acute disease showed alveolitis, and in some, fibrosis and emphysema (possibly from a Type III reaction). New protective measures were therefore instituted to overcome the problems. The incidence of sensitisation amongst exposed workers was estimated to be 40 to 50%, including non-atopics as well as atopics.
>
> (Interestingly hypersensitivity to porcine trypsin powder was also found amongst workers weighing it out and transferring it to a reactor in a plastics manufacturing plant.[45])

Exposures are not, however, restricted to those engaged in manufacture but may extend to users such as health workers (page 222 *et seq.*) and particularly pharmacists.

> 'Indeed if we questioned closely those who work . . . in the shops of apothecaries . . . as to whether they have at a time contracted some ailment

while compounding remedies that would restore others to health, they would admit that they have very often been seriously affected.' (B. Ramazzini, 1713)

Some drugs which have been reported to have caused various types of asthma, in many cases of late or dual type, and the occupations in which exposure occurred are summarised in Table 29.5 (after ref. 4).

Table 29.5 Some drugs reported to have caused occupational asthma

Sensitising agent	Occupation
Psyllium	Laxative manufacturer
Methyl dopa	Manufacturer
Salbutamol intermediate	Manufacturer
Amprolium HCl	Poultry feed mixer
Dichloramine	Manufacturer
Piperazine dihydrochloride	Process worker and chemist
Spiramycin	Manufacturing engineer
Penicillins	Manufacturer
Phenylglycine acid chloride	Ampicillin manufacture
Tetracycline	Encapsulator
Sulphathiozole	Manufacturer
Sulphonechloramides, chloramine T and halazone	Manufacturers and neighbourhood
Gentian powder	Pharmacists
Phosdrin	Manufacturers

Table 29.6 Principal types of pharmaceuticals reported to cause dermatitis[16]

Pharmaceutical type	Examples
Local anaesthetics	Benzocaine
	Amethocaine
	Procaine
	Cinchocaine
Antibiotics and anti-bacterial compounds	Chloramphenicol
	Neomycin
	Penicillin
	Quinolines
	Sulphonamides
Antifungal compounds	Nystatin
Antihistamines	Promethazine hydrochloride
Antimitotic compounds	5-Fluoracil
Dyes	Gentian violet
Corticosteroids	Hydrocortisone
Vitamins	Vitamins B, E and K
Stabiliser	Ethylene diamine

Occupational dermatitis may occur during either the manufacture or use of pharmaceuticals; it may accompany respiratory symptoms. Principal reported causes are listed in Table 29.6.[16]

Health impairments arising from exposure to hormones may resemble those resulting from disease of the endocrine system, or the side effects of hormone therapy.[43] Industrial exposure of male workers to oestrogens may give rise to all the symptoms of monolateral or bilateral true gynaecomastia; in female workers effects observed include menstrual disorders. Males exposed to progestogens may suffer diminution of libido, testicular pain, and reduced urinary excretion. Females exposed to androgens may experience menstrual disorders, reduced fertility and spontaneous abortions. Corticosteroids effect the skin and cause increases in body weight.

An investigation of a factory which manufactures oral contraceptives revealed that during the previous 12 months 20 per cent of male employees and 40 per cent of the company's female workers experienced hyperestrogenism. The affected males all worked in the area of greatest risk of contact and had gynecomastia and some reported decreased libido or impotence. The females exhibited menstrual disorders.[45b]

A man at a factory manufacturing a potent corticosteroid complained for years of general ill-health during periods away from work, and of nausea and vomiting during the second week of his holidays.[46] Upon investigation of his case, and those of other workers from different parts of the processing plant, clinical examinations including urine and blood analyses were undertaken. It was concluded that the glucocorticoid was absorbed when workers were accidentally exposed, resulting in suppressed adrenocortical levels. In view of the plethora in several workers it appeared that they had experienced prolonged facial contact with the drug, although ingestion or inhalation of fine powder was probably more important.

It was subsequently recommended that workers engaged on the manufacture of potent steroids should have adrenocortical function tests and regular screening, by measuring their morning plasma cortisol concentrations. It was also proposed that they have regular spells on tasks concerned with non-steroid, or non-physiologically active, corticosteroid processes.

Occupational disease in workers exposed to antibiotics such as penicillin, streptomycin and tetracyclin produced symptoms such as asthma, modifications to intestinal bacterial flora and antibiotic resistance.

In one investigation 12.2% of workers in a penicillin manufacturing plant had allergic symptoms and three had asthma. There was a strong suggestion of sensitisation due to inhalation of penicillin dust.[47] Work with penicillin has been implicated in other allergic reactions.[48]

Workers engaged on the production of penicillin antibiotics developed asthma.[49] Exposure to a range of chemicals was by skin contact or inhalation. One worker developed sneezing and wheezing 15–30 minutes after exposure; another developed a skin rash on hands and face. Following transfer to another part of the factory with less exposure to penicillin dust he developed respiratory problems, on one occasion merely as a result of contact over the tea break with colleagues whose clothes were contaminated with penicillin dusts. A third worker developed conjunctivitis after two years' production experience. The symptoms abated on wearing a protective hood but two months later he developed eczema on his hands and then rhinitis and asthma. After a period off sick he returned to a 'penicillin-free' building; he remained symptom-free for a month but then developed attacks of coughing and wheezing. A fourth worker who came into contact with benzyl penicillin and 6-amino penicillanic acid dusts after four years developed shortness of breath and coughs, the severity of which appeared to relate to dust exposure levels.

(Penicillin antibiotics are a well-known cause of allergic reaction during treatment. Development of such a reaction in workers exposed whilst engaged on their manufacture is not therefore surprising, although the chemicals to which they reacted varied, e.g. ingestion of ampicillin in one patient and benzyl pencillin in another in therapeutic doses led to asthma and skin and gastrointestinal problems.)

Sources of exposure

Milling, granulation, coating, weighing, blending, conveying, tableting and manual handling/transfer (e.g. scooping, tipping kegs, filling containers) pose the greatest opportunity for dust generation and dispersion.

Operators involved in the manufacture of pharmaceuticals developed rhinitic reactions.[50] Spirometry and skin prick tests using five common allergens plus a 10% cimetidine solution confirmed that inhalation of cimetidine tableting dust caused non-atopy-related specific acquired hypersensitivity, the intervals ranging from weeks to months. No skin reactions were illicited by cimetidine.

Similarly, whilst preventative maintenance is necessary to avoid loss of containment from valves, glands, gaskets, etc., the maintenance and cleaning operations can themselves give rise to dust exposure, e.g. changing dust filters, breaking powder transfer lines, dismantling and cleaning capsule fillers and dry sweeping/brushing filling machines, capping machines and intermediate container conveyers. High efficiency vacuum cleaning is therefore required.

Control measures

Potential dust-producing operations are normally segregated from the rest of the factory, in areas with controlled access. If an aseptic environment is required the whole area is air-conditioned with recirculation via appropriate filters. Open manual transfer into bulk filling machine reservoirs requires particularly careful containment.

As with control of any industrial use of toxic solids the emphasis should first be on containment and ventilation, with an extraction hood, placed as near as possible to the point of dust escape. To illustrate, the precautions for antibiotic production include avoidance of direct contact; keeping atmospheric contamination to a minimum; conducting the dustiest processes in automated and enclosed systems; carrying out less dusty operations in sealed chambers, boxes or rooms under negative pressure or on benches fitted with effective local exhaust ventilation; collection of any spillage using an appropriate vacuum cleaner; washing down surfaces with hypochlorite solution or other bacterial detergents to prevent contamination by antibiotic-resistant microbes or mycetes; correct use of personal protective equipment; adoption of a high standard of personal hygiene; and prohibition of eating, drinking or smoking in the workplace.

Provisions during a detraying operation and subsequent -harmaceutical powder transfer are given in ref. 42. Reliance on personal protection is only a secondary line of defence. The cleaning of dust collectors is one example where an engineering solution is not reasonably practicable to protect operators. Resort must then be made to appropriate protective equipment. Further advice on ventilation and personal protection is given in Chapters 6 and 12 and in refs. 41, 51 and 52.

Individual plant items/unit operations such as tableting machines are often isolated in air-conditioned booths. A laminar-flow cabinet, with the working space continually purged with clean air moving in a non-turbulent flow pattern, is shown in Fig. 29.4. This has a completely open front for

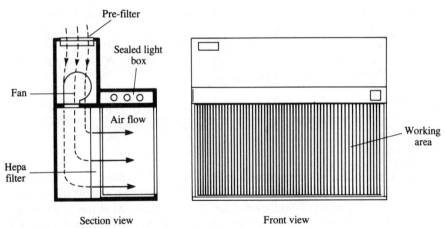

Fig. 29.4 Operating principle of horizontal laminar flow air cabinet

improved accessibility.[53] Environmental monitoring, particularly at the point of emission, is important for therapeutically active substances. However, whilst monitoring may indicate the level of airborne material in operators' breathing zones to be low, any settled material, if disturbed, may create critical levels or cause risk from direct skin contact. Swab testing of surfaces may be appropriate. Guidance on sampling hardware and strategies is given in Chapter 7.

Hygiene standards

There are few therapeutically active materials such as aspirin, glycerol trinitrate, and warfarin for which official environmental or biological standards have been set. However, this industry more than most may be in a strong position to set in-house provisional hygiene standards from the data obtained from animal experiments (e.g. oral, dermal, intravenal toxicity; eye and skin irritancy; skin, and possibly respiratory, sensitisation; mutagenicity; target organ toxicity; reproductive toxicity; along with studies on absorption, metabolism, toxicokinetics) and human clinical trials designed for assessment of product safety and efficacy, etc. Side-reactions need serious considerations, e.g. the sedative effect of certain analgesics, anticonvulsants, antihistamines, antidepressants or anxiolytics may be of greater significance than their therapeutic intended effect when setting occupational hygiene standards. Cytotoxic substances developed for the treatment of cancer may themselves possess irritant or toxic potential.[54] The influence of mode of input into the body also needs to be accounted for; e.g. when respirable dust consisting of a soluble therapeutic component such as codeine is inhaled it may be readily absorbed into the systemic circulation, the implications of which in setting occupational exposure limits are that its bioavailability will be more akin to the intravenous route than to oral administration, when absorption may be slow with subsequent metabolism in the liver.[54] The key aspects for consideration when setting in-house standards for pharmaceutical substances, summarised in Table 29.7,[39] are amplified in refs. 52 and 55. Table 29.8 provides provisional hygiene standards set by one company for a selection of therapeutic substances.[56] These are derived using the equation,

$$\text{ECL (mg/m}^3\text{)} = \frac{\text{NOEL (mg/kg/day)} \times \text{BW (kg)} \times \alpha \times \text{SF}}{V \text{ (m}^3\text{/day)} \times S \text{ (days)}}$$

where:

ECL = Exposure Control Limit. Like TLVs (page 257) these are defined as the time-weighted average concentration for a normal 8-hour workday and 40-hour workweek to which nearly all workers may be repeatedly exposed day after day without adverse effects. Unlike TLVs, ECLs are pharmacology-based rather than toxicity-based.

NOEL = The No-Observable-Effect-Level derived from clinical studies. Some are based on local effects, some on nuisance value, but most are based on either the therapeutic effect for which the drug was intended (e.g. diuresis with ethacrynic acid) or clinically recognisable side-effects observable at or below the therapeutic dose (e.g. drowsiness with amitriptyline).

Table 29.7 Items to be considered when deriving a Provisional Hygiene Standard (PHS)

1. Review background:	• therapeutic area/mode of action • types of drug used/structures • known effects of these compounds/structures
2. Review candidate drug:	• experimental toxicology • experimental pharmacology • experimental toxicokinetics • theoretical effects • anticipated human therapeutic dose • possible effects of repeated subclinical doses
3. Review assumptions in light of:	• experimental toxicokinetics • known species differences • local/systemic effects • route/absorption differences • observed dose–response relationships • specially susceptible populations
4. Review:	• severity of possible effects • severity of risk • need for safety factors • need for special restrictions • availability of analytical methods • need/availability of surveillance methods • need/availability of biological monitoring
5. Leads to:	• proposal for PHS with any special proviso regarding applicability • identification of further studies needed/desirable
6. Review proposal for PHS:	• technical arguments • ability to meet in workplace • impact on operations
7. Limitations:	• lack of data – especially human low dose chronic studies • relevance of regulatory/therapeutic safety studies • (in)validity of assumptions

BW = The average human body weight (i.e. 70 kg for males and 50 kg for females).

α = Absorption. Assumed to be 100% absorption ($\alpha = 1.0$) of some known atmospheric level.

SF = A Safety Factor, usually 10 when variability within a species is known to be large, and 100–1000 when inconclusive human or animal data have been used.

V = Volume of air breathed in an 8-hour day (10 m^3/day).

S = Time to achieve a plasma steady state. For such intermittent, chronic exposures as experienced industrially, the steady-state

Table 29.8 Exposure Control Limits (ECLs) for human health drugs

Generic name	Pharmacological classification	ECL[a] (mg/m)	Pharmacological effect	NOEL (mg/kg/day)	Plasma $t_{1/2}$ (hours)
Amiloride	diuretic	0.1	diuresis	0.04	6–9
Amitriptyline	antidepressant	0.1[b]	sedation	0.02	20–50
Benzotropine mesylate	anticholinergic	0.01	decreased blood pressure	0.002	unknown
Carbidopa	anti-Parkinson's agent	5	nausea	6.0	2
Cefoxitin	antibiotic	2	antibiotic tolerance	0.4	1
Chlorothiazide	diuretic	5	diuresis	2.0	1.5
Cilastatin	antibiotic	5	nuisance effect	not established	1
Cocaine	opioid analgesic	0.1	sedation	0.08	3.5
Codeine	opioid analgesic	0.5	sedation	0.4	2.9
Cortisone	anti-inflammatory	0.7	adreno-pituitary suppression	0.35	unavailable[c]
Cyclobenzoprine	muscle relaxant	0.08	sedation	0.05	6–8
Cyproheptadine	antihistaminic	0.2	sedation	0.04	13
Dexamethasone	anti-inflammatory	0.02	adreno-pituitary suppression	0.01	2.8–6.1
Diflunisal	analgesic	5	URT irritation	not established	8–12
Enalapril	antihypertensive	0.1	decreased blood pressure	0.02	11
Ethacrynic acid	diuretic	1	diuresis	0.2	unavailable[d]
Hexylcaine	local anaesthetic	1	local anaesthesia	0.2	unknown
Hydrochlorothiazide	diuretic	5	diuresis	1.0	2.5
Hydrocortisone	anti-inflammatory	0.5	adreno-pituitary suppression	0.27	4
Imipenem	antibiotic	5	antibiotic tolerance	not established	unavailable
Indomethacin	anti-inflammatory	0.5	headache	0.1	4.5
Levodopa	anti-Parkinson's agent	2	nausea	1.0	2–3
Lisinopril	antihypertensive	0.1	decreased blood pressure	0.02	unknown
Methyldopa	antihypertensive	5	sedation	1.0	2
Morphine	opioid analgesic	0.1	sedation	0.1	2.9
Penicillamine	chelating agent	5	chelation	21.0	unknown
Prednisolone	anti-inflammatory	0.1	adreno-pituitary suppression	0.07	unknown
Probenecid	uricosuric	5	uricosuria	2.0	6–12
Protriptyline	antidepressant	0.08[b]	sedation	0.05	74
Pyrazinamide	antituberculosis agent	5	nuisance effect	2.0	unknown
Sulindac	anti-inflammatory	1	dizziness	0.2	16[e]
Timolol maleate	antihypertensive	0.1	decreased heart rate	0.02	3–5

[a] ECLs given as 8-hour time-weighted average concentrations.
[b] Employees restricted from handling material for more than 8 hours within any 48-hour period.
[c] Plasma half-life unavailable; however, compound has a short (8–12 hour) biological half-life.
[d] Plasma half-life unavailable; however, pharmacological (diuresis) lasts 6–8 hours.
[e] Mean half-life of the sulphide metabolite.

level is directly proportional to the exposure concentration and the biological half-life assuming elimination processes are first-order.

Clearly, like TLVs, any such provisional standards are not fine dividing lines between safety and danger, but rather indicators of whether immediate investigation of control procedures is warranted. They do not guarantee 100 per cent protection for all individuals, e.g. pregnant women or hypersensitive individuals.

The effectiveness of improved protective measures for workers engaged in steroid manufacture (such as self-contained breathing apparatus, environmental monitoring, exclusion of unprotected personnel, modification to the ventilation system, and more efficient decontamination of protective clothing) has been elegantly demonstrated by a programme of biological monitoring.[57]

Information, training and supervision

The objectives of providing information, training and supervision are to[42]
- communicate effectively information on potential hazards and precautions when handling therapeutic materials, and ensure these are fully understood;
- summarise relevant properties of, and precautions to be taken in handling, therapeutic substances; and to instruct personnel on the immediate action to be taken to protect persons and the environment if accidental spillage occurs;
- ensure, so far as is reasonably practicable, compliance with safe operating procedures by instruction, supervision and example.

In general these aims do not differ from considerations covered on page 1628. Provision of information on health hazards is, however, particularly specialised. The nature of therapeutic action should be given, with a description of likely effects and associated symptoms which may be expected to arise in the event of acute or chronic overexposure.[41] Account should be taken of the intended pharmacological effects and the spectrum and severity of effects observed in toxicological studies and during clinical use. Examples of substances with potent pharmacological activity include cytotoxic agents, steroids, anticoagulants and digoxin. A range of toxicological, pharmacological and clinical data are, of course, available for most therapeutic substances; this is more valuable than acute toxicity data from experimental animal studies, e.g. LD_{50} values, which may be used for substance classification (e.g. in the UK under the Classification, Packaging and Labelling Regulations, 1984). It is necessary to identify any potential for absorption via intact skin, and any indications of irritancy or sensitisation. The degree of any mutagenic, teratogenic or carcinogenic potential and the nature of any such hazard should be indicated.

Reference is required to any groups who because of their medical, or other, status may be particularly vulnerable from pharmacological effects of the substance. Potential exposure routes need identifying with emphasis given to the most significant hazards. All relevant effects should be described concisely. The status (i.e. statutory or in-house) of OELs, and numerical value of any exposure limits, should be stated together with any qualifications. Standard reference sources provide a useful initial source of information for products already in commercial use.[58]

Health surveillance

Occupational health surveillance is particularly appropriate to therapeutic substance workers for the following purposes (after ref. 41):
- to protect the health of employees by identifying
  —the foreseeable effects of the substance under conditions of use, and to set up a system capable of identifying them,
  —at an early stage any adverse variations in employees' health which may be work-related, and to seek to remove the cause of any variation and advise employees of appropriate actions;
- to assist in the evaluation of control measures; and
- to ensure compliance with any statutory requirements, e.g. COSHH Regulations.

Because the process of standards setting is imprecise, health monitoring may also be of value as a check on the adequacy of the standards.

The methods available are clinical examination, biological (e.g. concentration of substance or its metabolites in blood or urine, or biological effect monitoring), physiological, and psychological testing, and review of medical records plus a knowledge of the job and working conditions. An occupational health physician will need to advise on the most appropriate surveillance for each employee, having regard to the working conditions, the processes involved, the substances contacted and any statutory requirements. Thus lung function tests may be advisable for workers exposed to respiratory sensitisers (e.g. antibiotics) or substances with potential for direct broncho-constricting effects (e.g. β-blockers, prostaglandins). Biological effect monitoring such as prothrombin time may be more appropriate for workers exposed to anticoagulants, and psychological tests may be suited to those handling psychotropic drugs. General advice on procedures, possible actions if occupational health problems do arise, and the keeping of records is provided in ref. 42.

Like many industrial activities, safety within the pharmaceutical industry is governed by the myriad legislation described in Chapter 16. Key examples include the HASAWA 1974, the COSHH Regulations 1988 (Chapter 28), the Factories Act 1961 and the Environmental Protection Act. Pollution control legislation may be particularly relevant since large-scale antibiotic fermentations result in huge volumes of aqueous effluent in the form of spent broth from which the active substances have been extracted. Process liquors from purification stages may also arise in large quantities and are high in biochemical oxygen demand.[59,60]

An example of a process for treatment of 60 tonne/day of aqueous waste from a pharmaceutical plant is shown in Fig. 29.5.[61] The water content of the waste is

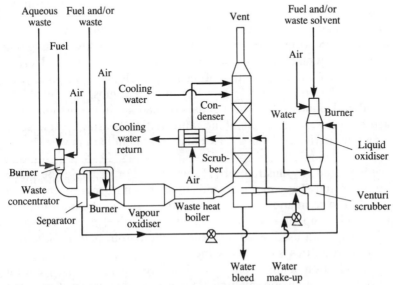

Fig. 29.5 Incineration process for pharmaceutical wastes[61]

80–90% and it contains both organic and inorganic impurities and has a high COD value which precludes dumping. The feed is split into two phases using a fuel-oil-fired concentrator to give a vapour phase containing 75% by weight of the water and some organic solvents, and a smaller liquid phase containing phosphoric acid and hydrochloric acid. Heat recovered from the hot gases is used to generate steam at 1140 kN/m^2 (150 psi).

The liquid phase from the concentrator is burnt off in a ZTO fuel-oil-fired incinerator from which gases are cooled and scrubbed in a high-energy venturi to remove inorganic compounds and then in a packed bed for hydrochloric acid removal.

The Medicines Act 1968 is of specific relevance. It licenses the industry on two fronts, viz.
- a Product Licence issued to allow the product to be marketed once its safety and efficacy has been adequately established
- a Manufacturer's Licence issued once the adequacy of the operations, premises, equipment, documentation and personnel are considered by the regulatory authorities as satisfactory

Further activity in this area is likely to arise from extensive EC legislation aimed at harmonisation of controls for the licensing, manufacture and sale of pharmaceuticals.[62] Product quality in this industry sector is assured by Good Manufacturing Practice.[63]

Lead and lead compounds

Introduction

Lead is a soft bluish/silvery-grey lustrous metal. It is one of the most versatile metals, ranking next to iron, copper, aluminium and zinc in consumption. Important properties of the metal include low melting point, high density, low strength, inertness to attack by acids, water and air, ease of casting, etc. With the exception of the nitrate, chlorate, and chloride the salts tend to be poorly soluble in water, although most are dissolved by acids and body fluids. The principal uses of lead are in the manufacture of storage batteries, solders, plumbing, anti-friction bearings, cable sheathing, as a filler in the car industry, as a heat treatment bath in wire drawing, for sound attenuation, and for radiation shielding. Lead compounds are also widely used, as illustrated by Table 29.9. For example the oxide is used in glass and glazes in the ceramic industry, in pigments for paint and as a catalyst in resin and varnish manufacture. One of the most important organic lead compounds is the tetraethyl derivative, a water insoluble, volatile compound used as an anti-knock additive in petrol, although its use in some countries is on the decline.

Lead production entails separation of ore from gangue by dry crushing, wet grinding, flotation and smelting of the liberated minerals. This involves blast sintering and blast furnace reduction. Metals such as copper, tin, arsenic, etc., are removed from the blast furnace bullion by refining.

Lead toxicity

The first report of lead poisoning was by Hippocrates around 500 BC, i.e.:

> '. . . the worker who was employed in the extraction of metals, suffered a constriction in the stomach area, his abdomen hardened and became stiff, filled with gas and the tissues became discoloured. The affliction moved to his left knee, then back to his abdomen and the attack finally ended in crisis.'

Lead is a microconstituent of foodstuffs and is normally present in human tissue and fluids. However, if a worker inhales lead fume it is absorbed; consequently an amount greater than normal will be found in the blood and urine. This excess is not necessarily indicative of lead poisoning, since individuals vary in their sensitivity to lead, i.e. some may exhibit symptoms when the blood level is only slightly raised whilst others may have no symptoms in the presence of a grossly raised blood lead level.[64]

Lead ingested by mouth is mostly excreted in the faeces; the fraction absorbed is trapped by the liver and largely returned to the intestine in the

Table 29.9 Selected uses of lead compounds

Compound	Uses
Acetates	Raw material for preparation of other lead compounds, rubber antioxidants, processing agent for cosmetics and toiletries, mordant in dyes, water repellant, printing
Arsenate	Insecticides
Azide	Detonator, electrophotography, information storage on resins
Benzoate	Antioxidant in engine lubricants, catalyst, fluorescence quenching
Borate	Pottery glaze, enamels for cast iron, thermistors
Bromide	Catalyst, photographic/duplicating processes, fire-retardant
Carbonates	Catalyst, friction liners, bonding agent, high-pressure lubricants, dielectric component for plastics, photoconductor in electrophotography, thermistors, white pigment, ceramic glaze, temperature-sensitive inks, for red and ultraviolet light
Chloride	Preparation of organolead compounds, catalyst, flame retardant, manufacture of printed circuits
Chromates	Pigments, e.g. chrome yellow, chrome orange, chrome red and chrome green
Fluoride	Low-power fuses, catalyst, glass coatings for infrared reflection, low-melting glass, in phosphors for television tubes
Hydroxide	Batteries, oxidation catalyst, electrical insulating paper, lubricating grease, photothermography
Iodide	Optical imaging and photography, asbestos brake linings, mercury vapour arc lamps, lubricating greases, filters for far-infrared astronomy
Naphthenate	Dryer
Nitrate	Ore processing, pyrotechnics, photothermography, catalyst
Oxides:	
Monoxide (lithage)	Plates for electric storage batteries; optical, electrical and electronic glasses and fine tableware
Dioxide	Dyes, chemicals, matches, pyrotechnics
Sesquioxide	Catalyst, manufacture of high-purity diamonds, radiation detectors, vulcanisation accelerator
Tetroxide	Red pigment in anti-corrosion paints, batteries, glass ceramics, catalyst, sealant for television tubes, radiation shields, waterproofing putty
Phosphite	Polymer additive
Phthalates	Heat stabilisers for PVC
Silicates	Glass and ceramics, paint pigment, rubber additive
Sulphate	Photography, stabilisation of clay soils, white pigment in paint, rubber filler, textile dyes
Sulphide	Photoconductive cells, infrared detectors, transistors, catalysts, humidity detectors, mirror coatings, blue lead pigments
Telluride	Pyrometry, catalyst, thermoelectric elements, electrical contacts for vacuum switches, lead-ion selective electrodes
Tetraethyl	Anti-knock additives for petroleum
Zirconate	Mixed lead titanate–zirconate materials have high piezoelectric properties and are used in acoustic-radiating transducers and hydrophones

bile. Conversely lead absorbed via the pulmonary tract passes directly and almost completely into the circulation. Hence much smaller doses of inhaled lead may rapidly produce symptoms.

The cause of industrial lead poisoning is almost invariably the inhalation of lead-bearing dust and/or fumes, i.e. of exposure to airborne lead or inorganic lead compounds.

However as with toxic solids generally,

'. . . poor personal hygiene (including bad habits such as nail biting) and general carelessness on the part of the worker can result in the ingestion of lead and may be a contributory cause of poisoning.'[65]

The organic lead compounds, e.g. tetraethyl lead, are absorbable via unbroken skin, but potential occupational exposures to these are more restricted.

When lead is rapidly absorbed into the body it becomes widely distributed, and whether or not it causes symptoms of illness depends largely on the ratio of the rate of absorption to the rate of excretion. It is a cumulative poison which in cases of chronic absorption is deposited in the calcareous portion of the bones and may cause no symptoms. However, the metabolism of lead parallels that of calcium; therefore stored lead is remobilised and returned to circulation by conditions, e.g. depletion of the alkali reserve, which alter the reaction of the body fluids.

In lead works in Great Britain the practice over many years was to provide workmen with a free glass of milk each morning. This anticipated the discovery that a high calcium intake assists the storage of lead in a harmless form in the bones.[66]

Inorganic lead poisoning

In cases of inorganic lead poisoning symptoms occur in various combinations, usually referable to either the gastro-intestinal tract or the nervous system. In acute cases, e.g. after heavy exposure to fumes, after some vague prodromal symptoms such as a sweetish taste in the mouth with anorexia, nausea, possibly vomiting and headache, there may be persistent constipation and intermittent colic. In other cases symptoms focus on the nervous system with a sudden attack of meningoencephalitis. In more chronic poisoning the classical signs are headache, blue line on the gums (as may also occur with bismuth), myalgia and arthralgia, with pallor, anaemia and palsy with or without intestinal disturbance and encephalopathy.[64] The commonest form of palsy is the well-known, but now rare, 'wrist-drop'.

Tetraethyl lead poisoning

Tetraethyl lead is not a solid but a liquid but is discussed briefly in order to differentiate it from the inorganic lead solids. It is still used as an antiknock compound in gasoline and has caused acute poisoning of a severe and even fatal type amongst persons handling it. It is volatile at ambient temperatures and is lipo-soluble; hence it can readily be inhaled and may, on absorption, affect the central nervous system. It is more easily absorbed than any other lead compound and even a few days' exposure may cause a rapidly fatal illness.[64] Although rare nowadays amongst persons exposed to metallic lead, encephalopathy may occur after exposure to organic lead compounds; it is characterised by mental dullness, inability to concentrate, tremors, deafness, faulty memory, aphasia, coma and convulsions.[64]

Assessment of nature and degree of exposure

The dangers of excessive exposure to lead have become increasingly appreciated over the years and ever-stringent controls have been introduced. Nowadays deaths and serious illness from lead poisoning are rare. However, lead poisoning in the UK is a reportable disease under The Reporting of Injuries, Diseases and Dangerous Occurrences Regulations, 1985.

An assessment of the nature and degree of exposure to lead should be made by a competent person, e.g. an occupational hygienist, medical adviser or safety officer. This should cover any work which exposes persons to lead which is liable to be inhaled, ingested or absorbed through the skin. The exposures to be considered are those to employees, non-employees, e.g. outside contractors maintaining lead plant, and other persons who may be affected by the work activity, e.g. neighbours who may be affected by emissions of lead dust and fume and the families of lead workers who may be exposed to lead carried home on clothing or footwear. (This is avoidable by following the measures on p. 1681.)

Hygiene standard

The current 8-hour time-weighted average threshold limit and occupational exposure standard for airborne lead of 0.15 mg lead per m^3 (0.10 mg lead per m^3 for tetraethyl lead) is designed to prevent overt illness and long-term renal damage. Various published techniques are available for collection and analysis of dust/fume for lead, e.g. atomic absorption spectrometry,[67-72] colorimetry, and X-ray fluorescence spectrometry, covering a range of analytical sensitivity and interfering species.

Since lead absorbed through the lungs and gastrointestinal tract first passes into the bloodstream prior to distribution to various organs with partial excretion, the level of lead in blood (PbB) provides a good indicator of total exposure. Lead in urine (PbU) may also be used but values tend to be more susceptible to fluctuation; however, it is of use in monitoring exposure to lead alkyls and is 'non-invasive'.

People not occupationally exposed to lead tend to have PbB levels <40 μg per 100 ml and usually <30. Lead in urine (PbU) levels are normally <65 μg per litre. Symptoms of lead poisoning rarely develop if PbB levels are <100 μg per 100 ml.[73] The frequency of checks will be influenced by findings as illustrated by Table 29.10.

Table 29.10 Guide to the frequency of monitoring of lead levels[74]

Lead level	Frequency
PbB (per 100 ml blood)	
<40 μg	12 months
40–59 μg	6 months
60–69 μg	3 months
≥70 μg	<3 months
PbU (per litre urine)	
<120 μg	6 weeks
120–149 μg	1 week
≥150 μg	at doctor's discretion

If a blood level exceeds 70 μg per 100 ml then it is likely that a doctor would consider the person unfit for work with lead until the level fell. Further guidance on interpretation of UK hygiene standards for lead is given in the Code of Practice.[74] The ACGIH Biological Exposure Index is 50 μg/100 ml for PbB and 150 μg/g creatinine for PbU.[75] The OSHA standard requires removal of an employee from the worksite if PbB levels exceed 60 μg per 100 ml or an average of 50 μg per 100 ml on three separate occasions in a 6-month period. Clearly, if work patterns change which may alter exposures then the frequency of biological monitoring should be adjusted.

> In the lead crystal glass industry replacement of composite cutting wheels with more efficient, high-speed diamond wheels resulted in an increase in blood lead levels of workers in some cutting rooms. This was attributed to greater production of fine glass/water aerosol particles coupled with inadequate local ventilation. Exposure in some locations was in excess of 20 times the hygiene standard for airborne lead levels.[76]

Since lead in the blood of a pregnant woman can pass to the blood of the foetus/baby and possibly affect its development, the level of lead in the blood of any woman of child-bearing capacity should be kept as low as possible. Blood levels in excess of 40 μg per 100 ml in such women are likely to be interpreted as rendering them unsuitable for work with lead. Any woman known to be pregnant should be excluded from work with lead.

Unlike environmental monitoring, however, biological monitoring cannot usually identify the source of exposure, and a combined monitoring strategy is required for meaningful assessment of risk. The following cases, along with that on page 262, illustrate the importance of a monitoring programme to protect workers likely to be exposed to lead.

> Following several violations of the lead standard, an investigation of a small foundry by OSHA in the USA revealed that six of the thirteen workers had blood lead levels in excess of 60 μg per 100 ml, the highest being 277 μg per 100 ml. All six should have been immediately removed from further exposure.

Workers reported a variety of complaints including irritability, fatigue, memory loss, headaches, pain in the muscles or joints, etc., and one employee had pigmented gums, all consistent with lead poisoning. Measured airborne 8-hour time-weighted concentrations of lead proved to be 66 μg/m^3 with peaks at 330 μg/m^3. Advice from the regulatory authorities included use of personal protective equipment, engineering controls, and medical screening.[77]

In one US study it transpired that foundries and metal-working industries are the primary cause of overexposure to lead.[77]

A company which manufactures car batteries admitted three offences under the Control of Lead at Work Regulations and was fined for exposing employees to health risks. Lead dust was produced during the process and blood-lead levels in its female workers were higher than that at which female employees should be suspended. The company was prosecuted for not providing proper training or regular medical examinations, and they had failed to ensure that the workplaces were kept clean.[78]

In a similar incident two workers at a minerals company suffered with lead poisoning when lead dust levels exceeded 200 times the permitted value. Twenty of the firm's 23 employees had high levels of lead in their blood and the firm was prosecuted under the Control of Lead at Work Regulations.[79]

Sources of exposure

Metallic lead readily becomes coated with a film of easily removed, absorbable oxide, especially when heated. Volatilisation occurs at high temperatures. Therefore industrial processes which may result in lead poisoning include those generating vapour, aerosol or dust containing lead, e.g. lead smelting, melting and burning; vitrous enamelling on glass or metal; manufacture of lead compounds and lead colours; glazing of pottery; lead battery manufacture, ship-building and ship-breaking, painting, plumbing and soldering, and demolition (e.g. of old lead-covered roofs which have become covered in dust 'age scale'). For example, in the metal industries exposure to airborne lead dust can be encountered
- during sampling and unloading of concentrates from copper smelting
- during arsenic recovery from copper refining, particularly from
  —leaking seals between rotating drum and housing
  —bins below the blender
  —hoppers over roasters
- in lead production and purification, e.g.
  —during crushing and grinding of ore prior to floatation
  —during sampling and unloading of concentrates.

Potential for lead fume exposure exists
- during copper roasting
- during smelting of lead bullion, pouring crucibles and cupelling (and whenever molten lead is disturbed, e.g. to add fresh ingots)

- during repair of lead liners, e.g. of electrolytic tanks
- while welding lead battery plates.

In the pottery industry lead exposure can arise
- in the manufacture of frits, glazes or colours
- during colour or glaze spraying
- during colour grinding, sieving, or mixing
- while mixing dry powdered glaze with water
- while drying after glaze application
- during maintenance.

Exposure to lead oxide dust can be encountered during dry weighing litharge and in the batch production of paste (lead oxide/sulphuric acid/water) in the manufacture of storage batteries. In addition, whenever lead or its compounds are handled the opportunity exists for contamination of body surfaces with subsequent ingestion. However, as exemplified by Tables 29.11 and 29.12 there is a wide diversity between possible levels of exposure.[74]

Precautions

Within the UK the Control of Lead at Work Regulations apply to all work which exposes persons to lead in any form such that it may be ingested, inhaled or otherwise absorbed.[78] In association with an Approved Code of Practice,[74] these aim to control the exposure to lead of people at work and, where such controls cannot reduce exposure to an acceptable low level, to monitor the lead absorption of each individual so that they may be removed from work involving exposure to lead before their health is affected. A summary of the Regulations is given in Table 29.13.

Control measures will generally involve some combination of those listed in Table 29.12. As in other cases, substitution is a primary consideration (see page 278). This has, for example, included replacement of white lead by titanium oxide, zinc oxide or zinc sulphide, or barium sulphate in the manufacture of paints and colours, and the use of lead-free glazes in pottery.

When the use of lead or its compounds is unavoidable, plant and systems of work should clearly be designed and operated such that personal exposures to lead are eliminated or reduced. (This contol must not, however, result in unacceptable emissions of dust, fume, or vapour to the general environment.)

Typical control measures include use of enclosure and exhaust ventilation over material transfer points, feeders, hammer mills, grinders, sieves, crushing rolls, mixers, chutes, etc.; extracts are exhausted via, e.g. electrostatic precipitators. Leaks may occur unless seals are effective and plant is well maintained. Respirators may still be required and these may need to cope with lead dust and fume and with gases such as oxides of sulphur. Enclosure and ventilation is required whenever lead is heated,

Table 29.11 Work that can involve significant exposure to lead

Types of lead work where there is liable to be significant exposure to lead (unless adequate controls are provided)	Examples of industries and processes where such work could be carried out
Lead dust and fumes	
1. High-temperature lead work (above 500°C), e.g. lead smelting, melting, refining, casting and recovery processes; lead burning, welding and cutting	Lead smelting and refining; casting of certain non-ferrous metals, e.g. gunmetal; leaded steels manufacture; scrap metal and wire-patening processes of lead coated and painted plant and surfaces in demolition work; ship-building, breaking and repairing; chemical industry; miscellaneous industries
2. Work with lead compounds which gives rise to lead dust in air (e.g. any work activity involving a wide variety of lead compounds) (other than low solubility lead compounds)	Manufacture of lead batteries, paints and colours, lead compounds, rubber products; certain mixing and melting processes in the glass industry, certain colour preparation and glazing processes in the pottery industry battery breaking manufacture of detonators (explosives industry)
3. Abrasion of lead giving rise to lead dust in air, e.g. dry discing, grinding, cutting by power tools	Miscellaneous industries, e.g. motor vehicle body manufacture and repair of leaded car bodies Firing of small firearms on indoor ranges
4. Spraying of lead paint and lead compounds other than paint conforming to BS 4310/68 and low solubility lead compounds	Painting of bridges, buildings, etc., with lead paint
Lead alkyls	
1. Production of concentrated lead alkyls	Lead alkyl manufacture
2. Inspection, cleaning and maintenance work inside tanks which have contained leaded gasoline, e.g. road, rail and sea tankers and fixed storage tanks	Oil refineries, oil transport terminals and certain works where tank cars are inspected or repaired

e.g. above 550 °C. Flue ducts, e.g. from smelters, should be equipped with access for cleaning to remove accumulated dust which can reduce the ventilation efficiency. Gases, vapours and fumes containing lead should be treated prior to exhausting to air. Cyclones and impingement scrubbers are acceptable for relatively coarse particles, but finer particles, e.g. from smelting or refining, require fabric filters or high-efficiency wet scrubbers. Electrostatic precipitators are effective but depend upon the power supply being maintained and the electrodes being clean.

Use of water sprays may also be effective in suppressing dust generated, e.g. during processing or by traffic. However, if 'wet working' is adopted the wetting needs to be sufficient to prevent dust formation, and the wetted materials, or surfaces, cannot be allowed to dry out because of the associated risk of dry lead dust generation. Water sprays are unlikely to be

Toxic hazards

Table 29.12 Work that is unlikely to involve significant exposure to lead

Types of lead work where there is not liable to be significant exposure to lead	Examples of industries and processes where such work could be carried out
Lead dust and fumes	
1. Work with galena (lead sulphide)	Mining and working of galena when its character or composition is not changed
2. Low-temperature melting of lead (below 500°C) (such low temperatures control the fume but some care is still required in controlling any dust from dross)	Plumbing; soldering; linotype and monotype casting processes in printing industry
3. Work with low-solubility inorganic lead compounds	Painting with low solubility paints
4. Work with materials which contain less than 1% total lead	
5. Work with lead in emulsion or paste form where the moisture content is such and is maintained so that lead dust and fume cannot be given off throughout the work duration	Brush painting with lead paint and the use of some stabilisers for plastics
6. Handling of clean solid metallic lead, e.g. ingots, pipes, sheets, etc.	Miscellaneous metal industries, stock-holding, general plumbing with sheet lead
7. Lead emissions from petrol-driven vehicles (other products of combustion such as carbon monoxide are the major risk)	Testing of petrol-driven engines
Lead alkyls	
1. Any exposure to lead alkyl vapours from leaded gasoline where the lead content is limited under the Motor Fuel (Lead Content of Petrol) Regulations	Work with such leaded gasoline including, e.g., the filling of petrol vehicles on garage forecourts (except for work inside tanks which have contained fuel)

Table 29.13 Control measures for lead materials, plant and processes

- Substitution by lead-free materials or low-solubility lead compounds (i.e. a lead compound which does not yield to dilute hydrochloric acid >5% of its dry weight as soluble lead compound calculated as lead monoxide in a standard test)
- Use in emulsion or paste form
- Control of molten lead temperature to <500°C at which fume emission levels are insignificant, although PbO formation and dust emission remain a hazard
- Containment of lead in totally enclosed, well-maintained plant, and in enclosed containers, e.g. drums and bags. Opening of the enclosure under exhaust ventilation where reasonably practicable
- Effective exhaust ventilation (page 676), terminating in a dust and/or fume collection unit with appropriate filtration or arrestment equipment. Lead-contaminated air from the exhaust system should not be discharged into work areas unless the concentration of lead has been reduced to an acceptable standard or into other areas such as the general environment unless it has been adequately treated
- Wet methods including the wetting of lead and lead materials (e.g. wet grinding and pasting processes), the wet rubbing or scraping down of lead-painted surfaces, and the wetting of floors and work benches whilst work is in progress.

effective in controlling an airborne dust cloud. Moreover, wetting is not practicable in some instances, e.g.
- with lead materials containing arsenides or antimonides due to the possibility of toxic gas evolution (page 495)
- where furnaces could create a risk of a 'steam-explosion' (page 84).

In other cases, however, 'wetting' may be an inherent part of the process, e.g. in the conversion of pulp white lead into oil paint by the use of oil in automatic closed-in machines without dry grinding.

Feeder rooms/control rooms and overhead crane cabs may be pressurised with HEPA filtered air. Walls, equipment and floors need routine water wash-downs. Environmental and biological monitoring programmes should be implemented as a check on adequacy of controls.

Similarly in lead battery manufacture, potential sources of exposure should be enclosed and ventilated with exhaust being discharged via a cyclone or wet scrubber as appropriate. 'Shaker' or 'pulse' type fabric filters are sometimes employed. Respirators are usually required for access to the paste mixer. Attention is needed to avoid paste drying out and leaving dusty deposits in trolleys, on protective clothing, on floors after spillage, etc.

External ledges on plant on which dust can settle should be avoided and plant surfaces should be as smooth and impervious as reasonably practicable to facilitate cleaning. Equipment should have good joints and seals to prevent leakages. Consideration should cover all stages of the work cycle, e.g. when exhaust booths are used for weighing-out lead material it is important to assess how the material is transfered to the booth and how it is handled subsequently.

If a substantial increase in exposure to lead is likely as a result of an incident, non-essential workers should be excluded from the affected area; workers who remain should wear appropriate protective clothing and respiratory equipment.

In general, lead and its compounds should be purchased where possible in a non-dusty form, i.e. as paste, crystals, etc. Food, drink and smoking materials should be prohibited from work areas with provision for changing clothes and showering: protective clothing should be laundered regularly, e.g. at least weekly. Workers should be encouraged to wash thoroughly on leaving the work area, with detailed attention to folds of the skin.

Provision of information, instruction and training

The requirement to provide employees with appropriate information, instruction and training is emphasised elsewhere, e.g. pages 1013 and 1628. In the present context the objectives are to familiarise them with
- health hazards of lead and its compounds
- reasons for the nature of the general control measures, etc., to protect

them, and other persons including their families who may be affected by their work with lead, and for the specific job-related control measures
- how to use the control measures effectively, e.g. respiratory protection (page 295)
- significance of air and biological monitoring
- role of medical surveillance
- duties under appropriate legislation, e.g. in the UK involving particular emphasis on the need
  —for correct use of control measures
  —for cleanliness and to practise a high standard of personal hygiene
  —to attend for medical examinations and biological tests
  —to report any defects/inadequacies in control measures which could affect people's health.

Environmental impact

Best practicable means for lead works are summarised elsewhere.[80] The environmental impact of lead has been reviewed.[81]

Asbestos

Brief reference to the occupational health problems with asbestos is made on page 221 and micro-photographs of typical fibre types are shown in Fig. 7.19.

The disease first identified as being due to occupational exposure to asbestos was asbestosis, a fibrosis of the lung which is irreversible and which may progress even after cessation of exposure. In the UK this risk was first recognised in the asbestos industries in 1930[82] but the regulation introduced in 1931[83] did not cover many activities, particularly insulation application and removal which generated dust.[84] The observation in 1938 that,[85]

> We are but on the threshold of knowledge to the effects on the lungs of dust generally . . . It is not many years ago when the dust of asbestos was regarded as innocuous while today it is regarded as highly dangerous . . .'

was prophetic. In the early 1950s,[86] an increased incidence of lung cancer was established, the increase in risk depending upon the degree of exposure and being very much greater for smokers than for non-smokers. Mesothelioma, a cancer of the inner lining of the chest wall or of the abdominal wall, may be caused by much lower exposures. The incidence in the general population is very low but currently it accounts for some 1000 deaths per year in the UK.

For asbestos-related diseases there is commonly a latency period of 10 to 20 years between first exposure and the onset of symptoms but for cancer this period may be 40 years of more. In any event, the three diseases are considered to be related to the cumulative dose an individual has experienced and the time lag. The evidence for any difference in risk of contracting asbestosis arising from different types of fibre is slight, but crocidolite and amosite are believed to be more hazardous than chrysotile in relation to increased risk of lung cancer. Certainly for mesothelioma crocidolite appears to be more hazardous than chrysotile.

Asbestos minerals may occur in natural mineral products and in manufactured or compounded materials. Their presence may not be identifiable from appearance, or the trade name, or trade literature, but in the UK all asbestos materials or products containing asbestos must now be labelled as such.[87,88] However, since older materials or products may be unlabelled, advice should be sought from suppliers if the presence of asbestos is suspected, otherwise testing should be instituted.

Asbestos was widely used, e.g. in construction work, buildings, industry, laboratories and in the home. The major uses were:

- asbestos cement products, e.g. for roofing and pipes. These generally contain 10–15 per cent of chrysotile bound in a matrix of hydrated Portland cement.[89]
- insulation boards, e.g. for fire protection of buildings. These contained 2–40 per cent asbestos, viz. amosite with a small percentage of chrysotile in some cases.[89] Thus strict procedures are necessary when working with asbestos insulating board[90] and appropriate precautions should be taken with asbestos cement fabrications.[91]
- as a constituent of thermal insulation, i.e. lagging on pipes, equipment in heat and steam-raising plant and to insulate storage vessels in industrial plant. In factories, power stations, hospitals and similar large buildings this often consisted of a layer of 85 per cent magnesia reinforced with 15 per cent asbestos with a surface coat of asbestos composition.
- in asbestos resins, e.g. for clutch faces and brake linings.
- as yarn or cloth, e.g. for protective clothing.
- in floor tiles, roofing felts, textured paints and plasters.
- in engineering products, e.g. filters, gaskets, pump packings and electrical insulation.
- for acoustic insulation of buildings and plant.
- for insulation of steelwork against fire.

Whilst only chrysotile asbestos is now used in the manufacture of articles, e.g. mainly asbestos cement products such as sheets, tiles and decorative finishes, atmospheric contamination may arise during the removal or repair of older materials. Slow disintegration of asbestos-containing materials, e.g. pipework or boiler insulation, may also cause contamination.

Some indication of the dust concentrations probably generated by various construction activities, albeit obviously dependent upon individual circumstances (e.g. the exact composition of the material, the mechanics of the operation, the degree of confinement and the extent of general ventilation), is given in Table 29.14.[92]

Table 29.14 Representative asbestos dust concentrations from construction activities[92]

Activity	Details	Concentration (fibres/ml)
Asbestos spraying*	Using recommended pre-damping equipment	5–10
	Without the use of this equipment	>100
Demolition (de-lagging)†	With thorough soaking	1–5
	With use of water sprays	5–40
	Carried out dry	>20
Use of asbestos/cement sheet and pipes	Machine drilling	<2
	Hand sawing	2–4
	Machine sawing without effective local exhaust ventilation:	
	jig saw	2–10
	circular saw	10–20
	Machine sawing with effective exhaust ventilation	<2
Use of asbestos insulation board	Drilling vertical structures, e.g. column casing	2–5
	Drilling overhead, e.g. suspended ceilings	4–10
	Sanding and surforming	6–20
	Scribing and breaking	1–5
	Hand sawing	5–12
	Machine sawing without effective local exhaust ventilation:‡	
	jig saw	5–20
	circular saw	≥20
	Unloading deliveries of board (short-term sampling):	
	cut pieces	5–15
	manufacturers' standard-size sheets	1–5

* Concentrations 6–10 m away are roughly one-tenth of these.
† Airborne asbestos concentrations depend on the constitution of the individual lagging.
‡ Dust concentrations evolved by machine drilling or sawing can be reduced to 2–4 fibres/ml by the use of dust control equipment.

Because of the serious potential risk associated with asbestos exposures there has been an increase in the stringency of legislative controls in developed countries.

For example, in the UK The Control of Asbestos at Work Regulations 1987 and Approved Codes of Practice apply to all work with asbestos products which may give rise to dust, including processing, manufacturing, repairing, maintenance, construction, demolition, removal and disposal. The requirements are summarised below.

Assessment

Before starting any work which is liable to expose employees to asbestos dust, an assessment of the work is required to help decide the measures necessary to control exposure. This should

- identify the type of asbestos (or assume that it is crocidolite or amosite, to which stricter controls are applicable than to chrysotile)
- determine the nature and degree of exposure
- set out steps to be taken to prevent that exposure, or reduce it to the lowest level reasonably practicable.

The assessment should be in writing unless the work involves low-level exposure and is simple, so that the assessment can be easily repeated and explained.

Control limits

Maximum exposures to dust levels are set by control limits (Chapter 28) expressed as fibres per millilitre, measured or averaged over 4 hours or 10 minutes: see Table 29.15.

Table 29.15 Control limits for exposure to asbestos dust (fibres/ml)

Type of asbestos		4 hours	10 minutes
Chrysotile	(white asbestos)	0.5	1.5
Amosite	(brown asbestos)	0.2	0.6
Crocidolite	(blue asbestos)	0.2	0.6

Employees should never breathe air containing a level of asbestos which exceeds these limits. Moreover the level should always be reduced so far as it reasonably can be. Use should be made of
- suitable systems of work
- exhaust ventilation equipment
- other technical measures
- all of these techniques if reasonably practicable.

If the dust level is, or could be, above the control limit an employer must
- provide suitable respiratory protective equipment and ensure that it is used properly; and
- post warning notices that the area is a 'respirator zone'.

Action levels

Action levels are a measure of the total amount of asbestos to which a person is exposed within a 12-week period. These are set in fibres/hours per millilitre, as in Table 29.16.

When these are, or may be, exceeded the employer must ensure the Enforcing Authority has been notified, maintain a health record of exposed workers and make sure that they receive regular medical examinations, and identify work areas where the action level is liable to be exceeded as 'asbestos areas'.

1686   Toxic hazards

Table 29.16 Action levels for asbestos dust exposure (fibres/hour per ml)

Type of asbestos	Level over 12 weeks
Chrysotile	120
Amosite	48
Crocidolite	48

Other provisions

There are also requirements for an employer to
- monitor the exposure of employees to asbestos where appropriate
- ensure that employees liable to be exposed to asbestos receive adequate information, instruction and training – so that they are aware of the risks and the precautions which should be observed
- provide protective clothing for workers when a significant quantity of asbestos is liable to be deposited on their clothes
- check that the plant or premises where work with asbestos is carried out is kept clean
- make sure that there are adequate washing and changing facilities
- provide separate storage areas for any protective clothing and respiratory protective equipment required, and for personal clothing
- make sure that all asbestos articles, substances and products for use at work are specially labelled
- keep raw asbestos and asbestos waste sealed and labelled.

Sodium hydroxide

Sodium hydroxide finds wide use in, e.g. manufacture of organic and inorganic compounds, polymers and pulp and paper, cleaning (soap, detergents, household bleaches, polishes), petroleum and natural gas industries, cellulosics, and cotton mercerisation. It is manufactured by reacting sodium carbonate solution with calcium hydroxide, or by electrolysis of solutions of sodium salts (mainly the chloride) in mercury or diaphragm-type cells. The product is usually an aqueous solution varying in concentration from 45–75%. This is converted into anhydrous caustic soda by dehydration at $>400\,°C$ in, e.g. tubular flash-type nickel or Inconel evaporators. After the water has been driven off, the molten caustic is pumped either to rotary drum flakers or to drum loading stations and then poured into thin steel drums for shipment. Flaked solid caustic is also shipped in drums.

Sodium hydroxide is a brittle, white translucent crystalline solid whose physical properties are summarised in Table 29.17. It is deliquescent on exposure to air, resolidifying because of sodium carbonate formation upon absorption of carbon dioxide. It is very soluble in water, liberating considerable heat on dissolution which can cause the solution to boil. Solid

Table 29.17 Physical properties of pure sodium hydroxide

Property	Value
Molecular weight	39.998
Specific gravity at 20°C	2.130
Melting point, °C	318
Boiling point, °C at 101.3 kPa	1388
Specific heat, J/g°C at 20°C	1.48
Refractive index at 589.4 nm	
320°C	1.433
420°C	1.421
Latent heat of fusion, J/g	167.4
Latice energy, kJ/mol	737.2
Entropy, J/(mol·K) at 25°C and 101.3 kPa	64.45
Heat of formation ΔH_f, kJ/mol	
α form	422.46
β form	426.60
Solubility in water g/100 ml at 0°C	42
Bulk density	1.175

caustic soda should, therefore, be added gradually to water (not vice versa) with cooling. It also reacts with acids and certain metals and may release hydrogen.

Sodium hydroxide is corrosive to many materials of construction: steel is acceptable for <50% solution below 40 °C, e.g. as liners for tank cars or tanks. Nickel is suitable for handling material at all concentrations including molten caustic at up to 480 °C. Advice should be sought from the supplier and open literature.

Caustic soda in both liquid and solid form is extremely corrosive to tissue, causing severe burns (even dilute solutions can have marked effects after prolonged contact) with subsequent scaring. Inhalation of the dust or mist can damage the upper respiratory tract and even prove fatal as a result of spasm. Symptoms of exposure may include burning sensations, coughing, wheezing, laryngitis, shortness of breath, headache, nausea and vomiting. Whilst prolonged exposure to high concentrations may cause discomfort and even ulceration of nasal passages, subjective symptoms are usually indicative of the need for control, and short-term exposure levels of 2 mg/m^3 have been set in the UK and USA. Ingestion causes damage to the mucous membranes and can injure the digestive system. All forms of the compound are dangerous to the eyes and may cause corneal ulcers and permanent impairment of vision.

Clearly, handling precautions must ensure correct materials of construction are used, contact with incompatible chemicals (e.g. keep away from aluminium, zinc, lead, tin, acids and chlorinated solvents) is avoided, and operator contact or exposure to the compound in airborne particulate form is prevented. As with any hazardous chemical, workers should be trained in the hazards and the correct handling procedures both for normal operations and emergencies. Containment is the prime strategy with reliance on snuggly fitting personal protective equipment for eyes, skin and respiratory system. Since sodium hydroxide is a common constituent of domestic/commercial oven cleaners, appropriate hand and forearm protection, without gaps, is essential for their safe use. Protective clothing/footwear should be of rubber, chlorobutadiene or other caustic-resistant material. Adequate ventilation must be provided where caustic soda mist or dust are present. Showers and eye wash solutions must be available for emergencies.

First-aid measures tend to rely on copious washing of the contaminated area for prolonged periods and removing contaminated clothing: speed is of the essence and medical attention should be sought. Provide resuscitation and oxygen if breathing has stopped. In the case of ingestion provide ample water to drink but do not induce vomiting. Spillages should be scooped up and residues flushed with water by operatives wearing appropriate protection (including respiratory protection), avoiding dust/aerosol formation. Spillages to drains or sewers must be immediately reported to the appropriate regulatory water authority.

Toxic solids in the rubber industry[95–98]

The natural and synthetic rubber industry manufactures a vast range of products. This requires a wide variety of processes over a range of temperatures, e.g. blending, mastication (breaking down the long polymer chains to create a softer, more plastic material into which other ingredients can be incorporated), compounding, milling, extrusion, buffing, etc. together with a catalogue of over 500 raw materials, many of which are supplied as powders. Official hygiene standards are available but for a small percentage of the solids used, although recommended values are available from, e.g. The British Rubber Manufacturers' Association.[93] However, it cannot be assumed that hazards are associated with exposure to raw materials alone, since the rubber processes cause many chemical reactions which can give rise to more toxic materials such as nitrosamines. The working conditions in the industry have improved significantly during recent years but excess incidence of certain occupational diseases has been associated with previous exposures to toxic solids, e.g.

- lung cancer which may be linked to exposure to raw material during storage, weighing, mixing, compounding, extruding and calendering, although the exact causative agents are unknown
- bladder cancer linked to β-naphthylamine in antioxidants
- stomach cancer in tyre manufacture
- skin irritation, dermatitis, blood disorders, nasal and eye irritation associated with fillers and other powders ('drugs')
- respiratory distress linked to general dust exposure

Compounding ingredients for vulcanising the rubber (cross-linking or curing agents) include sulphur, zinc oxide, stearic acid and organic accelerators. Carbon black, amorphous silica or clays are used for reinforcement. Antioxidants, antiozonants or clays are used to prolong service life. Oils, talc, waxes and resins are compounded to aid processing/rubber extension. Other additives include fillers and colorants.

Hazards arise during storage, weighing and milling. Fumes can be liberated during extrusion of mixed rubber compound and from moulds. Dusts are formed during buffing of finished rubber.

Techniques for controlling exposure include:
- providing dust removal systems for mixing mills, internal mixers and weighing plant, sack disposal, etc. (see Figs. 29.6 and 29.7)[97]

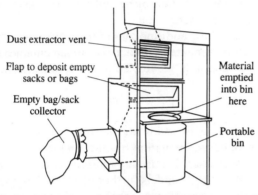

Fig. 29.6 Sack emptying and disposal under local exhaust ventilation

Fig. 29.7 Manual weighing in local exhaust booth

- use of enclosed or automated bag and powder handling facilities
- bulk delivery of high throughput items to reduce the number of containers that need to be handled
- weighing directly into the container to be used rather than into a weighing scale pan for subsequent transfer

- arranging for minimum distance between bins and weigh scales for manual scooping
- using solids in least dusty form, i.e. granules, pellets, lozenges, etc., rather than powder
- where powder form is unavoidable use dust-reduced powders containing binders wherever available, or wet powder with liquid softeners, e.g. carbon black with mineral oil
- providing efficient ventilation for mixing highly toxic solids, e.g. lead compounds (see also page 1680)
- purchasing batches of ready-mixed powders from suppliers equipped with high-grade facilities
- enforcing good housekeeping, particularly in stacking, transporting, bag opening and bag disposal, and ensuring spillages are cleaned promptly using suitable vacuum cleaners
- prohibiting eating, drinking and smoking in the work area
- avoiding prolonged or repeated contact of rubber chemicals with the skin. Protective clothing, e.g. gloves, aprons, overalls, should be worn as appropriate
- encouraging high standards of personal hygiene and providing adequate washing facilities

With such complex mixtures of solids no one precaution is adequate for prevention of dust formation, inhalation hazards and skin contact. Thus a combination of control measures is usually required even for dust control as summarised in Table 29.18,[96] and as further illustrated by the following selected examples. The concepts are equally applicable to other toxic solids.

Reinforcing agents

During compounding all the key ingredients are blended with the raw rubber to provide the finished product with the desired properties. The solid additives are received as powder, flake, pellets, etc., in bags, drums, or small containers that can be emptied into the hopper bins or weighed. Selected toxic materials can be obtained in batch-sized quantities in plastic bags that can be added to the mixer without opening.

Exposure during handling arises from over-dusted rubber stocks, leaks from conveying or piping systems, container damage, spillage from rail or road tankers, loss during hook-up of bulk delivery vehicles for discharge, poor handling practices, and from ineffective ventilation, e.g. during weighing.

Carbon black can be particularly troublesome. This compound is added to enhance the stiffness and tensile strength of the rubber. Problems arise from its fine particle size, greasy nature and possible contamination by polyaromatic hydrocarbons (PAHs), some of which when extracted and isolated are known to be carcinogenic (e.g. benzo-α-pyrene). Handling this

Table 29.18 Combination of measures for dust and fume control in the rubber industry

Factory process	Health hazard	Control measures
Drug room	Dust from 'small drugs' (complex organic compounds)	Substitution Master batches Preweighed, sealed bags Dust-suppressed chemicals Local exhaust ventilation Care in handling
	Dust from bulk fillers and whitings	Local exhaust ventilation Care in handling
	Dust from carbon black	Master batches Local exhaust ventilation Totally enclosed systems Not by 'careful handling' alone
	Skin contact with process oils	Direct metering into mixer Care in handling and protective clothing
Compounding	Dust	Local exhaust ventilation Master batches Preweighed, sealed bags Dust-suppressed chemicals Care in handling
	Fume	Local exhaust ventilation Removal of hot product from workroom – cool before handling
	Skin contact with process oils	Direct metering Care in handling and protective clothing
Moulding	Fume	Local exhaust ventilation Removal of hot product from workroom – cool before handling Deflection by shields
Calendering and extruding	Fume	Local exhaust ventilation Water cooling of extrudate
	Dust from release agents (chalk stearate or talc)	Substitution of wet methods Enclosure and local exhaust ventilation
Curing	Fume	Local exhaust ventilation at autoclave door and storage racks Allow autoclave to cool before opening
Spreading	Fume	Local exhaust ventilation Care in handling mixes

material therefore poses cancer, pulmonary fibrosis and dermatitis hazards, although the latter is generally associated with indirect factors such as vigorous washing. Precautions include containment, ventilation, personal protection and, most effectively, use of the material in pellet form and supply in small bags for unit dosing directly into the mixer, or in tote bins, which need effective seals between bin and silo. Bulk deliveries are effectively transported by pneumatic conveying. Figure 29.8 illustrates bulk handling facilities based on a less-effective bucket elevator system. Pellets can be easily broken but this damage can be minimised by:

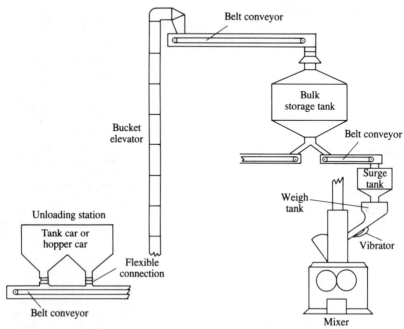

Fig. 29.8 Typical carbon black bulk-handling system

- reducing the height of fall
- shielding the material from external air flows
- reducing the conveying velocities
- arranging for continuous flow rates rather than dumping
- reducing the ratio of surface area to volume of the flowing stream
- using gentle conveyors or processes

Amorphous silica powder is not generally considered dangerous, but special handling procedures are justified to avoid respiratory problems. Usually the material can be supplied in prepacked bags of appropriate sizes to enable the entire contents to be incorporated directly without further weighing/dispensing.

Processing aids

Talc is often employed to aid processing such as milling. This material is a hydrous magnesium silicate which may be contaminated by other silicates, including asbestos (see pages 1634 and 1682). Because of this, its close similarity to asbestos, and the fact that long exposures to high concentrations can lead to 'talc pneumoconiosis', the careful handling that is required to prevent dust formation is often supplemented with effective ventilation and personal protection.

Vulcanising agents

Curing is a final step in the process to convert raw polymer into finished rubber with the desired properties. The terms 'vulcanising' and 'curing' are used interchangeably and refer to the introduction of cross-links between polymer chains to give the polymer 'memory' when stretched and released, i.e. elastomeric properties. This process was originally achieved using sulphur but faster curing and superior products can be obtained using a range of organic peroxides or organic sulphur-containing substances.

The acute oral and systemic toxicity of organic peroxides is not high, although they are irritating to skin and mucous membranes. Non-dust forms are available. During curing these compounds degrade in some cases into more toxic compounds. Key precautions include: store below 40 °C, isolate from sources of ignition, avoid contamination, protect eyes with goggles and skin with gloves, use ventilation to regulate dust/fumes, provide adequate water for firefighting and washing contaminated skin, and provide eye wash solutions.

Whilst many of the accelerators are skin irritants only a few, e.g.

$$\text{thioureas } (>N-\overset{\overset{\displaystyle S}{\|}}{C}-N<),$$

$$\text{dithiocarbamates } (>N-\overset{\overset{\displaystyle S}{\|}}{C}-S-) \text{ and}$$

$$\text{guanidines } (>N-\overset{\overset{\displaystyle NH}{\|}}{C}-N<)$$

show dermatitic activity/sensitisation potential. Thus the opportunity theoretically exists to select safe alternatives. Where this is not feasible the risk can be minimised by careful handling techniques, use of barrier creams and cleansing agents, protective clothing and education in personal hygiene. Whilst incidents of skin irritation can thereby be eliminated there is always the chance that susceptible operators will show allergic responses; the only effective measure may then be to change their work station.

Autoclave or dry heat curing releases very little fume. Moulding techniques using higher temperatures can, however, discharge considerable fumes, e.g. in tyre curing, although the composition of these is little under-

Antioxidants

Because of the carcinogenic risk associated with β-naphthylamine-derived antioxidants this has been substituted by, e.g. phenyl-β-naphthylamine (PBN), the specification of which is carefully controlled so that the free β-naphthylamine content is less than 1 ppm. Even so, the risk of occupational cancer, albeit low, may still exist in this industry because of the complex mixture of hazardous substances to which operators may be exposed.

PBN

General principles for protection against toxic solids

The general principles for protection when storing, handling and disposing of toxic solids summarised on pages 323–30 are amply illustrated by the foregoing examples. However, the safe design and operation of facilities will only result in acceptable occupational hygiene with the training, instruction and co-operation of the workforce. Typical duties of employees include the following.
- Maintaining the workplace in a clean and tidy condition.
- Complying with no-smoking regulations.
- Complying with the prohibition of eating or drinking in the work area.

> Inorganic pigments, e.g. lead chromate, basic copper carbonate, basic copper ethanoate and copper arsenite, were used in the mid-19th century in wallpaper. Boys were employed to spread the colours on the printing blocks and, since it was considered uneconomic to stop the machines for mealbreaks, they ate their food as they worked.[99]

- Co-operating with health surveillance requirements, e.g. regular medical check-ups.
- Following working practices which minimise dust generation or redispersion, e.g. avoiding the use of airlines to blow dust away, or dry sweeping, but using approved vacuum cleaning or wet cleaning methods.
- Wearing all protective clothing provided and replacing it as necessary. Ensuring that overalls are disposed of, or laundered and not taken home.

- Disposing of waste strictly in accordance with instructions.
- Using ventilation equipment in the recommended manner and reporting any apparent defaults promptly.
- Wearing approved respiratory protection as instructed, e.g. for cleaning or maintenance jobs. Taking care of respiratory protective equipment, e.g. correct fitting, cleaning, filter-changing and storage.
- Washing thoroughly, and if necessary taking a shower, before leaving the workplace.

The boys who filled the moulds at colour factories with the highly toxic inorganic pigments referred to on page 1694 and ran with them to the drying ovens were identified as 'blue boys', 'yellow boys', and 'green boys'.[99]

- Co-operating with atmospheric monitoring exercises, e.g. by wearing a sampling device (Chapter 7).

References

1. Ross, J. G., *Chrysotile Asbestos in Canada*, Canada Dept. of Mines, Ottawa, 1931, p. 128.
2. Figley, K. D. & Elrod, R. H., *J. Am. Med. Assoc.*, 1928, **90**, 79.
3. Wegman, D. H. & Giusti, R., in *Occupational Health, Recognising and Preventing Work-related Disease*, eds. B. S. Levy and D. H. Wegman, Little Brown & Co., Boston, 1983, p. 55.
4. Parkes, W. R., *Occupational Lung Disorders*, 2nd edn., Butterworths, 1982.
5. Butterworth, D. J., MSc Thesis, *Health and Safety in the Agricultural Industry*, Aston University, 1976.
6. Flindt, M. L. H., *The Lancet*, 1969 (14 June), 1177.
7. Pepys, J., et al., *The Lancet*, 1969 (14 June), 1181.
8. Newhouse, M. L., et al., *The Lancet*, 1970 (4 April), 689.
9. Greenberg, M., et al., *Brit. Med. J.*, 1970 (13 June), 629.
10. Pepys, J., et al., *Clinical Allergy*, 1973, **3**, 143.
11. Juniper, C. P. & Roberts, D. M., *J. Occup. Med.*, 1984, **34**, 127.
12. Juniper, C. P., et al., *Brit. J. Ind. Med.*, 1985, **42**, 43.
13. Report by the Industrial Diseases Sub-Committee of the Industrial Injuries Advisory Council on Occupational Asthma, 1981, HMSO.
14. Zachariae, H., et al., *Acta Dermatovenereologica*, 1973, **53**, 145.
15. How, M., et al., *Clinical Allergy*, 1978, **8**, 347.
16. Cambridge, G. W. & Goodwin, B. F. J., *Allergy to Chemicals and Organic Substances in the Workplace*, Science Reviews Ltd, 1984.
17. Report of the Symposium on Biological Effects of Proteolytic Enzyme Detergents on 4th to 5th May, 1976: MRC Pneumoconiosis Unit, Cardiff, in *Thorax*, 1976, **31**, 621.
18. The Standing Committee on Enzymatic Washing Products, Fifth Report, 1990, The Soap and Detergent Industry Association.
19. Markussen, E. K. & Schmidt, A. W., *Ger. Offen.*, 2, 730, 481/1978 (patent).
20. Health & Safety Executive, Guidance Note EH40, HMSO, 1993.
21. Bruce, C. F., et al., *Ann. Occup. Hyg.*, 1978, **21**, 1.

22. Little, D. C., in *Occupational Asthma*, ed. C. A. Frazier, Van Nostrand Reinhold, 1980, p. 186.
23. Recommendations of the Medical Sub-committee, 1989 (November), Soap and Detergent Industry Association.
24. Ransome, J. H. & Schuster, M., *J. Allergy Clin. Immunol.*, 1981, **67**(5), 412.
25. How, C., Erlanger, B. F., Beiser, S. M., Ellison, S. A. & Cohen, W., *New England J. Med.*, 1961 (17 April), 332.
26. Hayes, J. P. & Newman Taylor, A. J., *Brit. J. Ind. Med.*, 1991, **48**, 355.
27. Grime, K. & Clauss, A., *Chem. & Ind.*, 1990 (15 October), 647.
28. Hendrick, D. J., et al., *Thorax*, 1988, **43**, 501.
29. Stenton, S. C., et al., *Brit. J. Ind. Med.*, 1990, **47**, 405.
30. Ferguson, S., et al., *Human and Experimental Toxicology*, 1990, **9**, 83.
31. Health & Safety Executive, Guidance Note MS22, 1983.
32. Health & Safety Executive, Guidance Note EH16, 1984.
33. Health & Safety Commission Newsletter, 1991 (77), June, 9.
34. Health & Safety Executive, Guidance Notes MDHS 25, 1987; MDHS 49, 1985.
35. Chan-Yeung, M. & Grybowski, S., Occupational Asthma due to Western Red Cedar (*Thuja plicata*), in *Occupational Asthma*, ed. C. A. Frazier, Van Nostrand Reinhold, New York, 1980, p. 5.
36. Health & Safety Executive Woodworking Sheet No. 1, 1990: IND (S) 21L, *Control of Hardwood Dust*, 1987.
37. Guidance Notes with Regard to Hardwood Dust Control Limit, British Woodworking Federation, London, 1987.
38. *Allergy to Chemicals at Work*, The Chemical Industries Association, London, 1983.
39. McHattie, G. V., Rackham, M. & Teasdale, E. L., *J. Soc. Occup. Med.*, 1988, **38**, 105.
40. Teichman, R. F., Fleming, F. & Brandt-Rauf, P. W., *J. Soc. Occup. Med.*, 1988, **38**, 55.
41. Harrington, J. M. & Goldblatt, P., *Brit. J. Ind. Med.*, 1986, **43**, 206.
42. ABPI, *Guidelines for the Control of Occupational Exposure to Therapeutic Substances*, The Association of the British Pharmaceutical Industry, 1985.
43. Farina, G. & Alessio, L., Pharmaceutical industry, in *Encyclopaedia of Occupational Health*, International Labour Organisation, 3rd edn, 1983.
44. Wiessemann, K. J. & Baur, X., *Eur. J. Respir. Dis.*, 1985, **66**, 13.
45. a) Colten, H. R. & Strieder, D. J., Occupational asthma: hypersensitivity to porcine trypsin, in *Occupational Asthma*, ed. C. A. Frazier, Van Nostrand Reinhold, New York, 1980, p. 266.
    b) Harrington, J. M. & Stein, G. F., *Arch. Environ. Health*, 1978 (Jan), 12.
46. Newton, R. W., Browning, M. C. K., Iqbal, J. & Adamson, D. G., *Brit. Med. J.*, 1978 (14 January), 73.
47. Maffei, R. & Napolitane, L., *Minerva Med.*, 1955, **46**, 1785.
48. Roberts, A. C., *Arch. Ind. Hyg. Occup. Med.*, 1953, **8**, 340.
49. Davies, R. J., Hendrick, D. J. & Pepys, J., *Clinical Allergy*, 1974, **4**, 227.
50. Dally, M. B., Coutts, I. I., Burge, P. S., Hunter, J. V., Page, R. C. & Newman Taylor, A. J., *Thorax*, 1980, **35**, 716.
51. HSE, *Respiratory Protective Equipment: A Practical Guide for Users*, HS(G) 53, 1991, HMSO.
52. Llewelyn, A. E., *Manuf. Chemist and Aerosol News*, 1975 (September), 53.

53. Yalden, M. J., Laminar flow as applied to bacteriology, in *Safety in Microbiology*, eds. D. A. Shapton and R. G. Board, Academic Press, New York, 1972, p. 48.
54. Agius, R., *Ann. Occup. Hyg.*, 1989, **33**(4), 555.
55. U.K. Chemical Industries Association, London.
56. Sargent, E. & Kirk, G. D., *Am. Ind. Hyg. Assoc. J.*, 1988, **49**(6), 309.
57. Newton, R. W., et al., *British J. of Industrial Medicine*, 1982, **39**, 179.
58. Martindale, *The Extra Pharmacopoeia*, 30th edn, Pharmaceutical Press, London, 1993.
59. Hurran, W. J., *Int. J. Environmental Studies*, 1975, **6**, 193.
60. DOE, *Wastes from the Manufacture of Pharmaceuticals, Toiletries and Cosmetics*, Waste Management Paper No. 10, Department of the Environment, 1978.
61. Anon., *Process Engineering*, 1974 (April), 68.
62. a) Murray, M., *A Guide to the European Community Legislation on Health and Safety at Work*, The Association of the British Pharmaceutical Industry, 1990.
b) Charlesworth, F. A. & Griffin, J. P., *A Brief Guide to the European Directives Concerning Medicines*, 2nd edn., The Association of the British Pharmaceutical Industry, 1988.
63. Moss, G. F., QA in Good Manufacturing Practice, in *Good Laboratory and Clinical Practices*, eds. P. A. Carson and N. J. Dent, Heinemann Newnes, Oxford, 1990.
64. Zielhuis, R. L., in *Encyclopedia of Occupational Health and Safety*, 3rd edn, ILO, 1983, **2**, 1200.
65. H.M. Factory Inspectorate, *Prevention of Industrial Lead Poisoning*, Technical Data Note 16(Rev), Department of Employment, 1971.
66. Hunter, D., *The Diseases of Occupations*, 5th edn, The English Universities Press, 1975, p. 287.
67. Health & Safety Executive, *Methods for the Determination of Hazardous Substances, Lead and Inorganic Compounds of Lead in Air*, MDHS 6 (Laboratory Method using Atomic Absorption Spectrometry), 1987.
68. Health & Safety Executive, *Control of Lead in Air at Work, Air Sampling Techniques and Strategies*, Guidance Note EH28, 1981.
69. Health & Safety Executive, *Methods for the Determination of Hazardous Substances, Lead and Inorganic Compounds of Lead in Air*, MDHS 7 (Laboratory Method using X-Ray Fluorescence Spectroscopy), 1987.
70. Health & Safety Executive, *ibid.*, MDHS 8 (Colorimetric Field Method Using *sym*-Diphenyl thiocarbazone), 1987.
71. Health & Safety Executive, *ibid.*, MDHS 9 (Tetra-alkyl lead Compounds in Air. Personal Monitoring Method), 1981.
72. Health & Safety Executive, *ibid.*, MDHS 10 (Tetra-alkyl lead Compounds in Air. Continuous Monitoring), 1981.
73. Health & Safety Executive, *Lead and You*, Health & Safety Executive, London, 1989.
74. Health & Safety Commission, Approved Code of Practice, *Control of Lead at Work*, HMSO, 1985.
75. American Conference of Governmental Industrial Hygienists, Documentation of Threshold Limit Values, 5th edn, American Conference of Governmental Industrial Hygienists, Cincinnnati, OH, 1986.
76. Health & Safety Executive Technology and Pollution Division, Specialist

Inspector Reports, *Dust Control at Lead Crystal Glass Cutting*, Report No. 2, October 1986.
77. Anon., *Morbidity and Mortality Weekly Report*, 1991, **40**(47).
78. The Control of Lead at Work Regulations, 1980.
79. British Safety Council, *Safety*, 1982 (Sept), 14.
80. Health & Safety Executive, *Lead Works, Best Practicable Means, BPM 16*, 1985.
81. World Health Organisation, *Environmental Health Criteria 85: Lead–Environmental Aspects*, WHO, Geneva, 1989.
82. Merewether, E. R. A. & Price, C. W., *Report on Effects of Asbestos Dust on the Lungs and Dust Suppression in the Asbestos Industry*, Home Office, 17 March 1930.
83. The Asbestos Industry Regulations, 1931.
84. Ministry of Labour, *Problems arising from the use of asbestos*, 1967.
85. Report of Chief Inspector of Factories and Workshops, 1938.
86. Reports of Chief Inspector of Factories 1947 and 1954.
87. The Control of Asbestos at Work Regulations, 1987.
88. The Asbestos Product (Safety) Regulations, 1985.
89. *Asbestos Materials in Buildings*, Department of the Environment, 1983.
90. Health & Safety Executive, *Work with asbestos insulating board*, Guidance Note EH37, October 1984.
91. Health & Safety Executive, *Work with asbestos cement*, Guidance Note EH36 (revised) 1989.
92. Health & Safety Executive, *Probable asbestos dust concentrations at construction processes*, Technical Data Note 42, 1975.
93. Code of Practice on the *Toxicity and safe handling of rubber chemicals*, British Rubber Manufacturer's Association, 3rd edn., 1990.
94. Health & Safety Executive, *Rubber Health and Safety 1976–80*, HMSO, London, 1981.
95. Health & Safety Commission Rubber Industry Advisory Committee, *Control of dust and fume at two roll mills*, HMSO, London, 1986.
96. Health & Safety Executive, *Cleaning the air-dust and fume control in the rubber industry*, HMSO, London, 1987.
97. Health & Safety Executive Rubber Industry Advisory Committee, *Dust Control in Powder Handling and Weighing*, HMSO, London, 1989.
98. Cralley, L. J. & Cralley, L. V., *Industrial Hygiene Aspects of Plant Operations*, Vol. 2, Collier Macmillan, London, 1984.
99. Campbell, W. A., *Chemistry in Britain*, 1990 (June), 558.

CHAPTER 30

Toxic gases and vapours

Introduction

Fifty years ago, gas poisoning was considered a 'frequent hazard to industrial life'.[1] Conditions in early steel strip pickling plants using hydrochloric acid were, by modern standards, primitive:[2]

> '... the bad atmosphere that was prevalent in all steel pickling and galvanising plant buildings which were always open and draughty for the purpose of dispersing the acid fumes on heavy and foggy days. The visibility in such buildings was often as low as 20 ft. with a choking atmosphere of HCl fumes. It was claimed that men who worked in these atmospheres very seldom developed the common cold, but it was true to say they had no teeth either.'

Some 25 years ago four substances – hydrogen sulphide, oxides of nitrogen, benzol and carbon tetrachloride – were identified as 'chemical criminals' on the basis that they were at that time frequently encountered in laboratories, because of familiarity they were treated with indifference and used carelessly, and because – with the notable exception of carbon monoxide – they had resulted in more injuries and fatalities than any other compounds.[3]

An up-to-date review would probably exclude the last two – because of their substitution by less toxic alternatives – but include carbon monoxide, common 'oxygen-deficiency' producing gases in general, and asbestos dust. This exemplifies how toxic chemical risks, and perceptions of them, change with increasing knowledge.

> A review in the USA revealed that of 4,756 deaths investigated during 1984–86, 233 (5.0%) resulted from asphyxiation and poisoning, with the highest rate in the oil and gas industries.[4] Toxic gases were the largest group, followed by simple asphyxiants (Table 30.1).
>
> Of the 25 fatalities from carbon monoxide, six resulted from faulty pumps and compressors, five from running vehicles indoors, four from faulty space heaters, and two from industrial furnaces. Of the 30 deaths involving hydrogen sulphide exposure, eight were in sewers or manholes and nine in tanks or pits containing crude oil. Of the 35 deaths associated with solvent exposure, 24 were attributed to chlorinated hydrocarbons, and eleven to naphtha, toluene,

Toxic hazards

Table 30.1 Causes of asphyxiation and poisoning in death investigations in the OSHA IMIS Database 1984–86.*

Classification	No. of cases	Chemical
Simple asphyxiants	1	Propane
	10	Methane
	7	Argon
	18	Nitrogen
	12	Carbon dioxide
Toxic gases	3	Ammonia
	25	Carbon monoxide
	3	Chlorine
	30	Hydrogen sulphide
	4	Nitrous oxide
Oxygen deficient atmosphere (substance not reported)	27	Total
Solvents	1	Chlorodifluoromethane
	1	Coal tar pitch volatiles
	2	Dichlorodifluoromethane
	1	Diesel fuel
	3	Gasoline
	1	Isopropanol
	1	Methyl chloride
	4	Methyl chloroform
	8	Methylene chloride
	3	Naphtha
	1	Perchloroethylene
	2	Toluene
	4	Trichlorotrifluoroethane (F-113)
	2	Trichloroethylene
	1	Trichloroethane
Unidentified or report incomplete	37	Total
Mechanical	42	Total
Other	1	Dimethyl sulphate
	1	Maleic anhydride
	1	Sulphuryl fluoride
	1	Cyanide
	4	Hydrogen fluoride
	1	Phosphorus (yellow)
Table total	233	

* This table does not include the 190 asphyxiations due to trench cave-ins.

gasoline, diesel fuel, isopropanol and coal tar pitch volatiles. Nine of the 35 fatalities were in degreasing plant. Hydrofluoric acid burns have been reported to be frequent but rarely with fatal consequences. Of the four deaths from hydrogen fluoride exposure in the OSHA study, three involved exposure to large areas of the body during maintenance operations.

The present chapter supplements Chapters 5 and 6 by considering the properties, toxicology, sources and handling precautions for selected common gases and vapours encountered industrially. Clearly the general principles discussed for the handling of, and emergency procedures with, these specific chemicals are also applicable to less common materials.

It is apparent from earlier chapters that many different combinations of circumstances may result in undesirable instantaneous, intermittent short-term or continuous long-term emissions of toxic materials into the workplace.

The physical properties of the toxic gas or vapour, e.g. its water solubiity and density relative to air (page 95), are clearly important factors.

Any release of hydrogen fluoride will result in a dense plume which can hug the ground for considerable distances. Its effects on people include severe skin and lung burns.

An accidental release of HF from a refinery in Texas in 1987 resulted in the hospitalisation of 1000 people.[5]

Lists of occupational processes and operations giving rise to a wide range of environmental toxic contaminants are given in Tables 5.1, 6.3 and 6.4, whilst Table 5.15 identifies principal industrial toxicants giving rise to lung disease. The commonest potential sources are summarised in Table 30.2. In attempting to control exposures of personnel or neighbours, or of the environment, it is often important to allow for any given pollutant not being present in isolation. For example, potentially lethal concentrations of a mixture of gases may build up in silage towers or storage bins due to chemical reactions in harvested crops.

Very high concentrations of both carbon dioxide and oxides of nitrogen can build up following filling a tower silo with a green crop.[6] The evolution of a mixture of fumes and gases in welding, mixed gases in combustion, and vapours from mixed effluent streams are other common examples. 'Permissible' exposure levels and hence control measures must be modified accordingly. Furthermore all routes of entry require consideration.

Occupational leucoderma, involving chemical depigmentation of the skin, can be caused by phenolic compounds damaging or destroying the melanocytes. Workers engaged in the production of *p-tert*-butyl phenol suffered from depigmentation on both exposed skin and in strikingly symmetrical distribution on covered areas; this indicated absorption not only through the skin but also by inhalation and possibly by ingestion.[7]

Substances which are gaseous at room temperature are usually kept in closed systems and for workplace contamination to arise loss of containment, such as leaks or accidental venting, must occur. High value or highly toxic substances are tightly controlled whereas substantial leaks may be tolerated for gases such as carbon dioxide or nitrogen. Even here, however, asphyxiation could result in confined spaces as a result of oxygen displacement.

Toxic hazards

Table 30.2 Potential sources of toxic atmospheres in industry

- From improper storage, handling, use or disposal of specific chemicals
  e.g. leakages*
      improper venting or draining*
      'open' handling*
      incorrect notification on disposal
      'use' of wrong material
- 'Accidental' releases, spillages
  e.g. overfilling of containers
      transport incidents
      equipment failure
      unexpected reactions
      runaway reactions
- Admixture of chemicals
  e.g. by mistake (e.g. wrongly identified)
      in wrong proportions
      in wrong circumstances*
      in wrong sequence*
- In fires
  e.g. pyrolysis products
      combustion products*
      vaporisation
      through 'domino' effects*
- In 'confined spaces'
  e.g. improper isolation from residues
      oxygen deficiency (inherent or from rusting or from 'purging')
- Maintenance or cleaning of equipment
  e.g. residues
      loss of containment (breaking lines)
      stripping insulation
      burning-off paint/flame heating components
      reaction/vaporisation of cleaning agents
- From 'wastes'
  e.g. anaerobic break-down
      admixture of effluents or 'wastes'*
      open-handling of effluents
      from atmospheric venting
      'solid' wastes
      uncontrolled incineration
- From fabrication, manufacturing or machining operations, etc.
  e.g. welding fumes*
      spray painting*
      curing of paints
      use of adhesives*
      curing of adhesives
      cutting/grinding/fettling, shot-blasting*
      electroplating*
      degreasing/cleaning/etching/pickling*
      plastics 'forming' or overheating

* May result in 'long-term' exposures (i.e. throughout period of operation or in adjacent areas of workplace).

Oxygen deficiency

Although it is of considerable industrial importance, problems associated with oxygen deficiency due to depression of an atmosphere's oxygen content by an 'inert' gas (page 242) are not strictly classifiable as 'toxic gas' situations. Nevertheless since some of the situations, and precautions necessary, have some commonality they are considered here.

Oxygen deficiency resulting in asphyxia can be simply caused by consumption of oxygen by a chemical reaction in a confined space.

> A large steel evaporator used for magnesium chloride was shut down, drained and washed. Two men entered the vessel to perform maintenance and were ovecome due to lack of oxygen. A reconstruction of the accident revealed that, unexpectedly, the oxygen content of the atmosphere in the closed evaporator decreased from normal to about 1% in just under 24 hours. The oxygen was deduced to have been consumed by rapid rusting of the steel under humid warm conditions with traces of magnesium chloride.[8]

More often, however, it is caused by an excess of the common 'inert' gases, particularly in confined spaces. Responses at given concentrations of oxygen in the atmosphere are given on page 242.

The inert gases (e.g. nitrogen, helium, argon) have no colour, odour or taste and deficiency of oxygen, the effects of which are often rapid and insidious, can arise in numerous ways and incidents are commonplace.

> A fitter was replacing the manway cover on a vessel which contained nitrogen. He fell into the vessel and was killed.

> A man knelt by an open manway to recover a rope which was half inside the vessel and caught on something. The vessel contained nitrogen and he was overcome. Fortunately he survived.

> A plant manager was witnessing the testing of a water-sealed gasholder. Nitrogen was used to raise the bell. During this the water-sealed lute pot on the gas outlet blew in the pit around the base of the gasholder. He climbed down the ladder into the pit to turn on the water supply to seal the lute. Halfway down the ladder he felt weak, realised why, and was fortunately able to climb out.

> An experienced process supervisor was found unconscious with his head in the manway of a column, containing an oxygen-deficient atmosphere. Despite the application of artificial respiration, he was declared dead shortly afterwards.

> In a somewhat similar incident, a man was overcome when he put his head into the manway of a tank to simply observe the condition of the pump suction line which had been nitrogen-purged.

These incidents all involved nitrogen, which has probably been responsible for more gassing accidents than any other chemical. Simple rules for its safe use as an inert gas include:
- Provision of good ventilation in any room in which inert gases are handled or can enter.

- Prohibition of the use of nitrogen to drive power tools, etc., designed to work on compressed air.
- Clear recognition and identification of confined spaces.
- Prohibition on entry into a confined space until the atmosphere has been tested and confirmed to contain at least 21 per cent oxygen.
- When entry into an oxygen-deficient atmosphere is essential this should be only when wearing approved breathing apparatus and a harness with a full-time watcher outside, also provided with breathing apparatus.

The last two precautions are covered on page 983 but the full procedures, e.g. to comply with S.30 of the UK Factories Act, are given in ref. 9.

Unfortunately, however, accidents still occur following the misuse of nitrogen.

> A fitter put on a mask supposedly supplied from a breathing air main around the factory but in reality attached to a nitrogen line; he collapsed immediately. Fortunately, a colleague immediately removed the mask and the fitter survived.

> A contract employee was overcome when working inside a vessel. He had inadvertently connected a portable grinder to a nitrogen line.

> Two inspectors suffered from the effects of oxygen deficiency when they entered a reactor after a break in the work, during which an air hose had been used as a purge. The hose had mistakenly been connected to a nitrogen line.

This hazard can be overcome by the use of different plug-in connections for air and nitrogen lines and clear identification of lines, together with training and instruction never to use an air connector on a nitrogen line or vice versa. Whilst the above cases referred to exposure to nitrogen, similar incidents are repeatedly reported for the other 'inert' gases, for example:

> A welder was using electric arc welding gear inside a compartment of an 18 000-litre road tanker. The compartment was ventilated by an extractor fan. During the lunch break he switched off the fan and left the welding gear inside the compartment. The inerting gas, argon, leaked during this time and displaced the air. The welder returned to work but was soon seen to have collapsed and was lying face upwards displaying signs of sickness. He was hauled from the tank by the rescue team and eventually recovered. With the fan off it would have taken only 10 minutes for the air to become unbreathable.[10]

Unfortunately it is characteristic of such incidents that they commonly occur in 'confined spaces' (page 981) and tend to result in multiple casualties – often fatalities. For example, following an incident in a disused mineshaft at a London hospital in which one workman died and another was rendered seriously ill it was observed that it appeared to be,[11]

> '... yet another example of the classic case of entry into confined spaces without adequate precautions. In this case only one other person went in unprotected to attempt a rescue. On other occasions, there have been multiple fatalities as one man or more has been overcome in rescue attempts.'

Incidents frequently involve contractors or maintenance personnel:

> A contracts manager and a foreman welder were asphyxiated whilst attending to a pipe freezing job in a 1.8 m deep excavation. Two pipe freezing jackets had been clamped 1.8 m apart around a 15 cm hot water pipe in the excavation; these jackets were fed with liquid nitrogen from two 500-litre tanks. Cold, and therefore relatively dense, nitrogen gas was vented into the excavation near its base. The contracts manager apparently entered the excavation several times during the filling operation; the only precautions involved shutting off the supply of nitrogen and waiting 10 minutes until most of the mist had cleared in the excavation.
>
> Later he and a colleague who entered the excavation were overcome. The foreman welder jumped in to attempt a rescue; he was thrown a lifeline by a fourth man but collapsed before he could tie it on. The second man was then successfully rescued.
>
> (In this case the company were found guilty of a breach of S.2 of The Health and Safety at Work Act, it being found that an oxygen meter did not work, there was no breathing apparatus, no mechanical ventilation and no written system of work or rescue procedure.)

The manner in which oxygen-deficient atmospheres arise is not always so obvious and may be unexpected.

> Three men were allowed to enter a 4,000 gallon, empty lager tanker about to be MOT-tested at a haulage contractors. The tankers apparently usually arrived from the brewery full of water but on this occasion it was filled with nitrogen. The first man who entered to remove a sight glass, so that a hosepipe could be led in to add cold water, collapsed on the floor. A second man who went to help him also collapsed. Their foreman then entered to attempt a rescue but had to get out quickly. Luckily the men were rescued.[12]

It is precisely because the nature of the risk may not be immediately apparent that a formal system for entry, including inspection and testing, as outlined on page 981 is essential.

Carbon dioxide

Carbon dioxide is a colourless gas which is about 1.57 times denser than air. It is present in the atmosphere at a concentration of 300 ppm and it is a normal body constituent arising from cellular respiration. However, in high concentrations it is toxic, acting directly on the respiratory centres in the brain.

> A horizontal cylindrical vessel of 6,800-litre capacity was pressure-tested with carbon dioxide. A charge-hole cover joint was found to be leaking and this was therefore removed to remake the joint. In so doing a fitter's mate slipped and fell into the vessel which contained residual gas from the pressure test. Although rescued soon afterwards, he was found to be dead.[13]

Sometimes carbon dioxide may be evolved from chemical reactions, including fermentation processes, and accumulate at low level, e.g. in pits, sumps or mash tanks, because it is heavier than air (page 95).

An 8 m deep concrete pit was under construction on an old site beneath which lay disused acid lines and drains. The excavation was 'sweetened' with compressed air prior to entry but a man who entered it felt unwell. A concentration of 3% CO_2 was determined within 0.6 m of the surface and 20% at 1.8 m from the surface. The pit was pumped clear but the CO_2 had returned by the following day. It was attributed to reaction between leaked acid and carbonates in the ground.[14]

Whilst small quantities are transported as a gas in cylinders (page 627), liquid carbon dioxide may be stored in bulk for a variety of industrial uses, e.g.[15]
- heat transfer medium at nuclear power stations
- carbonation of beer and soft drinks
- raw material for the manufacture of chemical products
- inert gas for blanketing and purging
- binding agent in the sodium silicate process for the production of cores and moulds
- inert atmosphere for the welding of steel
- enrichment of the carbon dioxide concentration in glasshouse atmospheres for increasing the crop yield and advancing the cropping date
- fire extinguisher
- refrigerant for the freezing and chilling of foodstuffs
- tobacco processing
- pipe freezing
- solvent in certain solvent extraction processes.

Transportation and storage as a liquefied gas is generally at a temperature of $-17\,°C$ and a pressure of 20.7 bar gauge.

Toxicity

Exposure to an atmosphere containing approximately 5 per cent carbon dioxide results in laboured breathing and headache. At concentrations >5 per cent in air it has a slightly pungent odour and the above symptoms become more severe at concentrations of 7–10 per cent.[15] Exposure over 10 per cent for some time results in headache, visual disturbances and tremors, and may be followed by loss of consciousness. Death from respiratory failure may follow further prolonged exposure. This demonstrates that a dangerously high atmospheric concentration may be attained before displacement results in a sufficiently low oxygen concentration to cause collapse and death due to oxygen deficiency (page 242). Whilst it is possible for a person to withstand exposure to high levels of carbon dioxide, e.g. in submarines, the Occupational Exposure Limit of 5,000 ppm as an 8 h time-weighted average value[16] should normally be observed.

A van salesman entered a vehicle refrigerated by dry ice to recover some stock. Whilst he was inside, the door closed and latched shut. Despite the internal release mechanism being in perfect working order, he was overcome before he could reach it.[17]

(A rake was provided for stock recovery and this example serves to illustrate the need to check on the observance of procedures intended to reduce the risk of being affected by toxic gas, or oxygen deficiency in a confined space.)

Concentrations of the order of 20–30 per cent carbon dioxide are immediately hazardous to life.[15]

Fatalities have occurred amongst persons entering process tanks without appropriate checks after cleaning with solutions containing sodium hydroxide. This reacts with sugar residues and whey solids to generate carbon dioxide.[8]

Hydrogen sulphide

Hydrogen sulphide is used in the preparation of metal sulphides, oil additives, etc., in the purification and separation of metals, as an analytical reagent and as raw material in organic synthesis. Physical properties and effects of temperature on vapour pressure are summarised in ref. 18. Hydrogen sulphide is a dense, colourless, highly flammable water-soluble gas with an offensive odour of rotten eggs.

Its acute toxicity was described on page 248. In effect when the amount of hydrogen sulphide absorbed into the bloodstream exceeds that which is readily oxidised systemic poisoning occurs and respiratory paralysis may follow immediately. This situation corresponds to an atmospheric concentration of approximately 700 to 1000 ppm.[19]

Poisoning can occur almost without warning since odour may no longer be detectable due to olfactory fatigue. Unless breathing is stimulated or induced by artificial respiration death is likely to follow.

An operator was gassed and rendered unconscious whilst filling liquid hydrogen sulphide into cylinders. He had disconnected a filling line with the supply valve apparently still open due to an oversight and inhaled the gas whilst attempting to isolate the supply line valve. Fortunately this man, who had not been using respiratory protection, was rescued by a nearby foreman.[20]

'Massive' releases of hydrogen sulphide, e.g. any which could result in atmospheric concentrations >500 ppm, should obviously be avoided. Therefore amongst the situations to be avoided – by a combination of equipment design and an appropriate system of work – are the following (after ref. 19).

- Gas release from a cylinder due to clumsy manipulation.

    A laboratory assistant required 500 ml of sulphuretted water. She succeeded, after some difficulty, in opening the cylinder valve but unfortunately turned it too far; the gas pressure blew the rubber tube off the nozzle. Inhalation of gas resulted in loss of consciousness and she was found lying on the ground.[19]

- Addition of sulphides to acidic liquor.

    A workman inhaled hydrogen sulphide whilst processing scrap batteries for recycling at a precious metal recovery plant. During partial processing of the batteries at a different company, and unknown to the victim, they had already been treated with iron sulphide. When he added nitric acid, as part of the routine processing, H_2S was evolved. He was overcome and remained in a coma for several months; he was left totally paralysed, speechless and partially deaf. (The suppliers of the batteries were held to be responsible for the accident on the basis that the employers had a contract with them for the supply of materials of known composition with which their process could cope.)

    When sulphuric acid was added to a tank at a beauty products plant toxic gases were liberated from sulphur-containing contents which caused the evacuation of the workforce, many of whom had to be rushed to hospital as they choked and vomited. More than 90 were admitted and kept in for over 24 hours; three of the worst affected were taken to an intensive therapy unit with respiratory problems.[21]

    Six tannery workers were killed when a tank of sodium hydrosulphide was mistakenly emptied into a vat containing chromic acid. The first man collapsed and the others were exposed when they ran to his assistance.[19]

    A different, albeit somewhat similar, incident is described on page 245 involving a new batch of hides which had been wet-salted leading to the reaction

    $$Na_2S + 2H^+ \rightarrow H_2S + 2Na^+$$

    (There the system of work was changed subsequently so that every consignment of hides was examined. If acidity levels were high, sodium hydroxide was used to supplement the sodium sulphide solution.)

    A cleaner working in a storm water settlement pit removing silt was overcome by hydrogen sulphide. Having completed the job he removed a bung from a 0.3 m diameter drain leading into the pit from a water treatment plant. Some 10 000 litres of liquid containing hydrochloric acid flowed into the pit and reacted with sulphides in the silt to generate hydrogen sulphide. The man collapsed and subsequently drowned. (There was a written permit-to-work system for entry into confined spaces (page 983) but it was not implemented.)

    A further example is given on page 249. This problem may arise from the mixing of effluents, e.g. in drainage systems, and examples during liquid waste disposal are given on pages 848–9.

- Loss of containment on small-scale work due to breakage of glass equipment, hose failures or maloperation.

    Two laboratory chemists were affected by hydrogen sulphide whilst performing sulphide determinations on soil samples using a standard method, albeit with a 10 g sample compared with the 1 g recommended. The test, which involved the generation of hydrogen sulphide from the soil by the addition of hydrochloric acid in a flask and estimation of the gas by passing it through a Draeger tube (see page 362), was carried out on an open bench rather than within a fume cupboard. A violent reaction occurred and blew out the bottle top but the chemists only noticed the effects of hydrogen sulphide after some minutes.[21]

- Loss of containment on a larger scale, e.g. due to failure of glands, gaskets or valves.

    At Poza Rica, Mexico in 1950 hydrogen sulphide escaped for 20 to 25 minutes from a new plant designed to remove it, and produce sulphur, from natural gas; the cause was a valve problem. The gas did not disperse due to heavy fog, absence of wind and an atmospheric temperature inversion, but spread to the town. Some people were awakened by the characteristic stench but became unconscious whilst attempting to escape. In total 22 persons died and 320 were hospitalised; the survivors developed headaches, eye irritation, coughs, shortness of breath and temporarily lost their sense of smell.[22]

- An unsecured cylinder falling and shearing its valve.
- Agitation of decomposing materials liberating hydrogen sulphide gas by anaerobic bacterial action.

    In slurry stores on farms bacterial decomposition of slurry produces gases including carbon dioxide, methane and hydrogen sulphide.[23] Inhalation of such fumes in confined spaces resulted in eight fatalities in the UK over a period of only three years. Therefore slurry should be stored and emptied with no need for persons to enter the store.

- Unexpected by-products of reaction (see page 491) possibly due to changes in the proportions of reactants or their order, or rate, of addition.
- Unexpected desorption.

Fortunately major releases involving this gas appear to have been rare.

In October 1973 an oil well near Camrose, Alberta erupted and released natural gas containing 5% hydrogen sulphide into the atmosphere. The accident occurred at 15.45 and by 16.30 a large cloud had accumulated and was drifting towards the town at about 1.5 m/s. Some 800 people were evacuated from their homes before the well was capped at 18.00 on the next day.

Where there is some potential for hydrogen sulphide escape or formation it is inadvisable for single persons to work in isolated situations. If a person collapses then, even if the air clears, breathing may not restart voluntarily. Rapid help and artificial respiration are essential. Adequate training and instruction is crucial since in a 'massive release' persons tend not to appreciate the danger, because they either lose their sense of smell or simply overestimate their ability to cope. Therefore operators attending to leaks and rescuers must wear respiratory protection. Experience has shown that, as with incidents involving oxygen deficiency in confined spaces (see page 982), would-be rescuers may make up a very high proportion of the casualties.

> A manager was seen lying on the floor near a light oil tank. One man who rushed to his aid was overcome by the fumes. A further six people were subsequently overcome during their attempts to rescue the first two victims. (This incident was precipitated by the tank manhole being left open; the oil contained mercaptans and emitted traces of hydrogen sulphide.)[19]

Cylinders are typically protected from overpressurisation by frangible gold-plated discs and fusible plugs.

Important precautions include, in addition to those in Table 20.9,

- Use in well-ventilated conditions and eliminate sources of ignition.
- Operators should work in pairs.
- Do not rely on the sense of smell to detect hydrogen sulphide. Leak-strips of wet lead acetate paper turn black on exposure to hydrogen sulphide and offer a simple indicator, as do colour indicator tubes. For plant-scale operations instrumental multi-point detectors and alarms are likely to be more appropriate.
- Segregate cylinders of hydrogen sulphide from oxygen or other highly oxidising or combustible materials.
- Ground all lines and equipment used with hydrogen sulphide.
- Insert traps in the line to prevent suck-back of liquid into the cylinder.
- Provide respiratory protection for emergencies.
- In the event of exposure apply first-aid as indicated in Table 30.3.

Table 30.3 First-aid measures for toxic gas exposure

OBTAIN MEDICAL HELP IMMEDIATELY	
Inhalation	Remove victim to uncontaminated area and effect artificial respiration if breathing has stopped. In the case of *hydrogen sulphide*, ensure that the victim rests and refrains from exercise for 24 hours. For *chlorine* gassing, lay victim on stomach with head and shoulder slighly lowered. Discourage from coughing
Skin contact	Use emergency shower, removing contaminated clothing and shoes at the same time
Eye contact	Wash promptly with copious amounts of water for at least 15 minutes

Carbon monoxide

Carbon monoxide is produced by steam reforming or partial oxidation of carbonaceous materials. It is primarily produced for on-site process applications but it is available commercially in cylinders in a range of purities and finds use as a metallurgical reducing agent, a feedstock in the manufacture of a range of key chemicals, as described later, and an ingredient of gaseous fuels (producer gas and water gas, which have been largely replaced by natural gas). More recently, carbon monoxide emission from vehicle exhaust has been recognised as an important air pollutant. It is notorious for its acute toxicity (see page 243).

> In 1982 a barge on the River Moselle collided with the support for an overhead pipeline carrying carbon monoxide. The line was severed and the released gas caused five fatalities, possibly because of the lack of warning properties.[24]

Important physical properties are listed in Table 30.4. Carbon monoxide

Table 30.4 Physical properties of carbon monoxide

Property	Value
mol wt	28.011
m.p.	68.09 K
b.p.	81.65 K
ΔH, fusion (68 K)[a]	0.837 kJ/mol
ΔH, vaporisation (81 K)[a]	6.042 kJ/mol
density (273 K, 101.33 kPa (1 atm))	1.2501 g/litre
s.g., liquid, 79 K[b]	0.814
s.g., gas, 298 K[c]	0.968
critical temperature	132.9 K
critical pressure	3.496 MPa (34.5 atm)
critical density	0.3010 g/cm^3
triple point	68.1 K/15.39 kPa ($-205°$C/115.4 torr)
$\Delta G°$ formation (298 K)[a]	-137.16 kJ/mol
$\Delta H°$ formation (298 K)[a]	-110.53 kJ/mol
$S°$ formation (298 K)[a]	0.1975 kJ/(mol.K)
$C_p°$ (298 K)[a]	29.1 J/(mol.K)
$C_v°$ (298 K)[a]	20.8 J/(mol.K)
auto-ignition temperature	925 K
flammability limits in air[d]	
upper limit, %	74.2
lower limit, %	12.5

[a] To convert J to cal, divide by 4.184.
[b] With respect to water at 277 K.
[c] With respect to air at 298 K.
[d] Saturated with water vapour at 290 K.

is a colourless, odourless, tasteless, flammable gas burning readily in air (with a flammable range of 12.5–74 per cent) or oxygen and with ignition temperatures of 644–658 °C and 637–658 °C, respectively.

Because the gas is undetectable to man, it has resulted in multiple casualties in situations when its presence is not anticipated.

> Two construction workers were going to continue repairs to the roof of a refinery compressor hall at around 8.00 a.m. A refinery employee showed them in, gave them the works labour pass, and moved the bridge crane into position. The oxygen and nitrogen compressors had been shut down two days previously.[25]
> One worker climbed up onto the crane and at 8.20 a.m. was seen by his colleague to have collapsed. The works fire brigade were called and arrived

four minutes later; the works doctor arrived at 8.27 a.m. Some of the firemen rushed directly to the victim, on the assumption that he had suffered an electric shock; others made up a pulley and apparatus ready for the rescue.

At around 8.32 a.m. several firemen collapsed on the bridge crane. The gas alarm was sounded, the available firemen recalled from the bridge crane, the hall ventilators switched on and all the compressed air connections opened. At 8.40 a.m., 600 ppm CO was recorded by test gauges at 4 m above the floor. Minutes later the rescue operations were continued by firemen wearing breathing apparatus. Assistance was provided by the local voluntary fire brigade and two neighbouring works brigades and rescue was completed by 9.30 a.m.

Unfortunately the construction worker subsequently died in hospital. Four firemen were released after medical treatment on the day of the accident but five were hospitalised for two days.

Opposite the hall were several plant units in which synthesis gas (28% CO and 70% H_2) was produced; all were shut down except the nearest unit. The only connection to this unit was via a two-sided cooling water outlet. The synthesis gas from the outlet valve of a standing cylindrical condensate separator entered the funnel-shaped outlet lying beneath it. The synthesis gas therefore penetrated the cooling water pipe system with injector action as a gas–air mixture. It was able to leave from the funnel-shaped outlets in the hall because (a) the cooling water was not flowing against it and (b) the high hydrogen content gave it natural lift. (In fact the CO and H_2 concentration was particularly concentrated close to the hall ceiling.) (Subsequently the cooling water systems of the plants were physically separated to avoid outbreaks of gas from the adjacent unit.)

Toxicity

The acute toxicity of carbon monoxide is well established. It is absorbed into the bloodstream via the lungs and replaces oxygen by forming carboxyhaemoglobin. This reduces the oxygen-carrying capacity of the blood; the dissociation of oxyhaemoglobin is also affected so that oxygen supply to the tissues is further reduced. The symptoms of exposure depend upon the degree of saturation of the haemoglobin with carbon monoxide. Symptoms at various saturated levels are summarised on page 243. The relationship between the atmospheric concentration of carbon monoxide and blood carboxyhaemoglobin level in exposed workers is summarised in Table 30.5.[26]

This demonstrates, for example, that a headache is likely to be associated with one or two hours of heavy work in an atmosphere containing 200 pm carbon monoxide or four to eight hours at rest. The body may be rendered more susceptible to the effects of oxygen deprivation associated with even low carbon monoxide levels due to pre-existing severe respiratory or cardiovascular disease. Other populations, e.g. those living at high altitudes, young children or pregnant women, may also have an increased susceptibility.

Table 30.5 Per cent carboxyhaemoglobin levels in the blood of exposed workers versus time at various carbon monoxide in air concentrations[26]

Time	Air concentration											
	220 ppm			100 ppm			75 ppm			50 ppm		
	S*	L*	H*	S	L	H	S	L	H	S	L	H
15 min	1.8	3.5	5.2	1.2	2.0	2.8	1.0	1.6	2.2	0.82	1.2	1.6
30 min	3.1	6.2	9.2	1.8	3.3	4.8	1.5	2.6	3.7	1.1	1.9	2.6
45 min	4.3	8.7	12.6	2.4	4.6	6.5	1.9	3.5	4.9	1.4	2.5	3.4
60 min	5.5	11.0	15.5	3.0	5.7	7.9	2.3	4.3	6.0	1.7	3.0	4.1
90 min	7.7	14.9	20.2	4.0	7.6	10.2	3.1	5.8	7.7	2.2	4.0	5.2
2 h	9.7	18.1	23.7	5.0	9.2	11.9	3.9	7.0	9.0	2.7	4.7	6.1
4 h	16.3	26.2	30.4	8.3	13.2	15.3	6.3	10.0	11.5	4.4	6.9	7.7
6 h	21.1	30.0	32.4	10.7	15.1	16.2	8.1	11.3	12.2	5.5	7.6	8.2
8 h	24.5	31.7	32.9	12.4	15.9	16.5	9.4	12.0	12.4	6.4	8.0	8.3
24 h	32.7	33.2	33.2	16.5	16.7	16.6	12.4	12.5	12.5	8.4	8.4	8.3
∞	33.4	33.2	33.2	16.8	16.7	16.6	12.7	12.5	12.5	8.5	8.4	8.3

Time	35 ppm			25 ppm			10 ppm			5 ppm		
	S	L	H	S	L	H	S	L	H	S	L	H
15 min	0.72	1.0	1.3	0.66	0.84	1.0	0.55	0.61	0.67	0.52	0.54	0.56
30 min	0.93	1.4	1.9	0.80	1.2	1.5	0.61	0.72	0.82	0.54	0.57	0.60
45 min	1.1	1.9	2.5	0.95	1.4	1.9	0.66	0.81	0.95	0.56	0.61	0.64
60 min	1.3	2.2	3.0	1.1	1.7	2.2	0.71	0.90	1.1	0.58	0.63	0.68
90 min	1.7	2.9	3.7	1.3	2.1	2.7	0.80	1.1	1.2	0.62	0.69	0.74
2 h	2.0	3.4	4.3	1.6	2.5	3.1	0.89	1.2	1.4	0.66	0.73	0.78
4 h	3.2	4.7	5.4	2.4	3.4	3.9	1.2	1.5	1.6	0.77	0.84	0.86
6 h	4.0	5.4	5.7	2.9	3.9	4.1	1.4	1.6	1.7	0.85	0.83	0.88
8 h	4.5	5.7	5.8	3.3	4.1	4.2	1.5	1.7	1.7	0.91	0.91	0.89
24 h	5.9	5.9	5.9	4.3	4.2	4.2	1.9	1.8	1.7	1.05	0.93	0.89
∞	6.0	5.9	5.9	4.4	4.2	4.2	1.9	1.8	1.7	1.06	0.93	0.80

* S = sedentary subjects, L = light physical work, H = heavy physical work.

The amount of carboxyhaemoglobin in the blood depends upon
- concentration of carbon monoxide in inspired air;
- duration of exposure;
- respiration rate, i.e. degree of activity of exposed person;
- individual susceptibility.

In an uncontaminated environment carbon monoxide is eliminated solely via the respiratory system. Elimination rates on average vary between 30 per cent and 50 per cent per hour, depending on the amount of uncontaminated air which is breathed.

It has been suggested that long-term exposure to elevated levels of carbon monoxide may be associated with atherosclerotic cardiovascular disease.[26]

The approved Occupational Exposure Standard for carbon monoxide in the UK is 50 ppm (55 mg/m^3) calculated as an 8-hour time-weighted average concentration. The Short-term Exposure Limit is 300 ppm (340 mg/m^3) calculated as a 10-minute time-weighted average concentration.

Correlation of biological levels with atmospheric exposures is, as mentioned earlier, complicated and other factors, e.g. smoking habits, may increase the carboxyhaemoglobin base line.

It has been suggested by the American Conference of Governmental Industrial Hygienists that a carboxyhaemoglobin level of up to 10 per cent is acceptable for healthy workers.[26] However, exposure should be limited to a maximum of 4 per cent for workers with pre-existing disease, e.g. cardiovascular effects.

Sources

Inhalation of methylene chloride, a common constituent of paint strippers, may serve as an unexpected source of carbon monoxide exposure. This solvent is metabolised to CO; a 3-hour exposure even in a 'well-ventilated room' can produce a carboxyhaemoglobin saturation of 16%.[27]

Carbon monoxide is frequently encountered as a combustion product; it is also used in some chemical and metallurgical processes.[28] It is also produced when coal smoulders deep inside large tips or bunkers.

Most plastics produce carbon monoxide on decomposition by heating (plus other complex compounds) and minor amounts can be produced during plastic moulding. The use of certain explosives in, e.g. tunnels and quarries, can also result in carbon monoxide generation.

Chemical processes

Carbon monoxide is used in the manufacture of ammonia from nitrogen and hydrogen. When steam and methane are catalytically converted under high temperatures in the reformer, carbon monoxide, carbon dioxide and hydrogen are produced. Tight containment is crucial to avoid worker exposure and sensors and warning devices are required to indicate leaks, when self-contained breathing apparatus should be worn until it can be demonstrated that areas are safe. When reactors are opened nickel carbonyl exposure is a possibility and purging and respiratory protection become important.

Carbon monoxide finds use in the industrial synthesis of several important organics such as (1) methanol, (2) oxo alcohols and aldehydes, and (3) acrylic acid, thus:

(1) $CO + 2H_2 \xrightarrow[3,000 \text{ lb/in}^2]{Cr_2O_3 + ZnO, 350-400 \,°C} CH_3OH$ (almost quantitative)

(2) $RCH_2\overset{R}{\underset{|}{C}}=CH_2 + CO \xrightarrow[125\,°C, 3,000 \text{ lb/in}^2]{H_2, [Co(CO)_4]_2} RCH_2\overset{R}{\underset{|}{C}}HCH_2CH_2OH$

(3) $HC \equiv CH + CH_3OH + CO \rightarrow CH_2 = CHCOOCH_3$

Also, its catalytic reaction with water is used in the preparation of high-grade hydrogen:

$$CO + H_2O \rightarrow H_2 + CO_2$$

whilst the product of its reaction with chlorine is phosgene, a valuable, albeit toxic, intermediate used in the industrial manufacture of, e.g. isocyanates (see page 883):

$$CO + Cl_2 \xrightarrow{\text{activated carbon}} COCl_2$$
(quantitative)

To illustrate the precautions, in the oxo-process the entire process is conducted in a closed system and employee exposure to carbon monoxide is therefore limited to leaks around pump seals, sampling and maintenance (e.g. line breaking, vessel entry, pump repairs, etc.). The use of respiratory protection is important when breaking lines unless after purging with air, steam or inert gas, environmental monitoring indicates the absence of toxic material.

Carbon monoxide may be generated in other processes as a by-product, e.g. in the catalytic cracking of hydrocarbons. In such situations potential for exposure may arise near gas generating and purification plant, reactors, gas discharge treatment and venting equipment, etc.

A chemist collapsed whilst attempting to carry out tests on gas cleaning equipment. He subsequently died.[29] Together with a colleague, this man had climbed up to a sampling platform where there was a bleeder pipe discharging 'carbon monoxide' about 1.2 m above their heads. Respirators were at hand but there was no procedure requiring their use. Both men began to feel unwell and it was as they were getting down that the one collapsed due to inhaling the toxic fumes. (At this time a document had been published advising that certain operations, including the purging of gas, be done under strict safety standards.)

Inert gas for blanketing (page 132) may be generated by combustion. Carbon monoxide may be present in such gas in significant concentrations; hazards may hence arise at the gas generator or from the distribution and venting system.

Significant quantities of carbon monoxide may be produced in the manufacture of carbon black and in certain graphitising operations. Hazards may arise mainly near filter units and venting equipment.

Metallurgical processes

Carbon monoxide has proved invaluable in the purification of certain metals. Thus in the production of nickel the metal is converted to the carbonyl derivative for purification by distillation and dissociation back to pure metal (the 'Mond' process):

$$Ni + 4CO \underset{180\,°C}{\overset{50\,°C}{\rightleftarrows}} Ni(CO)_4$$

In the production of lead, galena is first oxidised and the resulting oxide reduced by carbon monoxide to elemental lead:

$$2PbS + 3O_2 = 2PbO + 2SO_2$$
$$PbO + CO = Pb + CO_2$$

Employees around the blast furnaces must wear appropriate respiratory protection. Typical engineering controls include automation of furnace charging operations, ventilation at the top of the furnace and enclosing and ventilating the settler, lead pot and slag pot. Some blast furnaces have automatic carbon monoxide sensors linked to warning alarms and activation of ventilation systems.

Aluminium is produced electrochemically in a carbon-lined cell by the Hall–Heroult process which liberates carbon monoxide at the anode. Pot emissions of carbon monoxide (and particulate fluorides) are controlled by ventilated hoods. Control against exposure is required during maintenance of the fume collection system.

Carbon monoxide is formed as a process by-product in the steel industry, and is subsequently used as a fuel.

> During a test run of a new electric-arc furnace and combustion gas tank, gas containing 80% to 90% carbon monoxide was fed from the tank to a boiler at the heating plant. Dirt caused the burner's shut-off valve to cut off the gas supply resulting in a temporary increase in pressure, up to 0.6 bar, in the pipeline. A water lock had mistakenly been left in the line between the tank and burner without being blanked off. The water lock, with an initial level of 500 cm, was emptied by the pressure surge and gas entered the sewerage

system. As a result eight persons sustained carbon monoxide poisoning in little more than two hours. The first victim, who was worst affected, remained unconscious for three days but was able to return to work within two months.[30]

A crucial mitigatory measure in this case was the ability of workers who found the first victim to administer mouth-to-mouth resuscitation and the availability of an emergency oxygen apparatus within a couple of minutes.

The main hazard areas include blast furnaces, basic oxygen steelmaking converters, coke ovens, gas cleaning and distribution plant, power generating plant, soaking pits and electric arc furnaces. In the latter carbon is removed from the melt by burning off as carbon monoxide above the slag. Because of the scale, and specialist processes, procedures involved for hazard reduction are dealt with by Industry Codes of Practice.

Hazards may arise in iron foundries around cupola furnaces and in associated equipment. Of the gases emitted from cupolas 20-30 per cent comprises carbon monoxide. Long-term average CO exposures of approximately 10 ppm are typically found at charge platforms but short-term exposures >500 ppm have arisen on charge platforms and spark arrester platforms. When cupolas are operated in pairs the CO produced by one of them may enter the other if it is out of service; this must be considered when arranging for inspection and maintenance work.

> During normal operation in iron foundries the cupola stack gases are likely to contain 13% to 17% carbon monoxide but in certain conditions the concentration may be as high as 20%. During lighting up and before the blast is turned on concentrations in the well of the cupola may be as high as 35%. Several fatalities have arisen during maintenance and operation in the past and safe working practices and emergency procedures have been published to avoid a recurrence.[31]

Combustion of carbonaceous materials in moulds and cores may result in typical concentrations of 25-60 ppm carbon monoxide in the atmosphere of the casting floor during metal casting; this is in addition to other pollutants, e.g. sulphur dioxide and sulphur trioxide.

Carbon monoxide may be used as a component in reducing or inert atmospheres in metal treatment furnaces. Potentially hazardous atmospheres may exist at the gas generators, at the furnaces, and at the furnace venting stacks. Ventilation ducts should be inspected regularly for obstructions, and metering equipment, piping, and burners need attention to prevent leaks. The decomposition of graphite electrodes in aluminium refining also generates carbon monoxide.

Heating operations

All heaters fuelled by liquid petroleum gas (LPG) or natural gas require an adequate supply of fresh air to ensure complete combustion and to minimise carbon monoxide formation. Operation of flueless equipment, with which the products of combustion are recirculated into the general atmosphere of a room, also need an adequate supply of make-up fresh air; this is to ensure an adequate supply of oxygen for efficient combustion and to dilute the products of combustion to a satisfactory, safe level.

Flueless heaters have presented problems in ill-ventilated rooms, e.g. domestic premises and construction site huts and caravans.

> A security guard was asphyxiated by carbon monoxide emitted from an LPG-fuelled gas brazier used for heating a construction site office. The brazier was of a type unsuitable for use indoors and was poorly maintained in that the gas input jet was partly blocked with mud; this resulted in excessive generation of carbon monoxide.

Some water heaters may permit flame impingement on cold surfaces, e.g. water tubes, which can cause excessive carbon monoxide formation.

If flueless heaters are used in sealed rooms, carbon monoxide concentrations of several hundred ppm – with high levels of carbon dioxide and depleted levels of oxygen – can soon be generated. Inefficient burner design or poor maintenance may result in excessive CO formation under 'normal' operation of gas-fired appliances. Leaks from flues, or re-entry of combustion products into buildings, may result in a toxic hazard as exemplified on page 243.

Thus in the UK there are stringent duties imposed upon those who install and maintain gas appliances, except in mines or factories, under The Gas Safety (Installation and Use) Regulations, to ensure that the flue and supply of combustion air are adequate, that the appliance can be used without constituting a danger, and that there is adequate room ventilation. (Reference should be made to the complete statutory instrument which imposes other comprehensive duties.)[32]

Clearly flue gases from any carbonaceous fuels may result in a hazard – either in the stack or at its outlet, or as a result of blow-back through any boiler or furnace – whether in hot water, steam or electricity generating plant or in refuse disposal incineration.

Engines

Internal combustion engines fuelled by gasoline, LPG or diesel oil may generate significant quantities of CO. This depends upon engine design, carburation, engine load, operating temperature, level of maintenance, etc. Exhaust gases of petrol-driven combustion engines contain 4–10 per cent CO whilst a diesel engine exhaust contains about 0.1 per cent when the engine is operating properly. (Other toxic fumes and gases include carbon dioxide, oxides of nitrogen and possibly hydrocarbon particulates and lead compounds.)

Care is clearly necessary in locating engines in factory premises, to ensure adequate removal of exhaust gases, and indeed in the open air.

In commercial air range diving associated with the offshore petroleum industry sources of compressed air are used to supply breathing systems through umbilicals or from cylinders. The compressors, and other engines, will be exhausting in the vicinity; the most common cause of CO contamination is therefore simply the positioning of air intakes downwind of exhausts.[33]

The problem of vehicle emissions is likely to be particularly acute in any confined spaces.

> Two sewage workers died whilst attempting to use a petrol-driven pump to empty a flooded well at a pumping station. An electric submersible pump which would have been perfectly safe to use was out of commission. The accident was caused by the emission of carbon monoxide within the confined space.

When it is essential to run petrol engines indoors, e.g. during garage repairs, the garage doors should be opened to provide dilution ventilation and the vehicle exhaust should be connected directly to an exhaust ventilation manifold. Dangerous exhaust gases from fork-lift trucks in warehouses can be minimised or reduced by the use of trucks powered by electricity or propane.

Hidden dangers

Care must be taken to avoid insidious sources of carbon monoxide which could pose health or explosion risks. For example, carbon monoxide may be liberated during the storage of carbon black (used in the rubber industry, see Chapter 29) in confined spaces. Before entry into storage areas therefore, levels of carbon monoxide should be demonstrated to be below the Occupational Exposure Limit (50 ppm) and the oxygen level should not have been depleted. Dimethyl formamide, a common dipolar aprotic solvent, undergoes disproportionation to carbon monoxide and dimethylamine particularly at its boiling point.[34]

$$HCON(CH_3)_2 \xrightarrow{\Delta} CO + HN(CH_3)_2$$

Similarly formic acid decomposes to carbon monoxide and water

$$HCOOH \rightarrow CO + H_2O$$

The rate of decomposition, K_v is given by

$$K_v = x/t$$

where x is the mole fraction which decomposes and t is the time. At 25 °C K_v for formic acid is approximately 13×10^{-10} min^{-1}.

> Slow decomposition in storage of 98–100% formic acid with liberation of carbon monoxide led to the rupture of the sealed glass containers. (Without venting a full Winchester of formic aid could develop a pressure of 7 bar after a year at 25 °C). A full bottle of 96% formic acid burst when the ambient

1720 Toxic hazards

temperature fell to $-6\,°C$ because of expansion of the contents on freezing and the pressure build-up from previous partial decomposition to carbon monoxide.[35]

Two tanks of 930 m³ and 570 m³ capacity aboard a ship were loaded with 900 tonnes and 600 tonnes 98% formic acid, respectively. After the 35-day journey the cargo was off-loaded and the tank washed with sea water at about 70 °C for 1 hr followed by a cold 30 minute rinse. The tank manhole was opened and a crew member entered to remove residual water. He quickly got into difficulty and a companion wearing a demand valve self-contained breathing apparatus entered the tank, but he too soon experienced difficulties. When both men were eventually removed they were found to be dead.[36]

With between 68–81% of the saturation level of carbon monoxide in the haemoglobin of the blood (see page 243), death was attributed to carbon monoxide poisoning. Using K_v from above, 1,282 moles of carbon monoxide could have formed in the 930 m³ tank. After off-loading this would give rise to an atmosphere containing up to 3.1% by volume of carbon monoxide. Similarly the small tank could have contained up to 3.4% carbon monoxide. (Levels of 0.5–1% are considered fatal.)

The respiratory protection was not of the positive pressure type and the wearer had a heavy growth of beard. This allowed the contaminated atmosphere to be drawn into the face-piece rather than fresh air through the demand valve of his breathing set. Besides use of adequate personal protection, the tanks should have been vented after washing and the tank atmosphere analysed for carbon monoxide before operators were permitted to enter, following the correct procedures for entry into confined spaces as discussed in Chapter 17.

General

Wherever CO is handled or stored the area should be well ventilated. Where exposure risks are high, levels should be monitored by frequent grab samples or continuous background multipoint samplers as appropriate. Biological monitoring of blood carboxyhaemoglobin levels, taking account of smoking, may also be valuable. Data from monitoring and inspection schedules should be recorded. On filling or on receipt of CO containers the vessel should be checked for leaks, and thereafter say quarterly. Areas should be prominently signed to warn of dangers, and self-contained breathing apparatus should be available for emergencies. Staff should be forewarned of the dangers and trained in proper handling and emergency procedures.

Sampling and analysis

The range of detectors for continuous atmospheric monitoring include infra-red spectrophotometers (page 366), semiconductor devices (page 379) and electrochemical sensors (page 381). Simple colorimetic detector tubes (page 362) are available for assessment of short-period and time-weighted average exposures. Alarm and monitoring instrumentation is commercially available.

Standard methods of biological monitoring, spectrophotometry, gas chromatography, etc., may be used to determine carboxyhaemoglobin levels in blood samples (page 243).

Suitable short-term sampling strategies may be used to identify peak exposures and to help assess acute gassing incidents. However, it may be difficult to use such results to assess time-weighted average long-term exposures unless a sufficient number of grab samples are taken (page 401).

Precautionary measures

The prime aim when a carbon monoxide hazard has been positively identified is to ensure that personnel are exposed to the lowest concentration that is reasonably practicable; in any case exposure concentrations should not exceed the recommended Exposure Limit. The effectiveness of the engineering control measures adopted for this objective can be checked by atmospheric and/or biological monitoring.

Substitution

One control strategy (pages 278 and 1736) is to alter equipment or processes so that no carbon monoxide is generated or used, and without increasing risks from other hazards. For example, flueless gas-fired heaters may be replaced by electric heaters, or vehicles driven by internal combustion engines (e.g. fork-lift trucks) may be replaced by battery-driven designs. This may be the optimal strategy when fume-containing carbon monoxide is produced in confined spaces and ventilation is difficult or prohibitively expensive to engineer (e.g. in certain warehouses).

Total enclosure

Total enclosure of a process or equipment (page 286) can avoid the emission of carbon monoxide into a working area. This would include, for example, the use of gas-fired indirect heaters – including balanced flue-types – which prevent combustion products venting internally. Exhausts from stationary internal combustion engines should always be vented to the open air.

Many chemical and metallurgical processes are designed as closed systems. The effectiveness of the enclosure does depend, however, on the reliability of, and quality of maintenance upon, the plant, e.g. mechanical seals.

Local exhaust ventilation/treatment

If a source of carbon monoxide cannot be effectively eliminated or contained, gas may be removed and prevented from entering a working environment using local exhaust ventilation situated close to the source (pages 285 and 676). The carbon monoxide and other pollutants may then be treated or vented safely into the open air.

Motor vehicle engine testing facilities may be equipped with exhaust fume extraction equipment preferably coupled directly to the engine exhaust pipe. In certain circumstances, when a vehicle is mobile, it may be possible to reduce emissions by means of proprietary emission control units. However, as carbon monoxide is not the only product of engine exhaust fumes, care must be taken to ensure that the concentration of the other contaminants will also be reduced to a safe level.

General ventilation

General, i.e. dilution, ventilation may be practicable in some situations to reduce an atmospheric carbon monoxide concentration to well below the appropriate hygiene standard. Hence, flueless gas-fired heaters or mobile internal combustion engines may be used safely inside buildings with effective ventilation. This is a common economic solution to some problems, e.g. in vehicle garages working with the garage doors open, but routine monitoring of contaminant levels at peak emission rates is recommended to check the adequacy of the ventilation system and, e.g. during winter when the temptation is to close the doors for comfort.

Gas monitoring

Audible and visual warnings of carbon monoxide gas leaks, or accumulations, may be provided by detectors. These alarms can be interlocked so as to isolate or shut down plant or to initiate evacuation/emergency procedures.

Oxygen depletion sensors or flame vitiation detectors can be fitted to gas-fired appliances. These automatically shut off the gas supply if the oxygen content of the inlet air is significantly reduced, hence preventing the production of dangerous amounts of carbon monoxide. Such devices should be fitted to all LPG-fired space heaters, excluding catalytic heaters, for domestic use. Other flueless heating appliances for indoor use should preferably also have oxygen depletion sensors.

Systems of work

As with other toxic gas hazards, satisfactory safe systems of work need to be established and maintained, e.g. through training and supervision. As discussed on page 984 a rigorous permit-to-work system is required for entry into confined spaces.

Personal protection

If control of a carbon monoxide hazard cannot be assured so far as is reasonably practicable, respiratory protective equipment may be required, e.g. for occasional access to certain areas in steel works, in some circumstances until an atmosphere has been tested and isolated, and during firefighting. The equipment should be selected to provide an adequate degree of protection and be suitable for the circumstances of use (page 294). As for personal respiratory equipment generally, arrangements should be made for servicing and inspection, and employees should be properly trained in its use.

First-aid

Any person affected by inhalation of carbon monoxide-contaminated air should be examined by a doctor as soon as possible. Until then the treatment summarised in Table 30.6 is recommended.

Table 30.6 First-aid treatment for any person affected by carbon monoxide

- Carry the patient to a warm uncontaminated atmosphere and loosen clothing at the neck and waist
- *If unconscious:* begin artificial respiration at once, preferably by the mouth-to-mouth method. If breathing has ceased continue artificial respiration until breathing recommences or until told to stop by a doctor
- *In other cases:* keep the patient at rest, avoiding unnecessary exertion. After recovery, the patient should be taken home, and not allowed to walk there

Hydrogen cyanide

Hydrogen cyanide is a colourless volatile liquid (b.p. 26.5 °C); the vapour density is 0.93 so that it tends to rise in air. The characteristic odour of bitter almonds does not necessarily provide good warning of its presence since it is one of the most rapidly acting poisons known. The vapour is also flammable and the compound can polymerise explosively (page 1484).

The vapour has been widely used in the fumigation of ships, buildings, orchards and various foods. (Within the UK use for fumigation is now controlled by COSHH Regulation 13, with the exception of fumigations carried out for research or in fumigation chambers or in the open air to control or kill mammal pests.) Hydrogen cyanide is also used in the production of nitriles and certain monomers, e.g. acrylates and methacrylates, and as a chemical intermediate. Together with traces of isocyanates, and of course carbon monoxide, it is evolved in fires involving foamed polyurethane plastics.

> In some factory fires employees and firemen appear to have suffered long-term effects from inhaling smoke and fumes not characteristic solely of CO exposure. This may be explained by the additional hazard from hydrogen cyanide and isocyanates.[37]

Absorption of hydrogen cyanide vapour via the respiratory tract is very rapid and the liquid and concentrated vapour are absorbed directly through intact skin. Therefore the current Maximum Exposure Limit of 10 ppm in the UK is afforded on 'Sk' notation, explained on page 260.

> A process supervisor was preparing a pipeline for maintenance by clearing a small amount of hydrocyanic acid from it and into a storage tank. The acid drained back into the area where he was working and 50 kg was accidentally released. He was overcome by the fumes but most of the 200 people on site reached toxic refuges, i.e. any building which can be enclosed and containing a telephone.[38]

> A leak of liquid HCN occurred to atmosphere due to failure of a bonnet gasket on a 20 mm drain valve whilst an operator was preparing to drain it from a drum to a closed-loop drainage system. He retreated rapidly to the adjacent control room, without isolating the leak, and subsequently collapsed. It was deduced that an oversized round mechanical pump seal had been used as the bonnet gasket instead of cutting one from polytetrafluoroethylene (PTFE) sheet. Moreover it was brittle apparently due to polymerisation of HCN. (Specific recommendations were subsequently made regarding the use of PTFE and the manufacture of gaskets.)[39]

Typical physiological effects of exposure to hydrogen cyanide are summarised in Table 30.7.[18] However, concern over the toxicity of hydrogen cyanide should not divert attention from its other hazards, e.g. exothermic polymerisation of the anhydrous liquid under certain conditions[40] (see page 1484).

Table 30.7 Physiological effects of exposure to hydrogen cyanide[18]

Effect	Concentration	
	mg/l	ppm
Immediately fatal	0.3	270
Estimated human LC_{50} after 10 min	0.61	546
Fatal after 10 min	0.2	181
Fatal after 30 min	0.15	135
Fatal after ½–1 hour or later, or dangerous to life	0.12–0.15	110–135
Tolerated for ½–1 hour without immediate or late effects	0.05–0.06	45–54
Slight symptoms after several hours	0.02–0.04	18–36

Complete advice on the industrial handling and use of anhydrous hydrogen cyanide and aqueous solutions is given in ref. 41.

Cyanide salts are used in electroplating, in the case hardening of metals and in the extraction of gold from mineral ores. Hydrogen cyanide is evolved as a gas when cyanide salts react with an acid (Table 10.3a). Hence a serious hazard may be created in metal treatment factories, particularly since acid pickling or phosphating may also be employed.

Therefore the crucial precautions with all such salts are as follows.[42]
- Maintain strict segregation of them well away from all acids.
- Store in labelled, closed containers in a place accessible only to authorised persons. Replace container lids immediately after use.
- Provide, and insist upon the wearing of, impervious gloves when handling cyanides or cyanide-contaminated articles or implements, e.g. scoops or shovels.
- Instruct and encourage all workers to wash thoroughly before eating, drinking and smoking and at the end of each period of work.
- Prohibit smoking in areas where cyanides are stored, used or disposed of.
- Prohibit the bringing of food, drink or utensils into all such areas.
- Ensure that contaminated clothing is removed, handled safely, and never taken home.

Oxides of nitrogen

The oxides of nitrogen are significant air pollutants. Smog is the result of photochemical reaction of NO_x and hydrocarbons; they may also contribute to ozone formation.

Oxides of nitrogen comprise nitrous oxide (N_2O), nitric oxide (NO), nitrogen dioxide (NO_2), dinitrogen tetroxide (N_2O_4) and dinitrogen pentoxide (N_2O_5). The latter is a low-melting solid which rapidly decomposes in air to NO_2/N_2O_4.

Nitrous oxide has no irritating properties and is used extensively as an anaesthetic. Patients recovering from it, or those exposed to low concentrations such as students in early lecture demonstrations, are occasionally hysterical; hence the name laughing gas.[43] A higher than expected incidence of spontaneous abortions among female workers exposed directly to anaesthetic gases has been reported but the current 8-hour time-weighted average TLV of 50 ppm is believed to be sufficiently low to prevent embryofoetal toxicity in humans. Nitric oxide is a colourless gas which is oxidised in air to the dioxide.

Nitrogen dioxide can cause lung damage due to either acute or chronic exposures. Its natural background level over land areas is generally in the range 0.0002–0.0005 ppm but concentrations in urban areas can be 1–2 orders of magnitude higher, e.g. with a global annual mean of 0.03–0.06 ppm.[44]

Effects of given concentrations of nitrogen oxides are listed in Table 30.8. The margin between concentrations which will provoke mild symp-

Table 30.8 Effects of nitrogen oxides

Concentration (ppm in air)	Effect
<60	No warning effect (although the odour threshold is below 0.5 ppm)
60–150	Can cause irritation and burning in nose and throat
100–150	Dangerous in 30–60 minutes
200–700	Fatal on short exposure (1 hour or less)

toms and those proving to be fatal is small. Evidence suggests that humans with normal respiratory function may be affected by exposures as low as 5 ppm and workers with diseases such as bronchitis may be aggravated by such exposures.

Moreover an absence of a warning effect at levels which are hazardous to health arises because nitrogen dioxide has such a low solubility in water that it penetrates deep into the lung, producing alveolar damage rather than affecting the airways.

Thus a 2-hour exposure to NO_2 at a concentration of 5 ppm causes an increase in airway resistance and may render normal persons more sensitive to broncho-constrictive agents that produce pneumonia. Similar effects may occur at considerably lower concentrations, e.g. 0.1 ppm, with some asthma sufferers. Exposure to a concentration of 25 ppm for 1 hour causes respiratory irritation and bronchitis.[45]

'Nitrous fume' exposure generally involves the inhalation of airborne NO_2/N_2O_4 mixtures, usually in the equilibrium ratio of approximately 3:7, which at high concentrations exist as a reddish brown gas. It may arise from the following:

- Fuming nitric acid.

    Sixty people were moved out from near the centre of a town when 'orange' fumes of oxides of nitrogen were released from a drum of nitric acid at a garage. The drum was apparently used for a pickling process for clean steel and the incident arose when a metal tool was accidentally dropped into it. The emergency was dealt with by firemen wearing protective clothing and breathing apparatus.

- Chemical reactions, including the firing of organic, nitrogen-based explosives including ethylene glycol dinitrate, nitroglycerine and trinitrotoluene (Chapter 26).

    Characteristically explosives are used for blasting during mining, quarry or construction work. The amount of NO_2 produced depends upon the quantity and type of explosive, the method of detonation and the work-

ing volume; thus highest concentrations occur when blasting is in confined spaces, e.g. coal mines.
- When metals are welded electrically or cut with oxy-acetylene, propane or butane flames, or whenever such flames burn in air.

    A slinger and two other men were required to work in a 16.1 m long, 5.8 m diameter cylindrical steel vessel with hemispherical ends; two internal bulkheads divided the tank into thirds with 0.46 m by 0.3 m manholes between and in each top part. Stress relieving had resulted in a kink in the bottom of the tank and in the centre section and an oxy-propane burner was therefore used to heat the plate for half-an-hour. The foreman felt unwell and noticed the slinger looked ill so got the gang out of the tank; they had been in the tank for a total of one hour. Exhaust ventilation and compressed air were available but whether or not they were used is subject to conflicting evidence.

    All the men finished the shift on other jobs but on returning home the slinger suffered from tightness of the chest. He was hospitalised for ten days due to pulmonary oedema which appeared to settle satisfactorily under treatment but died forty-three days later due to acute viral pneumonia.[46]

- Forage tower silos. Accidental exposures have occurred due to entry into improperly vented agricultural silos, in which NO_x has been produced from enzymatic action on nitrates in crops during silage production.
- The exhaust of metal dissolution or cleaning processes.

    An operator inhaled NO_x fumes whilst attempting to remove a basket from the manhole of a copper dissolver. He left the area, donned a Chemox mask and returned to complete the task. However, because he did not have the mask correctly positioned, he again had to leave the area. Eventually with the mask on properly he returned and finished removing the basket after which the fumes subsided. He did not report the incident until the following day; a substitute doctor was then reluctant to admit him to hospital.[47]

    Operators on the plant were not familiar with the symptoms and effects of NO_x exposure – since normally the basket was removed quickly without wearing a mask. Subsequently such information was included in verbal instructions and written operating procedures. The need for prompt reporting of all injuries was emphasised. The exhaust system was improved and operators retrained periodically in the correct use of Chemox masks.

- Fires, e.g. involving ammonium nitrate, Nitrogen dioxide is also formed by combustion of organic-nitro compounds, e.g. in soft furnishings, wallcoverings and veneers.

    A fire at the Cleveland Clinic involved burning of radiographic film coated with nitrocellulose and resulted in over 100 fatalities.[48]

- Exhausts from diesel vehicles.

As with other toxic gases, mistakes in processing can also lead to NO_x emissions.

> Standard practice at a fertiliser plant was for reject material to be mixed with the usual fertiliser components. A process change resulted in only material that had been rejected for use in the blending plant going through the factory. When chemicals in the reject batch of 50 tons went through the drier a chemical reaction led to the formation of nitrous oxide, ammonia and hydrogen chloride which spread over the surrounding area. Two workers and several local residents who inhaled the toxic mixture were hospitalised. One of the employees affected died eleven days later.[49]

This serves to emphasise the importance of a proper process evaluation as discussed in Chapter 20. Piloting this change in the laboratory would have revealed the hazards associated with the new procedures, particularly the possibility of generating toxic gases. The effects of this mixture of gases are insidious: several hours may elapse before lung irritation develops. (It is feebly irritant to the upper respiratory tract due to its relatively low solubility.)

> To accelerate the cleaning of a large copper fermenter a brewery worker used a nitric acid-based solution instead of the usual cleaning materials comprising dilute alkali and tartaric acid solvents. He finished work early and went home but subsequently had to be admitted to hospital suffering from delayed pulmonary oedema due to the inhalation of nitrous fumes.[50]

Ozone

Ozone, O_3, is a bluish gas with a slightly pungent odour. It occurs naturally through the action of ultraviolet radiation on oxygen at high altitudes and from natural electrical discharges. It is also generated by photochemically initiated reactions between hydrocarbons and oxides of nitrogen as illustrated in Fig. 21.4.

Occupational exposures may occur due to:
- its formation during arcing in high voltage electrical equipment;
- generation during electric arc welding;
- leakage from processes in which it is used, e.g. for water or sewage treatment or the stabilisation of industrial wastes;
- emissions from photocopying machines in poorly ventilated rooms (in particular when cleaning is required).

A concentration of 0.015 ppm in air produces a barely detectable odour but 0.1 ppm produces a disagreeable odour and may result in irritation of the eyes and upper respiratory tract and headache. As with many chemicals, however, detection by odour varies between individuals from 0.02 ppm up to 0.13 ppm compared with the current UK approved Occupational Exposure Standards of 0.1 ppm (8-hour TWA value) and 0.3 ppm (10-minute STEL).

Welders have exhibited no effects from short exposure to 0.25 ppm, irritation of the throat and a sensation of chest constriction at 0.3–0.18 ppm, irritation of the eyes and nose at 0.7–1.7 ppm and severe respiratory irritation and headaches with signs of pulmonary oedema or pneumonia at 9.2 ppm, but some exposures included other gases. Healthy persons engaged in light exercise may experience chest discomfort and cough after <2 hours' exposure to <0.37 ppm. Inhalation of 50 ppm for 30 minutes is potentially fatal.[51]

Ozone is extremely chemically reactive as an oxidiser, e.g. it can react with alkenes to form ozonides, some of which are explosive.

$$>C=C< + O_2 \rightarrow$$
Alkene

$$>C\underset{O}{\overset{O-O}{\diagup\hspace{-0.5em}\diagdown}}C<$$
Ozonide

With certain aromatics it can form gelatinous explosive ozonides.

Ammonia

At ambient temperature and atmospheric pressure ammonia is a colourless gas with a characteristic pungent odour. It is alkaline and dissolves readily in water to produce the aqueous ammonia solutions commonly used in laboratories, e.g. as a 35 per cent solution with a specific gravity of 0.88. The solution causes severe burns and hence, if ingested, severe internal injuries.

> A factory cleaner died due to cardiac arrest following inhalation and ingestion of ammonia. He had kept proprietary disinfectant and cleaning materials in a small locked cupboard together with two lemonade bottles. One of these bottles contained lemonade and the other ammonia, which had not been officially issued to him. On the day of the accident he inadvertently drank from the wrong bottle and although he spat out the liquid and subsequently vomited he died about one hour later.
>
> (The risk associated with improper identification of chemicals was described on page 774.)

The solubility of ammonia gas in water is illustrated in Fig. 3.2 and the incident involving sudden discharge from concentrated 0.88 ammonia solution due to increased temperature described on page 82 is not an isolated case.[52] 'Boiling-off' usually commences as soon as the cap is loosened, so that this should be done in a fume cupboard whilst wearing appropriate gloves and eye-protection.[53] When technically acceptable, use of less concentrated 0.99 solution avoids this hazard.[52]

Ammonia is transported as a liquefied gas under its own vapour pressure of 114 psig at 21 °C. The gas is irritating to the eyes, mucous membranes and all parts of the respiratory system. As a result of its odour and irritant properties, persons are unlikely to be unwittingly overexposed for a prolonged period. A summary of the effects at various concentrations is given in Table 15.7. Thus compared with a current UK Occupational Exposure

Standard of 25 ppm as an 8-hour TWA value, a concentration of 700 ppm causes severe eye irritation and less than half-an-hour's exposure to 1700 ppm may be fatal.

> A fitter's labourer was required to cut a hole in a brick wall in the basement of a cold storage plant. The ceiling height was only 2 m so that he had to work in a crouched position. He leaned against an adjacent pipe which came adrift causing him to be enveloped in a cloud of gaseous ammonia. He was rescued by a colleague wearing a respirator but subsequently died from bronchial pneumonia.[50]

On contact with the skin liquid ammonia produces severe burns compounded by frostbite.

Ammonia gas is flammable, the limits in air being 16–25 per cent and the ignition temperature 651 °C. Its use as a refrigerant, from which liquid leakage can produce an 850-fold increase in volume on vaporisation, has therefore resulted in some fires and explosions within buildings.

> A 5 cm nipple failed on an ammonia compressor at a milk-bottling factory. Ignition resulted in multiple explosions which spread fire throughout the building.[54]

Because of its corrosive, irritant properties the following precautions are necessary, in addition to those summarised in Table 19.9.
- Avoid heating cylinders directly with steam or a flame to accelerate gas discharge.
- Protective equipment should include rubber gloves, chemical goggles and – on any scale beyond the laboratory – a rubber apron or full chemical suit.
- Ensure that the gas cannot be ignited (the flammable limits are fairly narrow, 16–25 per cent, and the auto-ignition temperature 651 °C).
- Use under well-ventilated conditions.
- Provide convenient safety shower and eye-wash facility.
- Ensure appropriate first-aid provisions.

Any leaks can be detected with moist litmus paper or carefully with concentrated hydrochloric acid, the fumes from which are converted into dense white fumes of ammonium chloride.

Chlorine

In liquified form chlorine is a clear amber dense liquid. The gas is greenish-yellow, about 2.5 times as dense as air, and non-flammable. Liquid chlorine causes severe irritation and blistering of skin. The gas has a pungent suffocating odour and is irritant to the nose and throat. It is an extremely powerful vesicant and respiratory irritant. Physical properties are summarised in ref. 18.

Chlorine is used for the preparation of organic chemicals, e.g. trichloroethylene, carbon tetrachloride, vinyl chloride plastics; pesticides; propellants and refrigerants. A range of inorganic chemicals, e.g. chlorides and metal compounds, are also based on chlorine. It is also used for bleaching and deodorisation in the pulp and paper industries and in water and sewage treatment processes.

Typically exposure to chlorine concentrations of 3–6 ppm results in a stinging and burning sensation in the eyes. Exposures for 0.5–1 hour to concentrations of 14–21 ppm cause pulmonary oedema, pneumonitis emphysema and bronchitis. This is usually associated with marked bronchospasm, muscular soreness and headache. A summary of effects at various concentrations is given in Table 15.5. Whilst there is inevitably a variation in individual susceptibility, typically 4 ppm is the maximum concentration that can be breathed for one hour without damage, 40–60 ppm is dangerous for a 30-minute exposure and a concentration of 1000 ppm is likely to be fatal after a few breaths.

Moist chlorine is corrosive to skin and to most common materials of construction. Thus corrosion in the presence of moisture has caused localised thinning of steel vessels.[55] Wet chlorine at low pressure can be handled in chemical stoneware, glass or porcelain and in certain alloys and plastics.

Chlorine is shipped in bulk in liquid form, either by road or by rail. The type of facility for bulk storage is shown in Fig. 13.9. These facilities are, of necessity, designed with great care and generally subjected to HAZOP studies; strict operating procedures are established.[56]

> A leak occurred in a liquid chlorine storage tank in Cleveland. Twenty-seven people required hospital treatment. Of 18 studied for pulmonary function, airways obstruction and hypoxemia disappeared after 3 months but five still had persistent reduction in air inflow after 14 months.[57]
>
> (The consequences of a massive escape following an explosion are described on page 886.)

Chlorine is also shipped in cylinders which are normally protected from overpressurisation by a fusible metal plug which melts at about 85 °C. It is in connecting up and bringing on stream, and in disconnecting, cylinders that problems may sometimes arise.

> During change-over of two 860 kg chlorine drums in a water treatment plant gas seeped from a connecting pipe despite closure of the main safety valve. A slight leak was usually tolerated but this was larger so a fitter fetched canister-type respirators. Attempts to rectify the leak resulted in a large emission of chlorine (eventually traced to a faulty valve). The respirators were inadequate for the duty imposed upon them and the fitter and several colleagues were badly gassed.[50] (Air-fed respiratory apparatus, page 303, would have provided appropriate protection in this case.)
>
> After changing a cylinder of chlorine two men opened valves to check that

there were no leaks on the lines to the vaporiser. They did not expect to find any leaks and therefore did not wear breathing apparatus. There were, however, leaks and they were affected by the gas.[58]

Therefore the precautions summarised in Table 30.9 are essential when dealing with bulk or cylinder supplies.

Table 30.9 Additional precautions with chlorine cylinders (to supplement those in Table 19.9)

- Provide appropriate respiratory protection, conveniently situated eye-wash facilities and showers for emergency use.
- Handle in well ventilated areas whilst wearing appropriate skin protection and respirator.
- Check for leaks, e.g. with aqueous ammonia. If appropriate provide detection and alarm systems.
- Establish a procedure for local evacuation and to deal with leaks.
- Avoid connection of a cylinder directly to a vessel containing liquid. (Suck-back into the cylinder could cause a violent reaction. A trap should be inserted with a capacity to accommodate all the liquid.)
- Avoid heating of the cylinder.
- Segregate from cylinders of, e.g., ammonia, hydrogen, acetylene or fuels and ensure mixing cannot occur in the cylinder.
- Ensure accidental contact cannot occur with hydrocarbons, ethers and other organic liquids or with finely divided metals.

Hypochlorites

As explained on page 495 accidental admixture of acids with hypochlorites will generate chlorine. Therefore a proper system of work is required irrespective of the scale of operation.

> Eight workers were detained in hospital following a mistake at a small chemical factory. Instructions were issued to place phosphoric acid into a 4,000-litre tank which had contained bleach. A cloud of chlorine gas was evolved as soon as the acid was introduced. (The owners subsequently admitted a failure to ensure the health and safety of workers.)[59]

As with other toxic gases, as exemplified on page 494, undesirable concentrations of chlorine are sometimes generated from what may appear to be relatively innocuous operations.

> A company expected a road tanker delivery of sodium hypochlorite. It arrived bearing warning panels for sodium hypochlorite but, in fact, contained ferric chloride. As the driver commenced unloading into a bleach storage tank chemical reaction resulted in the evolution of a large cloud of chlorine. Twenty-nine people, mostly employees, were taken to hospital; 16 were detained for between one and three days.

> A bulldozer driver and two other men were affected by fumes on a waste disposal site; all suffered respiratory congestion. Drums containing sodium

hypochlorite were emptied from one skip and drums containing an acid solution from another. The bulldozer crushed the drums and admixture of the chemicals resulted in chlorine evolution.

$$4NaClO + 2H_2SO_4 \rightarrow 2Na_2SO_4 + 2H_2O + O_2 + 2Cl_2$$

These examples re-emphasise the need for great care in labelling (page 780) and segregation of incompatible chemicals (page 491).

A large dye filter pan was cleaned using sodium hypochlorite solution instead of the established method of washing with water and with sulphuric acid. Reaction between a film of sulphuric acid in the pan and the bleach solution liberated chlorine. The chargehand concerned and a colleague suffered from the effects of inhalation; fortunately this necessitated only two weeks off work.[50]

Transportation

As with all chemicals, transportation incidents may pose a special problem simply because the circumstances and location are unpredictable.

One of twenty-seven 50 kg chlorine cylinders developed a leak whilst on a truck passing over the Manhattan Bridge into Brooklyn, USA. The driver was alerted by the smell; when he pulled into the kerb to investigate, the cylinder was leaking steadily from its base. Dense fumes caused him to retreat. The bulk of the escaping gas flowed into ventilator gratings linked to a nearby subway station; the remainder drifted down a busy street causing some 40 to 50 people to be overcome within minutes. The subway continued in operation and persons on the station and within the compartments of trains became casualties. There were 418 treated in hospitals but no immediate fatalities.[60]

Supposedly empty cylinders of liquid chlorine were being unloaded from a freighter when the main valve was snapped off one cylinder. Some 156 people sustained acute respiratory irritation; some suffered haemorrhages from the lungs whilst others experienced asthma-like wheezing and pneumonia-like lung infiltration, resulting in long-term impairment of lung function.[33]

In the past acute epidemics have resulted from accidents involving bulk chlorine tanks in transit.

A rail-tankcar was ruptured in a wreck in Louisiana in 1961 releasing 6,000 gallons of liquid chlorine.[61] The resulting gas cloud covered 6 square miles. Some seven hours later concentrations were still as high as 400 ppm at 70 m from the site and 10 ppm at the fringe of the cloud. One thousand people were evacuated; 100 required treatment for varying degrees of respiratory illness and an infant died from the exposure.

(A bulk toxic release following derailment is also described on page 808 and one in which derailment was followed by a fire and explosions on page 806.)

Phosgene

Phosgene, i.e. carbonyl chloride, is a pale yellow liquid with a boiling point of 8.3 °C. The colourless vapour has a musty odour and is a potent respiratory irritant.

Phosgene is used in the production of specific organic chemicals including isocyanates and acid chlorides. However, the more frequent, suspected or actual, occupational exposures arise following the thermal degradation of volatile chlorine compounds. For example chlorinated hydrocarbon solvents, e.g. trichloroethylene or 1,1,1-trichloroethane which are used for metal degreasing, may decompose in contact with hot objects or naked flames; the main gas produced is hydrogen chloride but phosgene and other acid gases are also present. Therefore such solvents should not be used in locations where vapour is likely to disperse and reach welding flames or welding arcs or to enter gas or other flame-heated equipment. Alternatively if this is not practicable adequate screening and local exhaust ventilation are essential.

A hazardous concentration of phosgene can be encountered without warning from its irritant properties. Thus whilst 0.5 ppm can be detected by odour, by persons familiar with it, a gradually increasing concentration may pass unnoticed due to adaptation or olfactory fatigue. The least concentration causing immediate throat irritation is 3 ppm, whilst 4 ppm causes immediate eye irritation and 4.8 ppm causes coughing.[62]

When inhaled, since it is relatively insoluble, phosgene reaches the lungs. There may then be a delay of some hours without symptoms; acute pulmonary oedema may then occur suddenly accompanied by profound circulatory collapse. Typical physiological effects of various exposures are summarised in Table 30.10.[63]

Table 30.10 Physiological effects of exposure to phosgene[60]

Effect	Concentration (ppm)
Maximum amount for prolonged exposure	1
Dangerous to life, for prolonged exposure	1.25–2.5
Cough or other subjective symptoms in 1 min	5
Irritation of eyes and respiratory tract in <1 min	10
Dangerous to life in 30–60 min	12.5
Severe lung injury in 1–2 min	20
Dangerous to life in 30 min	25
Rapidly fatal ≤30 min	90

Radon

Radon is a naturally occurring radioactive gas from uranium which is present in all soils and rocks to some extent. It seeps from the ground and can collect in enclosed spaces, e.g. workplaces and homes. Radon levels are inherently higher in some parts of the UK than others, because of variations in the amount of uranium in the ground and because some ground types allow easier air movement.

Health hazard

The decay of radon forms minute radioactive particles which may be inhaled. Radiation from these particles can cause lung cancer after an extended latency period. The effect is synergistic with smoking. In the UK, where the workplace is occupied for a normal working day, no further action is likely to be required if extended measurements confirm radon levels <400 Bq m^{-3} (becquerels per cubic metre, refer to Chapter 8). Employers should seek expert advice from the Health and Safety Executive or Local Environmental Health Department for levels >400 Bq m^{-3}. (The corresponding Action Level for homes is 200 Bq m^{-3}, to account for the difference in exposure periods and those at risk).

Workplace characteristics

Most workplaces in the UK are not considered to have significant radon levels.[64] However, premises built on certain areas, e.g. in Cornwall, Devon and parts of Derbyshire, Northamptonshire, Somerset, Grampian and the Scottish Highlands, are more likely to have high indoor radon levels.

Building construction is an important factor, since radon seepage from the ground is drawn into buildings via cracks in floors and gaps around drains, pipes and cables, etc. Radon levels are generally low in well-ventilated workshops, but problems have arisen in buildings with relatively low ventilation rates, e.g. shops, offices and public buildings.

Precautions

Within the UK, the Ionising Radiation Regulations 1985 require employers to take action where radon is present above the level defined above. Therefore, if location, construction and ventilation of a workplace render elevated radon levels likely, the premises should be tested. Successful and relatively inexpensive remedial techniques may be applied if appropriate.[65]

General principles for protection against toxic gases and vapours

From the previous discussion, and from common experience, it is possible to summarise general principles for protection of workers, and others who may be exposed, against toxic gases and vapours. These are not restricted in application to factories, for example toxic gases may accumulate in slurry pits and tanks on farms, so that a strict procedure is advisable for safe emptying.

The first requirement is obviously to assess the risk as discussed in Chapter 28. In so doing it must be realised that, as discussed on page 1583, toxic gases are frequently encountered in admixture with others.

> A librarian used a common chlorine-containing cleanser to remove a stain from a laminated plastic counter top. To achieve better effect she added vinegar and rubbing alcohol to the cleansing agent. Almost immediately she experienced extreme irritation to the eyes and upper respiratory tract and was subsequently hospitalised for 14 days, much of the time in an oxygen tent. Chlorine was presumed to have removed hydrogen from surface and tissue water releasing nascent oxygen and ozone: the latter plus hydrochloric acid were believed to have been the main respiratory irritants.[66] (The practice of mixing cleansing agents was advised against on page 774.)

Substitution

A basic principle is that it is better to eliminate or reduce risks by substitution rather than deal with the, often difficult, problems of adequately controlling these risks. Guidance has therefore been published on substitution for use by manufacturers or suppliers of substances, customers and others who select products or set specifications.[67] In addition to reducing health and safety risks, benefits may include increased efficiency in production, reduced inventories of substances held, or the adoption of different processes based upon less hazardous processes. It is, however, a proviso that other risks are not allowed to increase to unacceptable levels. Thus from the early 1970s 1,1,1-trichloroethane was actively promoted as a safer substitute for trichloroethylene for degreasing and cleaning applications. This substitute compound has now been implicated along with chlorofluorocarbons (CFCs) in the depletion of the earth's ozone layer. Trichloroethane will therefore be phased out completely by the end of 1995 under EC Regulations introduced to implement the Montreal Protocol (see Chapter 24). Conversely, substitutes introduced for environmental purposes, so far as is reasonably practicable, should not pose unnecessary or unacceptable risks in the workplace. Many of the alternative cleaning chemicals are either more toxic than 1,1,1-trichloroethane or are flammable.

If, following assessment, substitution appears to be the most suitable control strategy then the recommended steps to ensure that all the consequences of the change have been considered are:[67]

- seek to identify the relevant hazards and evaluate the foreseeable risks involved in storage, use and disposal of the existing substance or process;
- identify possible alternatives;

- consider the consequences of the alternatives;
- compare the range of alternatives with each other, and the original;
- decide, if possible, on the most suitable substitute;
- implement the change;
- evaluate the solution.

Assessment of the risks likely to arise in both the workplace and the environment involves collecting information about the hazards associated with the substances, and looking at how the substances are used and how they may interact in the workplace to create risks. All possible routes of exposure need to be considered.[68]

> Substitution involving a pesticide may simply be the replacement of a broad spectrum pesticide with one that is pest specific and no more hazardous. (Alternatively the use of a pesticide may be avoided altogether by the use of alternative cultural techniques.[67])

The basic aim of identifying alternatives is that their use should result in a reduction in overall risk. Consideration is needed of

- effectiveness of the alternative, e.g. quality of the product, and customer acceptability
- acute health effects, e.g. the immediate consequences of exposure to corrosive substances
- chronic health effects from longer-term exposure to hazardous substances, e.g. risk of cancer from carcinogens
- risks of fire and explosion from flammable liquids and gases, e.g. solvents
- risk of adverse effects on the environment
- waste disposal, e.g. quantity and toxicity of waste produced
- cost impact
- continuing availability of the substance in the marketplace, e.g. the phasing out of 1,1,1-trichloroethane
- physical form, e.g. solid, dust or vapour
- physical and chemical properties, e.g. volatility
- impending legislation.

Hence full information on possible alternative substances should be obtained, e.g. following the procedure summarised on page 1158.

In order to determine the consequences of using alternatives, assessment is necessary of the risks from exposure both inside and outside the workplace and of effects on the environment. The way in which a proposed substitution may affect other processes also requires consideration. An indication of whether a substance is likely to damage the environment may be obtainable by reference to lists compiled for the purposes of statutory legislation, e.g. in the UK ref. 69.

However 'substitution', although bringing long-term benefits, can be a costly and time-consuming exercise. Thus, public concern in the UK arising in connection with conversions of domestic appliances to burn natural gas – which is 'non-toxic' compared to manufactured town gas because of the absence of carbon monoxide – resulted in two inquiries.[70,71] Over 150 years previously the introduction of coal gas met with considerable opposition.

> This revolutionary transition from rushlights, tallow candles and oil lamps to coal gas was not accomplished without much opposition, distrust and ill-will particularly from those with vested interests in tallow and whale-oil. The objections were based mainly upon the alleged dangers of fire and explosion, which were seized upon and distorted by the virile caricaturists of the day.[43]

In comparing alternatives a balance is sought between the effectiveness of any potential alternative and the level of risk created by its use. Factors involved will include known or suspected toxicity, flammability, volatility, droplet or particle size and the number of people at risk. Whilst it may be straightforward to compare alternatives with similar properties, some comparisons may be difficult, e.g. comparing the risks between a highly flammable substance and a highly toxic one, or between acute and chronic toxic effects (e.g. mutagenicity). The recommended procedure is to consider each risk in turn, in the context of other risks, and to decide whether it can be adequately controlled or is unacceptable.

If a substitute is determined to be appropriate, a decision is required as to how, and when, it should be introduced. It may be advisable to substitute on a small scale to identify any unforeseen problems;[67] whether equipment to be used will cope, or will need adaptation or replacement. Staff will need training in implementing the use of the substitute. Assuming that proper practice has been followed to make them aware of the risks and the precautions with the original material, they must be instructed in the different risks associated with the substitute and given appropriate training. Finally the substitution should be evaluated in operation to check that the anticipated benefits actually accrue. Obviously any unforeseen problems impinging upon health and safety should be dealt with promptly.

On-demand generation

If substitution is impracticable, it may be feasible to develop an on-demand process in which a toxic gas or vapour is generated only in the quantity needed for immediate use.

> Arsine is utilised in the manufacture of electronic components.[72] Compressed gas cylinders must be stored on-site in specially-designed containment facilities to limit the risk of accidental releases. An on-demand generator has been developed in which arsine is electrochemically synthesised at an arsenic cathode containing 1N potassium hydroxide. This has eliminated the need for expensive storage facilities.[73]

Toxic gases and vapours 1739

On-demand generators are also under development for phosphine and silane for semiconductors manufacture.

Containment

The primary control method with all toxic gases and vapours is containment, the general features of which are summarised on page 286 *et seq*. This includes measures to cope with loss of containment, emergency relief, etc.

Important examples of ways to prevent escape of toxic gases include use of all-welded pipework, pumps with double mechanical seals and a pressurised liquid between each seal so that any gas leakage is retained in the system, canned pumps, and special valves with no glands, e.g. bellow valves.

> A process worker was fatally gassed when a sightglass failed in a pipeline containing phosphorous oxychloride. The line was steam-traced and the valves at each end were closed. The possible cause of this incident was an increase in temperature of the phosphorous oxychloride, due to steam tracing, resulting in expansion: the pressure increased until the sightglass, i.e. the weakest section, failed.[74]
>
> (The phenomenon of pressure generation due to thermal expansion of a trapped liquid is discussed on page 99. If a hazardous liquid in a pipeline with a sightglass and a valve at each end can be heated even slightly, e.g. from the surroundings, a pressure relief system must be included. Otherwise the pipeline should be designed without a sightglass and to withstand the maximum pressure which can be generated.)

As already noted on page 288, and exemplified on page 698 and 883, a scrubbing system is required for pressure relief discharges. The merits of using stronger equipment – capable of withstanding the highest pressure that can be reached – so that relief valves and subsequent, expensive scrubbing installations may be dispensed with, are often overlooked.[75]

It is important to subject all tasks including toxic gases to detailed scrutiny and analysis. The majority of operations are designed to incorporate two safety features; if there is a greater risk, three safety features may be desirable.[76]

Local exhaust ventilation

The general principles for effective local exhaust ventilation are well established and are summarised in Chapter 12, pages 675–89. As described on page 285 the need is for fumes/vapour removal with dilution being a back-up.

For example, control of exposure to the wide range of fumes and gases generated in industrial welding processes may be achieved by capturing

them at, or near, the point of generation, i.e. the arc or gas flame. The types of exhaust ventilation are:
- fixed installations, e.g. side draught or down-draught tables and benches; or partially, or completely, enclosed booths.
- portable installation, e.g. mobile hoods attached to flexible ducts.
- fume extractors attached to the welding gun.

Examples are shown in Fig. 30.1.[77] Since the object is to remove fumes

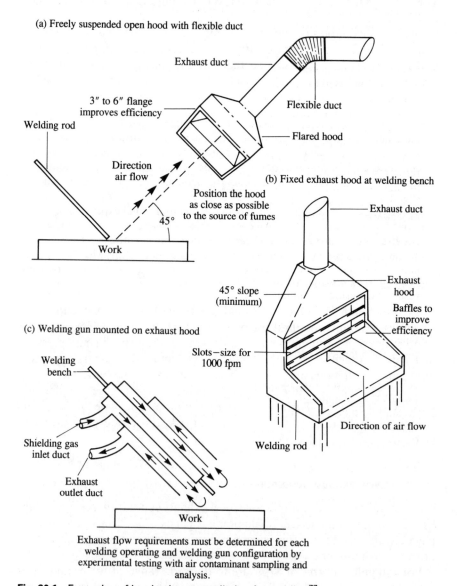

Fig. 30.1 Examples of local exhaust ventilation for welding[77]

and gases before they reach the welder's breathing zone, installations such as overhead canopy hoods are not generally recommended. Mere recirculation is unlikely to be satisfactory on its own.

> A boilermaker used two oxyacetylene torches with multiple jets to heat-shrink a tiller on a rudder post in a 93 m^3 volume compartment. The heated part was beneath a circular manhole containing an extraction fan and there was an open hatch approximately 0.6 m square in one corner of the compartment. A powerful air blower was placed behind the operator but served only for recirculation, i.e. it was not fed with air from outside.[78]
>
> The boilermaker was assisted by other men, who entered for short periods only, and operated the torches for about two hours. He apparently suffered little or no discomfort and continued with other work for several hours without complaint. Subsequently that evening he developed a cough; he was admitted to hospital but died the following day.
>
> Tests were performed with identical equipment except that a steel block was used tto receive the heat. The powerful blower was found to disperse fumes fairly evenly throughout the compartment, instead of allowing them to rise through the manhole. Within three minutes of lighting the torches the concentration of nitrogen dioxide was >20 ppm. They were then operated for a further two minutes. Thirteen minutes were then required for the concentration to fall to 10 ppm. It was estimated that the boilermaker – who suffered from a lack of taste and smell – had been exposed to >100 ppm nitrogen dioxide with probably a higher concentration still of nitrous fumes.

As the following cases demonstrate sometimes there has simply been a failure to provide any, or adequate, local exhaust ventilation.

> Women using brazing rods developed hoarseness and dry, sore throats as a result of exposure to cadmium fumes. The company were subsequently prosecuted and the Court ruled that they were aware of the risks but the exhaust ventilation provided at the bench and at the bath were inadequate as it failed to capture the fumes and prevent them entering the workroom. Hence, there was a failure to provide and maintain adequate ventilation.[79]

> A man poured several gallons of solvent into a 500-gallon paint pan to clean it using a long-handled brush. After inhaling vapours he collapsed and required urgent hospitalisation. Subsequently management changed the procedures so that the paint pans were cleaned elsewhere using special equipment and always under extract ventilation.[80]

In other cases exhaust ventilation, although provided, has not been made use of; sometimes it has been turned off to conserve power and at other times to reduce the noise level in the factory. Alternatively as illustrated below and by the ethyleneimine example on page 333[29] it has been used but has not been sufficiently effective.

> A research worker was recovering dimethylhexanol from an unsuccessful esterification experiment. Some 200 g had been stored overnight in a fume cupboard. Fume extraction was not complete and a strong smell was present in

the laboratory on the following morning. After two hours the researcher felt sick and left the laboratory for a few minutes. On recovery he returned and told a colleague who shortly after also felt sick; other staff were symptom-free. The researcher again suffered nausea and slight giddiness. On reporting to the first-aid room he was very pale with a pulse of 60 and a temperature of 36.3 °C. After twenty minutes he felt normal but the following day suffered from a severe headache.[81]

In less common situations the local exhaust provisions have failed in service.

A student was diluting the cholinesterase inhibitor diisopropyl fluorophosphate (DFP) and had taken all necessary precautions. Manipulation was in a standard laboratory fume cupboard with adequate ventilation. During the operation some debris came down the flue of the fume cupboard and caused the fan to stop. Within seconds he began to feel the effects of the compound; he was admitted to hospital but fortunately was discharged on the same day with no ill-effects.[82] (This illustrates the short period required for compounds of this type to produce ill-effects on the human body and emphasises the need for stringent precautions. Absorption via the skin is rapid and effects upon health are often irreversible. Medical assistance, i.e. atropine injection plus a doctor, should be readily available.)

In all situations in which local exhaust ventilation is relied upon, a properly designed plant needs to be reinforced by a proper system of work, including inspection and maintenance and compliance with an established operating procedure. For example, in vapour degreasing processes, commonly used in the engineering industry, components are immersed in hot solvent vapour in plant shown, in the simplest outline, in Fig. 30.2. The

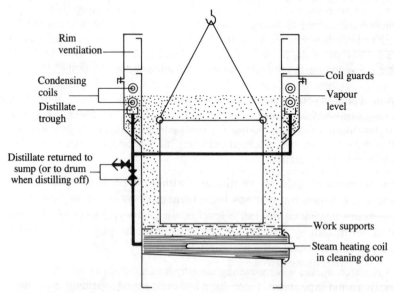

Fig. 30.2 Simple vapour degreasing tank arrangement

Toxic gases and vapours 1743

open-topped tank is fitted with heaters and a water-cooled condensing coil; rim ventilation is fitted as standard. Solvent is boiled in the sump and components are suspended in the vapour zone which is held down by the cooling coils. Vapour, which condenses on the components, dissolves off soluble oil, grease, etc. Safe operation requires that

- plant is installed in a draught-free area to avoid undue disturbance of the vapour level;
- work should be moved in, and out, at slow uniform rates, e.g. 1.5 to 3 metres/min, with adequate time for liquor and vapour draining beneath the level of rim ventilation;
- transfers of work within the plant should preferably be below the vapour level;
- the plant should be closed with a lid when not in use;
- plant should not be sited near furnaces, and flame-cutting, welding or similar operations should not be permitted in the vicinity, as explained on page 1734.

Additional design requirements are that the flow of water, or other coolant, through the coils should be adequate to ensure that vapour does not escape when the tank is at its normal operating temperature. Also the freeboard zone, i.e. the distance between the top of the coils and the top of the tank, should be at least 0.6 times the width of the tank, or more in very large tanks. Moreover mechanised plants can be enclosed and solvent losses reduced by linking them to activated carbon recovery plants.

Within the UK thorough examination and tests of local exhaust ventilation plant are required at least once every 14 months or in special cases (listed in Table 30.11) at lesser intervals.[83]

Table 30.11 Frequency of thorough examination and testing of local exhaust ventilation plant used in certain processes: UK requirements[80]

Process	Minimum frequency
Processes in which blasting is carried out in or incidental to the cleaning of metal castings, in connection with their manufacture	Every month
Processes, other than wet processes, in which metal articles (other than of gold, platinum or iridium) are ground, abraded or polished using mechanical power, in any room for more than 12 hours in any week	Every 6 months
Processes giving off dust or fume in which non-ferrous metal castings are produced	Every 6 months
Jute cloth manufacture	Every month

System of work

Permits-to-work to control, e.g. clean-up and repair of plant, or entry into confined spaces, were discussed on page 975. They are intended to predetermine a safe procedure and to record clearly that all 'foreseeable' hazards have been considered and that all appropriate precautions are defined and followed in the correct sequence. The principles are well established but the work involved will only in fact be performed safely if all the participants follow the requirements of the system at work conscientiously, i.e. without short cuts or 'improvements' to avoid inconvenience or to save time or through lack of knowledge.

> A cleaner entered an autoclave used for PVC production in order to check the need for cleaning. He was overcome by vinyl chloride vapour. This vessel was entered twice a week. A permit system had once been operated but had fallen into disuse.[29] (Whilst this illustrates the short-term narcotic effects of vinyl chloride, exposure to lower concentrations of the compound may result in a host of long-term effects such as angiosarcoma (Table 5.11).)

This is an example of failure to document and enforce established procedures (the requirements of s.30 of the Factories Act were evidently applicable in the UK).

> When a valve was accidentally opened during the cleaning of a reactor ethylenediamine spilt onto two operators inside it. The valve was opened due to a misunderstanding on adjacent plant.[29]

This is a simple case of inadequate isolation.

> A cleaner was overcome by fumes whilst removing residues from inside a vessel using a high-pressure water jet. Entry was regulated by a Permit-to-Work; this had been issued in the belief that the vessel was gas-free. The possibility of 'fumes' being evolved as the residues were disturbed had been overlooked.[29]

This is an example of the need to consider all possible hazards.

'Systems of Work' also encompass procedures for normal operation, transport and waste disposal activities.

> A technician obtained a Winchester bottle of nitrobenzene from the stores in the basement. He departed from his usual practice of using an approved carrier, because he was in a hurry. The storeman also departed from a standing rule that no full Winchester should be handed over without a carrier.[84]
>
> The technician found the goods lift occupied and therefore, contrary to regulations, went to use a passenger lift. Whilst waiting for the lift cab he placed the bottle on the floor but unfortunately kicked it over when the doors opened.
>
> The odour of nitrobenzene permeated all fifteen floors of the building within a few minutes despite the smoke extract fans being used. It was several weeks before it could no longer be detected in the basement.

Respiratory protection

Strategies based on a 'Safe Place' are preferred to those relying on a 'Safe Person' approach as discussed on page 294. As an example, respiratory protection against fumes and gases f............................ing operations should only be considered as a m................... if other control measures are impracticable. Eve............, however, as discussed on page 307, protection depends heavily upo................. 'pe of respirator – and upon its conscientious and .... Most welding operations do not require the us............................ion provided a high standard of ventilation is .... or general ventilation is not available, and in co................... movement, respiratory protection is usually .... General recommendations for respirators for welders are given in Table 20.12.[77]

Thus personal protection should not normally be used as the primary safety precaution during operation or maintenance (with obvious exceptions, e.g. work in oxygen-deficient atmospheres, welding in restricted areas, stripping asbestos-containing insulation). Its use is sometimes, however, required by law and can supplement other control measures and make for safer working. Incidents frequently occur in which correct, properly maintained, protection was either not provided or not specified as being necessary, which serves to highlight the weakness of this strategy.

> A senior operator worked for most of his shift cleaning a troublesome batch of crude intermediate products from a plant. Fumes of sulphur dioxide were being evolved from the product so he wore a respirator. The next morning he had a particularly tight chest as a result of constant inhalation of low concentrations of sulphur dioxide. This arose because he was bearded which prevented the respirator from forming a good seal with his face thus allowing gas to pass.[85]

> Paint-sprayers at a firm of metal finishers were provided with ordinary mouth masks rather than approved respiratory protection. As a result many were exposed to hazardous concentrations of isocyanates which damaged their health.[86]

> To replace the mixer blades in a resin tank the trichlorofluoromethane solvent was drained to a level of 0.23 m from the tank bottom when a departmental manager, with the knowledge of one of the company directors, arranged for a maintenance fitter to enter the tank. He came out immediately complaining about fumes and was given a cartridge respirator but found it difficult to breathe when wearing it. The black cover for the filter of the respirator had been left in place. A plastic hose was connected to the compressed air supply and the end poked inside the respirator. The fitter re-entered the tank, had difficulty breathing and was asphyxiated.[87]

In these cases protective equipment was used but was inadequate and the need for training in the use of personal protection is clearly highlighted.

Whilst, as with other types of personal protection, there is a limit to what is expected of an employer, when there is a statutory duty to provide respiratory protection it is a very onerous duty.

> A man sustained an eye injury whilst working at a grinding machine. Some

Table 30.12 Minimum respiratory protection recommended for different welding processes[73]

		Shop welding		Field welding	
		Ventilation good: exhaust vent is used to capture fumes and gases	Ventilation poor: vent cannot be used due to physical or process restrictions	Ventilation good:* open area spark enclosure, or inside vessel with excellent air movement	Ventilation poor:† spark enclosure or inside vessel with poor air movement
Shielded metal arc welding	Carbon steel	Not required	Not required	Not required, except for galvanised	Fume mask
	Other alloys	Not required	Fume mask	Not required, except for galvanised	Fume mask
Arc cutting or gouging		Not recommended for shop welding: see field welding requirements		Fume mask except for open plant areas	Air-supplied respirator Helpers to wear fume mask (minium)
Oxy-acetylene torch cutting		Not required	Not required except for all plasma arc cutting	Not required except for galvanised steel	Air-supplied respirator
Plasma arc cutting		Air-supplied respirator required for all plasma arc cutting			
Gas metal arc welding		Not required	Air-supplied respirator	Not required	Air-supplied respirator
Gas tungsten arc welding		Not required	Air-supplied respirator	Not required	Air-supplied respirator

* General criteria to be met:
Spark enclosures: At least two open sides; no fume accumulation; no more than one welder.
Inside vessels: Directional air flow moving fume away from welder; no more than one welder; no major structural or scaffolding barriers to air flow.
† Examples:
Spark enclosure with all sides enclosed
Vessel with welders working at different elevations
Visible accumulation of fume
Short circuiting of air flow by open manways.

time previously a pair of goggles had hung on a nail near the machine but, finding they were not used, a foreman removed them to his office a few yards away. A practice had grown up by which anyone wanting goggles was left to ask for them. In a civil action brought against the employers, it was held that goggles had not been put in a place 'where they came easily and obviously to the hand of the workman' nor had the workman been given 'clear objectives as to where they could be obtained'.[88]

Personal protection must be worn continuously throughout the duration of exposure to a hazard. For example, respiratory protection keeping all contamination out of the operatives' air supply must be worn for 99 per cent of the time to be effective, and if worn for only 95 per cent of the exposure period then effective protection drops significantly. Clearly, then, detailed and specific information and training are necessary if respiratory protection is to be relied upon. This should encompass instructions as to when it must be used, e.g. in emergency situations.

It is a not uncommon fallacy that a person without a breathing apparatus can rescue someone who has collapsed by holding their breath. It has been pointed out that anyone who believes this should try to move an 80 kg weight for 10 m without breathing.[89]

In other situations respiratory protection may not be provided because the requirement for it is simply not recognised.

Following a start-up, a pump, which has been incorrectly repaired, released 1,000 gallons of monochlorobenzene into an open bund. Whilst cleaning up this spillage two men were overcome by the effects of the vapour.[90] (The bund was in the open; however, monochlorobenzene has a vapour density of about 4 and a vapour pressure of 10 mm at 22 °C, with strong narcotic effects at concentrations around 1,000 ppm and fatal consequences at exposures approaching 3,700 ppm.)

A teenager on a youth training scheme used paint stripper to clean a large pot of one metre depth. He was overcome by the vapour and was later found collapsed at the bottom of the vessel. The OEL for the stripper was only 100 ppm but the youth had been exposed to more than 50,000 ppm.[91] (The cleaning operation had been carried out in a well-ventilated area next to double doors and the pot was not enclosed. However, the solvent used in the stripper was known to be highly toxic and, being much denser than air, would have accumulated in the base of the pot.)

Sometimes respiratory protection has let down the wearer because of inadequate maintenance, including checking.

Two operators had been called to change a filter on a plant. One donned an air-fed hood and the other self-contained breathing apparatus. Both felt the air supply to be only just adequate. Shortly, the operator wearing the hood felt uncomfortable and removed it. He immediately inhaled a very high concentration of methylene chloride and methanol vapour and collapsed. His companion removed his own breathing set to drag him away. A third man was summoned and first-aid was given. The affected man recovered consciousness and was given medical attention.

It was subsequently found that the hood had not received the normal cleaning treatment after use, had not been disinfected and had not been inspected and placed in a protective polythene bag. The silencer in the air supply was blocked with rust and there was a defective regulator on the compressed air system.

Appropriate measures should have included:
- identification of all foreseeable risks;
- checking that the protective equipment or clothing would provide adequate protection and was adequately maintained;
- checking that the protective equipment was compatible with any other working clothing and the system of work to be adopted.

For highly toxic gases, such as hydrogen cyanide or phosgene it is common to operate a 'buddy system'. Thus an operator will be equipped with full protective equipment and be supervised from an appropriate distance by a second person, who is also provided with full protective equipment and a radio communication to a control room, from which emergency aid can be summoned. Speed is vital when dealing with a toxic emergency, e.g. in providing oxygen to anyone over-exposed to hydrogen cyanide. Hence adequate training and regular practice exercises are crucial.

Atmosphere monitoring

Monitoring of workplace atmospheres, using procedures summarised in Chapter 7, is a necessary precaution to ensure that established occupational exposure limits are not being exceeded and to check that particular control measures are working properly, e.g. with the use of isocyanates in polyurethane foam manufacture or near ovens in which polyurethane paint films are cured, when a serious risk to health could arise if any control measure failed or deteriorated. In the UK monitoring is a statutory requirement under the Approved Code of Practice[92] for the substances or processes listed in Table 30.13. Other materials for which monitoring is likely to be required include asbestos and lead.

Table 6.1 identified emission rates from a variety of process equipment and fittings. More recent data on fugitive emissions from equipment leaks have been published by the Environmental Protection Agency in the USA.[93] Examples were included in the programmes needed to reduce emissions through more rigorous field testing, data management, maintenance and emission estimation methods. Whilst the data primarily related to

Table 30.13 Specific substances and processes for which monitoring is required in the UK[92]

Substance or process	Minimum frequency
Vinyl chloride monomer	Continuous or in accordance with a procedure aproved by the Health and Safety Commission
Vapour or spray given off from vessels at which an electrolytic chromium process is carried on, except trivalent chromium	Every 14 days

volatile organic compounds, the principles are applicable to the control of any toxic materials.

Health surveillance

With many toxic gases or vapours a known adverse health effect can be readily observed. If this can reasonably be anticipated under the circumstances of work then some form of medical surveillance is desirable.

Within the UK health checks are a statutory requirement for the agents, operations and processes summarised in Tables 30.14 to 30.16. (Exposures

Table 30.14 Substances and processes for which medical surveillance is required unless exposure is insignificant[94]

Substances for which medical surveillance is appropriate	Processes
Vinyl chloride monomer (VCM)	In manufacture, production, reclamation, storage, discharge, transport, use or polymerisation
Nitro or amino derivatives of phenol and of benzene or its homologues	In the manufacture of nitro or amino derivatives of phenol and of benzene or its homologues and the making of explosives with the use of any of these substances
Potassium or sodium chromate or dichromate	In manufacture
1-Naphthylamine and its salts Orthotoluidine and its salts Dianisidine and its salts Dichlorbenzidine and its salts	In manufacture, formation or use of these substances
Aurumine. Magenta	In manufacture
Carbon disulphide Disulphur dichloride Benzene, including benzol Carbon tetrachloride Trichlorethylene	Processes in which these substances are used, or given off as vapour, in the manufacture of indiarubber or of articles or goods made wholly or partially of indiarubber
Pitch	In manufacture of blocks of fuel consisting of coal, coal dust, coke or slurry with pitch as a binding substance

1750 Toxic hazards

Table 30.15 Other substances and processes to which the definition of carcinogen relates and requiring health surveillance[94]

Aflatoxins
Arsenic and inorganic arsenic compounds
Beryllium and beryllium compounds
Bichromate manufacture involving the roasting of chromite ore
Electrolytic chromium processes, excluding passivation, which involve hexavalent chromium compounds
Mustard gas (B,B'Dichlorodiethyl sulphide)
Calcining, sintering or smelting of nickel copper matte or acid leaching or electrorefining of roasted matte
Ortho-toluidine
Coal soots, coal tar, pitch and coal tar fumes

The following mineral oils:
   (i) unrefined and mildly refined vacuum distillates;
   (ii) catalytically cracked petroleum oils with final boiling points above 320°C;
   (iii) used engine oils;

Auramine manufacture*
Leather dust in boot and shoe manufacture, arising during preparation and finishing
Hard wood dusts
Isopropyl alcohol manufacture (strong acid process)
Rubber manufacturing and processing giving rise to rubber process dust and rubber fume
Magenta manufacture
4-Nitrobiphenyl

* See also Table 30.14.

Table 30.16 Other agents and operations for which health surveillance is required in the UK

Agent/operation	UK legislation
Asbestos	Control of Asbestos at Work Regulations 1987
Compressed air (other than diving operations	The Work in Compressed Air Special Regulations 1958
Diving operations	Diving Operations at Work Regulations 1981
Ionising radiations	Ionising Radiations Regulations 1985
Lead	Control of Lead at Work Regulations 1980 (HSE Guidance Note EH 29)
Mine dusts	The Coal Mines (Respirable Dust) Regulations 1975

to some of these agents are clearly relevant in forms other than the gaseous phase but the complete list is given for reference purposes.) Health surveillance is also advised for the agents listed in Table 30.17.[94-111]

Health surveillance includes the keeping of an individual health record.

Table 30.17 Agents for which health surveillance is advised

Agent	Ref.
Agents liable to cause skin disease	92
Antimony	93
Arsenic*	94
Beryllium*	95
Cotton dust	96
[Genetic modification]	97
Isocyanates	98
Mineral wool	99
Platinum	100
Talc dust	101
Biological monitoring	
Cadmium	102
Mercury	103
Trichloroethylene	104
Biological effect monitoring	
Organo phosphorus pesticides	105

* See also Table 30.15.

The range of procedures will depend upon the specific chemical, or mixture of chemicals, to which exposure is possible but may include:
- Medical surveillance
  Under a qualified medical practitioner and possibly encompassing clinical examinations and measurements of physiological and psychological effects of exposure as indicated by alterations in body function or constituents.
- Biological monitoring
  Measurement and assessment of chemicals or their metabolites in tissues, secreta, excreta or expired air, in exposed persons.
- Biological effect monitoring
  Measurement and assessment of early biological effects of exposure.
- Routine medical enquiries/examinations
  Inspection or examination, or enquiries about symptoms, by e.g. an occupational health nurse. For example, enquiries seeking evidence of respiratory symptoms related to work are appropriate with chemicals known to cause occupational asthma, i.e. including those in Table 28.14.
- Routine inspections
  Inspection by e.g. a supervisor or manager.
- Review of records and occupational history
  Review of personal records and medical history during and after exposure.

Emergency procedures

It is generally found that predictions of casualties following the accidental release of a toxic gas tend to be pessimistic compared to the results of

1752 *Toxic hazards*

known incidents. (The Bhopal incident described on page 883 was an exception to this.) In such a release the workforce and neighbours normally take actions to minimise the toxic effects, e.g. stay indoors, and other mitigative measures may be appropriate; neither of these possibilities is properly accounted for in models based simply upon dispersion and toxicity estimations.[58]

> A maintenance man mistakenly removed the bonnet on a valve in a chlorine line. About 200 construction workers in a neighbouring company 0.4 to 0.8 km downwind were enveloped in the cloud and required hospitalisation but were released the next day after treatment. Emergency procedures involving a water curtain and foam blanket were put into effect after the leak had been isolated. The local emergency plan was put into action; incoming traffic to the area was halted – except for emergency vehicles – and outward-moving traffic was accelerated.

Pre-planning and training, to the extent discussed in Volume 4, is crucial to cope with any foreseeable emergency involving toxic gas release since – apart from donning an appropriate respirator immediately – survival may depend upon which of the strategies shown in Fig. 30.3 each individual adopts. (This is confirmed by reference to the Potchesftroom incident described on page 888.) The technical background to emergency procedures is summarised below.

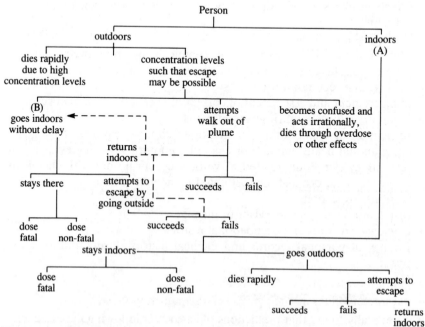

Fig. 30.3 Possible actions of persons exposed to a toxic gas[108]

Escape from a gas cloud

Mathematical models have been derived with which to assess the chances of an individual engulfed in a toxic gas cloud being able to escape by walking out of it by proceeding in a crosswind direction. The basis is that successful escape would occur if the time to walk across the wind is less than the 'lethal exposure time' at the cloud axial concentration.[25]

However, the predictions from such models may be over-optimistic if the assumptions are made that the individual perceives the release and sets off walking at right-angles to the cloud without panic, i.e. at an even pace to ensure that his respiration rate does not increase substantially, and without deviation.[58] In fact an escapee may adopt a tortuous route out of a cloud – thus receiving a dose greater than predicted – and have to cope with physiological reactions to exposure to gas levels below those considered fatal. Thus it may be concluded that escape is only likely to be successful for a person at such a distance from release that his theoretical straight-line cross wind path at walking pace would accumulate less than 0.1 of the 'fatal dose'.[84]

Numerous modelling techniques also exist to help predict off-site risk from toxic releases.

> The mean release rate of MIC from the Bhopal plant was 7.4 kg per second over a 90-minute period. Experimental animal studies indicate LC_{50} values for 1 and 2 hour exposures of 30 and 20 ppm respectively. A computer model estimated downwind distances from the Union Carbide plant for these concentrations to be approximately 4.2 and 5.7 km and the angle of emergency sector was 68 °. The calculated emergency sector had a width of 7.7 km (which included all reported casualties).[112]

However, uncertainties arise from e.g.[113]
- the selection of failure cases from the range of possibilities
- failure probabilities for each failure case
- scale release rates and durations
- conversion of a failure case to a source term for use in further calculations
- the validity of the dispersion model
- meteorological inputs
- topographical inputs
- human and environmental response to toxic, pressure, or thermal burdens
- ameliorating factors
- enhancing factors
- ignition probabilities
- parameter values in many of the mathematical models

> One analysis of risk associated with a 15,000 tonnes ammonia tank farm near Thessaloniki concluded that the outcome of gas release could range from tens

of fatalities (mainly staff working at, or near to, the terminal) in unstable conditions and a north wind, to 200,000 fatalities in stable conditions and a northwest or southwest wind.[112]

Staying indoors

It can be shown by calculation that staying, or going, indoors will provide substantial protection against short to medium duration continuous and quasi-instantaneous toxic gas releases.

> A cloud of ammonia gas from a ruptured tank truck enveloped a building in Houston but the workers inside remained unaffected. Some 94 people caught out-of-doors were, however, injured, four of them fatally.[48]

The correct procedure therefore is to escape indoors and to reduce the ventilation rate in the building as soon as possible. For example, at least one order of magnitude attenuation of the outdoor dose may be achieved at ventilation rates of 2.0 per hour or less, a range obtainable in naturally ventilated housing with the doors and windows closed.

Once a cloud has passed, the gas inside the building only leaks out, and is replaced by fresh air, at a rate proportional to the ventilation rate. Therefore people who escape, or remain indoors, will continue to add to their dose if they remain there *after* the cloud has passed by. They should either increase ventilation by opening windows and/or switching on ventilation systems or go outdoors. They must, however, in no circumstances go outside until the cloud has completely passed.[114]

> A family were caught in their home very close to a rail car accident in which there was a quasi-instantaneous release of 30 tonnes of chlorine. They received a high dose and after 15–20 minutes the father carried his eleven-month-old child, who was 'choking' outside. After 30 minutes' exposure the family were rescued but the critical dose received by the child outside led to his death.[15]

Evacuation

Complete evacuation is the best solution for the protection of the public but only providing that it can be confidently predicted that there is sufficient time before an incident occurs or escalates. Evacuation is time-consuming and difficult, and during it people are vulnerable to the effects of an incident which occurs before it is expected.

> With low windspeeds of 2 m/s the front of a chlorine gas cloud will travel 1.2 km within the first ten minutes. Hence there is essentially no scope for outside intervention or assistance during the incident for a community within 2 km of the release.[114]

Furthermore the probabilities are that[114]

- most serious toxic incidents which involve rupture of a vessel or pipeline will occur quasi-instantaneously or escalate rapidly;
- a significant time will elapse before the emergency services can arrive in nearby communities and a problem may arise in persuading people to move at all.

Hence pre-incident evacuation may be a strategy of limited value. Conversely once an incident is over, provided there is no risk of recurrence, post-incident evacuation may be valuable to bring people out of their homes into the fresh air.

Post-incident evacuation is also, of course, a consideration if a persistent toxic chemical has been released into the environment.

Following the accident at Seveso (page 510 and 892) it was 27 hours before the local authorities were informed that, as a result of an escape of trichlorophenol, a cloud of herbicide had been released which had caused damage to surrounding crops. Production was resumed but was halted a few days later and the plant closed.[109] Some five days later the authorities were told that a hyper-toxic material might have escaped along with other by-products. On the weekend following the release the company revealed that TCDD had been released and subsequently advised the authorities to evacuate an area two miles 'downwind' and 0.5 miles wide.

Although it was on 23 July that advice to evacuate was passed to the authorities, the official order to proceed was not given until 2 August. The forbidden zone was not fenced or policed.[115]

An emergency commission specified a contamination level of 5 µg/m^3 in soil samples as acceptably safe. As a result three zones were drawn on the map, as in Fig. 30.4. Zone A of 267 acres was where 95 per cent of the TCDD was

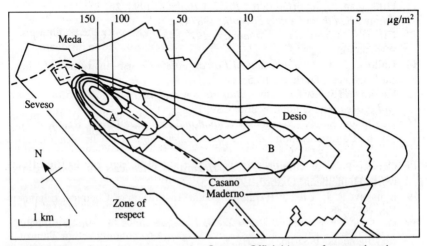

Fig. 30.4 Plan of zones contaminated at Seveso. Official 'zone of respect' and estimated contours of the cloud for 150, 100, 50, 10 and 5 µg/m^3 assuming a 2 kg TCDD release

thought to have fallen and was therefore evacuated. The inhabitants of Zone B, where the average level of TCDD in the soil was 3 µg/m^3, were allowed to stay. However, they were advised to wash regularly, and children and pregnant women were evacuated by day and allowed to return at night.

It was later concluded[115] that Zone A should have been extended another 200 acres across the Milan motorway, where, soil samples were found containing 40 µg/m^3, and that the authorities should have evacuated this zone whilst leaving the motorway open. Instead the inhabitants of this area were dealt with as for Zone B.

References

1. Clark, I. and Drinker, P., *Industrial Medicine*, National Medical Book Co., New York, 1935, 199.
2. Mulcahy, E. W., *The Pickling of Steels*, 1973, 4.
3. Foulger, J. H., in *Safety and Accident Prevention in Chemical Engineering Operations*, Fawcett, H. W. & Wood, W. S., eds., Ch. 16, John Wiley & Sons, New York, 1965.
4. Surda, A. & Agnew, J., *British J. Industrial Medicine*, 1989, **46**, 541.
5. Anon., *Chem. Eng.*, 17 January 1991, 16.
6. MAFF, *Current Topics*, Ministry of Agriculture Fisheries and Food, July 1970, UK.
7. Cronin, E., *Contact Dermatitis*, 1980, 869.
8. Private Communication, *Chemical Safety Summary*, Chem. Ind. Assoc., 1984, **66**, 55.
9. Health and Safety Executive, *Entry into Confined Spaces*, Guidance Note GS5.
10. *Eastern Daily Press*, 28 October 1989.
11. Hinksman, J., reported in the *RoSPA Bulletin*, 1990, **20**, 4.
12. Anon., *Health and Safety at Work*, 1987 (Nov).
13. CISHEC, *Chem. Safety Summary, 1977.* **48**, 190, 22, Chemical Industries Association.
14. Bullock, C. J., *Fundamentals of Process Safety Course*, Institution of Chemical Engineers, Durham, 1985.
15. Health and Safety Executive, *Bulk storage and use of carbon dioxide: hazards and procedures*, Guidance Note CS9, 1985.
16. Health and Safety Executive, *Occupational Exposure Limits*, EH40, 1993.
17. Private Communication 1981, *Chemical Safety Summary*, Chemical Industries Association, 1981, 29.
18. Carson, P. A. & Mumford, C. J., *Hazardous Chemicals Handbook*, Butterworth-Heinemann, Oxford, 1994.
19. Lynskey, P. J., *Loss Prevention Bulletin*, Institution of Chemical Engineers, 1985(063), 9.
20. Health and Safety Executive, *Loss Prevention Bulletin*, Institution of Chemical Engineers, 1985(063), 13.
21. Anon., *RoSPA Bulletin*, 1986 (Jan), 2.
22. Katz, M., *Effects of the contaminants other than sulphur dioxide on vegetation*

and animals, National Conference on Pollution and Our Environment, Canadian Council of Resource Ministers, Montreal, Oct–Nov, 1966.
23. HSE, *Farmwise, your Guide to Health and Safety*, HMSO, 1992.
24. Marshall, V. C., *Major Chemical Hazards*, Ellis Harwood Ltd., Chichester, 1987, 341.
25. Anon., *Loss Prevention Bulletin*, Institution of Chemical Engineers, 1990(091), 19.
26. Health and Safety Executive, *Carbon Monoxide*, Guidance Note EH43, 1986.
27. Steward, R. D., *Annu. Rev. Pharmacol.* 1975, **15**, 409.
28. *Carbon Monoxide Poisoning Causes and Prevention*, Ministry of Labour SHW 29, 1965, HMSO.
29. Annu. Report of HM Chief Inspector of Factories 1974, 61, HMSO.
30. Huuinen, M., *Loss Prevention Bulletin*, Institution of Chemical Engineers, 1990(092), 17.
31. CFA Health and Safety Committee, *Carbon monoxide hazards in cupola operation and maintenance*, Council of Ironfoundry Associations, December, 1969.
32. *The Gas Safety (Installation and Use) Regulations* 1984, SI 1984 No.1358.
33. Kowitz, T. A., Reba, R. C., Parker, R. T., et. al., *Arch. Environ. Health*, 1967, **14**, 545.
34. Coppinger, G. M., *J. Am. Chem. Soc.*, 1954, **76**, 1372.
35. Bretherick, L., *Handbook of Reactive Chemical Hazards*, 4th edn, Butterworths, London, 1990, 150.
36. Anon., *Loss Prevention Bulletin*, Institution of Chemical Engineers, 1984(056), 24.
37. HM Factory Inspectorate, *Fire Risk in the Storage and Industrial Use of Cellular Plastics*, Technical Data Note 29 (Rev), Department of Employment, 1972.
38. Anon., *Chem. Eng.*, 13 September 1990.
39. Anon., *Loss Prevention Bulletin*, Institution of Chemical Engineers, 1985(066), 23.
40. Bretherick, L. *Handbook of Reactive Chemical Hazards*, Butterworths, 1990, 136.
41. Manufacturing Chemists Association, Safety Data Sheet SD-67, 1961.
42. Health and Safety Executive, *What you should know about cyanide poisoning. A guide for employers*, MS(A) 9, 6/90.
43. Read, J., *Humour and Humanism in Chemistry*, G. Bell & Sons Ltd., 1947.
44. World Health Organisation, *Oxides of Nitrogen*, Environmental Health Criteria, WHO, Geneva, 1977, 4.
45. Orehek, J., Massari, J. P., Gayrard, P., *J. Clin. Invest.*, 1976, **57**, 301.
46. Morley, R. & Silk, S. J., *Ann. Occup. Hyg.*, 1970, **13**, 101.
47. Manufacturing Chemists Association, MCA Case History 1903.
48. Gray, E., *Arch. Ind. Health.*, 1959, **196**, 479.
49. *RoSPA Bulletin*, 1985, 3.
50. Atherley, G. R. C., *Occupational Health and Safety Concepts, Chemical and Processing Hazards*, Applied Science Publishers, 1979.
51. Baxter, P. J., 'Gases' in *Hunter's Diseases of Occupations*, 8th edn, Edward Arnold, 1994, p. 245.

52. Bretherick, L., *Handbook of Reactive Chemical Hazards*, Butterworths, 1990, 1229.
53. Henderson, D., *Ammonia Plant Safety*, 1975, **17**, 132.
54. Meidl, J. H., *Flammable Hazardous Materials*, 2nd edn, Glencoe, 1978, 140.
55. Anon., *Loss Prevention Bulletin*, Institution of Chemical Engineers, 1989(086), 27.
56. Institution of Chemical Engineers, Rugby, *Preventing Emergencies in the Process Industries*, Training Package.
57. Kaufman, J. & Burkons, D., *Arch. Environ. Health*, 1971, **23**, 79.
58. Purdy, G. & Davies, P. C., *Loss Prevention Bulletin*, Institution of Chemical Engineers, 1985(062) 1.
59. Anon., *RoSPA Bulletin*, April 1990, **20**, 4, 3.
60. Anon., *Loss Prevention Bulletin*, Institution of Chemical Engineers, 1989(086), 1.
61. Joyner, R. E. & Durel, E. G., *J. Occup. Med.*, **4**, 1962, 152.
62. Beard, R. R., in *Patty's Industrial Hygiene and Toxicology*, 3rd Rev. Edn., Vol. 2C, G. D. Clayton and F. E. Clayton (eds), 1982, 4127.
63. Flury, F. & Zernik, F., *Schadliche Gase*, Springer, Berlin, 1931.
64. Health and Safety Executive, *Radon in the Workplace*, Leaflet INE (G) 138 L, December 1992.
65. Department of the Environment, *The Householders Guide to Radon*, UK.
66. Gafney, E. T., *Harper Hosp. Bull.*, 1966, **24**, 262.
67. Health and Safety Executive, *The Principles of Substitution for Hazardous Substances*, Draft Guidance Note, April 1993.
68. Health and Safety Executive, *A Step by Step Guide to COSHH Assessment*.
69. Schedules 4, 5 & 6 of the Environmental Protection (Prescribed Processes and Substances) Regulations 1991.
70. Morton, F., *Report of the Inquiry into the Safety of Natural Gas as a Fuel*, Ministry of Technology, July 1970.
71. King, P. J., Clegg, G. T. & Walters, W. J., *Report of the Inquiry into serious gas explosions*, Department of Energy, June 1977, UK.
72. Department of Employment, *Arsine: Health and safety precautions*, Technical Data Note 6 (Rev.), 1972.
73. Ember, L. R., *Chemical & Engineering News*, 8 July 1991, 7.
74. Anon., *Chemical Safety Summary*, Chemical Industries Association, 1982, **53**, 364.
75. Kletz, T. A., *Cheaper Safer Plants or Wealth and Safety at Work*, Institution of Chemical Engineers, Rugby, 1984, 77.
76. Bond, J. & Bryans, J. W., *Loss Prevention Bulletin*, Institution of Chemical Engineers, 1987(075), 27.
77. Sampara, P., *Control of Exposure to Welding Fumes and Gases*, Canadian Centre for Occupational Health and Safety, Ontario, 1985.
78. Maddocks, J. G., *Annals Occup. Hyg.*, 1970, **13**, 247.
79. *RoSPA Bull.* 1985 (May), 3.
80. *RoSPA Bull.*, 1986 (March), 2.
81. Anon., *Occupational Safety Bulletin*, Royal Society for the Prevention of Accidents, February 1969.
82. Anon., *Safety News*, Universities Safety Association, November 1978, **10**, 10.
83. Control of Substances Hazardous to Health Regulations 1988, Reg. 9.2.(a).

84. Anon., *Safety News*, Universities Safety Association, January, 1972, 8.
85. a) Health and Safety Executive, *Entry into confined spaces*, Guidance Note GS5.
    b) CISHEC, *Chem. Safety Summary*, Chemical Industries Association, 1978, **49**(194), 30.
86. *RoSPA Bull.*, 1986 (Oct), 3.
87. Health and Safety Commission Newsletter, 1988 (Aug), 5.
88. Finch v. *Telegraph Construction & Maintenance Co. Ltd.*, 1949, 1 All E.R. 452.
89. Hempseed, J., *Loss Prevention Bulletin*, Institution of Chemical Engineers, 1991(097), 1.
90. Health and Safety Executive. *Dangerous Maintenance*
91. *RoSPA Bull.* 1987 (Dec), 3.
92. Health and Safety Commission, *Approved Code of Practice, Control of Substances Hazardous to Health (General ACOP – 4th edn)*.
93. Scagtaglia, A., *Fugitive Emissions Estimation Methods and Reduction Strategies Utilising Leak Detection and Repair Programmes For Equipment Leaks*, paper presented at *Management of Environmental Protection and Safety* Conference, Institution of Chemical Engineers, London, 10–11 February 1993.
94. Health and Safety Commission, *Approved Code of Practice, Control of carcinogenic substances (carcinogens ACOP – 4th edn)*.
95. Health and Safety Executive, *Occupational skin diseases: health and safety precautions*, Guidance Note EH 26, HMSO, 1981.
96. Health and Safety Executive, *Health surveillance of occupational skin disease*, Guidance Note MS 24, HMSO, 1990.
97. Health and Safety Executive, *Antimony – health and safety precautions*, Guidance Note EH 19, HMSO, 1978.
98. Health and Safety Executive, *Arsenic-toxic hazards and precautions*, Guidance Note EH8 (Rev), HMSO, 1990.
99. Health and Safety Executive, *Beryllium – health and safety precautions*, Guidance Note EH 13, HMSO, 1977.
100. Health and Safety Executive, *Byssinosis*, Guidance Note MS9, HMSO, 1977.
101. Advisory Committee on Genetic Manipulation & Health and Safety Executive, *Guidelines for the health surveillance of those involved in genetic manipulation at laboratory and large-scale*, ACGM/HSE Note 4.
102. Health and Safety Executive, *Isocyanates: toxic hazards and precautions*, Guidance Note EH16, HMSO, 1984.
103. Health and Safety Executive, *Isocyanates: medical surveillance*, Guidance Note MS8, HMSO, 1983.
104. Health and Safety Executive, *Man-made mineral fibres*, Guidance Note EH46 (Rev), HMSO, 1990.
105. Health and Safety Executive, *Medical monitoring of workers exposed to platinum salts*, Guidance Note MS22, HMSO, 1983.
106 Health and Safety Executive, *Control of exposure to talc dust*, Guidance Note EH32, HMSO, 1982.
107. Health and Safety Executive, *Cadmium – health and safety precautions*, Guidance Note EH1, HMSO, 1986.
108. Health and Safety Executive, *Mercury – health and safety precautions*, Guidance Note EH17, HMSO, 1977.

109. Health and Safety Executive, *Mercury – medical surveillance*, Guidance Note MS12, HMSO, 1978.
110. Health and Safety Executive, *Trichloroethylene – health and safety precautions*, Guidance Note EH5, HMSO, 1985.
111. Health and Safety Executive, *Biological monitoring of workers exposed to organo-phosphorous pesticides*, Guidance Note MS17 (Rev.), HMSO, 1987.
112. Psomes, S., *Disasters*, 1990, **14**(4), 301.
113. Cassidy, K. & Pantony, M. F., *Institution of Chemical Engineers, Symposium Series*, 1988, No. 10, 75.
114. Purdy, G., and Davies, P. C., *Loss Prevention Bulletin*, Institution of Chemical Engineers, 1985(062), 1.
115. Lihou, D. A., *Loss Prevention Bulletin*, Institution of Chemical Engineers, 1982(042), 25.

CHAPTER 31

Toxic liquids

Introduction

Evidently a common source of toxic vapours of the kind discussed in Chapter 30 is spillage or leakage of volatile liquids. If the liquid is 'obviously' volatile the hazard is likely to be appreciated; the potential problem from less volatile liquids, particularly in confined spaces or rooms, is less well understood.

> Four power-station workers were exposed to high concentrations of mercury vapour when a manometer broke inside a condenser where they were working. After an hour they exhibited symptoms of acute pneumonitis with coughs, chest pains, shortness of breath, shivering, loss of appetite, weakness and joint pains. The following day the symptoms had largely abated and there were no serious consequences.[1]

(A brief discussion of acute and chronic poisoning due to mercury is given on pages 78 and 79. Toxic effects may follow inhalation of mercury vapour, or absorption via unbroken skin, or ingestion of mercury or its inorganic compounds. Organo-mercury compounds produce different effects as noted on page 249. With metallic mercury, ingestion is not an expected route of absorption, since if swallowed it is poorly absorbed; by comparison approximately 90% of mercury vapour which is inhaled is absorbed through the lungs into the bloodstream. Direct exposure to mercury vapour can be significantly added to by the local environment created if there is contamination of clothing and exposed parts of the body; this 'microenvironment' may contain vapour concentrations several times greater than in the workplace itself.)

Some liquids are inherently 'toxic', e.g. with the potential to cause damage to the skin or mucous membranes, or to be absorbed via them, or to cause severe systemic effects on ingestion.

The risks associated with accidental ingestion of chemicals are fortunately now, in general, well appreciated. However, this was not always the case.

> A 1925 survey of employees in a watchmaking factory revealed an abnormally

high incidence of cancer of the jaw. Ladies engaged in painting using luminous radium paint had developed the habit of shaping the small paint brushes to points with their lips (see page 1588). This resulted in cumulative ingestion of radium.[2]

Conversely the potential for skin absorption, identified in published Occupational Exposure Limits[3] or Threshold Limit Values by an 'Sk' notation as described on page 260 is often underestimated.

A contract pipefitter was splashed with acetone cyanohydrin whilst working on a line from a 7 m high scaffold. This line had previously been opened, drained, depressurised and purged. A plumber's plug popped out of the line, probably due to a build-up of nitrogen pressure, and residual acetone cyanohydrin escaped. The man was able to climb down and was driven to the shower-room, since the nearest emergency safety shower had not been identified and no portable shower had been provided. He stripped off his clothes and showered; no symptoms of cyanide poisoning were observed over the next 55 minutes.[4]

He was provided with clean overalls and boots and put his contaminated socks back on. Shortly after being turned over to the contractor's safety supervisor he exhibited symptoms of cyanide poisoning, i.e. reddening of eyes, trembling, respiratory difficulty. The supervisor did not recognise the symptoms and did not know that antidote kits were available on site.

The man was driven to hospital, given the correct intravenous antidote and fortunately recovered rapidly, such that he returned to work three days later. (Because of the hazardous nature of acetone cyanohydrin, full protective clothing, i.e. neoprene suit and full breathing apparatus, should be considered).

The potential for absorption is naturally increased if the skin is damaged, e.g. due to a cut or puncture.

A research student was performing an experiment with 83×10^{10} Bq (22 Ci) of tritiated water contained in a glass ampoule previously sealed by a colleague. The work was carried out in a fume cupboard but on scoring the ampoule with a glass knife it fractured into a number of pieces; tritiated water splashed onto the student's face and clothing and he sustained a small cut on the thumb of the right hand from broken glass.[5]

The student washed his hands and face copiously with cold water and removed contaminated clothing as rapidly as possible.

Tritium as tritiated water is easily absorbed throughout the whole body, behaving like ordinary water with the exception of a small fraction of tritium which becomes fixed to proteins. It is a pure beta emitter with a radioactive half-life of 12.3 years; the biological half-life is typically ten days but closely related to body fluids. In this case assessment of external activity on the skin etc. was carried out by swabbing the areas and liquid scintillation counting the swabs. A urine sample was taken four hours after the accident and early morning samples were taken daily afterwards for radioactivity measurements using liquid scintillation counting.

To increase the body water cycle the student was advised to increase his liquid intake and promote diuresis.

Internal contamination from inhalation and absorption via the skin and open wound resulted in a dose commitment of 3.2×10^9 Bq (81 μCi) compared with an annual limit intake of 3×10^9 Bq. The increase in liquid intake over the initial period resulted in an observed effective biological half-life of 7.3 days, demonstrating the usefulness of the increased liquid intake.

For prevention purposes, it is advisable when using resealed ampoules to cool to -10 °C to freeze the contents before opening but an alternative form of storage is strongly recommended.[5]

Increased hazards are, of course, associated with any toxic liquid if it is dispersed as an aerosol.

An employer was prosecuted for breaches of the UK COSHH Regulations (page 1617) and the Control of Pesticides Regulations in a case in which a nineteen-year-old apprentice became ill due to occupational exposure to a fungicide. She was permitted to spray it onto the ceiling of a property rather than brush it on; this resulted in her being exposed to high levels of the fungicide.[6]

A technician developed a serious lung condition after working for seven years with diazotised sulphanilic acid (Paul's reagent) which was sprayed as an aerosol onto thin layer chromatography papers to detect phenolic and imidazole type amino acids. Previously a dipping technique had been used.

Spraying as an aerosol should hence only be carried out, as a minimum standard, in a fume cupboard of known high efficiency.[7]

Thus the substitution of methylene bisphenyl for toluene diisocyanate as a constituent of polyurethane paints resulted in a dramatic reduction in vapour pressure, which is highly desirable having regard to the very low Occupational Exposure Limits established for isocyanates in general. Thus,

	Vapour pressure (mm Hg at 25 °C)
Toluene diisocyanate	0.025
Methylene bisphenyl diisocyanate	0.00009

However, cases of sensitisation still occurred amongst spray painters in the absence of appropriate precautions against paint mists. More sophisticated polyurethane paint and lacquer formulations now usually contain <0.5 per cent free volatile isocyanate monomer. However,

'... when such paints are applied by spray (as in car refinishing) the aerosol mist droplets produced are mainly in the respirable size range and precautions are necessary to avoid an inhalation hazard. Even if spraying is carried out in an enclosure under exhaust ventilation, adequate and suitable respiratory protective equipment will be needed if employees need to work inside the enclosure'.[8] [Adequate ventilation is also essential on any ovens used for heat curing.]

1764 Toxic hazards

The general principles for protection against toxic liquids, and emergency procedures, are discussed in this chapter. This is followed by reference to some common toxic liquids, their manner of use, and procedures involved in handling them.

General principles for protection against toxic liquids

As indicated on pages 278 and 1623, certain general principles can be applied to the control of toxic liquid hazards. These are discussed below.

Substitution

Whenever practicable a less toxic liquid should be used (see page 278). However, other relevant physico-chemical properties of the liquid must be considered in conjunction with its toxicological properties, e.g. its Occupational Exposure Limit. As discussed on page 76, the vapour pressure is especially important.

> As a result of animal and human exposure data, 1,1,1-trichloroethane has earned a reputation of being one of the less toxic chlorinated solvents. Because of this its use is often preferred to that of the comparable but more toxic trichloroethylene which has a TLV of 100 ppm. However, the greater volatility at room temperature of the 'safer' solvent in some cases offsets the advantage of higher TLV.[9]

Furthermore, substitution should take into account a proper hazard evaluation of the type discussed in Chapter 20 in order to avoid inadvertently introducing other hazards.

> In the interests of safety, less toxic solvents were substituted for benzene in a standard laboratory process to reduce the water content of a polymer. This process relied on the removal of benzene and water by azeotropic distillation of a benzene solution of polymer; because temperature was important solvents of similar boiling points were used as a replacement for benzene. Uncontrollable boiling was followed by a fume-cupboard fire. The polymer was only partially soluble in the new solvent so that a two-phase mixture was heated during the distillation.[10]
>
> (The difference in vapour pressure associated with an immiscible liquid mixture is referred to on page 97.)

Clearly the more 'open-handling' associated with any particular liquid the greater the benefits achievable by substitution. Thus, following a survey in 1991 which found that over 90 per cent of painters and decorators had suffered 'ill-health' after using solvent-based decorative paints, relevant associations in the UK agreed on efforts to promote the substitution wherever possible of water-based products. (However, these currently give less than 80 per cent of the gloss of solvent-based paints.)[11]

Change of process may also be appropriate to reduce potential escape of liquid or to use it in a manner which minimises dispersion.

Processing, transfer, application and cleaning operations which generate liquid aerosols require careful assessment. Not only may the droplet size distribution be in the respirable range, but given the enormous increase in surface area to volume ratio noted on page 114, the evaporation rate can be surprisingly rapid. Any process that imparts energy to a liquid which overcomes the surface tension can produce an aerosol. The following are common operations which may require specific control measures.

- All liquid spray coating/treatment processes, e.g. timber treatment; application of paints, lacquers or adhesives; application of surface finishes as onto leather; application of mould release agents.
- Condensation of exhausted vapour on contact with air (page 1105).
- Leaks of liquid under pressure (page 1105).
- Venting gas via liquid traps or blowing through of equipment or pipelines containing liquid.
- Heating of liquids in open vats.
- Centrifugation.
- High speed machinery subject to lubrication by liquids, e.g. generating mineral oil mists (page 1596).
- High speed agitation.
- Splash filling or impingement of liquid jets upon solid surfaces.
- Bursting of bubbles in open liquid tanks, e.g. in electrolytic processes (page 282).

Reduction/elimination of storage

Minimisation of inventory/concentration was mentioned on page 275 as an important control measure and there is a tendency to reduce storages and in-process inventories of toxic liquids (page 889). For example, chlorination at water treatment works using gas drawn from liquefied storage and injected into carrier water in chlorinators is likely to be increasingly replaced by on-site electrolytic sodium hypochlorite generation.[12,13] This avoids the requirement for emergency procedures relating to toxic releases[14] but does require control measures for the hydrogen which is co-produced.[15]

Appropriate storage

Storage should preferably be in the open air or other well-ventilated place. Liquids which generate vapours heavier than air should not be stored below ground level, unless there is no access to the storage tanks, since any vapour will naturally collect in pits and basements.

There will of course be other requirements specific to the particular liquid, e.g. some analogies may be drawn from the need for cylinders and drums of chlorine to be[16]

- stored on the ground floor under cover, preferably in an annexe or separate room which is not in a main building and not near to any exit;
- protected from any source of heat or damp;
- kept away from the neighbourhood of any flammable material or of any plant in which fire/explosion is liable to occur; and
- kept so as to allow for ready removal in case of fire, e.g. on wheeled racks to facilitate handling.

Temperature control may be an important feature to avoid overpressurisation.

> An unopened 2.5 litre bottle of fuming nitric acid burst during the night after being stored for approximately 9 months in an unheated store, albeit within a heated building subject to normal ambient temperature fluctuations.[17] An extremely hazardous situation developed because of the toxic fumes emitted and impregnation of adjacent wooden doors and shelves in the store, resulting in spontaneous combustion several hours later (page 1361).
>
> The label on the bottle correctly advised 'pressure may develop on storage, store in a cool place and open with care'. Rupture probably occurred, however, due to pressure build-up during prolonged storage. Subsequently glass bottles with an external plastic coating were introduced for fuming nitric acid.

(An incident involving 0.88 ammonia solution is described on page 81.) The precautions with storage of such liquids in laboratory quantities therefore include the following (after ref. 17).

- Familiarisation with, and observance of, hazard information and storage instructions provided on labels.
- Maintaining stocks of chemicals which may develop pressure on storage to a minimum.
- Using stocks of chemicals on a first-in, first-out basis, e.g. with a system of date stamping.
- Occasionally loosening caps on bottles of chemicals prone to pressure build-up, ensuring adequate provision for venting any hazardous fumes likely to be evolved.

Enclosure

Where possible liquids should obviously be handled in enclosed plant. Excessive escape of mist or vapour from essential openings, e.g. during loading or discharge, may be controlled by either:

- maintaining the plant under sufficient negative pressure, or
- maintaining local exhaust ventilation of adequate capacity at these points. This should be designed to cope with the greatest foreseeable evolution of vapour during normal operation.

For example, the basic precautions for handling liquid isocyanates include the following (after refs. 8 and 18).
- Avoid handling in open vessels as much as possible.
- Transfer by pumps or by vacuum techniques. (Dip-leg transfer by the application of air pressure to supplier's drums is unacceptable because of the potential for vapour dispersion.)
- Use fully enclosed systems whenever practicable.
- Perform all mixing, weighing and dispensing operations under local exhaust ventilation or in suitable booth enclosures, under effective inward flow conditions.
- Perform sampling so as to prevent exposure to vapour (e.g. as on page 290) or wearing suitable respiratory protective equipment.
- Avoid vapour displacement from vessels during filling.
- Clear up spillages immediately using a suitable decontaminant liquid.

Leaks have occurred in the past during transfers of various liquids from mobile tankers to storage tanks during movement of the tanker whilst still coupled by flexible piping to fixed pipework.[19] In some cases interlinkage between the filling system and a physical barrier to prevent movement of the tanker may be advisable.

Appropriate bunding/control of drainage

Bunding with provision for surface water drainage is necessary to control spillages.

An operator had to transfer potassium hydroxide solution from a tank farm to a head tank elsewhere on the site. He knew that the pump outlet valve had been changed and to check for leaks immediately after pump start-up.[20] He commenced solution transfer, checked there were no leaks near the pump outlet valve, and telephoned to confirm there was an increase in level in the head tank. On return to the stock tank he found solution overflowing from a drum outside the bund wall. He immediately stopped the pump and discovered the drain valve was open; he closed this valve and alerted the foreman.

Instructions were issued to hose down the affected area and a call made to the Fire Brigade to assist with this. It was realised that a significant quantity of solution may have been spilled and entered the storm water drain. Action therefore had to be taken to prevent contamination of a nearby brook into which this discharged.

Amongst the list of recommendations following this incident were the need to ensure that any future solution draining was to drums inside the bund, and to reroute drainage to the effluent drains. Drains were also recommended across vehicle access gates to prevent liquid leaving the area.

Bunds should be kept clear of debris and a regular check made that bund drain valves are shut.

About 5 tons of process liquid escaped through an open bund drain valve during a transfer; it had entered the bund via a pump drain valve which had been left open. The bund drain valve, a gate valve, had been inadvertently opened during the transfer and a piece of debris trapped in the seat kept it partly open.

Provision should also be made for fire water catchment and, if necessary, treatment prior to disposal.

Systems of work

The importance of proper systems of work has been emphasised on pages 320, 967 and 1623. This must include the supply of appropriate information to users of toxic liquids, even if they are commonly used for routine purposes.

> A supermarket assistant was pricing cans using an ink pad and rubber pad and rubber stamp. To apply more marking ink he decided to slash off the top of an unopened polythene bottle using a Stanley knife. As he was sawing at the bottle ink suddenly spurted out into his eye causing severe pain. No first-aid instructions were displayed on the bottle or anywhere in the store, and since irrigation with water proved ineffective, the casualty was taken to hospital. Treatment there was delayed until next morning when the solvent base of the ink was identified in the hospital laboratory. (The bottles were actually imported in cartons containing information on hazards and emergency treatment but these were not passed on from the distribution depot.)[21]

With highly toxic chemicals, e.g. solutions of cyanides, systems of work will entail written instructions, two-man operation, and the correct use of safety equipment and full personal protection to guard against mishap. Thus when handling hydrogen cyanide or acrylonitrile, full neoprene clothing is worn including gloves, boots and overalls with hoods, together with breathing apparatus. Back-up emergency procedures must also be provided.

The need for permit-to work systems when dealing with toxic liquids in confined spaces was explained on page 981 and is illustrated by the following example.

> A sixteen-year-old plant operator collapsed and died within a 1.5 m deep by 1.75 m wide tank containing sludge from an alkaline paint stripper containing methylene chloride and methanol. The tank had been used to recover aluminium from powder-coated aluminium components. It had been drained but still contained 13 cm of sludge; without authority, the young person entered the tank and began to drain liquid from the sludge into a bucket. He was told to leave but this instruction was not apparently enforced and shortly after he was found in a collapsed state. (The correct procedure for entry into such a confined space, involving control by a permit-to-work system, is summarised on page 981 and covered in full in ref. 22.)

The significance of the vapour densities of such solvents being much heavier than air, and therefore being likely to accumulate at low level, was previewed on page 95. In such circumstances a confined space may be anywhere in which the worker necessarily has to breathe a localised atmosphere.

An eighteen-year-old trainee used a proprietary paint cleaner containing methylene chloride to clean out a large paint pot, about 2 m diameter by 1 m deep. He was found unconscious within it but fortunately recovered. It was subsequently estimated, by simulating the conditions, that the exposure level was 500 times the Occupational Exposure Limit of 100 ppm. Although the work was done in a well-ventilated area, near some open double doors, proper ventilation or breathing apparatus was needed.[23]

The requirement for controlled work procedures extends to the loading/unloading, transportation, and disposal of toxic liquids as well as to their normal usage.

Following a fatal accident, a company analysed the off-loading of road tankers of nitric acid and hydrofluoric acid and initiated the following actions:[24]

- All valves at the bottom of storage tanks were so positioned that personnel could open them without getting under, or in close proximity to, the tanks. A number of valves were fitted with stalks.
- A second means of escape was ensured from the top of every storage tank.
- For nitric acid only, special stainless steel snap couplings with Viton A seals were fitted for off-loading. These required only a half-turn to engage – thus reducing the time to make a joint – and totally eliminated fuming when seals were broken.
- Additional emergency showers were provided at storage sites. These were subjected to weekly, recorded, checks.
- Compressed air lines were re-sited away from the off-loading area.
- All off-loading of acid was restricted to the hours of daylight. During these operations the areas were sealed-off from road, rail and pedestrian traffic.
- Electric lighting was repositioned away from the area and floodlighting installed, so eliminating any need for electricians to work on top of the tanks.
- Special storage lockers were positioned in the area, for protective clothing and engineering items, e.g. flanges, bolts, off-loading pipes.
- All main acid discharge valves were fitted with a special locking-off facility.

Personal protection

The type and nature of personal protection will depend upon the chemical and process concerned. (As noted for toxic vapours and solids, however, this should always be considered a last line of defence.)

Impervious gloves should be provided and worn where hand contact is likely.

Impervious aprons, armlets or suits may also be necessary as will suitable eye protection. Any respiratory protection equipment or clothing must clearly be appropriate to the specific risk.

> A supervisor sustained burns to both feet whilst attempting to stop a leak between a storage tank, containing several thousand litres of acid, and a valve. He was wearing protective boots but they were too short and acid was able to enter via the tops.[25] The sealing of impervious gloves at the wrists, e.g. using armlets or simply elastic bands, may be an important precaution.

A common injury with phenol occurs when it drips inside a glove and remains in contact with the skin for e.g. an hour or more, producing a severe burn.

(Pure phenol deposited on the skin does not give rise to an immediate burning sensation. The skin will often go white and become anaesthetised and insensible to touch. The severity of the burn, which depends upon the contact time, is only evident many hours afterwards.)

Also all the protective apparatus/clothing should be maintained in a sound condition.

> An operator was cyanosed whilst removing monochloroaniline from the base of a still. He was wearing protective clothing but one of the gloves had a hole in it.[25]

> A fitter was changing a valve on an anhydrous hydrofluoric acid receipt manifold. Both he and a standby operator were dressed correctly for the job in protective clothing comprising a two-piece PVC suit, PVC gloves and Vistavision combination air-line mask/hood with their wrists and ankles taped.
>
> Some anhydrous hydrofluoric acid fume and liquid was evolved when changing the valve and this was treated with water. The fitter felt an irritation on his left forearm; he removed the PVC suit and recognising that he had sustained a burn placed his arm under cold water and cooled it with ice. He was then treated at the Works Surgery and fortunately the burn proved to be of a very mild nature. (Hydrofluoric acid can produce deep, painful blisters after a delay period so prompt action is essential.)
>
> Examination of the suit jacket subsequently revealed a 5 cm tear in the left outer sleeve and a smaller tear, approximately 1 cm long, in the inner sleeve of which the fitter was not aware. (The recommended procedure for opening up the manifold involved steaming of its outer surface with low pressure steam during purging to assist removal of anhydrous HF; sodium carbonate solution was provided to neutralise leaks.)

Continuity of protection

As explained in Chapter 30 (page 1747) personal protection must be worn continuously throughout the duration of exposure to a hazard. There are numerous examples where misfortune has resulted in an incident coincident with the removal of face, head or other protection.

An operator had been working energetically. He took off his safety helmet to wipe his brow and sustained acid burns on his head when acidic liquid dripped onto him.

An operator was assisting with off-loading of a phenol road tanker. Because of difficulty in seeing clearly what he was doing in one operation he removed his RED hood. Liquid splashed in his face; fortunately it was not phenol.

A fitter was instructed to fit a blank flange downstream of a caustic solution valve. The plant had been shut down but still contained hazardous materials; no clearance certificate or permit-to-work was issued. He was unable to move the bolts downstream of the valve which he also found to be blocked solid and inoperable.
Wearing safety goggles, he therefore began to break the line upstream of the valve. When two top bolts were removed from the four-bolt flange, liquid dribbled out and then stopped. He removed his goggles – believing the job was safe – and began to remove the remaining bolts. Caustic solution squirted into his face resulting in burns to his eye.[26] (This was also, of course, a communications breakdown, where a clearly-worded permit-to-work was desirable.)

It may be that management has been at fault in some of these incidents in that adequate training, instruction and admonition to use the personal protection has not been ensured.

A workman sustained burns to his face whilst he was lagging a pipe adjoining a caustic soda filter containment area in which transfer pumps were leaking. He stood in liquid, which he assumed was rainwater, wearing canvas shoes instead of the protective footwear provided.[4]

A Plant Engineer had occasion to check whether a caustic soda line, in an Eye Protection Area, had been cleared by steam heating. He was not wearing the eye protection provided and, unfortunately, as he cracked open the valve to the line a leak from the flange sprayed into his eyes.[25]

Clearly in such situations the measures necessary are:
- training and instruction in the correct use of protective equipment; and
- admonishing the workforce to use protective equipment when necessary, and checking that they do so.

In some cases indeed the failure to use protective clothing has followed directly from a faulty system of work.

A process operator sustained burns to his head, back and legs from hydrofluoric acid when removing a spade from a pipe. The plant was thought to be freed of acid at the commencement of the work and checks were not repeated. Hence adequate protective clothing was not worn.[4]

This example illustrates the importance of ensuring that all the conditions stated on a permit are satisfied, including checking what isolation has been carried out and for freedom from process materials.[27]

Once personal protection has been 'provided' and proper instructions given as to its use, maintenance, and replacement, some onus is placed on the individual to follow the system.

> An employer provided certain workmen with special Wellington boots, incorporating non-slip material on the soles, because where they worked was very slippery at times. When the soles of the boots became worn they tended to lose their non-slip character, so the employer made it clear that at the slightest sign or suspicion of wear an individual should request new boots which would be provided. When an employee who failed to renew his boots slipped on the floor and was injured his civil claim was unsuccessful, it being held that there was no further obligation to instruct the workers nor to instigate a system of inspection of boots to assess whether they needed replacement.[28]

> A fitter's mate, who lost an eye when a piece of metal flew into it while trying to remove a steel sprocket by hitting it with a heavy hammer, was unable to recover damages from his employer. This was due to his own failure to wear safety glasses. (It was concluded that the man had been told to wear glasses the day before the accident and that he was aware that such glasses were available to use.)[29]

There remain isolated cases in which, despite the correct types of personal protection being worn, it has not provided full protection when a system of work or engineering failure has resulted in discharge of a jet of liquid, due to displacement or reflection off the inside of a visor. Mechanical displacement also increases vulnerability.

> A process worker sustained a superficial eye burn requiring hospital treatment as a result of being splashed with 32% sodium hydroxide solution. He was working near to a transfer system under pressure when a flexible hose whipped out of a valve and knocked his glasses off. The emergency shower was starved of water but prompt action was taken by his colleagues.[30]

Emergency provisions

The requirement for appropriate antidotes, first-aid provisions and eye-wash provisions is described on page 323. Emergency water showers may also be necessary.

The water supply to such systems should be protected from cross-contamination, e.g. by the provision of dedicated supplies.

> During its routine daily test a safety drench shower system on an acid purification plant was found to be charged with phosphoric acid.
> Some time previously an operator had connected a water line to the pressure side of the acid feed pumps, by means of a temporary hose, in order to dilute acid via a feed tank. This hose was coupled to a wash-down point without the

operator realising that it was fed from the drench shower system. When the valves were opened acid flowed into the system because the acid pressure was higher.[31]

Subsequently a dedicated water supply was provided for all shower systems, an alternative wash-down system was fitted, all extraneous wash-down points or connections removed from the shower system, and specific instructions issued on connection of water lines into pressurised acid lines.

Rapid, and completely thorough, decontamination is essential in most cases. Percutaneous absorption is of particular concern with chemicals which damage the skin, especially if a diffuse dermatosis occurs.

A female worker was drenched with triphenyltin chloride from a 1500-gallon mixing vessel. She was led from the accident area by two other workers; her hard hat, goggles and coveralls – but not clothing – were removed and she was placed under a water shower. She was transferred to hospital some distance away, by which time she had sustained first-degree thermal and chemical burns to 10% of the body. Unfortunately renal failure proved fatal 12 days after exposure.[32]

Conclusion

The importance of applying these general principles, and indeed common sense, when dealing with toxic liquids is exemplified by the following incident in which a combination of errors resulted in a serious exposure to a very corrosive acid during what should have been a simple transfer operation.[33]

An operator half-filled an 18.9-litre plastic bucket with glacial acetic acid, placed a lid on it, and began to carry it from one building to another. After some 9 metres he noticed that the lid was beginning to fall off. As he set the bucket down it bumped a pallet causing it to tip and acid splashed in his face and eyes. There were no eye baths in the area but he immediately washed his eyes and face under a safety shower 12 metres away. Assistance arrived in about five minutes and following first-aid he was transported to hospital. Fortunately the acid burns did not result in permanent damage.

The following features of this accident are particularly relevant, although other recommendations were made subsequently.
- The bucket was too small and the lid could not be easily secured.
- The bucket lost its shape when lifted causing the lid to slide off.
- The area was congested, and cluttered due to construction work.
- Gloves and goggles were specified for the job but only gloves were in fact worn; this rule had been broken before, and since, but was apparently condoned.
- Eye baths had been ordered and received but not installed.

(Apart from adequate training and the provision of appropriate personal protective equipment, and ensuring its use, the provision and use of an adequate, labelled, 'closed' container and enforcing its use was the main corrective action.)

That the principles are widely applicable is exemplified by the case of mineral oil-based liquids which are widely used as lubricants, as cutting oils and in soluble oil coolants. Substitution or dilution is now used extensively and suitable refining, e.g. solvent refining and severe hydrogenation, has considerably reduced the carcinogenic activity of the mineral oil base stocks incorporated into the majority of lubricants. It is also recognised as being important to minimise exposure to lubricant mist or vapour of fume, e.g. generated due to pyrolysis on overheating by friction or quenching of hot components. The range of precautions then considered advisable to protect the worker's skin and mucous membranes are summarised in Table 31.1.

Table 31.1 Measures for protection against liquids used as lubricants

General
- Avoidance of all unnecessary contact by selection and control of working practices (e.g. on machine tools diversion or cut-off of lubricant flow whilst components are changed, or automatic ejection of components)
- Provision of enclosure, e.g. splash guards on machine tools
- Maintenance of a high standard of housekeeping, including absorption and removal of spillages
- Provision and use of appropriate protective clothing, e.g. a frequent change of overalls, impervious aprons (with detachable absorbent fronts if splashing is unavoidable) Short-sleeved overalls may be selected for machinists to avoid skin-friction from wetted cuffs. Impervious, elasticated armlets may be appropriate
- Provision and use of suitable gloves, i.e. which remain clean inside. (Unfortunately impervious gloves tend to promote excessive perspiration; porous gloves may prolong exposure) (page 311 *et seq.*)
- Provision and use of eye protection
- Provision and use of disposable rags. It is particularly important that used rags are not carried in overall or trouser pockets

Personal hygiene
- Provision of a high standard of washing and, where appropriate, showering facilities (page 319)
- Encourage, and allow time for, thorough washing of the skin at breaks and at the end of work to remove all traces of lubricant
- Encourage the wearing of clean work clothes and prompt changing of any underclothes that become wetted with lubricant
- Avoidance of the use of strong soaps or detergents, abrasive skin cleansers
- Provision and use of a suitable barrier cream before work or after washing hands during work (page 316)
- Provision and use of a skin-reconditioning cream after washing hands

Supervision and health surveillance
- Consider whether potential exposure justifies pre-employment medicals, routine medical enquiries/examinations, etc. (page 1627)
- Encourage (a) prompt attention for all cuts or abrasions; (b) prompt reporting of any skin irritation or other abnormality, via a clearly identified route to occupational health nursing/medical expertise. Encourage self-checks
- Use the issue of leaflets[34] and the display of notices (page 321) to promote good work practices and a good standard of personal hygiene.

Finally when devising safe plant and safe systems of work it is vital to consider the possibility of toxic vapour/gas generation (e.g. due to vaporisation, chemical reaction or thermal degradation) or aerosol formation. In such cases the factors outlined in Chapter 30 also apply. For example, the manner in which routine control of tetrachloroethylene exposure is achieved in dry cleaning operations by a combination of system of work, ventilation and protective clothing, with respiratory protection reserved for emergency use, is exemplified by the following recommendations.[35]

- Provide continuous mechanical ventilation in all workroom areas.
- Transfer solvent from the storage drum to the cleaning machine via an enclosed hose or piping system.
- Fully closed (single unit) system. The integral machine ventilation must be actuated automatically when the loading door is open and maintain an inward velocity of 0.5 to 0.75 m s^{-1}. The drying cycle should run long enough to remove as much residual solvent as possible from the clothing.
- Semi-closed (separate washer and dryer) system. Any machine situated in an enclosure should be mechanically ventilated with controls outside it. The washer spin cycle and the drying machine should both operate long enough to remove as much solvent as possible. Manual transfer of wet clothing from washer to dryer should be over as short a distance as possible. It requires mechanical general ventilation of the transfer area.
- Impermeable gloves and aprons should be worn when transferring wet clothes.
- All filters should be changed frequently. Spent filters and solvent reclamation residues should be disposed of as hazardous wastes, with suitable worker protection.
- Spills must be cleaned up immediately with all doors and windows open and all ventilation systems in operation. Suitable personal protection, including respirators, should be provided for all personnel engaged in clean-up. Floor drains and curbing around the machines are recommended.
- All ventilation systems should discharge to the outdoors so as to avoid re-entry into the workroom.

Examples of Toxic Liquids

The application of these general principles to specific chemicals is discussed below.

Sulphuric acid

Sulphuric acid is a colourless, viscous, dense liquid with a specific gravity of 1.86 and a boiling point of 270 °C. It is miscible with water (with which it reacts violently) in all proportions and it is a powerful oxidising agent. It finds widespread use in industry because of its properties and cost. Its anhydride, sulphur trioxide, is a liquid which fumes in air and which possesses a similar specific gravity but boils at 44.8 °C. The trioxide undergoes polymerisation in the presence of traces of water or sulphuric acid to form α, β or γ forms differing in crystal structure. Substances such as dimethyl sulphate are often added as a polymer inhibitor. (The latter is highly toxic and can accumulate through concentration, e.g. if the trioxide is distilled.) Oleum is concentrated sulphuric acid containing up to 60% sulphur trioxide. It is an oily liquid which emits fumes in moist air and which sometimes contains additives, e.g. nitric acid as an antifreeze. The acid, its anhydride and mixtures thereof are all classified as corrosives for labelling and freight purposes.

Correct selection of construction materials is essential for the safe handling of sulphuric acid. For example, in sulphuric acid production, PTFE, graphite or anodically-passivated stainless steel are used for coolers. Carbon steel is the traditional construction material for oleum but low oleum velocities should be maintained to avoid corrosion. Carbon steel is not recommended for oleum containing low levels (e.g. <5%) of trioxide. Cast iron fails catastrophically by cracking in oleum service. Special forms of cast iron, stainless steel or plastic are used for sulphuric acid piping, depending upon concentration, temperature, flow rates, pipe diameter, etc. Suppliers should be consulted for advice relating to the grade of sulphuric acid to be used and the conditions of use. When used in agriculture, e.g. a silage additive, it is usually supplied at 45% concentration, and at 70–77% w/w as a pesticide. For the former it is normally delivered in barrels made from high density polyethylene, PVC or polypropylene. It should not be stored in unlined metal containers unless they are manufactured from mild steel and the acid is >77% w/w concentration.

Sulphuric acid is supplied in concentrated form at concentrations in the range 70–100 per cent, as a colourless to dark brown oily liquid. It is a strong, corrosive acid and its toxic hazards derive from this. However, as with all toxic liquids, it is important not to overlook other – sometimes equally serious – hazards. For example, although it is non-flammable sulphuric acid will react with numerous metals and generate hydrogen as exemplified on page 102. The hazardous reactions in which it could be involved can be deduced from Chapter 10; some are listed in Table 31.2. These illustrate that, because of its reactivity, toxic gas generation is possible on contact with a wide variety of chemicals.

> A screen print employee was using a bleach solution to clean the rubber rollers of a film processor. A small amount of bleach from a cloth splashed into a solution of sulphuric and chromic acids in a sink. She inhaled the chlorine gas evolved and was subsequently off work for several days. (There was apparently a lack of adequate instructions on the hazards of darkroom chemicals and of precautions for handling.)[37]

Table 31.2 Some hazardous reactions of concentrated sulphuric acid*

Acids	With nitric acid, hydrochloric acids, formic acid, dilute acids generally – dangerous reactions with possibility of toxic gas evolution
Alkalis	Violent reaction
Oxidising agents	Violent reaction with hydrogen peroxide, permanganates, chlorine oxyacid salts. With highly concentrated or anhydrous acid the metal perchlorates form explosively unstable anhydrous perchloric acid.[36]
Water	Vigorously exothermic on dilution (acid must always be added to water to avoid local boiling)
Miscellaneous chemicals	Hazardous reactions are possible with halocarbons. Ketones, nitro-compounds, nitriles and aldehydes, e.g. acetaldehyde, undergo violent polymerisation with concentrated acid

* This list is not exhaustive; supplier's advice should be sought and/or a comprehensive hazard evaluation performed (Chapter 20).

Concentrated sulphuric acid is extremely irritating, corrosive and toxic to tissue.

An electrician sustained burns to his right foot when he stepped into a puddle of sulphuric acid in a depression caused by collapse of a gulley which contained a 27% concentration. Acid has spilled during off-loading and had not been diluted and the gulley collapsed due to corrosion and traffic loading, so that there was no means for effective drainage.[38]

Exposure results in rapid tissue destruction and severe burns; contamination of a significant area may culminate in shock, collapse and symptoms similar to those of severe thermal burns.

Sulphuric acid 'mist' may form through the reaction of sulphur trioxide with ambient moisture. Dispersion by gas or any of the other processes summarised on page 1765 may also generate a corrosive aerosol.

Precautions against exposure to liquid are as follows.
- Enclosed transfer.
- Correct selection and treatment of materials of construction. The lowest concentration which can be stored in steel without significant corrosion is 70 per cent. Surface dilution due to ingress of humid air or water can result in corrosion and hydrogen evolution or vigorous reaction. Flooding with large quantities of water after draining is practised with stainless steel tanks and a mild alkali wash with mild steel tanks.
- Shielding of pipe flanges, bunding of storage areas, etc.
- Provision of safety showers and eye-wash facilities in close proximity to areas where accidental exposure is possible.
- Provision, and conscientious wearing, of PVC plastic chemical suit, PVC plastic gloves with gauntlets, acid-resistant boots and chemical goggles/face shield.

A fitter sustained burns to the insides of his arms from drippage of sulphuric acid from a pressure gauge. [Although the system had been drained until apparently 'acid-free', failure of an internal connection had allowed leakage of a small quantity of acid into the gauge casing.] He was wearing goggles and gloves but not gauntlets or an acid-resistant suit. Subsequently the permit-to-work system was modified so as to require the wearing of full protective clothing on an acid system even when it was believed to have been drained.[39]

- Reduction of mist from acid vats may be achieved by covering the surface with plastic floats or by the addition of chemicals to alter the surface tension. However this tactic is not a primary control measure. Ventilation is always likely to be required, reinforced where necessary with personal protective equipment.
- Local exhaust ventilation is required if mist or vapour can arise. For open vats a typical arrangement comprises a lateral-slot type exhaust system positioned along a single edge, or with two slots – one on each of the long sides of the vat, capable of maintaining slot velocities of 1800–2000 f.p.m. and capture velocities ranging between 50–150 f.p.m.

In situations where exposure to mist or fumes cannot be avoided by other measures, full breathing apparatus is essential. Environmental monitoring is required (see chapter 7). Emergency procedures for concentrated sulphuric acid are summarised in Table 31.3.

Table 31.3 Emergency procedures for 70–100 per cent concentrated sulphuric acid

Eye contact	Immediately flood with copious quantities of water, holding the eye open if necessary Obtain urgent medical attention
Skin contact	Immediately wash with water, preferably under a shower; remove contaminated clothing while washing Obtain medical attention if blistering occurs or irritation persists Contaminated clothing should be washed or dry-cleaned
Inhalation	Remove from exposure. Keep warm and at rest. If there is respiratory distress give oxygen. If respiration stops or shows signs of failing, apply artificial respiration. Obtain urgent medical attention
Ingestion	Wash out mouth with water. Give plenty of water or other fluids to drink Obtain urgent medical attention; do not induce vomiting Treatment may be required for pain and shock
Spillage	Wear full protective clothing. Avoid contact with liquid. Flood with *copious* quantities of water (N.B. Reactivity with water) Contain spillages with sand or earth Inform appropriate authorities of any major spillage or entry into drains, watercourses, etc.

Emptied containers retain product residues and vapour. Therefore all labelled safeguards should be observed until they have been thoroughly cleaned. Containers should be carefully vented periodically.

Barrel storage yards should be correctly sited, secure and provided with warning notices. Stocks should be segregated from incompatible materials (see chapter 10). The floor area needs to be of acid-resistant construction and bunded to retain at least 10% by volume of the largest container.

Bulk containers for agricultural uses should be under the control of the operator. Construction must be from compatible materials; piping is normally of mild steel or polypropylene. Each container should be vented and be fitted with draining arrangements only for authorised usage. The storage area must be secure and bunded. Regular inspection is recommended for the containers and ancillaries.[40]

> A glass reinforced plastic storage tank containing dilute sulphuric acid failed catastrophically and 90,000 litres of acid were released. The violence of the failure caused the bund wall to be breached and a spillage occurred onto waste ground. The cause of the failure was acid environmental stress cracking; the tank had not been regularly inspected.

Dilution of acid requires gradual addition with agitation to the surface of the aqueous solution to avoid splashing, boiling and eruption; cooling is likely to be required (see page 100).

Organic solvents

Industrial organic solvents are used to dissolve reactants or extract oils, fats and medicinal products from vegetable and animal matter; for dry cleaning/degreasing; in paint varnishes, wood treatments, printing inks; in agricultural products; etc. Commonly these are classified according to chemical type, e.g. aromatic or aliphatic hydrocarbons, alcohols, aldehydes, ketones, amines, esters, ethers, glycols or halides. In addition to their flammability consideration must be given to their local and systemic toxic properties from skin absorption, inhalation or ingestion. Their affinity for fat-like material leads to defatting of skin and toxicity to nerve cells. Narcosis is a common outcome of over-exposure. Occasionally, the first member of an homologous series poses greatest danger (e.g. benzene, ethylene glycol monomethyl ether, ethylene oxide, formaldehyde, methyl chloride). Individual reaction to solvents is dictated by the toxicity of the specific solvent or mixture, the duration of exposure and solvent uptake, which depends upon partition between air/blood and blood/fat, pulmonary ventilation, individual stature, work practices, drug and alcohol intake, and genetic variability.

Commercial solvents often comprise mixtures, marketed under proprietary names, so that reference to individual Health & Safety Data Sheets and computation of hygiene standards (page 404) is necessary. Amongst the most common solvents in use are[41]

acetone	'petroleum spirits'
dichloromethane	toluene
hexane	1,1,1-trichloroethane
methanol	trichloroethylene
methyl ethyl ketone	white spirits
perchloroethylene	xylene

Hazards and precautions relating to this 'class' of chemical were mentioned in Chapters 5 and 6. Precautions against vapour hazards follow those described in Chapter 30. Particular care is necessary with any work in a confined space (page 981). For example, common degreasing solvents, comprising chlorinated solvents, are potent anaesthetics. A common feature of fatalities arising from their misuse, or lack of appreciation of the risks, has been loss of consciousness followed by unobserved collapse into a very high vapour concentration, e.g. within a confined space or into a vapour degreasing tank.

> A director of a small company, working alone, climbed into a standard trichloroethylene degreasing tank to check the bottom. A few gallons of liquid remained in the tank. He was overcome by the vapour, collapsed and died.[42]

The use of chlorinated hydrocarbon solvents as degreasing agents may result in significant vapour concentrations even without the application of heat.

> A young person died after using a very small quantity of solvent to remove marks from the seat of a car in a car valeting operation. The solvent vapour built up in the enclosed space of still air formed within the car.[43]

It is essential that personnel who carry out cold cleaning with solvents, even on a very small scale, should follow the supplier's instructions carefully; they should be properly instructed and supervised. The deliberate 'sniffing' of solvent has resulted in fatalities. Therefore local management need to be alert to signs that young employees may be 'sniffing'; the issue and use of solvents should be carefully controlled to avoid their misuse.

The potential for absorption of liquid via the skin or mucous membranes also requires consideration. Prolonged contact can also result in dermatitis, either through primary irritation or because action as a degreasing agent removes natural grease from the skin, rendering it prone to attack by other agents, e.g. other chemicals or metal swarf.

Precautions against excessive exposure to liquid include, in addition to appropriate education, training and supervision:
- seek solvent-free materials or 'safer' solvents;
- avoid skin contact; wear suitable protective clothing (e.g. gloves, apron) where necessary;
- avoid the use of solvents for removing paint, grease, etc. from the skin.

Use, hazards and appropriate control measures can be illustrated by reference to paints.

Paints

Paints and similar finishes are usually applied for protection or aesthetics. They comprise:

- binders such as natural oils (e.g. linseed oil) or synthetic resins made by, for example, copolymerisation of poly carboxylic acids with polyols, or from vinyl monomers, epoxy derivatives, amines or urethanes.
- pigments, usually finely ground water-insoluble dyes, e.g. white lead, titanium dioxide, lead sulphate, zinc oxide, iron oxide, lead and zinc chromates, compounds of cadmium, molybdenum, cobalt, and powdered metals including aluminium and zinc. In many countries lead pigments are rarely used nowadays because of their inherent toxicity.
- additives, e.g. to facilitate drying, reduce skin formation, etc. Examples include cobalt naphthenate, manganese soaps of naphthenic acid, oximes, substituted phenols, soaps and clays.
- solvents to dissolve and/or make the binder fluid, or to adjust viscosity. Selection is dictated by cost, solvent power, toxicity, volatility, odour, legislation, etc. Most are flammable. Common examples include white spirits, methylene chloride, toluene, xylene, acetone, butyl alcohol and acetate, ethyl alcohol and acetate, methyl ethyl ketone, paraffins, toluene, xylene. Benzene, tetrachloroethane and carbon tetrachloride have tended to be substituted on toxicity grounds.

Paints are usually applied as liquids to the substrate by brush, roller, dip wipe or as a spray using air driven or airless pressure, or electrostatic guns. In addition to the hazards associated with preparation of the substrate, hazards can arise as a result of direct contact with the liquid, or by inhalation of aerosol, or inhalation/skin absorption of volatile components as they evaporate.

Health hazards stem from over-exposure due to direct skin contact or inhalation of vapours. Provision of good ventilation, sanitation, personal protection and information are crucial controls, and operators must adopt high standards of personal hygiene.

> An operator fell off a stepladder whilst painting the second of two air heaters which were still warm. The paint was of a solvent-based aluminium type and it was deduced that inhalation of vapour whilst painting the first heater without wearing a respirator, and apparently with limited ventilation, had induced dizziness. (Clearly, painting on any warm surface will significantly increase the vaporisation rate.)

Biological monitoring may be appropriate to detect sub-clinical effects of over-exposure. The flammable hazards of paints must not be ignored and systems of work are needed wherever solvent vapour can produce hazardous concentrations.

> A paint-spray booth at a car manufacturing plant had a sloping floor down which excess paint and solvent ran to a sump below. One wall of the booth contained a projecting pipe supply of highly flammable thinners. The booth was equipped with mechanical ventilation which, when operating, effected rapid air changes in the sump. The booth and sump were cleaned by contractors at weekends when the plant was shut down. The arrangements were for the

ventilation to be switched on, for no one to be working in the booth at the same time, for the contractors to provide their own thinners, and for only approved inspection lamps to be used in the sump. One weekend a contractor entered the sump whilst another cleaned the booth. Contrary to instructions the cleaner in the booth rigged up a system to allow thinners to be run off from the protruding pipe into a drum from where it eventually overflowed into the sump, where his colleague was using an unapproved inspection lamp. A sudden flash fire in the sump killed the contractor.[44]

General precautions applicable to preparation and painting work are summarised in Table 31.4 (after ref. 45).

Table 31.4 Precautions in preparation and paintwork

Information (at least a safety data sheet and comprehensive container label) and training related to the hazards of handling and using the range of chemicals.

Use, where practicable, of less harmful chemicals, e.g. water-based paints.

Full use of any spray booth, enclosure, exhaust ventilation or dilution systems, and automatic handling equipment. Correct positioning of work.

Full use, where appropriate, of general ventilation, e.g. by opening doors, windows.

Prompt attention to any damaged or malfunctioning equipment.

Replacement of lids on containers, and solvent bottles.

Use, where appropriate, of a properly fitting respirator with correct filter or air-fed equipment.

Use of a vacuum cleaner or damping techniques to minimise dust generation.

Prompt disposal of impregnated rags etc.

Avoid skin contact and ingestion of chemicals by:
Use of protective clothing and eye protection.

Use of barrier cream and skin conditioning cream.

Removal of jewellery, etc. which can trap chemicals in contact with the skin.

Avoidance of excessive skin contact with solvents, e.g. when cleaning brushes and spray guns; not washing hands in solvents.

Avoidance of eating, drinking or smoking while painting.

A good standard of personal hygiene, i.e. washing hands before eating, and showering or bathing at the end of work.

Maintaining overalls and respiratory protection in a clean state.

Leaving protective clothing at work.

Phenol

Phenol is actually a white wax-like solid at temperatures below 40 °C; it is generally processed as a pinkish liquid at higher temperatures.

It is caustic so that contact causes softening and whitening of the skin followed by painful burns. The first skin reaction is tingling and local anaesthesia with a feeling of numbness which lasts some hours. Because of this the affected person may feel nothing until serious damage has been done (page 234).

If a large surface area is splashed there is a real danger of death from collapse and kidney damage.

Absorption by exposure of 412 cm^2 of skin can result in death in 30 minutes to several hours.

The symptoms following rapid absorption through the skin are headache, dizziness, rapid and difficult breathing, weakness and eventually collapse. (This is exemplified by the case described on page 116.) Several cases of acute phenol poisoning have been reported[46] including the following.

A chemist stepped into a pool of phenol waste and his leg became soaked. He complained of ringing in the ears, dizziness and dyspnoea. He became dazed and excited but left the building. He was found dead on the road next morning with his leg discoloured black/green to the knee.

Precautions mirror those for concentrated sulphuric acid. Transfer lines are insulated to avoid solidification but special care is required when freeing any blocked pipelines or valves, e.g. using indirect steam heating. The full range of protective clothing should be worn and provision made for coping with the full discharge of material in, and upstream of, the blockage. Enclosed/shielded sampling procedures are recommended (page 290). The extreme corrosivity and toxicity of phenol merit the display of appropriate information on hazards and precautions at the points of use.

Impervious plastic chemical suits should have elasticated wrists, and gloves should have extended gauntlets to preclude any ingress of phenol via the open tops. Suits and gloves should be inspected for soundness before use and washed after use.

The emergency and first-aid procedures following any phenol contamination of the skin are summarised in Table 31.5.

Table 31.5 Phenol – emergency and first-aid procedures

- Remove all contaminated clothing immediately (care being taken to avoid cross-contamination of unaffected skin)
- Swab contaminated area for >10 minutes with either glycerol, or a mixture of 30 parts ethanol and 70 parts liquid polyethylene glycol
- If these solvents are unavailable flush with water for >10 minutes
- Seek medical attention

References

1. *The Lancet*, 5 February 1977.
2. Castle, W. B., Drinker, K. R. & Drinker, C. K. *J. Industrial Hygiene*, 1925, **7**, 371.

3. Health & Safety Executive, *Occupational exposure limits* 1993, EH 40/93.
4. Anon, *Loss Prevention Bulletin*, Institution of Chemical Engineers, 1989(086), 17.
5. Anon., *Safety News*, Universities Safety Association, April 1980, **13**, 13.
6. Anon, *Health & Safety at Work*, December 1990, 5.
7. Anon, *Safety News*, Universities Safety Association, November 1978, **10**, 9.
8. Health & Safety Executive, *Isocyanates: toxic hazards and precautions*, Guidance Note EH16, 3rd edn., 1988.
9. *Health & Safety Information Bulletin*, March 1982, **75**, 11.
10. U.M.I.S.T. *Health and Safety Bulletin*, June 1983.
11. Anon, *The Chemical Engineer*, 15 August 1991, 10.
12. Strickland, J. E. T., *J. Institution of Water Engineers and Scientists*, **40**, 1986, 389f.
13. Condliffe, W. H., *Environmental Protection Bulletin*, Institution of Chemical Engineers, 1989(003), 8.
14. Health & Safety Executive, *Chlorine from drums and cylinders*, HSG40, October 1987.
15. Health & Safety Executive, *Fire and explosion hazards at electrochlorination plant*, Information document HSE 498/11, November 1987.
16. Department of Employment, *Liquid Chlorine*, Health & Safety at Work booklet 37, 1970, HMSO.
17. Anon, *Chemical Safety Summary*, Chemical Industries Association, 1979, **50**, 62.
18. Corbett, E., *Second Symposium on Chemical Process Hazards with Special Reference to Plant Design*, Institution of Chemical Engineers Symposium Series No. 15, 1963, 41.
19. Anon, *Loss Prevention Bulletin*, Institution of Chemical Engineers, April 1990(092), 26.
20. Anon, *Loss Prevention Bulletin*, Institution of Chemical Engineers, 1988(085), 27.
21. Anon, *Health & Safety at Work*, July 1980, 49.
22. Health & Safety Executive, *Entry into confined spaces*, Guidance Note GS5, 1980.
23. Anon, Health & Safety at Work, February 1988, 21.
24. Anon, *Loss Prevention Bulletin*, Institution of Chemical Engineers, 1977(013), 5.
25. Health & Safety Executive, *Dangerous Maintenance*.
26. Anon, *Loss Prevention Bulletin*, Institution of Chemical Engineers, April 1990(092), 18.
27. Institution of Chemical Engineers, Rugby. *Preparation for Maintenance*, Training Module 004.
28. Smith-v-Scott Bowyers Ltd., 1986, IRLR 315.
29. *Health & Safety Information Bulletin*, 1 September 1987, 141.
30. Anon, *Chemical Safety Summary*, Chemical Industries Association, 1980, **51**, 42.
31. Anon, *Chemical Safety Summary*, Chemical Industries Association, June 1981, **52**, 181.
32. Stokinger, H. E., *The Metals* in *Patty's Industrial Hygiene and Toxicology*, 3rd

rev. edn., Vol. 29, G. D. Clayton & F. E. Clayton (eds), John Wiley & Sons, New York, 1981, 1965.
33. Anon, *Loss Prevention Bulletin*, Institution of Chemical Engineers, December 1989(090), 13.
34. Health & Safety Executive, *Skin Cancer caused by Oil*, Leaflet MS(B)5.
35. Feiner, B., *Control of Workplace Carcinogens* in *Cancer Causing Chemicals*, Sax, N. I. (ed), Vol. 3, Van Nostrand Reinhold Co., New York, 1981, 61.
36. Bretherick, L., *Handbook of Reactive Chemical Hazards*, 4th edn., Butterworths, 1990, 1680.
37. Health & Safety Commission, *Monitoring for Health and Safety in Print*, Health & Safety Executive, 10/92.
38. Anon, *Chemical Safety Summary*, Chemical Industries Association, **51**, 1980, 78.
39. Anon, *Chemical Safety Summary*, Chemical Industries Association, **52**, 1981, 150.
40. Health & Safety Executive, *Sulphuric acid used in agriculture*, Guidance Note CS20, December 1990.
41. Health & Safety Executive, *Solvents & You*, Leaflet IND(G) 93(L), 1990.
42. Anon, *The Safety Practitioner*, March 1985.
43. U.M.I.S.T. *Health and Safety Bulletin*, December 1982.
44. R-v-Austin Rover Group, 1989, IRLR, 404.
45. Health & Safety Executive, *Health Hazards to Painters*, Leaflet IND(G) 72(L), 1990.
46. Hunter, D., *The Diseases of Occupations*, 5th edn., 1975, English Universities Press Ltd., 494.

CHAPTER 32

Microbiological hazards and preventative measures

Introduction

Those engaged in work involving chemicals are invariably exposed to a range of additional occupational hazards, e.g. electrical and mechanical energy, ionising and non-ionising radiations, and physical dangers. These are discussed in Chapters 2 and 8. Workers may also be exposed to biological hazards as a direct result of the working environment. These may include acute and chronic infections, parasitism, and toxic or allergic reactions to plant or animal matter. The causative agents may be bacteria, viruses, rickettsiae, chlamydiae, or fungi, depending on the nature of the work and the geographical location.

Detailed discussion of the risk with such a wide spectrum of hazards is beyond the scope of this book but some awareness of the dangers of, and precautions for, working with pathogenic micro-organisms in general, and more specifically in areas also associated with chemical usage, is crucial for an overall appreciation of the potential for occupational ill-health. Indeed, as discussed in Chapter 28, in the UK the scope of the Control of Substances Hazardous to Health Regulations, 1988 embraces 'a micro-organism which creates a hazard to the health of any persons'. In any event there are similarities between control measures with them and with chemicals as aerosols or particulates. Risks of exposure to chemicals used in conjunction with micro-organisms, e.g. disinfectants, must also be controlled.

Whilst many animal diseases do not infect man some are more suspect, e.g. bovine spongiform encephalopathy (BSE) and even though it may be unlikely that BSE can affect human health, because of the uncertainty it is reasonable to take hygiene precautions in handling carcases of infected cows.[1] There are many occupational diseases of biological origin, zoonoses, i.e. animal diseases which are easily communicable from animal to man. Those working with animals such as farmers, veterinary surgeons, and animal technicians may be at risk. These are only mentioned in passing. In the present context, however, the biological hazards of greatest relevance arise from work in a biochemistry laboratory, in bioprocessing, and from use of plant such as humidifiers and cooling systems, installed to control industrial chemical processes or working environments but which provide a potential source of microbiological diseases such a Humidifier Fever and Legionnaires' Disease. These hazards and their control are therefore discussed in more detail.

Definitions

Micro-organisms

Micro-organisms are any microscopic biological entity which are capable of replication.

Viruses

Viruses are microscopic agents containing genetic material coated in protein and which only replicate using the mechanisms of a host cell. They vary in size from 30 nm to 400 nm and are capable of being crystallised, which illustrates that they are not cells but simpler chemical entities.

Bacteria

Bacteria are normally considered as unicellular organisms that possess no nucleus (although recent techniques have revealed a nuclear area) but which can live and multiply outside living cells using their own genetic and metabolic resources provided that a suitable nutrient is available. They vary in shape, e.g. rods (bacilli), spheres (cocci), and helicals (spirochaetes) with diameters between 0.5 and 1 μm and lengths ranging from 1 to 30 μm. Bacteria also exist in the form of spores which may be highly resistant and long lived.

> A student developed a typical anthrax lesion on his chin at the site of a cut made from shaving. Fortunately the infection was mild and he recovered.
>
> He had been handling fixed and stained smears from anthrax (see page 1791) cultures. Each slide used by the student was subsequently tested for viable anthrax spores by making cultures of them. Nearly all proved to be positive. Even anthrax culture smears made some years previously proved upon examination to contain viable spores of *Bacillus anthracis*. Experiments also demonstrated that spores of *B. anthracis* in films on slides may remain viable even after heating in a Bunsen flame for five seconds or so. Methods and dyes used in preparing bacteriological slides do not destroy all the spores. Viable spores can become airborne if spore suspensions are heated in open containers.[2]

Most occupational bacterial infections arise from minor wounds/abrasions and are frequently caused by species such as staphylococci and streptococci. Rickettsiae, originally classed as viruses, are now recognised as a class of small bacteria. They multiply in arthropods which become the reservoir and transmit these organisms to man. Similarly chlamydiae, once considered as viruses, are now classed as bacteria.

Fungi

Fungi are complex cellular structures usually with rigid cell walls and lacking chlorophyll. They may replicate by the production of spores. Fungal cells are typically 5 to 10 μm in diameter.

Parasites

Parasites are multicellular organisms which live in, or on, a living host. Whilst often visible to the naked eye, their eggs and larvae may be microscopic. Parasitism may involve protozoa, helminths, or arthropods.

Biological diseases

The main routes by which micro-organisms enter the body are by inhalation, ingestion or through broken skin. Occupational infection can result in a variety of diseases such as those listed in Table 32.1 (after ref. 3) and exposure to certain chemicals may cause disease which predisposes to infections, e.g. silicosis and pulmonary tuberculosis.[2] Selected biological diseases are briefly discussed below, whilst those caused by biological agents as listed under the UK National Insurance (Industrial Injuries) (Prescribed Diseases) Regulations are identified in Table 16.7 on page 940.

Viral infections

Viral diseases likely to be encountered industrially include animal respiratory viruses and infections. Thus, orf is a viral skin infection of sheep and goats which can be transmitted to animal handlers or those who handle the meat or its products. The disease in man, contagious pustular dermatitis, can result in pus-containing skin lesions at the site of infection.

Viral hepatitis is an inflammatory condition of the liver due to the two hepatitis viruses A and B. Organs other than the liver may become affected. Serum hepatitits is now a well-known occupational hazard for staff working in renal dialysis units, as a result of contact with infected blood. Cleaners may become exposed when cleaning-up spills of body fluids after an accident; hence the need for training, provision of cleaning agents and protective clothing including gloves. Vaccinations are recommended for those at highest risk.

Hepatitis manifestation varies from subclinical to acute illness, and in particular inflammation of the liver. As the liver swells bile channels become compressed and jaundice appears. Between 5 and 10 per cent of

Table 32.1 Selected examples of occupational diseases resulting from exposure to biological agents

Class and examples	Route of entry	Principal occupations at risk
Viral		
Rabies	Bite	Veterinarians, wild animal handlers, cave explorers
Cat-scratch Disease	Skin lesion	Animal laboratory attendants, cat and dog handlers, veterinarians
Orf	Skin lesion	Shepherds, stockyard workers, shearers, veterinarians
Milkers' Nodules	Skin lesion	Milk producers, dairy farmers, veterinarians
Newcastle Disease	Inhalation (upper respiratory tract)	Poultry handlers, virologists, veterinarians
Viral serum hepatitis	Parenteral infection	Health workers, oral surgeons
Rickettsial and chlamydial		
Tick Fever	Bite	Laboratory workers, foresters, ranchers, construction workers
Q Fever	Inhalation/ingestion	Dairy farmers, slaughterhouse workers, hide and wool handlers, laboratory workers
Ornithosis	Inhalation	Petshop owners, taxidermists, zoo attendants, poultry workers
Bacterial		
Tetanus	Skin lesion	Farmers, gardeners, laboratory workers
Anthrax	Skin lesion/inhalation	Agricultural workers, occupations handling goat hair, wool, hides
Brucellosis	Skin lesion	Meat packers, livestock producers, veterinarians
Leptospirosis	Skin lesion/ingestion	Farmers, field workers, sewer workers, miners, military troops
Plague	Flea bite	Shepherds, farmers, hunters, geologists
Tuberculosis	Inhalation	Health workers
Fungal		
Candidiasis	Skin and mucous membranes	Dishwashers, poultry processors, bar staff, cooks, bakers, cannery workers
Aspergillosis	Inhalation	Farmers, mill workers, bird handlers
Histoplasmosis	Inhalation	Workers in barns and chicken houses
Mycetoma	Skin lesion	Farmers
Sporotrichosis	Skin lesion	Gardeners, horticulturists, florists, nursery workers
Chromoblastomycoses	Skin lesion	Farmers
Dermatophytoses	Local skin effect	Farmers, animal handlers, pet handlers, wool sorters, athletes, gymnasium workers
Parasitic		
Swimmer's itch	Wet skin	Workers around fresh lakes and ponds, dock workers
Creeping eruption	Skin lesion	Ditch diggers, gardeners, plumbers, lifeguards
Hookworm Disease	Skin lesion	Ditch diggers, lifeguards, sewer workers

those infected become carriers, a state which may persist for life. A link has been established between hepatitis B and primary cancer of the liver.

Bacterial infections

Examples of bacterial disease include tetanus, leptospirosis (and leptospira hardjo or bovine leptospirosis) and anthrax.

Tetanus

Tetanus is a wound infection caused by the release of exotoxins from *Clostridium tetani* requiring a broken skin for the passage of the organism. Neither the bacterium nor its toxin is harmful by mouth. The prevalence of the disease in man varies in different localities, depending upon the contamination of, e.g., soil by tetanus spores. Wholesale anti-tetanus immunisation, e.g. of farm workers, is not common practice; therefore any wound, particularly a puncture wound, acquired on agricultural land calls for urgent anti-tetanus measures.[4]

Leptospirosis

Leptospirosis, or Weil's Disease, is one of the zoonotic bacterial diseases found worldwide in developed and underdeveloped countries, in both rural and urban areas. Common signs and symptoms include sudden temperature and shivering, headache, severe myalgia of the legs and back, eye redness and prostration.

> Part of the normal duties of a maintenance worker employed in canal workshops was the repair of a canal bank. Rats were numerous on this bank, attracted by effluent from numerous nearby pig farms. He contracted Weil's disease, complaining on admission to hospital of nausea, head pains, intermittent abdominal pain and loss of appetite; fortunately he recovered after treatment.

Until recently it was believed that leptospirosis, could only be caused by ingestion of food or water contaminated by the urine of infected rats or dogs; alternatively the bacteria may enter via the skin or mucous membranes. It was assumed, therefore, that in the UK it was associated mainly with occupations with potential for contact with rat urine, e.g. farming, stock-rearing, sewer worker, casual work,[4] coal miners and bargemen.[1] For example, a farmhand contracted the disease through contact with water, hay, etc. contaminated with rat urine.[5] More recently, however, human infections have been linked with other animals, e.g. voles, field mice, cattle, and pigs.[6] Leptospires can live for days in slaughtered animals and therefore those working with pigs and cattle or their meat products are also at risk. Leptospires are particularly likely to enter the body through cuts, scratches or abrasions of the skin. Therefore in addition to eradication of rats, careful covering of all cuts and scratches, especially on the hands and arms, and wearing of gloves are crucial. Lacerations sustained at work should be immediately washed well with soap and water, rinsed, dried and dressed. Meticulous personal hygiene is required before eating, drinking and smoking. Other precautions include regular rubbish removal, especially that from food factories, canteens, dairies, restaurants, etc. to discourage vermin. Rubbish should be securely contained and waste areas should be kept scrupulously clean. Regular inspections for rodents should be carried out. Unnecessary water should be drained away and water supplies chlorinated. Workers must be alerted to the hazards and the precautions. Although not a common disease in the UK, cases are reportable to the Health & Safety Executive under the 'Reporting of Injuries, Diseases and Dangerous Occurances Regulations, 1985'.

> A council worker cut his hand whilst clearing a river bank. He was first treated for pneumonia five days later, then developed jaundice. Weil's disease was only diagnosed on his admission to hospital in a critical condition and unfortunately this proved fatal.[7]

Leptospira hardjo affects 60 per cent of cattle in the UK, posing a risk mainly to those exposed to cattle urine, e.g. during milking. At least an eighth of dairy farmers have experienced the disease, which includes flu-like symptoms and headaches. It responds to antibiotics but if untreated can lead to meningitis, jaundice, kidney failure and possibly death.[8] Clearly in making COSHH assessments it is important to involve employees who, if adequately instructed and experienced, may be able to contribute their observations of the condition of the herd. Other requirements to satisfy the Regulations would include establishing, with the aid of a vet, whether the herd was infected and identifying where infection occurs. Routine control measures include

- checks on all abortions and sudden drops in milk yields within the herd
- good personal hygiene
- use of rubber gloves and waterproof apron and face visor during milking
- use of respiratory protection if an epidemic exists within a herd
- use of suitable hand and body protection during assisted calvings
- consider use of vaccine.

Anthrax

Anthrax, or Wool Sorter's Disease, is a disease primarily of herbivores as a result of infection with *Bacillus anthracis*. The spores, which can be killed by boiling for 10 minutes, can survive for long periods in soil and in animal remains[6] and as discussed on page 1787 even on microscope slides. Infection may be external, i.e. painful skin lesions (cutaneous type of malignant pustule), or internal (pulmonary type). In addition to those handling animal carcases, cases of anthrax have occurred in workers employed in the following industries and processes.[9]

- Manufacture of glue and gelatine (contact with imported dried bones and sinews).
- Manufacture of bone charcoal used for sugar refining (contact with imported bones).
- Manufacture of fertilisers (contact with bone meal imported as such or derived from imported bones).
- Tanneries (contact with imported dry and dry-salted hides and skins).
- Brush making (contact with imported bristles and hair).
- Manufacture of hair cloth and production of stuffing for mattresses and upholstery (contact with imported horsehair, hog and cow hair).
- Wool, worsted and felt industries (contact with imported wool and hair).
- Plastering (where imported cow hair is used as a binder).
- Manufacture of manure (contact with imported hooves, horns, dried blood, bone and horn meal, hide cutting and skin waste, shoddy and wool waste).
- Docks (contact with various imported materials).
- Fur skin dressing (contact with imported furs).

> After two weeks working as a woolpicker at a carpet factory a 57-year-old man was admitted to hospital with a typical anthrax lesion on the upper lip. His temperature continued to rise and submaxillary glands swelled. Pain and bulging developed behind the eye, and twitching in the arms and feet developed during the night. Anthrax meningitis was diagnosed bacteriologically. The patient made an uneventful recovery after treatment with drugs.[2]

Within the UK active immunisation against anthrax is now available free to employees[9] which is probably the main reason why anthrax is fortunately now rare in man,[4] and the pulmonary form extremely rare[9] albeit usually fatal. With modern treatment the prognosis is now good in cases of malignant pustule, complete recovery being the rule. Control measures are specified in connection with various products by established legislation[9] and a summary of key precautions is given in Table 32.2.

Rickettsial and chlamydial infections

Diseases from these micro-organisms include Rocky Mountain Spotted Fever or Tick Fever (caused in man by the bite of ticks infected with *Rickettsia rickettsii*); Q Fever (resulting in man mainly from contact with infected animals or inhalation of dust infected with rickettsiae, usually around lambing pens and pastures); and Psittacosis.

Q Fever is predominantly occupational in origin, usually due to inhalation of contaminated aerosols and less often due to drinking infected

Table 32.2 General precautions against anthrax

- A high standard of cleanliness in premises where materials potentially infected with anthrax are handled.
- Mechanical handling techniques wherever possible to keep to a minimum physical contact with unprocessed material.
- Exhaust ventilation of a high standard for dusty processes (e.g. wool sorting) and dust collected in extraction plants should be burned.
- Washing facilities of a high standard.
- Separate cloakroom accommodation, each with drying facilities, for protective clothing and clothing taken off during working hours.
- First-aid equipment of a high standard so that all cuts and abrasions can be treated without delay.
- Messroom facilities so that no meals are taken in rooms where potentially infected materials are handled or stored.
- Suitable protective clothing and equipment for the handling of material, for the cleaning of machinery or premises and in other instances where contact with materials having an anthrax risk is unavoidable.
- Control over certain imported biological materials coming from areas where animal infection is common and control poor, exercised by means of compulsory disinfection.
- Persons exposed to the known risk advised of the importance of early recognition and prompt treatment, e.g. in the UK by means of the exhibition of the Anthrax Cautionary Notice (F410) and the distribution of the Anthrax Card (F1893), on which the name of the designated hospital for the area should be inserted.
- In all premises where materials with an anthrax risk are handled, there should be recognised procedures for referring all cases of suspected anthrax for immediate medical advice. In the UK a scheme has been instituted whereby in certain areas particular hospitals have been designated as ones to which suspect cases of anthrax should be referred.

milk.[10] Occupations at risk include farmers, shepherds and their families; veterinary workers; stockyard workers; abattoir and knacker's yard workers; workers engaged in fertiliser manufacture from animal products; medical, pathology and post-mortem technicians. However, as with other microbiological hazards the boundary is indistinct.

> Twenty-eight people in an art school contracted Q Fever due to exposure to dusty, contaminated straw used in packing cases.[11]

Whilst spread of disease via milk is preventable by pasteurisation, prevention is difficult since the health and milk production of infected animals remain unaffected.[10]

Psittacosis (ornithosis), once thought to be of viral origin, is a disease in birds infected with *Chlamydia psittaci*. The disease can be passed on to man through inhalation of dried, infected droppings, through contact with the feathers or tissues of infected birds, or even through the bite of an infected bird.

> A risk of psittacosis, a disease caused by the same organism as ornithosis (Table 32.1) but from the parrot family, may arise in readily identifiable occupations, e.g. pet shop staff; zoo attendants; aviaries and bird sanctuaries; veterinary surgeons and pathologists; laboratory workers handling infected materials; medical and nursing staff; and those exposed during transportation

of sick birds, e.g. railway guards. Cases have also arisen amongst employees in turkey farms and processing plants and amongst workers exposed in roof spaces of buildings heavily contaminated with excreta from pigeons (after ref. 12).

Symptoms include slow rise in temperature, headache and malaise, pronounced cough and in severe cases delirium, cyanosis, jaundice, nausea and vomiting. Although normally associated with diseased birds, it would seem that other sources for man also exist.

A bacteriology research worker was photographing viruses with an electron microscope. Preparative work entailed ultrasonic disintegration of animal specimens to rupture cells and release viruses. On one occasion he was working with crushed spleen from mice infected with psittacosis. He apparently became exposed to aerosol either during the ultrasonification (although the test tube was plugged with cotton wool at this stage) or by convection currents from the very hot tube as the plug was removed from pipetting.

He developed marked malaise, shivers, high fever, headache and vomiting. Symptoms abated but reappeared and a cough and sore throat developed. Eventually hospitalisation became necessary and he was diagnosed as suffering from psittacosis. After treatment with drugs he recovered and was eventually discharged symptom-free.[2]

Fungal infections

Table 32.1 includes examples of diseases resulting from infection by fungi via cuts in the skin or by inhalation. Transmission of fungal diseases from person to person is extremely rare and they are therefore not considered to be contagious.

As explained on page 192, the size of a micro-organism determines how far into the respiratory tract it can penetrate before deposition and hence the type of adverse effect it may provoke. Those exposed may often show an allergic response. Since spores of a given species vary only within a narrow size range, most will be deposited in the same region of the tract and hence provoke similar symptoms. Particles >10 μm which are trapped in the nose during normal breathing may cause rhinitis; conversely such particles inhaled via the mouth can enter deeper into the respiratory system and become deposited in the trachea and bronchi and cause asthma. Particles <5 μm mainly penetrate to the alveoli and may cause alveolitis; the optimum size for deposition is 2–4 μm. Particles of intermediate size, i.e. 5–10 μm, are mostly deposited in the bronchi and bronchioles and may cause asthma, or if in sufficient concentration penetrate to the alveoli and result in alveolitis.

The related occupational diseases are, in the main, most common amongst outdoor workers and animal handlers. There are, however, exceptions.

Workers engaged in citric acid production developed a respiratory complaint similar to Farmer's Lung. Trays of nutrient medium with surface growths of *Aspergillus niger* became contaminated by *Aspergillus fumigatus* and various *Pencillium* spp. Dense clouds of spores were released during harvesting at 3-daily intervals and the workers did not wear any respiratory protection.[13]

The types of extrinsic allergic alveolitis caused by inhaled spores are listed in Table 32.3 (after ref. 14.). Occupational dusts of vegetable origin

Table 32.3 Extrinsic allergic alveolitis due to inhaled spores

Source of dust	Organisms	Disease
Mouldy hay	*Micropolyspora faeni* *Thermoactinomyces vulgaris*	Farmer's Lung
Air-conditioning systems	*Micropolyspora faeni* *Thermoactinomyces vulgaris*	Hypersensitivity pneumonitis
Bagasse	*Thermoactinomyces sacchari*	Bagassosis
Redwood sawdust	*Aureobasidium pullulans* *Graphium* sp	Sequoiosis
Malting barley	*Aspergillus clavatus* *Aspergillus fumigatus*	Maltworker's Lung
Maple bark	*Cryptostroma corticale*	Maple bark pneumonitis Coniosporosis
Mouldy cork	*Penicillium frequentans*	Suberosis
Cheese	*Penicillium casei*	Cheese Washer's Lung

include bagasse, cork, cotton, flax, flour, hay, hemp, sisal, straw, tea and wood. Pollen and fungal spores from these may be inhaled into the nasal passages and upper respiratory tract, and many can produce inhalant allergy such as hay fever, urticaria and asthma.

> A 51-year-old maintenance fitter had been employed at a bakery for 20 years, without previous experience of asthma. During the past 15 years, however, he complained of increasing breathlessness, wheeze and cough. His eyes often became red and watery and he also suffered bouts of sneezing. His condition tended to improve when he was away from work at weekends or on holidays. This was confirmed by lung function tests and he was diagnosed as suffering from flour dust allergy.[15]
>
> The exhaust ventilation was improved and the man returned to work but in a less dusty part of the plant. He was also issued with respiratory protection. The man continues to work but only with the aid of medication to alleviate his symptoms and he continues to suffer from chronic asthma for which he receives disablement benefit. Further investigation revealed other cases of flour dust allergy, but because their sensitisation had been detected their condition was less severe. Even so they too rely on medication to relieve their symptoms, and they are likely to suffer respiratory distress for the rest of their lives.

Workers exposed to mill dust for the first time often develop dry throats, coughs, sneezing, headache, malaise, high temperatures, occasional nose bleeding, nausea and vomiting. The onset of the symptoms may be delayed until the worker arrives home and then reappear each night for a week or so until the individual becomes acclimatised.[2]

Exposure to heavy concentrations of mouldy hay or grain can result in severe respiratory distress, termed Farmer's Lung. This is a respiratory allergy to the inhaled spores of certain actinomycetes which are profuse in mouldered animal feedstuffs. Those at risk, therefore, include all those who,[16]

- handle hay, straw, grain or similar produce that has been harvested in damp conditions and stored in a way that leads to heating and drying out
- are exposed to dry dust from the harvesting and storage of grain or the milling, moving and handling of feedstuffs
- work with intensively kept livestock, e.g. pigs and poultry.

A variant of the disease, Mushroom Worker's Lung, arises due to the inhalation of spores when mushroom compost is disturbed; this contains many thermophilic actinomycetes so that a significant concentration of spores is generated in closed sheds.

> A man worked in the mushroom industry for over 20 years prior to becoming sensitised to fungal spores.[15] Initially he worked as a supervisor in the relatively clean growing area and then transferred to the composting facility of the plant where spores were released during the composting process.
>
> The employers had conducted a COSHH assessment and were aware of the hazards. Engineering controls were instituted where possible with reliance on personal protection to control exposures where spore levels were highest. However, the man was unconvinced of the value of a respirator and hence failed to wear the ventilated helmet provided. Within two months he began to feel cold in the afternoons, became lethargic and developed sweating bouts at home which lasted late into the night. His condition deteriorated with acute attacks of breathlessness and exhaustion, even when he used a disposable mask. He lost 5 kg in weight in one week. Allergic alveolitis was confirmed and on removal from exposure his condition settled down. Upon return to work he wore the ventilated helmet fitted with filters suitable for removing the spores and has experienced no further symptoms.
>
> It is possible that improvements to the engineering controls are warranted but the case serves to demonstrate that where reliance is placed on personal protection to be effective it must be of the correct type and must be worn. (In the UK this is specifically required by COSHH Regulations 7 and 8.)

A different respiratory allergy, Harvester's Lung, is attributable to the inhalation of clouds of spores released by combine harvesting from fungi growing on fields of cereals. Advice has been published on precautions to avoid dust inhalation, including recommended respirators and their use.[17]

In general breathing in these types of dust and spores should be avoided wherever possible by routine precautions, e.g.[16]

- not creating more dust than necessary during working
- ensuring that machinery is constructed, adapted and maintained so as to contain dust as far as possible

- providing adequate ventilation – either local exhaust ventilation or dilution ventilation of the whole area (where produce is handled indoors)
- observing good housekeeping where materials are handled. An industrial vacuum cleaner should be used to remove excess dust from floors, walls and rafters of buildings. Thorough wetting of dust is required before use of a brush
- where other methods of preventing exposure are not reasonably practicable, ensuring that appropriate, properly fitted and maintained, respiratory protective equipment is worn
- provision and use of a coverall and headgear
- cleaning working clothing and, if possible, leaving it on the premises at the end of the working day.

Parasitic infections

Parasitic diseases may have reservoirs in infected insects, animals or humans. Occupations in which the worker can be exposed to, e.g. infected animal (or human) faeces or to mites, chiggers and ticks are at greatest risk.

For example, Hookworm Disease results from infection by larvae of *Ankylostoma duodenale*, *Ankylostoma braziliense* or *Necator americanus*. The adult worm is *ca.* 1 cm long and ova hatch in the soil. The larvae enter the body through skin which has been exposed to damp earth or muddy water. The optimum temperature for the ova to hatch is *ca.* 25 °C. Parasite embryos invade the skin and travel through the veins to the lungs from where they are coughed up and swallowed to end up in the faeces. In the early stages skin eruption may occur characterised by itching and lesions. Interestingly, in Cornish tin miners it was the forearms that were commonly affected because these were generally bare and came into contact with the rungs of ladders which were contaminated with mud containing ova of the parasite. In plantation workers it is the feet that become affected.[2] Shortness of breath, weakness and gastro-intestinal difficulties develop within a few months of skin lesions appearing. This may be followed by ulcerations, stunting of physical and mental growth and anaemia.

> A worker accidentally spilled a culture of hookworm larvae upon his hand. Dermatitis developed at the site of contact and hookworm ova were found in his faeces.[2]

Outdoor workers such as farmers, gardeners and construction workers are at a risk from bites by a variety of mites which contaminate grain, animals, dairy produce, etc. Dermatitis often results and secondary infection is a possibility. In the case of the tick, for example (see tick fever), if the head is not completely removed or if the tick remains feeding on the human blood for days, weakness and even paralysis can develop in the host as a result of the neuotoxins injected into the host's bloodstream.

Micro-organisms in the laboratory

Risks

Laboratories handling both chemical and biological specimens (blood, urine, gastric contents, bile, faeces, cultures, etc.) include clinical biochemistry departments in universities and industry, forensic departments, hospitals, safety testing research facilities, veterinary practices, etc. Laboratory-acquired infection may result from working with the agent, from clinical or autopsy specimens, or from animals. Certain micro-organisms pose a threat by ingestion and by absorption via broken skin (or even intact skin in certain instances, e.g. Yellow Fever), or conjunctiva, or via inhalation. Airborne particulates may arise from a variety of laboratory equipment (e.g. shakers, homogenisers and centrifuges) or operations (such as pipetting, stirring, liquid dispensing, culture preparation, opening ampoules, and mechanical fragmentation of infected animal tissue). Table 32.4 lists spray factors for a range of laboratory operations. Lyophilised cultures, dried bacterial colonies, material on the caps of culture vessels, dusts from animal cages, etc. are also potential sources of hazardous particulate matter. Laboratory workers may also become infected by inhalation of aerosols from the cough of experimental animals or by bites.

> '. . . every infectious microbial agent which has been studied in the laboratory has at one time or another caused infection of operators. In some instances, laboratory infections outnumber natural infections and have been the only known human infections'.[18]

Thus workers in animal, microbiology or biochemistry institutions are at risk from laboratory acquired infections as illustrated by the cases of psittacosis and anthrax in bacteriologists described on pages 1794 and 1787 respectively, and by the following selected case histories.[2]

> A scientist at the Koch Institute for Infectious Diseases died from plague in 1903 as a result of infection when fluid material from a guinea-pig infected with bubonic plague was accidentally splashed over an agar plate.

> In 1909 a worker at a branch of the Lister Institute was working with pneumonic plague when he inhaled the organism and susbsequently died.

> Four workers were infected in 1905 when a centrifuge containing a glanders suspension broke, scattering the culture. Three of those infected died but the man operating the centrifuge recovered.

> A scientist who discovered the method of infecting animals with foot-and-mouth disease cut his thumb with a broken glass vessel whilst collecting virus

Table 32.4 Spray factors for various operations

Operation	Spray factors* Minimum	Spray factors* Maximum
Sonic oscillation	5×10^{-7}	9×10^{-5}
Blender (opened)	1×10^{-6} (tight cover on, open 1 min later)	2×10^{-3} (open)
Blender (closed)	9×10^{-9} (screw cap on)	2×10^{-8} (plastacap, loose)
Dropping liquids, 90 cm		2×10^{-6}
Mixing with pipette	2×10^{-6} (not blowout type)	1×10^{-4}
Centrifuge 'spill' (drop of culture on rotor)		2×10^{-6}
Vortex mixer (capped tube)		0
Shaking dilution bottle	7×10^{-9}	2×10^{-7}
Lyophile tube breakage:		
Milk and broth		3×10^{-9}
Nutrient broth		2×10^{-9}
With cotton pledget (wet)		0
With cotton pledget (dry)		6×10^{-11}
Transfer of lyophilised material (shaking tube)		2×10^{-10}
Removing cotton plugs from test tubes	6×10^{-8}	2×10^{-7}
Flaming loop	3×10^{-9}	3×10^{-7}

* When the spray factor is multiplied by the number of bacteria or virus per millilitre of solution the expected number of airborne particles released by the operation can be calculated.

from a guinea-pig. Within two days he had developed a headache and shivers. Vesicles containing clear fluid formed on his palm and soles on the third day. When later innoculated with foot-and-mouth virus the man was found to be immune.

In 1932 the death of a 29-year-old man led to the discovery of a new neurotropic virus. Whilst engaged in experimental work on poliomyelitis he was bitten on the hand by an apparently healthy monkey. Vesicular lesion and lymphangitis appeared and thirteen days later he developed myelitis (inflammation of the spinal cord) and died on the eighteenth day.

One review of some 3921 cases of laboratory-acquired infections revealed that bacterial and viral infections accounted for most incidents and for most of the 164 deaths which resulted, as summarised by Table 32.5 (after ref. 19).

Table 32.6[19] identifies the source of the infection whilst Table 32.7 provides an analysis of the 703 'accidents' by type.

The risk of infection is not confined to those actually working in the laboratory as the following two incidents involving smallpox demonstrate.

Toxic hazards

Table 32.5 Summary of 3,921 incidents involving laboratory-associated infections

Type of infection	% of total incidents	% of total deaths
Bacterial	42	42
Viral	27	33
Rickettsial	15	14
Fungal	9	3
Chlamydial	3	6
Parasitic	3	1
Unspecified	1	1

Table 32.6 Distribution of cases according to proved or probable source of infection

| Sources | Agents | | | | | | | |
	Bacteria	Viruses	Rickettsiae	Fungi	Chlamydiae	Parasites	Unspecified	Total
Accident	378	174	45	33	14	38	21	703
Animal or ectoparasite	149	249	66	151	32	11	1	659
Clinical specimen	90	175	2	1	0	19	0	287
Discarded glassware	34	10	2	0	0	0	0	46
Human autopsy	56	9	4	0	0	1	5	75
Intentional infection	14	1	0	0	0	4	0	19
Aerosol	101	92	217	88	22	2	0	522
Worked with the agent	381	213	100	62	43	28	0	827
Other	7	1	7	0	1	0	0	16
Unknown or not indicated	459	125	130	18	16	12	7	767
Total	1,669	1,049	573	353	128	115	34	3,921

Table 32.7 Number of laboratory-associated infections resulting from various types of accidents

| Type of accidents | Agents | | | | | | | |
	Bacteria	Viruses	Rickettsiae	Fungi	Chlamydiae	Parasites	Unspecified	Total
Accident involving needle and syringe	83	43	16	12	5	16	2	177
Contact with infectious material resulting from spills, sprays, etc.	82	72	11	6	8	9	0	188
Injury with broken glass or other sharp object	75	11	4	9	0	1	12	112
Aspiration through pipette	67	20	3	0	1	1	0	92
Bite or scratch of animal or ectoparasite	41	25	9	4	0	9	7	95
Other	3	0	0	0	0	0	0	3
Not indicated	27	3	2	2	0	2	0	36
Total	378	174	45	33	14	38	21	703

Because of pressure for laboratory space a spectrophotometer and a balance for general use were located in a pox viruses laboratory. On 28 February 1973 a laboratory technician was harvesting a number of strains of pox virus on the open bench in the pox laboratory. A young graduate zoologist, who often needed to visit the laboratory to use the communal equipment, watched the operation and nothing untoward occurred. By 11 March the graduate zoologist developed headache, backache, vomiting and high fever. By mid-March a rash appeared over her body and on 16 March she was hospitalised in a general ward. It was not until 31 March that smallpox was confirmed, probably contracted as an aerosol or by virus from a contaminated surface when watching the harvesting operation.

In the intervening period two visitors to the hospital to a patient in the next bed to the zoologist contracted the disease and eventually died, one on 6 April, the other on 16 April.

Problems highlighted in the inquiry report[20] were symptomatic of a lack of adequate handling facilities and procedures for work with pathogenic micro-organisms, shortcomings in the laboratory's organisation, poor communication, lack of appreciation of the magnitude of the risks, no recognition of the required urgency once the incident had occurred and had been reported, and inadequate emergency procedures for control of outbreaks of communicable disease, etc.

A medical photographer at a research facility became unwell with headache and muscular pain on 11 August 1978 but went to work. By 13 August she again felt unwell and developed a rash. She was admitted to hospital on 24 August but died of smallpox on the 11 September that year.[21]

The organisation was engaged on research work involving pox viruses in a laboratory housed in the same building in which the photographer worked. However the photographer never had reason to visit the pox virus laboratory and the likely source of exposure was either personal contact with a visitor to the laboratory or, more probably, airborne contamination in the corridors or the 'telephone room' which was directly above the laboratory. The opportunities for the virus to become airborne and to contaminate surfaces arose from poor design to ensure containment, poor laboratory procedures, a failure to use the safety cabinet for all open work with smallpox, the practice of entering and leaving the smallpox room during work without changing gowns or gloves, a failure to use sealed containers for infected material, and poor personal hygiene. Furthermore airflow tests revealed that with the pox room door closed the fan in the safety cabinet provided the necessary negative pressure within the pox room, but with the door open air flowed into the adjoining room whether the fan was on or off.

Staff from the pox virus laboratory were vaccinated annually and those with limited access, e.g. cleaners, security staff and maintenance personnel, were vaccinated every two years; those from other Departments within the medical school were not offered vaccination. The photographer had been vaccinated against smallpox in 1966. However, whilst immunity is never absolute, to provide effective protection from smallpox, vaccination would have to have been carried out less than two years before exposure. Vaccination within three days is also generally protective. When the photographer's illness was diagnosed her mother was vaccinated on 24 August, and although she developed smallpox on 7 September she recovered.

This case emphasises the need for complete containment when handling hazardous micro-organisms, backed up by adequate administrative arrangements. It was luck rather than skill that prevented a major outbreak of smallpox.

Since even 'safe' cultures such as those used in school biology laboratories may become contaminated when a sub-culture is made, it is good practice to consider any micro-organism as potentially pathogenic. Even cultures grown from field specimens such as air, soil or pond-water samples, though possibly safe, should be regarded as potentially hazardous because of the unknown identity of the micro-organism grown.[22]

Sterilisation

Established sterilisation procedures must be followed conscientiously. For example, when materials are sterilised in an autoclave it is essential that all the load reaches, and is maintained at, the required temperature for the appropriate length of time.[23] Suitable techniques are required to ensure that the middle of the load as well as the outside is fully sterilised. For example, when a large quantity of closely packed material of poor thermal conductivity is sterilised there is a substantial time lag between the autoclave chamber and the middle of the load reaching the requisite temperature. Load thermocouples should be considered and specialist tests are available for assessing sterility. The most definitive approach is to assay autoclaved material for the presence of surviving micro-organisms, but this tends to be impractical for routine use. Biological standards comprising strips or ampoules containing a known number of spores can be included with the autoclave load and subsequently incubated to provide a useful indicator of the efficiency of the autoclave operation. Chemical sterilisation indicators which respond only to exposure to steam and not to dry heat alone, whilst less reliable than biological indicators, are more convenient and provide a more immediate reading.

Disinfection

Disinfection is not sterilisation, i.e. no disinfectant should be expected to kill all micro-organisms, but all laboratories involved in the handling of micro-organisms or animal materials should have the correct disinfectant solutions readily available. Chemical disinfectants are useful where it is impractical to steam-sterilise, e.g. large surfaces or bulky apparatus. It is important, however, to appreciate the limitations of general-purpose disinfectants, e.g. all disinfectants are deactivated to some extent by organic matter, hard water, natural materials (cotton, wood, cellulose, sponge, rubber, cork) and synthetic substances (polystyrene, polyvinylchloride, polyvinyl acetate, polyurethane, polyethylene, nylon).

Of the many laboratory-acquired infections with no obvious cause, a number are likely to have resulted from the lack of use, or inappropriate use, of disinfectants. Misuse includes
- disinfectant not used when it should have been
- disinfectant provided at unknown dilutions
- disinfectant changed infrequently
- efficacy unknown for particular use
- articles not completely immersed
- failure to discard jars overloaded with paper, protein, etc. which rapidly deactivate the disinfectant.

Added detergent at a concentration of 1 per cent v/v will aid penetration of disinfectant but compatible agents should be selected. For example, phenolics and hypochlorites are both anionic and therefore compatible with anionic and non-ionic detergents; however, an anionic disinfectant can be inactivated by a cationic surfactant and hence allow micro-organisms to survive.[24]

The effectiveness of a disinfectant is governed by its activity, concentration and contact time. General recommendations are as follows (after ref. 25).
- A clear soluble phenolic type disinfectant is the best type for general swabbing of surfaces, washing of plant areas, and disposal of routine laboratory apparatus, e.g. pipettes. A 1 per cent solution may be used to disinfect bench tops and areas after spillage; a 0.5 or 0.75 per cent solution may be used in containers into which pipettes, swabs and slides are discarded.
- Where phenolic-type disinfectants may be too irritant to personnel handling them it may be necessary to use a disinfectant of lower activity, e.g. chloroxylenol types, ampholytic types or quaternary ammonium compounds.
- Hypochlorite solution is only recommended for special purposes, e.g. spillage of high-hazard fluids, since, in addition to its inherent hazards, stock solutions diminish in strength through loss of free chlorine.
- A 5 per cent solution of formalin may be used as a general disinfectant in discard containers or for bench tops instead of 1 per cent hypochlorite solution.
- A 2 per cent glutaraldehyde solution is suitable for use with equipment and instruments.

Table 32.8 (after ref. 29) summarises the properties of classes of disinfectant. The parameters need to be determined on a case-by-case basis and further advice is given in ref. 26. As a generalisation, steam sterilisation is preferred to disinfection, e.g. for treating used cultures prior to disposal. In all cases the disinfectant/sterilant should be chosen with the advice of a professional microbiologist. Once selected, it should be used strictly in

Table 32.8 Properties of classes of disinfectants

Property	Phenols	Halogens	Alcohols	Aldehydes	Surfactants
Active against					
Vegetative bacteria	+	+	+	+	+
Bacterial spores	−	+	−	+	−
Fungi	+	1	−	+	1
Lipid virus	+	+	+	+	2
Non-lipid virus	2	+	2	+	2
Mycobacteria	+	+	+	+	−
Inactivated by					
Organic matter	−	+	−	−	+
Hard water	+	−	−	−	+
Detergent	3	3	−	−	4
Corrosive to metals	−	+	−	−	−
Flammable	−	−	+	−	−
Hazard to human health	toxic	corrosive	harmful	toxic	—

1 = Limited anti-fungal activity.
2 = Depends on the virus.
3 = Inactivated by cationic detergents.
4 = Inactivated by anionic detergents.

accordance with the supplier's instructions as to required concentrations, compatibility and precautions against specific hazards (e.g. glutaraldehyde is a potential skin irritant and sensitiser;[27] chlorine gas can be liberated from hypochlorite solutions during autoclaving; concentrated phenolic and quaternary ammonium disinfectants are especially harmful to the eyes). Suitable tests should be applied to verify its effectiveness in use.[25]

Handling precautions

The main causes of laboratory-acquired infection are accidents, carelessness, poor experimental technique and generally inadequate containment. One study of haematology and biochemistry laboratories reported many lapses from ordinary hygiene such as the communal use of unplugged mouth pipettes, even by experienced technicians.[28] The enquiry report on the smallpox incident described on page 1801 concluded[21]

> 'We think the main lesson to be learnt for the future is that containment of dangerous pathogens within laboratories working with them depends on adopting safe methods of working with adequate training and supervision and the correct use of physical containment facilities.'

Risk assessment

Risk assessment often tends to be qualitative and concentrates on taking account of the medium in which the organism is suspended, the numbers of organisms involved, and their likely survival rates in the environment. Conclusions are also based on a knowledge of their exposure in the host, routes of entry into the body, organism potency, whether prophylaxis and treatment is available, resistance of the host and the possibility of secondary spread. The UK philosophy, recently reviewed,[29] categorises dangerous pathogens according to the following criteria:

(i) the pathogenicity of the organisms for man
(ii) the infective hazard to laboratory workers
(iii) the transmissibility of the organism in the community
(iv) the availability of effective prophylaxis and treatment

In general the approach should be to contain infective micro-organisms at source. The Advisory Committee on Dangerous Pathogens[30] identified four levels of containment to reflect the hazard, thus:

Class 1 Non-pathogenic organisms and low-risk pathogens. These may be handled on the open bench since they are unlikely to pose risk of human disease.

Class 2 These micro-organisms may pose a hazard for laboratory workers but laboratory exposure rarely produces infection and effective prophylaxis or effective treatment is usually available. Infection of the community is unlikely. Examples include *Legionella pneumophila*, *Herpes simplex* virus and some species of *Salmonella*. These organisms may be handled on the open bench but activities likely to generate significant aerosol must be undertaken in a microbiological cabinet.

Class 3 Pathogens such as the AIDS virus HIV, Hepatitis B, and *Mycobacterium tuberculosis* may cause severe human disease and thus pose high risk to laboratory workers and could spread to the community at large, although there is available effective prophylaxis or treatment. Higher levels of containment are therefore necessary. Laboratories should be under negative pressure, preferably equipped with an independent High Efficiency Particulate Air (HEPA) filtration system. Work with these micro-organisms must be undertaken in a microbiological cabinet conforming with the specification set out in British Standard BS 5726: 1979 for Class I or III (Fig. 32.1 and 32.3),[19] or of equivalent performance.

Class 4 Here standards of containment are most stringent and are for the most dangerous pathogens, i.e. micro-organisms which present a serious threat to those involved with their handling as well as to the community as a whole, and for which no effective treatment or prophylaxis exists. Examples include smallpox and Lassa fever. Laboratories should be under a negative pressure of 7 mm (70 Pa) and be provided with an airlock itself under a negative pressure of 3 mm (30 Pa). Extract ventilation should be double HEPA filtered and micro-organisms must be handled in a Class III glove box.

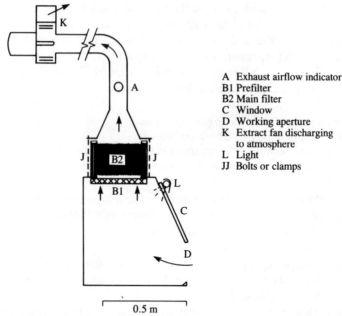

Fig. 32.1 Typical Class I cabinet

The Class I microbiological safety cabinet is an exhaust protective cabinet, where air is drawn in through the front aperture at a velocity of between 0.7 and 1.0 m/s. The cabinet must be capable of achieving a protection factor of at least 1×10^5 when subjected to a containment test specified in the British Standard. The Class I cabinet is fitted with an exhaust HEPA filter allowing not greater than 0.003% penetration as measured by dioctyl phthalate or sodium chloride testing, and should duct via independent ducting direct to atmosphere. Ducting should be fitted with appropriate anti blow-back devices, to guard against the possibility of contaminated particles being dislodged from pre-filters due to reverse air flows.

A Class II microbiological safety cabinet (Fig. 32.2) provides a degree of protection to the worker, whilst maintaining the sterility of work within the cabinet. The Class II cabinet has an inward airflow at the face of not less than 0.4 m/s, in addition to a downward airflow across the work of 0.25–0.5 m/s. The cabinet should also meet the British Standard containment test of not less than 10^5, and exhaust HEPA filter penetration of not more than 0.003%.

The Class III microbiological safety cabinet is essentially a glove box with HEPA filters on the air intake and extract from the cabinet. The cabinet should be capable of holding a positive pressure of 250 Pa for 30 minutes, have a minimum working negative pressure of 200 Pa, and have an air flow through the inlet filter of not less than 3 m^3/min when in use.

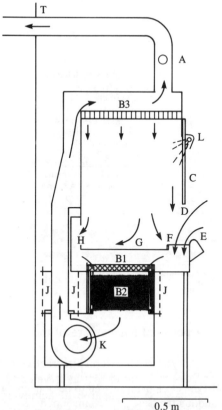

Fig. 32.2 Typical free-standing Class II cabinet

A Exhaust airflow indicator
B1 Prefilter in recess with lip all round to contain spills
B2 Main filter
B3 Diffuser
C Window
D Working aperture
E Edge aerofoil
F Front extract grille
G Working surface solid with lip all round
H Rear extract grille
JJJ Bolts or clamps
K Fan chamber
L Light
T Extract from room

Twin glove ports on the front of the cabinet allow access to the material to be handled.

The performance of microbiological safety cabinets requires periodic monitoring[31] and these arrangements for containment need to be backed up by a raft of general precautions.

General precautions

Anyone intending to conduct laboratory work with biological specimens or dangerous pathogens should in the first instance seek expert advice from a variety of sources including the regulatory authorities. They should consult the literature such as refs. 23 and 32–36. Some essential elementary precautions for handling micro-organisms in the laboratory are summarised by Table 32.9 (after ref. 36) but the reader is advised to consult original texts and experts for more comprehensive guidance. Chapter 11 provides detailed guidance for general laboratory safety.

It is clear that management commitment and training in appropriate

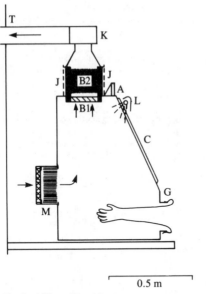

Fig. 32.3 Typical Class III cabinet

A Pressure indicator
B1 Prefilter
B2 Main filter
C Window
G Glove port, with glove
JJ Bolts or clamps
K Extract fan
L Light
M Inlet filter (shown on back but normally mounted on side)
T Extract from room

skills coupled with high standards of personal hygiene are of paramount importance in minimising risk.

> A laboratory technician became infected by a virus with which she was working and required hospitalisation.[37] Whilst the equipment proved to be of adequate standard the technician had been handling the Category 2 human pathogenic virus on the open bench with little previous experience in working with viruses. Her training during one-month's employment with the laboratory was also judged inadequate: she had not been given the usual induction training and was unaware of the dangers of the work. Furthermore her supervisor had a poor command of the English language and was oblivious of the technician's limited background.
>
> The laboratory had only recently expanded its work from plant to human pathogens and had not retrained staff to cope with the increased hazard. Further investigations revealed a history of infections together with the exposure route by which they arose. The laboratory was fined £600 under the Health and Safety at Work Act.

Precautions for hazardous pathogens

In addition to the foregoing minimum general provisions, more demanding precautions will be required for work with more hazardous biological micro-organisms. Requirements may include,

- the Biohazard sign displayed on the entrance, particularly for containment class 3 and 4 areas, together with an indication of the containment class.

Table 32.9 Summary of precautions for work in biochemistry laboratories

- Adequate locker facilities should be available, so that staff do not have to bring any personal belongings into the working areas.
- Any cuts or abrasions should be kept covered when in the laboratory.
- Vaccination of laboratory staff is recommended when satisfactory immunogenic preparations are available.
- No eating, drinking or smoking in the laboratory, except in rooms dedicated for the purpose.
- No food should be kept in laboratory refrigerators. The gum on envelopes, labels or forms should not be licked.
- Gloves should be worn when dealing with 'high-risk' specimens. Surgeons' gloves may be more generally suitable than the lighter disposable polythene type which are a very loose fit and puncture easily.
- Laboratory coats should be fully buttoned-up; other protective clothing should be worn when appropriate.
- The hands should be kept away from the face and neck.
- Broken glass should not be picked up with unprotected fingers and should be placed in covered tins for disposal.
- Staff should remove gloves and laboratory coats and leave them in a designated area, annex or lockers. They should wash their hands thoroughly before going for coffee, lunch or tea, and before going home.
- Safe pipetting practices should be adopted.
- If a specimen is spilled, the area should be swabbed down with a suitable disinfectant (e.g. 1 per cent hypochlorite solution; i.e. 1:10 dilution of commercial hypochlorite containing 10 g available chlorine per 100 ml and a suitable indicator – an example is 'Chloros').
- Techniques should avoid aerosol production; mixtures of infectious materials should not be prepared by bubbling expired air through liquids.
- Inspect centrifuge tubes for cracks prior to use, and exercise extreme care when removing them after centrifugation. Any broken tubes should be removed using e.g. tweezers.
- Syringes should be avoided whenever possible and handled with care so as to avoid injury.
- Disintegrators, sonicators and homogenisers should be operated in safety cabinets whenever aerosol formation is a possibility.
- Hazardous biological samples should be stored in labelled, leakproof containers. Freezers, cupboards, etc., should where relevant be labelled with the biohazard sign.
- Specimen containers should be transported in adequately labelled secondary vessels and packed with absorbent material if necessary.
- Hazardous microbiological materials should be rendered safe by autoclaving, incineration or chemical treatment prior to disposal.
- The laboratory and associated areas should be kept clean and tidy.
- Special arrangements are desirable for the collection and laundering of protective clothing.
- A procedure should be established for disposal of used syringes.
- Procedures should be established for decontamination, sterilisation and wash-up of all apparatus before removal from the laboratory.
- Provision should be made for sterilisation of hands before leaving the laboratory.
- There should be strict control over all 'open-bench' work.
- Safety cabinets should be provided with low airflow alarms and be tested regularly.

- restricted access to containment class 2–4 laboratories
- appropriate protective clothing to be worn, changed regularly and subsequently sterilised prior to laundering, particularly for containment class 3 and 4 areas. Suitable gloves to be worn in class 3 and 4 areas (and for class 2 if hands could become contaminated)

- laboratory design to reflect containment class
- transport of containment class 4 materials requires prior notification to the Health and Safety Executive
- procedures for dealing with mishaps and emergencies.

Biochemical processes

Micro-organisms and biochemical processes have been exploited industrially for many years, e.g. in bread and wine making, in limited production of pharmaceuticals by culturing and fermentation of naturally occurring organisms, and in sewage and effluent treatment. Each poses differing risks from microbiological hazards. Thus sewage treatment and certain effluent treatment workers face a range of hazards including exposure to pathogenic organisms, e.g. hepatitis A, resulting from contamination by human or other wastes.

In the UK contamination from airborne micro-organisms from sprays and splashes associated with effluent treatment plants, e.g. activated sludge plants, is not considered a source for concern.[38] However, if high levels of pathogenic organisms are likely to be present the risk should be reviewed and appropriate measures taken to reduce emissions. The risk of exposure to endotoxins from dead micro-organisms released with dust from sludge composting and drying processes requires the provision of adequate ventilation at sludge processing operations and the provision and use of appropriate personal protective equipment.[38]

In addition to the above, in recent years the scope of biochemical processing has been significantly expanded by genetic modification.

The main stages of a biochemical process include inoculum preparation, large-scale growth and subsequent separation of the organism, and product purification. Organisms are traditionally grown in dilute solution under moderate conditions of pH, temperature and pressure. Reactive, flammable and toxic raw materials and solvents are therefore usually avoided in biochemical processes. In most operations the micro-organism is intrinsically of low risk, although some may pose health hazards through inhalation, ingestion, or absorption through broken skin.

Control procedures

Control measures for handling micro-organisms and biochemicals should follow good occupational hygiene practice. In principle those general precautions listed in Table 32.9 are applicable. The detailed arrangements and measures will be influenced by the hazardous properties of the organisms coupled with the nature of the process. In performing risk assessments and devising handling strategies it is advisable to consult with experts such as microbiologists, hygienists and medical officers. Consideration should extend to disinfectant usage, sterilisation, storage, cleaning, waste disposal, laundering of protective clothing, etc. as well as to the bioprocess *per se*. It should embrace the need for monitoring, health screening and immunisation or vaccination. Any genetic manipulation work may present serious risks to personnel, and within the UK there is a requirement to notify the Health and Safety Executive of any intention to carry out such work.[39] Dangerous pathogens must be avoided and care taken to prevent the inadvertent culturing of animal and other pathogens. Advice on techniques and practice is given in ref. 23.

Contamination of equipment, room surfaces and air, and of the outside environment, by organisms can arise from releases during routine procedures or as a result of accidents. Whilst micro-organisms cultured on small volumes of solid agar media may pose little risk, when grown in larger quantities of liquid nutrient there is greater potential for accidents and aerosol formation. Furthermore even when using organisms which are unlikely to cause human disease, conditions could permit the growth of pathogens if contamination occurs, e.g. through faulty design or technique. It is preferable if possible to avoid ideal culture conditions for pathogens. Therefore, it is potentially safer to use an organism which requires for culture pH 3 or 4 at 25 °C rather than pH 7 at 37 °C.[25]

The risk can be minimised by substitution of the hazardous micro-organism by use of biologically disabled strains of the organism, e.g. use of the gene fragments of the AIDS virus rather than the whole virus, etc. Where substitution is not an option, containment by engineering methods is the preferred strategy, e.g. enclosed fermentation systems, Class III biological cabinets, etc. Containment may be important because of risk of occupational disease or threat to the environment from biochemical products, regulatory requirements, employee and public concern regarding release of micro-organisms, or process security. The level of containment should reflect the hazard. Thus, the majority of biochemical proceses warrant only minimal containment whereas high-risk organisms demand high levels of containment by engineering controls.

Physical containment consists of well-designed reactors fitted with appropriate seals and gas filters with secondary levels of containment provided by waste treatment and room air filtration. Where complete enclosure is impractical then local extract ventilation such as Class I and II biological cabinets may be appropriate, coupled with good operator technique. Personal protective equipment represents a secondary protective measure. It may be necessary to ensure the bioreactor and downstream processing areas remain aseptic whilst also protecting the areas outside this part of the facility from exposure to organisms or product.

As with any process involving hazardous substances the plant should be professionally designed, installed, commissioned, maintained and operated. Equipment should be easy to clean and sterilise internally, and to clean externally. Budgetary constraints may encourage use of 'home-made' devices such as fermenters, which often prove difficult to sterilise. Indeed, even some commercial designs are less than ideal and are too bulky to autoclave easily, employ 240 V aquarium heaters for temperature control, and produce excessive volumes of culture requiring disposal.[22] When work, or cleaning, is carried out on benches in the open workroom the surrounding area should be isolated and ventilated to the atmosphere. Personal hygiene is important and a sink must be available for hand washing and proper waste disposal facilities provided.[25] Obviously, the storage of any biological material must be segregated from any storage of food or drink.

Training and monitoring are of paramount importance depending upon the level of risk. For fermentation processes, pressure testing and air monitoring are widely carried out. Currently there are few hygiene standards and monitoring techniques for selected micro-organisms and biochemicals; typical details of standards and monitoring of detergent enzymes are given on page 1649. Control measures need reinforcement with in-house codes of safe working and permit-to-work procedures.

Legionnaires' disease and humidifier fever

Background

Building services generally incorporate mechanical systems for the supply and distribution of fresh air; this is sometimes filtered, heated, cooled or humidified. Micro-organisms from outside the building may enter the air-conditioning system where conditions prove ideal for their survival, leading to multiplication, and subsequent distribution throughout the workplace. The contaminated air has the potential to cause human respiratory distress by, e.g. either

- infection from an invasive growth of micro-organisms in the respiratory system, e.g. Legionnaires' disease, a rare form of pneumonia
- allergy from hypersensitivity to inhaled microbial particulates, e.g. humidifer fever.

In the UK, occupational exposure to hazardous microbial matter is subject to the Health and Safety at Work Act, 1974 (Chapter 16), and, as mentioned earlier, more specifically The Control of Substances Hazardous to Health Regulations, 1988 (Chapter 28).

> Inspectors from the Health and Safety Executive investigating an outbreak of Legionnaires' disease at a Blackburn wallpaper company in 1990 concluded that the company did not take all reasonable precautions to ensure that measures were properly carried out for controlling the legionella bacterium, by application of biocide to the water of a cooling tower system. The company had the required chemicals but had failed to ensure that they were used. They were fined for a breach of Regulation 8(1) of the Control of Substances Hazardous to Health Regulations, 1988. (This was the first prosecution under the COSHH Regulations.[40])

Legionnaires' disease

Two main types of infection are caused by legionellas, viz. Legionnaires' disease and Pontiac fever. The latter is less severe and is similar to influenza with symptoms appearing between five hours and three days after exposure. The illness lasts between two and five days and is rarely fatal. Legionnaires' disease, however, is a pneumonia-type illness. Symptoms, which include high fever, chills, headache, muscle pain, dry cough and breathing difficulties, usually manifest themselves within three to six days after exposure to the bacterium, but this can vary from two to ten days.

Legionnaires' disease was mentioned briefly on page 464. Outbreaks caused by emissions from evaporative cooling towers, e.g. those associated with air-conditioning plant, receive considerable publicity but more cases are actually caused by contaminated domestic hot-water systems.

The disease is an infection caused by the inhalation of water aerosols contaminated by *Legionella pneumophila*; it results in a form of pneumonia and may have other effects, e.g. diarrhoea, kidney failure and delirium. Several other species of *Legionella* can cause disease in man, as shown in Fig. 32.4,[41] each with varying ability to cause Legionnaires' disease; serogroup 1 is the most virulent.[42]

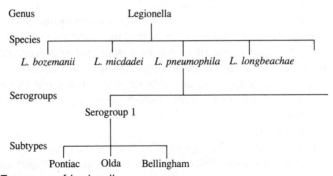

Fig. 32.4 Taxonomy of *Legionella*

About 10 per cent of cases have proved fatal; older people and the infirm are most at risk (hence the prevalence of cases in hospitals).

In July 1976 the Pennsylvanian Branch of the American Legion held its 58th convention at the Bellevue Stratford Hotel, Philadelphia. A few days after returning home many of the delegates became ill with pneumonia. In all over 200 were affected and 29 subsequently died.

This was the first occasion on which the disease was recognised as a specific entity, although serological work on stored materials showed that cases of Legionnaires' disease had occurred up to 30 years previously; sporadic cases were identified retrospectively in Britain back to 1972.[42] Further examples of cases recorded in the five years following the Philadelphia outbreak are given in Table 32.10.[43] Currently there are about 200 cases each year in England and Wales, about 10 per cent of which are fatal. Reporting is voluntary and the disease is not a Notifiable Disease, hence there may be unreported cases. Indeed, based on one survey which concluded that 2 per cent of all community-acquired pneumonias were Legionnaires' disease,[44] then between 1500 and 2000 cases of the disease would be expected annually.[42]

Microbiology

Legionellas are aquatic organisms found in a range of habitats including streams, rivers, the shores of lakes, thermally polluted water and natural thermal ponds.[45] It is inevitable, therefore, that legionellas will invade and colonise man-made water systems. This fastidious short Gram-negative flagellated bacillus is, unlike most fresh-water aquatic bacteria, slow growing and has exacting nutritional requirements. It is believed that legionellas live in association with other organisms such as protozoa (amoebae), cyanobacteria (blue algae) and flavobacteria, all of which are frequently present in water. Indeed legionellas are able to infect amoebae, which exist in a wide range of freshwater habitats including potable water, water tanks, showers and swimming pools. As depicted in Fig. 32.5,[42] one amoeba can contain 1000 legionellas. Whilst 0.4 mg/l free chlorine kills free-living legionellas, amoebal cysts produced from infected trophozoites protect legionellas from concentrations of at least 50 mg/l free chlorine. Effectively, therefore, legionellas can persist in chlorinated waters.

The legionella organism is susceptible to a range of antibiotics under laboratory conditions, but in the human body it thrives and multiplies within the lung alveolar macrophages, where it remains protected from many antibiotics.

The cause of the Philadelphia outbreak eluded scientists for many months, since, whilst not new, the infection had escaped recognition because the causative organism failed to grow on the conventional culture media used in hospital laboratories.

Humidifier fever

Humidifier fever is a flu-like illness caused by the inhalation of fine water droplets from humidifiers which have become contaminated.[46,47] Although various micro-organisms such as bacteria, fungi and protozoa have been found in contaminated water from humidifiers the relationship between these micro-organisms and the illness is unclear. It differs from Legionnaires' disease in terms of both the causative micro-organisms and the symptoms, although both illnesses are associated with exposure to contaminated water systems in buildings. Inhalation of contaminated water droplets has also, albeit rarely, caused extrinsic allergic alveolitis, which

Table 32.10 Numbers of cases and deaths for some Legionnaires' disease outbreaks in the USA and the UK

Location of outbreak	Year	Cases	Deaths
Washington, DC	1965	81	14
Pontiac	1968	144	0
Benidorm	1977/80	At least 3 fatalities	
Philadelphia (Oddfellows' convention)	1974	11	10
Philadelphia (Legionnaires' convention)	1976	182	29
Vermont	1977	56	19
Bloomington	1978	21	0
Memphis	1978	15	0
Atlanta	1978	8	0
Corby, UK	1978	4	0
Kingston, UK	1980	12	4

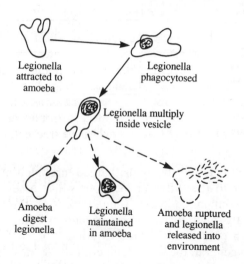

Fig. 32.5 Legionella and Amoeba

has similar acute symptoms but can lead to irreversible lung fibrosis if exposure is prolonged, or humidifier asthma, which although accompanied by wheezing has no flu-like symptoms.

The symptoms of humidifier fever range from mild fever with headache, malaise, and muscle ache, to acute illness with high fever, cough, chest tightness and breathlessness on exertion. The onset of symptoms is delayed, beginning 4–8 hours after the start of work. The illness is also known as 'Monday sickness' since symptoms usually appear on the first day back at work after a break and tend to resolve over 12–16 hours.

Air-conditioning systems

Figure 32.6 illustrates a typical hospital air-conditioning system installed in the 1960s.[43] External air is drawn into the system and heated or cooled as

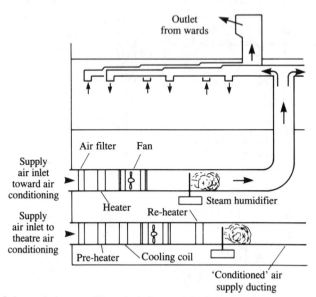

Fig. 32.6 Schematic layout of hospital air-conditioning system

necessary, the tempered air being discharged through a humidifier. Ductwork supplies air to the different locations whilst a separate extraction system removes contaminated air. Temperatures on the heating side are usually sufficiently high to kill micro-organisms. Air cooling involves refrigeration with effluent air vented externally via cooling towers (Fig. 32.7)[43] where, in general, the water temperatures are near optimum for the survival and growth of *Legionella pneumophila*. Water is continually recirculated in cooling towers which act as effective air scrubbers and which leads to contamination by airborne dusts, micro-organisms, scale and sludge. In a recent survey of 300 cooling towers, half of the samples taken were at between 30 and 35 °C. Most systems contain sufficient slime and sludge to provide ample nutrition for *Legionella*, and to nullify the effects of added chlorine or biocides.[48]

Humidifiers

Humidification of ambient air is achieved by water evaporation or steam injection. The temperatures associated with the latter are high enough to be biocidal. Evaporative humidifiers may function by discharging all the

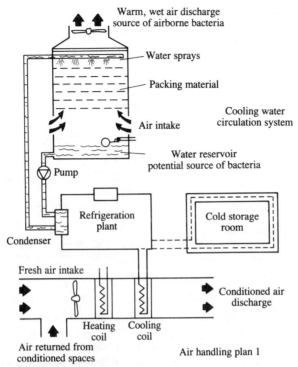

Fig. 32.7 Cooling system for a large public building, showing potential source and discharge of airborne bacteria

water drawn from the supply, or by evaporation of only a small percentage of the flow with the bulk of the water being recycled to the reservoir.

Several types of humidifier are available as illustrated by Fig. 32.8.[43] Spinning disc and spray humidifiers and air washers present a considerable risk once contaminated. With systems relying on atomised sprays baffles are fitted to prevent discharge of large droplets, but whilst highly effective they are not 100 per cent efficient.

In most systems water evaporation takes place at ambient temperatures, hence the reason why humidifiers tend not to harbour *Legionella pneumophila*, which prefers 35–40 °C. In plants requiring high levels of humidity, such as printing works and textile mills, there is the additional complication of organic dust which provides nutrition for bacteria, and the air inlet supply needs filtering.

Preventative measures against Legionnaires' disease

General

In the UK the duty imposed under S3(1) of the Health and Safety at Work Act 1974 for an employer to ensure, so far as is reasonably practicable, that persons not in his employment are not exposed to risks to their health, covers risks due

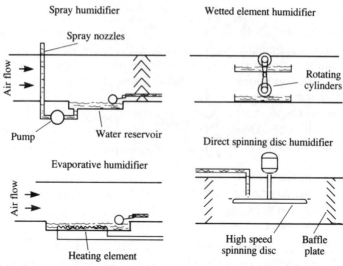

Fig. 32.8 Common types of humidifier

to *Legionella pneumophila*. It has been held that the public outside a museum had been exposed to such risks by reason of an inadequate system of maintenance of the air-conditioning system. It was not necessary to prove that members of the public had actually inhaled the bacterium or that it had been present, it was sufficient to show that there had been a risk of it being present.[49]

Prevention of Legionnaires' disease relies on avoidance of inhalation of contaminated respirable mists, i.e. aerosols. This entails controlling the growth of *Legionella pneumophila* in water systems and preventing the generation of aerosols. Colonies of bacteria develop on internal surfaces or pipes, on grommets and washers, etc. and in the aqueous environment generally.

Aerosols form by atomising or break-up of water streams, e.g. as they strike a surface or by bursting and dispersion of bubbles from the water surface. Such phenomena are common with running water taps, showers, fountains, humidifiers and cooling towers.

The risk is increased with the smallest droplets because

- the terminal velocity of the droplets decreases with droplet diameter so that the smallest droplets remain airborne for longer periods, as discussed on page 117
- the most dangerous droplets are those <5 μm in diameter since they can penetrate deeply into the lungs.

However, larger droplets may enter the respirable range due to evaporation. Small droplets can travel over 50 metres. Survival of bacteria within the aerosol is influenced by temperature, humidity, sunlight, wind speed, etc.

Detailed advice on control measures is given in refs. 50–52.

Design

Because of the ubiquitous nature of *Legionella pneumophila*, contamination of water systems is impossible to prevent. In order to grow, the bacteria require water, a suitable temperature and pH, time and nutrients. Control of these conditions is the basis for prevention and they should be addressed at the design stage. Since *Legionella* can only grow in water, pipe runs should be as short as possible and hot pipes should not be located too close to cold pipes. Deadlegs and other areas where water may become stagnant should be eliminated. Any disused plant or water system capable of generating an aerosol should be designed or modified so that it can be drained and kept dry when not in operation. If economics and technical factors permit it, air-cooled chillers are preferable to evaporative cooling towers. Shower heads should be designed to drain and not to retain a small pool of stagnant water.

In mid-1989 a routine check for bacterial infection revealed the presence of *Legionella pneumophila* serogroup 6 in one of the shower heads in an amenity building of a plant in Scotland. This was not the virulent strain but can cause flu-like symptoms when inhaled in droplet form. Since neither the hot water system, at 85 °C, nor the cold water, at 17 °C, would promote bacterial growth the infection was deduced to have been localised to the shower head. A system of routine sterilisation of the shower heads was subsequently introduced based on the methods summarised in Table 32.11.[53]

Table 32.11 Procedure for sterilisation of shower heads[53]

- For metal shower heads: remove from shower, place in a heated oven at 150°C for one hour. Switch off oven and leave shower head in to cool down.
- For plastic heads: remove and soak in 5–10% household bleach (hypochlorite) for one hour (or overnight if more convenient). Rinse with hot water and refit.
- When the head is removed, flush the piping feed to the head at full bore for a few minutes.
- This procedure can be repeated every 3 months.

The optimum temperature for growth of *Legionella pneumophila* is 37 °C but it will grow within the range 20 °C to 50 °C. Wherever practicable water supplies should, therefore, be stored either below 20 °C or above 55 °C. There is a risk of scalding above 60 °C. At 60 °C, 90 per cent of *Legionella* are killed within two minutes.

Wooden cooling-tower packings should be replaced if practicable and iron piping should be avoided. The major sources of aerosols are cooling towers, taps and showers. Even ordinary taps produce aerosols when the jet of water impinges on a bath, basin or vat. All forced and induced draught cooling towers need to be provided with drift and windage eliminators; these installations and their sump tanks should be inspected and maintained on a regular basis.

Water treatment

All water, except that used for drinking, should be chemically treated. Water treatments include scale or corrosion inhibitors, dispersants to prevent sediment from settling, biocides to kill bacteria and other organisms, and disinfectants. Their effectiveness depends on proper use, and in certain cases the components may react with one another or with the system, resulting in corrosion.[54] Many water treatment chemicals are themselves highly toxic and their use should be subjected to a risk assessment. Recommendations on suppliers' safety data sheets should be followed and the advice of a competent consultant or reputable water company may be required; indeed water treatment is commonly undertaken by specialist contractors.

Piped water supplies can be rendered safe by continuous chlorination of cold water: once contamainted water enters the building services, such as piping, tanks, cisterns, etc. it comes into contact with sediment and the temperature rises. Consequently, the concentration of any added chlorine falls rapidly and its disinfectant property is destroyed. Water treatment is not an alternative to the physical removal of solid matter from a system, since the efficiency of a biocide is severely impaired by accumulation of debris and slime.[45] Bacteria can also become resistant to biocides, and some corrosion inhibitors, e.g. polyphosphates and nitrites, can actually encourage the growth of micro-organisms. However, where water treatment to control bacterial and algal growth forms part of the preventative programme it is important to monitor its implementation, as illustrated by the Blackburn case on page 1812 and by the following example.[55]

> In 1988 an outbreak of Legionnaires' disease, traced to the cooling towers at a corporation's headquarters in Central London, culminated in three fatalities. In all, 58 people were affected, of whom 18 were employees and the remainder were members of the general public. The retirement of one man, under whose regime there had apparently been regular water treatment, was followed by a period of confusion over requirements and duties. The organism probably travelled 50 m downstream of the building and still caused an infection. (The organisation was prosecuted under the Health and Safety at Work Act and fined £3600.)
>
> (A taxi driver who contracted Legionnaires' disease due to exposure whilst driving past the headquarters was awarded £26 000 some three years after the incident and it has been suggested that the corporation could face up to £1 million in claims.[56])

The nutrients required by the bacteria comprise organic foodstuffs, trace elements and minerals, including soluble iron (usually from rust) and nitrogen (from nitrite descalers) or the organic 'baffle jelly' which accumulates in plumbing. Therefore all storage cisterns should be protected against ingress of plant and animal matter, and the growth of algae and other microbes, which will generate organic nutrients, should be controlled.

Water fittings, e.g. tap washers of plastic or rubber, or the thin film of slime which coats them can support bacterial growth. Hence only approved fittings and materials should be used in water systems.

If iron components are used in cooling towers or similar systems then corrosion should be prevented by water treatment.

All water systems should be inspected regularly, as a check on the control features. The frequency of inspection and cleaning will be dictated by specific circumstances. Ideally, however, cooling tower packing should be drained and cleaned on a regular basis (e.g. every two to three months) with a final rinse in disinfectant. Systems should certainly be overhauled before any air-conditioning plant is started up, and again when it is closed down; outbreaks of Legionnaires' disease have been associated with the start-up of cooling towers after lengthy shut-down. It is good engineering practice to inspect, and if necessary disinfect, at least annually. The degree of general microbiological contamination can be measured using simple dislides, as a useful check that conditions are not favourable for microbial growth.

Inspection and maintenance

If stagnation of water in deadlegs or in excessive storage capacity is avoided, the continuous water flow will allow insufficient time for dangerous levels of bacterial contamination to develop. Careful inspection is recommended to ensure there are no disused/little-used pipe sections, particularly if they are likely to equilibrate with building temperatures and reach 20 °C for considerable periods of time.[41] Visual inspection of condensate from cooling coils in air-conditioning ductwork is needed to ensure that water drains away through a trap which has an air-break to prevent back-siphonage. Glass traps are recommended in sensitive areas. Regular draining and thorough cleaning of water systems are the key preventative measures since the build-up of debris, e.g. scale or corrosion, may effectively produce tiny pockets of stagnant water.

It is during the cleaning process that health hazards are at their greatest for maintenance staff and other occupants in the area. Precautions to consider include those for entry into confined spaces (including disinfection before entry), controlling aerosol generation, and the provision of suitable protective clothing and respiratory equipment. Respiratory protective equipment should be worn for any operation where a spray of heavily contaminated water is foreseeable, e.g. during cleaning with high-pressure water jetting. Training is essential and, in the UK, is a requirement under the COSHH Regulations.

Regular testing for *Legionella pneumophila* is not considered necessary since low levels of contamination are likely to be confirmed and furthermore contamination may occur immediately after testing.[41] Checking the primary control measures is more important. Routine monitoring for pH and dissolved solids, routine bacterial counts, and the addition of disinfectant or biocide concentrations will prove useful for, e.g. cooling water systems, but this should not be confused with sampling for *Legionella*. The UK Health and Safety Executive have recently issued further advice in ref. 58; this includes contentious proposals that cooling towers should be registered with local authorities.

A checklist for air-conditioning equipment and associated systems is given in Table 32.12.[50]

Record keeping

Formal records should be kept to ensure good management of the system, to fulfil legal requirements such as the UK COSHH Regulations (Chapter 28), and to demonstrate that all reasonable steps have been taken in the event of an incident. Items to cover include
- cleanliness
- risk assessment
- altered usage resulting in excess capacity or deadlegs in the system
- temperatures
- displaced tank covers
- inspection programmes
- compliance with cleaning/maintenance schedules
- water treatment regimes, e.g. chlorination levels (if used as a biocide)
- water quality checks.

Emergencies

If an outbreak of Legionnaires' disease is suspected the local hospital's control of infection specialists will normally work in association with the Public Health Service, the Health and Safety Executive, and the local medical officers of environmental health. The important early steps are to rapidly identify the population likely to be affected, and to ensure that the contaminated water system is treated as early as possible.

The engineer's role is important in directing specialists to the various water systems within the buildings and in providing access to sampling points.

Preventative measures against humidifier fever

Guidance on the prevention of humidifier fever and other related health problems includes the following.

Equipment selection

Select humidification equipment that is least likely to become contaminated, i.e. in order of preference:[46]

Table 32.12 System checklists

Air conditioning equipment

Has water been found in the ductwork?	Yes	Search for cause and correct
	No	No action
Are the drains to the chiller coil trapped and with an air break?	Yes	No action
	No	Provide them and consider using glasstraps in sensitive buildings such as hospitals
Do the drains dry out in winter?	Yes	Provide a liquid seal
	No	No action
Are there sufficient inspection hatches to check on state of heat exchangers?	Yes	No action
	No	Provide them
Is there a recirculating water type humidifier?	Yes	Ensure it is kept scrupulously clean, and drained down when not in use, e.g. weekends
	No	No action

Cooling towers

Is the system water volume marked on the tower?	Yes	Check that it includes pipework volume
	No	Mark it on
Is there an operator's manual?	Yes	Ensure it is readable, readily available and describes the existing equipment
	No	Prepare one
Does it define normal operating conditions?	Yes	No further action
	No	Provide the data
Is there a maintenance manual?	Yes	Ensure it is readily available
	No	Consult the tower manufacturer and the cooling circuit designer to have one prepared
Is there a maintenance schedule?	Yes	Ensure that actions are clearly recorded and dated
	No	Draw one up
What is the normal maximum operating temperature of the water?	20°C	No action
	20–30°C	Caution: check biocide is used
	30°C	Check biocide used carefully
Is a water meter fitted to the mains supply?	Yes	Ensure it can easily be read and is regularly recorded
	No	Fit one
Is chemical and biocide use recorded?	Yes	Check that quantities used agree with expectations
	No	Record it
Is microbiological activity checked?	Yes	Ensure that it is steady and below 10 colony forming units/ml
	No	Consider providing regular monitoring
Can sunlight enter the pond?	Yes	Consider screening the tower
	No	No action
Is a drift eliminator fitted?	Yes	No action
	No	Fit one

Table 32.12 (continued)

Is the tower outlet within 10 m of an air inlet or window?	Yes	Consider moving them apart or converting to a dry tower at the end of the useful life of the present one
	No	No action
Is the cooling tower drain provided with an air break?	Yes	No action
	No	Provide one
Hot water supply		
What is the storage temperature in the calorifier?	>60 °C	Safe if left for 2 or more hours
	55–60 °C	Safe if left overnight
	50–54 °C	Caution
Does the hot water reach 46 °C at all taps?	Yes	No action
	No	Consider trace heating, pipe insulation or modifying the installation
Does the system contain materials not listed by the Water Research Centre, e.g. natural rubber gaskets or seals?	Yes	Remove as soon as convenient and replace with approved materials
	No	No action
Are there long deadlegs?	Yes	Consider modifying the installation to include a high-use fitting at the end of the line, or shortening the deadleg by rearranging secondary pipework
	No	No action
Do these deadlegs serve outlets?	Yes	No action
	No	Remove the redundant piping
Are there showers or emergency chemical or eye washes on site?	Yes	Discharge weekly
	No	No action
Does the water supply provide for susceptible people, e.g. elderly males, smokers, people who are ill?	Yes	Examine where aerosols are likely because of high pressure operation or close proximity of basin to tap. If likely, take action to minimise aerosol production
	No	No action
Cold water storage tanks		
What is the temperature of the water?	<20 °C	Safe
	20–25 °C	Caution
	26–30 °C	Take action to lower temperature, particularly in summer if in roof space, or in winter if a heating pipe runs alongside. Possible actions: (a) Replace with smaller capacity cistern (b) Insulate cistern, paint outside with reflecting paint (c) Improve local ventilation or reduce other heat gains (d) Link in with neighbouring tank to ensure series flow (e) Insulate cold feed supply
Is the cistern or inspections hatch covered?	Yes	No action
	No	Fit cover to protect tank from dirt

Table 32.12 (continued)

Is the inside of the tank clean?	Yes	No action
	No	Clear loose debris and where internal surface is marked paint with approved paint
Is it a solitary tank?	Yes	If the building is in continuous use such as an hotel or hospital consider the addition of a small break tank to facilitate cleaning
	No	Caution: ensure flow occurs through the tanks simultaneously, for example by coupling them in series
Does the overflow warning pipe have an insect guard?	Yes	No action
	No	Fit one

- Steam humidification, since droplets are not produced and the steam may kill organic growth.
- Compressed air atomisers that take water directly from the mains. These present some risk if supplied from a holding tank.[47]
- Evaporative-type humidifiers that do not create a water spray, although other measures may still be necessary to avoid contamination of any water reservoir.

Water supply

- The water supply to the humidifier should be clean and free from contamination. Where permitted by the water supplier, taking water directly from the mains greatly reduces the risk. If a holding tank or reservoir is used this should preferably be sited away from sources of contamination and fitted with a cover.
- Water tanks and reservoirs need to be emptied and flushed out regularly with fresh water.

Maintenance of cleanliness

- Regular and thorough cleaning and disinfection of humidifier and storage tanks is most important.
- The frequency of these will depend largely on environmental conditions. Frequent cleaning will be needed if the environment is dusty, especially with organic or cellulose dust: in this case they should be inspected on, e.g., a weekly basis. Frequent cleaning may also be necessary in warm weather, or if the water is warm. For office environments cleaning and disinfection may be required every two to three months, in some industrial environments it has been on a monthly or even weekly basis to control bacterial and organic growth.
- Cleaning and disinfection should be thorough. All accessible wetted parts should be cleaned and the humidifier should be washed through

with disinfectant followed by clean water. (Care should be exercised in cleaning, e.g. to provide good ventilation and avoid contact with disinfectant.)
- Humidifiers should be inspected weekly, especially any internal reservoirs and holding tanks. If there are any signs of organic growth, or if the water appears to be contaminated, cleaning and disinfection should be performed.
- Biocides or other water treatment chemicals should not normally be used when humidifiers are in operation.
- Humidifiers should be drained and kept dry during periods when they are not in use or cleaned and disinfected before reuse.

Maintenance

A good standard of maintenance is necessary to avoid malfunction that could result in water entering ductwork. Particular attention should be given to baffles and eliminators intended to minimise the release of water droplets.

In cases where humidifier fever is suspected, the humidifier should be turned off and advice sought from a medical practitioner and a competent ventilation engineer with specialised knowledge of humidifier systems. Exposure to contaminated humidifiers can be confirmed by tests on blood samples from the workforce with extracts from the residues or sludge in contaminated systems.[46] Once it has occurred remedial measures with an existing humidifier may be ineffective, because despite subsequent maintenance and cleaning the humidifier is liable to continue to release some organical material. If symptoms persist after such work, the alternatives include
- moving sufferers away from the influence of the humidification system, which may prove difficult (i.e. they need to be away from the humidifier building);
- dispensing with humidification, which may be impractical; or
- replacing the system with one that presents little risk, e.g. steam humidification.

Sewage workers

Sewage is the waste water from domestic and industrial washing, food preparation and industrial processes. It also contains urine, faeces and other body wastes. It is frequently quoted as having 300 ppm suspended solids with about 30 ppm ammoniacal nitrogen, which arises from breakdown of protein and urea, and about 250 g/m^3 of dissolved organic matter that can be oxidised by bacteria in the receiving water. As discussed in Chapter 21, this Biological Oxygen Demand (BOD) will consume some, or all, of the oxygen available in the water; there is only about 11.3 g of oxygen per m³ of river water at 10 °C.

Sewage treatment aims to reduce the potential for liquid-borne wastes to be harmful to humans or the environment. Techniques include:
- disinfection by chemicals, heat or radiation
- primary treatment to remove solids from the inflowing sewage, which settle to afford supernatant liquor which passes to secondary treatment. The majority of the sludge is degraded by anaerobic bacteria. In some cases sludge is dumped at sea or on land, or incinerated
- secondary treatment – aerobic oxidation

Hazards to sewage workers include exposure to micro-organisms arising mainly from sewage of domestic origin. Possible pathogenic organisms include typhoid, dysentery, cholera, polio and hepatitis (Table 32.13).[58]

Table 32.13 Pathogens in sewage and their fate in the sewage treatment process[58]

Pathogens[3] in crude sewage:
Total coliforms $(3-500) \times 10^6$ per 100 ml
Faecal coliforms $(1-30) \times 10^6$ per 100 ml
Viruses (PFU) 10^4-10^7 per litre

	Percentage reduction			
Process	Salmonella	Viruses	Protozoa and Metazoa	Coliforms
Activated sludge	70–90	80–99	0–99	
Trickle filters	84–99	40–80	60–99	
Anaerobic digestion	84–93	90–99	45–97	
Primary treatment		0–50		30–60
Secondary treatment		75–100*		90–99

* Depends on suspended solids concentration.

Airborne dust from sewage sludge can be heavily contaminated with a variety of micro-organisms including bacteria, fungi and protozoa but not necessarily thermophilic actinomycetes. Liming sludge reduces pH and can eliminate salmonella and inactive viruses. Other precautions against infection are mainly education, personal protection and inoculation against, e.g. tetanus, typhoid and poliomyelitis.

References

1. Health & Safety Executive, ND(G)85L, HSE, 1991.
2. Hunter, D., *The Diseases of Occupations*, 5th edn., English Universities Press, 1974.
3. Barber, T. E. & Hustings, E. L., in *Occupational Diseases: A Guide to Their Recognition*, eds. M. M. Key, A. F. Henschel, J. Butler, R. N. Ligo and I. R. Tabershaw, US Dept. of Health, Education, and Welfare, National Institute for Occupational Safety and Health, Washington, DC, 1977.

4. Butterworth, D. J., MSc Thesis, *Health & Safety in the Agricultural Industry*, Aston University.
5. *Tremain* v. *Pike & Another*, 1969 All E.R. 1303.
6. Department of Health and Social Security, *Notes on the Diagnosis of Occupational Diseases*, Rev. edn., 1983.
7. Kinnersley, P., *The Hazards of Work: How to Fight Them*, Pluto Press, 1973, p. 172.
8. Health & Safety Executive, *Leptospira hardjo (bovine leptospirosis and human health*, Agricultural Sheet No. 5, HSE, Nov. 1992.
9. HM Factory Inspectorate, *Anthrax*, Technical Data Note 20 (Rev.), Department of Employment, 1974.
10. Parkes, W. R., *Occupational Lung Disorders*, 2nd edn., Butterworth, 1982, p. 391.
11. Harvey, M. S., Forbes, G. B. & Marmion, B. P., *The Lancet*, 1951, **2**, 1152.
12. Parkes, W. R., *Occupational Lung Disorders*, 2nd edn., Butterworth, 1982, p. 389.
13. Horejsi, M., Sach, J., Tomsikova, A. & Mecl, A. *Thorax*, 1960, **15**, 212.
14. Lacey, J., Pepys, J. & Cross, T., Actinomycete and fungal spores in air as respiratory allergens, in *Safety and Microbiology*, eds. D. A. Shapton and R. G. Broad, Academic Press, New York, 1972.
15. *Health & Safety Commission Newsletter*, HSC 1991 (No. 77) June, p. 9.
16. *Farmer's Lung*, Pamphlet AS5 Revised, Health & Safety Executive, 1992.
17. MAFF, *Farmer's Lung, Mushroom Workers' Lung*, Ministry of Agriculture Fisheries and Food.
18. Chatigny, M. A. & Clinger, D. I., Contamination control in aerobiology, in *An Introduction to Experimental Aerobiology*, ed. Dimmick, R. L., Chapter 10, John Wiley, New York, 1969, p. 194.
19. Pike, R. M., *Health and Laboratory Science*, 1976, **13**, 105.
20. *Report of the Committee of Enquiry into the Smallpox Outbreak in London in March and April 1973*, HMSO.
21. a) *Report of the Investigation into the Cause of the 1978 Birmingham Smallpox Occurrence*, HMSO, 1979.
    b) McGinty, L. *New Scientist*, 1979 (January), 8.
22. Tranter, J., *Safety & Health Practitioner*, 1991, **9**(6), 14.
23. Collins, C. N., *Laboratory Acquired Infections*, Butterworths, 1983.
24. Maurer, I. M., The management of laboratory discard jars, in *Safety in Microbiology*, eds. D. A. Shapton and R. G. Board, Academic Press, New York, 1972.
25. Institution of Chemical Engineers, Rugby, *Safety in Chemical Engineering Research and Development*, 1991, p. 92.
26. Cottam, A. N., The Selection and Use of Disinfectants, Specialist Inspectorate Report No. 17, Health & Safety Executive, 1989.
27. Health & Safety Commission, *Glutaraldehyde and You*, Leaflet IAC 64(L), 1992.
28. Public Health Services Working Party on Haemodialysis Units, *Brit. Med. J.*, 1968, **3**, 454.
29. Keddie, J. R., The UK approach to laboratory ventilation and the control of dangerous pathogens, in *Proc. Second Int. Symp. on Ventilation for Contaminant Control*, 20–23 September 1988, London.

30. a) Advisory Committee on Dangerous Pathogens, *Categorisation of Pathogens According to Hazard and Categories of Containment*, HMSO, 1984.
    b) Advisory Committee on Dangerous Pathogens, *Guidance on the Use, Testing and Maintenance of Laboratory and Animal Flexible Film Isolators*, HMSO, 1985.
31. British Standard BS 5726: 1979, *Specification for Microbiological Safety Cabinets*, British Standards Institution.
32. The Control of Substances Hazardous to Health Regulations, 1988.
33. Department of Health, *The Control of Substances Hazardous to Health: Guidance for the Initial Assessment in Hospitals*, HMSO, 1989.
34. Department of Health & Social Security, *Code of Practice for the Prevention of Infection in Clinical Laboratories and Post-mortem Rooms*, HMSO, 1978.
35. Health & Safety Commission, *Safety in Health Service Laboratories. Hepatitis B*, HMSO, 1985.
36. a) Neill, D. W. & Doggart, J. R., Safety in hospital biochemistry laboratories, in *Hazards in the Chemical Laboratory*, ed. G. D. Muir, 2nd edn., The Chemical Society, 1977.
    b) Rayburn, S. R., *The Foundations of Laboratory Safety: A Guide for the Biomedical Laboratory*, Springer-Verlag, New York, 1990.
    c) HSAC, *Safe Working and the Prevention of Infection in Clinical Laboratories*, HMSO, 1991.
37. *Report by HM Chief Inspector of Factories 1986–87*, Health & Safety Executive, 1987.
38. Institution of Chemical Engineers, Rugby, *Aqueous Effluents*, Training Package EO1, 1992.
39. Health & Safety Executive, *Genetic Manipulation Regulations and Guidance Notes*, HMSO, 1978.
40. Ballards, J., *The Safety and Health Practitioner*, 1991 (April), 2.
41. B.P. Health, Safety and Environment Bulletin, reproduced in *Loss Prevention Bulletin*, 1991(090), 23.
42. Marshall, J., *The Safety and Health Practitioner*, 1991 (January), 19.
43. Agner, B. P. & Tickner, J. A., *Ann. Occup. Hyg.*, 1983, **27**(4), 34.
44. Kurtz, J. B., *Ann. Occup. Hyg.*, 1988, **32**, 59.
45. Dennis, P. J. L., *Ann. Occup. Hyg.*, 1990, **34**, 189.
46. Health & Safety Commission, *Precautions Against Humidifier Fever in the Print Industry*, 1989.
47. Sykes, J. M., *Precautions Against Illness Associated with Humidifiers*, Specialist Inspector Report No. 11, Health and Safety Technol. Div., HSE, 1988.
48. Bateman, P., *The Safety and Health Practitioner*, 1989 (July), 36.
49. *Regina v. Board of Trustees of the Science Museum*, Times Law Report, 15 March 1993.
50. Chartered Institute of Building Services Engineers, *Minimising the Risk of Legionnaires' Disease*, Technical Memorandum, 1987, p. 13.
51. Department of Health and Social Security and the Welsh Office, *The Control of Legionellae in Health Care Premises. A Code of Practice*, HMSO, 1989.
52. Health & Safety Executive, *Legionnaires' Disease*, Guidance Note EH48, HMSO, 1987.
53. Anon., *Loss Prevention Bulletin*, 1990(091), 25.
54. Sykes, J. M. & Brazier, A. M., *Ann. Occup. Hyg.*, 1988, **32**, 63.

55. Employment Committee, *Legionnaires' Disease at the BBC*, HMSO, 1989.
56. National Press, e.g. *Today*, 1991 (18 May), 11.
57. Health & Safety Executive, HS(G)70.
58. Lacey, A., *The Safety and Health Practitioner*, 1994 (April), 12.

INDEX

(for individual chemicals see 'Chemicals', for individual cases of major disasters see 'Major hazards', for individual cases of pollution see 'Pollution disasters')

Acid rain 1221
Aerosols 1103, 1343
Aerosol filling plant 1343
Air conditioning 1616, 1816
Air pollution 1232
Allergens (see respiratory sensitisers, sensitisers)
Angiosarcoma 1598, 1602
Anhydrides 1656
Antibiotics 1663
Anti knock additives 1375
Anthrax 1787, 1791
Asthma (see also respiratory sensitisers) 1661, 1795
Atmosphere control 1508
Audits 1273

Bacteria 1787
Bagging operations 1461
Batch reactors 1173, 1190
Best practicable environmental options (BPEO) 1259
Best practicable means not entailing excessive costs (BATNEEC) 1260, 1305, 1309
Biological agents 1636
Biological diseases 1788
Biological oxygen demand 1301
Bladder cancer, 1584, 1589, 1602, 1688
Boiling Liquid expanding Vapour Explosion 1091, 1322
Bucket elevators 1442
Building design 1543
Bunding 1331, 1767

Carcinogens 1585, 1591, 1635, 1750
CENELEC 1629
Chemical reactions 1111
Chemicals
  Acetic acid 1773
  Acetone 1112, 1145
  Acetone cyanohydrin 1762
  Acetonitrile 1116
  Acetylene 1133, 1150, 1518

Acrylic acid 1106
Acrylonitrile 1175, 1207
Adipic Acid 1149
Alkyl aluminiums 1375, 1400, 1404, 1410
Alkyl boranes 1375
Alkyl lithiums 1383, 1400
Alkyl sodiums 1383
Alkyl zincs 1382
Aluminium 1221, 1362, 1389
Aluminium Chloride 1190
4-Aminodiphenyl 1589
Ammonia 1729, 1730, 1736, 1753
Ammonium nitrate 1500, 1507
Ammonium Salts 1150, 1172
Aniline 1584
Argon 1128, 1704
Arsenic 1593, 1598, 1602
Arsenic trioxide 1247
Arsine 1092
Asbestos 1241, 1242, 1363, 1531, 1573, 1603, 1634, 1642, 1682
Azides 1152
Azo diisobutyronitrile 1542
Barium peroxide 1320
Benzene 1633, 1764
1, 3-Benzodithiolylium perchlorate 1152
Benzo-*a*-pyrene 1603
Benzoyl peroxide 1481
Biphenyls 1449
Bis chloromethylether 1572, 1587
Bitumen 1362, 1533
Bleach 1494, 1776
Boron tribromide 1101
Bromine 1100
m-Bromophenyllithium 1387
Butadiene 1115, 1187
Butane (see LPG)
Butyl lithium (see alkyl lithiums) 1383, 1388
p-tert-Butyl phenol 1701
BZ 1162
Cadmium 1220, 1242, 1740

Chemicals – *cont.*
  Calcium 1372
  Calcium sulphide 1205
  Carbides 1367
  Carbon dioxide 1092, 1101, 1128, 1176, 1705, 1706
  Carbon monoxide 1235, 1699, 1710
  Castor beans 1642
  Chloramphenicol 1661
  Chloraniline 1770
  Chlorine 1100, 1156, 1162, 1730–32, 1752, 1754, 1776
  Chlorobenzene 1747
  Chlorofluorocarbons 1243, 1344
  Chloroform 1112, 1179
  Chloronitro compounds 1172
  5-Chloro-1,2,3-thiadiazole 1156
  Chromium 1220
  Cimetidine 1664
  Citric acid 1794
  Coal 1364
  Colophony 1604
  Copper 1097, 1150, 1218
  CS gas 1163
  Cyanides 1222
  Detergent enzymes 1644
  Detergents 1359
  Diazonium compounds 1095
  Dibromochloropropane 1586
  Dichloronitrobenzene 1102
  Difluoronitrobenzene 1102
  Dimethylacetamide 1102
  Dimethylformamide 1719
  Dimethyl hexanol 1742
  Enzymes (see also Detergent enzymes) 1661, 1718
  Epoxy resins 1609
  Ethylene 1123, 1484, 1541
  Ethylenimine 1485
  Ethylene oxide 1145, 1172, 1174, 1486, 1488
  Flour 1417
  Fluorides 1276
  Formaldehyde 1171
  Formic acid 1098, 1719
  Fuel oil 1120
  Glucocorticoid 1663
  Grignard reagents 1382
  Hay 1362
  Hormones 1662
  Hydrazine and hydrazides 1153, 1154
  Hydrides 1367, 1393
  Hydrochloric acid 1223, 1273, 1699
  Hydrogen 1097, 1132, 1136, 1197, 1366
  Hydrogen cyanide 1161, 1484, 1573, 1723–25
  Hydrogen fluoride 1700, 1701, 1770, 1771
  Hydrogen peroxide 1098, 1144
  Hydrogen sulphide 1699, 1707–09
  Hypochlorites (see Bleaches) 1732
  Imines 1173
  Iminobisacetonitrile 1389
  Inert gases 1702
  Iron 1097, 1100, 1148, 1362, 1371, 1717
  Isopropanol 1173
  Isocyanates 1093–95, 1115, 1148, 1156, 1643, 1655, 1724, 1745
  Kerosene 1508
  Lead 1220, 1592, 1635, 1643, 1671, 1682, 1716
  Lead azides 1501
  Linseed oil 1364
  Liquefied Gases 1124
  Liquefied Natural Gas (LNG) 1510, 1535
  Liquid Propane Gas (LPG)
    Aerosols 1334
    BLEVE 1322, 1332, 1333
    Bulk Storage 1326, 1330
    Cylinders 1333
    Fire appliances 1334
    Fire ball 1322
    Flammability 1321
    Hygiene standards 1320
    Physico-chemistry 1313
    Toxicity 1320
    Transport 1339
  Lithium 1370, 1397
  Lithium compounds 1383, 1397, 1400
  Magnesium 1390
  Manganese 1243
  Mercury and its compounds 1220, 1572, 1574
  Metals 1218, 1242, 1368, 1389
  Methacrylic acid 1096
  Methylamine 1604
  Methylene chloride 1768, 1769
  Methyl ethyl ketone peroxide 1481
  Methyl isocyanate 1094
  Milk powder 1447
  Mineral oils 1363, 1596, 1774
  Mustard gases 1162
  β-Naphthylamine 1602
  Nerve gas 1162
  Nickel 1370, 1393, 1716
  Nickel carbonyl 1716
  Nitrates 1223
  Nitration 1148, 1174, 1180
  Nitric acid 1096–1100, 1766
  Nitrides 1368, 1370
  Nitrites 1173
  Nitrobenzene 1744
  Nitrocellulose 1500
  Nitro explosives 1495
  Nitrogen 1126, 1128, 1508, 1509, 1703–05

Nitrogen oxides 1725, 1740
Nitrogen tri-iodide 1474
Nitroglycerine 1502
Nitrosamines 1587
Nitrotoluene 1388
Nitrous fumes 1578
N-methyl-N-nitrosotoluene-4-sulphonamide 1149
Oil 1359
Organic acid anhydrides 1656
Organic peroxides 1190, 1476
Organic solvents 1779
Organometallics 1374, 1397
Oxides of nitrogen (see nitrogen oxides)
Oxygen 1106, 1115, 1126, 1171, 1532, 1702
Ozone 1728
Paint 1245, 1263, 1273, 1740, 1763, 1780, 1781
Pentafluorophenyl lithium 1387
Pentane 1531
Peracetic acid
Peracids 1653
Perchlorates 1152, 1170, 1172, 1180
Perchlorethylene 1269
Petroleum jelly 1363
Phenacetin 1524
Phenols 1225, 1770, 1771, 1782, 1783
Phosgene 1146, 1162, 1174, 1715, 1734
Phosphates 1223
Phosphides 1367
Phosphine 1368, 1394, 1395
Phosphoric acid 1772
Phosphorus 1373, 1395
Phosphorus oxychloride 1094
Phosphorus trichloride 1094
Platinum salts 1654
Polyacetate/butyrate 1525
Polychlorinated biphenyls 1224
Polyethylene 1099, 1511, 1525
Polymethyl methacrylate 1447
Potassium 1369, 1372, 1391
Potassium dichromate 1095
Potassium hydroxide 1767
Potassium isocyanate 1095
Potassium permanganate 1146
Propane (see LPG) 1533
iso-Propyl ether peroxide 1112
Pyrites 1531
Pyruvic acid 1101
Radium 1588
Radon 1735
Sarin 1162
Sawdust 1361, 1365, 1548
Silanes 1368, 1396
Silicon tetrachloride 1147
Silver sulphate 1197
Smog 1244

Sodium 1369, 1373, 1391
Sodium chlorate 1507
Sodium cyanide 1507
Sodium hydride 1091, 1367
Sodium hydroxide 1153, 1686, 1771, 1772
Sodium hydrosulphite 1092
Sodium iso-nonanoyl oxybenzene sulphonate 1653
Sodium nitrite 1359
Sodium persulphate 1542
Solvents (see organic solvents) 1742
Steel 1362
Styrene 1171, 1524
Sugar 1445
Sulphur 1359
Sulphur oxides 1240, 1745
Sulphur dioxide 1745
Sulphuric acid 1197, 1775, 1776
Tetrachlorodibenzo-p-dioxin 1161, 1247, 1755
Tetra ethyl lead 1674
Tetrahydrofuran 1571
Thionyl chloride 1093
Titanium 1371, 1391
Toluene 1277, 1526, 1527
o-Toluidine 1584
1,1,1-Trichloroethane 1764
Trichloroethylene 1780
Trichlorofluoromethane 1745
Triethanolamine 1148
Triglycidyl isocyanate 1176
Triphenyl tin chloride 1773
Tritium 1762
Vinyl acetate 1527
Vinyl chloride 1576, 1595, 1597, 1633, 1744
Vinyl ethyl ether 1147
Welding fume (see also Hot work) 1584, 1727
Wood dust (see also sawdust) 1657
Zinc 1097, 1218
Zirconium 1371
Chimney sweepers cancer 1586
Combustion 1121, 1239
Commissioning plant 1191
Compressed gases 1129, 1132
Compression 1532
Confined Vapour Cloud Explosions 1345
Contact allergic dermatitis 1605
Containment 1739, 1766
Corrosion 1097, 1776
Cost of pollution 1206
Critical temperature 1128

Deflagration 1118
Detonations 1118
Detonators 1493, 1502
Disinfection 1802

1834 Index

Disposal
  diisocyanate processes 1309
  food industry 1301
  metal processing 1303
  metal working fluids 1301
  paint industry 1309
  pharmaceutical industry 1309
  rubber industry 1305
Drugs (see therapeutic substances)
Dryers 1468
Dust collection 1468
Dust control 1646, 1657, 1664, 1678
Dust conveyance 1461
Dust explosibility testing 1419, 1438
Dust explosions 1416
Dust generation 1441

EEC directives 1629
Electrical ignition sources 1535
Electrochemical series 1369
Emergency procedures 1751, 1772, 1822
Enthalpy 1475
Environmental assessment 1267
Environmental monitoring 1649, 1660
Environmental pollutants (see also 'Pollutants') 1163
Environmental Protection Agency (EPA) 1225, 1241, 1246, 1259, 1263, 1600
Evacuations 1754
Expected environmental concentration 1268
Explosimeters 1509
Explosion isolation 1452
Explosion relief 1454
Explosion resistant design 1452
Explosions 1091, 1121
Explosion tests 1158
Explosives 1149, 1150, 1473
Exposure limits (see also 'Hygiene standards') 1624, 1632

Farmers' lung 1794, 1796
Fire 1117
Fire ball 1324
Fire detection 1550, 1552
Fire drills 1562
Fire extinguishers 1554
Fire fighting 1411
Fire prevention 1507
Fire protection 1542
First aid 1723
Flame cutting 1514
Flames (see Fire) 1533
Flammability 1358
Flash point 1104, 1508
Fractionation 1106
Friction 1444
Fungi 1788, 1794

Gas clouds 1753
Gas detectors 1551
Gaseous waste 1286
Gas oil 1527
Grain 1469
Grinders 1469

Harvesters lung 1796
Hazard (health) evaluation 1573
HAZOPS 1145, 1169
Health control measures 1591
Health surveillance 1627, 1651, 1669, 1749
Heat detectors 1552
Henry's constant 1278
Hepatitis 1788
Hookworm 1797
Hot work 1578
Housekeeping 1457
Hygiene standards 1600, 1624, 1632, 1649, 1665, 1675, 1685, 1687
Humidifiers and humidifier fever 1814, 1816

Ignition sources 1119, 1442, 1512
Incineration 1260
Inerting 1449, 1454, 1702
Information 1628, 1681, 1768
Integrated Pollution Control (IPC), 1259

Joule Thompson Effect 1529

Latent heats 1122
Legionaires' Disease 1613, 1813
Legislation
  The Air Quality Standards Regulations 1259
  Classification and Labelling of Explosives Regulations (1983) 1503
  Clean Air Act 1259
  The Comprehensive Environmental Response, Compensation and Liability Act 1256
  The Control of Asbestos at Work Regulations (1987) 1684
  The Control of Explosives Regulations (1991) 1504
  The Control of Industrial Air Pollution (Registration of Works) Regulations 1259
  The Control of Pollution Act 1259
  The Control of Substances Hazardous to Health Regulations (1988) 1591, 1617, 1644, 1652
  Dangerous Substances in Harbour Areas Regulations (1987) 1505
  The Environmental Protection Act 1259
  Fire Precautions Act (1971)

The Health and Safety (Emissions to the Atmosphere) Regulations 1259
Management of Health and Safety at Work Regulations (1992) 1563
The Reporting of Injuries, Diseases and Dangerous Occurrences Regulations (1985) 1985
The Resources Conservation and Recovery Act 1256
The Road Traffic (Carriage of Explosives) Regulations (1989) 1504
Town and Country Planning (Assessment of Environmental Effects) Regulations 1267
The Toxic Substances Control Act 1241, 1256
Trade Effluents (Prescribed Processes and Substances) Regulations 1263
Water Act 1263
Leptospirosis 1790
Lightning strikes 1537
Limiting oxygen concentration 1440
Liquid solid phase change 1107
Lung cancer 1572, 1579, 1603, 1688

Maintenance 1624
Major hazards (see also Pollution disasters)
  Bantry Bay 1531
  Bhopal 1753
  Brazil 1313
  Bremen 1416
  Feyzin 1320
  Grangemouth 1131
  Houston 1754
  Indiana 1345
  Kings Cross 1544
  Louisiana 1733
  Manhattan Bridge 1733
  Mexico City 1322
  Nebraska 1373
  Pepcon 1563
  Peterborough 1504
  Piper Alpha 1511
  Portishead 1374
  Port Hudson 1321
  Poza Rica 1709
  Sandos 1560
  Seveso 1755
  Silvertown 1488
  Texas 1511
  Urals 1325
Management controls 1270
Maximum explosive pressure 1440
Mechanical sources of ignition 1530
Mesothelioma 1682
Metal fume fever 1380
Metallurgical processes 1716
Microbiological hazards 1104

Microbiology laboratories 1798
Microbiology safety cabinets 1806
Micro-organisms 1786
Minimum explosible concentration 1439
Minimum ignition energy 1419, 1440, 1521
Minimum ignition temperature 1439
Mixtures of chemicals 1576, 1591
Monitoring 1625, 1722, 1748
Monomers 1483
Multiple chemical sensitivity 1579
Mutagens 1586

Nasal cancer 1657
No observed effect concentration 1268

Occupational Exposure Limit (see Hygiene Standards) 1649, 1655, 1656
Odour 1231, 1237, 1290
Orf 1788
Overpressure 1091
Oxygen deficiency 1702
Ozone layer 1243, 1344

Parasites 1788, 1797
Particulates 1103, 1241
Penicillin 1663
Permit to work 1194, 1199
Personal protective equipment 1623, 1624, 1632, 1633, 1650, 1723, 1769
Pharmaceuticals (see Therapeutic substances)
Physico-chemical properties 1091, 1313
Pipe accidents 1109
Pilot plants 1142, 1176
Pollutants
  acids 1222
  arsenic trioxide 1247
  asbestos 1242, 1241
  beryllium 1249
  carbon monoxide 1235
  chlorofluorocarbons 1243
  combustion products 1239
  cyanides 1222
  cypermethrin 1269
  foam 1231
  metals 1218, 1242
  nitrates 1223
  paint 1245, 1263, 1273
  particulates 1241
  PCBs 1224
  perchlorethylene 1267
  pesticides 1224, 1229
  phenols 1225
  phosphates 1223
  oil 1226
  smog 1244
  smoke 1235
  sulphur dioxide 1235, 1239

Pollutants – *cont.*
  suspended matter 1230
  temperature 1230
  tetrachlorodibenzo-*p*-dioxin (TCDD) 1247
  turbidity 1231
  volatile organic chemicals 1237
Pollution and its control 1256
Pollution audits 1273
Pollution disasters
  Amoco Cadiz 1227
  California 1246
  Camelford 1221
  Coode Island 1207
  Denver 1206
  Donora 1235
  Exxon Valdez 1227
  Ixtoc-1 1228
  Lake Baikal 1207
  Lekkerkerk 1244
  Lincolnshire 1242
  London 1239
  Love Canal 1205
  Manfedonia 1247
  Mouse River Valley 1235
  MV Braer 1230
  New York 1207
  Perentis 1225
  River Mersey 1226
  Sandos 1225
  Sauda 1243
  Strathclyde 1226
  Seveso 1247
  Tucson Airport 1246
  Ust-Kamenogorsk 1249
  Woodkirk 1225
Pollution legislation (see also Legislation) 1256
Polymer fume fever 1643
Polymerisation inhibitors 1106
Powder coating 1470
Process development 1142, 1728
Process physics 1173
Pyrophoric chemicals 1366
Pyrotechnics 1492
Psittacosis 1793

Q fever 1792

Reaction hazard index 1156
Reaction kinetics 1111
Reformatsky reagents 1382
Refrigeration 1330, 1366
Respiratory protection (see also Personal protection) 1745
Respiratory sensitisers 1606, 1645, 1653
Rickettsiae 1787, 1792
Risk assessment 1574, 1583, 1624, 1684, 1804

Risk control 1574
Rubber 1419, 1444, 1688

Scale of operation 1180
Scrotal cancer 1586
Self-initiated reactions 1100
Sensitisers 1604, 1644
Sewage works 1827
Sick building syndrome 1613
Siderosis 1579
Smallpox 1801
Smoke 1235
Smoke detectors 1551
Smokeless powder 1495
Smoking 1532
Solidification problems 1107
Solid waste 1278
Spontaneous combustion 1357, 1443, 1447
Sprinklers 1558
Static electricity 1448, 1520
Steam 1186, 1187
Sterilisation 1802
Storage 1175, 1765
Strict liability 1264
Substitution 1592, 1623, 1737, 1764, 1774, 1811
Suppression 1453
Synergistic substances 1583
Systems of work 1723, 1744, 1768

Taste 1231
Temperature changes 1109
Teratogens 1586
Tetanus 1790
Therapeutic substances 1659
Thermal expansion/contraction 1108
Thermal inhibitors 1450
Thermite reaction 1531
Threshold Limit Value (see also 'Hygiene Standards') 1277, 1320
Threshold odour number 1290
Tick fever 1792
Toilet cleaner 1571
Toxic gases 1699
Toxic liquids 1761
Toxic solids 1642
Training 1628
Transport 1503, 1733, 1769
Typewriter fluid 1571

Underwater pipelines 1226
Unplanned chemistry 1091, 1171
Unstable chemicals 1474

Vegetable oil 1107
Vehicles 1518
Ventilation 1722, 1739
Viruses 1787, 1788, 1801, 1809

Waste chemicals 1205, 1278, 1286, 1294, 1670
Water hammer 1138
Water pollution 1207
Water treatment 1820

Weil's disease 1790
Welding gas (see also Hot work) 1514, 1584, 1727
Wood dust 1657